Chemical Carcinogenesis

Models and Mechanisms

Chemical Carcinogenesis

Models and Mechanisms

Edited by

Francesco Feo

University of Sassari
Sassari, Italy

Paolo Pani and
Amedeo Columbano

University of Cagliari
Cagliari, Italy

and

Renato Garcea

University of Sassari
Sassari, Italy

SPRINGER SCIENCE+BUSINESS MEDIA, LLC

Library of Congress Cataloging in Publication Data

Sardinian International Meeting on Models and Mechanisms in Chemical Carcinogenesis
(4th: 1987: Alghero, Sardinia)
    Chemical carcinogenesis: models and mechanisms / edited by Francesco Feo . . . [et al.].
    p.   cm.
    "Proceedings of the Fourth Sardinian International Meeting on Models and Mechanisms in
Chemical Carcinogenesis, held October 23–27, 1987, in Alghero (Sassari), Italy"—T.p. verso.
    Includes bibliographies and index.
    ISBN 978-1-4757-9642-1   ISBN 978-1-4757-9640-7 (eBook)
    DOI 10.1007/978-1-4757-9640-7

    1. Carcinogenesis—Congresses. 2. Carcinogens—Metabolism—Congresses. I. Feo,
Francesco. II. Title.
    [DNLM: 1. Carcinogens—metabolism—congresses. 2. Neoplasms—chemically induced—
congresses. QZ 202 S244c]
RC268.5.S27   1987
616.99'4071—dc19
DNLM/DLC   88-29013
for Library of Congress   CIP

SCIENTIFIC COMMITTEE

F. Baccino, *Torino, Italy*
P. Bannasch, *Heidelberg, Federal Republic of Germany*
A. Columbano, *Cagliari, Italy*
F. Feo, *Sassari, Italy*
P. Pani, *Cagliari, Italy*
G. Prodi, *Bologna, Italy*
M. Roberfroid, *Brussels, Belgium*
D. S. R. Sarma, *Toronto, Canada*
R. Schulte-Hermann, *Vienna, Austria*

Proceedings of the Fourth Sardinian International Meeting on
Models and Mechanisms in Chemical Carcinogenesis,
held October 23–27, 1987, in Alghero (Sassari), Italy

© Springer Science+Business Media New York 1998
Originally published by Plenum Press, New York in 1998
Softcover reprint of the hardcover 1st edition 1998

FOREWORD

About two centuries after the communication by Sir Percival Pott that the "chimney sweeper disease" was a cancer and its suggestion that active compounds of soot were the causative agents, and about one century after the description of urinary bladder cancer in dye workers, an enormous number of substances have been synthesized and have probably come into contact with man. Research in cancer prevention is of primary importance, and may receive continuous support from new discoveries on cancer etiology and pathogenesis.

If one accepts the multistage model of chemical carcinogenesis, one has also to accept that many events occur between the contact of carcinogenic compounds and their specific targets and the development of a clinically recognizable neoplasm. Thus, animal studies become essential to elucidate the different steps by which chemical carcinogens induce neoplasia. The analysis of these steps and the comparative evaluation of experimental models is essential to an understanding of pathogenesis.

Serial meetings are an appropriate forum to analyze successively various topics related to chemical carcinogenesis. While not aiming at being comprehensive, each meeting of the series highlights particular areas of interest and tries to give a good indication of further research possibilities. This series of biannual meetings, on Chemical Carcinogenesis, organized by the "Istituto di Patologia generale, Università di Sassari", and the "Istituto di Farmacologia e Patologia biochimica, Università di Cagliari", was started in 1981. In the Fourth Sardinian International Meetings, held in Alghero (Italy) between October 23 and 27, 1987, a number of topics dealing principally with the analysis of different experimental models to study the main steps of the carcinogenic process from the biochemical, biological and molecular aspects were presented. The topics ranged from the recent progress on the metabolism of carcinogenic substances and their interaction with target molecules to some basic mechanisms in human and experimental nutritional carcinogenesis, through an analysis of developmental stages of carcinogenesis and growth control mechanisms. Different experimental models for the study of chemical carcinogenesis were analyzed comparatively. Developmental stages and growth regulation were considered in some detail from the molecular and biochemical points of view and modulatory mechanisms were considered.

The meeting program included five keynote addresses, sixty-five plenary lectures, and forty posters, some of which were selected for oral presentation. The speakers chosen represented individuals of international acclaim who are very active in the area of chemical carcinogenesis. This meeting brought together scientists from sixteen countries including: Austria, Australia, Belgium, Canada, Denmark, France, Holland, Japan, Norway, Poland, Sweden, Switzerland, The United Kingdom, The United States, West Germany, and, of course, Italy. The presentations were of the highest quality and were justly appreciated by those who had the

opportunity to attend. We trust that the readers of this volume,
published as the Proceedings of the Fourth Sardinian International
Meeting, will also be rewarded.

We would like to thank the Members of the Scientific Committee who
provided us guidance and advice, the Chairpeople and Reviewers, who worked
together to insure the success of the meeting. Special thanks and ap-
preciation are given to the staff of the "Istituto di Patologia generale",
University of Sassari, for the many functions that they expertly and
carefully conducted under the guidance of Miss Lucia Daino. We would like
to express our deep gratitude to the "European Association for Cancer
Research", the "International Agency for Research on Cancer", the "Società
Italiana di Cancerologia" and the "Società Italiana di Patologia" for the
sponsorship of the "Fourth Sardinian International Meeting" and to the
Members of the National and Regional Governments and to all other govern-
mental and private institutions which, by their financial support, made
possible the organization of the meeting.

Francesco Feo
Paolo Pani
Amedeo Columbano
Renato Garcea

Sassari, May 1988

During the preparation of this book, the Editors received the unhappy news of the untimely death of Professor Giorgio Prodi.

Professor Prodi was the chairman of the Institute of Cancerology, University of Bologna (Italy). He taught Experimental Oncology at this University and was involved in the research on interaction of chemical carcinogens with DNA for several years. He was a member of the Organizing Committee of the Sardinian International Meetings and actively contributed to the success of these meetings. His wide experience, enthusiasm and criticism were highly stimulatory for us.

We are profoundly convinced that the death of Professor Prodi represents a heavy loss. Indeed, we have lost an invaluable collaborator and a friend.

Francesco Feo
Paolo Pani
Amedeo Columbano
Renato Garcea

CONTENTS

SECTION 1
XENOBIOTICS: METABOLISM AND INTERACTION
WITH CELLULAR MACROMOLECULES

SECTION II
DEVELOPMENTAL STAGES OF CARCINOGENESIS

SECTION III
REGULATION OF CELL GROWTH IN CHEMICAL CARCINOGENESIS

SECTION IV
NUTRITIONAL CARCINOGENESIS AND ANTICARCINOGENESIS

SECTION V
MECHANISMS IN HUMAN CARCINOGENEIS

SECTION I

XENOBIOTICS: METABOLISM AND INTERACTION

WITH CELLULAR MACROMOLECULES

CYTOCHROME P-450 FUNCTION ANALYSES WITH MONOCLONAL ANTIBODIES AND cDNA EXPRESSION VECTORS [1]

Harry V. Gelboin, Frank J. Gonzalez, Sang S. Park, Junji Sagara and Narayana Battula

Laboratory of Molecular Carcinogenesis, National Cancer Institute, National Institutes of Health Bethesda, Maryland, USA

INTRODUCTION

It is quite fitting and proper to initiate this International Conference devoted to the problems of chemical carcinogenesis with a discussion of the role and function of cytochrome P-450. Cytochrome P-450 is the enzymatic interface between mammalian organisms, including man, and a large variety of foreign chemicals, i.e. xenobiotics, which include almost all chemical carcinogens.

Table 1 shows the classes of xenobiotics, that is, compounds foreign to the organism as well as endobiotics, those compounds which are endogenous to normal metabolism which are substrates for the cytochrome P-450 containing mixed function oxidases. Among the xenobiotics, more than 90% of the various classes of carcinogens, including polycyclic aromatic hydrocarbons, amino azo-dyes, aflatoxins, nitrosamines, aromatic amines as well as numerous other synthetic and natural occurring carcinogens are substrates for the cytochrome P-450 system. Other environmental pollutants such as polychlorinated biphen-

Table 1. Classes of xenobiotic and endobiotic substrates of cytochrome P-450

Xenobiotics	Endobiotics
Carcinogens > 90%	Steroids (all classes)
Drugs > 90%	Fatty acids
Environmental	Prostaglandins
Pollutants	Leukotrienes
Pesticides	
Herbicides	

---

[1] This paper is dedicated to the memory of Professor Elizabeth C. Miller.

yls (PCBs), a large variety of pesticides and herbicides, as well as the vast majority of drugs which are of therapeutic and clinical use.

In addition to these xenobiotics, some very important endobiotics concerned in regulation of normal metabolism and regulation of physiological function are substrates for the cytochrome P-450 system. Thus all classes of steroids, ranging from vitamin D and sex steroids to cortisone, progesterone and cholesterol are substrates for P-450. Fatty acids, prostaglandins, leukotrienes, all very important regulatory substances, are P-450 substrates.

Aspects of P-450 action on xenobiotics are shown in Table 2. The cytochrome P-450s are essential for the detoxification of a whole variety of drugs, carcinogens and environmental chemicals. Although the P-450s have largely beneficial value, the P-450 system is also responsible for the activation of many compounds to toxic forms, to mutagenic compounds inducing mutations in DNA, to the formation of teratogens, and finally to the formation of active carcinogens. Thus there is a dichotomy between the results of P-450 action, on the one hand, the beneficial effects of detoxification and the opposite detrimental effects of activation of inactive compounds to active toxins and carcinogens. Thus the cytochrome P-450 puzzle is both important and complex and is at the core of understanding man's interaction with the chemicals in his environment.

Table 3 shows the nature of the P-450 system. First, there are a large multiplicity of xenobiotic and endobiotic P-450 substrates. Secondly, there are a multiplicity of cytochrome P-450 forms. There are at least 30 described forms and likely there are considerably more. An added feature of this complex puzzle is the fact that cytochrome P-450 distribution, that is, the isozyme make-up of a tissue, changes with induction, hormonal, nutritional and developmental state and differs with different tissues, species, strains, sex and age. Thus, we have a very complex puzzle for which some of the solutions we seek are shown in Table 4. First of all, we need to define the contribution of each form of cytochrome P-450 to the metabolism of single and specific xenobiotics and endobiotics in a crude cell extract or mixture like microsomes or cell homogenates. Secondly, we need to define the specificity of each cytochrome P-450 form for each single substrate. These two goals are complementary; in other words, we need to know the substrate specificity for each P-450, and then, we need to be able to determine the contribution of each specific P-450 form to the metabolism of any given substrate in a tissue. In many cases many more than one P-450 will contribute to the metabolism of a single substrate. In other cases, there may be a high specificity of a substrate for only one or a few cytochrome P-450s. Thirdly, we seek to define the molecular, genetic and structural basis for cytochrome P-450 differences in different individuals. We then need to determine human individual differences in cytochrome P-450-related carcinogen and drug sensitivity. Finally we want to determine the relationship of the individual differences in sensitivity and the molecular genetic differences exhibited by

Table 2. Biological activities resulting
from P-450 action on xenobiotics

1) Detoxification	(drugs, carcinogens environ. chemicals)
2) Activation to:	Toxic compounds Mutagens Teratogens Carcinogens

Table 3. Cytochrome P-450

---

1) Multiplicity of xenobiotic and endobiotic P-450 substrates

2) Multiplicity (> 30) of cytochrome P-450 forms

3) Cytochrome P-450 distribution changes with induction, hormonal, nutritional and developmental state and differs with tissues, species strain, sex and age

4) Responsible for both detoxification and activation to toxins, mutagenic and carcinogenic compounds

---

the gene structure of different individuals for different P-450 genes.

The strategic tools towards these major goals are shown in Table 5. First, we have partially constructed a monoclonal antibody library to individual and classes of cytochrome P-450 [1-5]. The inhibitory Mabs can be used for reaction phenotyping [6,7] and the addition of one of these inhibitory monoclonals to a reaction mixture can determine the contribution to the reaction by the cytochrome P-450 to which the monoclonal was directed. Secondly, we have cloned and sequenced a large number of cytochrome P-450 cDNAs which can be constructed into expression vectors such as vaccinia or expressed in COS cells via SV40-based expression vectors. Since a cloned cDNA is expressed as enzyme activity we can precisely define the enzymatic character of that P-450 derived from a single cDNA. In this way, we can define precisely the substrate and product specificity of individual P-450s. We are continuing to clone and sequence additional P-450 cDNAs and plan to have a large, complete as possible, library. These cloned and sequenced genes can be used to compare relatedness of the different P-450s by site-directed mutagenesis studies in which active sites of P-450 genes will be studied. The cDNA will also be used as probes for restriction fragment length polymorphism analysis and should be very useful for detecting potential polymorphisms in the human population that relate to specific sensitivity to certain classes of carcinogen, or to specific abnormalities in the metabolism of different classes of drugs that are therapeutically effective.

RESULTS AND DISCUSSION

Table 6 shows a brief summary of our library of monoclonal antibodies. We have panels of Mabs to seven different forms of cytochrome P-450. In some cases we have large numbers of positive clones ranging from 6 to 31. In the

Table 4. Major goals of cytochrome P-450 research

---

1. Define contribution of each form of cytochrome P-450 to specific xenobiotic and endobiotic metabolism

2. Define specificity of each cytochrome P-450 form

3. Define the molecular genetic basis for cytochrome P-450 differences

4. Determine human individual differences in gene structure and cytochrome P-450 related carcinogen and drug sensitivity

---

Table 5. Strategic tools of cytochrome P-450 research

---

1. Monoclonal antibody library to individual and classes of cytochrome P-450 (inhibitory Mabs for reaction phenotyping)

2. Cloning and sequencing of cytochrome P-450 genes

3. Expression of cloned single P-450 cDNAs by vectors (Vaccinia,
   COS cell, etc.)

4. Restriction fragment length polymorphism (RFLP) analysis

---

fourth column you see the number of inhibitory clones. The positive clones, independent of their inhibitory activity, can be used to determine the amount of specific forms or classes of P-450 in a given tissue or individual, and the inhibitory clones can be used for "reaction phenotyping". The latter is done by measuring the inhibitory effect of a monoclonal in a microsomal preparation incubated with a specific P-450 substrate. This defines the contribution of that particular P-450 form to the total reaction. The mono-clonals we have used for this purpose all inhibit the enzymatic activity of the pure P-450 for the substrate studied by more than 90%. In one case, the monoclonals made to PCN inducible P-450 did not yield inhibitory clones.

With Mab 1-7-1, a monoclonal which inhibits two forms, P-450c and P-450d, we find that in the rat liver microsomes from uninduced rats and from phenobarbital(PB)-treated rats there is essentially no inhibition of AHH . If, however, the rat is treated with methylcholanthrene, there is a 70% inhi-bition of AHH. This experiment tells us that the forms of P-450 which are active for benzopyrene hydroxylation (AHH) in the untreated and in the pheno-barbital(PB)-treated rat are insensitive to the monoclonal antibody. In the MC-induced animal, 70% of the total AHH activity of liver is due to forms of P-450 recognized by the monoclonal, and 30% is due to P-450 which is differ-ent than that recognized by the monoclonal. Thus 70% of the AHH in MC-treated rats is the result of a P-450 different than the P-450 responsible for AHH in control and PB rats. In a similar experiment with ethoxycoumarin deethylase activity the Mab 1-7-1 distinguished ECD activity due to the sensitive and insensitive forms of P-450. In similar experiments done with two strains of mice the 1-7-1 detected P-450 differences in the two strains (Table 7). In the C57bl mice, none of the AHH in control mice and very little in the PB mice are inhibited by the monoclonal. On the other hand, 85% of the AHH in the MC-treated rat liver is sensitive to the 1-7-1. With the DBA strain, however, there is only a small inhibitory effect ranging from

Table 6. Library of monoclonal antibodies to cytochrome P-450

P-450	Inducer	Positive Clones(RIA)	Inhibitory Clones	%Inhibition of Pure P-450
P-450 MC(forms c,d)	PAH	10	3	90%
P-450 Pb	PB	10	4	30-90%
P-450 Et	Ethanol	31	1	92%
P-450 SCUP	PAH	8	4	87-94%
P-450 PCN	PCN	11	0	0
P-450 RLM-5	CONSTIT.	6	2	90%

Table 7. Inhibition (%) of AHH and ECD of mouse tissues by Mab 1-7-1

| Strain | Aryl hydrocarbon hydroxylase | | | | | |
| | Liver | | Lung | | Kidney | |
	Control	Mab	Control	Mab	Control	Mab
C57B1						
Control	186	(0)	14	(50)	1.2	(0)
PB	854	(4)	18	(43)	1.7	(6)
MC	4882	(85)	141	(83)	57.0	(86)
DBA/2						
Control	327	(11)	18	(17)	2.2	(27)
PB	610	(17)	19	(0)	2.9	(7)
MC	283	(2)	21	(51)	1.5	(27)
	Ethoxycourmarin deethylase					
C57B1						
Control	2.5	(15)	0.18	(6)	0.12	(0)
PB	10.8	(8)	0.22	(5)	0.14	(7)
MC	14.0	(55)	0.22	(45)	0.12	(0)
DBA/2						
Control	4.8	(8)	0.67	(6)	0.34	(0)
PB	16.0	(2)	0.45	(0)	0.65	(9)
MC	4.0	(17)	0.48	(0)	0.22	(5)

2-17% by the Mab 1-7-1 which suggests that all of the AHH in this strain is due to a different P-450. We examined the effect of the Mab 1-7-1 in mouse liver, lung and kidney. In the lung tissue there is a significant amount of AHH present which is due to the Mab 1-7-1 sensitive to P-450, in the control 42%, PB-treated 24% and the MC-treated rats 78%. Likewise in the kidney, there is 11%, 39% and 87% inhibition by Mab 1-7-1 respectively. Thus in the extra-hepatic tissues a higher percentage of the AHH is due to the 1-7-1 defined P-450 in the control and PB animals than in the livers of the same animals. The Mabs can define the contribution of epitope specific single or classes of P-450 to any reaction in respect to either substrate utilization or product formation. With the proper library of monoclonals one can construct an atlas of reactions in different tissues and define the amount of each reaction catalyzed by particular forms of P-450s.

Table 8 shows reaction phenotyping with a new monoclonal antibody we have recently prepared to ethanol-induced cytochrome P-450. This P-450 is active in aniline hydroxylation and in the demethylation of nitrosodimethyl-amine demethylase is inhibited by 92%. With microsomes from acetone-treated rats, acetone is an inducer of this enzyme, we see that 54% of the aniline hydroxylase is inhibited by this monoclonal and 77% of the nitrosodimethyl-amine demethylase is inhibited, indicating the percent contribution by that particular P-450 to total microsomal enzyme activity. This also indicates

Table 8. Effect of Mab 1-91-3 on aniline hydroxylase and nitrosodimethylamine demethylase of cytochrome P-450$_{et}$ and rat liver microsomes

| Mab | Purified P-450$_{et}$ | | | |
	Aniline hydroxylase	Inhibition	NOMAD	Inhibition
Control	2.50	-	4.5	-
Mab 1-91-3	0.25	90%	0.4	92%
	Microsomes (acetone-treated rats)			
Control	1.36	-	3.5	-
Mab 1-91-3	0.63	54%	0.8	77%

that the remaining activity, 46% in the case of aniline hydroxylase and 23% in the case of the nitrosamine demethylation is contributed by P-450s different than those recognized by the Mab to ethanol-induced P-450. Thus with a proper and complete library of monoclonal antibodies, one could determine precisely the contribution of each of the P-450s to any reaction occurring in a biological mixture. This method can be used not only for substrate disappearance and product formation but also to determine quantitatively the role of different P-450s in biological effects due to P-450 action.

We performed a collaborative study with Drs. Bartsch and Hietanen of Lyon in which they used the monoclonals to assess the contribution of specific P-450s to aflatoxin and nitrosomorpholine mutagenicity[8]. Fig. 1 shows the effect of two Mabs, 1-7-1 and 2-66-3 an Mab to PB-induced P-450 in different strains of mice treated with different inducers: control, MC, PB and PCN. In the B6 mice, inducible strains, the monoclonal directed to PB-induced P-450s had a significant inhibitory effect on the control mice as well as the PB- and PCN-treated mice, in some cases inhibiting 50% of the mutagen activation of aflatoxin. In the D2 mice we observed similar effects in the PB-treated mice but a lesser effect in the PCN and control mice. The monoclonal to the MC inducible form of P-450 had very little inhibitory effect; in some cases, it actually stimulated to some extent the activation. This stimulatory effect we view as due to an inhibitory effect on detoxification which permits more of the substrate to be activated to the mutagen form. This is vividly shown in Fig. 2 where the antibody to PB induced P-450, Mab 2-66-3, had a major stimulatory effect on nitrosomorpholine on mutagenesis in both strains of mice and in all mice after treatment with MC, PB, or PCN. We interpret this to mean that the detoxification pathway has been largely inhibited by the Mab, making more substrate available for activation to the mutagen form. As you can see, in contrast to the phenobarbital having a major stimulatory effect, the monoclonal to the MC-inducedP-450 has essentially no effect.

We have used the monoclonals in a variety of systems examining a number of different drugs - aminopyrine, theophyllin, antipyrine, prostaglandins, and in steroid metabolism. I have shown you an example of what can be done with inhibitory monoclonals. Thus a full library of these reagents would permit us to construct a precise map of the contribution of each P-450 to the

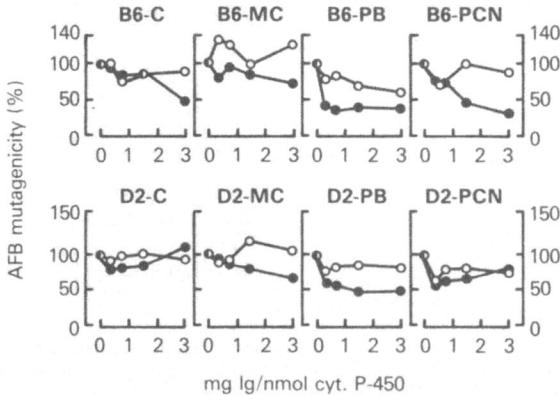

Fig. 1. Effect of Mab 1-7-1 (MC) (O—O) and Mab 2-66-3 (PB) (●—●) on
aflatoxin B-1 mutagenicity in B6 and D2 mice.

total activity of a tissue toward any compound that is metabolized by P-450s,
be it carcinogen, drug or endobiotic.

CLONING AND SEQUENCING OF P-450 GENES

A second major approach we have taken has been based on the new advances
in molecular biology, and we have applied some of these techniques to the
problems of cytochrome P-450. Dr. Gonzalez has accomplished the cloning and
sequencing of a variety of P-450 cDNAs, and currently we have 13 cloned and
sequenced cDNAs for different P-450 forms. Thirteen of these are rat cDNAs
and four are human. A listing of these is shown in Table 9.

The rat cDNAs include three forms which we believe are debrisoquine
metabolizing P-450s. This is a form of major interest since there is a polymorphism of debrisoquine metabolism in the human population, with approximately 8% to 10% of individuals being deficient metabolizers[9]. Another
interesting form is P-450j which is known to metabolize nitrosamines and is
induced by ethanol. In addition to forms metabolizing drugs and carcinogens,
we have also isolated and sequenced the forms that are involved in steroid

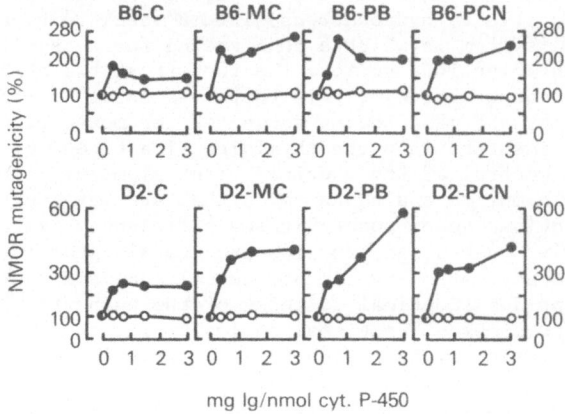

Fig. 2. Effect of Mab 1-7-1 (MC) (O—O) and Mab 2-66-3 (PB) (●—●) on
mutagenicity of nitrosomorpholine.

Table 9. Cloned and sequenced cytochrome P-450 cDNAs

Cytochrome P-450	Major substrate
Rat	
P-450a	Testosterone
P-450a	Testosterone
P-450b	Unknown
P-450e	Benzphetamine
P-450f	-
P-450 PB1	-
P-450j	Nitrosamine, aniline
P-450 db1	Debrisoquine, bufaralol
P-450 db2	Unknown
P-450 db3	Unknown
P-450 PCN1	Testosterone
P-450 PCN2	Testosterone
P-450 LA omega	Lauric acid, palmitic acid, arachidonic acid
Human	
P-450j	Nitrosamine, aniline
P-450 db	Debrisoquine, other drugs
P-450 PCN1	Nifedipine, testosterone
P-450 PCN2	Unknown

and fatty acid metabolism and are likely to be very instrumental in regulatory processes. These cDNAs and genes can be used for a variety of purposes.

VACCINIA VIRUS EXPRESSION VECTORS FOR P-450 cDNAs

One major thrust is to determine the precise specificity of single forms of P-450. This has been very difficult since purified forms often contain contaminating P-450s, and there are now a number of reports in the literature that purified reconstituted mixed-function oxidases containing P-450 do not exhibit the same substrate specificity that they do in intact microsomes. Dr. Battula in our lab has inserted two cDNAs, P1-450 and P3-450 (these are forms that are inducible by hydrocarbons) into a vaccinia virus vector[11]. We sought to develop a system in which a cDNA coding for a specific P-450 could be inserted into an infectious vector. We took these two cDNA clones, and the diagram of the construction of this vector is shown in Fig. 3. These cDNA clones represent the mRNA coding sequences for mouse P1-450 and mouse P3-450. They were inserted into the thymidine kinase gene of the wild type vaccinia under the control of the vaccinia virus promoter. The construct was successful, and both murine and human cells that are infected with each of the resulting infectious recombinant viruses efficiently expressed the P-450 proteins. The newly synthesized proteins are translocated into the microsomes. Their characterization by immunochemical analysis that they are the size of the polypeptides identical to those of the authentic P-450s found in mammalian liver microsomes. Functional analysis of each of the proteins both spectrally and enzymatically indicates that the proteins have incorporated heme and display the appropriate P-450 enzymatic activity. Fig. 4 shows the Western blot of the proteins made by the cells infected by this infectious P-450 vaccinia vector. Fig. 5 shows that the P1-450 expresses AHH activity but does not express any aniline hydroxylase activity. The control Vaccinia containing no P-450 cDNA shows no enzyme activity. Fig. 6 shows that the

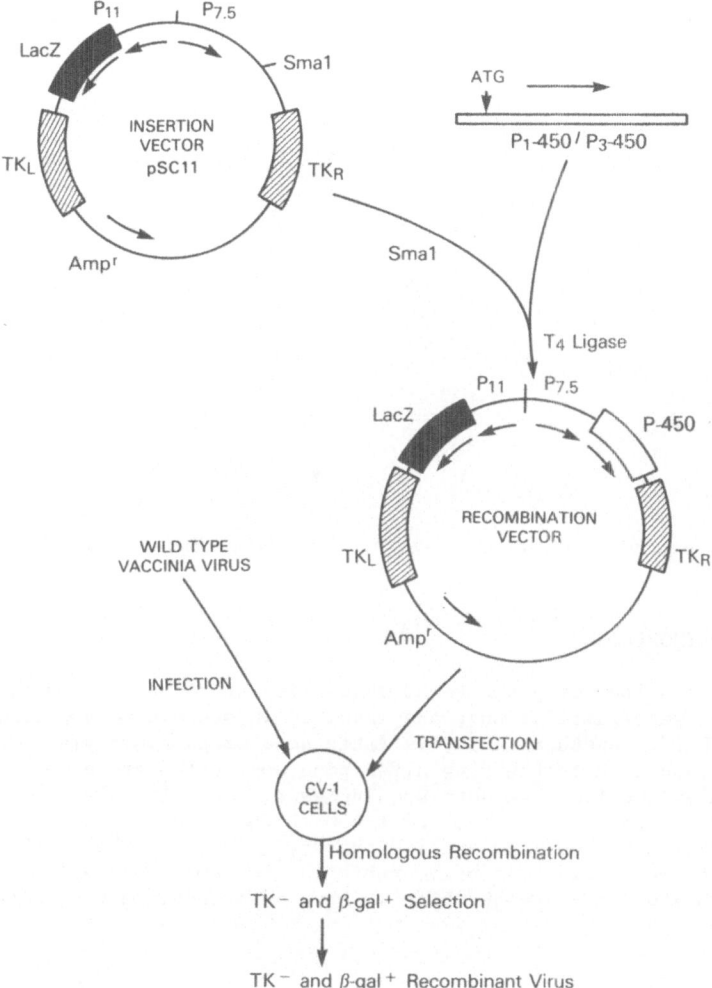

Fig. 3. Construction of vaccinia virus P-450-cDNA for cytochrome P-450
expression.

vaccinia constructed with the P3-450 shows (this is a chromatographic assay
for aniline hydroxylase) that the P3-450 exhibits good activity, but there is
no activity observed for the AHH. Table 10 shows two other enzymes activities often associated with AHH. These were ethoxycoumarin deethylase (ECD)
and ethoxyresofurin deethylase (ERDE). As is seen, the AHH is specific for
P1-450 and the aniline hydroxylase is specific for P3-450. However, the ECD
and the ERDE are enzymatic functions of both of these P-450s which was suggested in a less definitive way by our monoclonal antibody work. Thus the
vaccinia-type expression system can be of extraordinary use in the analysis
of the precise specificity of different forms of P-450. Our current plan is
to construct all of the human cDNAs and rat cDNAs into these vectors to
determine their specificity. In addition to enzyme activity one can measure
of the precise specificity of different forms of P-450. Our current plan is
to construct all of the human cDNAs and rat cDNAs into these vectors to
determine their specificity. In addition to enzyme activity one can measure
any biological activity which depends on P-450 enzyme function, such as
toxicity or mutagenicity.

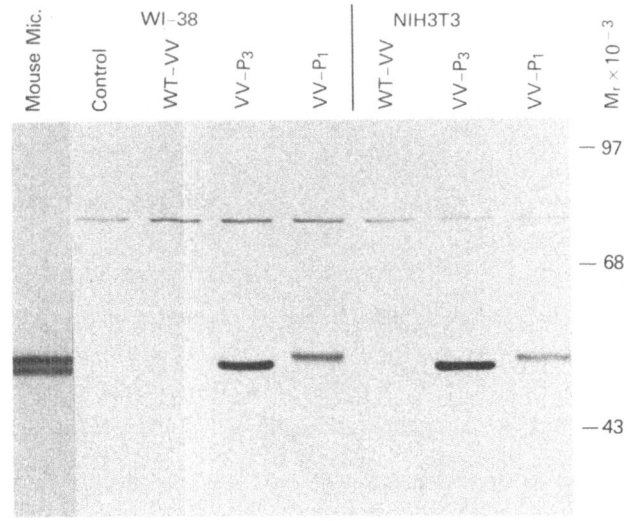

Fig. 4. Western blot of vaccinia vector expressed P1-450 and P3-450.

P-450 HUMAN POLYMORPHISMS

There are a number of human P-450 genes with which we are concerned. There are three gene families that are involved in steroid metabolism, C-21, aromatase and C-17. Another family is fatty acid omega hydroxylase. PCN induced P-450 is active in the drug nifedipine and testosterone metabolism. In another family, we find the debrisoquine gene. This is polymorphic in humans. P-450j is the alcohol-induced P-450 which is active in dimethylni-trosamine metabolism as well as aniline hydroxylation. Another P-450, for mephenytoin drug metabolism, is polymorphic in the human population. Finally, there is the P3-450 and P1-P-450 which we have inserted in a vaccinia virus

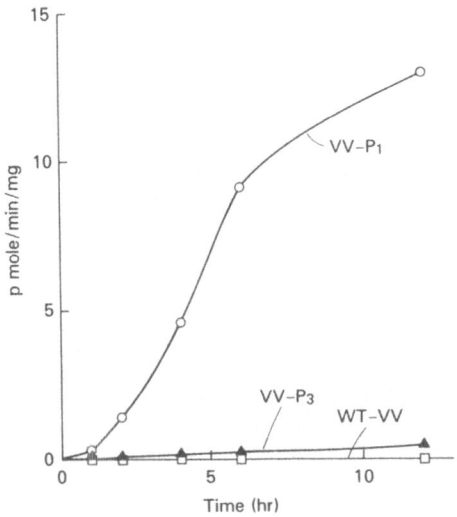

Fig. 5. Expression of AHH activity by VV-P1-450 and VV-P3-450.

Table 10. Enzymatic activities of P1-450 and P3-450 in
recombinant virus infected HeLa cells

Enzyme	AHH	EROD	ECD	A-OH
P1-450	61	15.5	23.5	3
P3-450	0.9	5.4	44.0	74

Activities expressed are pmoles of product/min/mg protein of cell lysates. AHH = Aryl hydrocarbon hydroxylase. EROD = Ethoxy resorufin deethylase. ECD = 7-Eth-
oxy coumarin deethylase. A-OH = Acetanilide hydroxylase

and which is active in carcinogen metabolism. We believe the P3-450 is active in the metabolism of different carcinogenic agents, including aflatoxin,
acetylaminofluorene, and other aromatic amines.

An important substrate for the P-450j is dimethylnitrosamine. This
P-450 is inducible by a variety of agents, including ethanol, acetone, and
pyrazole. One of the aims of our studies is to determine individual differences in human populations. We have looked at a number of different human
livers for the content of P-450j, the alcohol-nitrosamine metabolizing
P-450[12,13]. Fig. 7 is a Western blot showing that there is little or no difference between the livers in respect to antibody recognizable P-450 protein.
We have not, however, analyzed for functional activity, and although the protein content may be identical, it is quite possible that enzymatic activity

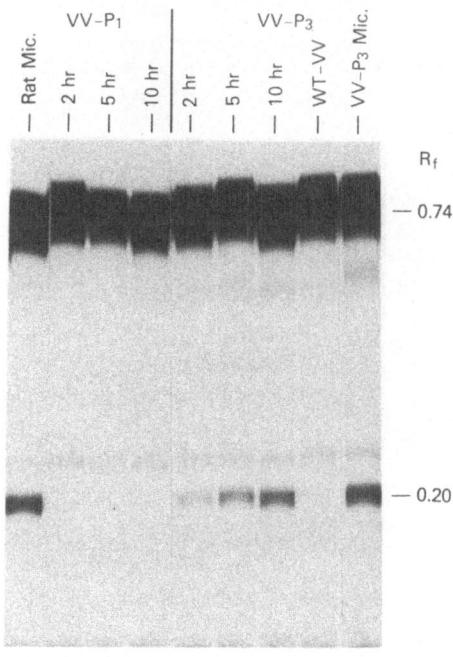

Fig. 6. Expression of aniline hydroxylase from VV-P1-450 and VV-P3-450.

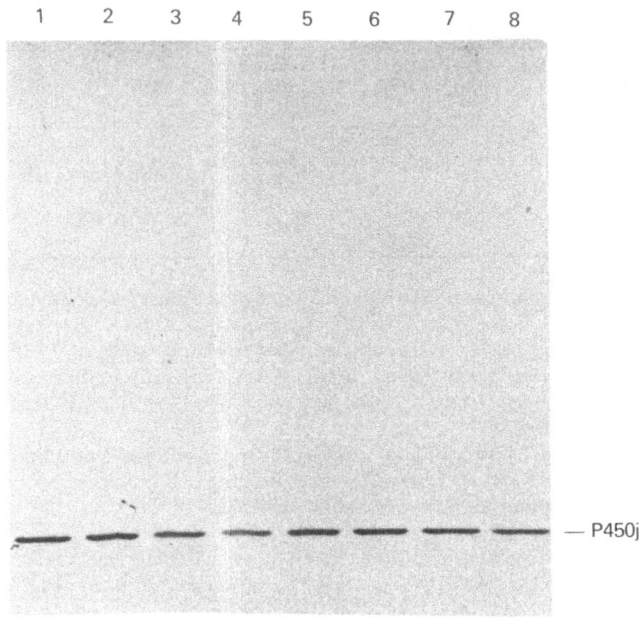

Fig. 7. Western blot analysis of P-450j in different human livers.

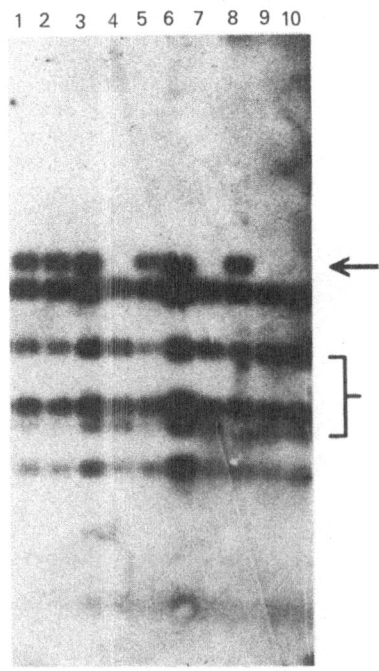

Fig. 8. Restriction fragment length polymorphism of P-450j gene in humans.

14

varies among different individuals[12,13]. With respect to the P-450j, potential polymorphisms were analyzed with an RFLP analysis. Here, clearly, about half of the subjects examined show differences in their restriction map patterns.

Fig. 8 shows that a restriction fragment length polymorphism (RFLP) was found linked to the P-450j gene. The human lymphocyte DNA from these individuals was digested with Taq-1 which showed a RFLP with a high allelic frequency. Twenty other restriction enzymes did not show any RFLPs[14].

REFERENCES

1. H. V. Gelboin and F. K. Friedman, Monoclonal antibodies for studies on xenobiotic and endobiotic metabolism. Cytochromes P-450 as paradigm, Biochem. Pharmacol. 34:2225 (1985).
2. S. S. Park, T. Fujino, D. K. West, F. P. Guengerich and H. V. Gelboin, Monoclonal Antibodies that Inhibit Enzyme Activity of 3-Methylcholanthrene-induced Cytochrome P-450, Cancer Res. 42:1798 (1982).
3. S. S. Park, T. Fujino, H. Miller, F. P. Guengerich and H. V. Gelboin, Monoclonal antibodies to phenobarbital-induced rat liver cytochrome P-450, Biochem. Pharmacol. 33:2071 (1984).
4. I. Y. Ko, S. S. Park, B. J. Song, C. Patten, Y. Tan, Y. C. Hah, C. C. Yang and H. V. Gelboin, Monoclonal Antibodies to Ethanol-induced Rat liver Cytochrome P-450 That Metabolized Aniline and Nitrosamines, Cancer Res. 47:3101 (1987).
5. S. S. Park, D. J. Waxman, H. Miller, R. Robinson, C. Attisano, F. P. Guengerich and H. V. Gelboin, Preparation and characterization of monoclonal antibodies to pregnenolone 16-alpha-carbonitrile inducible rat liver cytochrome, Biochem. Pharmacol. 35:2859 (1986).
6. T. Fujino, S. S. Park, D. West and H. V. Gelboin, Phenotyping of cytochromes P-450 in human tissues with monoclonal antibodies, Proc. Natl. Acad. Sci. 70:3682 (1982).
7. T. Fujino, D. West, S. S. Park and H. V. Gelboin, Monoclonal antibody-directed phenotyping of cytochrome P-450-dependent aryl hydrocarbon hydroxylase and 7-ethoxycoumarin deethylase in mammalian tissues, J. Biol. Chem. 259:9044 (1984).
8. E. Hietanen, C. Malaveille, F. K. Friedman, S. S. Park, J. C. Bereziat, G. Brun, H. Bartsch and H. V. Gelboin, Monoclonal Antibody-directed Analysis of Cytochrome P-450-dependent Monooxygenases and Mutagen Activation in the Livers of DBA/2 and C57BL/6 Mice, Cancer Res. 46:524 (1986).
9. J. R. Idle and R. L. Smith, Polymorphisms of oxidation at carbon centers of drugs and their clinical significance, Drug Metabol. Rev. 9:301 (1979).
10. R. P. Peng, P. Tennant, N. A. Lorr and C. S. Yang, Alterations of microsomal monooxygenase system and carcinogen metabolism by streptozotocin-induced diabetes in rats, Carcinogenesis 4:703 (1983).
11. N. Battula, J. Sagara and H. V. Gelboin, Expression of P1-450 and P3-450 coding sequences as enzymatically active cytochromes P-450 in mammalian cells, Proc. Natl. Acad. Sci. 84:4073 (1987).
12. B. J. Song, H. V. Gelboin, S. S. Park, C. S. Yang and F. J. Gonzalez, Complementary DNA and protein sequences of ethanol-inducible rat and human cytochrome P-450s. Transcriptional and post-transcriptional regulation of the rat enzyme, J. Biol. Chem. 261:16689 (1986).
13. B. J. Song, T. Matsunaga, J. P. Hardwick, S. S. Park, R. L. Veech, C. S. Yang, H. V. Gelboin and F. J. Gonzalez, Stabilization of cytochrome P450J mRNA in teh diabetic rat, Molec. Endocrinol. in press (1987).
14. O. W. McBride, M. V. Umeno, H. V. Gelboin and F. J. Gonzalez, A TaqI polymorphysm in the human P450 IIE1 gene on chromosome 10, Nucleic Ac. Res. in press (1987).

CYTOCHROMES P-450 AS DETERMINANTS OF SUSCEPTIBILITY TO CARCINOGENESIS BY AROMATIC AMINES AND NITROAROMATIC HYDROCARBONS

Fred F. Kadlubar, F. Peter Guengerich[1], Mary Ann Butler and K. Barry Delclos

National Center for Toxicological Research, Jefferson AR 72079 USA and [1]Vanderbilt University, Nashville TN 37232, USA

INTRODUCTION

Aromatic amines and nitroaromatic hydrocarbons represent two major classes of occupational and environmental carcinogens. Aromatic amines are known to be present in the workplace, in cigarette smoke, synthetic fuels, and as pyrolysis products formed during the cooking of foods[1-5]; while nitroaromatic hydrocarbons have been used as dye intermediates, anti-microbial drugs, and food additives or have been detected as pollutants in coal fly ash, urban air samples, and emissions from auto, diesel, and airplane engines[6]. Epidemiological studies in humans and carcinogenicity experiments in laboratory animals have clearly indicated that these chemical carcinogens can induce cancers in several tissues, especially the urinary bladder, colon, liver, and lung[7]. For both aromatic amines and nitro-aromatics, metabolic activation is a necessary prerequisite to the formation of carcinogen-DNA adducts; and cytochromes P-450 have been shown to play a major role in this process (reviewed in ref. 8). In this paper, the catalytic activity of specific cytochrome P-450 isozymes for the activation of these carcinogens will be discussed in relation to species and individual differences in cancer susceptibility.

AROMATIC AMINES

For aromatic amines, hepatic N-oxidation is generally regarded as a critical metabolic activation pathway. Cytochromes P-450 have been shown to be primarily responsible for the N-oxidation of primary arylamines and arylamides, while the flavin-containing monooxygenase is known to catalyze the N-oxidation of secondary N-alkylarylamines[8-9]. The N-hydroxy arylamine and N-hydroxy arylamide metabolites are then subject to a variety of con-jugation reactions which can result in their transport to extra-hepatic tissues and/or their conversion to electrophilic derivatives that form covalent aromatic amine-DNA adducts in the carcinogen-target tissue (review-ed in ref. 10). These electrophilic reactants may include N-acetoxy or N-sulfonyloxy esters, O-glucuronides, or O-aminoacyl derivatives[11]. N-Hydroxy arylamines, which undergo protonation under acidic to neutral conditions, are also electrophilic per se and can react directly with DNA, in a concentration-dependent manner[11].

The catalytic involvement of cytochromes P-450 in the metabolic activation of carcinogenic aromatic amines was first established in 1973[12] for the arylamide, 2-acetylaminofluorene (2-AAF) and in 1976 for the arylamine, 4-aminoazobenzene[13]. Since that time, the availability of purified P-450 isozymes has allowed a detailed examination of aromatic amine substrate specificity and has provided an insight into the basis of species and individual differences in aromatic amine susceptibility as well as the effects of enzyme inducers on experimental carcinogenesis. For example, 2-AAF N-oxidation in the rat is catalyzed primarily by the major aromatic hydrocarbon-inducible hepatic cytochromes P-450, P-450$_{BNF-B}$ (see ref. 14 for P-450 nomenclature) and P-450$_{ISF-G}$, while detoxification by ring-oxidation is catalyzed preferentially by P-450$_{BNF-B}$ and several other isozymes. Upon pretreatment of rats with hydrocarbon inducers, there is a selective increase in the P-450's catalyzing 2-AAF ring-oxidation that results in the inhibition of 2-AAF carcinogenesis. In contrast, chronic 2-AAF treatment which induces tumor formation results in induction of P-450$_{ISF-G}$, P-450$_{BNF-B}$, P-450$_{PB-B}$, and P-450$_{PB-D}$, but in such a manner that the ratio of N-/ring-oxidation is unaltered while overall metabolism increases 4-fold. Similar findings have been also reported for hamsters. In rabbits and humans, however, there appears to be a single inducible P-450, corresponding to rat P-450$_{ISF-G}$, that is responsible for 2-AAF N-oxidation (reviewed in ref. 8).

Studies with purified rat hepatic cytochrome P-450 isozymes have also shown that 4-aminobiphenyl (4-ABP), 2-naphthylamine (2-NA) and the heterocyclic amines 3-amino-1-methyl-5H-pyrido[4,3-b]indole (Trp-2), 2-amino-6-methyl-dipyrido[1,2-a:3',2'-d]imidazole (Glu-P-1), and 2-amino-3-methylimidazo[4,5-f]quinoline (IQ) are selectively N-oxidized by major isosafrole-inducible isozyme P-450$_{ISF-G}$ (Table 1). In contrast, 4,4'-methylene-bis(2-chloroaniline) (MOCA), is oxidized preferentially by the phenobarbital-inducible enzymes P-450$_{PB-B}$ and P-450$_{PB-D}$, and 6-aminochrysene is activated to DNA-reactive metabolites by P-450$_{PCN-E}$ and P-450$_{UT-A}$. Hepatic ring-oxidation, an important detoxification pathway for these arylamines, also varied greatly between the substrates with ring-oxidation representing >60% of total metabolism for 2-NA, <2% for ABP, and 14-80% with MOCA with different isozymes[15-17].

Human hepatic microsomal preparations from different individuals have

Table 1. Rates of N-oxidation of aromatic amines by purified rat hepatic cytochromes P-450

	Rates (nmol/min/nmol P450)					
	2-NA[a]	4-ABP[b]	MOCA[b]	Trp-P-2[c]	Glu-P-1[c]	IQ[c]
P450 ISF-G	2.52	13.56	2.74	2.09	3.208	0.628
P450 BNF-B	<0.05	4.40	1.83	0.70	0.102	0.048
P450 PB-B	<0.05	2.05	8.98	0.05	0.004	0.012
P450 PB-D	<0.05	0.95	6.62			

[a]Taken from ref. 15; [b]taken from refs. 16 and 17; [c]taken from ref. 18.

been shown to vary considerably in their oxidative ability to activate (N-oxidation) and detoxify (ring-oxidation) several aromatic amines of significant potential human exposure (Table 2). In one set of microsomal preparations from ten individuals, rates of N-oxidation of 2-NA varied at least 30-fold, when expressed per nmol P-450, while total P-450 levels varied only 3-fold. Hepatic microsomal preparations from 19 additional individuals showed a 43-fold variation in rates of N-oxidation of 4-ABP and an 8-fold variation in MOCA N-oxidation, while total cytochrome P-450 content varied only 5-fold. In regard to metabolic detoxification, the ring-oxidation of 2-NA comprised from 42 to 100% of total metabolism; whereas less than 6% of total metabolism resulted in ring-oxidation of 4-ABP, which is considered a more potent carcinogen[1,7]. Ring-oxidation of MOCA varied from 8-19% of total metabolism in the different individuals[17]. Such variability in activation and detoxification proficiency in different individuals is considered to be due to expression of specific P-450 monooxygenases and their control by genetic polymorphisms and/or by environmental factors such as diet, lifestyle, or drug exposure. Accordingly, these predispositions may be significant determinants in individual susceptibility to carcinogenesis by aromatic amines.

Because of immunochemical reactivity and amino-terminal sequence similarity between rat hepatic cytochrome P-450$_{ISF-G}$ and human hepatic cytochrome P-450$_{PA}$, which catalyzes the O-deethylation of phenacetin[20-21], a comparison was made between N-oxidation of 4-ABP and O-deethylation of phenacetin in the 19 hepatic microsomal preparations. There was an excellent correlation ($r = 0.91$, $P < 0.001$) between these activities, suggesting that a single cytochrome P-450 is involved in both catalyses. Since a genetic polymorphism for phenacetin O-deethylation exists in the human population[22], interindividual variability in metabolism of 4-ABP and perhaps other carcinogenic arylamines may also be due to an absent, defective, or low level of this particular cytochrome P-450, in addition to different levels of cytochromes P-450 controlled by environmental factors.

NITROAROMATIC HYDROCARBONS

For nitroaromatic hydrocarbons, metabolic activation can involve nitroreduction, aromatic ring-oxidation, or a combination of these pathways[6,23]. Nitroreduction results in the stepwise formation of nitrosoarenes, N-hydroxyarylamines, and arylamines, each of which are subject to metabolic oxidation, reduction, and/or conjugation reactions that can lead to carcinogen-DNA adduct formation (vide supra). Although cytochromes P-450 can serve as nitroreductases under hypoxic conditions, at least nine other

Table 2. Human hepatic cytochrome P-450 levels and rates of aromatic amine N-oxidation

Number of individuals	Cytochrome P-450 (nmol/mg protein)	N-Oxidation (nmol product/min/nmol P-450)		
		2-NA[a]	ABP[b]	MOCA[c]
10	0.36-0.92	<0.03-0.86		
19	0.14-0.74		0.10-4.37	1.23-10.22

[a]Taken from ref. 15; [b]taken from refs. 16 and 17; [c]taken from ref. 17.

19

enzymes, primarily flavo-proteins, have been shown to mediate the reduction of nitro-, nitroso-, and N-hydroxy-derivatives (reviewed in ref. 8). Work with purified P-450's has been limited to the major phenobarbital- and hydrocarbon-inducible rat liver isozymes; and each P-450 was found to catalyze the reduction of 1-nitropyrene[24]. Aromatic ring-oxidation has been studied primarily with the polycyclic nitroaromatic hydrocarbons (nitro-PAHs); and except for 6-nitrochrysene (vide infra), experiments with purified cytochromes P-450 have not yet been undertaken. Nevertheless, since the stereoselectivity of oxidative metabolism of several nitro-PAHs by microsomal monoxygenases is very similar to that of their parent hydrocarbons, it seems likely that the major hydro- carbon-metabolizing enzyme, cytochrome P-450$_{BNF-B}$, is the predominant isozyme involved in the ring-oxidation of nitro-PAHs[23]. Interestingly, some of the ring-oxidized metabolites of 5-nitroacenaphthene, 6-nitrochrysene, 1-nitro- pyrene and 3- and 6-nitrobenzo(a)pyrene show high mutagenic activity. Thus, their metabolic activation is thought to arise from the combination of an initial ring-oxidation and a subsequent nitroreduction to form a reactive phenolic N-hydroxy arylamine derivative.

Recent bioassays have shown that 6-nitrochrysene (6-NC) is perhaps the strongest carcinogen of this class, inducing multiple malignant neoplasms in the lung, liver and lymphatic system of mice treated neonatally[25,26]. Our studies on the metabolic activation 6-NC to electrophilic species that form DNA adducts have indicated that there are at least two distinct metabolic activation pathways which are operative in vivo. Furthermore, the pathway that predominates in a given tissue appears to be determined by constitutive or induced levels of specific cytochrome P-450's. The effect of P-450 induction can best be seen in studies that we have carried out with hepa-tocytes isolated from induced or uninduced adult mice and rats. After 6 h of exposure to [^{3}H]6-NC, DNA was isolated from hepatocyte cultures, enzy-matically hydrolyzed, and the carcinogen-nucleoside adducts analyzed by HPLC. In hepatocytes isolated from mice or rats which either were untreated or pretreated with phenobarbital, two major adducts were detected in the DNA. These adducts cochromatographed with N-(deoxyguanosin-8-yl)-6-aminochrysene and N-(deoxyinosin-8-yl)-6-aminochrysene, products which are formed in vitro from the reaction of N-hydroxy-6-aminochrysene with calf thymus DNA[27].

In contrast, DNA from [^{3}H]6-NC-treated mouse or rat hepatocytes isolated from animals which had been pretreated with Aroclor contained a single major adduct which is not derived from N-hydroxy-6-aminochrysene. This adduct can be isolated from in vitro incubations containing 6-aminochrysene-1,2-dihydro-diol, calf thymus DNA, liver microsomes from 3-methylcholanthrene-pretreated rats and an NADPH-generating system and, to a lesser extent, in identical incubations that contain 6-aminochrysene (6-AC) as substrate. Incubations of 6-AC with microsomes from phenobarbital-pretreated rats in the presence of DNA leads to formation of adducts derived from N-hydroxy-6-AC. Preliminary spectral characterization of the adduct obtained from hydrocarbon-induced microsomal incubations containing 6-AC-1,2-dihydrodiol and DNA have thus far indicated that this adduct contains an aminophenanthrene nucleus. Additional experiments with purified P-450 isozymes have shown that the formation of the DNA-binding metabolite from 6-AC-1,2-dihydrodiol is catalyzed by the major hydrocarbon-inducible, P-450$_{BNF-B}$, but not by the isosafrole-inducible form P-450$_{ISF-G}$, nor the phenobarbital-induced form, P-450$_{PB-B}$. Since P-450$_{BNF-B}$ has been shown to be the major P-450 isozyme responsible for the conversion of polycyclic aromatic hydrocarbons to reactive diol-epoxides, these results indicate that 6-aminochrysene-1,2-diol-3,4-epoxide is likely to be the ultimate reactive metabolite formed from 6-NC and 6-AC in these experimental systems.

The results described above suggest that carcinogen-DNA adducts derived from at least two independent activation pathways might be expected to be

found in tissues of animals or humans exposed to 6-NC. In addition, it
might be expected that other activation pathways, such as nitroreduction of
the two metabolically formed dihydrodiols of 6-NC, 6-NC-1,2-dihydrodiol and
6-NC-9,10-dihydrodiol, to the corresponding N-hydroxy derivatives would also
lead to DNA adducts in vivo. The adduct pattern seen in hepatocytes from
Aroclor-pretreated rats and mice suggests that in situations where P-450$_{BNF-B}$
is present along with other induced P-450 isozymes, the predominant pathway
(Fig. 1) involves the metabolism of NC to NC-1,2-dihydrodiol, followed by
metabolic reduction to AC-1,2-dihydrodiol, and a subsequent P-450$_{BNF-B}^{-}$
mediated activation of the AC-1,2-dihydrodiol to form a reactive arylamine
diol-epoxide. Such a situation apparently exists in the target tissues of
preweanling mice treated with [^{3}H]6-NC where we find a single major adduct
(up to 90% of the total carcinogen-DNA adducts) that is chromatographically
and chemically identical to the adduct derived from the further microsomal
metabolism of 6-AC-1,2-dihydrodiol[28].

The major metabolic pathway for 6-NC in humans might also be dependent
on the specific constitutive or induced forms of cytochromes P-450 present.
In the DNA isolated from two explants of human bronchial epithelium treated
with [H]NC, two distinct adduct patterns were observed. In one case, adducts derived from N-hydroxy-6-AC were present; whereas in the second sample,
the only detectable adduct was that derived from 6-AC-1,2-dihydrodiol.
Further experiments with human bronchial epithelial explants are in progress,
and an attempt to determine whether the observed adduct pattern can be related to the smoking history of the donors will be made.

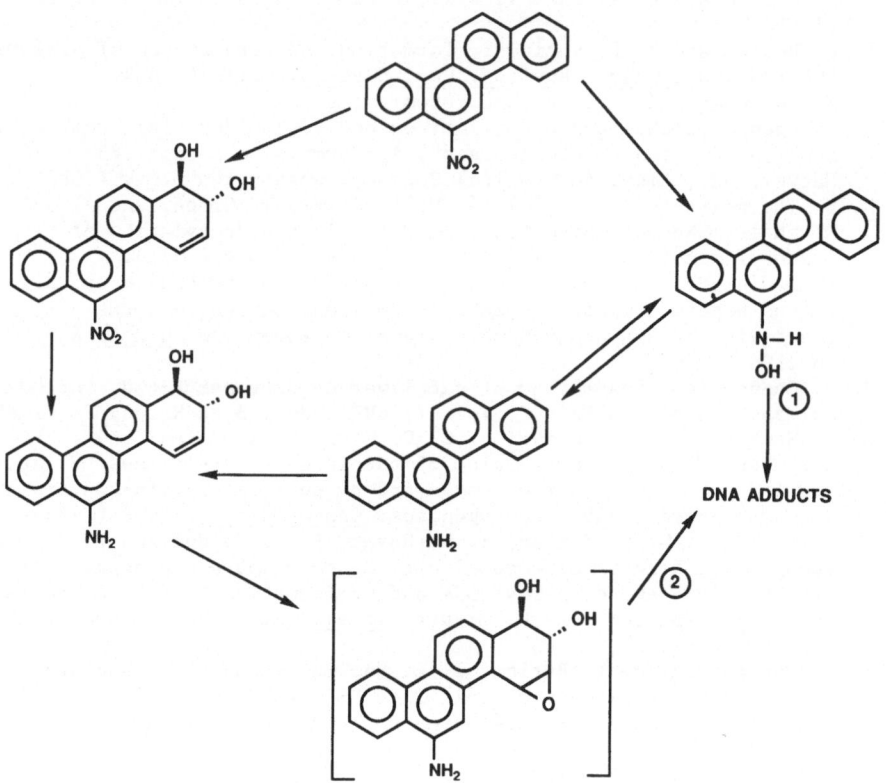

Fig. 1. Metabolic Activation Pathways for 6-NC and 6-AC.

REFERENCES

1. H. G. Parkes and A. E. J. Evans, Epidemiology of aromatic amine cancers, in: "Chemical Carcinogens", C. E. Searle, ed., American Chemical Society, 2nd edition, Washington D.C., (1984).
2. C. Patrianakos and D. Hoffmann, Chemical studies on tobacco smoke, LXIV, On the analysis of aromatic amines in cigarette smoke, J. Anal. Toxicol. 3:150 (1979).
3. D. Haugen, M. J. Peak, K. M. Suhrbler and V. C. Stamoudis, Isolation of mutagenic aromatic amines from coal conversion oil by cation exchange chromatography, Anal. Chem. 54:32 (1982).
4. M. S. Bryant, P. S. Skipper, S. R. Tannenbaum and M. Mclure, Hemoglobin adducts of 4-aminobiphenyl in smokers and non-smokers, Cancer Res. 47:602 (1987).
5. R. Kato, Metabolic activation of heterocyclic aromatic amines from protein pyrolysates, CRC Crit. Rev. Toxicol. 16:307 (1986).
6. H. S. Rosenkranz and R. Mermelstein, The genotoxicity, metabolism and carcinogenicity of nitrated polycyclic aromatic hydrocarbons, J. Environ. Sci. Health 2:221 (1986).
7. R. C. Garner, C. N. Martin and D. B. Clayson, Carcinogenic aromatic amines and related compounds, in: "Chemical Carcinogens", C. E. Searle, ed., American Chemical Society, 2nd edition, Washington D.C. (1984).
8. F. F. Kadlubar and G. J. Hammons, The role of cytochrome P-450 in the metabolism of chemical carcinogens, in: "Mammalian Cytochromes P-450" Vol. II, CRC Press, Boca Raton, FL, (1987).
9. D. M. Zeigler, Molecular basis for N-oxygenation of sec- and tert-amines, in: "Biological Oxidation of Nitrogen in Organic Molecules", J. Gorrod and L. A. Damani, eds., Ellis Horwood Ltd., Chichester, (1985).
10. F. A. Beland and F. F. Kadlubar, Formation and persistence of arylamine DNA adducts in vivo, Environ. Health Persp. 62:19 (1985).
11. F. F. Kadlubar and F. A. Beland, Chemical properties of ultimate carcinogenic metabolites of arylamines and arylamides, in: "Polycyclic Hydrocarbons and Carcinogenesis", ACS Symposium Series 283, R. G. Harvey, ed., American Chemical Society, Washington D.C., (1985).
12. S. S. Thorgeirsson, D. J. Jollow, H. A. Sasame I. Green, and J. R. Mitchell, The role of cytochrome P-450 in N-hydroxylation of 2-acetylaminofluorene, Mol. Pharmacol. 9:398-404 (1973).
13. F. F. Kadlubar, J. A. Miller and E. C. Miller, Microsomal N-oxidation of the hepatocarcinogen N-methyl-4-aminoazobenzene and the reactivity of N-hydroxy-N-methyl-4-aminoazobenzene, Cancer Res. 36:1196 (1976).
14. F. P. Guengerich, Enzymology of rat liver cytochromes P-450, in: "Mammalian Cytochromes P-450", Vol. I, CRC Press, Boca Raton, FL (1985).
15. G. J. Hammons, F. P. Guengerich, C. C. Weis, F. A. Beland and F. F. Kadlubar, Metabolic activation of carcinogenic arylamines by rat, dog and human hepatic microsomes and by purified flavin-containing and cytochrome P-450 monooxygenases, Cancer Res. 45:3578 (1985).
16. F. F. Kadlubar, M. A. Butler, B. W. Hayes, F. A. Beland and F. P. Guengerich, Role of microsomal cytochrome P-450 and prostaglandin H synthase in 4-aminobiphenyl-DNA adduct formation, in: "Microsomes and Drug Oxidation", D. W. Birkett et al., eds., Taylor and Francis, London, in press.
17. F. P. Guengerich, M. A. Butler, T. L. MacDonald and F. F. Kadlubar, Oxidation of carcinogenic arylamines by cytochrome P-450, in: "Carcinogenic and Mutagenic Responses to Aromatic Amines and Nitroarenes", C. M. King, L. J. Romano, and D. Schuetzle eds., Elsevier, New York (1987).
18. R. Kato and Y. Yamazoe, Metabolic activation and covalent binding to nucleic acids of carcinogenic heterocyclic amines from cooked foods

and amino acid pyrolysates, Jpn. J. Cancer Res. (Gann) 78:297 (1987).

19. M. A. Butler, K. B. Delclos, F. P. Guengerich and F. F. Kadlubar, unpublished data (1987).

20. F. P. Guengerich, D. R. Umbenhauer, P. F. Churchill, P. H. Beaune, R. Bocker, R. G. Knodell, M. V. Martin and R. S. Lloyd, Polymorphism of human cytochrome P-450, Xenobiotica 17:311 (1987).

21. S. A. Wrighton, C. Campanile, P. E. Thomas, S. L. Maines, P. B. Watkins, G. Parker, G. Mendez-Picon, M. Haniu, J. E. Shively, W. Levin and P. S. Guzelian, Identification of a human liver cytochrome P-450 homologous to the major isosafrole-inducible cytochrome P-450 in the rat, Mol. Pharm. 29:405 (1986).

22. H. Devonshire, I. Kong, M. Cooper, T. Sloan, J. Idle and R. Smith, Contribution of genetically-determined oxidation status to inter-individual variation in phenacetin disposition, Brit. J. Clin. Pharm. 16:157 (1983).

23. F. A. Beland, R. H. Heflich, P. C. Howard and P. P. Fu, The in vitro metabolic activation of nitro polycyclic aromatic hydrocarbons, in: "Polycyclic Hydrocarbons and Carcinogenesis", R. G. Harvey, ed., ACS Symposium Series Number 283, American Chemical Society, Washington D.C., (1985).

24. K. Saito, T. Kamataki and R. Kato, Participation of cytochrome P-450 in reductive metabolism of 1-nitropyrene by rat liver microsomes, Cancer Res. 44:3169 (1984).

25. P. G. Wislocki, E. S. Bagan, A. Y. H. Lu, K. L. Dooley, P. P. Fu, H. Han-Hsu, F. A. Beland and F. F. Kadlubar, Tumorigenicity of nitrated derivatives of pyrene, benzo[a]anthracene, chrysene and benzo[a]pyrene in the newborn mouse assay, Carcinogenesis 7:1317 (1986).

26. W. F. Busby, Jr., R. C. Garner, F.-L. Chow, C. N. Martin, E. K. Stevens, P. M. Newberne and G. N. Wogan, 6-Nitrochrysene is a potent tumorigen in newborn mice, Carcinogenesis 6:801 (1985).

27. K. B. Delclos, D. W. Miller, J. O. Lay, Jr., D. A. Casciano, R. P. Walker, P. P. Fu and F. F. Kadlubar, Identification of C8-modified deoxyinosine and N2- and C8-modified deoxyguanosine as major products of the in vitro reaction of N-hydroxy-6-aminochrysene with DNA and the formation of these adducts in isolated rat hepatocytes, Carcinogenesis 8:1703 (1987).

28. K. B. Delclos, R. P. Walker, K. L. Dooley, P. P. Fu and F. F. Kadlubar, Carcinogen-DNA adduct formation in the lungs and livers of preweanling CD-1 male mice following administration of [^{3}H]6-nitro-chrysene, [^{3}H]6-aminochrysene and [^{3}H]1,6-dinitropyrene, Cancer Res. (1987), in press.

SPECIES-DEPENDENT DIFFERENCES IN THE METABOLIC ACTIVATION

OF POLYCYCLIC AROMATIC HYDROCARBONS IN CELLS IN CULTURE

William M. Baird, Teresa A. Smolarek[1], Said M. Sebti[2]
and Donna Pruess-Schwartz[3]

Department of Medicinal Chemistry and Pharmacognosy
School of Pharmacy and Pharmacal Sciences
Purdue University
West Lafayette, IN 47907 U.S.A.

Polycyclic aromatic hydrocarbons are widespread environmental contaminants that require metabolic activation in order to induce biological effects[1]. One of the most widely studied carcinogenic hydrocarbons is benzo-(a)pyrene (BaP). Most pathways of BaP metabolism result in the production of metabolites that are detoxification products. However, a small proportion of the BaP metabolites are reactive derivatives that bind to DNA in cells and these DNA interactions are involved in the initiation of the cancer induction process[1]. Although these reactive metabolites cannot be isolated from cells so that their production can be quantitated, it is possible to measure their formation through detection of the DNA adducts they produce. The DNA serves as both a critical target for the "ultimate carcinogenic metabolites" and as a nucleophilic trapping agent for detection and measurement of these reactive electrophiles.

The major pathway of activation of BaP to DNA-binding metabolites in many tissues and cells in culture involves formation of a diol-epoxide on the "bay region"[2] of the molecule[3], reviewed in 1. The bay-region diol-epoxide of BaP exists as a pair of diastereomers, anti-BaP-7,8-diol-9,10-epoxide (anti-BaPDE: benzylic hydroxyl and epoxide on opposite faces of the plane of the molecule) and syn-BaPDE (benzylic hydroxyl and epoxide on the same face of the plane of the molecule). Each diastereomer consists of two enantiomers (plus and minus). The tumor initiating activity of the four optical isomers of BaPDE differs: (+)-anti-BaPDE is much more tumorigenic in the newborn mice and mouse skin assays than the other three optical isomers[4,5]. The structures of these four optical isomers and their relative carcinogenic potency based upon the review of Dipple[6] are given in Fig. 1. The major differences in carcinogenic activity of these optical isomers indicated that studies of metabolic activation of BaP to reactive derivatives in vivo required the development of techniques for analyzing the BaP-DNA interaction products

---

[1] Present address: Pfizer Chemical Company, Drug & Safety Evaluation, Groton, CT 06340. [2] Present address: University of Pittsburgh, School of Medicine, Department of Pharmacology, Pittsburgh, PA 15261. [3] Present address: Department of Chemistry, Wayne State University, Detroit, MI 48201.

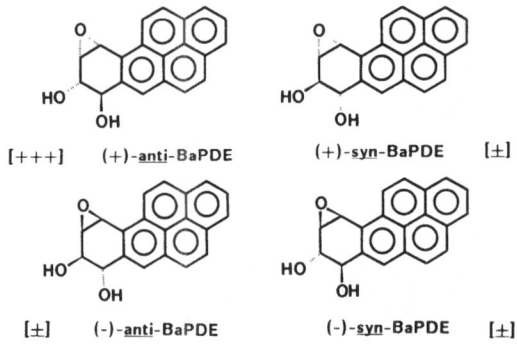

[+++]   (+)-anti-BaPDE   (+)-syn-BaPDE   [±]

[±]   (-)-anti-BaPDE   (-)-syn-BaPDE   [±]

Relative Carcinogenic potency [   ]

Fig. 1. Structures of the four optical isomers of BaPDE. The relative
carcinogenic activities are based upon the data summarized by
Dipple[6].

which would allow the optical isomer of BaPDE involved in their formation to
be identified. Our laboratory has developed techniques for the analysis of
BaP-DNA adducts formed in cells and applied these to studies of the DNA
adducts formed from BaP in cell cultures derived from different species.

The low extent of binding of BaP to DNA in cells in culture and in animals, typically less than one BaP per 10^5 bases, necessitates the use of tri-
tium-labeled high specific radioactivity BaP to measure these DNA interactions. It also prevented use of standard analytical techniques for identification of these adducts and required the development of chromatographic techniques capable of resolving these [3H]BaP-DNA adducts and allowing their
characterization by comparison. of their chromatographic behavior with chemically synthesized marker compounds. The development of these techniques and
the contributions made by a number of laboratories has recently been reviewed
by Baird and Pruess-Schwartz[7]. This chapter will describe the recent results
from our laboratory on the mechanism of formation of BaP-DNA adducts in cells
from different species.

Cell cultures were prepared from Wistar or Fisher 344 rat embryos,
Sencar or Balb/c mouse embryos, or Syrian hamster embryos as described previously[8]. Third passage cultures were exposed to [3H]BaP (0.5 µgm/ml medium)
for 5 to 48 h, the cells were harvested and the nuclei prepared[8]. After
isolation of the DNA, analyses of BaP-DNA adducts were carried out by enzymatic degradation of the DNA and separation of the normal deoxyribonucleosides from those containing bound BaP by use of a Sep-Pak column[9,10]. The
BaP-deoxyribonucleoside adducts were then analyzed by high-performance liquid
chromatography (HPLC) on a reverse-phase column eluted with a methanol-water
gradient[11]. Fig. 2 (top) shows the profile of adducts obtained after HPLC
of a DNA sample obtained from Wistar rat embryo cell cultures. The peaks
were not well resolved and the peak that chromatographed with the marker of
[14C]-(+)-anti-BaPDE-deoxyguanosine (dG) contained two components. A procedure was developed for separation of the syn- and anti-BaPDE-deoxyribonu-
cleoside adducts by passing the BaP-deoxyribonucleoside adducts through a
column of immobilized boronate[11]. The syn-BaPDE-deoxyribonucleoside adducts
pass through the column in the morpholine buffer and the anti-BaPDE-deoxyri-
bonucleoside adducts are retained and subsequently eluted with a morpholine
buffer containing 10% sorbitol. The HPLC profiles of the BaP-deoxyribonucleoside adducts in the morpholine and the morpholine-sorbitol fractions are
shown in the middle and bottom segments of Fig. 2. The adducts that eluted
in the morpholine buffer were separated into three distinct peaks as were the

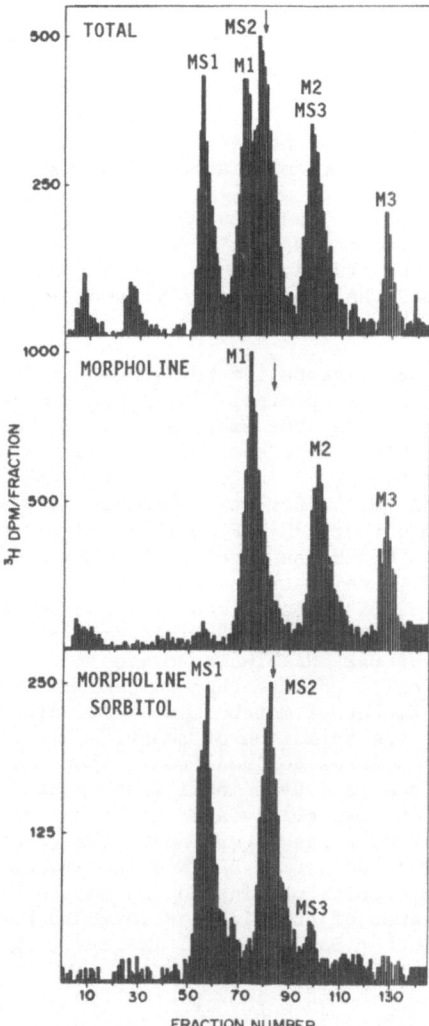

Fig. 2. HPLC analysis of [³H]BaP-deoxyribonucleoside adducts from the
DNA of third passage Wistar rat embryo cells exposed to 0.5 μgm
[³H]BaP per ml medium for 24 h. The samples are the total
adducts (top), the adducts eluted from an immobilized boronate
column in morpholine buffer (middle) and morpholine-10% sorbitol
buffer (bottom). The identities of the peaks are given in Table
1. The methods used are described in ref. 11. ↓ indicates the
elution position of a [¹⁴C]-(+)-anti-BaPDE-dG marker.

adducts eluting in the morpholine-sorbitol buffer. The peaks which previously overlapped were completely resolved permitting quantitation of these
adducts. Peak MS2 coelutes with the adduct formed by reaction of (+)-anti-
BaPDE with the 2-amino group of deoxyguanosine[12].

In order to provide further characterization of these adducts rather
than simply cochromatography in a single solvent system, acid hydrolysis
techniques were developed[10]. By treatment of the (+)-anti- and (-)-anti-
BaPDE-deoxyguanosine adducts with weak acid at 37°C it was possible to remove
the deoxyribose. This results in conversion of these two diastereomers into

27

enantiomers which cochromatograph. Fig. 3 demonstrates the application of this technique for the analysis of the morpholine-sorbitol buffer eluted peaks obtained from[^{3}H]BaP-treated cells. Peak MS2 cochromatographed with a [^{14}C]-(+)-anti-BaPDE-dG marker. As can be seen in the lower portion of Fig. 3 after acid hydrolysis the [^{14}C] marker peak and the [^{3}H] from the cells continued to coelute. In contrast peak MS1 had eluted with a (-)-anti-BaPDE-dG adduct marker. However, mild acid hydrolysis of this peak from cells demonstrated that the [^{14}C] and [^{3}H] did not coelute: thus adduct MS1 in cells was not formed from (-)-anti-BaPDE [10]. Similar conclusions were reached using a different acid hydrolysis procedure which resulted in removal of both the deoxyribose and the guanine to give a BaPDE-tetraol[10]. Based upon these studies the BaP-DNA adduct peaks present in rat embryo cells were identified as shown in Table 1.

The relationship of BaP metabolism to BaP-DNA binding was examined in cell cultures from a number of species. Third passage cell cultures were prepared from mouse, rat and hamster embryos as described above. In addition, cell lines derived from three species of fish, rainbow trout embryonic gonad (RTG-2), brown bullhead (BB) and bluegill fry (BF-2) were tested[13]. All cultures were exposed to [^{3}H]BaP at a concentration of 0.5 µgm/ml medium for 24 h and then the amount of [^{3}H]BaP metabolized was determined by organic solvent extraction and HPLC and the level of BaP-DNA binding was determined. The results shown in Fig. 4 demonstrated that there was no simple relationship between the amount of BaP metabolized by a culture and the amount of BaP bound to DNA. All of the cultures metabolized from 50 to 90% of the benzo-(a)pyrene but the level of BaP-DNA binding ranged from less than 5 pmol/mg DNA in the bluegill fry cells to more than 80 pmol/mg DNA in the Sencar mouse embryo cells. Thus the amount of metabolism of BaP alone is unlikely to be an adequate predictor of the formation of reactive ultimate carcinogenic metabolites in the cell cultures derived from a species. The BaP-DNA adducts from these samples were analyzed by immobilized boronate chromatography and high-performance liquid chromatography and the amount of each individual BaP-deoxyribonucleoside adduct was determined. The relationship between the amount of (+)-anti-BaPDE-dG adduct/mg DNA and the amount of BaP metabolized is shown in Fig. 5. The results obtained on formation of this particular adduct are presented because of the high carcinogenic potential of (+)-anti-BaPDE compared with the other BaPDE optical isomers. Again, there was great variation between cultures from different species in the proportion of the BaP metabolized to (+)-anti-BaPDE. This proportion was higher in mouse embryo cells, especially those from Sencar mice, than in the other species examined. In general, however, it represented a very small proportion of the total BaP metabolized.

Comparison of the proportion of BaP bound to DNA through (+)-anti-BaPDE in the above cultures demonstrated that a different relationship exists (Fig. 6). In several types of cell cultures including those derived from the brown bullhead, the Sencar mouse embryos and both normal human mammary epithelial cells and the T47D human mammary epithelial cell line[14], the (+)-anti-BaPDE-dG adduct accounted for more than 70% of the total BaP-DNA adducts formed. In cells from other species it accounted for a much smaller percentage. In the bluegill fry, rainbow trout and Syrian hamster embryo cells this adduct accounted for approximately 30 to 40% of the BaP-DNA adducts and in the Wistar rat embryo cells it accounted for less than 10% (Fig. 6).

These complex relationships between BaP metabolism and the formation of the (+)-anti-BaP diol-epoxide indicate the importance of various metabolic pathways in determining the metabolic fate of the BaP applied to cells. For example, comparison of the Syrian hamster embryo cell cultures with the bluegill fry cell line demonstrates that the proportion of (+)-anti-BaPDE adducts formed was around 40% in both (Fig. 6). However, although the hamster embryo

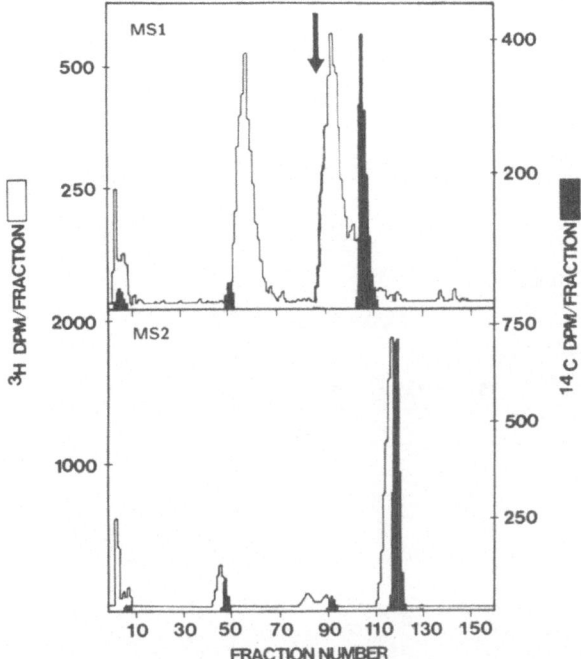

Fig. 3. Mild acid hydrolysis of peaks MS1 and MS2 from Wistar rat embryo
cell DNA isolated as described in Fig. 2. The [³H] is from
adducts from [³H]BaP-treated cells, the [¹⁴C] from [¹⁴C]-(+)-
anti-BaPDE-dG. The procedures are described in ref. 10.

cells metabolized less than twice as much as BaP as the bluegill fry cells,
they bound nine times as much as BaP to DNA as did the bluegill fry cells,
(Fig. 4). This difference can be explained by examination of the major pathways of metabolism of BaP in these cell cultures. Whereas formation of a
glucuronic acid conjugate of the 7,8-diol of BaP (a proximate carcinogenic
metabolite of BaP) was not detected in hamster embryo cell cultures, formation of the 7,8-diol glucuronide was a major pathway of BaP metabolism in the
bluegill fry cell cultures[15]. In the bluegill fry cells even though a high
proportion of the BaP is converted to this proximate carcinogenic metabolite,
BaP-7,8-diol, the majority of this diol is conjugated with glucuronic acid
rather than converted to the diol-epoxide.

The proportion of BaP bound to DNA through (+)-anti-BaPDE also changes
with time in cells in culture. Baird and Diamond[16] observed that the propor-

Table 1. Identification of BaP-DNA Adducts Formed in Cell Cultures

MS1 = Unidentified BaP-metabolite-adduct; not (-)-anti-BaPDE-dG
M1 = (-)-Syn-BaPDE-dG
MS2 = (+)-Anti-BaPDE-dG
M2 = (+)-Syn-BaPDE-dG
MS3 = Syn-BaPDE-adduct breakdown product (7,9,10/8-tetraol)

The BaP-deoxyribonucleoside adduct profiles and the methods used
for their analysis are described in Fig. 2.

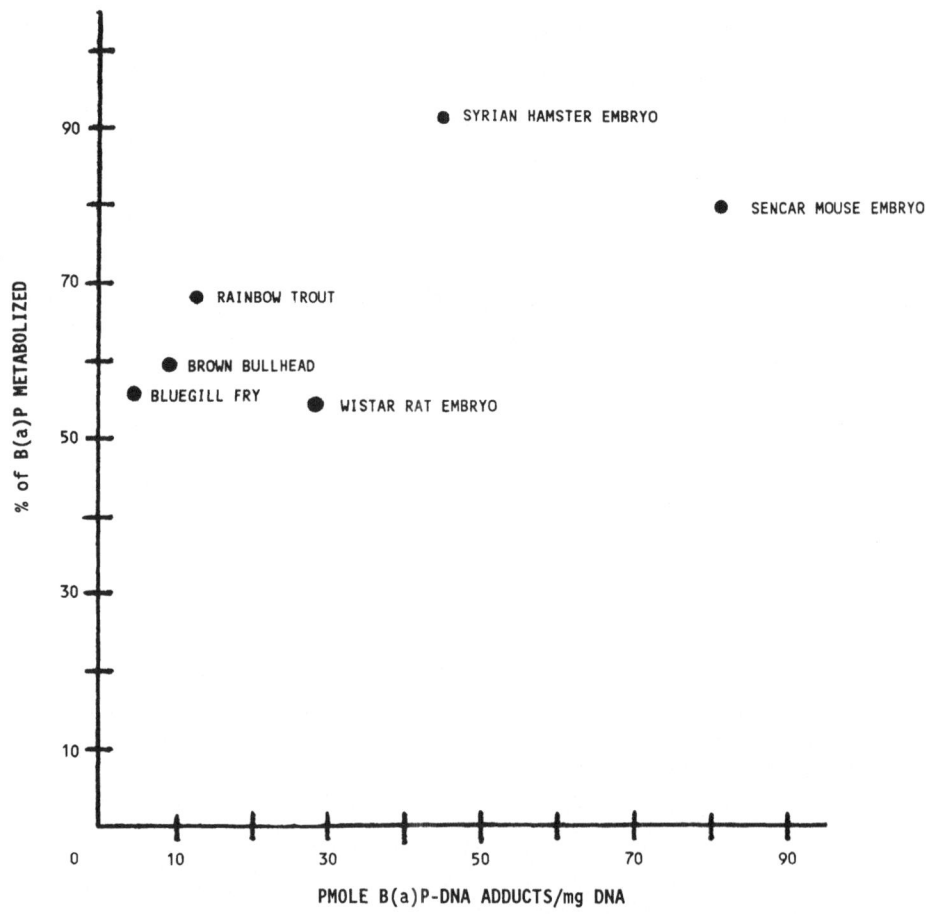

Fig. 4. The relationship between metabolism of [^{3}H]BaP and the amount
bound to DNA in cell cultures derived from various species.
All cultures were exposed to 0.5 μgm [^{3}H]BaP per ml medium for
24 h. The methods used to analyze metabolites and BaP-DNA
binding are given in ref. 19 and 9.

tion of anti-BaPDE adducts formed in BaP-treated hamster embryo cells increased with the length of time of exposure of the cultures to the BaP. However,
the LH20 chromatography technique used did not allow the identification of
individual adducts. The effects of time of exposure to BaP on the morpholine-sorbitol eluted BaP-DNA adducts formed in BaP-treated Wistar rat and
Syrian hamster embryo cell cultures are shown in Fig. 7. In the rat embryo
cell cultures (Fig. 7, left) adduct MS2, the (+)-anti-BaPDE-dG adduct, was
present in a tiny amount after 5 h of exposure. The major adducts were the
unidentified peak MS1 and the syn-BaPDE adduct breakdown product peak MS3.
After 24 h of exposure peak MS2 was clearly detectable and by 48 h of
exposure it had become one of the major adducts. In hamster embryo cell
cultures (Fig. 7, right) the change in the proportion of this adduct was much
greater. After 5 h of exposure the (+)-anti-BaPDE-dG adduct (MS2) was one of
three major peaks. After 24 h this adduct had become the major peak and
remained the major adduct at 48 h. Thus in both the rat and hamster embryo
cell cultures there was a time-dependent increase in the proportion of BaP
activated to a DNA binding intermediate through (+)-anti-BaPDE.

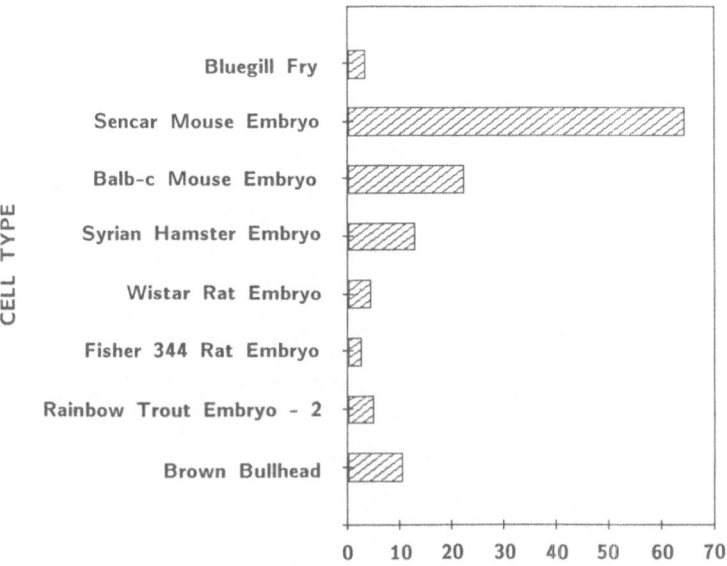

Fig. 5. The relationship between the amount of (+)-anti-BaPDE-dG
adduct in cellular DNA and the amount of [^{3}H]BaP metabolized in cell cultures from various species. All cultures
were exposed to 0.5 μgm [^{3}H]BaP per ml medium for 24 h.
The amounts of BaP metabolized and (+)-anti-BaPDE-dG were
measured as described in Fig. 4.

The above results suggested that BaP treatment itself may increase the
proportion of a certain enzyme(s), probably a cytochrome P_{450} isozyme(s),
which selectively forms (+)-anti-BaPDE. In studies of the metabolic activation of 7,12-dimethylbenz(a)anthracene (DMBA) to DNA-binding metabolites in
mouse embryo cells and rat mammary cells, Milner et al.[17], and Singletary and
Milner[18] demonstrated that DMBA treatment induced an enzyme which increased
the proportion of DMBA activated to the anti-diol-epoxide. This induction
could be inhibited by Actinomycin D, an inhibitor of RNA synthesis. Examination of the effect of BaP pretreatment on the binding of BaP to DNA in Syrian
hamster embryo cell cultures demonstrated that BaP pretreatment does increase
the binding of [^{3}H]BaP to DNA[19]. In these studies the cell cultures were exposed to either acetone or BaP for 24 h, the medium was removed and the cultures were then refed with fresh medium containing [^{3}H]BaP for 5 h. The
BaP-pretreated cultures metabolized 34 nmol [^{3}H]BaP per flask as compared
with 20 nmol in the acetone-pretreated cultures. The BaP-pretreated cultures
contained about 30% more BaP-DNA adducts than the acetone pretreated cultures
(Fig. 8). The level of binding of BaP to DNA reported for the BaP-pretreated
cultures is an underestimate since it is impossible to remove all of the unlabeled BaP by changing the medium. The remaining unlabeled BaP would dilute
the specific activity of the [^{3}H]BaP and result in an underestimate of the
level of BaP-DNA binding in the BaP-pretreated cultures. Analysis of the
BaP-DNA adducts by HPLC demonstrated that there was an almost two-fold increase in the amount of the (+)-anti-BaPDE-dG adduct in the BaP-pretreated
group (Fig. 8). Thus BaP-pretreatment resulted in an increase in BaP binding
and this was due mainly to an increase in the amount of (+)-anti-BaPDE-dG
adduct formed.

When hamster embryo cell cultures were pretreated with BaP and then

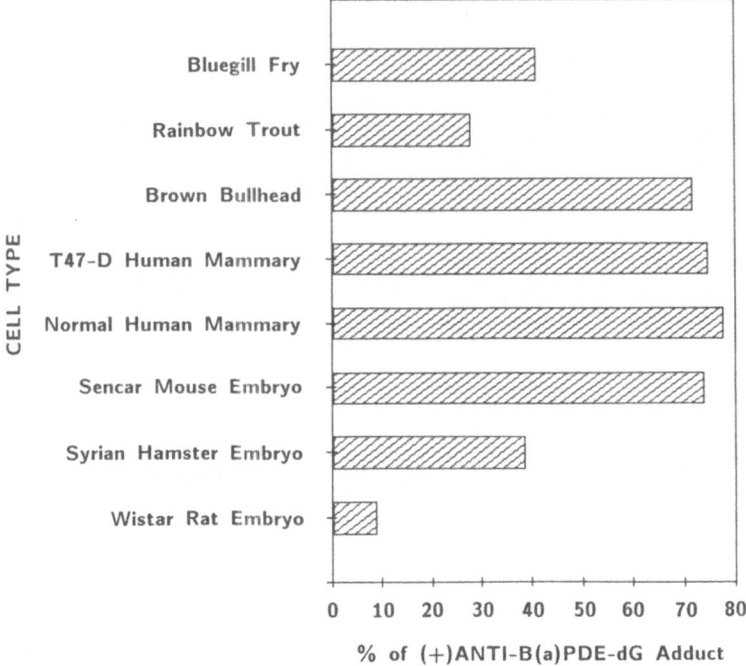

Fig. 6. The proportion of BaP-DNA binding due to formation of (+)-anti-
BaPDE in cell cultures from various species. All cultures were
exposed to 0.5 μgm [³H]BaP per ml medium for 24 h except the two
human mammary cell cultures (see ref. 14) which were exposed to
1 μgm per ml. The (+)-anti-BaPDE-dG adducts and BaP-DNA binding
were analyzed as described in ref. 11.

refed with [³H]BaP plus actinomycin D for 5 h, these increases were abol-,
ished (Fig. 8). The group with actinomycin D had a much lower level of binding of BaP to DNA than the group without Actinomycin D; the level was lower
than that of the acetone-pretreated group. Analysis of the BaP-DNA adducts
formed in the Actinomycin D treated group showed that the amount of (+)-anti-
BaPDE-dG adduct was similar to that found in the acetone control group
(Fig. 8). The large increase in this adduct observed in the BaP-pretreated
group was not observed when Actinomycin D was included with the [³H]BaP.
However, the total amount of BaP metabolized in the Actinomycin D group,
34 nmol/flask was identical to that in the BaP-pretreated group. Thus
Actinomycin D had no detectable effect on the overall metabolism of BaP but
greatly inhibited the binding of BaP to DNA and specifically the binding of
BaP to DNA through (+)-anti-BaPDE. These results are consistent with the
induction of specific isozymes of cytochrome P_{450} which metabolize a very high
proportion of the BaP to the (+)-anti form of BaPDE.

The above studies in cells in culture have demonstrated that different
species exhibit a wide divergence in the metabolism and activation of BaP to
DNA binding-metabolites. It is clear that overall BaP metabolism is a relatively poor measure of the efficiency of conversion of BaP to reactive DNA-
binding metabolites. Although the cultures from different species examined
all metabolized between 50 and 90% of the dose of BaP applied within 24 h,
the level of binding of the hydrocarbon differed by more than 16-fold between
cultures from various species (Fig. 4). The efficiency of conversion of the
BaP to an ultimate carcinogenic metabolite, (+)-anti-BaPDE, also differed
greatly between cultures from different species (Fig. 5). The proportion of

32

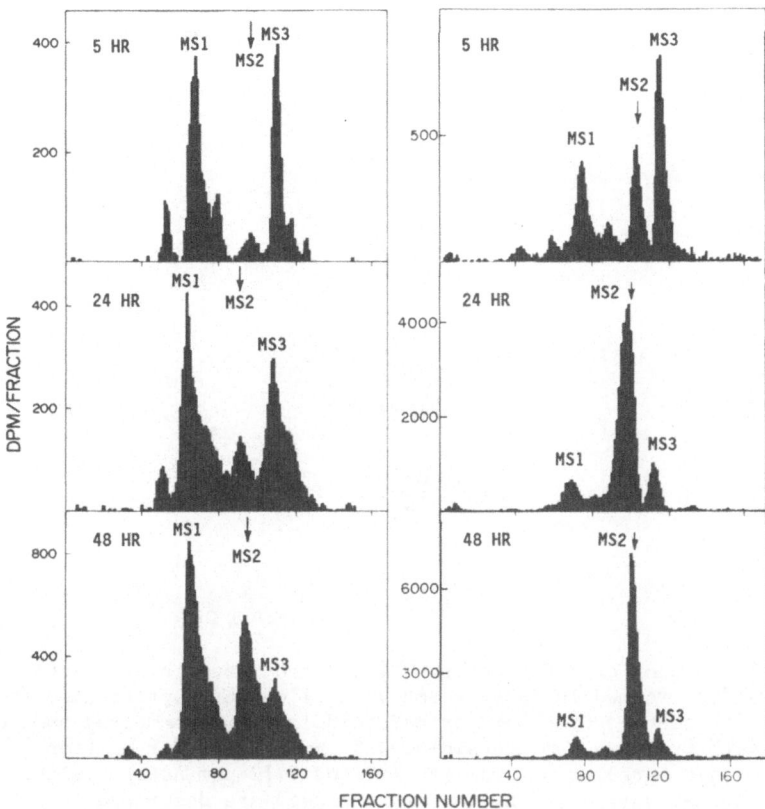

Fig. 7. Time-dependent differences in the proportion of (+)-anti-BaPDE-dG adduct present in [³H]BaP treated Wistar rat embryo cell cultures (left) and Syrian hamster embryo cell cultures. The cultures were exposed to 0.5 μgm [³H]BaP per ml medium for 5 (top), 24 (middle) or 48 (bottom) h and the adducts eluted in the morpholine-sorbitol fractions of an immobilized boronate column were analyzed by HPLC. ↓ indicates the elution position of a [¹⁴C]-(+)-anti-BaPDE-dG marker. The methods used are described in ref. 8.

the BaP-DNA binding due to the (+)-anti diol-epoxide also differed between cell cultures from different species (Fig. 6). Thus neither the amount of BaP metabolized nor the amount bound to DNA is sufficient to indicate the relative ability of the cell to activate BaP to (+)-anti-BaPDE.

Despite the above differences, the BaP-DNA adduct profiles from all of the cultures tended to converge with time to an adduct profile containing a very high proportion of the(+)-anti-BaPDE-dG adduct. This time dependent increase in the proportion of DNA binding due to the (+)-anti-BaPDE adduct is illustrated in Fig. 9. In mouse embryo cell cultures from both strains the proportion of DNA binding due to (+)-anti-BaPDE was relatively high at all times examined. A similar result was observed in the bluegill fry cells at 24 and 48 h. In the Syrian hamster embryo cells the proportion of binding through (+)-anti-BaPDE was low at 5 h but it increased three-fold by 24 and 48 h. In the Fisher and Wistar rat embryo cells the proportion was very low at 5 h and increased very slowly over the period examined. Thus after a

33

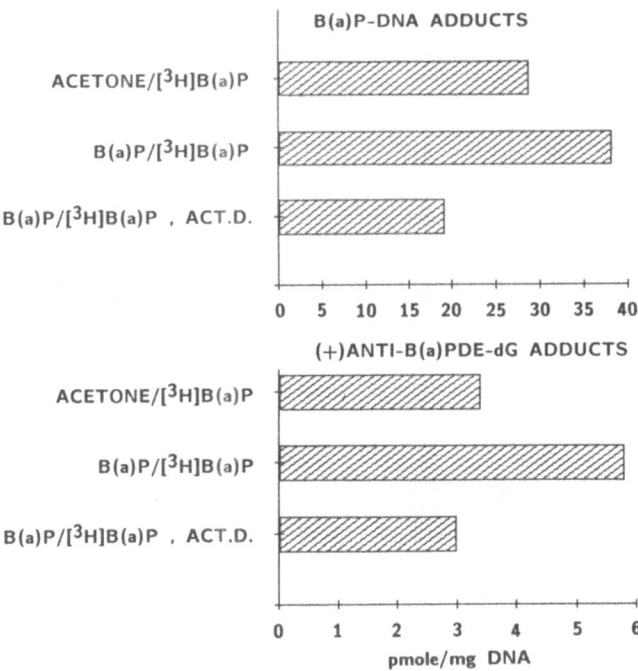

Fig. 8. The amount of BaP bound to DNA and the amount of (+)-anti-BaPDE-dG
adduct formed in hamster embryo cell cultures pretreated for 24 h
with acetone (top bar) or BaP (middle and lower bars) and refed
with fresh medium containing 0.5 μgm [^{3}H]BaP per ml (top and
middle bars) or 0.5 μgm [^{3}H]BaP and 0.12 μgm Actinomycin D per ml
(bottom bar) for 5 h. The methods used are described in ref. 19.

at 5 h and increased very slowly over the period examined. Thus after a
period of time of BaP treatment most of the cell cultures examined would
produce a very similar BaP-DNA adduct profile with the major adduct due to
reaction of (+)-anti-BaPDE with DNA. In the hamster embryo cells this appears to result from induction of certain forms of cytochrome P_{450} which selectively activate the BaP or the BaP-7,8-diol to (+)-anti-BaPDE. Pretreatment with BaP increased the proportion of (+)-anti-BaPDE formed suggesting
that the BaP dose itself may induce the formation of this cytochrome P_{450}(s).
These results in cells in culture would suggest the different species or tissues may respond very differently to low doses of hydrocarbons than to doses
which are capable of inducing this type of change in cytochrome P_{450}. They
would also suggest that conditions which result in induction of cytochrome
P_{450} such as continual exposure to cigarette smoke might increase the proportion of BaP activated to an ultimate carcinogenic metabolite (+)-anti-BaPDE.
By the use of cell cultures from different species which show different rates
and degrees of induction of this type of cytochrome P_{450} it should be possible
to establish the role of induction of specific isizymes of cytochrome P_{450} in
the activation of BaP to DNA binding metabolites in cells.

ACKNOWLEDGEMENTS

The authors wish to acknowledge the technical assistance provided by
Stephanie Morgan, Mark Ferin, Connie Moynihan and Cyndy Salmon and thank
Marilyn Hines for typing the manuscript. This work was supported by PHS
grants CA28825 and CA40228 from the National Cancer Institute, DHHS.

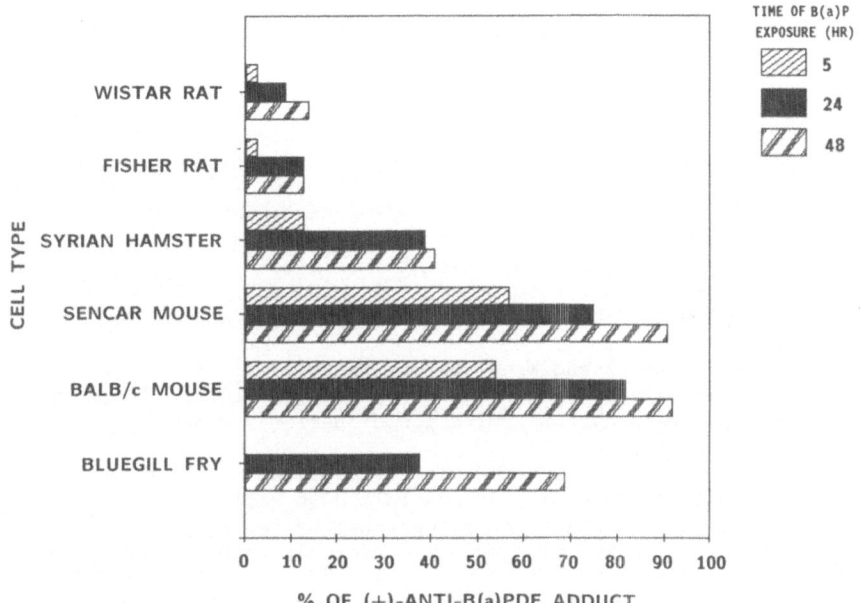

Fig. 9. Effect of time of treatment on the proportion of BaP bound to DNA through (+)-anti-BaPDE in cell cultures derived from various species. The cultures were exposed to 0.5 μgm [³H]BaP per ml medium for 5, 24 or 48 h and the level of binding of BaP to DNA and amount of (+)-anti-BaPDE-dG adduct determined as described in ref. 8.

REFERENCES

1. R. G. Harvey, "Polycyclic Hydrocarbons and Carcinogenesis", American Chemical Society Symposium Series 283, American Chemical Society, Washington D.C. (1985).
2. R. E. Lehr, S. Kumar, W. Levin, A. W. Wood, R. L. Chang, A. H. Conney, H. Yagi, J. M. Sayer and D. M. Jerina, The bay region theory of polycyclic aromatic hydrocarbon carcinogenesis, in: "Polycyclic Hydrocarbons and Carcinogenesis", R. G. Harvey, ed., American Chemical Society Symposium Series 283, American Chemical Society, Washington, D.C. (1985).
3. P. Sims, P. L. Grover, A. Swaisland, K. Pal and A. Hewer, Metabolic activation of benzo(a)pyrene proceeds by a diol-epoxide, Nature 252:326 (1974).
4. M. D. Buening, P. G. Wislocki, W. Levin, H. Yagi, D. R. Thakker, H. Akagi, M. Koreeda, D. M. Jerina and A. H. Conney, Tumorigenicity of the optical enantiomers of the diastereomeric benzo(a)pyrene-7,8-diol-9,10-epoxide in newborn mice: Exceptional activity of (+)-7ß,8α-dihydroxy-9α,10α-epoxy-7,8,9,10-tetrahydro-benzo(a)pyrene, Proc. Natl. Acad. Sci. USA 75:5358 (1978).
5. T. J. Slaga, W. J. Bracken, G. Gleason, W. Levin, H. Yagi, D. M. Jerina and A. H. Conney, Marked differences in the skin tumor-initiating activities of the optical enantiomers of the diastereometic benzo-(a)pyrene 7,8-diol-9,10-epoxides, Cancer Res. 39:67 (1979).
6. A. Dipple, Polycyclic aromatic hydrocarbon carcinogenesis: An Introduction, in: "Polycyclic Hydrocarbons and Carcinogenesis", R. G. Harvey, ed., American Chemical Society Symposium Series 283, American Chemical Society, Washington, D.C. (1985).

7. W. M. Baird and D. Pruess-Schwartz, Polycyclic aromatic hydrocarbon-DNA adducts and their analysis: A powerful technique for characterization of pathways of metabolic activation of hydrocarbons to ultimate carcinogenic metabolites, in: "Polycyclic Aromatic Hydrocarbon Carcinogenesis: Structure-Activity Relationships", B. D. Silverman and S. K. Yang, eds., CRC Press, Boca Raton, FL (1988).

8. S. M. Sebti, D. Pruess-Schwartz and W. M. Baird, Species- and length of exposure-dependent differences in the benzo(a)pyrene:DNA adducts formed in embryo cell cultures from mice, rats and hamsters, Cancer Res. 45:1594 (1985).

9. W. M. Baird and P. Brookes, Isolation of the hydrocarbon-deoxyribonucleoside products from the DNA of mouse embryo cells treated in culture with 7-methylbenz(a)anthracene-^{3}H, Cancer Res. 33:2378 (1973).

10. D. Pruess-Schwartz and W. M. Baird, Benzo(a)pyrene:DNA adduct formation in early-passage Wistar rat embryo cell cultures: Evidence of multiple pathways of activation of benzo(a)pyrene, Cancer Res. 46:545 (1986).

11. D. Pruess-Schwartz, S. M. Sebti, P. T. Gilham and W. M. Baird, Analysis of benzo(a)pyrene:DNA adducts formed in cells in culture by immobilized boronate chromatography, Cancer Res. 44:4104 (1984).

12. A. M. Jeffrey, Polycyclic aromatic hydrocarbon-DNA adducts: Formation, detection and characterization, in: "Polycyclic Hydrocarbons and Carcinogenesis", R. G. Harvey, ed., American Chemical Society Symposium Series 283, American Chemical Society, Washington, D.C. (1985).

13. T. A. Smolarek, S. L. Morgan, C. G. Moynihan, H. Lee, R. G. Harvey and W. M. Baird, Metabolism and DNA adduct formation of benzo(a)pyrene and 7,12-dimethylbenz(a)anthracene in fish cell lines in culture, Carcinogenesis 8:1501 (1987).

14. D. Pruess-Schwartz, W. M. Baird, A. Nikbakht, B. A. Merrick and J. K. Selkirk, Benzo(a)pyrene:DNA adduct formation in normal human mammary epithelial cell cultures and the human mammary carcinoma T47D cell line, Cancer Res. 46:2697 (1986).

15. I. Plakunov, T. A. Smolarek, D. L. Fischer, J. C. Wiley Jr. and W. M. Baird, Separation by ion-pair high-performance liquid chromatography of the glucuronide, sulfate and glutathione conjugates formed from benzo[a]pyrene in cell cultures from rodents, fish and humans, Carcinogenesis 8:59 (1987).

16. W. M. Baird and L. Diamond, The nature of benzo(a)pyrene-DNA adducts formed in hamster embryo cells depends on the length of time of exposure to benzo(a)pyrene, Biochem. and Biophy. Res. Comm. 77:162 (1977).

17. J. A. Milner, M. A. Piggot and A. Dipple, Selective effects of selenium selenite on 7,12-dimethylbenz(a)anthracene-DNA binding in fetal mouse cell cultures, Cancer Res. 45:6347 (1985).

18. K. W. Singletary and J. A. Milner, DNA binding and adduct formation of 7,12-dimethylbenz(a)anthracene by rat mammary epithelial cell aggregates in vitro, Carcinogenesis 7:95 (1986).

19. T. A. Smolarek, S. L. Morgan, J. Kelley and W. M. Baird, Effects of pretreatment with benzo(a)pyrene on the stereochemical selectivity of metabolic activation of benzo(a)pyrene to DNA-binding metabolites in hamster embryo cell cultures, Chem. -Biol. Interactions 64:71 (1987).

COMPLEX METABOLIC ACTIVATION PATHWAYS OF POLYCYCLIC AROMATIC HYDROCARBONS:
3-HYDROXY-trans-7,8-DIHYDROXY-7,8-DIHYDROBENZO[a]PYRENE AS A PROXIMATE
MUTAGEN OF 3-HYDROXYBENZO[a]PYRENE

Hansruedi Glatt, Paul C. Hirom[1], Charles A. Kirkby[1],
Odartey Ribeiro[1], Albrecht Seidel and Franz Oesch

Institute of Toxicology, University of Mainz, Obere
Zahlbacher Strasse 67, D-6500 Mainz, Federal Republic of
Germany, and [1]Department of Biochemistry, St. Mary's Hospital
Medical School, Norfolk Place, London W2 1PG, United Kingdom

INTRODUCTION

3-Hydroxybenzo[a]pyrene (3-OH-BP) is a major metabolite of benzo[a]py-
rene (BP) in various systems. Metabolites of 3-OH-BP, formed by liver enzymes, bind to DNA[1,2] and are mutagenic[3,4]. However, the active species have
not yet been identified. Administration of 3-OH-BP to rats results in the
excretion of sulfate and glucuronic acid conjugates of 3-hydroxy-trans-7,8-
dihydroxy-7,8-dihydrobenzo[a]pyrene (3-OH-BP-7,8-diol) (Fig. 1) as major
metabolites in the bile[5]. The hydroxyl groups of this triol are structurally
superimposable to those of 9-hydroxy-trans-1,2-dihydroxy-1,2-dihydrochrysene
(9-hydroxychrysene-1,2-diol, Fig. 1), which is a metabolite of chrysene[6,7] and
a potent promutagen[8,9]. 9-Hydroxychrysene-1,2-diol is activated by mammalian
enzymes to anti-chrysene-1,2-diol-3,4-oxide[7], which is chemically more reactive, is more mutagenic in bacterial and mammalian cells and is more potent
in the malignant transformation of C3H10T1/2 cells than is the simple bay-
region diol-epoxide of chrysene[8,9]. Furthermore, after application of
chrysene to mouse skin, a target organ for carcinogenicity, this triol-epoxide forms a major type of DNA-adduct[7,10]. The higher chemical reactivity and
biological activity of 9-hydroxychrysene-1,2-diol-3,4-oxides as compared to
the chrysene-1,2-diol-3,4-oxides may be explained, and were predicted, on
electronic grounds[11]: the hydroxyl group may resonance-stabilize the bay-region carbonium ion that results from opening of the oxirane ring. Analogous
resonance stabilization is expected to occur with the carbonium ions formed
from 3-OH-BP-7,8-diol-9,10-oxides[11]. Therefore, it was of interest to investigate the novel, secondary BP metabolite, 3-OH-BP-7,8-diol, for mutagenity.

MATERIALS AND METHODS

Chemicals

BP was obtained from Sigma and was purified by flash chromatography.
3-OH-BP was obtained from the NCI Chemical Carcinogen Reference Standard
Repository, a function of the Division of Cancer Cause and Prevention, NCI,
NIH, Bethesda, Md. 20205. Trans-7,8-dihydroxy-7,8-dihydrobenzo[a]pyrene
(BP-7,8-diol) was synthesized as described[12]. 3-OH-BP-7,8-diol was prepared
biochemically. Female Lewis rats (250-300 g) were anaesthetized with sodium

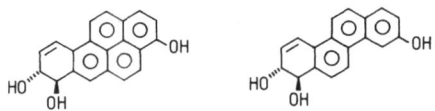

Fig. 1.   Structures of 3-OH-BP-7,8-diol (left) and
9-hydroxychrysene-1,2-diol (right).

pentobarbitone and the bile ducts cannulated. 3-OH-BP (50 mg/kg), dissolved
in 0.2 ml of Tween 80:ethanol (1:2, v/v), was injected intraperitoneally.
Bile was collected for 5 h. Samples (15 ml) of the bile were added to an
equal volume of Sigma H-2 ß-glucuronidase/arylsulphatase and incubated at
37°C overnight in the dark. Incubates were then diluted with an equal volume
of 0.1 M sodium acetate buffer (pH 5) and extracted with an equal volume of
ethyl acetate (4 times). The combined organic fractions were dried over sodium sulfate and the solvent evaporated. The residue was dissolved in methanol. The crude aglycones were applied to preparative layer chromatography
plates. The plates were developed with toluene:ethanol (9:1, v/v), dried
under nitrogen and rerun in the same solvent system. The band with an R_f
value of 0.2 was scraped off and eluted with methanol. Analysis by high
pressure liquid chromatography showed a single product without any detectable
impurities. The structure of this product was deduced by analysis of its
u.v., p.m.r. and mass spectra[5] and later confirmed by comparison with
chemically synthesized 3-OH-BP-7,8-diol (R. Schrode and A. Seidel, unpublished result).

Preparation of S9 Mix

Male Sprague-Dawley rats (200-300 g) were given a single intraperitoneal
injection of Aroclor 1254 (500 mg per kg body weight; Aroclor was diluted
with sunflower oil 1:5, v/v) 6 days before they were killed. The livers were
homogenized in 3 volumes of sterile, cold KCl (150 mM, containing 10 mM sodium phosphate buffer, pH 7.4) in a Potter-Elvehjem glass/teflon homogenizer.
The homogenate was centrifuged at 9,000 g for 10 min. The resulting supernatant fraction was diluted 1:3 with homogenization buffer and was then mixed
with two volumes of a solution that contained 12 mM $MgCl_2$, 50 mM KCl, 6 mM
$NADP^+$, 7.5 mM glucose 6-phosphate and 75 mM sodium phosphate buffer, pH 7.4.
This preparation, which is termed S9 mix and was always freshly prepared from
rats killed on the day of the mutagenicity experiment, was employed in the
experiments with bacteria. For use in the tests with V79 cells, the preparation was modified. The livers from animals treated as above were homogenized
in 3 volumes of Dulbecco's phosphate-buffered saline additionally containing
10 mM of 4-(2-hydroxyethyl)-1-piperazine ethane sulfonic acid, pH 7.4 (PBS).
The 9,000 x g supernatant, in aliquots of 20 ml, was frozen and stored at
-70°C. One volume of this fraction and 3 volumes of a cofactor solution
(197 mM glucose 6-phosphate, 28 mM $NADP^+$, 26 mM NADH and 11 mM NADPH in PBS)
were mixed to yield the S9 mix that was used for the V79 mutagenicity test.

Mutagenicity in Bacteria

His⁻ strains TA 97, TA 98, TA 100 and TA 1537 of Salmonella typhimurium
were grown overnight in nutrient broth (25 g Oxoid Nutrient Broth No. 2 per
liter). For inoculation, stock cultures which were stored at -70°C, were
used. Before the experiment, bacteria were centrifuged, resuspended in medium B (1.6 g Bacto Nutrient Broth and 5 g NaCl per liter) and adjusted neph-
elometrically to a titer of 1 to 2 x 10⁹ bacteria (colony forming units per
milliliter. The test compound (in 10 µl methanol), 500 µl S9 mix (or 150 mM
KCl in 10 mM sodium phosphate buffer pH 7.4), 100 µl of the bacterial suspension and 2 ml of 45°C warm top agar (0.55% agar, 0.55% NaCl, 50 µM his-

tidine, 50 μM biotin, 25 mM sodium phosphate buffer, pH 7.4) were mixed in a test tube and poured into a petri dish containing 22 ml of minimal agar (1.5% agar, Vogel-Bonner E medium with 2% glucose). After incubation of the plates for 2 to 3 days at 37°C in the dark, the colonies (his$^+$ revertants) were counted. All incubations were performed in duplicate or triplicate. Generally, individual values deviated by less than 10 colonies or by less than 10% from the mean.

Mutagenicity in V79 Cells

Chinese hamster V79 cells were maintained in Dulbecco's modified minimum essential medium supplemented with fetal calf serum (5%), penicillin (100 units/ml), and streptomycin (100 μg/ml). The cells were grown at 37°C in a humidified atmosphere containing 5% CO_2. For determination of mutagenicity a total of 1.5×10^6 cells and 30 ml medium were added to each 150-cm^2 dish. After 18 h, the medium was replaced by 18 ml of S9 mix or PBS, respectively, and the test compound (dissolved in 36 to 240 μl methanol) was added. S9 mix or buffer, and the test compound were removed 2 h later. Medium (30 ml) was added to the cultures after washing with PBS. After 3 days the cells were detached by treatment with trypsin. Solvent control cultures yielded $6-10 \times 10^7$ cells in all experiments in this study. As a measure for toxicity, the number of cells of the treatment group was expressed as a percent of the solvent control. The cells were then subcultured at a density of 3×10^6 cells per 150-cm^2 petri dish. Three days later, they were again detached and replated at a density of 10^6 cells per 150-cm^2 petri dish in medium containing 6-thioguanine (7 μg/ml) to determine the number of mutants (6 replicate plates) and, at a density of 100 cells per 22-cm^2 dish in medium without 6-thioguanine, for the determination of the cloning efficiency (3 replicate plates). The plates were fixed and stained, and the colonies were counted after about 7 days in the case of the cloning efficiency plates and 10 days in the case of the selection plates.

RESULTS

Mutagenicity in Salmonella typhimurium

3-OH-BP-7,8-diol, BP and BP-7,8-diol showed no direct mutagenicity in any bacterial strain. The doses used were the same as in the S9 mix-mediated experiment (see Fig. 2). At the dose of 4 μg per plate, BP-7,8-diol began to be toxic. No toxicity was seen with 3-OH-BP-7,8-diol up to the highest dose used, 16 μg per plate with strain TA 1537 and 8 μg with the other strains. Higher doses were not tested due to the limited amount of compound available. In contrast to the above compounds, 3-OH-BP was directly mutagenic (Table 1, detailed data not shown). Effects were seen in all four bacterial strains, but were relatively weak (2- to 4-fold increases in the number of colonies above control value). TA 98 was the most responsive strain.

In the presence of liver S9 mix, 3-OH-BP-7,8-diol, BP-7,8-diol and BP were activated to mutagens and the mutagenicity of 3-OH-BP was potentiated (Fig. 2, Table 1). Each compound reverted all four bacterial strains, except that BP-7,8-diol showed no effects with strain TA 1537. In the other three strains, however, the diol was the most potent mutagen per nmole test compound. In strain TA 1537, the triol and BP were most mutagenic, both showing about equal potencies. The weakest mutagenic activity in all strains was noticed for 3-OH-BP (apart from the negative result of the diol in TA 1537). 3-OH-BP was marginally (TA 98) to 18 times (TA 1537, TA 97) less mutagenic than the next weakest mutagen, which always was its potential metabolite, 3-OH-BP-7,8-diol. Therefore, these two compounds substantially differed in the pattern of mutagenic effects they evoked.

Table 1. Summary of the results of the bacterial mutagenicity experiments

Compound	S9 mix	Revertants per nmol[a]			
		TA 98	TA 97	TA 100	TA 1537
BP	−	<0.2[b]	<0.5	<0.3	<0.1
	+	33	88	78	9
3-OH-BP	−	2	2	1	0.2
	+	9	3	6	0.5
BP-7,8-diol	−	<1	<2	<3	<0.2
	+	66	110	310[c]	<3
3-OH-BP-7,8-diol	−	<0.5	<1	<1	<0.1
	+	10	54	31[c]	9

[a] Initial slope of the dose-response curve or, in cases in which the effect increased more than linearly with increasing doses, maximal ratio of effect (above control) and dose. [b] Estimated detection limit. [c] In modified assays (30 min incubation at 37°C before addition of the top agar), the mutagenic potencies of BP-7,8-diol and 3-OH-BP-7,8-diol differed by a factor of only 2.5.

Mutagenicity and cytotoxicity in V79 cells

3-OH-BP-7,8-diol and BP-7,8-diol did not show mutagenicity in V79 cells, when they were tested in the absence of an exogenous xenobiotic-metabolizing system (data not shown). However, in the presence of liver S9 mix, the diol elicited strong mutagenic effects, while the triol regularly showed weak effects (Fig. 3). The concentration-response curve for the diol was non-linear with the relative effects being stronger at high concentrations. Appreciable effects of the triol were observed only at the highest concentration used. In the experiment in which several concentrations of the diol were tested, the triol at a concentration of 5 μg per ml was equimutagenic to one fourth concentration of the diol, but was only one fifteenth as active as the diol at equal concentration. Determination of mutagenicity of the triol at higher concentrations than 5 μg/ml was not possible due to cytotoxicity, which was even stronger than that of the diol. Moreover, it is possible that the cytotoxicity of the triol was not due to metabolites formed by the S9 mix, since similar (slightly stronger) toxicity was observed in the direct test (data not shown).

DISCUSSION

Mutagenic Potencies of BP Derivatives in Salmonella typhimurium

In the present study we found that bioactivated 3-OH-BP-7,8-diol is a potent mutagen in Salmonella typhimurium. Its potency is similar to those of BP and BP-7,8-diol and is higher than that of 3-OH-BP. The differences are too small to allow predictions about the relative mutagenic or carcinogenic activities in mammalian organisms.

The higher activity of 3-OH-BP-7,8-diol, as compared to 3-OH-BP, is

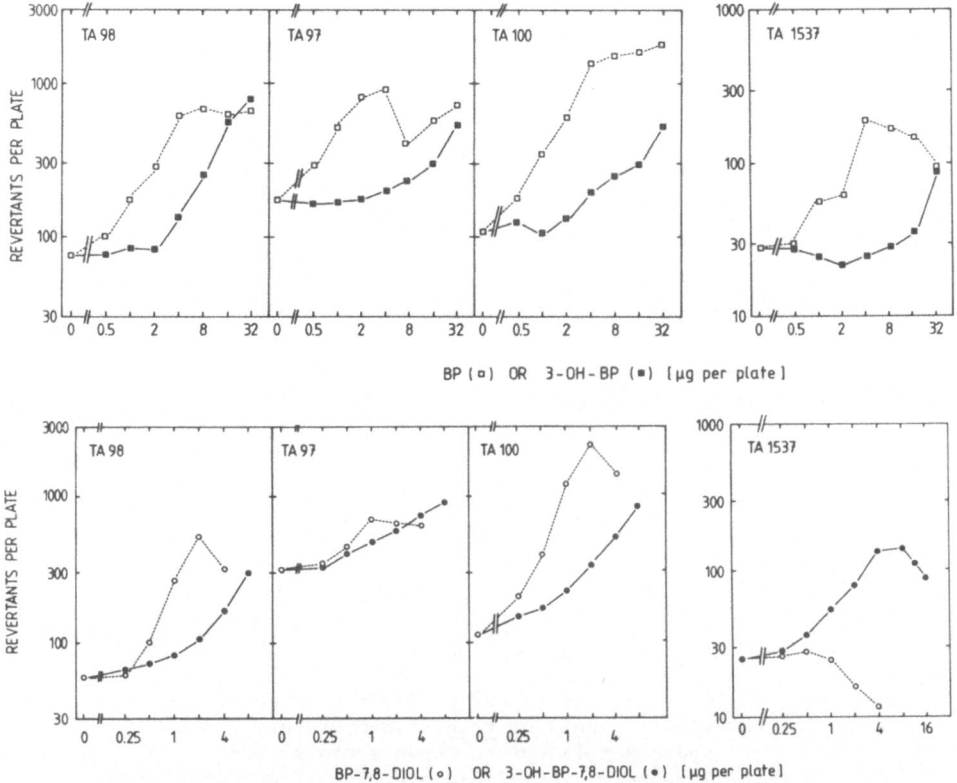

Fig. 2. Mutagenicity of benzo[a]pyrene (□), 3-hydroxybenzo[a]pyrene (■),
trans-7,8-dihydroxy-7,8-dihydrobenzo[a]pyrene (O) and 3-hydroxy-
trans-7,8-dihydroxy-7,8-dihydrobenzo[a]pyrene (●) in the presence
of rat liver S9 mix in four his-strains of Salmonella typhimurium.
Similar results, including the multiphasic dose-response curve for
benzo[a]pyrene in strain TA 97, were observed in repeat experiments.

compatible with the assumption that the triol is a proximate mutagen of the
phenol. However, since the metabolically activated triol has a preference
for strains TA 97 and TA 1537 (18 times the activity of metabolically activated 3-OH-BP), but shows relatively weak activity in strain TA 98, additional active species formed from 3-OH-BP have to be postulated. The hypothesis that 3-OH-BP can be activated to several mutagenic metabolites and
that one of them may be formed via 3-OH-BP-7,8-diol is supported by previous
findings: (a) inhibition of microsomal epoxide hydrolase (probably implying
inhibition of formation of 3-OH-BP-7,8-diol) reduced the S9 mix-mediated mutagenicity of 3-OH-BP in strain TA 98 under some experimental conditions but
had no effect under other conditions[13]; (b) incubation of 3-OH-BP with mouse
liver microsomes and DNA resulted in the formation of nine distinguishable
nucleoside adducts, whose frequency was differentially affected with modification of the metabolizing system[1].

Mutagenicity and Cytotoxicity in V79 Mammalian Cells

Anti-BP-7,8-diol-9,10-oxide was by far the strongest direct-acting mutagen in V79 cells among all investigated BP metabolites (reviewed in reference 13). Several other metabolites, including 3-OH-BP, showed weak direct
mutagenicity in V79 cells. In tests in which exogenous xenobiotic-metaboli-

41

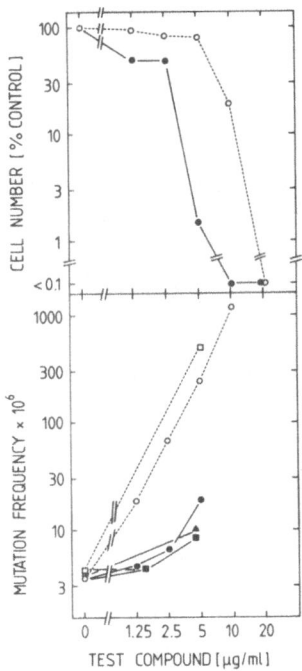

Fig. 3. Cytotoxicity (upper panel) and mutagenicity
(lower panel) of trans-7,8-dihydroxy-7,8-di-
hydrobenzo[a]pyrene (open symbols) and
3-hydroxy-trans-7,8-dihydroxy-7,8-dihydro-
benzo[a]pyrene (closed symbols), in the
presence of rat liver S9 mix, in V79 Chinese
hamster cells. Acquisition of resistance to
6-thioguanine served as the marker for
mutagenicity. Circles, triangles and squares
represent separate experiments.

zing systems, intact cells or liver S9 mix, were added, only BP and BP-7,8-
diol gave positive results. Efficient activation of BP-7,8-diol by liver S9
mix in this test was confirmed in the present study, and weak S9 mix-mediated mutagenicity was demonstrated for 3-OH-BP-7,8-diol. Depending on whether comparison was made at equal effects or at equal concentrations of the
test compound, the mutagenic potencies of the diol and the triol differed by
a factor of 4 or 15. The efficacies, i.e. the maximal increases in mutation
frequency, differed however by a factor of 70, since the triol was more cytotoxic than the diol. Interestingly, the cytotoxicity of the triol did not
require the presence of liver S9 mix; it appeared to be linked to the presence of the phenolic hydroxyl group. Indeed, it has been shown previously
that some phenols, e.g. 5-hydroxychrysene, are strongly cytotoxic to V79
cells (whilst others are not)[9].

Comparison of the Observed Biological Activities with those of Corresponding Chrysene Derivatives

Apart from 3-OH-BP-7,8-diol, 9-hydroxychrysene-1,2-diol is the only
triol for which mutagenicity data are available. This compound was not directly mutagenic, but activated by liver S9 mix[9]. The numbers of mutants induced in Salmonella typhimurium TA 98, Salmonella typhimurium TA 100 and V79
cells per nmole 9-hydroxychrysene-1,2-diol were 1, 39 and 23% of the corres-

42

ponding values for 3-OH-BP-7,8-diol. However, since the chrysene triol was virtually non-toxic and therefore could be tested at high concentrations, its mutagenic efficacy was much higher than that of 3-OH-BP-7,8-diol. Strong evidence exists that anti-9-hydroxychrysene-1,2-diol-3,4-oxide is a major active metabolite of 9-hydroxychrysene-1,2-diol[6,7]. The high structural similarity of 9-hydroxychrysene-1,2-diol and 3-OH-BP-7,8-diol (Fig. 1) leads to the speculation that 3-OH-BP-7,8-diol-9,10-oxide(s) may be the mutagenic metabolite(s) of the BP triol. Anti-3-OH-BP-7,8-diol-9,10-oxide, detected as a metabolite of anti-BP-7,8-diol-9,10-oxide, is chemically more reactive than its precursor diol-epoxide[14]. Anti-BP-7,8-diol-9,10-oxide is much more reactive than anti-chrysene-1,2-diol-3,4-oxide, as indicated by half-lives in buffers used for mutagenicity experiments of 6 to 12 min[15] and 74 min[9], respectively. Hence, it is difficult to predict whether additional enhancement of reactivity by introduction of the hydroxyl group at the 3-position of anti-BP-7,8-diol-9,10-oxide will increase the mutagenic activity analogous to the effect of the 9-hydroxyl group in the chrysene triol-epoxide. It is also possible that the mutagenicity of the BP triol-epoxides versus the diol-epoxides could be diminished because of enhanced competitive reactions with cellular structures which are not part of the genetic apparatus. Different reactivity could also result in different patterns of mutagenic effects, as observed in the present study for bioactivated BP-7,8-diol and 3-OH-BP-7,8-diol.

Metabolism of a polycyclic aromatic hydrocarbon to a triol and its further activation to a mutagen involves many metabolic steps, giving possibilities for competitive metabolism. The appearance of high concentrations of conjugates of 3-OH-BP-7,8-diol in the bile of rats supports this notion[5]. The observation that anti-9-hydroxychrysene-1,2-diol-3,4-oxide forms a major nucleoside adduct after application of chrysene to mouse skin[7,10] demonstrates that competitive reactions do no always prevent the formation of significant amounts of triol-epoxides.

ACKNOWLEDGMENTS

We thank Ms. Andrea Piée and Ms. Karin Pauly for expert technical assistance and Ms. S. Pollok for excellent help in preparing this manuscript. This work was supported in part by a grant (no. 0704851/4) of the Bundesministerium für Forschung und Technologie. The responsibility for the content of the publication is with the authors.

REFERENCES

1. I. S. Owens, C. Legraverend and O. Pelkonen, Deoxyribonucleic acid binding of 3-hydroxy- and 9-hydroxybenzo[a]pyrene following further metabolism by mouse liver microsomal cytochrome P$_1$-450, Biochem. Pharmacol. 28:1623 (1979).
2. J. Capdevila, B. Jernström, H. Vadi and S. Orrenius, Cytochrome P-450-linked activation of 3-hydroxybenzo[a]pyrene, Biochem. Biophys. Res. Commun. 65:894 (1975).
3. I. S. Owens, G. M. Koteen and C. Legraverend, Mutagenesis of certain benzo[a]pyrene phenols in vitro following further metabolism by mouse liver, Biochem. Pharmacol. 28:1615 (1979).
4. H. R. Glatt, R. Billings, K. L. Platt and F. Oesch, Improvement of the correlation of bacterial mutagenicity with carcinogenicity of benzo-[a]pyrene and four of its major metabolites by activation with intact liver cells instead of cell homogenate, Cancer Res. 41:270 (1981).
5. O. Ribeiro, C. A. Kirkby, P. C. Hirom and P. Millburn, Secondary metabolites of benzo[a]pyrene: 3-hydroxy-trans-7,8-dihydro-7,8-di-

hydroxybenzo[a]pyrene, a biliary metabolite of 3-hydroxybenzo[a]-pyrene in the rat, <u>Carcinogenesis</u> 6:1507 (1985).

6. R. M. Hodgson, A. Seidel, W. Bochnitschek, H. R. Glatt, F. Oesch and P. L. Grover, The formation of 9-hydroxychrysene-1,2-diol as an intermediate in the metabolic activation of chrysene, <u>Carcinogenesis</u> 6:1507 (1985).

7. R. M. Hodgson, A. Weston, A. Seidel, W. Bochnitschek, H. R. Glatt, F. Oesch and P.L. Grover, Metabolism of chrysene to triols and a triol-epoxide in mouse skin and rat liver preparations, <u>in</u>: " Polynuclear Aromatic Hydrocarbons: Chemistry, Characterization and Carcinogenesis", M. Cooke and A. J. Dennis, eds., Battelle, Columbus (Ohio) (1986).

8. H. R. Glatt, A. Seidel, W. Bochnitschek, H. Marquardt, H. Marquardt, R. M. Hodgson, P. L. Grover and F. Oesch, Mutagenicity in bacterial and mammalian cells of diol-epoxides, triol-epoxides and other metabolites of chrysene, <u>in</u>: "Polynuclear Aromatic Hydrocarbons: Chemistry, Characterization and Carcinogenesis", M. Cooke and A. J. Dennis, eds., Battelle, Columbus (Ohio) (1986).

9. H. R. Glatt, A. Seidel, W. Bochnitschek, H. Marquardt, H. Marquardt, R. M. Hodgson, P. L. Grover and F. Oesch, Mutagenic and cell-transforming activities of triol-epoxides as compared to other chrysene metabolites, <u>Cancer</u> <u>Res.</u> 46:4556 (1986).

10. R. M. Hodgson, A. Weston and P. L. Grover, Metabolic activation of chrysene in mouse skin: evidence for the involvement of a triol-epoxide, <u>Carcinogenesis</u> 4:1639 (1983).

11. P. B. Hulbert and P. L. Grover, Chemical rearrangement of phenol-epoxide metabolites of polycyclic aromatic hydrocarbons to quinone-methides, <u>Biochem.</u> <u>Biophys.</u> <u>Res.</u> <u>Commun.</u> 117:129 (1983).

12. P. P. Fu and R. G. Harvey, Synthesis of the diols and diol-epoxides of carcinogenic hydrocarbons, <u>Tetrahedron</u> <u>Lett.</u> 2059 (1977).

13. H. R. Glatt and F. Oesch, Structural and metabolic parameters governing the mutagenicity of polycyclic aromatic hydrocarbons, <u>in</u>: "Chemical Mutagens: Principles and Methods for Their Detection", Vol. 10, F. J. de Serres, ed., Plenum, New York. (1986).

14. A. Gräslund, F. Waern and B. Jernström, Studies by fluorescence and n.m.r. spectroscopy of the major product formed after further metabolism of (±)-7ß,8α-dihydroxy-9α,10α-epoxy-7,8,9,10-tetrahydro-benzo[a]pyrene, <u>Carcinogenesis</u> 7:167 (1986).

15. A. W. Wood, P. G. Wislocki, R. L. Chang, W. Levin, A. Y. H. Lu, H. Yagi, O. Hernandez, D. M. Jerina and A. H. Conney, Mutagenicity of benzo[a]pyrene benzo-ring epoxides, <u>Cancer</u> <u>Res.</u> 36:3358 (1976).

CARCINOGENICITY OF METHYL HALIDES:

CURRENT PROBLEMS CONCERNING CHLOROMETHANE

Hermann M. Bolt, Hans Peter and Rainer Jäger

Institut für Arbeitsphysiologie an der Universität Dortmund
Ardeystrasse 67,D-4600 Dortmund 1
F.R.G.

INTRODUCTION

Methyl halides (methyl chloride, bromide, iodide) comprise a group of compounds of "limited" or "insufficient" evidence of carcinogenicity; short-term tests with these materials, however, have revealed "sufficient evidence" of genetic activity[1].

Methyl iodide, when given to rats subcutaneously, produces local sarcomas[2]. Methyl bromide, when applied orally to the same species, induces squameous cell carcinomas of the forestomach[3,4]. This may be consistent with a direct alkylation of DNA by these methylating agents, as described by Djalali-Behzad et al.[5] for methyl bromide.

By contrast, methyl chloride, at very high concentrations in vitro, is an only very weak direct-acting mutagen for bacteria and cultured cells[6]. A DNA binding assay of methyl chloride in mice (B6C3F1) and rats (F-344), using ^{14}C-labeled material of high specific activity, did not show any DNA-methylation at the expected molecular targets (0^6 of adenine, 7-N of guanine) in livers and kidneys[7]. A carcinogenicity bioassay of methyl chloride, using the same strains of mice and rats, had resulted in an increased incidence of renal adenomas and cystadenomas in male mice exposed to the highest concentration (1,000 ppm) tested[8].

Based on investigations of the metabolism of methyl chloride to formate[9] it has been suggested that formaldehyde, formed intermediately within the tissues, could probably be responsible for such a tumor formation[10]. However, no significant increase of formaldehyde levels in target tissues could be found after exposures of mice to 1,000 ppm methyl chloride[11]. Moreover, other compounds giving rise to endogenous formaldehyde formation (methanol, aminopyrine) did not produce DNA-protein crosslinks in vivo which are a DNA damage characteristic for formaldehyde[12].

The determination of DNA-lesions, using the alkaline elution technique, also revealed no DNA-protein crosslinks in kidneys of male B6C3F1 mice after exposure to methyl chloride (1,000 ppm, 4 h/d, 4d). However, a faster elution of kidney DNA from exposed animals compared to untreated controls was taken as a possible indicator for single-strand breaks. Lipid peroxidation (production of thiobarbituric acid reactive material) was induced by a single exposure to methyl chloride (1,000 ppm, 6 h) in livers of (male and female) mice. Smaller increases of peroxidation were observed in kidneys of exposed mice as compared to the livers[11]. This was in line with earlier suggestions[13] that glutathione depletion and associated lipid peroxidation might be important for methyl chloride toxicity.

The importance of a glutathione-dependent pathway for metabolism of methyl chloride has also been demonstrated for humans by identification of the urinary metabolite S-methylcysteine[14]. However, there are indications for individual genetic differences as two discrete groups of men are observed which metabolise methyl chloride[15] and excrete S-methylcysteine[14] to widely different extents. Differences in susceptibility of mice (especially males) of different strains to chloroform are also known[16], and similarities in susceptibility to chloroform as compared with methyl chloride have been noted[17]. Chloroform is believed to be oxidatively biotransformed to phosgene, its toxic intermediate[18]. In mouse kidney, sex differences in cytochrome P-450 dependent enzyme activities are known in that renal microsomes from males oxidatively biotransform xenobiotics and steroids faster (8-14 fold) than those from females[19].

We investigated possible sex specifities of the glutathione dependent pathway of methyl chloride. The mouse hybrids (C3B6F1 and B6C3F1, males and females) where therefore exposed to a single high concentration (1,000 ppm, 6 h) of methyl chloride, and the time-course of depletion of glutathione in liver and kidneys was assessed.

MATERIALS AND METHODS

Male and female C3B6F1 and B6C3F1 mice were commercially bred by Lippische Versuchstierzucht, Extertal, F.R.G.. Prior to use, the animals were housed in humidity and temperature-controlled rooms with 12 h light-dark cycles and were allowed free access to food (Altromin[R] 1324, Altrogge GmbH, Lage/Lippe, F.R.G.) and water ad libitum, except during exposures.

Test groups of mice were exposed to 1,000 ppm methyl chloride for up to 6 h (see time points in Figs. 1-4) in a 50-liter plexiglass[R] chamber (open system, Bohlender, Lauda-Tauber, F.R.G.). Methyl chloride was metered from a gas bag by means of a peristaltic pump and was diluted with room air (provided by an oil-free compressor), using a mixing jet. Air flow into the chamber was maintained at 4 liters/min. Chamber concentrations were monitored at regular intervals by a gas chromatograph (Varian, series 1400, Darmstadt, F.R.G.) equipped with a flame ionization detector. Air samples from the exposure system were introduced onto the GC column via a gas-sampling loop (2 ml). Separation was performed at 100°C on a 3 m (2 mm i.d.) glass column packed with 35 - 60 mesh Tenax GC[R] (Chrompack GmbH, Mühlheim, F.R.G.). Nitrogen was used as the carrier gas at a flow rate of 60 ml/min. For calibration a desiccator was used and filled to a 1,000 ppm concentration by addition of pure methyl chloride (6.4 ml methyl chloride gas introduced into a 6.4 desiccator).

Animals were sacrificed by decapitation and exsanguinated. Livers and kidneys were immediately removed and chilled on ice. The excised organs were weight and homogenized in 5 vol. of ice-cold 0.1 M sodium phosphate buffer (pH 7.4). Determinations of glutathione (GSH and GSSG) were done enzymatically, according to a modification of Eyer and Podhradsky[20] of the method published by Griffith[21].

RESULTS

Control values of the glutathione content in liver and kidney of mice are shown in Table 1.

The exposure of mice to methyl chloride (1,000 ppm) leads to a severe reduction of glutathione in mouse livers (Figs. 1 and 2) and kidneys (Figs. 3 and 4). Compared with data published by others[13,22] the effect of a 6 h exposure, in these experiments, is even more pronounced. This difference may be due to different methodologies used. While others have used the chemical method of Sedlak and Lindsay[23] which uses Ellman's reagent to determine "nonprotein sulfhydryl compounds", we have employed a glutathione-specific enzymatic assay.

Upon exposure to 1,000 ppm methyl chloride, the depletion of glutathione in mouse liver is nearly complete (about 97 % depletion) within 2 h. Within these initial 2 h, there are minor sex differences in that GSH levels in females decline somewhat more rapidly, but this minor difference is probably not of biological significance. A similar tendency is visible in GSH levels of kidneys.

Although the time-course of GSH depletion in the kidneys is much slower and protracted than in livers, the six hours of exposure are sufficient to evoke a final depletion similar to that seen in the livers (although in the latter tissue this final degree of depletion is reached much earlier). This would mean that protection by GSH against oxidative stress, under methyl chloride exposure (1,000 ppm), is rapidly lost in mouse liver (within 1-2 h), but is principally also diminished in mouse kidneys although more time is required to induce this effect in this particular tissue.

Major differences between GSH depletion in B6C3F1 mice (Figs. 1 and 3) and C3B6F1 mice (Figs. 2 and 4) do not occur.

DISCUSSION

Possible strain differences in susceptibility of mice to the renal toxicity of methyl chloride have been discussed by Morgan et al.[17] on the background of data on chloroform.

It has been demonstrated[24] that mouse strains sensitive and insensitive to chloroform (renal) toxicity exist. Male C57BL6 mice are insensitive, male C3H mice are sensitive. After crossing female C3H mice with male C57BL6 mice

Table 1. Total glutathione (GSH + GSSG) in livers and kidneys of untreated C3B6F1 and B6C3F1 mice

Mouse strain	μmol/g wet wt.			
	Liver		Kidney	
	Male	Female	Male	Female
C3B6F1	7.58 ± 0.11	7.55 ± 0.16	3.66 ± 0.09	2.96 ± 0.13
B6C3F1	7.52 ± 0.14	7.43 ± 0.10	3.76 ± 0.11	3.12 ± 0.14

Means ± S.D.; n = 3

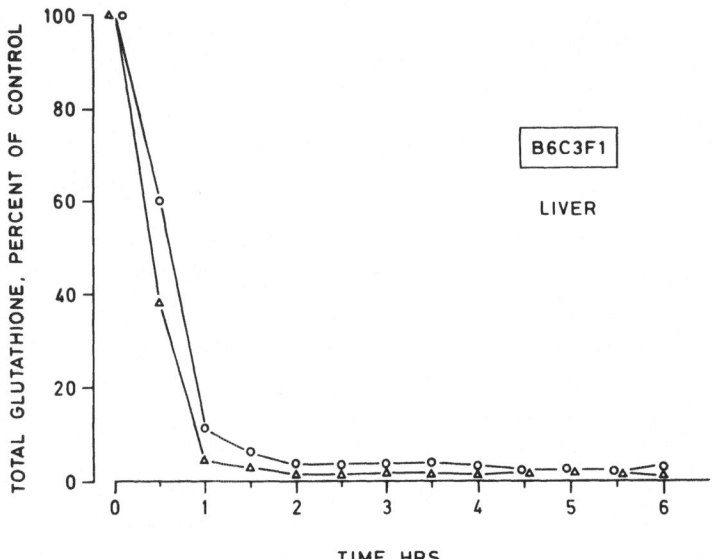

Fig. 1. Effect of exposure to methyl chloride (1000 ppm) on glutathione content of B6C3F1 mouse liver.

the offspring (C3B6F1 and B6C3F1) is insensitive[25]. It has been concluded[24] that the renal susceptibility of mice to chloroform should be inherited in an autosomal recessive mode. As outlined in the Introduction, chloroform is biotransformed to its ultimate toxic intermediate by the oxidative cytochrome P-450 dependent pathway. As the glutathione metabolic pathway may be relevant for cellular methyl chloride toxicity, either by an indirect effect of glutathione depletion[11;13], or by formation of an intermediate with ultimate

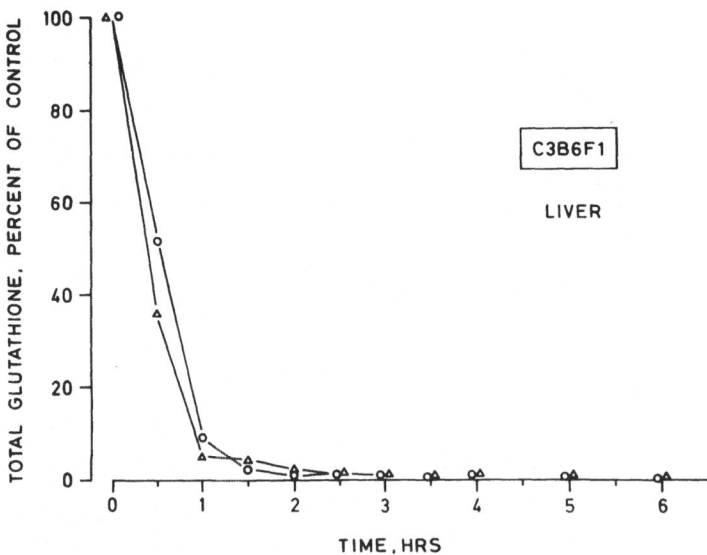

Fig. 2. Effect of exposure to methyl chloride (1000 ppm) on glutathione content of C3B6F1 mouse liver.

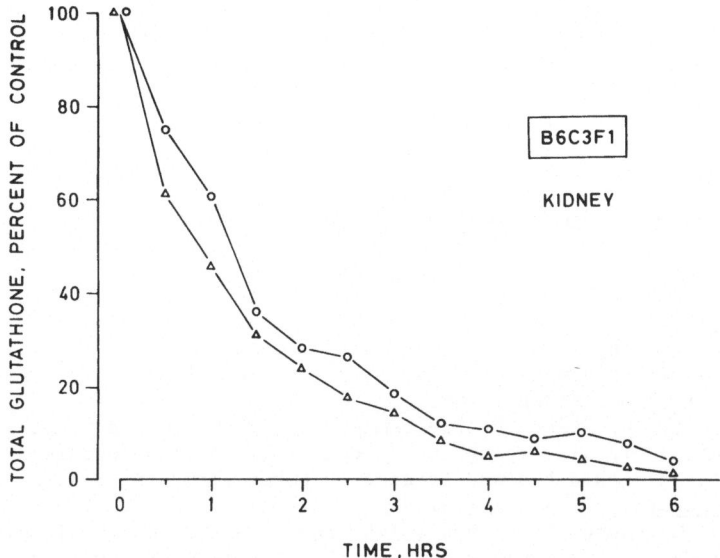

Fig. 3. Effect of exposure to methyl chloride (1000 ppm) on glutathione content of B6C3F1 mouse kidney.

target organ toxicity[22], we have investigated both parallel mouse hybrids (B6C3F1 and C3B6F1) as to differences in the GSH depletion pattern in liver and kidney during methyl chloride exposure. The results (Figs. 1-4) demonstrate that no such differences occur. The preponderant glutathione-dependent pathway in methyl chloride metabolism (which is confirmed by the observed extent of GSH depletion) must lead to a toxication mechanism different from that of chloroform.

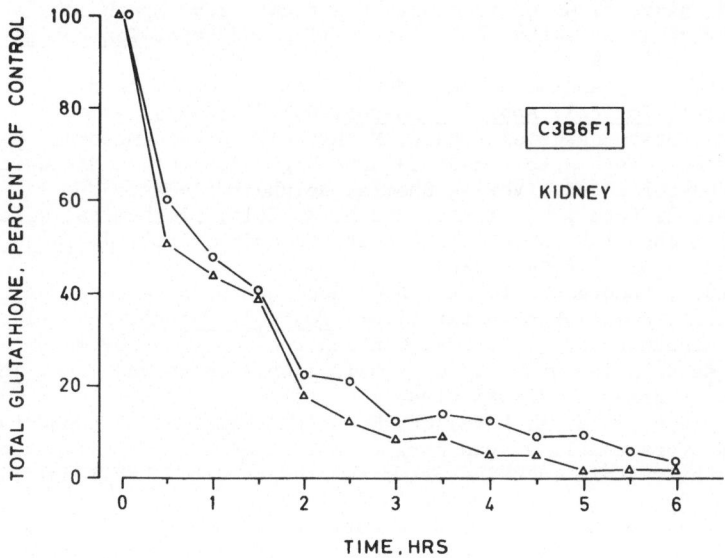

Fig. 4. Effect of exposure to methyl chloride (1000 ppm) on glutathione content of C3B6F1 mouse kidney.

Interpretation of the ultimate toxication mechanism is further hampered by obvious further species differences of mice[22] as opposed to rats[26]. This is visualized by the effectivity of antidotes: in mice, the GSH synthesis inhibitor L-buthionine-S,R-sulfoximine counteracts lethality due to methyl chloride[22], but in rats N-acetylcystein is an effective antidot after methyl bromide poisoning[27].

Therefore, further investigations will be necessary to elucidate the reasons of sex and strain specificity of methyl halide toxicity.

REFERENCES

1. IARC, IARC Monographs on the Evaluation of the Carcinogenic Risk of Chemicals to Humans, Vol. 41, International Agency for Research on Cancer, Lyon, France (1986).
2. H. Druckrey, H. Kruse, R. Preussmann, S. Ivankovic and C. Landschütz, Carcinogenic alkylating substances. III. Alkyl-halogenides, -sulphates, sulphonates and strained heterocyclic compounds, Z. Krebsforsch 74:241 (1970).
3. L. H. J. C. Danse, F. L. van Velsen and C. A. van der Heijden, Methyl bromide: carcinogenic effects in the rat forestomach, Toxicol. Appl. Pharmacol. 72:262 (1984).
4. G. A. Boorman, H. L. Hong, C. W. Jameson, K. Yoshitomi and R. R. Maronpot, Regression of methyl bromide-induced forestomach lesions in the rat, Toxicol. Appl. Pharmacol. 86:131 (1986).
5. G. Djalali-Behzad, S. Hussain, S. Osterman-Golkar and D. Segerbäck, Estimation of genetic risks of alkylating agents. VI. Exposure of mice and bacteria to methyl bromide, Mutat. Res. 84:1 (1981).
6. J. Fostel, P. F. Allen, E. Bermudez, A. D. Kligerman, J. L. Wilmer and T. R. Skopek, Assessment of the genotoxic effects of methyl chloride to human lymphoblasts, Mutat. Res. 155:75 (1985).
7. H. Peter, R. J. Laib, H. Ottenwälder, H. Topp, N. Rupprich and H. M. Bolt, DNA-binding assay of methyl chloride, Arch. Toxicol. 57:84 (1985).
8. K. L. Pavkov, W. D. Kerns, C. E. Chrisp, D. C. Thake, R. L. Persing and H. H. Harroff, Major findings in a twenty-four month inhalation toxicity study of methyl chloride in mice and rats, Toxicologist 2:161 (abstract) (1982).
9. D. J. Kornbrust and J. S. Bus, Metabolism of methyl chloride to formate in rats, Toxicol. Appl. Pharmacol. 65:135 (1982).
10. Deutsche Forschungsgemeinschaft, Methylchlorid/Chlormethan, in: "Toxikologisch-arbeitsmedizinische Begründungen von MAK-Werten", D. Henschler, ed., Verlag Chemie, Weinheim (1984).
11. R. Jager, H. Peter, W. Sterzel and H. M. Bolt, Biochemical effects of methyl chloride in relation to its tumorigenicity, J. Cancer Res. Clin. Oncol. 114:64 (1988).
12. W. K. Lutz, Endogenous formaldehyde does not produce detectable DNA-protein crosslinkes in rat liver, Toxicol. Pathol. 14:4 (1986).
13. D. J. Kornbrust and J. S. Bus, Glutathione depletion by methyl chloride and association with lipid peroxidation in mice and rats, Toxicol. Appl. Pharmacol. 72:388 (1984).
14. R. van Doorn, P. J. A. Borm, Ch.-M. Leigdeckers, P. Th. Henderson, J. Reuvers and T. J. van Bergen, Detection and identification of S-methylcystein in urine of workers exposed to methyl chloride, Int. Arch. Occup. Environ. Hlth. 46:99 (1980).
15. R. J. Nolan, D. L. Rick, T. D. Landry, L. P. McCarty, G. L. Agin and J. H. Saunders, Pharmacokinetics of inhaled methyl chloride in male volunteers, Fund. Appl. Toxicol. 5:361 (1985).
16. H. W. Casey, K. M. Ayers and F. R. Robinson, The urinary system, in: "Pathology of Laboratory Animals", K. Benirschke, F. M. Garner,

T. C. Jones, eds., Springer-Verlag, New York, Heidelberg, Berlin (1978).

17. K. T. Morgan, J. A. Swenberg, T. E. Hamm, R. Wolkowsky-Tyl and M. Phelps, Histopathology of acute toxic response in rats and mice exposed to methyl chloride by inhalation, Fund. Appl. Toxicol. 2:293 (1982).

18. L. R. Pohl, B. Bhooshan, N. K. Whittaker and G. Krishna, Phosgene, a metabolite of chloroform, Biochem. Biophys. Res. Commun. 79:684 (1977).

19. R. L. Hawke and R. M. Welch, Major differences in the specificity and regulation of mouse renal cytochrome P-450 dependent monooxygenases, Molec. Pharmacol. 27:283 (1985).

20. P. Eyer and D. Podhradsky, Evaluation of a micromethod for determination of glutathione using enzymatic cycling and Ellman's Reagent, Anal. Biochem. 153:57 (1986).

21. O. W. Griffith, Determination of glutathione and glutathione disulfide using glutathione reductase and 2-vinylpyridine, Anal. Biochem. 106:207 (1980).

22. G. J. Chellman, R. D. White, R. M. Norton and J. S. Bus, Inhibition of the acute toxicity of methyl chloride in male B6C3F1 mice by glutathione depletion, Toxicol. Appl. Pharmacol. 86:93 (1986).

23. J. Sedlak and R. H. Lindsay, Estimation of total, protein-bound and nonprotein sulfhydryl groups in tissue with Ellman's reagent, Anal. Biochem. 25:192 (1968).

24. W. M. Kluwe, The nephropathy of low molecular weight halogenated alkane solvents, pesticides, and chemical intermediates, in: "Toxicology of the Kidney", J. B. Hook, ed., Raven Press, New York (1981).

25. Z. Zaleska-Rutczynska and S. Krus, Effect of chloroform on the mouse kidney. II. Resistance of the F1 generation of a susceptible C3H/He strain and a resistant C57BL6/6JN strain, Pathol. Pol. 23:185 (1972).

26. G. J. Chellman, K. T. Morgan, J. S. Bus and P. K. Working, Inhibition of methyl chloride toxicity in male F-344 rats by the anti-inflammatory agent BW755C, Toxicol. Appl. Pharmacol. 85:367 (1986).

27. H. Peter, D. Hopp, U. Huhndorf and H. M. Bolt, Untersuchungen zur Methylbromid-Vergiftung und ihrer Behandlung mit N-Acetyl-Cystein, Verh. Dtsch. Ges. Arbeitsmedizin (Gentner, Stuttgart) 25:535 (1985).

RELATIONSHIP BETWEEN STRUCTURE AND MUTAGENIC/CARCINOGENIC ACTIVITY OF

SHORT CHAIN ALIPHATIC HALOCOMPOUNDS: A COLLABORATIVE PROJECT

Giorgio Cantelli-Forti[1], G. L. Biagi[1], P. A. Borea[2], G. Bronzetti[3], M. C. Guerra[1], P. Hrelia[4], M. Paolini[1], S. Simi[3] and Sandro Grilli[5]

[1]Istituto di Farmacologia, Università degli Studi
40126 Bologna, Italy
[2]Istituto di Farmacologia, Università degli Studi
44100 Ferrara, Italy
[3]Istituto di Mutagenesi e Differenziamento del CNR
56100 Pisa, Italy
[4]Division of Environmental Toxicology, The University
of Texas Medical Branch, Galveston, TX 77550 USA
[5]Istituto di Cancerologia, Università degli Studi
40126 Bologna, Italy

The halogenated hydrocarbons represent one of the most important categories of industrial chemicals owing to their use, production volume, environmental and toxicological activity and, hence most important, potential population risk. They are probably the most ubiquitous in occurrence. A number of these, because of their use as pesticides and aerosol propellents and their high chemical stabilities, have become distributed throughout the biosphere. Occupational exposure, seems to be the major risk[1]. In fact, several of these compounds are carcinogens in test animals[2,3,4] and their acute and chronic toxic effects on liver, kidney and central nervous system have been demonstrated[5,6]. However, in most cases, no adequate human data are available for estimating risk by halocompounds. They react with nucleophilic substrates by direct attack or after bioactivation by enzymatic systems. The direct reactivity of halocompounds decreases from iodo- to bromo- and chloro-substituted compounds. More generally, halocompounds are enzymatically "activated" to interact with macromolecules. The first metabolic step is an oxidation due to the mixed function oxidase system (MFO) in the presence of cytochrome P-450 with the exception of carbon tetrachloride and halothane metabolism where initial reductive reactions by MFO occur. Oxidation by P-450 dependent-MFO activity results in the addition of radical oxene to the carbon-carbon bond of substrate that leads to epoxide formation. Epoxides can undergo enzymatic or non-enzymatic hydrolysis or can isomerize to acyl-halides and haloaldehydes in aqueous environment (Fig. 1). Detoxification of most halocompounds proceeds via conjugation with glutathione in the presence of cytosolic glutathione-S-transferase with the exception of 1,2-dichloro-ethane (1,2-DCE) and 1,2-dibromoethane (1,2-DBE), whose conjugates with glutathione are known as mutagenic and carcinogenic species. Correlations between structure and activity have been postulated in order to explain the carcinogenicity of metabolites from halocompounds and particularly from halo-ethylenes[7]. On this subject _in vitro_ and _in vivo_ studies on vinyl chloride have demonstrated that the carcinogenicity of haloethylenes is determined by the rate of epoxidation and by stability and reactivity of epoxides. Epoxides require an "optimum" of stability to cause genotoxic response. Indeed, an epoxide, whose stability is too low, has too short life to reach distant targets. The stability of the epoxide ring is determined by the number of halosubstitutions. The epoxide C-O bond that has less substituted carbon in the halogenated oxirane ring is the weaker of the two bonds.

Haloethylenes with epoxides that fall within defined "threshold bond" limit should be considered potentially hazardous. Thus, potential carcinogenic activity is higher when asymmetric halosubstitution is lesser. With respect to haloethanes it has been postulated that covalent binding to DNA depends on energetic conditions of molecular orbitals which can accept the electron density from nucleophilic bases of DNA. Higher electron density of two carbon atoms leads to major affinity to covalent binding. The electron affinity increases with enhancement of the number of halosubstitutions. Since results in the literature are often contradictory for some halogenated compounds and incomplete for others, the aim of this collaborative work is to investigate, by using different end-points, the mutagenic/carcinogenic potential of short-chain aliphatic halocompounds in order to provide a quantitative structure-activity relationship study to predict the potential human risk. For this reason a collection of highly pure compounds ($\geq$ 95%) has been provided and stored at their respective appropriate temperatures. Before carrying out each experiment the purity of the test compound is controlled and strong attention is given in particular to the volatile property of these compounds from which may depend some of these aforementioned contradictory results.

In this paper the results obtained up to now with five haloethanes tested in microorganisms and in V79 cells are described. An in vivo study of the effects of these chemicals on drug-metabolizing enzymes is also considered in this project, as shown by the results obtained with 1,1,2,2-tetraclhoroethane (1,1,2,2-TTCE). As an expression of lipophilic character the chromatographic log k' and log P values were already determined for the 19 compounds considered. At the end of this collaborative project (1987-1989) more physico-chemical parameters will be determined and the above biological activities will be completed to provide a quantitative structure-activity relationship (QSAR) analysis.

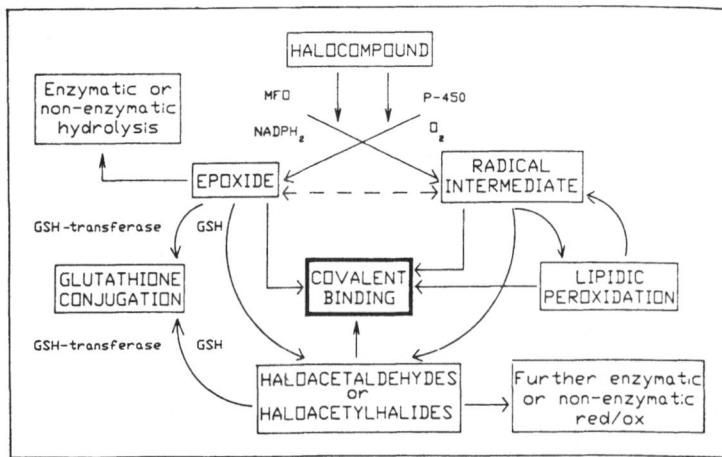

Fig. 1. General scheme of metabolism of halocompounds. MFO = Mixed Function Oxidase(s).

MATERIALS AND METHODS

Chemicals

The 19 short-chain aliphatic halocompounds under study are listed in Table 1. They were obtained from different sources and were at the highest rate of purity available. The test compounds were dissolved in DMSO before use. The other chemicals and solvents used were reagent grade.

Microorganisms Systems

Yeast test. The diploid D7 strain of Saccharomyces cerevisiae carries trp5-12/trp5-27 for detecting conversion events and a pair of alleles, ade2-40/ade2-119 for evaluating induction of mitotic reciprocal recombination. In addition, the strain is homoallelic ilv-92/ilv-92 for studying induced reverse mutation[8]. Cells from stationary phase (density 2 x 10^8 cells/ml) and from logarithmic growth phase (from a 20% glucose medium; density 5 x 10^7 cells/ml) were used in the experiments[9]. In the experiments with cells in the stationary phase, the metabolic activation was provided by 1 ml of S9 fraction from Aroclor 1254-induced rats and relative cofactors were added to the incubation mixtures[10]. Chemical inducing dose-related increases in the genetic effects and at least a doubling of the frequencies observed over the concurrent controls were considered as positive.

Salmonella test. The test procedure was performed as described by Ames et al.[11] and Maron and Ames[12]. The S9 fraction was used as metabolic activation system[10]. Where more than twice the number of revertants than the control were attained, the results were considered as positive only.

Cytogenetic Analysis.

V79 Cells Test. V79 cell line was established from a male Chinese

Table 1. Halogenated hydrocarbons under study

Chemicals	Molecular weight	Purity	Source
1,1-Dichloroethane	98.96	97%	Fluka-Aldrich
1,2-Dichloroethane	98.96	99.5%	Aldrich
1,1,1-Trichloroethane	133.42	99%	Fluka
1,1,2-Trichloroethane	133.42	99.5%	Fluka
1,1,2,2-Tetrachloroethane	167.86	98%	Fluka-Inalco
1,1,1,2-Tetrachloroethane	167.86	98.5%	Inalco
Pentachloroethane	202.31	95%	Aldrich
Monobromoethane	108.98	99%	Inalco
1,2-Dibromoethane	187.88	99.5%	Aldrich
1-Bromo-2-chloroethane	145.43	98%	Inalco
1,1,2,2-Tetrabromoethane	345.70	98%	Inalco
Monoiodoethane	155.98	99%	Inalco
1,1-Dichloroethylene	96.95	99%	Inalco
Cis-1,2-dichloroethylene	96.95	98%	Inalco
Trans-1,2-dichloroethylene	96.95	98%	Inalco
1,1,2-Trichloroethylene	131.40	98%	Inalco
1,1,2,2-Tetrachloroethylene	165.85	94%	Inalco
1,2-Dibromoethylene	185.87	98%	Inalco
Methylene dichloride	84.94	99%	Aldrich

hamster (cricetulus grisens) by Ford and Yerganian[13], and it has a model chromosome number of 22, which is also the diploid number of the Chinese hamster. V79 cells were routinely grown in Dulbecco's Eagle's Minimal Essential Medium, containing 100 IU of penicillin and 100 μg of streptomycin per ml, supplemented with 5% of fetal calf serum. In these conditions, the average cell cycle lasts 12 h. For each experiment, cultures were set up the day before treatment at a density of 8 x 10^5 cells per 100 mm Petri disk, allowing exponential growth at the time of treatment with agents[14]. After 16 h of treatment (time 0), cells were trypsinized, washed and re-fed with complete fresh medium. Cells were fixed at time 0 and 24 h later (time 24) for chromosome analysis. Cells were treated with 0.025 μg/ml of Colcemide for 1.5 h, then the cells were trypsinized, centrifuged and incubated with 10 ml of 0.075 M KCl at room temperature for 8 minutes, fixed in 3:1 methanol acetic acid 3 times, then spread on slides which were allowed to dry. For G-banding preparations, the trypsin digestion procedure was used[15].

Effects on MFO system (P-450 and P-448). Hepatic microsomes (105,000 x g) from uninduced Na-phenobarbital (500 ppm drinking water, ad libitum, for 12 days) ß-naphthoflavone (80 mg/kg in corn oil 48 h before sacrifice) and 1,1,2,2-TTCE (300 mg/kg) treated animals for each group were prepared as previously reported[16]. Pentoxyresorufin O-dealkylase and ethoxyresorufin in O-deethylase were determined as reported by Lubet et al.[17] and Klotz et al.[18], respectively. The protein concentration was determined according to Bailey[19].

Determination of log K' values. Chromatography was performed on a Waters 6000 A chromatograph using a μBondapak C 18 column (300 x 3.9 i.d.) (Waters), packed with silica gel (particle size 10 μm) with a C18 chemically bonded non-polar stationary phase. A UV detector (Waters Model 480) at 254 nm and Hamilton 802 chromatographic syringes (25 μl) were also used. The short chain aliphatic halocompounds were separated using methanol-water in various mixtures as the mobile phase at a flow-rate of 1 ml. Samples were dissolved in methanol (1 mg/ml) and applied to the column in 5 μl volumes. All solutions were before filtered to reduce contamination. The experiments were performed at room temperature (20-22°C). The retention times were expressed as log capacity factor (k'), where

$$k = \frac{(tx - to)}{to}$$

Statistics. The statistical analysis was made by the rank method of Wilcoxon as reported by Box and Hunter[20].

RESULTS AND DISCUSSION

Mutagenicity in Microorganisms

The test compounds were assayed in both yeast and Salmonella tests at four levels of concentration, proportionally decreasing from the dose in which a toxic effect was recorded. In Table 2 are reported the results from Salmonella typhimurium assay in presence and in absence of mammalian microsomal activation system. All the compounds under study showed positive results not only in the presence but also in the absence of S9 activation system. The most sensitive strains were TA100, TA98 and TA102. For all compounds no positive results were obtained in TA97, which seems to be the most liable strain to toxic effect. 1,1-Dichloroethane (1,1-DCE) was the compound less effective in this test showing a positive activity in TA100 strain with and without S9 fraction and in TA102 strain with S9 activation system. 1,2-Dichloroethane (1,2-DCE) was detected in base-pair strains TA1535 and TA100 and frame-shift strain TA98 without metabolic activation. On the contrary, a positive response in TA102 was observed only with the S9 fraction. A wide

Table 2 Evaluation of the effect of some short-chain aliphatic halocompounds under study when these chemicals were incubated with different **Salmonella typhimurium** strains, in presence and absence of S9 liver fraction from Aroclor 1254-induced rats. The results are the mean ± SD of three to five independent experiments

		R_t/R_c [a]									
Compound	Concentration μmol/plate	TA-1535		TA-100		TA-97		TA-98		TA-102	
		-S9	+S9	-S9	+S9	-S9	+S9	-S9	+S9	-S9	+S9
1,1 DCE	20	1.47±0.13	1.04±0.22	1.23±0.17	1.96±0.12	1.21±0.37	1.04±0.18	1.68±0.17	1.03±0.31	0.94±0.28	1.25±0.34
	40	1.94±0.18	0.99±0.20	2.61±0.29	2.06±0.37	0.88±0.15	0.65±0.10	1.35±0.21	1.73±0.24	1.33±0.20	1.17±0.25
	80	1.65±0.11	TOX	0.86±0.15	0.85±0.18	TOX	TOX	1.08±0.16	1.56±0.23	1.15±0.27	1.64±0.16
	160	1.21±0.09	TOX	1.19±0.16	TOX	TOX	TOX	1.51±0.26	TOX	TOX	2.15±0.28
1,2 DCE	10	1.83±0.18	1.27±0.20	1.46±0.24	0.92±0.16	0.89±0.16	0.95±0.09	1.12±0.22	1.32±0.21	0.88±0.15	1.56±0.16
	20	2.06±0.21	0.94±0.31	2.11±0.15	1.09±0.23	TOX	TOX	2.02±0.42	1.61±0.27	1.11±0.12	2.28±0.13
	40	2.47±0.27	1.12±0.23	2.82±0.40	1.73±0.31	TOX	TOX	1.27±0.33	1.55±0.16	1.15±0.21	1.16±0.08
	80	1.63±0.19	TOX	0.85±0.21	TOX	TOX	TOX	1.48±0.23	TOX	0.85±0.27	TOX
1,1,1 TCE	20	0.93±0.19	1.21±0.26	1.77±0.31	1.04±0.12	1.14±0.31	0.82±0.18	2.22±0.36	1.78±0.28	1.45±0.10	2.41±0.18
	40	2.12±0.25	1.15±0.44	2.40±0.24	1.87±0.31	1.76±0.18	1.11±0.25	3.11±0.64	3.23±0.36	2.19±0.17	3.58±0.36
	80	1.92±0.29	0.83±0.28	3.15±0.37	2.16±0.27	1.21±0.13	1.05±0.16	3.47±1.06	2.56±0.18	2.87±0.35	1.76±0.20
	160	1.47±0.12	TOX	1.26±0.15	0.89±0.20	TOX	TOX	1.31±0.29	TOX	3.26±0.24	TOX
1,1,2 TCE	10	0.86±0.23	1.11±0.26	1.04±0.37	0.90±0.18	0.79±0.21	0.71±0.16	2.69±0.41	2.05±0.29	1.27±0.21	1.17±0.19
	20	0.82±0.18	1.02±0.31	1.78±0.21	1.04±0.09	0.94±0.18	TOX	2.27±0.53	3.23±0.52	1.19±0.13	1.32±0.19
	40	0.75±0.16	1.08±0.18	1.02±0.18	0.94±0.17	1.02±0.32	TOX	0.70±0.18	3.38±0.38	1.84±0.24	TOX
	80	1.18±0.20	TOX	0.72±0.26	TOX	1.36±0.27	TOX	1.41±0.24	TOX	2.63±0.19	TOX
1,1,2,2 TTCE	5	1.86±0.40	1.64±0.12	2.42±0.32	1.98±0.20	0.74±0.05	0.91±0.09	2.71±0.22	3.50±0.44	1.87±0.16	1.16±0.11
	10	1.17±0.29	0.91±0.11	1.96±0.16	1.85±0.14	TOX	TOX	2.14±0.18	2.85±0.13	1.04±0.08	0.84±0.06
	20	0.70±0.15	0.98±0.08	0.87±0.07	0.97±0.13	TOX	TOX	1.21±0.15	0.81±0.08	1.21±0.12	0.61±0.16
	40	TOX	TOX	TOX	TOX	TOX	TOX	TOX	TOX	TOX	TOX
Hycanthone	40 μg/plate	7.86±0.94		5.99±0.91		2.73±0.51		3.72±0.69		19.38±1.15	
Methylmethansulphonate	1 μl/plate										
Sodium azide	2 μg/plate										
2 - Aminofluorene	1 μg/plate		7.12±0.55		6.57±0.30		15.43±0.84		12.37±0.41		2.97±0.65
Cyclophosphamide	200 μg/plate										

[a] R_t/R_c is the ratio of the number of revertants on treated plates to the number of spontaneous revertants on control plates The spontaneous number of revertants for the five strains were (mean ±SD). TA-1535,25±6; TA-100,131±32; TA-97,140±26; TA-98,49±4; TA-102,245±37.

spectrum of activity was observed with 1,1,1-trichloroethane (1,1,1-TCE), which appeared to be the most mutagenic compound in this test. 1,1,2-Tri-chloroethane (1,1,2-TCE) was detected by TA98 strain both with and without metabolic activation system and without S9 fraction in TA102 strain at the upper doses. 1,1,2,2-TTCE was assayed at four-fold lower doses than the other chemicals under study. However, a general toxic effect appears to be exerted on all strains. Significative increases in revertants were recorded with TA98 at 5 and 10 µmol/plate with and without S9 fraction and with TA100 strain at 5 µmol/plate in absence of S9 fraction. By using cells from sta-tionary phase, the short-chain aliphatic halocompounds under study did not induce either point mutation or mitotic gene conversion in Saccharomyces cerevisiae D7 strain in suspension test with and without metabolic activation system (Table 3). On the contrary, genotoxic results were obtained in cells of the same strain of yeast from logarithmic growth phase with the exception of 1,1-DCE which did not show any genetic event. 1,2-DCE was a weak mutagen giving a significant (P < 0.05) increase in convertants frequency at 60 mM dose only. In addition, 1,1,1-TCE increases only the gene conversion at two doses (20 and 30 mM). Genetic activity was found for both 1,1,2-TCE and 1,1,2,2-TTCE in the frequencies of trp5 conversion and ilvl-92 reversion. In particular 1,1,2,2- TTCE in the yeast test was the most effective compound showing genotoxicity at dose levels 4 or 8 times lower than the other test compounds. However, the mutagenic potency was very low; but the lack of dose-response relationship could depend on the high toxicity of this com-pound. The general aspect of the results obtained from the tests on micro-organisms, seems to confirm the poor sensibility of these systems (2-3 fold increase) in detecting the halogenated compounds. However, positive results were obtained and in both systems these are related. In particular, 1,1,1-TCE and 1,1,2-TCE are, in microorganism systems, the most mutagenic haloalkanes considered in this part of the study. Regarding the metabolic activation the literature reports that these compounds can act as direct mutagens or after metabolic activation[6]. In the Salmonella system the posi-tive results in the experiments with S9 metabolic fraction can depend on the balance between glutathione-conjugation and/or P-450 system (activation pro-cesses) and the detoxification processes by enzymatic pathways and/or unspec-ific binding with proteins. In Saccharomyces cerevisiae test the positive results obtained only in cells from logarithmic growth phase, can depend on different conditions of cell permeability and on enzyme content in these culture conditions.

Cytogenetic Analysis in V79 Cells

Of the five compounds tested, 1,1,1-TCE and 1,1,2-TCE were found to in-duce no chromosomal aberrations, while both dichloroethane and tetrachloro-ethane were able to induce them, even if without a dose-response relationship (Table 4). The analysis of data showed that chromosomal breaks are mostly induced, with a marked excess of centromeric breaks, giving rise to acrocen-tric "marker" chromosomes. In order to identify the chromosome bearing the centromeric breaks occurred preferentially on chromosome 1 and 3, producing acrocentric chromosomes which are the short arm of chromosome 1 and the long arm of chromosome 3. The frequency observed in cells treated with dichloro-ethanes is far more than can be observed in control cultures (Table 5). These results show that certain agents are able to induce chromosomal aberra-tions at specific sites, or make certain sites more prone to breakage. Cen-tromeric break on chromosome 3 is particularly interesting since it is also the location of a folate-sensitive fragile site found in this cell line[21]. Data from humans suggest that fragile sites may act as predisposing factors to cancer because of their association with break points involved in non-random chromosomal aberrations associated with leukemia[22]. In Chinese ham-ster cells, there has been reported a strong correlation between trisomy of of the long arm of chromosome 3 and tumorigenicity[23]. Recently two oncogenes (ras and fes) have been mapped on chromosome 3 [24]. Moreover, it has been

58

Table 3 Induction of gene conversion and point mutation in **Saccharomyces cerevisiae** D7 strain by short-chain aliphatic halocompounds under study
The results are the mean ± SD of three to five independent experiments ([a] p < 0.05; [b] p < 0.01).
The S9 fraction was prepared from Aroclor 1254 - induced rats

Compound	Concentration (mM)	CELLS FROM STATIONARY GROWTH PHASE						CELLS FROM LOGARITHMIC GROWTH PHASE		
		Survival %		Convertants/10^5 surv.		Revertants/10^6 surv.		Survival %	Convertants/10^5 surv.	Revertants/10^6 surv.
		-S9	+S9	-S9	+S9	-S9	+S9			
CONTROL	—	100	100	0.83±0.44	1.08±0.31	0.43±0.17	0.36±0.21	100	0.87±0.32	0.47±0.20
1,1 - DCE	20	98	94	1.18±0.28	1.16±0.14	0.32±0.24	0.32±0.19	98	0.86±0.06	0.32±0.17
	40	89	87	0.90±0.31	0.86±0.15	0.34±0.13	0.29±0.10	94	0.83±0.10	0.33±0.12
	60	83	82	1.06±0.33	0.84±0.26	0.45±0.27	0.38±0.09	83	0.88±0.19	0.28±0.16
	80	73	79	1.11±0.47	0.91±0.10	0.51±0.12	0.45±0.24	71	0.99±0.13	0.42±0.07
1,2 - DCE	30	73	71	1.18±0.37	1.11±0.11	0.37±0.16	0.39±0.11	86	1.08±0.18	0.48±0.13
	60	65	55	0.97±0.15	1.06±0.18	0.28±0.09	0.26±0.07	74	1.42±0.29[a]	0.64±0.10
	90	57	55	0.81±0.09	1.03±0.12	0.36±0.11	0.27±0.10	58	1.34±0.27	0.54±0.16
	120	TOX	TOX	TOX	TOX	TOX	TOX	34	1.32±0.42	0.47±0.21
1,1,1 - TCE	10	92	91	1.26±0.39	1.34±0.36	0.50±0.18	0.22±0.06	76	0.85±0.14	0.37±0.14
	20	91	65	1.08±0.21	1.50±0.47	0.48±0.19	0.33±0.11	70	1.45±0.26[b]	0.54±0.18
	30	39	53	0.55±0.23	1.10±0.22	0.26±0.10	0.34±0.12	22	1.90±0.32[b]	0.75±0.26
	40	TOX	TOX	TOX	TOX	TOX	TOX	TOX	TOX	TOX
1,1,2 - TCE	10	81	75	0.90±0.15	0.25±0.05	0.49±0.15	0.29±0.14	96	0.72±0.18	0.32±0.09
	20	58	65	0.94±0.17	0.29±0.19	0.73±0.24	0.32±0.09	77	1.03±0.27	0.32±0.12
	30	28	TOX	0.72±0.08	TOX	0.29±0.05	TOX	42	2.50±0.83[a]	1.87±0.28[b]
	40	TOX	TOX	TOX	TOX	TOX	TOX	TOX	TOX	TOX
1,1,2,2 - TTCE	2.5	96	95	1.16±0.38	1.19±0.29	0.48±0.27	0.48±0.21	85	1.30±0.27	0.50±0.13
	5.0	65	93	1.47±0.51	1.11±0.37	0.72±0.33	0.48±0.15	83	1.08±0.33[b]	0.34±0.10
	7.5	TOX	57	TOX	0.53±0.18	TOX	0.26±0.11	65	2.13±0.38[b]	1.03±0.39[a]
	10.0	TOX	TOX	TOX	TOX	TOX	TOX	59	1.85±0.55	0.69±0.26
Methylmethansulphonate	2	96	94	20.85±3.11[b]	0.98±0.27[b]	6.32±0.83[b]	0.51±0.11[b]	57	2.81±0.49[b]	1.13±0.27[a]
Dimethylnitrosamine	300	81	91	0.94±0.39	3.16±0.34	0.41±0.09	1.53±0.25			

Table 4 Cytogenetic effects of five chlorinated ethanes in V79 Chinese hamster cells in culture. For each dose, the first row shows the data at t = 0, the second row shows the data at t = 24

TREATMENT		CHROMATID ABERRATIONS[a]		CHROMOSOMAL ABERRATIONS[b]							N° ANALYSED CELLS	ABNORMAL CELLS[b]		% OF CELLS WITH Bc
		G	B	G	B	Bc	End	Pol	Quad	Dic		N	%	
CONTROL	V79/AP4	1		6	3	3					100	6	6.0	3.0
	DMSO	4		6	1	22	6			1	270	30	11.1	8.1
1,1 DCE	8 mM			3 / 3	5 / 1	20 / 24	3 / 3	2 / 1			115 / 150	30 / 29	26.1 / 19.3	17.3 / 16.0
	30 mM	2		3 / 3	9 / 4	31 / 24	3 / 2	1 / 4			150 / 150	40 / 37	26.6 / 24.6	20.6 / 16.0
	60 mM	1	5		6 / 8	17 / 6	1 / 5	3	2		80 / 50	23 / 23	28.7 / 46.0	21.2 / 12.0
	80 mM	1	1	2	22 / 4	10 / 6	2 / 3	2			50 / 32	22 / 15	44.0 / 46.8	20.0 / 18.7
1,2 DCE	8 mM	1	5	3	7	28	6				140	42	30.0	20.0
	15 mM	2	2	1	6	14	5	1			66	27	40.1	21.2
	30 mM	1	1	2	7	18	7	6		1	122	41	33.6	14.7
	60 mM		2		1 / 4	11 / 11	3				50 / 54	15 / 15	30.0 / 27.7	22.0 / 20.3
1,1,1 TCE	10 mM	1			1	9 / 13	3	5			100 / 110	12 / 14	12.0 / 12.7	9.0 / 11.8
	20 mM			1	1	7 / 11					60 / 110	8 / 11	13.3 / 10.0	11.6 / 10.0
1,1,2 TCE	10 mM	2	1	3	5	23 / 18	3 / 3	3			190 / 150	27 / 25	14.2 / 16.6	12.1 / 8.6
	20 mM	1	2		6	25 / 28	4 / 3	2			150 / 150	31 / 29	20.6 / 19.3	16.6 / 18.6
1,1,2,2 TTCE	2 mM	3			11	18	6	5		2	150	41	27.3	12.0
	4 mM				12	4		2			50	9	18.0	8.0
	6 mM	1		1	4	6	3	1			52	14	26.9	11.5

[a] G: gap, B: breack, Bc: centromeric break; End: endoreduplicated cells, Pol: polyploid; Quad: quadriradials; Dic: dicentrics,

[b] gaps are excluded

Table 5. G-Banding analysis of V79 cells treated with dichloroethanes

Treatment	Chromosomal number	No. of analyzed cells	Chromosomal Aberrations		Others
			Centromeric break of chromosome 3	Centromeric break of chromosome 1	
Control	22	37			
	21	11			
	20	2			
Total		50	2 (4.0%)	0 (0.0%)	3 (6.0%)
1,1-DCE 30 mM	22	23	7	3	0
	21	7	2	1	0
Total		30	9 (30.0%)	4 (13.3%)	0 (0.0%)
1,2-DCE 15 mM	23	2	1	0	1
	22	24	3	1	1
	21	4	1	0	0
Total		30	5 (16.7%)	1 (3.3%)	2 (6.6%)

reported[25] that mutagenic as well as carcinogenic compounds are able to induce chromosomal gaps and breaks at, or near, fragile sites on human chromosomes. This suggests that fragile sites can be general targets of mutagenic action.

Effects on MFO System (P-450 and P-448)

The in vivo effect of 1,1,2,2-TTCE on the hepatic microsomal monooxygenase activity from uninduced and phenobarbital (PB) or ß-naphthoflavone (ß-NF)-induced mice, are reported in Fig. 2. Male and female animals were used in the experiments. In uninduced mice both ethoxyresorufin O-deethylase

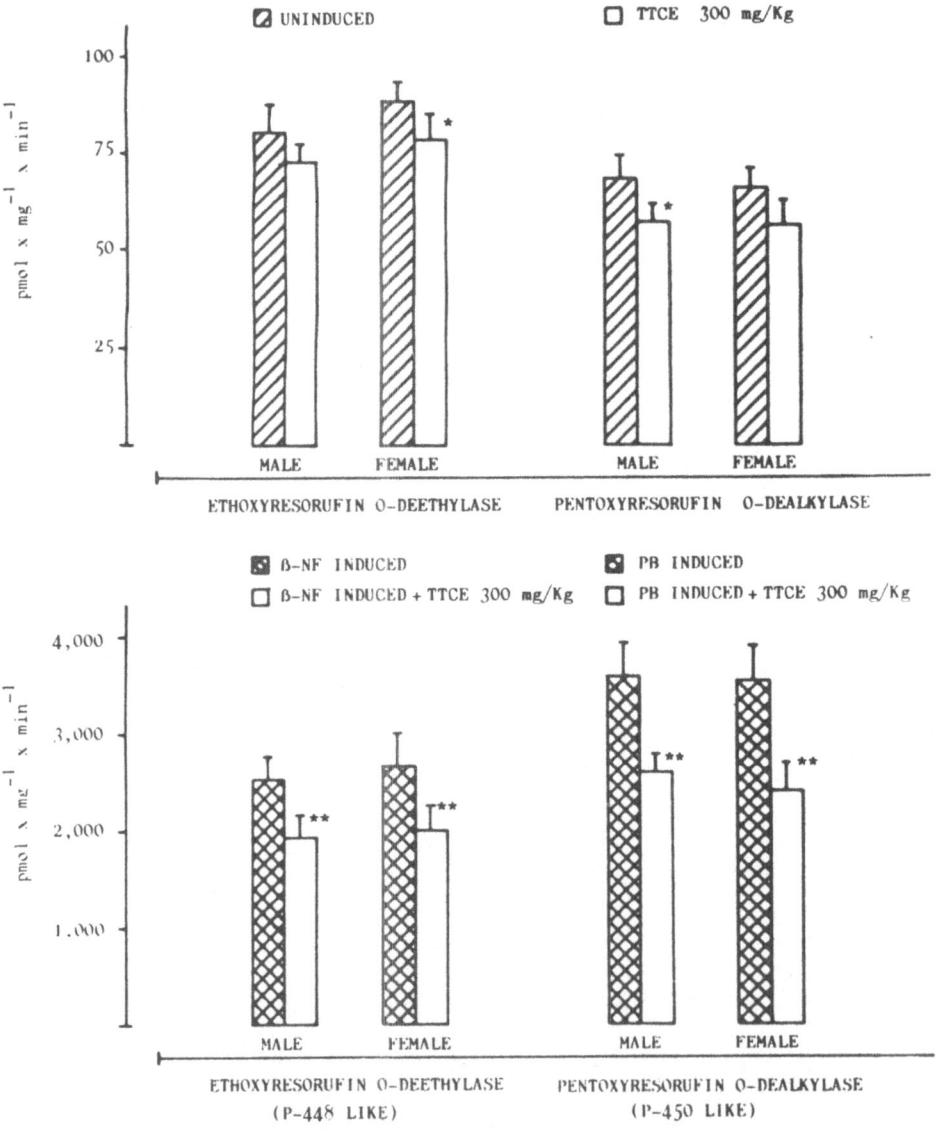

Fig. 2. Effect of 1,1,2,2-tetrachloroethane (TTCE) on microsomal enzymatic activities of liver from uninduced and ß-naphthoflavone (ß-NF) or phenobarbital (PB)-induced mice.

(P-448-like activity) and pentoxyresorufin O-dealkylase (P-450-like activity) were decreased by 1,1,2,2-TTCE treatment (P < 0.05) in, respectively, female and male animals. A significant decrease in the monooxygenase activities was achieved in both ß-NF and PB-preinduced mice (P < 0.01); the toxic effect of 1,1,2,2-TTCE is related to the isoenzymes of the "P-450 area", as shown by the decrease in pentoxyresorufin dealkylase (approximately 33% loss), as well as to the isoenzymes of the "P-448 area" as demonstrated by the approximately 28% loss of ethoxyresorufin O-deethylase activity.

QSAR Study

In this section of the research, only the chromatographic log k' values,

Table 6. Log k' (70% methanol) and log P values of short-chain aliphatic halocompounds

No.	Compound	Log k'	Log P
1.	1,1-Dichloroethane	-0.03	1.79
2.	1,2-Dichloroethane	-0.20	1.48
3.	1,1,1-Trichloroethane	0.18	2.49
4.	1,1,2-Trichloroethane	0.00	2.49
5.	1,1,1,2-Tetrachloroethane	0.31	2.88
6.	1,1,2,2-Tetrachloroethane	0.19	2.88
7.	Pentachloroethane	0.30	3.27
8.	1-Bromo-2-chloroethane	-0.08	2.00
9.	1,2-Dibromoethane	0.06	2.21
10.	1,1,2,2-Tetrabromoethane	0.41	3.41
11.	Bromoethane	-0.10	1.61
12.	Iodoethane	0.10	2.00
13.	Dichloromethane	-0.31	1.25
14.	1,1-Dichloroethylene	0.00	1.95
15.	1,2-Dichloroethylene (cis)	-0.04	1.93
16.	1,2-Dichloroethylene (trans)	-0.03	1.93
17.	1,1,2-Trichloroethylene	0.17	2.32
18.	1,2-Dibromoethylene	-0.01	2.35
19.	1,1,2,2-Tetrachloroethylene	0.20	2.71

determined by means of reversed-phase HPLC, and the log P values are reported in Table 6. A linear regression analysis showed that log k' values are well correlated with the log P values, as an expression of the lipophilic character of molecules, described by the following equation:

$$\log k' = -0.611 \ (\pm 0.064) + 0.295 \ (\pm 0.028) \ \log P$$

$$n = 19; \ r = 0.933; \ s = 0.068; \ f = 114.41; \ P < 0.005$$

The next step of this work will be a quantitative structure activity relationship (QSAR) between the considered biological activities and log k' and/or log P values of the present series of compounds [26,27].

CONCLUSIONS

The QSAR can be used as an experimental tool to predict the genotoxic risk of this class of halogenated compounds. The validity of this type of study depends on the broadness and variety of the biological data (mutagenic, cytogenetic and biochemical tests) and on the uniformity of data collection (high level of purity of compounds, uniform research methods on standard protocols, etc.). This paper is a presentation of our research goal and at present it is not possible to draw definitive conclusions and it should be considered as only a current collaborative research report.

ACKNOWLEDGEMENTS

This work was supported by CNR (National Research Council of Italy) grants applied projects "Preventive and Rehabilitative Medicine" (contract No. 86.01707.56) and "Oncology" (contract No. 86.00340.44) and by AIRC (Italian Association for Cancer Research) Milan.

REFERENCES

1. IARC 1979, Monograph on the Evaluation of the Carcinogenic Risk of Chemicals to humans, IARC, Lyon, Vol. 20.
2. E. K. Weisburger, Carcinogenicity studies on halogenated hydrocarbons, Environ. Health Perspec. 21:7 (1977).
3. L. Fishbein, Potential halogenated industrial carcinogenic and mutagenic chemicals, II. Halogenated saturated hydrocarbons, Sci. Total Environ. 11:163 (1979).
4. L. Fishbein, Potential halogenated industrial carcinogenic and mutagenic chemicals, III. Alkane, Sci. Total Environ. 12:170 (1980).
5. W. F. Von Ottigen, "The Halogenated Hydrocarbons of Industrial and Toxicological Importance", Elsevier, Amsterdam (1964).
6. W. M. Kluwe and J. B. Hook, Metabolism of nephrotoxic haloalkanes, Fed. Proc. 39:3129 (1980).
7. G. Cantelli-Forti and G. Bronzetti, Mutagenesis and carcinogenesis of halogenated ethylenes, Ann. N. Y. Acad. Sci. in press (1986).
8. F. K. Zimmermann, R. Kern and H. Rasenberger, A yeast strain for simultaneous detection of induced mitotic crossing-over, mitotic gene conversion and reverse mutation, Mutat. Res. 28:381 (1975).
9. G. Cantelli-Forti and G. Bronzetti, Improvements in in vitro genotoxicity tests, in: "In Vitro Short-Term Assays", N. Loprieno and C. Pantarotto, eds., Plenum Press, New York, in press (1987).
10. G. Cantelli-Forti, M. Paolini, P. Hrelia, C. Corsi, G. L. Biagi and G. Bronzetti, NADPH-generating system: influence on microsomal mono-oxygenase stability during incubations for liver microsomal assay with rat and mouse S9 fractions, Mutat. Res. 129:291 (1984).

11. B. N. Ames, J. McCann and E. Yamasaki, Methods for detecting carcino-
    gens and mutagens with the Salmonella/mammalian microsome muta-
    genicity test, Mutat. Res. 31:347 (1975).
12. D. M. Maron and B. N. Ames, Revised methods for the Salmonella muta-
    genicity test, in: "Handbook of Mutagenicity test procedures", B. J.
    Kilbery, M. Legator, W. Nichols and C. Ramel, eds., 2nd Ed.,
    Elsevier Science Publisher, Amsterdam (1984).
13. D. K. Ford and G. Yerganian, Observation on the chromosomes of Chinese
    hamster cells in tissue cultures, J. Natl. Cancer Inst. 21:393
    (1958).
14. C. M. Colella, G. Rainaldi and A. Piras, 8-Azaguanine versus 6-thiogua-
    nine: influence on frequency and expression time of induced HGPRT-
    mutations in Chinese hamster V79 cells, Mutat. Res. 107:397 (1983).
15. M. A. Seabright, A rapid banding technique for human chromosomes,
    Lancet 2:971 (1971).
16. G. Cantelli-Forti, M. Paolini, P. Hrelia, E. Sapigni and G. L. Biagi,
    Effects of Metronidazole, Azanidazole and Azathioprine on cytochrome
    P-450 and various mono-oxygenase activities in hepatic microsomes
    from control and induced mice, Arch. Toxicol. suppl. 11:264 (1987).
17. R. A. Lubet, R. T. Mayer, J. W. Cameron, W. N. Raymond, D. M. Burke,
    T. Wolff and F. P. Guengerich, Dealkylation of pentoxyresorufin: a
    rapid and sensitive assay for measuring induction of cytochrome(s)
    P-450 by phenobarbital and other xenobiotics in the rat, Arch.
    Biochem. Biophys. 238:43 (1985).
18. A. V. Klotz, J. J. Stageman and C. Walsh, An alternative 7-ethoxyreso-
    rufin O-deethylase activity assay: a continuous visible spectro-
    photometric method for measurement of cytochrome P-450 mono-
    oxygenase activity, Anal. Biochem. 140:138 (1984).
19. J. L. Bailey, "Techniques in Protein Chemistry", Elsevier, Amsterdam
    (1967).
20. G. E. P. Box and W. G. Hunter, "Statistics for Experiments", Wiley,
    New York (1978).
21. S. Simi and L. Vatteroni, Methotrexate induced fragile sites in
    different Chinese hamster cell lines, Proc. IX Meeting of the
    European Association for Cancer Research, Helsinki (1987).
22. M. Le Beau and J. D. Rowley, Heritable fragile site in cancer,
    Nature 308:607 (1984).
23. I. K. Gadi, J. J. Harrison and R. Sagzer, Genetic analysis of tumori-
    genesis, XVI. Chromosome changes in azacytidine and insulin induced
    tumorigenesis, Somat. Cell and Molec. Genet. 10:521 (1984).
24. R. L. Stalling, B. D. Grawford, R. J. Black and E. H. Chang, Assignment
    of Ras proto-oncogenes in Chinese hamster: implications
    for mammalian gene linkage conservation and neoplasia, Cytogenet.
    and Cell Genet. 43:2 (1986).
25. J. J. Yunis, A. L. Soreng and A. E. Bowe, Fragile sites are targets of
    diverse mutagens and carcinogens, Oncogene 1:59 (1987).
26. C. Hansch and A. J. Leo, "Substituent Constants for Correlation
    Analysis in Chemistry and Biology", John Wiley, New York (1979).
27. G. L. Biagi, A. M. Barbaro, M. C. Guerra, G. Cantelli-Forti, G. Aicardi
    and P. A. Borea, Quantitative relationship between structure
    and mutagenic activity in a series of 5 nitroimidazoles, Teratogen.
    Carcin. Mut.. 3:429 (1983).

ENZYMATIC DETOXICATION OF TUMORIGENIC BAY-REGION DIOL-EPOXIDES OF

POLYCYCLIC AROMATIC HYDROCARBONS BY CONJUGATION WITH GLUTATHIONE

Bengt Jernström[1], Iain G. C. Robertson[2],
Bengt Mannervik[3] and Lennart Dock[4]

[1] Department of Toxicology, Karolinska Institutet
Stockholm, Sweden
[2] Department of Pharmacology and Clinical Pharmacology
University of Auckland
Auckland, New Zealand
[3] Department of Biochemistry, Arrhenius Laboratory
University of Stockholm
Stockholm, Sweden
[4] National Institute of Environmental Medicine
Stockholm, Sweden

INTRODUCTION

Polycyclic aromatic hydrocarbons (PAH), such as benzo(a)pyrene (BP), benz(a)anthracene (BA) and chrysene (C), are widely distributed contaminants in the environment and proven tumorigens in experimental animals. Epidemiological data indicate a role for PAH also in the etiology of certain human tumors[1]. PAH require metabolic transformation to electrophilic intermediates to exert their toxic effects, most probably through covalent binding of these intermediates to critical targets in DNA. BP is activated through the action of cytochrome P-450-linked monooxygenases and epoxide hydrolase to diastereomeric trans-7,8-dihydroxy-9,10-epoxy-7,8,9,10-tetrahydro-BP (anti- and syn-BPDE). The (+)-enantiomer of anti-BPDE, with R,S,S,R-absolute configuration, is the most tumorigenic one of the isomers in animals[2]. Similar results have been obtained with the equivalent diol-epoxides of BA and C[3]. Furthermore, covalent binding of (+)-anti-BPDE to the exocyclic nitrogen of deoxyguanosine in DNA of target tissues is closely correlated with tumor formation[4]. Thus, it is likely that enzymatic and non-enzymatic processes that prevent intracellular accumulation of PAH diol-epoxides will counteract DNA-damage and resulting consequences.

Several possible routes of diol-epoxide detoxication can be suggested, such as enzymatic or spontaneous hydrolysis to tetrols, or conjugation with nucleophilic cell constituents such as glutathione (GSH), nucleotides, lipids and proteins. However, the most important cellular defence system towards the accumulation and subsequent action of diol-epoxides seems to be conjugation with GSH[5,6].

The present paper reviews some of our findings on GSH-transferase (GST) catalyzed conjugation of tumorigenic anti-diol-epoxides of BP, BA (anti-BADE) and C (anti-CDE). Several GSTs have been isolated from human and rat tissues and their activities towards the different diol-epoxides determined.

MATERIALS AND METHODS

GST isoenzymes from rat liver and lung as well as from human liver and placenta were isolated as described previously[7-9]. The nomenclature according to Jakoby et al.[10] has been adopted for the rat GSTs. The purified GSTs were incubated with 2.5-110 μM diol-epoxide substrate and 1-2 mM GSH at 37°C for 30 or 60 sec[11-13]. Freshly isolated and viable hepatocytes were incubated at 37°C with 20 μM (+)-anti-BPDE. Aliquots of the incubate were removed at various time points and the reaction terminated by addition of alkaline mercaptoethanol and methanol[14]. The GSH conjugates of anti-BPDE, anti-BADE and anti-CDE were analyzed by ion-exchange HPLC with fluorescence detection as described[5,15]. The (+)- and (-)-enantiomer conjugates of the diol-epoxides were further resolved by HPLC on a reversed phase analytical column using an ammonium acetate/methanol solvent system[11,12]. This system was further improved to enable rapid and convenient analysis of the various products of anti-BPDE formed in isolated cells. A Waters Z-module with a 5μ C_{18} Radial-Pak analytical cartridge and a solvent system of 25 mM ammonium acetate, pH 3.5, (solvent A) in acetonitrile (solvent B) delivered at a flow rate of 6 ml/min was used to separate the GSH-conjugates of (+)- and (-)-anti-BPDE as well as tetrols and the mercaptoethanol conjugates of anti-BPDE in a single run (Fig. 1A). The gradient elution system employed was 15-20% solvent B, linear gradient for 10 min; 20-30% B, linear gradient for 10 min and 30-100% B, linear gradient for 3 min.

RESULTS AND DISCUSSION

In rat liver, about 5% of the soluble proteins are accounted for by various GST isoenzymes whereas the corresponding figure for rat lung is about 0.5%. Based on isoelectric points the GSTs may be described as basic, near-neutral and acidic. In both liver and lung, the basic forms greatly predominate (about 95% and 90% of total GST, respectively). Table 1 shows the relative amounts of the GSTs in rat liver and lung. Both qualitative and quantitative differences are evident. For example, subunit 1 appears in liver but not in lung and subunit 7 appears in lung but not in liver. GST

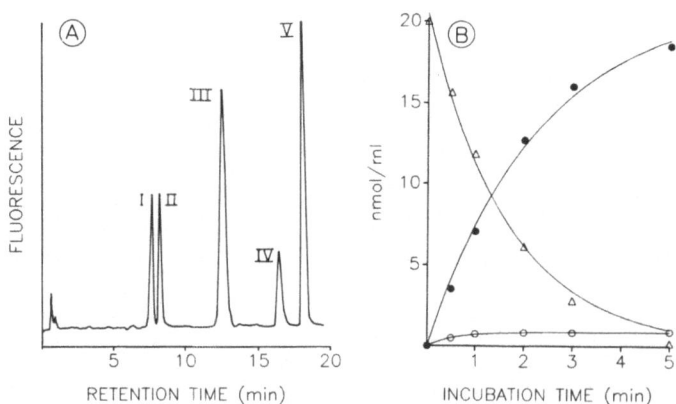

Fig. 1. HPLC separation of products derived from (±)-anti-BPDE and time dependent dispostion of (+)-anti-BPDE in isolated rat hepatocytes. A; Peaks I and II: GSH conjugates of (+)- and (-)-anti-BPDE, respectively; III: 7,10/8,9-tetrol; IV: 7/8,9,10-tetrol; V: BPDE-mercaptoethanol conjugate. B; -△-: BPDE-mercaptoethanol conjugate; -●-: BPDE-GSH conjugate; -O-: tetrols.

Table 1. Relative amounts of GSTs in rat
liver and lung

| GST | Relative amounts (%) | |
	Liver	Lung
"Acidic"	5	10
1-1	16	n.d.
1-2	17	n.d.
2-2	4	2
3-3	22	7
3-4	18	4
3-?	4	4
4-4	14	38
7-7	n.d.	35

n.d.: not detectable

5-5 is not included in Table 1 due to difficulties in its isolation and thus quantitation.

Kinetic parameters for the different GST isoenzymes with (±)-anti-BPDE as substrate are compiled in Table 2 together with the estimated contributions of the different forms to total GSH-conjugation activity. Transferases 4-4 and 7-7 are the most important forms in the lung; their combined activity is about 95% of the total calculated activity. Similar estimations using data for the liver isoenzymes demonstrate that the transferases containing subunit 4 account for approximately 90% of the total expected activity. These activities, calculated from the estimated distributions of the GST isoenzymes and the specific activities with anti-BPDE, agrees very well with the activities measured experimentally with cytosolic fractions (15-25 nmol conjugate/min x mg protein for rat liver and 10-15 nmol/min x mg protein for rat lung). Although the specific activity of lung cytosolic fraction expressed per mg GST protein is about 5-10 fold higher than in the liver, the total capacity of the liver can be estimated to be about 15-30 fold higher due to the larger tissue mass.

HPLC analysis of GSH conjugates derived from (±)-anti-BPDE by the rat GSTs showing high activities revealed almost exclusive conjugation (≥ 97%) of the (+)-enantiomer. It is interesting to note that the metabolism of BP to anti-BPDE in the rat is highly stereoselective and yields almost exclusively the (+)-enantiomer[16]. The rat is usually resistent to tumor formation by PAH and one factor is probably the very efficient and highly stereoselective GSH-conjugation of tumorigenic diolepoxides.

Table 3 shows data obtained with human GSTs and anti-BPDE. The basic forms (α-ε) exhibit low activity towards (±)-anti-BPDE, but the near-neutral (μ) and the acidic form (π) demonstrate high activity. However, both μ and π have lower catalytic efficiencies than the most active rat enzymes (GST 4-4 and 7-7). In contrast to GSTs α-ε and μ, GST π is highly selective for (+)-anti-BPDE. Incubation of μ or π with the individual enantiomers of anti-BPDE demonstrate that μ conjugates either enantiomer with equal efficiency. In contrast, transferase π exclusively conjugates (+)-anti-BPDE; the (-)-enantiomer is a competitive inhibitor of this reaction.

GST π is the dominant or exclusive form in adult extrahepatic and most fetal tissues[17]. GST μ on the other hand is absent or present in low amounts in fetal tissues and several extrahepatic adult tissues but present

Table 2. Kinetic parameters for purified rat GSTs with (±)-anti-BPDE as substrate and per cent contribution of individual isoenzymes to total conjugating activity

GST	V_{max}[a]	k_{cat}/K_m[b]	Contribution (%) Liver	Lung
"Acidic"			≤ 1	<3 [c]
1-1	9	0.5	≤ 0.5	n.d.
1-2	9	0.5	≤ 0.5	n.d.
2-2	8	0.5	≤ 0.5	$<<0.1$
3-3	16	0.9	2	$<<0.1$
3-4	440	24	35	1
3-?	340	19	6	1
4-4	880	46	55	14
7-7	5500	287	n.d.	. 81

[a] nmol x min^{-1} x mg^{-1}; [b] catalytic efficiency, mM^{-1} x s^{-1}; [c] not detectable

in adult liver in about 60% of the human population [18]. Therefore, in fetal and several extrahepatic adult tissues, including target organs such as lung and skin, the most important transferase in GSH-conjugation of (+)-anti-BPDE would be transferase π. In contrast, in the adult liver the predominant form would be transferase μ in those individuals having this form. Lack of transferase μ may increase the susceptibility of an individual to the tumorigenic action of PAH due to reduced capacity of elimination of reactive and harmful diol-epoxides by GSH-conjugation [18]. In fact, recent data obtained with human leukocytes indicate that in cigarette-smoking individuals, susceptibility to tumors in the lung is related to the polymorphic distribution of GST μ; individuals lacking GST μ have a significantly higher incidence of lung cancer than those expressing GST μ activity [19].

GST-catalyzed conjugation of GSH with PAH diol-epoxides is not re-

Table 3. Kinetic parameters and enantioselectivity for human GSTs with anti-BPDE as substrate

GST	BPDE[a]	V_{max}[b]	k_{cat}/K_m[c]	(+)-Enantiomer conjugate (%)
α - ε	racemic	40	<0.5	60
μ	racemic	600	18	60
μ	(+)	700	11	
μ	(-)	600	9	
π	racemic	900	13	>90
π	(+)	3000	28	
π	(-)	n.d.[d]		

[a] enantiomer of anti-BPDE used ; [b] nmol x min^{-1} x mg^{-1}; [c] catalytic efficiency, mM^{-1} x s^{-1}; [d] no activity detected under the assay conditions used

Table 4. Kinetic parameters and enantioselectivity for
rat GST 4-4 with racemic <u>anti</u>-diastereomers of
BPDE, BADE and CDE as substrates

Substrate	V_{max}^a	k_{cat}/K_m^b	(+)-Enantiomer conjugate (%)
(±)-<u>anti</u>-BPDE	600	42	98
(±)-<u>anti</u>-BADE	2100	14	95
(±)-<u>anti</u>-CDE	1500	12	97

^{a}nmol x min^{-1} x mg^{-1}; ^{b}catalytic efficiency, mM^{-1} x s^{-1}

stricted to <u>anti</u>-BPDE. Table 4 shows results obtained with rat liver GST 4-4
and (±)-<u>anti</u>-BADE and (±)-<u>anti</u>-CDE; results obtained with (±)-<u>anti</u>-BPDE are
included for comparison. It is evident that the catalytic efficiencies with
(±)-<u>anti</u>-BADE and (±)-<u>anti</u>-CDE are lower than with (±)-<u>anti</u>-BPDE. High
selectivity towards (+)-enantiomers of <u>anti</u>-diol-epoxides (R,S,S,R-absolute
configuration) is observed.

Lipophilic compounds such as PAH diol-epoxides are considered poor substrates for soluble GSTs in whole cells due to their partition into cellular
membranes [20]. The greatly extended half-life of <u>anti</u>-BPDE in the presence
of lipid-containing subcellular fractions is an expression of the stabilization of this substrate in the membrance phase [14]. However, this distribution is not a restriction for efficient and rapid conjugation of <u>anti</u>-BPDE
with GSH. Fig. 1B shows the disposition of (+)-<u>anti</u>-BPDE in isolated rat
hepatocytes and demonstrates GSH-conjugation as the most significant route of
elimination in an intact cell system with hydrolysis being of no or little
quantitative importance. The specific rate of GSH-conjugation observed with
isolated hepatocytes is very close to the rate obtained with cytosolic
fractions or the rate calculated from the contributions of the different
GSTs. Thus, it is unlikely that GST isoenzymes which significantly contribute to the total activity have not been accounted for.

In conclusion, mammalian tissues contain GST isoenzymes that are
highly efficient and selective in conjugating tumorigenic PAH diol-epoxides
with GSH. The GST/GSH system is probably the most important cellular defence against crucial DNA-damage caused by diol-epoxides. Polymorphism and
variations in GST distribution between tissues, species and individuals may,
in conjunction with cytochrome P-450-dependend monooxygenases and epoxide
hydrolase, determine the susceptibility to PAH-induced tumorigenesis.

ACKNOWLEDGEMENTS

This work was supported by grants from the Swedish Cancer Society, the
Swedish Natural Science Research Council and the Swedish Tobacco Company.

REFERENCES

1. A. Dipple, Polyaromatic hydrocarbon carcinogenesis, <u>in</u>: "Polycyclic
   Hydrocarbons and Carcinogenesis", R. G. Harvey, ed., American
   Chemical Society, Washington, D. C. (1985).
2. C. S. Cooper, P. L. Grover and P. Sims, The metabolism and activation of benzo(a)pyrene, <u>Prog</u>. <u>Drug</u> <u>Metab</u>. 7:295 (1983).

3. R. E. Lehr, S. Kumar, W. Levin, A. W. Wood, R. L. Chang, A. H. Conney, H. Yagi, J. M. Sayer and D. M. Jerina, The bay region theory of polycyclic aromatic hydrocarbon carcinogenesis, in: "Polycyclic Hydrocarbons and Carcinogenesis", R. G. Harvey, ed., American Chemical Society, Washington, D. C. (1985).

4. J. C. Pelling, T. J. Slaga and J. DiGiovanni, Formation and persistence of DNA, RNA and protein adducts in mouse skin exposed to pure enantiomers of 7ß,8α-dihydroxy-9α,10α-epoxy-7,8,9,10-tetrahydrobenzo(a)pyrene in vivo, Cancer Res. 44:1081 (1984).

5. B. Jernström, J. R. Babson, P. Moldéus, A. Holmgren and D. J. Reed, Glutathione conjugation and DNA-binding of (±)-trans-7,8-dihydroxy-7,8-dihydrobenzo(a)pyrene and (±)-7ß,8α-dihydroxy-9α, 10α-epoxy-7,8,9,10-tetrahydrobenzo(a)pyrene in isolated rat hepatocytes, Carcinogenesis 3:861 (1982).

6. B. Jernström, M. Martinez, S.-Å. Svensson and L Dock, Metabolism of benzo(a)pyrene-7,8-dihydrodiol and benzo(a)pyrene-7,8-dihydrodiol-9,10-epoxide to protein-binding products and glutathione conjugates in isolated hepatocytes, Carcinogenesis 5:1079 (1984).

7. I. G. C. Robertson, H. Jensson, C. Guthenberg, M. K. Tahir, B. Jernström and B. Mannervik, Differences in the occurrence of glutathione transferase isoenzymes in rat lung and liver, Biochem. Biophys. Res. Commun. 127:80 (1985).

8. M. Warholm C. Guthenberg, B. Mannervik and C. von Bahr, Purification of a new glutathione S-transferase (transferase μ) from human liver having high activity with benzo(a)pyrene-4,5-oxide, Biochem. Biophys. Res. Commun. 98:512 (1981).

9. B. Mannervik and C. Guthenberg, Glutathione transferase (human placenta), Methods Enzymol. 77:231 (1981).

10. W. B. Jakoby, B. Ketterer and B. Mannervik, Glutathione transferases: nomenclature, Biochem. Pharmacol. 33:2539 (1984).

11. I. G. C Robertson and B. Jernström, The enzymatic conjugation of glutathione with bay-region diol-epoxides of benzo(a)pyrene, benz(a)anthracene, and chrysene, Carcinogenesis 7:1633 (1986).

12. I. G. C. Robertson, H. Jensson, B. Mannervik and B. Jernström, Glutathione transferases in rat lung: the presence of transferase 7-7, highly efficient in the conjugation of glutathione with the carcinogenic (+)-7ß,-8α-dihydroxy-9α,10α-oxy-7,8,9,10-tetrahydrobenzo(a)pyrene, Carcinogenesis 7:295 (1986).

13. I. G. C. Robertson, C. Guthenberg, B. Mannervik and B. Jernström, Differences in stereoselectivity and catalytic efficiency of three human glutathione transferases in the conjugation of glutathione with 7ß,8α-dihydroxy-9α,10α-oxy-7,8,9,10-tetrahydrobenzo-(a)pyrene, Cancer Res. 46:2220 (1986).

14. L. Dock, M. Martinez and B. Jernström, Increased stability of (±)-7β,8α-dihydroxy-9α,10α-epoxy-7,8,9,10-tetrahydrobenzo(a)pyrene through interaction with subcellular fraction of rat liver, Chem.-Biol. Interact. 61:31 (1987).

15. B. Jernström, M. Martinez, D. J. Meyer and B. Ketterer, Glutathione conjugation of the carcinogenic and mutagenic electrophile (±)-7ß,8α-dihydroxy-9α,10α-oxy-7,8,9,10-tetrahydrobenzo(a)pyrene catalyzed by purified rat liver glutathione transferases, Carcinogenesis 6:85 (1985).

16. S. K. Yang, D. W. McCourt, P. P. Roeler and H. V. Gelboin, Enzymatic conversion of benzo(a)pyrene leading predominantly to the diol-epoxide r-7,t-8-dihydroxy-t-9,10-oxy-7,8,9,10-tetrahydrobenzo(a)-pyrene through a single enantiomer of r-7,t-8-dihydroxy-7,8-dihydrobenzo(a)pyrene, Proc. Natl. Acad. Sci. USA 73:2594 (1976).

17. B. Mannervik, The isoenzymes of glutathione transferase, Adv. Enzymol. 57:357 (1985).

18. M. Warholm, C. Guthenberg and B. Mannervik, Molecular and catalytic properties of glutathione transferase μ from human liver: an

enzyme efficiently conjugating epoxides, <u>Biochemistry</u> 22:3610 (1983).

19. J.-E. Seidegård, R. W. Pero, D. G. Miller and E. J. Beattie, A glutathione transferase in human leukocytes as a marker for the susceptibility to lung cancer, <u>Carcinogenesis</u> 7:751 (1986).

20. T. D. Boyer, D. Zakim and D. A. Vessey, Do soluble glutathione S-transferase have direct access to membrane-bound substrates? <u>Biochem. Pharmacol.</u> 32:29 (1983).

ANTIGENOTOXIC AND ANTICARCINOGENIC EFFECTS OF THIOLS. IN VITRO
INHIBITION OF THE MUTAGENICITY OF DRUG NITROSATION PRODUCTS AND
PROTECTION OF RAT LIVER ADP-RIBOSYL TRANSFERASE ACTIVITY

Silvio De Flora, Carmelo F. Cesarone[1], Carlo Bennicelli, Anna
Camoirano, Domizio Serra, Monia Bagnasco, Anna I. Scovassi[2],
Linda Scarabelli[1] and Umberto Bertazzoni[2]

Institute of Hygiene and Preventive Medicine and [1]Institute of
General Physiology, University of Genoa, 16132 Genoa, Italy
[2]CNR Institute of Biochemical and Evolutionary Genetics
27100 Pavia, Italy

INTRODUCTION

Reduced glutathione (GSH) is well-known to play a fundamental role in
the protection of the organism against toxic, mutagenic and/or carcinogenic
agents (see e.g. refs.1-3 for reviews). Among synthetic aminothiols, acting
as analogs and precursors of GSH, N-acetyl-L-cysteine (NAC) is of particular
interest, because this molecule is already extensively used in the treatment
of chronic respiratory diseases, and is extremely well tolerated in humans[4].
In addition, NAC is known to possess various antitoxic and antioxidant properties[3-5].

Previous in vitro and in vivo studies carried out in our laboratories
investigated a variety of antimutagenic and anticarcinogenic properties of
GSH and/or NAC, and aimed at elucidating the mechanisms underlying the observed inhibitory effects. Both thiols specifically inhibited the high
spontaneous mutability of strain TA104 of S. typhimurium, sensitive to oxidative damage. Moreover, NAC decreased the potency, in various S. typhimurium strains, of direct-acting mutagens, including, e.g., Cr (VI) compounds
4-nitroquinoline N-oxide, ICR 191, ICR 170, captan, folpet, epichlorohydrin,
sodium nitrite, formaldehyde, glutaraldehyde and hydrogen peroxide[6,7].
Antigenotoxic effects were also detected in a DNA-repair test system in E.
coli[8]. Inhibition of direct-acting genotoxic compounds could be ascribed to
the ability of NAC to act as a reducing agent, as a scavenger of reactive
oxygen species and as a blocker of electrophilic molecules[6]. Conversely,
thiols did not affect the mutagenicity of noncarcingenic compounds which are
known to be activated inside bacterial cells, such as nitrofurantoin and
sodium azide[7]. In the case of N-methyl-N-nitro-N-nitrosoguanidine (MNNG),
thiols yielded a marked inhibition of genotoxicity outside target cells,
while their intracellular reaction led to MNNG activation, as shown in
bacteria pretreated with either NAC, GSH or the GSH depletor diethyl maleate
(DEM)[9].

The in vitro influence of NAC on the mutagenicity of promutagens, including, e.g., benzo(a)pyrene, 2-aminofluorene, cyclophosphamide, aflatoxin
B1, a tryptophan pyrolysate product (Trp-P-2) and a cigarette smoke condensate, was variable depending on the dose of thiol and on the state of induc-

75

tion of liver metabolizing enzymes[6]. The use of liver and lung preparation from rats pretreated with either NAC, DEM, Aroclor 1254 or their combinations confirmed that NAC enhances the detoxification of direct-acting mutagens and that, in the case of promutagens, the effect of NAC depends on the balance between metabolic activation processes and trapping of electrophilic metabolites[10].

In vivo metabolic studies[10] showed that NAC, when administered i.p. to rats, increased GSH in erythrocytes and in liver and lung cells, and generally replenished its stores following depletion. NAC did not affect cytochromes P-450 in hepatic or pulmonary microsomes, whereas it stimulated, especially in combination with Aroclor, several enzyme activities, such as glucose-6-phosphate dehydrogenase, 6-phosphogluconate dehydrogenase, GSSG reductase and NAD(P)H-dependent diaphorases, in the liver and lung cytosol, as well as in isolated pulmonary alveolar macrophages[10,11]. A stimulation of various liver detoxifying activities, also including GSH S-transferase, was also detected following dietary administration of NAC or GSH to rats[12] or to mice[13].

In rats subjected to a discontinuous feeding regimen with N-2-acetyl-aminofluorene (2AAF), the oral administration of NAC or GSH resulted, at least during early treatment stages, in a delay in the development of gamma-glutamyl transpeptidase (GGT)-positive foci in the liver, and in a decreased damage to DNA, associated with a normal repair synthesis[14,15]. In the same animals, both NAC and GSH prevented the formation of ear-duct tumors (Zymbal gland carcinomas) induced by 2AAF feeding[15]. Moreover, administration of NAC with the diet efficiently prevented in mice the induction of lung tumors by the carcinogen urethan[13].

Therefore, on the whole, thiols appear to exert antimutagenic and anti-carcinogenic effects in various experimental test systems, and multiple mechanisms appear to be involved in their protective action. In the present paper we describe two additional mechanisms, detected by testing GSH and NAC in two different experimental models. In the first one, we compared the ability and mechanisms of these thiols with those of ascorbic acid, in inhibiting the bacterial mutagenicity of the nitrosation products of the two histamine H_2-receptor antagonists ranitidine and famotidine. These anti-ulcer drugs, like other H_2-blockers (e.g. cimetidine), react in vitro with high amounts of nitrite to form genotoxic derivatives, which can be inhibited by ascorbic acid[16-19]. Although such a reaction is unlikely to be important in vivo, as also shown by the lack of mutagenicity in the gastric juice of famotidine-treated patients[20], it can provide a model for investigating in vitro the nitrosation reaction and its prevention by inhibitors. In order to clarify the mechanisms involved, GSH, NAC and ascorbic acid were also compared in their ability to decrease the mutagenicity of sodium nitrite, MNNG and sodium dichromate.

The in vivo study was performed using a rat hepatocarcinogenesis model[21], and aimed at assessing the effect of the dietary administration of the carcinogen 2AAF and of GSH or NAC on liver ADP-ribosyl transferase (ADPRT) activity. This enzyme catalyzes the ADP-ribosylation of nuclear proteins, and is involved in DNA repair, possibly through modifications of the chromatin structure[22,23]. ADPRT is known to be affected in cells treated with DNA damaging agents[22,24,25].

MATERIALS AND METHODS

Chemicals

GSH (Sigma Chemical Co., St. Louis, MO) and NAC (gift from Zambon S.p.A

Bresso, Milano, Italy) were either dissolved in distilled water for in vitro studies, or incorporated into pelleted meal (Laboratori Piccioni, Brescia, Italy) for in vivo studies. MNNG (gift from IIT Research Institute, Chicago, ILL) sodium dichromate (Merck-Schuchardt, Münich, FRG), sodium nitrite and ascorbic acid (Carlo Erba, Milano, Italy) were dissolved in distilled water. Famotidine (gift from Merck & Co., Rahway, NJ) was dissolved in 0.3% glacial acetic acid, and ranitidine (gift from Glaxo S.p.A., Verona, Italy) was dissolved in distilled water. 2AAF (Sigma) was incorporated into pelleted meal (Laboratori Piccioni).

The antiprotease reagents phenyl-methyl-sulfonyl fluoride (PMSF), pepstatin (from Sigma), dithiothreitol and sodium bisulfite (from BDH Chemicals Ltd., Poole, UK) were prepared as previously described[26]. Nitrocellulose was from Schleicher and Schuell GmbH (Dassel, FRG). Newborn bovine serum was purchased from Flow Labs. Inc. (McLean, VA). Alkaline phosphatase-conjugated goat anti-rabbit IgG (H + L), and the color development reagents p-nitro blue tetrazolium chloride (NBT) and 5-bromo-4-chloro-3-indolyl-phosphatase toluidine salt (BCIP) were obtained from Bio-Rad Labs. (Richmond, CA). A rabbit antiserum against highly purified calf thymus ADPRT was a gift from Drs. M. E. Ittel and C. Niedergang (CNRS, Strasbourg, France). DE 81 paper was obtained from Whatman Ltd. (Springfield Mill, UK). Prestained protein molecular weight standards were from BRL Inc. (Gaithersburg, MD). Adenine-2,8-[^{3}H]NAD (25 Ci/mmol) and adenine-2,8-[^{32}P]-NAD (1000 Ci/mmol) were from New England Nuclear (DuPont Co., Wilmington DE), adenosine-8-[^{14}C] (56 mCi/mmol) and [^{3}H]dTTP (50 Ci/mmol) from Amersham Int. (Amersham, UK). Calf thymus DNA was activated to about 3-5% solubility by DNase I.

In vitro Assays

The mutagenicity of mixtures of nitrite with either famotidine or ranitidine was assessed as previously described[14,18]. Briefly, solutions of nitrite and of each one of the two drugs (or of their solvents) were mixed, acidified and incubated for 60 min (unless otherwise specified) at 37°C. After incubation, 0.5 ml of each mixture were neutralized with 200 µl phosphate-buffered saline (PBS), pH 7.4, prior to adding 2.5 ml of molten top agar and 50 µl of bacterial broth cultures, according to the procedure of the plate incorporation test[27]. The conditions used in most assays, i.e. doses of nitrite or of drugs, and the his- S. typhimurium tester strain (TA100 or TA102), were those indicated in Fig. 1, and corresponded to the optimal conditions leading to the detection of mutagenic nitrosoderivatives of these drugs in the test system used[19].

Inhibition of the above phenomenon was investigated by adding 100 µl of distilled water (controls) or of varying amounts of ascorbic acid, GSH or NAC, either at the start or at the end of the preincubation step of nitrite with drugs, before neutralization of the mixture. The effect of the 3 inhibitors on the mutagenicity of nitrite and MNNG in strains TA1535 and TA100 of S. typhimurium, respectively, was investigated under the above conditions, except that the drugs were omitted and replaced with distilled water. The chromium-reducing ability of thiols and of ascorbic acid was assessed by testing the mutagenicity of mixtures of sodium dichromate with varying amounts of inhibitors, using strain TA102, which is the most sensitive of the primary S. typhimurium tester strains in revealing the mutagenicity of hexavalent chromium[28].

All mutagenicity assays were carried out in triplicate plates. The results are expressed as mean ± SD values.

In vivo Studies

Male Wistar rats (Morini, S. Polo d'Enza, Reggio Emilia, Italy), weigh-

77

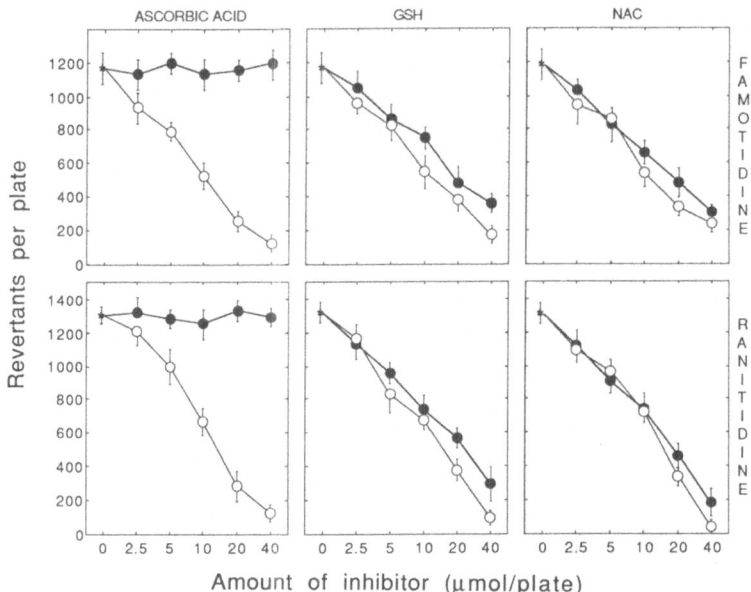

Fig. 1.   Inhibition by varying amounts of ascorbic acid, GSH or NAC, of the
formation of mutagenic derivatives from the reaction between sodium
nitrite (40 µmol/plate) and either famotidine (12 µmol/plate) or
ranitidine (0.6 µmol/plate). The inhibitors were added either at the
start (empty circles) or at the end (full circles) of a preincubation step (60 min at 37°C) in acidic environment (pH 2.0 in the
case of famotidine and pH 2.5 in the case of ranitidine). Mutagenicity was investigated in strains TA100 (famotidine) or TA102 (ranitidine) of S. typhimurium, in the absence of metabolic systems.

ing 80-100 g, were housed 3-4 per cage, given tap water and standard Vogt-
Möller diet (Laboratori Piccioni) ad libitum, and subjected to a 4-cycle discontinuous regimen, according to the Teebor and Becker model[21].   Each cycle
consisted of a 3-week feeding with a standard diet supplemented with 0.05%
2AAF and/or 0.1% NAC, followed by one week of standard diet.   At the end of
the third week of each cycle, 4-5 rats from each one of the 4 experimental
groups (i.e. untreated rats or rats fed with either NAC or 2AAF, or 2AAF plus
NAC) were sacrificed under slight ether anaesthesia.   The livers were rapidly
removed, washed in cold physiological saline, frozen in liquid nitrogen and
stored at -80°C.

Extracts were prepared by mincing 0.2 g of thawed liver in 10 volumes of
0.6 M NaCl, 50 mM Tris-HCl, pH 8.0, containing 1 mM EDTA, 0.5 mM dithio-
threytol, 10 mM $NaHSO_3$, 1mM PMSF and 1 µM pepstatin.   The suspension was
sonicated twice for 20 sec in ice at 50 W and centrifugated at 4°C for 30 min
at 105,000 x g.   The supernatant was used as enzyme extract for activity gel
analysis and biochemical assays.

Catalytic peptides of ADPRT were detected by the activity gel technique.
The sequential steps of this procedure included: SDS-polyacrylamide gel
electrophoresis; in situ renaturation of proteins; incubation of the intact
gel with $[^{32}P]$-NAD; removal of non-incorporated NAD by trichloroacetic acid
and autoradiography[24].   The first lane of the gel usually contained a mixture
of prestained molecular weight markers.

Protein blotting was carried out by Western blot procedure[29].   After

SDS-polyacrylamide gel electrophoresis, the proteins were transferred to nitrocellulose by applying 90 Volt, 350 mA for 90 min. The blot was incubated overnight with 10% newborn bovine serum-0.1% tween-20 in PBS, 3 h with rabbit antiserum against ADPRT, and 2 h with a 1:3000 dilution of goat anti-rabbit conjugated with alkaline phosphatase. Visualization of immunoreactive peptides was obtained by BCIP-NBT color development substrate.

RESULTS

In vitro Effects of Thiols and Ascorbic Acid on Drug Nitrosation and on the Mutagenicity of N-nitrosocompounds

Fig. 1 shows the results of an experiment aiming at assessing the effects of varying amounts of ascorbic acid, GSH and NAC on the mutagenicity of the nitrosation products of famotidine and ranitidine. Both drugs, following preincubation with nitrite in acidic environment, yielded an evident mutagenic response in S. typhimurium (dose 0 of inhibitors in the Figure). As confirmed in repeated assays, all 3 inhibitors decreased, with a dose-related effect, the mutagenicity of nitrite-drug mixtures when added at the start of the reaction (empty circles in Fig. 1). Mutagenicity was virtually eliminated at equimolar concentrations of nitrite and inhibitors (40 µmol/plate).

On the other hand, when the inhibitors were added at the end of the preincubation steps between nitrite and drugs (full circles in Fig. 1), both thiols were almost as effective as when added at the start of the reaction, while ascorbic acid did not affect at all the mutagenic response.

Other assays were carried out in order to check the effect of the 3 inhibitors on the mutagenicity of a preformed N-nitroso compound, i.e. MNNG, and on a reducible metal salt, i.e. sodium dichromate. As shown in Fig. 2, GSH and NAC eliminated the mutagenicity of this N-nitroso compound in dose-

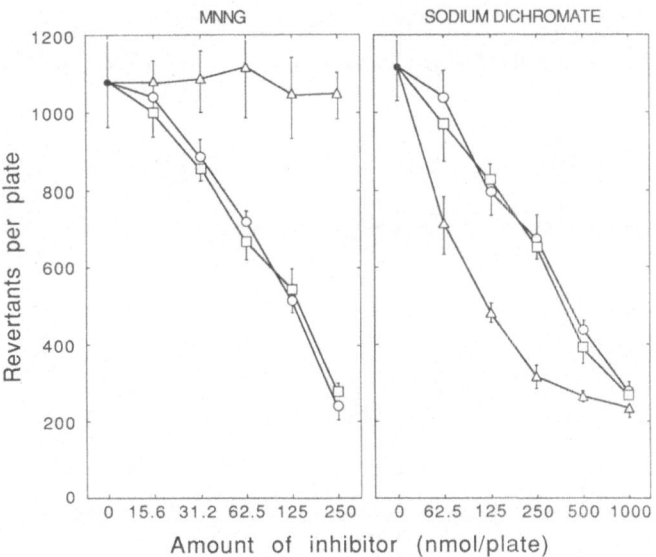

Fig. 2. Inhibition by varying amounts of ascorbic acid (triangles), GSH (circles) and NAC (squares) of the mutagenicity of MNNG (10 nmol/plate) and sodium dichromate (100 nmol/plate) in strains TA100 and TA102 of S. typhimurium, respectively.

related fashion. At the same doses (i.e. up to 250 nmol/plate), ascorbic acid failed to decrease MNNG mutagenicity (Fig. 2), some inhibitory effect being only observed over 2000 nmol/plate (data not shown). Similar results were obtained when the preincubation of thiols with MNNG was carried out in acidic (pH 1.5) rather than in neutral environment (data not shown). In contrast, ascorbic acid was even more efficient than GSH and NAC in lowering the mutagenicity of sodium dichromate (Fig. 2).

The ability of the 3 inhibitors to reduce the mutagenicity of sodium nitrite, after 60 min of preincubation, was rather similar, in both neutral (Fig. 3) and acidic (not shown) environments. The two thiols, and especially NAC, tended to produce toxic effects in bacteria, but only when mixed with nitrite (Fig. 3).

However, the time-dependence monitoring of nitrite reduction provided evidence that ascorbic acid was faster than thiols in producing this effect. In fact, as shown in Fig. 4, the drop of nitrite mutagenicity in the presence of equimolar amounts of ascorbic acid was not appreciably enhanced by prolonging the time of contact between these two compounds. The two thiols were less effective than ascorbic acid at time 0, but thereafter GSH produced a time-dependent decrease of mutagenicity, and NAC addition resulted in toxic effects. Again, toxicity could not be ascribed to NAC itself but to its mixture with sodium nitrite, and was only detected at concentrations of the two reactants exceeding 30 μmol/plate.

Protection by NAC of Liver ADPRT Activity in 2AAF-treated Rats

ADPRT activity and molecular weight were analyzed in the liver extracts of rats subjected to the 4-cycle discontinuous feeding regimen with 2AAF, at the end of each treatment cycle, i.e. after a week of withdrawal of 2AAF administration. As shown in Fig. 5A, a catalytic band of 116 KDa was clearly detectable in liver extracts of untreated rats (lane 1). In 2AAF-treated rats, the active band was no longer detectable at the end of the first cycle (lane 2), and progressively reappeared only in the subsequent cycles (lanes 3 to 5).

The situation was modified in rats fed with 2AAF plus NAC (Fig. 5B). In

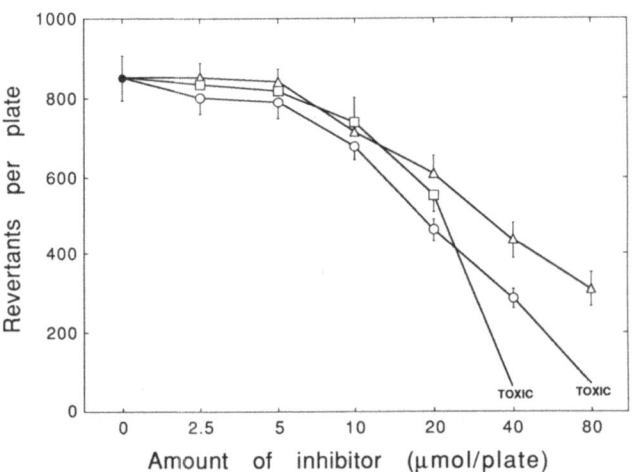

Fig. 3. Dose-dependence of the inhibition by ascorbic acid (triangles), GSH (circles) and NAC (squares) of the mutagenicity of sodium nitrite (40 μmol/plate) in strain TA1535 of S. typhimurium.

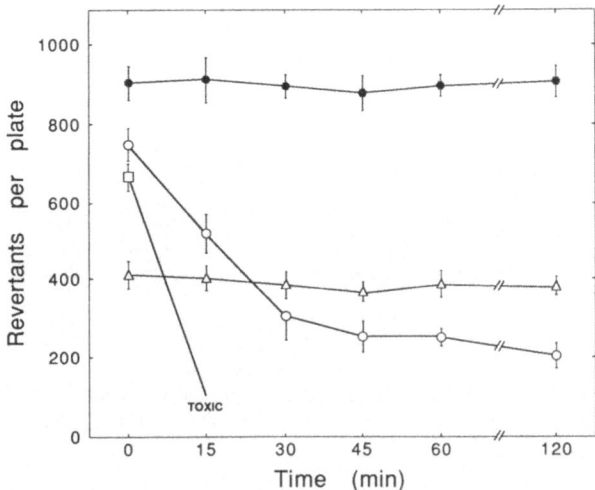

Fig. 4. Time-dependence of the inhibition by ascorbic acid (triangles), GSH (circles) and NAC (squares) of the mutagenicity of sodium nitrite in strain TA1535 of S. typhimurium. All reactants were tested at 40 μmol/plate.

fact, the active band of ADPRT detected in control rats (lane 1) was still evident at the end of the first cycle (lane 2), became barely detectable only at the end of the second cycle (lane 3) and returned to a normal appearance after the third and fourth cycle (lanes 4 and 5). Such a protective effect, delaying the suppression of the ADPRT band, was specifically produced by NAC, since its replacement with GSH did not modify the activity gel patterns observed in the carcinogen-treated rats.

The liver extracts of rats fed with 2AAF or with 2AAF plus NAC were analyzed by the Western blot technique, using a polyclonal anti-ADPRT antibody. As shown by the examples reported in Figs. 6A and 6B, respectively, there was a perfect agreement between appearance (or disappearance) of an immunoreactive peptide of 116 KDa in Western blot analysis and patterns of ADPRT bands observed by activity gel technique. This indicates that the loss in the specific catalytic band depends on a lack of ADPRT protein.

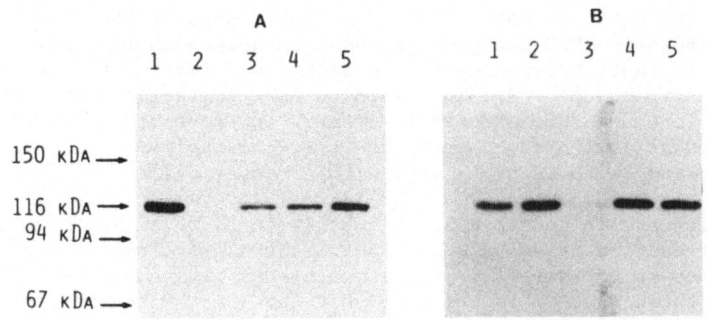

Fig. 5. Activity gel analysis of ADPRT in rat liver. Extracts from either untreated animals (lane 1 of panels A and B), rats fed with 2AAF (panel A), or with 2AAF plus NAC (panel B), were analyzed at the end of the 1st, 2nd, 3rd and 4th cycle of treatment (lanes 2, 3, 4 and 5, respectively). Molecular mass markers were: ß-galactosidase (116 KDa), phosphorylase b (94 KDa) and bovine serum albumin (67 KDa).

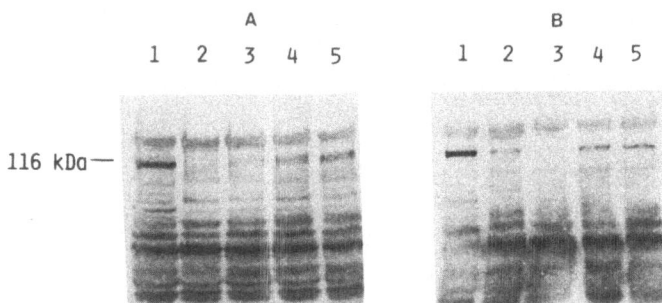

Fig. 6. Western blot analysis of ADPRT in rat liver. Extracts from either untreated animals (lane 1 of panels A and B) or rats fed with 2AAF (panel A) or 2AAF plus NAC (panel B), were analyzed at the end of 1st, 2nd, 3rd and 4th treatment cycle (lanes 2, 3, 4 and 5, respectively).

DISCUSSION

The results herein presented provide evidence for two independent mechanisms of protection afforded by thiols in mutagenesis and carcinogenesis.

The first one is the inhibition of the mutagenicity of N-nitroso compounds, either preformed (i.e. MNNG) or resulting from the nitrosation of in vitro nitrosatable drugs (i.e. ranitidine and famotidine). In such effect, it was of interest to compare the mechanisms of thiols with those of the most typical inhibitor of the nitrosation reaction, i.e. ascorbic acid[30,31]. Vitamin C was quite efficient in preventing the formation of mutagenic derivatives of the two drugs, while no inhibition was produced on the mutagenicity of the nitrosation products, once formed. Some loss of MNNG mutagenicity was produced only in the presence of a large excess of ascorbic acid. Occurrence of this reaction has been already described and interpreted as a Cu(II)-catalyzed decomposition of this N-nitroso compound, possessing an electronegative nitro group in its molecule[32].

In contrast, both thiols efficiently eliminated the mutagenicity of MNNG and of the nitrosoderivatives of ranitidine and famotidine. Such effect was maintained both in neutral and acidic pH ranges, which is a prerequisite for inhibiting those mutagens and carcinogens which act in the stomach environment, as is the case for MNNG. Several studies support the view that thiols play a dual role in MNNG mutagenesis and carcinogenesis (see e.g. ref. 9 for a review). In fact, the decomposition products formed by the reaction between sulfhydryl groups and MNNG are inactive when they are generated outside cells (e.g. in the gastrointestinal lumen or in the blood), whereas the same intermediates are expected to methylate critical nucleophilic molecules, e.g. DNA, when formed in their proximity, within target cells.

At variance with ascorbic acid, inhibition by thiols of mutagenicity of the drug nitrosation products was of a similar order of magnitude when GSH or NAC were added at the start or at the end of the preincubation step between nitrite and the drugs. This suggests that, at least with these precursors, thiols act more effectively by blocking the resulting nitrosoderivatives rather than by preventing the nitrosation reaction. Since thiols were only slightly less powerful than ascorbic acid in reducing the mutagenicity of hexavalent chromium, further assays were carried out in order to justify the apparently poor ability of thiols to prevent the reaction between nitrite and the two anti-ulcer molecules. Indeed, the decrease of nitrite mutagenicity was similar in the presence of ascorbic acid and of thiols, with only a more pronounced trend to form toxic derivatives in the case of thiols, and especially of NAC. Thiols, like ascorbate, α-tocopherol, 1,4-dihydroxyphenol and other sulfur compounds, have been reported to reduce nitrous acid to nitric oxide, but the reaction of thiols with nitrous acid may also result in the formation of thionitrite esters[33]. In addition, it is well known that, on contact with air, nitric oxide very readily oxidises to nitrogen dioxide and/or nitrogen tetroxide.

However, a time-dependence analysis of inhibition of nitrite mutagenicity provided evidence that the effect of ascorbic acid was virtually immediate, while the conversion into less mutagenic or toxic derivatives produced by thiols was slower. Taking into account that the in vitro formation of mutagenic nitrosoderivatives from both ranitidine[18] and famotidine[19] is very rapid yet not immediate, the findings of these assays may explain why ascorbic acid could successfully compete with these drugs in reacting with nitrite. Thiols probably reacted with nitrite at a slower rate compared with drugs, but on the other hand they successfully blocked the nitrosation products. Since ascorbic acid and thiols appear to prevent the formation of mutagenic nitrosoderivatives by exerting their main mechanisms at different levels, in theory a combination of these inhibitors would be expected to produce synergistic effects, which warrants further studies.

The in vivo study was performed in the same hepatocarcinogenesis model in the rat, based on a 4-cycle discontinuous feeding regimen with 2AAF[21], that we had already used in order to investigate various protective effects of dietary GSH or NAC, concerning, e.g., damage and repair of liver DNA, development of liver hyperplastic foci and Zymbal gland tumors[15], various biochemical activities related to GSH cycle and to liver cytosolic enzyme activities[12], and influence of treatments on the metabolic activation of aromatic amines[34].

Activity gel analyses of liver extracts from control rats showed an evident catalytic band of 116 KDa, corresponding to ADPRT activity, which was no longer detectable in 2AAF treated animals at the end of the first cycle. Similar patterns were observed in rats fed with 2AAF plus GSH, while co-administration of NAC resulted in a one-cycle delay of ADPRT inhibition. Western blot analyses indicated that the loss of catalytic band observed in activity gels was due to a lack of ADPRT protein and not to an inhibition of this enzyme activity.

In any case, this phenomenon was followed, after one additional cycle, by the reappearance of ADPRT protein, which could suggest a transient inhibition of ADPRT de novo synthesis produced by the carcinogen. The progressive reappearance of the catalytic peptide is likely to depend, rather than on reversal of the inhibitory effect in normal hepatocytes, on a selection of a population of "resistant" cells, which typically occurs in various hepatocarcinogenesis models[35].

The protective effect on ADPRT activity exerted by NAC, at least during the early stages of 2AAF-induced hepatocarcinogenesis, correlates with the previously demonstrated ability of thiols to counteract the DNA damage in liver hepatocytes, as assessed by the alkaline elution technique, and to antagonize the development of gamma-glutamyl transpeptidase (GGT)-positive foci[15]. The failure of GSH to delay ADPRT inhibition in 2AAF-treated rats may be ascribed to a slower penetration of this tripeptide into hepatocytes, compared to NAC. In fact, synthesis from exogenously supplied GSH requires a preliminary GGT-catalyzed hydrolysis of the tripeptide on the hepatocyte membrane[36]. On the contrary, plasmatic NAC is easily uptaken by liver cells and deacetylated to form cysteine[37], which, of the 3 amino acids forming the GSH molecule, has the lowest intracellular concentrations and therefore becomes rate limiting in its synthesis[38].

On the whole, the findings of the present study provide additional evidence that the natural thiol GSH and especially its synthetic analog and precursor NAC, which is more suitable for administration to humans[4], possess a wide array of mechanisms by which they can inhibit the mutagenesis and carcinogenesis processes. In fact, based on this study and on previous studies carried out in our and other laboratories (see the Introduction), the possible intervention levels of thiols cover a variety of events involved not only in mutagenesis and cancer initiation, but also in subsequent steps of carcinogenesis. The known protective mechanisms of thiols include, e.g., the herein described effects on the nitrosation reaction and on its products, their ability to act as antioxidants and nucleophiles, scavenging free radicals and blocking electrophilic compounds and metabolites, as well as the ability to induce multiple detoxifying enzyme activities in the cell cytosol, and to modulate DNA repair. Therefore, an induced enhancement of the so-called "GSH threshold"[1] by means of synthetic thiols seems to represent one of the most promising approaches in mutation and cancer chemoprevention.

ACKNOWLEDGEMENTS

This study was supported by Italian National Research Council (CNR) Special Project "Oncology" (grants Nos. 86.00382.44 and 86.00357.44), and by the Radiation Programme of the Commission of the European Communities (contract B16-158).

REFERENCES

1. D. J. Jollow, Glutathione thresholds in reactive metabolite toxicity, in: "Quantitative Aspects of Risk Assessment in Chemical Carcinogenesis", J. Clemmesen, D. M. Conning and F. Oesch, eds., Springer-Verlag, Berlin/Heidelberg/New York, Arch. Toxicol. suppl. 3:95 (1980).
2. A. Meister, Selective modification of glutathione metabolism, Science 220:470 (1983).
3. S. De Flora, C. Bennicelli, D. Serra, A. Izzotti and C. F. Cesarone, Role of glutathione and N-acetylcysteine as inhibitor of mutagenesis and carcinogenesis, in: "Absorption and Utilization of Amino Acids", M. Friedman, ed., CRC Press, Boca Raton, Florida, in press (1988).
4. J. W. Yarbro, R. S. Bornstein and M. J. Mastrangelo, eds., N-acetylcysteine (NAC): a significant chemoprotective adjunct, Seminars in Oncology 10:1 (1983).
5. P. Moldéus, I. A. Cotgreave and M. Berggren, Lung protection by thiol-containing antioxidant: N-acetylcysteine, Respiration 50:31 (1986).
6. S. De Flora, C. Bennicelli, P. Zanacchi, A. Camoirano, A. Morelli and A. De Flora, In vitro effects of N-acetylcysteine on the mutagenicity of direct-acting compounds and procarcinogens, Carcinogenesis 5:505 (1984).
7. S. De Flora, C. Bennicelli, A. Camoirano, D. Serra, C. Basso, P. Zanacchi and C. F. Cesarone, Inhibition of mutagenesis and carcinogenesis by N-acetylcysteine, in: "Anticarcinogenesis and Radiation Protection", P. A. Cerutti, O. Nygaard and M. G. Simic, eds., Plenum Press, New York, in press (1988).
8. S. De Flora, Detoxification of genotoxic compounds as a threshold mechanism limiting their carcinogenicity, Toxicol. Pathol. 12:337 (1984).
9. A. Camoirano, G. S. Badolati, P. Zannachi, M. Bagnasco and S. De Flora, Dual role of thiols in N-methyl-N-nitro-N-nitrosoguanidine genotoxicity, Life Science Advances- Exp. Cell Genetics 6:in press (1987).
10. S. De Flora, C. Bennicelli, A. Camoirano, D. Serra, M. Romano,

G. A. Rossi, A. Morelli and A. De Flora, In vivo effects of N-acetylcysteine on glutathione metabolism and on the biotransformation of carcinogenic and/or mutagenic compounds, <u>Carcinogenesis</u> 6:1735 (1985).

11. S. De Flora, M. Romano, C. Basso, M. Bagnasco, C. F. Cesarone, G. A. Rossi and A. Morelli, Detoxifying activities in alveolar macrophages of rats treated with acetylcysteine, diethyl maleate and/or Aroclor, <u>Anticancer</u> <u>Res.</u> 6:1009 (1986).

12. C. F. Cesarone, M. Romano, D. Serra, L. Scarabelli and S. De Flora, Effects of aminothiols in 2-acetylaminofluorene-treated rats. II. Glutathione cycle and liver cytosolic activities, <u>In</u> <u>Vivo</u> 1:93 (1987).

13. S. De Flora, M. Astengo, D. Serra and C. Bennicelli, Inhibition of urethan-induced lung tumors in mice by dietary N-acetyl-cysteine, <u>Cancer</u> <u>Lett.</u> 32:235 (1986).

14. C. F. Cesarone, L. Scarabelli and M. Orunesu, Effect of glutathione on alterations of liver DNA structure and metabolic activities induced <u>in</u> <u>vivo</u> by 2-acetylaminofluorene, <u>Anticancer</u> <u>Res.</u> 6:1283 (1986).

15. C. F. Cesarone, L. Scarabelli, M. Orunesu, M. Bagnasco and S. De Flora, Effects of aminothiols in 2-acetylaminofluorene-treated rats. I. Damage and repair of liver DNA, hyperplastic foci, and Zymbal gland tumors, <u>In</u> <u>Vivo</u> 1:85 (1987).

16. S. De Flora and A. Picciotto, Mutagenicity of cimetidine in nitrite-enriched human gastric juice, <u>Carcinogenesis</u> 1:925 (1980).

17. S. De Flora, Cimetidine, ranitidine and their mutagenic nitroso-derivatives, <u>Lancet</u> ii:993 (1981).

18. S. De Flora, C. Bennicelli, A. Camoirano and P. Zanacchi, Genotoxicity of nitrosated ranitidine, <u>Carcinogenesis</u> 4:255 (1983).

19. S. De Flora, A. Camoirano, C. Basso, M. Astengo, P. Zanacchi and C. Bennicelli, Bacterial genotoxicity of nitrosated famotidine, <u>Mutagenesis</u> 1:125 (1986).

20. S. De Flora, A. Picciotto, V. Savarino, C. Bennicelli, A. Camoirano, G. Garibotto and G. Celle, Circadian monitoring of gastric juice mutagenicity, <u>Mutagenesis</u> 2:115 (1987).

21. G. W. Teebor and F. F. Becker, Regression and persistence of hyperplastic nodules induced by N-2-fluorenylacetamide and their relationship to hepatocarcinogenesis, <u>Cancer</u> <u>Res.</u> 31:1 (1971).

22. S. Shall, ADP-ribose in DNA repair: A new component of DNA excision repair, <u>Adv.</u> <u>Rad.</u> <u>Biol.</u> 11:1 (1984).

23. K. Ueda, ADP-ribosylation, <u>Ann.</u> <u>Rev.</u> <u>Biochem.</u> 54:73 (1985).

24. A. I. Scovassi, M. Stefanini, P. Lagomarsini, R. Izzo and U. Bertazzoni, Response of mammalian ADP-ribosyl transferase to lymphocyte stimulation, mutagen treatment and cell cycling, <u>Carcinogenesis</u> 8:1295 (1987).

25. C. F. Cesarone, A. I. Scovassi, L. Scarabelli, R. Izzo, M. Orunesu and U. Bertazzoni, Loss of ADP-ribosyl transferase activity in liver of rats treated with 2-acetylaminofluorene, <u>in</u>: "Proceedings of the 8th International Symposium on ADP-Ribosylation: Niacin Nutrition, ADP-Ribosylation and Cancer", Ft. Worth, Texas, USA, May 30-June 3 (1987).

26. A. I. Scovassi, M. Stefanini and U. Bertazzoni, Catalytic activities of human poly(ADP-ribose) polymerase from normal and mutagenized cells detected after sodium dodecyl sulfate polyacrylamide gel electrophoresis, <u>J.</u> <u>Biol.</u> <u>Chem.</u> 259:10973 (1984).

27. D. M. Maron and B. N. Ames, Revised methods for the Salmonella mutagenicity test, <u>Mutat.</u> <u>Res.</u> 113:173 (1983).

28. C. Bennicelli, A. Camoirano, S. Petruzzelli, P. Zanacchi and S. De Flora, High sensitivity of Salmonella TA102 in detecting hexavalent chromium mutagenicity and its reversal by liver and lung preparations, <u>Mutat.</u> <u>Res.</u> 122:1 (1983).

29. H. Towbin, T. Staehelin and J. Gordon, Electrophoretic transfer of

proteins from polyacrylamide gels to nitrocellulose sheets: procedure and some applications, <u>Proc. Natl. Acad. Sci. USA</u> 76:4350 (1979).

30. S. S. Mirvish, Ascorbic acid inhibition of N-nitroso compound formation in chemical, food and biological systems, <u>in</u>: "Inhibition of Tumor Induction and Development", M. S. Zedek and M. Lipken, eds., Plenum Press, New York (1981).

31. H. Bartsh, H. Ohshima, J. Nair, B. Pignatelli and S. Calmels, Modifiers of endogenous nitrosamine synthesis and metabolism, <u>in</u>: "Antimutagenesis and Anticarcinogenesis Mechanisms", D. M. Shankel, P. E. Hartman, T. Kada and A. Hollaender, eds., Plenum Press, New York (1986).

32. J. B. Guttenplan, Mechanisms of inhibition by ascorbate of microbial mutagenesis induced by N-nitroso compounds, <u>Cancer Res.</u> 38:2018 (1978).

33. P. Sen, Formation and occurrence of nitrosamines in food, <u>in</u>: "Diet, Nutrition and Cancer: a Critical Evaluation", Vol.II. Micronutrients, Nonnutritive Dietary Factors, and Cancer, B. S. Reddy and L. A. Cohen, eds., CRC Press, Boca Raton, Florida (1986).

34. S. De Flora, A. Camoirano, C. Bennicelli, M. Orunesu and C. F. Cesarone, Effects of aminothiols in 2-acetylaminofluorene-treated rats. III. Metabolic activation of aromatic amines, <u>In Vivo</u> 1:101 (1987).

35. E. Farber, S. Parker and M. Gruenstein, The resistance of putative premalignant liver cell populations, hyperplastic nodules, to the acute cytotoxic effects of some hepatocarcinogens, <u>Cancer Res.</u> 36:3879 (1976).

36. M. H. Hanigan and H. C. Pitot, Gamma-glutamyl transpeptidase. Its role in hepatocarcinogenesis, <u>Carcinogenesis</u> 6:165 (1985).

37. L. Roy Morgan, M. R. Holdiness and L. E. Gillen, N-acetylcysteine: its bioavailability and interaction with isofosfamide metabolites, <u>Seminars in Oncology</u>, suppl. 1, 10:56 (1983).

38. D. Reed, A. Brodie and M. Meredith, Cellular heterogeneity in the status and function of cysteine and glutathione, <u>in</u>: "Function of Glutathione", A. Larsson, A. Holmgren, S. Orrenius and B. Mannervik, eds., Raven Press, New York (1983).

RELATIONSHIPS BETWEEN NADPH CONTENT AND BENZO(a)PYRENE METABOLISM IN NORMAL

AND GLUCOSE-6-PHOSPHATE DEHYDROGENASE-DEFICIENT HUMAN FIBROBLASTS

Rosa Pascale, Lucia Daino, Maria E. Ruggiu, Maria G. Vannini,
Renato Garcea, Serenella Frassetto, Luciano Lenzerini,
Leonardo Gaspa, Maria M. Simile, Marco Puddu and Francesco Feo

Istituto di Patologia generale, Università di Sassari
07100 Sassari, Italy

INTRODUCTION

Previous work in our laboratory has shown that human skin fibroblasts (HSF) and human lymphocytes carrying the Mediterranean variant of glucose-6-phosphate dehydrogenase (G6PD) exhibit a great decrease in hexose monophosphate shunt and in NADP /NADPH ratio[1,2]. G6PD deficiency protects in vitro growing HSF and lymphocytes from benzo(a)pyrene (BaP) toxicity. G6PD-deficient HSF give rise in soft agar, after incubation with BaP, to a lower number of colonies than normal HSF[1]. Aryl hydrocarbon hydroxylase (AHH) activities are lower in G6PD-deficient cells, when tested in the absence of exogenous NADPH[2]. G6PD-deficient cells, incubated in vitro with BaP, produce low amounts of organic- and water-soluble BaP metabolites and show a decreased ability to form BaP-7,8-diol-9,10-epoxide and BaP-DNA adducts[3,4].

It has been hypothesized[3,4] that NADPH content of G6PD-deficient cells, even though sufficient for steady state NADPH-dependent functions, could become insufficient for active NADPH-dependent reactions, as for instance the metabolism of relatively high BaP amounts. If this hypothesis is correct, the difference between normal and G6PD-deficient cells, as concerns BaP metabolism, should increase with carcinogen concentration. The study of the influence of BaP concentration on BaP metabolism by HSF could be a good tool to assess the modulatory role of NADPH concentration on BaP metabolism.

MATERIALS AND METHODS

HSF were obtained from normal males or males carrying the Mediterranean G6PD variant, aged between 20 and 50. They were subcultured from primary cultures in MEM Eagle's medium, containing 20% calf serum, penicillin (100 U/ml), and streptomycin (100 μg/ml), in atmosphere of 5% CO_2 in air, at 37°C. The cells were used between the 2nd and the 5th subculture generation. No differences between normal and G6PD-deficient cells with respect to morphology and growth rate (doubling time, 96 h) were observed. After 9 days in vitro, new medium containing 1 μg/ml of benzo(a)anthracene was added to the cultures and, after 24 h, the medium was renewed with medium containing 0.25, 0.4, 1.2 or 2.5 μM [^{3}H]BaP (Amersham, 19 Ci/mmol). Twenty-four hours later the cells were detached by trypsinization and the BaP metabolites were extracted from HSF and medium into ethylacetate/acetone (2:1 by vol.) containing 0.01% butylated hydroxytoluene. After evaporation of the organic phase under vacuum the residue was dissolved in methanol and the organic-soluble metabolites were determined by hplc as published[4]. The radioactivity of total organic-soluble metabolites was determined in the extracts after removal of BaP by tlc[4]. For determination of water-soluble metabolites, HSF suspensions were homogenized, buffered at pH 5.5 and incubated with ß-glucuronidase and arylsulfatase[4]. Metabolites were then extracted and chromatographed. Hplc profiles of organic-soluble metabolites were subtracted from those of enzyme-treated samples. Pyrimidine nucleotide cellular content, G6PD and AHH activities, and protein content were determined as described[1-3].

RESULTS

G6PD activity of HSF after 11 days in culture was 12.89 ± 1.84 and 1.63 ± 0.06 µmol of NADPH produced/min,mg protein for, respectively, normal and deficient HSF (n = 5, SD, P < 0.001). Specific activity of benzo(a)anthracene-induced AHH, measured in the presence of NADPH generating system was: 11.92 ± 2.55 and 12.03 ± 3.07 pmol of BaP hydroxylated/30 min,mg protein for, respectively, normal and deficient HSF (n = 5, SD).

Total metabolites, produced by normal and G6PD-deficient HSF, incubated with 2.5 µM BaP were: 9,10-diol, 7,8-diol, quinones, 9-hydroxy, 3-hydroxy and a more polar fraction which is probably a mixture of triols and tetrols. Organic-soluble metabolites represented 41-56% in both normal and deficient HSF, incubated with 0.25-2.5 µM BaP. With normal HSF, the relative percentages of organic-soluble metabolites, at all BaP concentrations, were: 20-27.2% for more polar metabolites, 23.2-28% for 9,10-diol, 9.7-13.6% for 7,8-diol, 17.8-23.2% for quinones and 15-22.2% for 9-hydroxy. With G6PD-deficient HSF these figures were: 23.4-26.9% for more polar fraction, 18.8-25.7% for 9,10-diol, 2.9-13.6% for 7,8-diol, 24.6-29.9% for quinones, and 14.8-28.9% for 9-hydroxy. Very low amounts of 3-hydroxy, belonging to the organic-soluble fraction, in both types of HSF, were not considered for percentage calculation. It thus appears that the percent distribution of organic-soluble metabolites, produced by normal HSF, did not exhibit any major change between the two types of HSF or, for both types of HSF, with the increase of BaP concentration except for 7,8-diol. At 1.2 µM BaP this metabolite represented only 3-4% of total metabolites in the deficient cells, against 10-14% in the same cells at lower BaP concentrations and about 10-14% in the control HSF at all BaP concentrations.

As shown in Fig. 1 total organic-soluble metabolites were 2-2.5 times as low in the deficient cells as in controls, at all BaP concentrations. The absolute concentration of each BaP metabolite produced by both types of HSF, increased with that of BaP in the culture medium. The comparison between the two types of HSF for the relative amounts of each BaP metabolite, showed that decreases in the G6PD-deficient cells, in comparison to controls at all BaP concentrations, ranged: 33-53% for more polar fraction, 40-56% for 9,10-diol, 32-87% for 7,8-diol, 28-43% for quinones and 41-64% for 9-hydroxy. It clearly appears, from the data in Fig. 1 that differences between the two types of HSF, with regard to the production of BaP metabolites, increased with BaP concentration in the medium.

Similar behavior has been observed for water-soluble metabolites produced by normal and G6PD-deficient cells. At all BaP concentrations the major water-soluble metabolites produced by the two types of HSF were: 9,10-diol, 9-hydroxy, 3-hydroxy, and a more polar fraction. Very low amounts of 7,8-diol were formed. Percent distribution of BaP metabolites did not exhibit any major difference between normal and G6PD-deficient cells. However, there was a lower percentage of more polar metabolites in the deficient cells compared to controls (2-9% in deficient HSF vs 7-18% in controls). Moreover, the percentage of these metabolites decreased, in both types of

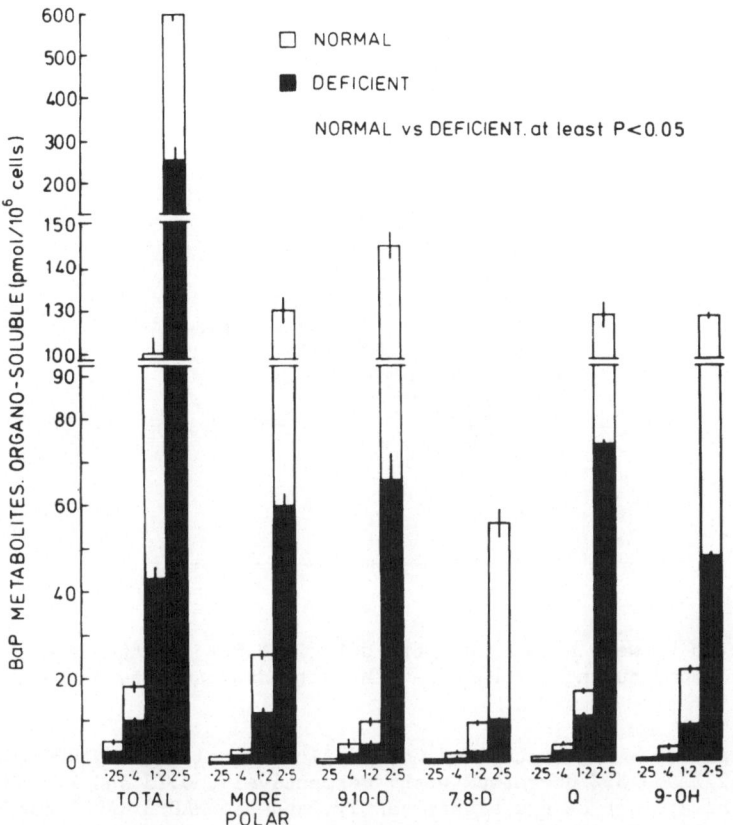

Fig. 1. BaP organic-soluble metabolites produced by in vitro growing normal
and G6PD-deficient human fibroblasts. Data are means ± SD of 6
different individuals. "t"-Test: deficient vs normal: at least
P < 0.01 at all BaP concentrations.

HSF, as BaP concentration increased. The percentages of the other water-soluble metabolites, produced by the two types of HSF, were 5-8% for 9,10-diol,
30-32% for 9-hydroxy and 37-51% for 3-hydroxy. The absolute amounts of
water-soluble metabolites underwent a great fall in the deficient cells at
all BaP concentrations, and the difference between the two types of metabolites increased with BaP concentration (data not shown).

Fig. 2 illustrates the variations of the normal/deficient ratio as
concerns the production of total (organic- plus water-soluble) metabolites.
The increase in BaP concentration was paralleled by a rise in the ratio
control/deficient for each metabolite. The increase proceeded slowly for
quinones, 9.10-diol and more polar metabolites. There was an about 2-fold
rise for 3- and 9-hydroxy, and a 4-fold rise for 7,8-diol.

These observations could indicate that a limiting factor may hinder
G6PD-deficient cells from metabolizing high BaP amounts. To investigate if a
low NADPH/NADP$^+$ ratio is the limiting factor, the pyrimidine nucleotide
content of fibroblasts growing in the presence of 0.25 and 2.5 μM BaP was
determined. The data in Fig. 3 show that NADPH content of normal HSF
underwent a 14% fall after incubation with 2.5 μM BaP. This was coupled with
a 15% rise in NADP$^+$ content and a 28% decrease in NADPH/NADP$^+$ ratio. No
great variations of the above parameters occurred in normal HSF incubated

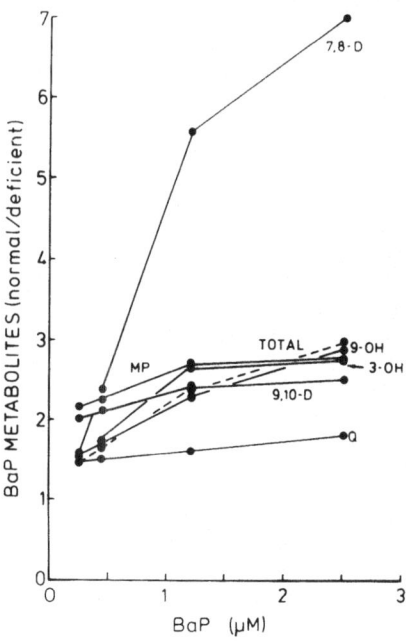

Fig. 2. Ratio between normal and G6PD-deficient HSF for the major metabolites produced during incubation with increasing amounts of BaP.

with 0.25 µM BaP. In the case of G6PD-deficient cells the NADPH content exhibited 27% and 35% decreases with, respectively, 0.25 and 2.5 µM BaP. The NADP content did not change significantly, and the NADPH/NADP$^+$ ratio exhibited a 37-39% decrease.

Thus, the NADPH/NADP$^+$ ratio varied with the BaP concentration only in normal HSF, whereas the deficient cell underwent a great fall at both concentrations used.

CONCLUSIONS

1. No major variations of BaP metabolic patterns take place, in normal and G6PD-deficient HSF, as a consequence of the increase in BaP concentration. Similar results have been obtained with mouse embryo fibroblasts, incubated with increasing BaP amounts[5].

2. HSF carrying the Mediterranean G6PD variant produce lower amounts of organic-soluble and water-soluble metabolites than normal HSF, when incubated with BaP concentrations varying from 0.25 to 2.5 µM.

3. The difference in the formation of BaP metabolites, between normal and G6PD-deficient cells, increases with BaP concentration. This is particularly evident for BaP-7,8-diol, a direct precursor of the ultimate carcinogen BaP-7,8-diol-9,10-epoxide. We have not determined this latter metabolite in the present research, but previous researches have indicated that G6PD-deficient cells are less prone than normal cells to transforming BaP-7,8-diol into 7,8-diol-9,10-epoxide. It thus seems that enhanced transformation of 7,8-diol to 7,8-diol-9,10-epoxide cannot account for the great fall in 7,8-diol in the deficient cells. On the other hand, our results also seem to exclude that high amounts of 7,8-diol are transformed into water-soluble metabolites in the deficient HSF. The possibility of a reduced 7,8-diol

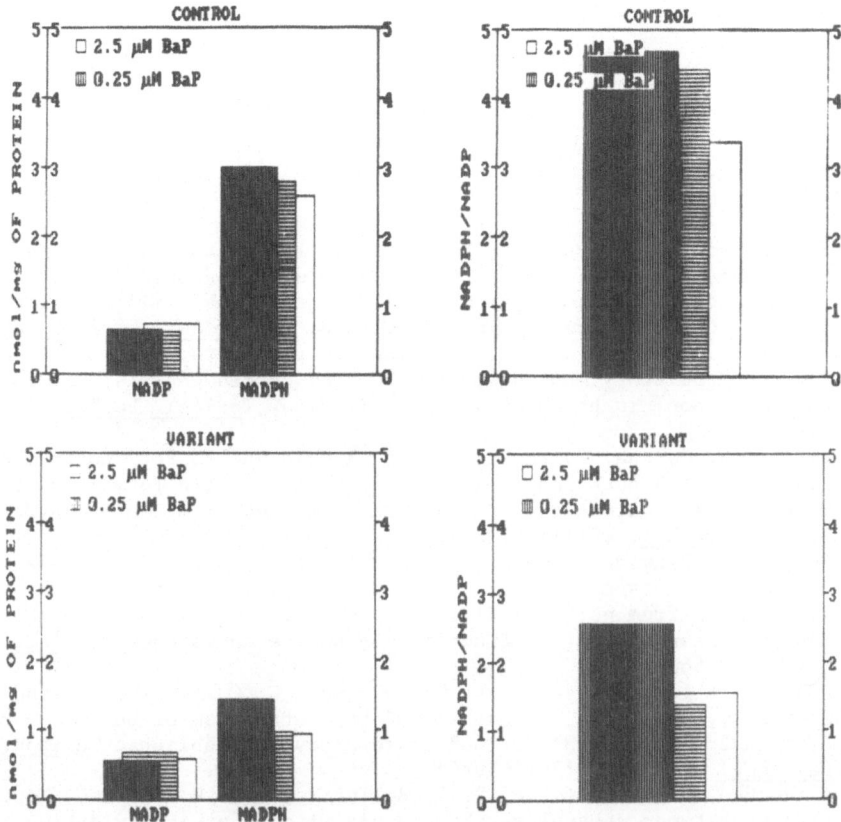

Fig. 3. Effect of incubation with varying BaP amounts on the NADPH and NADP content, and NADPH/NADP+ ratio of HSF carrying the normal (control) or Mediterranean variant (variant) of G6PD. Black histograms refer to untreated HSF. Data are means of 5 different individuals. Standard deviations were not higher than 10% of the mean. "t"-Test: variant vs normal: NADPH and NADPH/NADP+, P < 0.001. BaP vs untreated, control: 0.25 µM BaP, not significant; 2.5 µM BaP, P < 0.001. Variant: P < 0.001 at all BaP concentrations.

formation, in the G6PD-deficient cells, must be considered. This phenomenon is specific for 7,8-diol, since it does not involve other BaP-diols. An explanation of this phenomenon cannot be given at this stage of our research. It may be hypothesized that a fall in the reducing environment for the mixed function oxygenase system causes a shift in the activity of cytochrome P450 isozymes, leading to lower amounts of 7,8-diol precursors in the deficient cells. However, further work is necessary to substantiate this hypothesis.

4. Our data seem to indicate that the NADPH pool is a limiting factor for the metabolism of relatively high BaP amounts in G6PD-deficient cells. This agrees with the observation that ethanol infusion into rats, which causes NADPH oxidation [6], strongly inhibits BaP hydroxylation [7]. Moreover, recent studies with a whole-cell system permeabilized to NADPH suggest that this cofactor is rate-limiting in the mixed function oxidation of BaP [8]. These data, taken together, substantiate our previous hypothesis that the decrease in the NADPH pool, in G6PD-deficient cells, explains the reduction of BaP metabolism in these cells compared to normal cells.

ACKNOWLEDGEMENTS

Supported by grants from the "Progetto Finalizzato Ingegneria Genetica e Basi Molecolari delle Malattie Ereditarie del CNR", the "Ministero Pubblica Istruzione (programs 40% and 60%) and the "Regione Autonoma Sardegna".

REFERENCES

1. F. Feo, L. Pirisi, R. Pascale, L. Daino, S. Frassetto and R. Garcea, Modulatory effect of glucose-6-phosphate dehydrogenase deficiency on the benzo(a)pyrene toxicity and transforming activity for in vitro cultured human skin fibroblasts, Cancer Res. 38:3419 (1984).
2. F. Feo, L. Pirisi, R. Pascale, L. Daino, S. Frassetto, S. Zanetti and R. Garcea, Modulatory mechanisms of chemical carcinogenesis: the role of the NADPH pool in benzo(a)pyrene activation, Toxicol. Pathol. 12:262 (1984).
3. L. Pirisi, R. Garcea, R. Pascale, M. E. Ruggiu and F. Feo, Control of glucose-6-phosphate dehydrogenase deficiency on the formation of mutagenic and carcinogenic metabolites derived from benzo(a)pyrene, Toxicol. Pathol. 15:115 (1987).
4. F. Feo, M. E. Ruggiu, L. Lenzerini, R. Garcea, L. Daino, S. Frassetto, V. Addis, L. Gaspa and R. Pascale, Benzo(a)pyrene metabolism by lymphocytes from normal individuals and individuals carrying the Mediterranean variant of glucose-6-phosphate dehydrogenase, Int. J. Cancer 39:560 (1987).
5. K. Rudo, S. Ellis, B. J. Bryant, K. Lawrence, G. Curtis, H. Garland and S. Nesnow, Quantitative analysis of the metabolism of benzo(a)pyrene by transformable C3H10T1/2CL8 mouse embryo fibroblasts, Teratog. Carcinog. Mutagen. 6:307 (1986).
6. L. A. Reinke, F. C. Kauffman, S. A. Belinsky and R. G. Thurman, Interactions between ethanol metabolism and mixed function oxidation in perfused rat liver: inhibition of p-nitroanisole O-demethylation, J. Pharmacol. Exp. Ther. 213:70 (1980).
7. L. A. Reinke, P. McManus, F. C. Kauffman and R. G. Thurman, Benzo(a)pyrene phenol production by perfused rat liver and its inhibition by ethanol, Cancer Res. 4:1681 (1982).
8. I. J. Sadowski, J. A. Wright and L. G. Israels, A permeabilized cell system for studying regulation of aryl hydrocarbon hydroxylase: NADPH as rate limiting factor in benzo(a)pyrene metabolism, Int. J. Biochem. 17:1025 (1985).

COMPARISON OF THE COVALENT BINDING OF VARIOUS CHLOROETHANES WITH

NUCLEIC ACIDS

Giorgio Prodi, Annamaria Colacci, Sandro Grilli, Giovanna
Lattanzi, Mario Mazzullo and Paola Turina

Centro Interuniversitario per la Ricerca sul Cancro, Istituto
di Cancerologia, Università di Bologna, 40126 Bologna, Italy

INTRODUCTION

Chlorinated hydrocarbons are widely produced and utilized for various
purposes (as solvents, chemical intermediates, fumigants, vapor-pressure
depressants in aerosols, etc.)[1-3].

All of them share a central-nervous-system-depressing activity and can
cause damage to the liver, kidney and lung in mammals[3].

Several of these compounds have been found to possess mutagenic
properties in some short-term tests[1-3] or to cause cause carcinogenic effects
in animals[4].

However, haloalkanes differ greatly in their degree of toxicity or
oncogenic potency.

In recent years some investigators have tried to find a common mechanism of action for chlorinated compounds and have assumed that correlations
exist between the structure and activity of these compounds[5]. The genotoxic
activity is carried out through the interaction of electrophilic intermediates, such as chloroxirane and chloroacetaldehyde, with the nucleophilic
sites of nucleic acids.

In this paper we report the in vivo and in vitro covalent binding of
1,1-dichloroethane (1,1-DCE), 1,2-dichloroethane (1,2-DCE),1,1,1-trichloro-
ethane (1,1,1-TCE), 1,1,2-trichloroethane (1,1,2-TCE), 1,1,2,2-tetrachloro-
ethane (1,1,2,2-TTCE), pentachloroethane (PTCE), hexachloroethane (HCE) to
DNA and other macromolecules in an attempt to better define their genotoxic
activity whose evidence is still inadequate for some of them.

MATERIALS AND METHODS

[U-^{14}C]-1,1-dichloroethane (1,1-DCE), [U-^{14}C]-1,2-dichloroethane
(1,2-DCE), [U-^{14}C]-1,1,1-trichloroethane (1,1,1-TCE), [U-^{14}C]-1,1,2-tri-
chloroethane (1,1,2-TCE), [U-^{14}C]-1,1,2,2-tetrachloroethane (1,1,2,2-TTCE),
[U-^{14}C]-pentachloroethane (PTCE), [U-^{14}C]-hexachloroethane (14.6 mCi/mmol;
radiochemical purity: 98%) were purchased from the Radiochemical Centre,
Amersham, England.

DNA, polyribonucleotides, NADPH and GSH were obtained from Sigma
Chemical Co., St. Louis, Mo., USA; phenobarbitone (PB) was from Carlo Erba,
Milan, Italy.

Adult male Wistar rats and male BALB/c mice were obtained from Charles
River Italia, Calco, Italy.

In in vivo studies, each [U-^{14}C]-chlorocompound (127 µCi/kg b. wt.) was
injected i.p. to 6 male rats (250 g) and to 12 male mice (28 g). Fasted
animals were killed 22 h later and their organs (liver, kidney, lung and
stomach) were removed, pooled and processed in order to obtain DNA, RNA and
proteins. In the case of liver, individual binding values were also determined: Macromolecules were exhaustively washed with organic solvents until
no radioactivity was present in the supernatants. Recovery, purity and
labeling were assayed by ultraviolet absorption measurement, specific
colorimetric reactions and counting in a liquid scintillation spectrometer
Beckman LS-1801, as previously described[6-11].

Microsomal and cytosolic fractions used in in vitro studies were obtained from livers, kidneys, lungs and stomachs of male rats and mice. Some
of the animals were pretreated daily i.p. with 100 mg/kg b.w. of phenobarbitone for 2 days before sacrifice. Standard incubation mixture consisted of: 2.5 µCi [U-^{14}C]-chlorocompound 1.5 mg calf thymus DNA, 2 mg microsomal
proteins (+ 2 mg NADPH) and/or 6 mg cytosolic proteins (+ 9.2 mg GSH) to a
final volume of 3 ml 0.08 M potassium phosphate buffer, pH 7.7.

Reactions were carried out in triplicate at 37°C for 90 minutes in air
and in the dark. Controls were performed in the absence of coenzymes; in
addition some blanks were carried out in the absence of enzymes or with heat-
inactivated enzymes. 1.5 mM 2-diethylamino ethyl-2,2-diphenyl valerate. HCl
(SKF 525-A), an inhibitor of mixed function oxidase system (MFO), or 5 mM
diethylmaleate (DEM), an inhibitor of GSH-transferase(s), were added to
microsomal and/or to cytosolic standard incubation.

DNA, microsomal RNA and proteins, and cytosolic proteins were reisol-
ated, purified and their specific activity was determined as previously reported[6-12].

RESULTS

1,1-DCE, 1,2-DCE, 1,1,1-TCE, 1,1,2-TCE, 1,1,2,2-TTCE, PTCE, HCE were
bound covalently to macromolecules in vivo after i.p. injection. In Table 1
are shown the binding values to DNA of the liver, kidney, lung and stomach.
Unlike 1,1-DCE, chloroalkanes bound preferentially the DNA of mouse organs.
Other exceptions corresponded to the greater binding extent of 1,1,2,2-TTCE
to rat liver and kidney and of pentachloroethane to rat kidney DNA. Generally, DNA from liver was the most labeled. The Covalent Binding Index (CBI)
calculated according to Lutz[13] on the liver labelings shows this order of
reactivity: 1,1,2,2-TTCE > 1,1-DCE > Pentachloroethane > 1,2-DCE > Hexachloroethane > 1,1,1-TCE for rat liver labelings and 1,1,2,2-TTCE > Pentachloroethane ≥ Hexachloroethane > 1,2-DCE ≥ 1,1,2-TCE > 1,1-DCE > 1,1,1-TCE
for mouse. The other organs were labeled in various ways. The interaction
of 1,2-DCE with lung DNA from both species was slightly higher than that with
liver DNA.

Tables 2 and 3 and Figs. 1, 2 and 3 show the in vitro interactions of
chlorocompounds with DNA, mediated by enzymatic fractions from rat and mouse
organs. The efficiency of microsomal and cytosolic fractions in bioactivating the chlorocompounds was generally enhanced (1.5-20 fold) by the animals
pretreatment with PB. Moreover, the interaction resulted as linear up to 90

Table 1. In vivo binding (as pmol/mg) of chloroalkanes to DNA of rat and mouse organs

	Liver		Kidney		Lung		Stomach	
	Rat	Mouse	Rat	Mouse	Rat	Mouse	Rat	Mouse
1,1-DCE	3.10	2.54	1.81	0.65	2.24	1.51	4.78	2.33
1,2-DCE	1.32	2.12	1.32	2.99	1.70	2.59	0.34	0.49
1,1,1-TCE	0.22	0.44	0.84	0.91	0.73	0.63	0.37	0.56
1,1,2-TCE	0.74	2.07	1.11	1.30	0.49	1.82	0.96	1.88
1,1,2,2-TTCE	12.12	13.79	0.22	2.90	4.97	3.36	6.63	8.67
PTCE	1.29	4.35	0.71	0.43	1.11	2.49	2.31	2.81
HCE	0.50	3.39	0.42	0.50	0.14	0.35	0.26	0.37

Data refer to pooled organs from 6 male rats and 12 male mice.

Table 2. In vitro binding (as pmol/mg) of chlorocompounds mediated by enzymatic fractions from rat liver

	Microsome-mediated binding		Cytosol-mediated binding		Microsomes plus cytosol-mediated binding	
	Standard [a]	Control [b]	Standard [a]	Control [b]	Standard [a]	Control [b]
1,1-DCE	24.48 ± 1.77*	6.42 ± 0.39	UD [c]	UD [c]	UD [c]	UD [c]
1,2-DCE	154.66 [d]	9.66 [e]	166.90 [d]	19.90 [e]	282.20 [d]	19.20 [e]
1,1,1-TCE	10.28 ± 0.64*	3.29 ± 0.22	UD [c]	UD [c]	UD [c]	UD [c]
1,1,2-TCE	273.00 ± 21.64*	20.49 ± 1.44	UD [c]	UD [c]	UD [c]	UD [c]
1,1,2,2-TTCE	38.10 ± 3.30*	2.56 ± 0.3	62.11 ± 9.59*	6.76 ± 0.65	1,471.15 ± 8.42*	5.01 ± 0.90
PTCE	91.29 ± 7.00*	5.68 ± 0.86	139.33 ± 12.15	20.92 ± 0.83	398.93 ± 11.29*	51.68 ± 5.24
HCE	90.83 ± 5.31**	55.19 ± 4.90	195.51 ± 21.4**	92.96 ± 26.07	95.06 ± 6.29**	52.85 ± 12.93

[a] Incubation was carried out in the presence of coenzymes (NADPH and/or GSH). [b] Incubation was carried out in the absence of coenzymes (NADPH and/or GSH). [c] This compound was bioactivated by microsomal system only. [d] Data are reported as the mean of two values, each differing from the mean value in less than 3.5%. [e] Controls were carried out in the absence of specific enzymes. *Significantly different from control by Student's t test (p < 0.01). **Significantly different from control by Student's t test (P < 0.05).

Table 3. <u>In vitro</u> binding (as pmol/mg) of chlorocompounds mediated by enzymatic fractions from mouse liver

	Microsome-mediated binding		Cytosol-mediated binding		Microsomes plus cytosol-mediated binding	
	Standard[a]	Control[b]	Standard[a]	Control[b]	Standard[a]	Control[b]
1,1-DCE	27.89 ± 0.95*	6.25 ± 0.17	UD[c]	UD[c]	UD[c]	UD[c]
1,2-DCE	117.66[d]	9.66[e]	118.50[d]	19.90[e]	185.20[d]	19.20[e]
1,1,1-TCE	11.23 ± 2.87*	4.70 ± 0.46	UD[c]	UD[c]	UD[c]	UD[c]
1,1,2-TCE	179.81 ± 31.59*	20.49 ± 1.44	UD[c]	UD[c]	UD[c]	UD[c]
1,1,2,2-TTCE	48.62 ± 2.99*	2.87 ± 0.28	63.68 ± 7.90*	9.19 ± 0.83	1,765.83 ± 148.83*	4.75 ± 0.65
PTCE	97.80 ± 6.79*	5.21 ± 0.46	120.48 ± 26.10*	13.02 ± 0.43	215.32 ± 28.75*	7.65 ± 2.04
HCE	105.39 ± 7.80*	46.96 ± 4.19	346.17 ± 18.91*	128.56 ± 8.92	133.44 ± 2.42**	98.84 ± 8.06

[a]Incubation was carried out in the presence of coenzymes (NADPH and/or GSH). [b]Incubation was carried out in the absence of coenzymes (NADPH and/or GSH). [c]This compound was bioactivated by microsomal system only. [d]Data are reported as the mean of two values, each differing from the mean value in less than 3.5%. [e]Controls were carried out in the absence of specific enzymes. *Significantly different from control by Student's t test (P < 0.01). **Significantly different from control by Student's t test (P < 0.05).

97

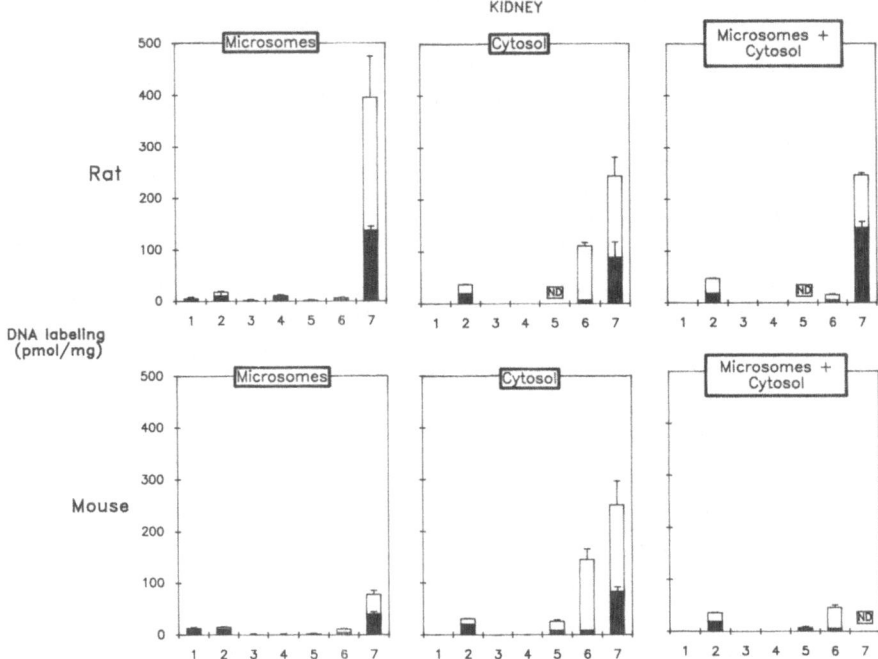

Fig. 1. In vitro binding of chlorocompounds mediated by enzymes from rat and mouse kidney. □ = total binding; ■ = controls; [1] = 1,1-DCE; [2] = 1,2-DCE; [3] = 1,1,1-TCE; [4] = 1,1,2-TCE; [5] = 1,1,2,2-TTCE; [6] = PTCE; [7] = HCE.

minutes incubation (data not shown). Thus, enzymatic fractions from organs of PB-treated animals and 90 minute incubation time were chosen as standard experimental parameters.

1,1-DCE, 1,1,1-TCE and 1,1,2-TCE were bioactivated by microsomal enzymatic system only, whereas cytosolic fractions from all assayed organs were capable of inducing the interaction of 1,2-DCE, 1,1,2,2-TTCE, pentachloroethane and hexachloroethane.

No particular species-specific differences were evidenced except for the higher efficiency of mouse lung enzymes in 1,2-DCE, 1,1,2,2-TTCE and pentachloroethane bioactivation, of mouse cytosolic fractions from liver, kidney and lung in hexachloroethane bioactivation and of rat kidney microsomes in activating hexachloroethane.

When both enzymatic systems were simultaneously present in the incubation mixture, the binding extents of 1,2-DCE, HCE and PTCE resulted equal or slightly less then those obtainable when adding values from microsomal incubation to those from cytosol-mediated binding. On the contrary, the interaction of 1,1,2,2-TTCE resulted strongly enhanced by the synergic effect of the presence of both activating systems.

The addition of 1.5 mM SKF 525-A to the microsomal standard incubation mixture inhibited the interaction of 1,1-DCE, 1,1,1-TCE, 1,1,2-TCE, and reduced that of 1,1,2,2-TTCE, hexachloroethane and, although slightly, that of pentachloroethane. The addition of 10 mM GSH to the NADPH-containing microsomal mixture inhibited binding of 1,1,2-TCE and hexachloroethane and increased the interaction of pentachloroethane and 1,1,2,2-TTCE. Even when

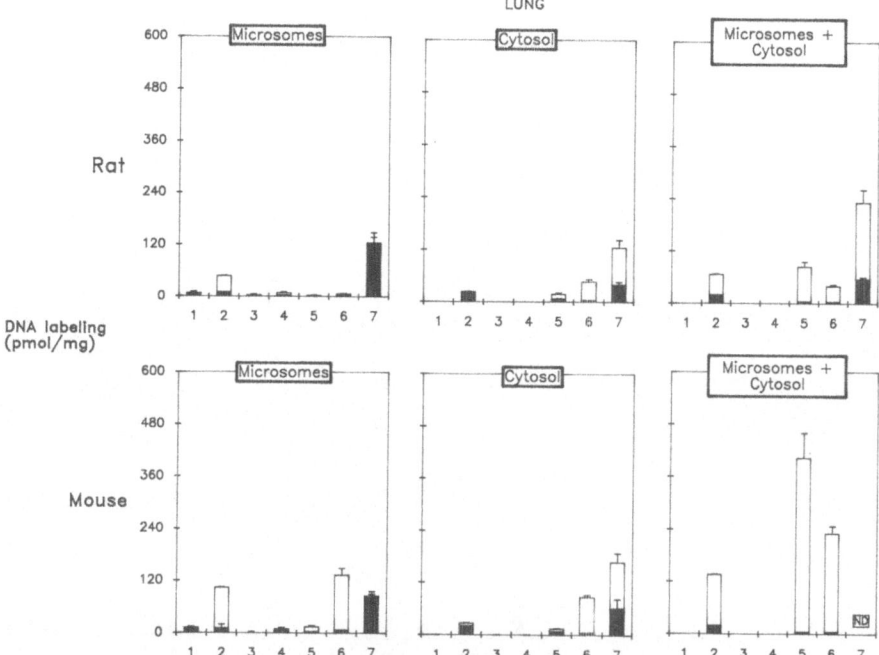

LUNG

Fig. 2. In vitro binding of chlorocompounds mediated by enzymes from rat and mouse lung. □ = total binding; ■ = controls; [1] = 1,1-DCE; [2] = 1,2-DCE; [3] = 1,1,1-TCE; [4] = 1,1,2-TCE; [5] = 1,1,2,2-TTCE; [6] = PTCE; [7] = HCE.

cytosolic fractions were added to the microsomal standard incubation the binding extent of 1,1,2,2-TTCE was enhanced (data not shown). The addition of 5 mM DEM to the GSH-containing cytosolic mixture suppressed the interaction of 1,1,2,2-TTCE and of PTCE. Also, the binding extent of HCE mediated by cytosolic fractions decreased (∿50%) in the presence of DEM (data not shown).

DISCUSSION

The Covalent Binding Index (CBI) values classify 1,1,1-TCE, 1,1,2-TCE (rat), HCE (rat) as weak initiators; 1,1-DCE, 1,2-DCE, 1,1,2-TCE (mouse), PTCE, HCE (mouse) as weak-moderate initiators and at least 1,1,2,2-TTCE as moderate initiator. These results parallel the available data on the muta-genicity and carcinogenicity of these compounds.

All of assayed chloroalkanes are bioactivated by P-450-dependent mixed function oxidase system to react covalently with DNA. The involvement of this enzymatic complex is confirmed by the higher efficiency of microsomes from PB-pretreated animals and by reduction or suppression of interaction when adding SKF 525-A, a noted inhibitor of MFO.

Some chloroethanes (1,2-DCE, 1,1,2,2-TTCE, PTCE and HCE) are metabolized even by cytosolic GSH-transferase(s) which exert a detoxificant role through metabolic pathways of other haloalkanes[14]. Indeed, the addition of GSH and/or cytosol to the NADPH-containing microsomal mixture reduces binding extents of 1,1-DCE and 1,1,2-TCE and suppresses that of 1,1,1-TCE. Also, cytosol-

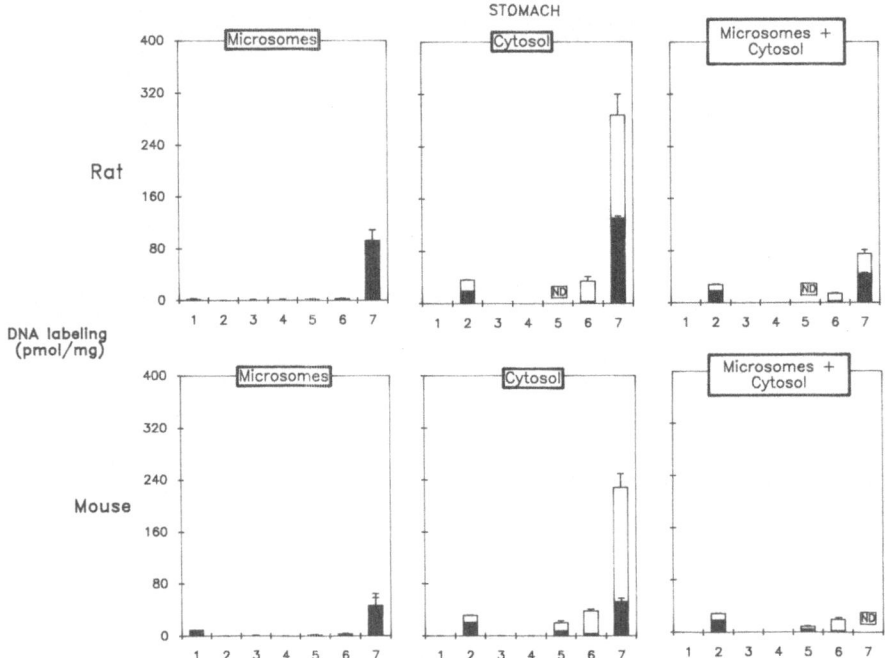

Fig. 3. <u>In vitro</u> binding of chlorocompounds mediated by enzymes from rat and mouse stomach. □ = total binding; ■ = controls; [1] = 1,1-DCE; [2] = 1,2-DCE; [3] = 1,1,1-TCE; [4] = 1,1,2-TCE; [5] = 1,1,2,2-TTCE; [6] = PTCE; [7] = HCE.

mediated binding extent is strongly affected by the presence of DEM, an inhibitor of GSH-transferase.

These reactions are an example of the ambiguity of metabolic processes[15]. Metabolism of chloroalkanes leads to the formation of reactive species such as chlorinated aldehydes, acyl chlorides or, under reductive conditions, their corresponding olefines which tend to be spontaneously transformed into epoxides. Chloroalkanes metabolism can also lead to free radicals formation[15]. All of these reactive forms can covalently bind the nucleophilic sites of nucleic acids and of other macromolecules giving rise to genotoxic effects.

It is generally held that precise correlations exist between the structure and the activity of halocompounds. It has been postulated that the covalent binding to DNA depends on the electron affinity, which increases with the number of non-symmetrical halosubstitution[5]. In this context, GSH-conjugation is a detoxifying step in unsymmetric chloroalkanes and, in contrast, has an activating function towards symmetric chlorocompounds.

These theories can be applied to the small chloroalkanes. As the number of halosubstitution increases more metabolic pathways come into play. They may compete with each other, rendering at times the identification of a general metabolic pathway difficult.

In spite of the well-known toxicity of chlorocompounds[3], informations on their mutagenic and carcinogenic activity are still rather scanty. Thus, the evidence of genotoxic effects from short-term tests is considered still limited or inadequate for 1,1,1-TCE, 1,1,2-TCE and 1,1,1,2-TTCE, whereas the mutagenicity of 1,2-DCE has been widely demonstrated[1,2].

Table 4. Oncogenic potency and DNA binding ability in vivo and in vitro of some chlorinated ethanes in mouse liver

Compound	OPI [a]	CBI	MCMBI [b]
1,1,1-Trichloroethane [c]	0.00049	16	0.11
1,1-Dichloroethane [d]	0.0096	65	0.27
1,2-Dichloroethane [e]	0.089	76	2.91
Hexachloroethane [f]	0.27	140	4.84
1,1,2-Trichloroethane [g]	0.994	73	2.84
Pentachloroethane [h]	1.96	143	3.64
1,1,2,2-Tetrachloroethane [i]	2.6	543	30.85

Correlation between Log_{10} CBI and Log_{10} OPI: ($r = 0.854$) ($P < 0.05$). Correlation between Log_{10} OPI and Log_{10} MCMBI: ($r = 0.914$) ($P < 0.005$). Correlation between Log_{10} CBI and Log_{10} MCMBI: ($r = 0.924$) ($P < 0.005$).
[a] Calculated according to Parodi et al.[16]
[b] Calculated according to Lutz[13].
[c] Turina et al.[11]
[d] Colacci et al.[7]
[e] Arfellini et al.[6]; Colacci et al.[12]
[f] Lattanzi et al.[9]
[g] Mazzullo et al.[10]
[h] Turina et al., unpublished data.
[i] Colacci et al.[8]

The comparison of Oncogenic Potency Index (OPI) in mouse liver with the DNA binding ability in vivo (CBI) and in vitro (MCMBI) of the assayed chloro-compounds (Table 4) shows a high degree ($p < 0.01$) of correlation between each binding index and OPI. We wish to emphasize that a simple in vitro cell-free system, such as that described in this report, leads to binding data which well correlate with in vivo binding. The high correlation among different end-points (DNA-adducts in vitro and in vivo and tumors) supports the hypothesis that a correlation exists between the structure and activity of carcinogenic compounds.

ACKNOWLEDGEMENTS

This study on structure-activity relationship is a part of a multi-center (Bologna, Genoa, Pisa and Rome) collaborative study on genotoxicity of halogenated hydrocarbons.

SKF 525-A was a kind gift of Smith Kline and French, Welwyn, U.K. This work was supported by grants from Consiglio Nazionale delle Ricerche, Progetto Finalizzato "Oncologia", contract no. 86.00444.44, Ministero della Pubblica Istruzione, Rome and Associazione Italiana per la Ricerca sul Cancro, Milan, Italy.

The authors thank Mr. Carlo Buttazzi for his technical assistance.

REFERENCES

1. IARC Monographs on the Evaluation of Carcinogenic Risk of Chemicals to

Humans, Vol. 20, "Some halogenated Hydrocarbons", International Agency for Research on Cancer, Lyon (1979).

2. IARC Monographs on the Evaluation of Carcinogenic Risk of Chemicals to Humans, Vol. 41, "Some halogenated hydrocarbons and pesticide exposure", International Agency for Research on Cancer, Lyon (1986).

3. T. R. Torkelson and V. K. Rowe, Halogenated haliphatic hydrocarbons, in: "Patty's Industrial Hygiene and Toxicology", G. D. Clayton, ed., Vol. 2, part B, John Wiley & Sons, Inc., New York (1981).

4. H. Greim and T. Wolff, Carcinogenicity of halogenated compounds, in: "Chemical Carcinogens", C. E. Searle, ed., American Chemical Society, Washington, D.C. (1984).

5. G. H. Loew, M. Rebagliati and M. Poulsen, Metabolism and relative carcinogenic potency of chloroethanes: a quantum chemical structure-activity study, Cancer Biochem. Biophys. 7:109 (1984).

6. G. Arfellini, S. Bartoli, A. Colacci, M. Mazzullo, M. C. Galli, G. Prodi and S. Grilli, In vivo and in vitro binding of 1,2-dibromo-ethane and 1,2-dichloroethane to macromolecules in rat and mouse organs, J. Cancer Res. Clin. Oncol. 108:204 (1984).

7. A. Colacci, G. Arfellini, M. Mazzullo, G. Prodi and S. Grilli, Geno-toxicity of 1,1-dichloroethane, Res. Comm. Chem. Pathol. Pharmacol. 49:243 (1985).

8. A. Colacci, S. Grilli, G. Lattanzi, G. Prodi, M. P. Turina and M. Mazzullo, The covalent binding of 1,1,2,2-tetrachloroethane to macromolecules of rat and mouse organs. Teratogen. Carcinog. and Mutagen. in press.

9. G. Lattanzi, A. Colacci, S. Grilli, M. Mazzullo, G. Prodi, G. and M. P. Turina, Binding of hexachloroethane to biological macromolecules from rat and mouse organs, J. Toxicol. Environ. Health submitted

10. M. Mazzullo, A. Colacci, S. Grilli, G. Prodi and G. Arfellini, 1,1,2-trichloroethane: evidence of genotoxicity from short-term tests, Jpn. J. Cancer Res. (Gann) 77:532 (1986).

11. M. P. Turina, A. Colacci, S. Grilli, M. Mazzullo, G. Prodi and G. Lattanzi, Short-term tests of genotoxicity for 1,1,1-trichloro-oethane, Res. Comm. Chem. Pathol. Pharmacol. 52:305 (1986).

12. A. Colacci, G. Arfellini, M. Mazzullo, G. Prodi and S. Grilli, In vitro microsome- and cytosol-mediated binding of 1,2-dichloroethane and 1,2-dibromoethane with DNA, Cell. Biol. Toxicol. 1:45 (1985).

13. W. K. Lutz, In vivo covalent binding of organic chemicals to DNA as a quantitative indicator in the process of chemical carcinogenesis, Mutat. Res. 65:289 (1979).

14. H. M. Bolt, Metabolism of genotoxic agents: halogenated compounds in: "Monitoring human exposure to carcinogenic and mutagenic agents", IARC Sci. Publ. No. 59, A. Berlin, M. Draper, K. Hemminki and H. Vainio, eds., International Agency for Research on Cancer, Lyon, (1984).

15. D. Henschler, Specific covalent binding and genotoxicity, in: Environ-mental carcinogens selected methods of analysis", Vol 7, "Some volatile halogenated hydrocarbons", IARC Sci. Publ. No. 68, L. Fishbein and I. K. O'Neill, eds., International Agency for Research on Cancer, Lyon (1985).

16. S. Parodi, M. Taningher, A. Zunino, L. Ottaggio, M. De Ferrari and L. Santi, Quantitative predictivity of carcinogenicity for sister chromatid exchanges in vivo, in: "Sister Chromatid Exchanges", part A, R. R. Tice and A. Hollaender, eds., Plenum Publishing Corporation, New York (1984).

IN SITU DETECTION OF HEPATIC AF-DNA ADDUCT FORMATION DURING CONTINUOUS FEEDING OF 0.02% ACETYLAMINOFLUORENE AND IN ENZYME-ALTERED FOCI INDUCED BY FOYR DIFFERENT PROTOCOLS

Henrik S. Huitfeldt[1], Henry C. Pitot[2], John M. Hunt[3], Jeffrey Baron[4] and Miriam C. Poirier[5]

[1] Laboratory for Immunohistochemistry and Immunopathology Institute of Pathology, The National Hospital, University of Oslo, Norway; [2] McArdle Laboratory for Cancer Research University of Wisconsin, Madison, WI 53706, USA [3] Department of Pathology and Laboratory Medicine University of Texas Medical School, Houston, TX 77225, USA [4] Department of Pharmacology, College of Medicine University of Iowa, Iowa City, IA 52242, USA [5] Laboratory of Cellular Carcinogenesis and Tumor Promotion National Cancer Institute, NIH, Bethesda, MD 20892, USA

INTRODUCTION

Formation of carcinogen-DNA adducts is thought to represent the critical event in the initiation of tumorigenesis in many carcinogenesis models. DNA-adduct formation and repair has therefore been profoundly studied, both for human monitoring of carcinogen exposure and in experimental carcinogenesis. Most studies on DNA-adducts are based upon extraction of DNA from whole tissue, with little possibility to distinguish between different cell populations. The ability to differentiate between subpopulations within the tissue is presumably important, both to assess adduct processing within the tumor progenitor cells, and to monitor biological alterations of initiated cell populations.

The aims of these studies has therefore been to establish a technique for in situ detection of carcinogen-DNA adduct formation within a liver carcinogenesis model. The technique should allow simultaneous demonstration of adducts in individual cells in combination with other cell markers. Also, we have been interested in quantifying adducts within single cells, so that accumulation and repair rates within cell populations could be studied.

METHODS AND RESULTS

Immunohistochemistry

To accomplish this, we have employed a paired immunofluorescence technique[1]. We have utilized a polyclonal rabbit antibody to N-(deoxyguanosine-8-yl)-aminofluorene[2]. This antibody recognizes the deacetylated aminofluorene-DNA adduct which is the major adduct formed during continuous feeding of 2-acetylaminofluorene (2-AAF). The antibody is visualized by a FITC-conjugated antibody to rabbit immunoglobulins, and a monoclonal antibody (PKK1) to cytokeratins[3] has been used for bile duct identification with a biotin-streptavidin-Texas Red staining sequence. By this method it has been possible to identify adduct formation within individual hepatocytes in different areas of the liver lobules during continuous feeding of 0.02% 2-AAF in the diet.

Adduct Formation and Removal during Continuous Feeding of 2-AAF

Young (200 g) male Fisher rats were fed 0.02% 2-AAF for up to 4 weeks. Paired immunofluorescence on frozen liver sections revealed adduct accumulation in periportal areas identified by cytokeratin staining of bile ducts. Adduct staining in midzonal and centrilobular nuclei was substantially weaker. Bile duct cells contained very low levels of AF-DNA adducts.

AF-DNA removal was studied in livers of young male Fisher rats fed 0.02% 2-AAF for 4 weeks, and then a control diet for additional 4 weeks[4]. Through this period, the overall AF-DNA adduct staining intensities diminished and the number of stained nuclei decreased. At all time points periportal adduct concentrations were the highest, and clearly positive after 1 month of control diet.

Microfluorometry

To allow semiquantitation of adducts within single cells in different regions of the liver lobule, we have used a microfluorometric technique[5]. Frozen liver sections from young male Fisher rats fed 0.02% 2-AAF for up to 4 weeks were stained by paired immunofluorescence as described. Fluorescence emitted at 525 nm from circular areas (6 μm in diameter) within nuclei was detected by a photomultiplier tube mounted on a fluorescence microscope. Emitted fluorescence was expressed as one minus absorbance (*100) so that a positive, linear relationship was obtained between the intensities of fluorescence staining and the microfluorometric measurements. 20-22 nuclei of each cell type from each rat at each time point were measured, and results expressed as mean fluorescence units ± SE (Fig. 1). The results demonstrate adduct accumulation in periportal, midzonal and centrilobular regions during the initial 12 days of feeding, followed by a plateau level where midzonal adduct concentrations were 70-80% and centrilobular adduct concentrations less than 20% of periportal concentrations. Bile duct cells also contained less than 20% of periportal adduct values. The profiles of adduct accumulation are similar to the profile obtained by RIA determination of whole liver AF-DNA and AAF-DNA adducts performed on the same livers.

AF-DNA Adduct Formation in Enzyme Altered Foci

To assess adduct formation in preneoplastic liver lesions induced by different carcinogenesis protocols, we have employed computer-aided image over-laying of histochemically stained serial sections to identify and score enzyme-altered foci[6,7]. Serial sections were stained for histochemical demonstration of gamma-glutamyl transpeptidase (GGT), canalicular ATPase and glucose-6-phosphatase (G6Pase), and for immunofluorescence-identification of AF-DNA adducts. Enzyme-altered foci were localized and the phenotypes scored by image overlaying of enzyme-stained slides. These overlays were carefully compared to fluorescence-stained slides for adduct identification, allowing recognition of most enzyme-altered foci on the adduct stained slide. Some foci were not successfully identified, either because they were damaged by the prolonged immunofluorescence staining procedure, or because they were too small for positive recognition.

Continuous Carcinogen Exposure Model

In the first model foci were scored in livers of young male Fisher rats

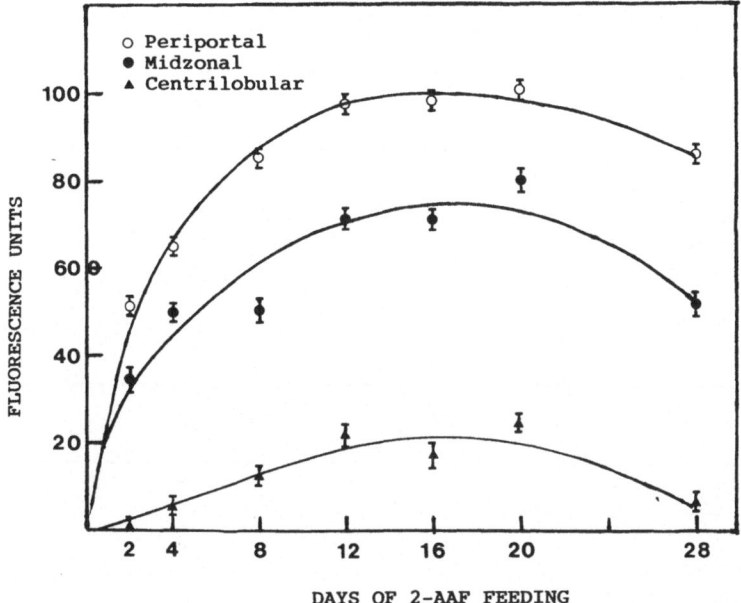

Fig. 1. Microfluorometric analysis of dG-8-AF immunofluorescence in nuclei
of periportal, midzonal and centrilobular hepatocytes of male
Fischer rats fed 0.02% 2-AAF for 2, 4, 8, 12, 16, 20 and 28 days
(<u>abscissa</u>). <u>Ordinate</u> <u>values</u>, relative fluorescence units ± SE for 40
to 44 nuclei in each cell type from two animals. Fluorescence values
from corresponding cells in livers of rats fed control diet have
been subtracted.

given 0.02% 2-AAF in the diet for 2 months. Altogether 102 foci of 7 different phenotypes were identified. 75 of these were recognized on the fluorescence-stained slide, and none of these showed adduct staining (Table 1).
The surrounding liver tissue showed bright adduct immunofluorescence, and
cytokeratin staining indicated pronounced bile duct proliferation.

<u>Initiation-Promotion Models</u>

Adduct formation was explored in foci induced by 70% partial hepatectomy
(PH) followed by diethylnitrosamine (DEN, 10 mg/kg) by intubation 24 h later.
Either phenobarbital (0.05% in the diet) for 9 months[8] or ciprofibrate (0.01%
in the diet) for 5 months followed by phenobarbital (0.05% in the diet) for 2
months[9] was used for promotion. The rats were given 0.02% 2-AAF in the diet
for 5 days prior to sacrifice. In phenobarbital-promoted livers, a total of
352 foci were identified by image overlaying of histochemically stained
slides. 304 of these were recognised on adduct immunofluorescence stained
slides (Table 2). All of these foci were adduct negative.

In livers from rats promoted by ciprofibrate and phenobarbital, 163
enzyme-altered foci were identified, and 125 of these detected on immunofluorescent-stained slides (Table 3). All these foci were also negative for
adducts. In these livers adduct formation in surrounding tissue was generally weaker than in the first model due to the shorter 2-AAF feeding period.
Also, bile duct proliferation was not noticed in these livers.

Table 1. AF-DNA Adduct localization in enzyme-altered foci in livers
of rats fed 0.02% 2-AAF in the diet for 56 days

Enzyme Phenotype	Number of Foci	AF-DNA Positive Foci
GGT+, ATPase-, G6Pase-	19	0
GGT+, ATPase-	13	0
GGT+, G6Pase-	5	0
ATPase-, G6Pase-	12	0
GGT+	16	0
ATPase-	4	0
G6Pase-	6	0
Total	75	0

Resistant Hepatocyte Model

Young male F344 (Fisher) donor rats received DEN (200 mg/kg, i.p.) and 2
weeks later a selective 2-AAF/PH regimen consisting of 2 weeks 0.02% 2-AAF in
the diet, with 70% partial hepatectomy after 1 week of 2-AAF feeding. Eight
days later donor rat liver cells were dissociated by in situ type I collagenase perfusion, and $2*10^6$ viable donor cells transplanted i.v. into livers of
(WF*F344) F host rats receiving the 2-AAF/PH regimen[10]. Host rats were maintained for 7.5, 16.5 or 19.5 months. The rats received 0.02% 2-AAF in the
diet for 6 days prior to sacrifice. In this model 83 enzyme altered foci
were identified, and 68 of these recognized on fluorescent-stained slides
(Table 4). All of these were generally adduct negative, but many of them
contained groups of weakly staining nuclei. It is interesting that this
phenomenon only was found in this model, since remodelling is known to occur
in many protocols where treatment is discontinued[11].

Table 2. AF-DNA Adduct localization in enzyme-altered foci in
livers of rats given partial hepatectomy, DEN (10 mg/kg),
0.05% phenobarbital for 9 month and 0.02% 2-AAf for 5 days

Enzyme Phenotype	Number of foci	AF-DNA Positive Foci
GGT+, ATPase-, G6Pase-	75	0
GGT+, ATPase-	126	0
GGT+, G6Pase-	0	0
ATPase-, G6Pase-	6	0
GGT+	49	0
ATPase-	44	0
G6Pase-	4	0
Total	304	0

Table 3. AF-DNA Adduct localization in enzyme-altered foci in livers of rats given partial hepatectomy, DEN (10 mg/kg), 0.01% ciprofibrate for 5 month, 0.05% phenobarbital for 2 month and 0.02% 2-AAF for 5 days

Enzyme Phenotype	Number of Foci	AF-DNA Positive Foci
GGT+, ATPase-, G6Pase-	10	0
GGT+, ATPase-	4	0
GGT+, G6Pase-	7	0
ATPase-, G6Pase-	15	0
GGT+	30	0
ATPase-	23	0
G6Pase-	36	0
Total	125	0

DISCUSSION

These studies demonstrate a paired immunofluorescence technique for in situ detection of AF-DNA adducts. Adduct staining can be combined with other cell markers to identify target cell for adduct formation. This has revealed a non-random distribution of adducts during continuous dietary exposure to 2-AAF within the liver lobule, where AF-DNA adducts are predominantly localized to periportal areas. This pattern is also maintained through 4 weeks on control diet following 4 weeks of 2-AAF feeding. Also, we have utilized microfluorometry, a technique for in situ measurement of AF-DNA adducts within individual cells. This has revealed an AF-DNA adduct gradient within the liver lobule, where midzonal adduct concentrations are 70-80% and centrilobular adduct concentrations less than 20% of periportal concentrations.

Table 4. AF-DNA adduct localization in enzyme-altered foci induced by resistant hepatocyte model

Enzyme Phenotype	Number of foci	AF-DNA Positive foci
GGT+, ATPase-, G6Pase-	24	0
GGT+, ATPase-	7	0
GGT+, G6Pase-	6	0
ATPase-, G6Pase-	8	0
GGT+	8	0
ATPase-	7	0
G6Pase-	8	0
Total	68	0

Donor rats received DEN (200 mg/kg), and 2 week of 0.02% 2-AAF with a 70% PH after 1 week of 2-AAF feeding. Livers were dissociated and donor cells transplanted into host rats receiving the 2-AAF/PH regimen. Host rats were maintained for 7.5, 16.5 and 19.5 months and received 0.02% 2-AAF for 6 days prior to sacrifice.

During the first 12 days of 2-AAF feeding adduct accumulation has been demonstrated in these zones, followed by a plateau phase. Differences in localization of enzymes or cofactors responsible for 2-AAF uptake, activation or detoxification within the liver lobule might explain this non random pattern. Alternatively, periportal hepatocytes may be exposed to the higher 2-AAF concentrations due to a drug concentration gradient across the liver lobule. So far we have not been able to identify differences in repair rates within the liver lobule.

A lack of AF-DNA adduct formation in all enzyme-altered foci of 4 different carcinogenesis models has been observed. Our findings demonstrate an apparently universal biochemical adaptation, common both to several phenotypes of different complexity levels, and to multiple carcinogenesis protocols. This suggests that an altered carcinogen metabolism is a very early and probably necessary event in the formation of preneoplastic lesions. In the resistant hepatocyte model, where no additional promotion is performed after the selection procedure, groups of nuclei within foci contain low levels of adducts, suggesting that cells may lose their protection from adduct formation as they remodel into normal hepatocytes. Our findings give support to the view that a common gene may be altered at an early time-point in these putatively premalignant lesions, conferring both protection from the toxic effects of xenobiotics and a growth advantage.

REFERENCES

1. H. S. Huitfeldt, E. F. Spangler, J. M. Hunt and M. C. Poirier, Immunohistochemical localization of DNA adducts in rat liver tissue and phenotypically altered foci during oral administration of 2-acetylaminofluorene, Carcinogenesis 7:123 (1986).
2. M. C. Poirier, J. Nakayama, F. P. Perera, I. B. Weinstein and S. H. Yuspa, Identification of carcinogen-DNA adducts by immunoassays, in: "Application of Biological Markers to Carcinogen Testing", H. A. Milman and S. Sell, eds., Plenum Press, New York (1983).
3. H. Holthofer, A. Miettinen, R. Paasivuo, V. P. Lehto, E. Linder, O. Alfthan and I. Virtanen, Cellular origin and differentiation of renal carcinomas. A fluorescence microscopic study with kidney-specific antibodies, anti-intermediate filament antibodies and lectins, Lab. Invest. 49:317 (1983) .
4. F. A. Beland, H. S. Huitfeldt and M. C. Poirier, DNA adduct formation and removal during chronic administration of a carcinogenic aromatic amine, Prog. exp. Tumor Res. 31:33 (1987).
5. H. S. Huitfeldt, E. F. Spangler, J. Baron and M. C. Poirier, Microfluorometric determination of DNA adducts in immunofluorescent-stained liver tissue from rats fed 2-acetylaminofluorene, Cancer Res. 47:2098 (1987).
6. H. C. Pitot, L. Barsness, T. Goldsworthy and T. Kitagawa, Biochemical characterization of stages of hepatocarcinogenesis after a single dose of diethylnitrosamine, Nature 272:456 (1978).
7. H. A. Campbell, H. C. Pitot, V. R. Potter and B. A. Laishes, Application of quantitative stereology to the evaluation of enzyme-altered foci in rat liver, Cancer Res. 41:465 (1982).
8. T. L. Goldsworthy and H. C. Pitot, The quantitative analysis and stability of histochemical markers of altered hepatic foci in rat liver following initiation by diethylnitrosamine and promotion with phenobarbital, Carcinogenesis 6:1261 (1985).
9. H. P. Glauert, D. Beer, M. Sambasiva Rao, M. Schwarz, Y-D. Xu, T. L. Goldsworthy and H. C. Pitot, Induction of altered hepatic foci in rats by the administration of hypolipidemic peroxisome proliferators alone or following a single dose of diethylnitrosamine, Cancer Res. 46:4601 (1986).

10. J. M. Hunt, M. T. Buckley, B. A. Laishes and H. A. Dunsford, Immuno-
    logical approaches to the purification of putative premalignant
    hepatocytes from genotypic mosaic rat livers, Cancer Res. 45:2226
    (1985).
11. S. Hendrich, H. P. Glauert and H. C. Pitot, The phenotypic stability of
    altered hepatic foci: effects of withdrawal and subsequent readmin-
    istration of phenobarbital, Carcinogenesis 7:2041 (1986).

INTERACTION OF ACTIVATED N-NITROSAMINES WITH DNA

Jamie R. Milligan, Laura Catz-Biro, Samim Hirani-Hojatti
and Michael C. Archer

Department of Medical Biophysics, University of Toronto
Ontario Cancer Institute
500 Sherbourne Street, Toronto
Ontario, Canada M4X 1K9

Some chemically induced tumors in rodents are associated with single point mutations in ras genes[1]. It is important to know whether these mutations arise at sites in DNA of unusually high reactivity for ultimate carcinogens, or whether carcinogen interaction is random. This information is not available for simple methylating agents such as those derived from N-nitroso compounds.

Metabolic activation of carcinogenic N-nitrosodialkylamines is by hydroxylation at an α-carbon atom[2]. Fragmentation of the α-hydroxynitrosamine produces an aldehyde and an alkyldiazohydroxide. The alkyldiazohydroxide then yields the alkyldiazonium ion as the ultimate electrophilic alkylating agent. There is evidence[3] that the α-hydroxynitrosamine may be the transport form of the activated carcinogen, and it may therefore be the species which first interacts with the DNA. If this is the case, α-hydroxylated derivatives of different nitrosamines could interact with DNA uniquely such that the sequence or conformation of nucleotides influences the extent of alkylation at specific sites.

For these various reasons, we have investigated the formation, in vitro, of 7-methylguanine and O^6-methylguanine in a variety of different DNA substrates following their reaction with three different N-nitroso compounds which are all methylating agents: N-nitroso(acetoxymethyl)methylamine (AcONDMA), N-nitroso(acetoxybenzyl)methylamine (AcONMBzA) and N-nitrosomethylurea (NMU). α-Acetoxy-N-nitrosamines are stable forms of activated nitrosamines that are readily hydrolyzed to the corresponding α-hydroxynitrosamines in the presence of a non-specific esterase[4]. NMU undergoes non-enzymatic, base-catalyzed decomposition to yield methyldiazotate[5]. In the second part of this study, we have investigated, using the plasmid pBR322, whether methylation of DNA by the three nitroso compounds displays any base sequence specificity.

We first showed, using calf thymus DNA, that below about 5mM, guanine methylation was proportional to the concentration of each of the three methylating agents. Above this concentration, guanine methylation levelled off. This saturation effect may be related to changes in DNA structure caused by the formation of a high proportion of positively charged 7-methylguanine residues.

We next measured methylated guanine formation in calf thymus DNA, a negatively supercoiled plasmid, pEC, and two synthetic polynucleotides of defined sequence, using the methylating agents at a concentration of 2 mM. Table 1 shows that all three N-nitroso compounds form similar amounts of O^6- methylguanine in the four DNA substrates (differences within and between columns are not significant). 7-methylguanine was formed in about ten times the yield of O^6-methylguanine in all of the DNA substrates except for poly(dGdC).poly(dGdC), for which the yield of 7-methylguanine was only about three times the yield of O^6-methylguanine. The ratios of the yields of the methylated guanines for the different DNA substrates and methylating agents are shown in Table 2.

Our results confirm and extend the observations of Briscoe and Cotter[6,7] who also observed a higher O^6-methylguanine:7-methylguanine ratio for poly(dGdC).poly(dGdC) ($\sim$0.2) than for other DNA substrates ($\sim$0.1) using NMU. Our somewhat higher value for poly(dGdC).poly(dGdC) (0.35) is probably due to different reaction conditions. This similarity of the results using calf thymus DNA and plasmid DNA, suggests that the supercoiling in the plasmids has no effect on the susceptibility of guanine to methylation. However, the results clearly show that the methylation of guanines is dependent in some manner on DNA structure.

Sequence specificity was studied by first reacting the plasmid pBR322[8] with 0.1 mM of each of the methylating agents. After isolating the 345bp Hind III - Bam HI restriction fragment, it was incubated at 65° at neutral pH to remove 7-methylguanine and 3-methyladenine residues, then end-labelled with ^{32}P. Treatment with spermidine cleaved the DNA at apurinic sites; piperidine treatment cleaved the DNA at both methyl phosphate triester sites and apurinic sites. Fragments were separated by polyacrylamide gel electrophoresis.

An autoradiograph of a sequencing gel is shown in Fig. 1. Unmethylated DNA was unchanged by the cleavage reactions (lanes 1-3); methylated DNA which was not treated with the cleavage reagents also remained intact (lanes 4, 7 and 10). Lanes 5, 8, and 11 illustrate that the methylating agents show no preference for particular guanines or adenines within the restriction fragment. Likewise, methyl phosphate triesters are formed in a random manner (lanes 6, 9 and 12). There is no difference between lanes in which the DNA was reacted with different methylating agents, but subsequently treated identically (lanes 5, 8 and 11, and lanes 6, 9 and 12).

Our sequencing results show that the influence of neighbouring bases is insignificant for methylation of DNA at the sites examined using three different methylating agents. It seems likely that the conformation of poly(dGdC).poly(dGdC) is responsible for its differential reactivity compared to other DNA substrates. Under conditions close to physiological pH and ionic strength[9], regions of DNA consisting of alternating purine and pyrimidine bases can assume the Z conformation. In our experiments, in order to maintain AcONMBzA in solution, it was necessary to use DMSO as co-solvent. So that results could be obtained under comparable conditions, reactions with NMU and AcONDMA were also carried out in this solvent. Although the ionic strength in our experiments and those of Briscoe and Cotter was much lower than in vivo, the effect of 20% DMSO in our case, or 15.5% ethanol in Briscoe and Cotter's system on DNA conformation, may be sufficiently large to drive the B-Z transition for poly(dGdC).poly(dGdC)[10,11].

Our results have a number of implications for our understanding of the molecular mechanisms involved in carcinogenesis by methylating agents.

Since we have shown that formation of 7-methylguanine, 3-methyladenine and methyl phosphate triesters is independent of sequence specific effects,

Table 1. Percentage yields (± SEM) of O^6-methylguanine and 7-methylguanine with respect to guanine in DNA substrates reacted with methylating agents

Adduct	Methylating agent	DNA substrate			
		Calf thymus DNA	pEC	poly(dG) poly(dC)	poly(dGdC) poly(dGdC)
O^6-Methylguanine	NMU	1.4 ± 0.1	1.4 ± 0.1	0.8 ± 0.1	1.0 ± 0.1
"	AcO-NDMA	1.0 ± 0.1	0.7 ± 0.1	0.8 ± 0.1	0.9 ± 0.1
"	AcO-NMBzA	0.8 ± 0.2	1.0 ± 0.1	0.6 ± 0.1	0.7 ± 0.1
7-Methylguanine	NMU	10.4 ± 0.3	9.3 ± 0.1	8.5 ± 0.6	2.9 ± 0.1
	AcO-NDMA	6.6 ± 0.1	6.7 ± 1.1	7.8 ± 0.3	2.9 ± 0.2
	AcO-NMBzA	7.8 ± 0.3	8.4 ± 0.5	7.2 ± 0.2	2.0 ± 0.1

Table 2. O^6-methylguanine: 7-methylguanine ratios ($\pm$ SEM) in DNA substrates reacted with methylating agents (to the nearest 0.05)

Methylating agent	DNA substrate			
	Calf thymus DNA	pEC	poly(dG) poly(dC)	poly(dGdC) poly(dGdC)
NMU	0.15 ± 0.001	0.15 ± 0.015	0.10 ± 0.0001	0.35 ± 0.040
AcO-NDMA	0.15 ± 0.011	0.10 ± 0.002	0.10 ± 0.002	0.30 ± 0.012
AcO-NMBzA	0.10 ± 0.016	0.10 ± 0.010	0.10 ± 0.005	0.30 ± 0.002

it is reasonable to assume that formation of O^6-methylguanine will also be independent of such effects. The c-Ha-<u>ras</u>-1 gene in rat breast tissue is activated by NMU treatment by mutation at the second rather than the first guanine of the 12th codon[12]. Our results suggest that sequence specificity of

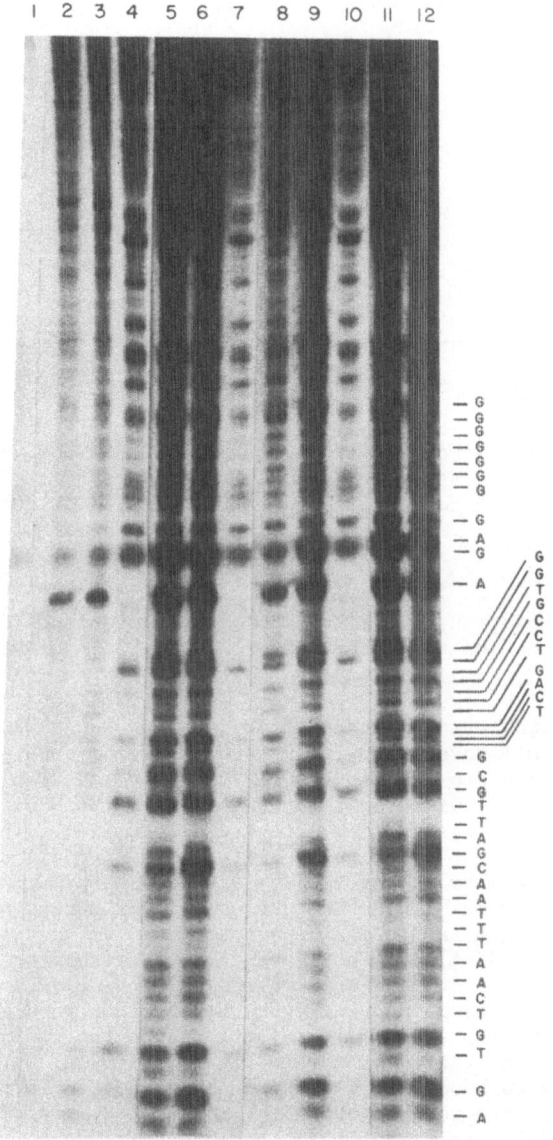

Fig. 1. Polyacrylamide gel electrophoresis of fragments derived from a 345 bp Hind III-Bam HI restriction fragment of pBR322 DNA. DNA methylation was as follows: lanes 1-3, unmethylated DNA; lanes 4-6, NMU treated DNA; lanes 7-9, AcONDMA treated DNA; lanes 10-12, AcONMBzA treated DNA. DNA treatment following modification was as follows: lanes 1, 4, 7, and 10, untreated; lanes 2, 5, 8 and 11, spermidine; lanes 3, 6, 9, and 12, piperidine.

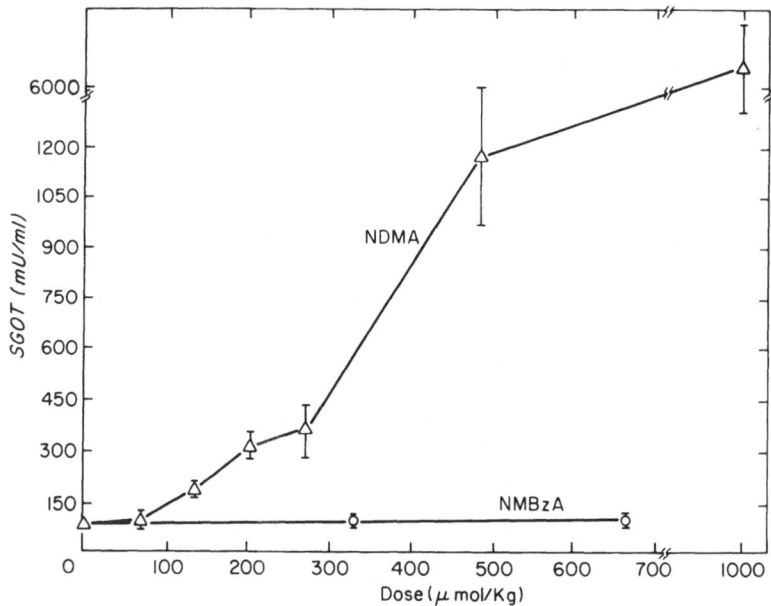

Fig. 2. Serum glutamic-oxalacetic transaminase (SGOT) levels 48 h after i.p.
administration to male Sprague-Dawley rats of various doses of NDMA
or NMBzA in normal saline containing 25% DMSO. Each point represents
the mean ± S.E.M. of the results from 4 animals.

the methylation reaction is unlikely to play a role in this transformation, a
more likely explanation being the differential repair of O^6-methyl-guanine at
various sites[13].

Since the methylation pattern of the various DNA substrates is indistinguishable for all three methylating agents, we have no evidence for interaction of α-hydroxynitrosamines with DNA prior to methylation. Rather, our
results are consistent with the reaction of the three N-nitroso compounds via
a common intermediate such as the methyl diazonium ion. This conclusion is
of interest with respect to the very different biological effects of the
three methylating agents in rat liver. Nitrosodimethylamine (NDMA) is a
potent hepatocarcinogen[14] while NMU produces liver tumors only after a promotional stimulus such as partial hepatectomy[15]. We have shown that nitroso-
methylbenzylamine (NMBzA) does not even produce preneoplastic foci in rat
liver when promotional stimuli are provided[16]. It is well established, however, that the three compounds methylate hepatic DNA in a grossly similar
fashion[17-19]. Furthermore, we have recently shown that NMBzA produces no
hepatotoxicity in the rat even at very high doses (Fig. 2). In contrast,
NDMA produces severe hepatotoxicity at much lower doses. In the light of the
results presented here which show that NDMA and NMBzA methylate DNA in a
qualitatively similar fashion, it is possible that DNA methylation is unrelated to the production of liver cell injury and, perhaps, carcinogenesis.

ACKNOWLEDGEMENTS

These investigations were supported by grant MT7025 from the Medical
Research Council of Canada, the National Cancer Institute of Canada, and
the Ontario Cancer Treatment and Research Foundation.

REFERENCES

1. M. Barbacid, Oncogenes in human cancer and chemically induced tumors, Prog. Med. Virol. 32:86 (1985).
2. M. C. Archer and G. E. Labuc, Nitrosamines, in: "Bioactivation of Foreign Compounds", M. W. Anders, ed., Academic Press, New York, NY, USA (1985).
3. B. Gold and W. B. Linder, α-Hydroxynitrosamines: transportable metabolites of dialkylnitrosamines, J. Am. Chem. Soc. 101:6772 (1979).
4. P. P. Roller, D. R. Shimp and L. K. Keefer, Synthesis and solvolysis of methyl(acetoxymethyl)nitrosamine. Solution chemistry of presumed carcinogenic metabolite of dimethylnitrosamine, Tetrahedron Lett. 25:2065 (1975).
5. J. K. Snyder and L. M. Stock, Reactions of alkylnitrosoureas in aqueous solution, J. Org. Chem. 45:1990 (1980).
6. W. T. Briscoe and L. E. Cotter, The effects of neighboring bases on N-methyl-N-nitrosourea alkylation of DNA, Chem. -Biol. Interact. 52:103 (1984).
7. W. T. Briscoe and L. E. Cotter, DNA sequence has an effect on the extent and kinds of alkylation of DNA by a potent carcinogen, Chem. -Biol. Interact. 56:321 (1985).
8. J. G. Sutcliffe, Complete nucleotide sequence of the Escherichia coli plasmid pBR322, Cold Spring Harbor Symp. Quant. Biol. 43:77 (1979).
9. J. H. Van de Sande, L. P. McIntosh and T. M. Jovin, Mn^{2+} and other transition metals at low concentration induce the right-to-left helical transformation of poly[d(G-C)], EMBO J. 1:777 (1982).
10. J. H. Van de Sande and T. M. Jovin, Z* DNA, the left-handed helical form of poly[d(G-C)] in $MgCl_2$-ethanol, is biologically active, EMBO J. 1:115 (1982).
11. W. Saenger, N. Hunter and O. Kennard, DNA conformation is determined by economics in the hydration of phosphate groups, Nature 324:385 (1986).
12. H. Zarbl, S. Sukamar, A. V. Arthur, D. Martin-Zanca and M. Barbacid, Direct mutagenesis of Ha-ras-1 oncogenes by N-nitroso-N-methylurea during initiation of mammary carcinogenesis in rats, Nature 315:382 (1985).
13. M. D. Topal, J. S. Eadie and M. Conrad, O^6-methylguanine mutation and repair is non-uniform, J. Biol. Chem. 261:9879 (1986).
14. R. C. Shank and P. N. Magee, Toxicity and carcinogenicity of N-nitroso compounds, in: "Mycotoxins and N-nitroso Compounds: Environmental Risks", R.C. Shank, ed., CRC press, Boca Raton, FL, USA (1981).
15. V. M. Craddock and J. V. Frei, J.V., Induction of liver cell adenomata in the rat by a single treatment with N-methyl-N-nitrosourea given at various times after partial hepatectomy, Br. J. Cancer 30:503 (1974).
16. G. E. Labuc and M. C. Archer, Comparative tumor initiating activities of N-nitrosomethylbenzylamine and N-nitrosodimethlamine in rat liver, Carcinogenesis 3:519 (1982).
17. K. C. Silinskas, P. F. Zucker and M. C. Archer, Formation of O^6-methylguanine in rat liver DNA by nitrosamines does not predict initiation of preneoplastic foci, Carcinogenesis 6:773 (1985).
18. P. F. Swann and P. N. Magee, Nitrosamine-induced carcinogenesis. The alkylation of nucleic acids of the rat by N-methyl-N-nitrosourea, dimethylnitrosamine, dimethylsulfate, and methylmethanesulfonate, Biochem. J. 110:39 (1968).
19. A. E. Pegg, Alkylation of rat liver DNA by dimethylnitrosamine: effect dosage on O^6-methylguanine levels, J. Natl. Cancer Inst. 58:681 (1977).

ANTIBODIES SPECIFIC FOR DNA COMPONENTS MODIFIED BY CHEMICAL CARCINOGENS AND THEIR BINDING EFFICIENCY TO DNA MODIFIED IN VIVO AND IN VITRO WITH THE CORRESPONDING CARCINOGENS

Erik Kriek, Frederik J. van Schooten, Michel J. X. Hillebrand and Maarten C. Welling

Division of Chemical Carcinogenesis
The Netherlands Cancer Institute
(Antoni van Leeuwenhoek Huis)
121 Plesmanlaan, 1066 CX Amsterdam, The Netherlands

INTRODUCTION

It is now well established that almost all carcinogens form a variety of reaction products with DNA, involving covalent binding to the various nucleophilic sites on all four DNA bases as well as the phosphate groups of DNA. Thus, even qualitative determination of all DNA adducts, even from a single carcinogen, presents a formidable analytical task. Although the major reaction products of many carcinogens have been characterized (reviewed by Singer and Grunberger[1]), the structures of a large number of carcinogen-DNA adducts, particularly those of minor products, have not yet been identified. The quantitative determination is further complicated by the extremely low levels usually found in vivo, the removal of adducts from DNA by repair processes at different rates, and the formation of secondary lesions from chemically unstable adducts. Also, unstable adducts may be released spontaneously, leaving apurinic sites. Despite these shortcomings, measurement of DNA adducts is important, since these are thought to represent initiating events leading to mutations and/or malignant transformation. Measurement of DNA adducts in situ in the DNA of cells should give the most direct evidence of genotoxic exposure. The potential value of measuring carcinogen-DNA adducts as a dosimeter of human exposure has been discussed in a number of recent reviews[2,3].

During the last ten years immunological methods have been developed for the detection of carcinogen-DNA adducts. A large number of polyclonal and monoclonal antibodies, specific to carcinogen-DNA adducts, are now available (reviewed by Strickland and Boyle[4]). The determination of carcinogen-DNA adducts by immunological procedure offers several advantages over other techniques. The sensitivity is often better than that obtained with radiolabeled carcinogens, which can be used only for experimental purposes. Antibodies have been prepared, which are specific for particular three-dimensional structures (see e.g. Adamkiewicz et al.[5]). These antibodies can be used for various purposes; they can be used to probe the conformation of unknown adducts in DNA, or they can be used to determine the chemical stability of specific adducts and measure their conversion into secondary lesions. As a typical example of the latter reaction, the opening of the guanine imidazole ring of N-(deoxyguanosin-8-yl)-2-aminofluorene (dGuo-8-AF) under certain conditions will be described in this report. Immunological assays are rapid, reproducible and obviate the necessity to use radiolabeled carcinogens. In addition, immunological techniques can be applied together with morphologic procedures (immunocytochemistry) to localize adducts in particular cells within a tissue[6]. The high degree of sensitivity of the enzyme immunoassays (0.1 - 10 fmol/μg DNA) has led a number of laboratories to investigate the feasibility of these assays for determination of DNA adducts in human populations exposed to carcinogens. Polyclonal and/or monoclonal antibodies have been raised to DNA modified with aflatoxin B_1, benzo(a)pyrene, N-nitrosamines, 1-nitropyrene, and chemotherapeutic agents like cisplatin and 8-methoxypsoralen. Typical applications of these antibodies for the detection of human exposure have been compiled in a monograph, which appeared recently[7]. The antibodies have been used in three types of immunoassays: radioimmunoassays (RIA), enzyme immunoassays (ELISA) and ultrasensitive enzymatic radioimmunoassays (USERIA). Immunoassays can be performed in a competitive or noncompetitive fashion. RIAs are competitive assays, but ELISA or USERIA can be performed in a competitive or noncompetitive way in so-called solid phase assays. In these assays the antigen (bound to DNA or protein) is coated to the wells of plastic microtiter plates. In competitive assays, a chemically identical hapten (inhibitor) competes with the antigen on the solid phase for the same antibody binding sites. When the amount of the first antigen coated to the plate is kept constant, the concentration of the inhibitor is reflected in the degree to which it inhibits the binding of the antibody to the antigen on the plate. Antibody binding is measured by binding of a second antibody, linked to an enzyme which cleaves a chromogen (p-nitrophenylphosphate), a fluorogen (4-methylumbelliferyl-phosphate) or [^{3}H]AMP. Bound enzymatic activity is then determined by spectrofotometric assay of p-nitrophenol, fluorometric assay of 4-methylumbelliferone, or radioactivity assay of [^{3}H]adenosine. The quantification is based on the comparison of unknowns with an inhibition curve obtained by competing increasing amounts of standard antigen (carcinogen-modified DNA). The most widely used method nowadays is the competitive ELISA, because the sensitivity is much greater than that of RIA, sometimes even more than 100-fold. The USERIA developed by Hsu et al.[8] is not generally applied, because the separation of [^{3}H]adenosine from unreacted [^{3}H]AMP is rather laborious and the sensitivity is not better than that of the highly sensitive ELISA developed in our laboratory by Van Der Laken et al.[9], employing a fluorogenic substrate. Immunocytochemical methods do not yet have the sensitivity of enzyme immunoassays.

The sensitivity of an assay depends on the antibody affinity, the amount of DNA assayed as well as the level of modification. As will be demonstrated in this report, the binding efficiency of antibodies strongly depends on the level of DNA modification, leading to 50% inhibition values that may differ more than 10-fold.

AFFINITY OF SPECIFIC ANTIBODIES

2-Aminofluorene and Derivatives

It is now well documented, that guanine-8-arylamine adducts in DNA are the major reaction products of aromatic amines and amides in vivo (see for reviews Kriek[10]; Kadlubar and Beland[11]). Moreover, it was shown that an important group of environmental contaminants, the nitro polycyclic aromatic hydrocarbons, generally exhibit the same target specificity as their amino analogues (reviewed by Beland et al.[12]). These compounds are metabolized to genotoxic agents through ring oxidation and/or nitroreduction. The nitro reduction pathway also leads to the formation of guanine-8-arylamine adducts in DNA, e.g. the structure of the DNA adduct formed when S. typhimurium[13] or rats[14] are treated with 1-nitropyrene,is N-(deoxyguanosin-8-yl)-1-aminopyrene. Therefore, we wanted to develop a sensitive immunoassay utilizing arylamine-modified denatured DNA, without the requirement of prior enzymatic degradation to the constituent nucleosides. A complicating factor in such an assay might be the instability of guanine-8-arylamines, because in previous studies we demonstrated that dGuo-8-AF is easily converted into its guanine imidazole ring opened form rodGuo-8-AF[15,16]. This appeared to be a general reaction of guanine-8-arylamines, in which the nature of the aromatic hydrocarbon moiety is an important determinant (Westra et al., unpublished observations). The ring opening can also occur at neutral pH, particularly under DNA denaturing conditions at 100° C, while dGuo-8-AAF does not give this reaction. The general reaction scheme is presented in Fig. 1. Thus, antibodies to guanine imidazole ring opened forms would offer certain advantages for the quantification of guanine-8-arylamines. The latter adducts are subject to depurination at slightly acidic pH (4-5), whereas the guanine imidazole ring opened forms are stable to hydrolytic removal from DNA. Additional interest in an anti rodGuo-8-AF antibody derives from the observation that rodGuo-8-AF is a lethal modification for the replication of ØX174 DNA[17].

The antibody, raised against the bovine serum albumin conjugate of roGuo-8-AF[15], was tested for its affinity to DNA modified to various levels with N-hydroxy-AF and converted into the guanine imidazole ring opened form by treatment with alkali. The 50% inhibition values, presented in Table 1, are strongly dependent on the level of modification. The lower detection limit of the competitive ELISA, utilizing DNA containing rodGuo-8-AF of low modification (1-50 fmol/μg) as standard, is 0.3 fmol/μg DNA, which is equivalent to 1 adduct per 10^7 nucleotides. The roGuo-8-AF antibody can be employed to monitor the stability of dGuo-8-AF under heat denaturing conditions, as shown in Table 2. Surprisingly, the conversion of dGuo-8-AF into rodGuo-8-AF appeared to be strongly dependent on the NaCl concentration, and proceeded much faster at higher NaCl concentrations. The 50% inhibition values presented in Table 1 also show that there is appreciable cross-reactivity of anti roGuo-8- AF antibody with AF-DNA and AAF-DNA (calculated as 9.7% and 1.4% respectively, from the data of Table 1). The 50% inhibition

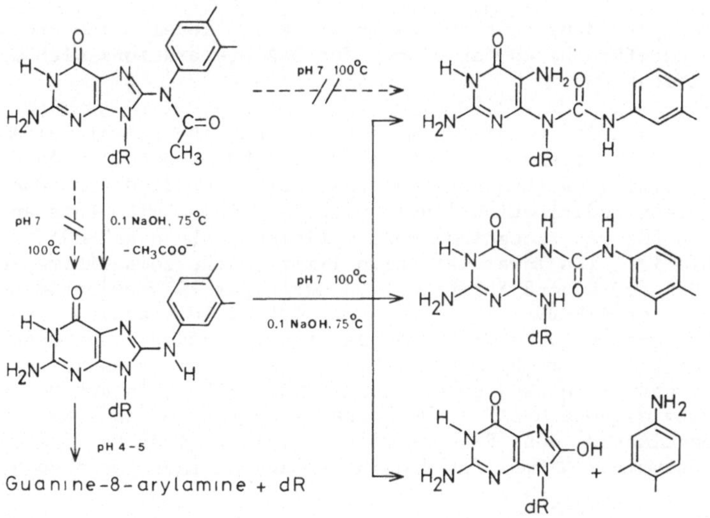

Fig. 1. Guanine imidazole ring opening of N-(deoxyguanosin-8-yl)arylamines at neutral and alkaline pH. N-(deoxyguanosin-8-yl)-N-acetylarylamines are stable upon heating at neutral pH and do not form guanine imidazole ring open forms under these conditions. The formation of 8-hydroxy-guanine as a result of arylamine modification at the C-8 position of guanine has not yet been demonstrated conclusively. dR, deoxyribose.

Table 1. Competitive inhibition of polyclonal antibody binding to AAF-DNA and roAF-DNA [a]

Competitor	Modification fmol/µg	Amount of competitor (fmol) causing 50% inhibition of antibody binding[b]	
		Anti roGuo-8-AF	Anti Guo-8-AAF
roAF-DNA	33600	6.7	n.d.
"	5000	7.8	n.d.
"	600	17	9200
"	43	74	n.d.
"	5	85	n.d.
ss AF-DNA	43	865	5500
ss AAF-DNA	5300	500	3.3
ds AAF-DNA	5300	4000	8.8
ss AAF-DNA	200	n.d.	12.0
ds AAF-DNA	200	n.d.	13.7
dGuo-8-AAF		n.d.	2.4

[a]AAF-DNA, DNA containing N-(deoxyguanosin-8-yl)-N-acetyl-2-aminofluorene; AF-DNA, DNA containing N-(deoxyguanosin-8-yl)-2-aminofluorene; Guo-8-AAF, N-(guanosin-8-yl)-N-acetyl-2-aminofluorene; roGuo-8-AF, 1-[6-(2,5-diamino-6-oxopyrimidinyl)-N[6]-riboside]-3-(2-fluorenyl)urea; roAF-DNA, DNA containing 1-[6-(2,5-diamino-6-oxopyrimidinyl-N)-deoxyriboside]-3-(2-fluorenyl)-urea; ds, double stranded; ss, single stranded; n.d. not determined. [b]Enzyme immunoassays were performed as described[15,24]. 10 fmol adduct/0.5 ng DNA was coated per microtiter well. The antibody dilutions were 1:10 .

values, however, are dependent on the modification level. The cross-reactivities can therefore be compared only for DNA preparations with similar devels of modification.

In previous studies we also demonstrated that the polyclonal antibody against Guo-8-AAF had its highest affinity for the mononucleoside dGuo-8-AAF[9,15]. The affinity of this antibody was further studied with DNA preparations of different modification levels with AAF. Also with this antibody, highly modified DNA was recognized more efficiently than AAF-DNA of low modification (Table 1). The lower detection limit in the competitive ELISA was 0.02 fmol/µg DNA for dGuo-8-AAF (i.e. 7 adducts per 10^9 nucleotides) and 0.1 fmol/µg DNA for AAF-DNA (35 adducts per 10^9 nucleotides). Anti Guo-8-AAF showed high cross-reactivity with the deacetylated adduct Guo-8-AF, but the affinity for AF-DNA was much lower (Table 1). For this reason, the anti Guo-8-AAF antibody cannot be used to determine dGuo-8-AF in intact DNA with the same sensitivity as dGuo-8-AAF. The cross-reactivity of anti Guo-8-AAF with the free mononucleoside dGuo-8-AF is 16%, calculated from the 50% inhibition values (respectively 2.4 fmol for dGuo-8-AAF and 15 fmol for dGuo-8-AF).

Benzo(a)pyrene

Benzo(a)pyrene (BP), like most polycyclic aromatic hydrocarbons (PAH), is a ubiquitous environmental pollutant produced by incomplete combustion of organic material. BP induces tumors in experimental animals of various species. Information of the mechanism(s) by which BP induces neoplasia is scant, however. In vivo BP is metabolized to epoxides and phenols by the cytochrome P-450-dependent mono-oxygenase and by epoxide hydratase. The majority of

Table 2. Conversion of AF-DNA into roAF-DNA under heat denaturing conditions measured by competitive ELISA with anti roGuo-8-AAF [a]

Concentration of NaCl (M)	Amount of competitor (fmol) causing 50% inhibition of antibody binding	roAF-DNA formed (%)
0	865	0
0.14	730	17.3
0.28	500	46.8
0.70	150	91.7
1.40	140	93.0
0	85	100

[a]See Table 1 for abbreviations. Solutions of AF-DNA (0.5 mg/ml, modification 43 pmol/mg) in 0.01 M sodium phosphate buffer pH 7.4, containing the indicated concentrations of NaCl, were heated at 100° C for 15 minutes and cooled rapidly by immersion in an ice bath. Enzyme immunoassays were performed as described[15,24]. 10 fmol adduct/0.5 ng DNA was coated per microtiter well. The anti roGuo-8-AF antibody dilution was $1 : 10^6$.

these metabolites is converted to water-soluble conjugates (glucuronides and sulfates), but a part of the parent hydrocarbon is activated to a reactive form capable to react covalently with DNA, RNA and proteins. The diol epoxide, (±)trans-7,8-dihydroxy-9,10-epoxy-7,8,9,10-tetrahydrobenzo(a)pyrene (BPDE), has been identified as the major enzymatic metabolite of BP7,8-diol (reviewed by Harvey[18]). The best documented DNA adduct of BPDE is that which results from the trans opening of the (+) enantiomer by reaction of the 2-amino group of guanine (see Jeffrey[19] for further references). A similar type of activation and reaction with DNA has been described for other PAH diol epoxides, e.g. chrysene[20], benzo(a)anthracene[21] and benzo(c)phenanthrene[22]. The diol epoxide of the latter compound binds extensively to adenine in DNA. The covalent binding of PAH diol epoxides to purine bases in DNA is considered to be an essential step in the initiation of PAH-induced neoplasia. A comprehensive book, dealing with various aspects of PAH carcinogenesis, has appeared recently[23].

A number of polyclonal and monoclonal antibodies against BPDE-modified DNA (1-2%) have been developed in our laboratory[24]. All antibodies showed a very high affinity for denatured BPDE-DNA, but had lower affinity for native BPDE-DNA. No affinity was detected for BPDE-tetrols or unmodified DNA. The affinity for the free mononucleoside BPDE-dGuo was much lower than that for BPDE-DNA. As expected, a high cross-reactivity was observed with DNA modified with the diol epoxide of chrysene, a hydrocarbon of closely related structure. Using four different antibodies, polyclonal as well as monoclonal, we found that these antibodies recognized highly modified (immunogen) DNA more efficiently than DNA modified to a low extent (Table 3). Even a monoclonal antibody, raised against the mononucleotide BPDE-dGMP (II.E.4) had a higher affinity for BPDE-DNA than for the free mononucleoside BPDE-dGuo. These observations, together with the expected low modification of DNA from biological samples (including human tissues or cells), led us to validate the competitive ELISA for BPDE-DNA with BP-DNA from animals treated with BP. Since Kulkarni and Anderson[25] showed that in mice, treated with [3H]BP the major adduct in DNA from various tissues was [3H]BPDE-dGuo, the mouse appeared to be a suitable experimental animal to test our competitive ELISA for measuring this adduct in DNA. Samples of [3H]BP-DNA, isolated from the

123

Table 3. Competitive inhibition of antibody binding to BPDE-DNA by
various inhibitors[a]

Competitor	Modification fmol/µg	Amount of competitor (fmol) causing 50% inhibition of antibody binding			
		Polyclonals		Monoclonals	
		F29	F30	41D3	II.E4
ss BPDE-DNA	42000	4	9	5	45
ds BPDE-DNA	42000	105	335	17	n.d.
ss BPDE-DNA	0.1-4	17	150	43	100
ss CDE-DNA	30000	10	17	6	n.d.
BPDE-dGuo		370	540	2000	90
BPDE-tetrols		$>10^5$	$>10^5$	$>10^4$	n.d.

[a]BPDE, (±)trans-7,8dihydroxy-anti9,10-epoxy-7,8,9,10-tetrahydro-
benzo(a)pyrene; BPDE-dGuo, trans-(7R)-N^2-[10-(7ß,8α,9α-trihydroxy-
7,8,9,10-tetrahydrobenzo(a)pyrenyl]-deoxyguanosine; BPDE-DNA, DNA modified by reaction with BPDE such that the major adduct is BPDE-dGuo;
BPDE-tetrols (±)7ß,8α,9α,10(α or ß)-tetrahydroxy-7,8,9,10-tetrahydro-
benzo(a)pyrene; CDE-DNA, DNA modified with (±)trans-1,2-dihydroxy-
anti-3,4-epoxy-1,2,3,4-tetrahydrochrysene such that the major adduct
is CDE-dGuo; ss, single stranded; ds, double stranded. Antibody II.E4
was raised against the bovine serum albumin conjugate of BPDE-dGMP[36];
all other antibodies were prepared against BPDE-DNA complexed with
methylated bovine serum albumin. The microtiter wells were coated with
denatured BPDE-DNA (15 fmol/-25 ng). Antibody dilutions were respectively 1:2 x 10^5 (F29 and F30), 1:10^5 (41D3) and 1:10^4 (II.E.4). The
50% inhibition values were taken partly from Van Schooten et al.[24]

livers of female BALB/c mice treated with different doses of [³H]BP, were
examined by competitive ELISA using [³H]BPDE-modified DNA of low modification
(1-10 fmol/µg). The binding levels of [³H]BP to mouse liver DNA, calculated
from ELISA, were in good agreement with those obtained from radioactivity
measurements[24].

Using the [³H]BP-DNA samples from mouse liver as standards, we also
examined a number of DNA preparations extracted from peripheral blood lymphocytes of coke oven workers, and unexposed controls. Among a group of 13 coke
oven workers, 10 (77%) were found to be carrying significant levels of PAH-
DNA adducts (corresponding to 0.1-1.2 fmol/µg DNA) in their blood lymphocytes
(Van Schooten et al., unpublished observations). No inhibition of anti BPDE-
DNA antibody binding was detected with DNA samples from the unexposed control
group. If the limit of detection of BPDE-dGuo in DNA by the competitive
ELISA is set at 20% inhibition, the lower detection limit of this method is
0.1 fmol/µg DNA. This is equivalent to the detection of 3 modified guanine
bases per 10^8 nucleotides in a 1 µg sample of DNA.

This value compares favorably with that reported by Haugen et al.[26] who
used a different antibody and another end-point in their immunoassay.

DISCUSSION

In this report we demonstrate that the guanine imidazole ring opening of dGuo-8-AF in DNA can also occur at neutral pH under heat denaturing conditions. The data of Table 2, obtained by competitive ELISA with anti roGuo-8-AF antibody, show that heating of AF-DNA at 100° C during 15 minutes in phosphate buffered saline (0.14 M NaCl) already results in partial formation of-roAF-DNA (15-20%). The cross-reactivity of anti roGuo-8-AF antibody with AF-DNA (calculated as 9.7% from the data of Table 1) does not allow the determination of small amounts of roAF-DNA (less than 20 %) in the presence of AF-DNA. We found also, that the conversion of AF-DNA into roAF-DNA is accelerated by increasing the concentration of NaCl. This effect may be related to changes in conformation of AF-modified DNA at higher salt concentrations. Based on various spectroscopical measurements of AF-modified DNA in solution, Van Houte et al.[27] reported that AF binding to the C-8 position of guanine introduces significant denaturation with strong interaction of AF with the neighbouring bases. Unlike the modification with AAF at the same position, AF binding does not induce the Z-structure in DNA[28]. High salt concentration, however, stabilizes the Z-structure in DNA[29] and may thus favour the conversion of local regions (containing dGuo-8-AF) from the B-form into the Z-structure. In this structure, the bases lie much more to the outer surface of the helix as compared to B-DNA[30]. Therefore, dGuo-8-AF is exposed on the outer surface of the molecule and more accessible to hydrolytic attack by water.

The instability of guanine-8-arylamines to heat denaturation has also been observed by other investigators. Hsieh et al.[31] reported the instability of DNA modified with 1-aminopyrene (AP), also measured by enzyme immunoassay. The AP modification is also at the C-8 position of guanine[13,14]. Therefore, it is very likely that in this case also a guanine imidazole ring-opened derivative is formed, although the authors did not identify such a product. Our results and those of Hsieh et al.[31] indicate that specific polyclonal or monoclonal antibodies to carcinogen-DNA adducts will be useful not only for detecting and quantifying carcinogen-DNA adducts, but also for monitoring adduct stability and even for probing local conformational changes in the DNA structure. In cases of adduct instability, like the guanine-8-arylamines, it may be advantageous to develop antibodies to the stable guanine imidazole ring open forms. Experiments are in progress in our laboratory to test this concept for a number of aromatic amines.

All antibodies described in this report recognize highly modified DNA more efficiently than carcinogen-DNA of low modification (biological samples). This difference in immunoreactivity has now been found for six antibodies, and may be common to all antibodies raised against highly modified carcinogen-DNA immunogens. Thus, DNA adduct values in human tissues and cells reported by other investigators may be underestimated several fold, because in all these studies serially diluted DNA of high modification was used as standard in the enzyme immunoassay. DNA preparations of low modification, obtained either in vitro by treating DNA with ultimate carcinogenic reactants, or preferably from animal tissues following exposure to the specific carcinogens, have to be used as reference in the enzyme immunoassays in order to obtain reliable quantification of DNA adducts in human tissues.

One of our polyclonal anti BP-DNA antibodies (F29 in Table 3) has been employed successfully for the immunochemical determination of PAH-DNA adducts in peripheral blood lymphocytes from coke oven workers. Although as yet only small groups of individuals have been examined, the data clearly show that levels of PAH-DNA adducts are significantly higher in groups of persons with a high exposure. Our data compare favorably with those reported by Perera et al.[32] for PAH-DNA adducts in white blood cells from foundry workers. The large variation in PAH-DNA adduct levels is probably a reflection of individual differences in.metabolic activation. In vitro studies with normal human tissues have shown large interindividual differences in the ability to metabolize PAH and other carcinogens[33]. These differences in metabolic capacity are due to genetic polymorphism within the human population[34].

The detection of carcinogen-DNA adducts at the level of individual cells has become possible by recent developments of immunocytochemical methods using specific antibodies[6]. The sensitivity of this method is reported to be better than 10^4 0^6-alkyldeoxyguanosine residues per diploid genome of 10^{10} nucleotides[35]. Using the same technique, we recently found that anti BPDE-DNA antibody can be used for the localization of BPDE-dGuo in lung tissue of mice following treatment with BP. The lower detection limit was 3×10^4 BPDE-dGuo adducts per diploid genome of 10^{10} nucleotides (Van Schooten et al., unpublished observations). The sensitivity of this method has to be increased to determine and quantify PAH-DNA adducts in human cells. A highly sensitive detection and quantification of carcinogen-DNA adducts, using computer-controlled scanning microscopy, is currently under investigation in our laboratory (Dr. E. Scherer, in collaboration with Dr. J. S. Ploem, Leiden University).

ACKNOWLEDGEMENTS

We wish to thank Dr. R. A. Baan (Medical Biological Laboratory TNO, Rijswijk, The Netherlands) for a gift of monoclonal anti BPDE-dGMP antibody, and Dr. P. L. Grover (Chester Beatty Laboratories, Institute of Cancer Research, London, UK) for a sample of CDE-modified DNA.

REFERENCES

1. B. Singer and D. Grunberger, "Molecular Biology of Mutagens and Carcinogens", Plenum Press, New York (1983).
2. G. N. Wogan and N. J. Gorelick, Chemical and biochemical dosimetry of exposure to genotoxic chemicals, Environ. Health Perspect. 62:5 (1985).
3. F. P. Perera, R. M. Santella and M. C. Poirier, Biomonitoring of workers exposed to carcinogens: immunoassays to benzo(a)pyrene-DNA adducts as a prototype, J. Occup. Med. 28:1117 (1986).
4. P. T. Strickland and J. M. Boyle, Immunoassay of carcinogen-modified DNA, in: "Progr. Nucleic Acid Res. Mol. Biol.", Vol. 31, W. E. Cohn and K. Moldave, eds., Academic Press, New York (1984).
5. J. Adamkiewicz, P. Nehls and M. F. Rajewski, Immunological methods for detection of carcinogen-DNA adducts, in: "Monitoring Human Exposure to Carcinogenic and Mutagenic Agents", A. Berlin, M. Draper, K. Hemminki and H. Vainio, eds., International Agency for Research on Cancer, Lyon (1984).
6. G. J. Menkveld, C. J. Van Der Laken, G. Hermsen, E. Kriek, E. Scherer and L. Den Engelse, Immunohistochemical localization of 0^6-ethyl-deoxyguanosine and deoxyguanosin-8-yl-(acetyl)aminofluorene in liver sections of rats treated with diethylnitrosamine, ethylnitrosourea or N-acetylaminofluorene, Carcinogenesis 6:263 (1985).
7. M. C. Poirier and F. A. Beland, eds., "Carcinogenesis and Adducts in Animals and Humans" (Progress in Experimental Tumor Research Vol. 31), Karger, Basel (1987).
8. I. C. Hsu, M. C. Poirier, S. H. Yuspa, R. J. Yolken and C. C. Harris, Ultrasensitive enzymatic radioimmunoassay (USERIA) detects femtomoles of acetylaminofluorene-DNA adducts, Carcinogenesis 1:455 (1980).
9. C. J. Van Der Laken, A. M. Hagenaars, G. Hermsen, E. Kriek, A. C. Kuipers, J. Nagel, E. Scherer and M. C. Welling, Measurement of 0 - ethyl-deoxyguanosine and N-(deoxyguanosin-8-yl)-N-acetyl-2-amino-fluorene in DNA by high-sensitive enzyme immunoassays, Carcinogenesis 3:569 (1982).

10. E. Kriek, Reactive forms of aromatic amines and amides: chemical and structural features in relation to carcinogenesis, in: "13th International Cancer Congress, Part B, Biology of Cancer (1)", E. A. Mirand, W. B. Hutchinson and E. Mihich, eds., Alan R. Liss, Inc., New York (1983).

11. F. F. Kadlubar and F. A. Beland, Chemical properties of ultimate carcinogenic metabolites of arylamines and arylamides, in: "Polycyclic Hydrocarbons and Carcinogenesis" (ACS Symposium Series 283), R. G. Harvey, ed., American Chemical Society, Washington, D.C. (1985).

12. F. A. Beland, R. H. Heflich, P. C. Howard and P. P. Fu, The in vitro metabolic activation of nitro polycyclic aromatic hydrocarbons, in: "Polycyclic Hydrocarbons and Carcinogenesis", (ACS Symposium Series 283) R. G. Harvey, ed., American Chemical Society, Washington, D.C. (1985).

13. P. C. Howard, R. H. Heflich, F. E. Evans and F. A. Beland, Formation of DNA adducts in vitro and in Salmonella typhimurium upon metabolic reduction of the environmental mutagen 1-nitropyrene, Cancer Res. 43:2052 (1983).

14. C. A. Stanton, F. L. Chow, D. H. Phillips, P. L. Grover, R. C. Garner and C. N. Martin, Evidence for N-(deoxyguanosin-8-yl)-1-aminopyrene as a major DNA adduct in female rats treated with 1-nitropyrene, Carcinogenesis 6:535 (1985).

15. E. Kriek, M. C. Welling and C. J. Van Der Laken, Quantitation of carcinogen-DNA adducts by a standardized high-sensitive enzyme immunoassay, in: "Monitoring Human Exposure to Carcinogenic and Mutagenic Agents", A. Berlin, M. Draper, K. Hemminki and H. Vainio, eds., International Agency for Research on Cancer, Lyon (1984).

16. E. Kriek and J. G. Westra, Structural identification of the pyrimidine derivatives formed from N-(deoxyguanosin-8-yl)-2-aminofluorene in aqueous solution at alkaline pH, Carcinogenesis 1:459 (1980).

17. J. T. Lutgerink, J. Retèl, J. G. Westra, M. C. Welling, H. Loman and E. Kriek, The biological activity of single-stranded ØX174 DNA, modified with N-hydroxy-2-aminofluorene, is inhibited by guanine imidazole ring opening of the major, non-lethal aminofluorene-DNA adduct, Carcinogenesis 7:1359 (1986).

18. R. G. Harvey, Activated metabolites of carcinogenic hydrocarbons, Acc. Chem. Res. 14:218 (1981).

19. A. M. Jeffrey, Polycyclic aromatic hydrocarbon-DNA adducts: formation, detection and characterization, in: "Polycyclic Hydrocarbons and Carcinogenesis" (ACS Symposium Series 283), R. G. Harvey, ed., American Chemical Society, Washington D.C. (1985).

20. D. H. Phillips, A. Hewer and P. L. Grover, Formation of DNA adducts in mouse skin treated with metabolites of chrysene, Cancer Lett. 35:207 (1987).

21. K. Hemminki, C. S. Cooper, O. Ribeiro, P. L. Grover and P. Sims, Reactions of "bay-region" and non-"bay-region" diol epoxides of benz(a)-anthracene with DNA: evidence indicating that the major products are hydrocarbon-N^2-guanine adducts, Carcinogenesis 1:277 (1980).

22. A. Dipple, M. A. Pigott, S. K. Agarwal, H. Yagi, J. M. Sayer and D. M. Jerina, Optically active benzo(c)phenanthrene diol epoxides bind extensively to adenine in DNA, Nature 327:535 (1987).

23. R. G. Harvey, ed., "Polycyclic Hydrocarbons and Carcinogenesis", (ACS Symposium Series 283), American Chemical Society, Washington, D.C. (1985).

24. F. J. Van Schooten, E. Kriek, M. J. S. T. Steenwinkel, H. P. J. M. Noteborn, M. J. X. Hillebrand and F. E. Van Leeuwen, The binding efficiency of polyclonal and monoclonal antibodies to DNA modified with benzo(a)pyrene diol epoxide is dependent on the level of modification. Implications for quantification of benzo(a)pyrene DNA adducts in vivo, Carcinogenesis 8:1263 (1987).

25. M. S. Kulkarni and M. W. Anderson, Persistence of benzo(a)pyrene metab-

olite-DNA adducts in lung and liver of mice, Cancer Res. 44:97 (1984).

26. A. Haugen, G. Becker, C. Benestad, K. Vähäkangas, G. E. Trivers, M. J. Newman and C. C. Harris, Determination of polycyclic aromatic hydrocarbons in the urine, benzo(a)pyrene diol epoxide-DNA adducts in lymphocyte DNA, and antibodies to the adducts in sera from coke oven workers exposed to measured amounts of polycyclic aromatic hydrocarbons in the work atmosphere, Cancer Res. 46:4178 (1986).

27. L. P. A. Van Houte, J. T. Bokma, J. T. Lutgerink, J. G. Westra, J. Retèl, R. Van Grondelle and J. Blok, An optical study of the conformation of the aminofluorene-DNA complex, Carcinogenesis 8:759 (1987a).

28. L. P. A. Van Houte, J. G. Westra, J. Retèl, R. Van Grondelle and J. Blok, A spectroscopic study of the conformation of poly d(GC).poly d(GC) modified with the carcinogen 2-aminofluorene, Abstracts 9th International Biophysical Congress, Jerusalem, August 23-28, 88 (1987b).

29. F. M. Pohl, Polymorphism of a synthetic DNA in solution, Nature 260:365 (1976).

30. A. H. J Wang, G. J. Quigley, F. J. Kolpak, G. Van Der Marel, J. H. Van Boom and A. Rich, Left-handed double helical DNA: variations in the backbone conformation, Science 211:171 (1981).

31. L. L. Hsieh, A. M. Jeffrey and R. M. Santella, Monoclonal antibodies to 1-aminopyrene-DNA, Carcinogenesis 6:1289 (1985).

32. F. P. Perera, K. Hemminki, R. M. Santella, D. Brenner and G. Kelly, DNA adducts in white blood cells of foundry workers, Proc. Amer. Assoc. Cancer Res. 28:373 (1987).

33. C. C. Harris, B. F. Trump, R. C. Grafstrom and H. Autrup, Differences in metabolism of chemical carcinogens in cultured human epithelial tissues and cells, J. Cell. Biochem. 18:285 (1982).

34. D. W. Nebert, S. Kimura and F. J. Gonzalez, Cytochrome P-450 genes and their regulation, in: "Molecular Biology of Development", E. H. Davidson and R. A. Firtel, eds., Alan R. Liss, Inc., New York (1984).

35. E. Scherer, A. A. J. Jenner and L. Den Engelse, Immunocytochemical studies on the formation and repair of O^6-alkylguanine in rat tissues, in: "Relevance of N-Nitroso Compounds to Human Cancer: Exposures and Mechanisms", H. Bartsch, I. K. O'Neill and R. Schulte-Hermann, eds., International Agency for Research on Cancer, Lyon (1987).

36. R. A. Baan, P. T. M. Van Den Berg, W. P. Watson and R. J. Smith, In situ detection of DNA adducts formed in cultured cells by benzo(a)-pyrene diol epoxide with monoclonal antibodies specific for the BP-deoxyguanosine adduct, Toxicol. Environ. Chem. 16:325 (1988).

LIVER DNA DAMAGE BY CHEMICAL CARCINOGENS: ROLE OF THYROID HORMONES

Marco Presta, Marco Rusnati, Jeanette A. M. Maier
and Giovanni Ragnotti

Department of Biomedical Sciences
Chair of General Pathology
Faculty of Medicine, University of Brescia
Via Valsabbina 19
25124 Brescia, Italy

INTRODUCTION

In 1948, Paschkis et al.[14] demonstrated the protective effect of thio-uracil on the induction of liver tumors by 2-acetylaminofluorene. Since then, several authors reported on the inhibition of hepatocarcinogenesis in hypothyroid rats [1,2,8,12,19] . These findings raise the point of the identification of the stage in chemical hepatocarcinogenesis which is inhibited by thyroid deficiency. Even though it has been demonstrated that thyroid digest increases the rate of growth of liver tumors in pituitary dwarf mice treated with aminofluorene[3], therefore suggesting a role for thyroid hormones in the promotion and/or progression of liver tumors, several data indicate that thyroid activity might influence liver carcinogenesis also at the stage of initiation. Bielschowsky and Hall[1], in fact, demonstrated that thyroidectomy, performed before, but not after, the administration of the carcinogen, prevents the development of liver tumors. These findings were confirmed by Goodall[8] who demonstrated that in thyroidectomized animals the treatment with thyroid digest performed after aminofluorene administration does not restore the susceptibility to the carcinogen. Moreover, thyroid hormones have been demonstrated to be necessary for the in vitro transformation of cultured cells by x-rays or chemical carcinogens[9,10].

The interaction between chemical carcinogens and cellular DNA is considered to be a crucial event in the induction of neoplastic transformation[7,13]. This interaction may result in DNA damage, demonstrable as alkaline labile sites by alkaline sucrose gradient analysis[4,5,26]. To clarify the role of thyroid activity on initiation by chemical carcinogens, we decided to compare the liver DNA damaging capacity of different carcinogens in intact, thyroidectomized (TDX) and TDX L-3,5,3'-triiodothyronine-treated rats. Because of sex differences in thyroid activity[20] the experiments were performed both in male and female rats. Two direct carcinogens that do not require metabolic activation, N-methyl-N-nitrosourea (MNU) and methyl-methanesulfonate (MMS)[13], two procarcinogens, N-2-fluorenyl-acetamide (2-AAF) and N-nitrosodimethyl-amine (DMNA)[13], and the sex-dependent hepatocarcinogen DL-1-(2-nitro-3-me-thylphenoxy)-3-tert-butylamino-propan-2-ol (DL-ZAMI 1305)[16], were used as genotoxic agents.

MATERIALS AND METHODS

Chemicals

All chemicals were of reagent grade and were purchased from E. Merck A. G. (Darmstadt, Federal Republic of Germany) or from British Drug Houses Ltd. (Poole, Dorset, England). T_3, sodium salt, was obtained from Calbiochem Behring Hoechst (San Diego, U.S.A.). 2-AAF, DMNA and MNU were purchased from Sigma Chemical Company (St. Louis, U.S.A.). MMS was obtained from Ega-Chemie (Steinheim, Albuch, West Germany). DL-ZAMI 1305 was a gift of Zambeletti S.p.A. (Baranzate, Milan, Italy).

Animals

Outbred Wistar rats (Nossan, Correzzana, Italy), weighing 120-200 g were used. They were fed a complete standard diet of laboratory chow (Piccioni, Brescia, Italy) and water ad libitum. A constant temperature (22-24°C) and alternating periods of 12 h dark and 12 h light were also maintained. Before experimentation animals were acclimatized to their environment for at least one week.

Treatments

Thyroidectomy was performed at Nossan breeding farm two weeks before the administration of the carcinogen. When the case, one week after thyroidectomy, T_3 was injected s.c. daily for five days at the dose of 200 µg/Kg body weight in 155 mmol NaCl/20 mmol NaOH. Animals which did not undergo thyroidectomy or T_3-treatment are named "intact".

2-AAF, dissolved in dimethylsufoxide, was injected i.p. at the dose of 25 mg/kg body weight, 20 h before sacrifice. DMNA, dissolved in redistilled water, was administered i.p. at the dose of 10 mg/Kg body weight, 4 h before experimentation. MNU and MMS were prepared freshly in 0.9% NaCl solution, and injected i.p. at the dose of 80 mg and 75 mg/Kg body weight respectively, 4 h before experimentation. DL-ZAMI 1305, dissolved in redistilled water, was administered i.p. at the dose of 75 mg/kg body weight, 14 h before experimentation. Control animals received an equal volume of the vehicle. For all carcinogens the time of sacrifice corresponding to the highest DNA damage was chosen[5,15].

Evaluation Hepatic DNA Damage

Animals were killed by cervical dislocation and their livers quickly removed. Liver nuclei were prepared according to the technique of Cox et al.[4]. DNA integrity was evaluated by alkaline sucrose gradient analysis as previously described[15], the only difference being that DNA content in the various fractions was determined by the fluorimetric procedure of Zubroff and Sarma[26]. The amount of DNA recovered from each tube ranged between 80 and 95% of that loaded onto the gradient.

The quantification of DNA fragmentation in each tube was obtained by calculating the value of the ratio light DNA/heavy DNA (R-DNA), where light DNA (mol. wt. $< 1 \times 10^9$) and heavy DNA (mol. wt. $> 1 \times 10^9$) are the amounts of DNA in fractions 6-16 and 1-5 of the gradient, respectively, fraction 1 being the first to be collected from the bottom of the tube.

Statistical Analysis

The data are expressed as the mean ± SEM. The significance of the difference of the means was evaluated by the analysis of variance (F test).

RESULTS AND DISCUSSION

Preliminary experiments were carried out to assay whether thyroidectomy, coupled or not with T_3-treatment, induces any change in the sedimentation properties of liver DNA in carcinogen-untreated animals. As shown in Fig. 1, a small but significant increase of DNA fragmentation is observed both in male and female thyroidectomized (TDX) rats in respect to intact animals. The damage is completely prevented by T_3-treatment (Fig. 1). Thyroid hormones act as trophic agents for liver cells[22,23]. It is then possible that thyroid hormones deprivation causes death of some hepatocytes with consequent DNA fragmentation. Since the increase of DNA fragmentation observed in TDX animals is very limited, it should not affect the interpretation of the data concerning the DNA damaging capacity of carcinogens like DMNA, MMS and MNU that cause a massive DNA fragmentation (see Figs. 2 and 3). On the contrary, it might affect the interpretation of the data on the genotoxic activity of carcinogens like 2-AAF and DL-ZAMI 1305 that cause a very limited DNA damage (see Fig. 2). To overcome this problem, for all experimental groups, DNA damage is expressed both as the absolute value "R-DNA" and as the relative value "R-DNA of carcinogen-treated animals minus R-DNA of untreated animals" (see Figs. 2 and 3).

In agreement with previous results[15], the sex-dependent hepatocarcinogen DL-ZAMI 1305 induces a small but significant DNA fragmentation in the liver of the intact female but not of the intact male rat (Fig. 2). 2-AAF and DMNA also exert a different genotoxic activity in the liver of the animals of the two sexes, the former being more potent in male and the latter in female rats (Fig. 2). This is in keeping with the different liver oncogenic activity of these two carcinogens in the two sexes[21,24].

Thyroidectomy affects the DNA damaging capacity of DL-ZAMI 1305, 2-AAF and DMNA (Fig. 2). A decrease of the amount of DNA damage is observed, in fact, in the liver of TDX female rats treated with DL-ZAMI 1305 or DMNA and in the liver of TDX male rats treated with 2-AAF in respect to carcinogen-treated intact animals of the same sex. The decrease is evident both when the absolute value (R-DNA) or the relative value (R-DNA treated minus R-DNA untreated) are considered. It must be noticed that thyroidectomy does not affect the DNA damaging capacity of these carcinogens in the sex less sensitive to their genotoxic activity, i.e. in males for DL-ZAMI 1305 and DMNA or in females for 2-AAF (Fig. 2).

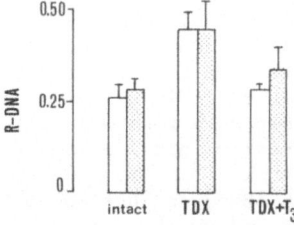

Fig. 1.   Effect of thyroidectomy and T_3-treatment on liver DNA integrity. For the details of surgery and T_3-treatment see Materials and Methods. DNA damage was evaluated by alkaline sucrose gradient analysis and expressed as R-DNA ratio. Each bar represents the mean ± SEM of 5-16 animals. TDX: thyroidectomized rats; TDX + T : thyroidectomized T_3-treated rats. ☐:males; ▦ :females.

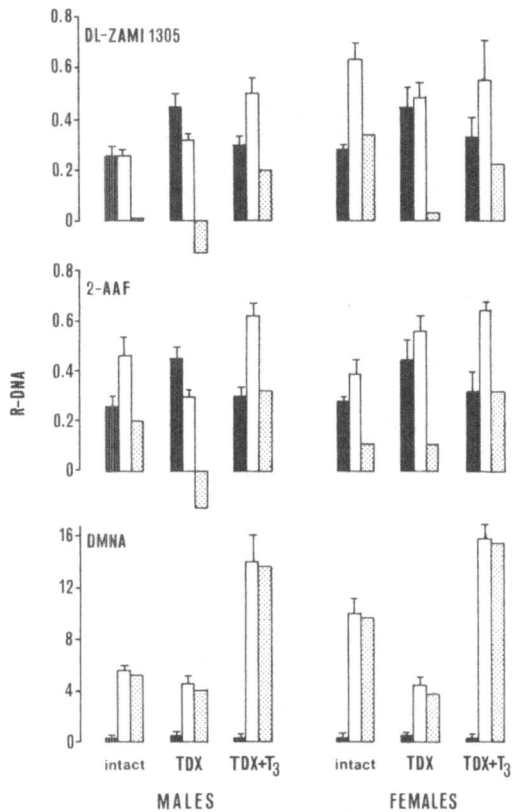

Fig. 2.   Effect of thyroidectomy and T_3-treatment on the DNA damaging
capacity of some chemical hepatocarcinogens in rat liver.
For details of surgery, T_3-treatment and carcinogen administration see Materials and Methods. DNA damage was evaluated
by alkaline sucrose gradient analysis and expressed as R-DNA
ratio. Note the different scales. Each bar represents the
mean ± SEM of 5-20 animals. TDX: thyroidectomized rats;
TDX + T_3: thyroidectomized T_3-treated rats. ■ : carcinogen-
untreated rats; □ : carcinogen-treated rats; ▨ : R-DNA of
carcinogen-treated minus R-DNA of carcinogen-untreated rats.

T_3 administration to TDX animals completely restores the DNA damaging
capacity of DL-ZAMI 1305, 2-AAF or DMNA.   The amount of DNA damage induced by
these carcinogens in TDX T_3-treated animals of both sexes is, in fact, equal
or even higher than that observed in intact animals.   Moreover, T_3 administration to TDX male rats evidences a genotoxic activity of DL-ZAMI 1305 in
this sex, which is not observed in intact animals (Fig. 2).

No differences have been observed instead in the DNA damaging capacity
of the direct carcinogens MMS and MNU both in respect to the sex or to the
thyroid status of the animals (Fig. 3).

Our data indicate that thyroid activity modulates the genotoxic and
then, by extension, the initiating activity of the hepatocarcinogens DL-ZAMI
1305, 2-AAF and DMNA.   These carcinogens require a metabolic activation to
exert their oncogenic potential, as demonstrated for 2-AAF and DMNA[13] and
hypothesized for DL-ZAMI 1305 [17].   They also show a different genotoxic (our
results) and oncogenic[16,21,24] activity in the two sexes, possibly because of

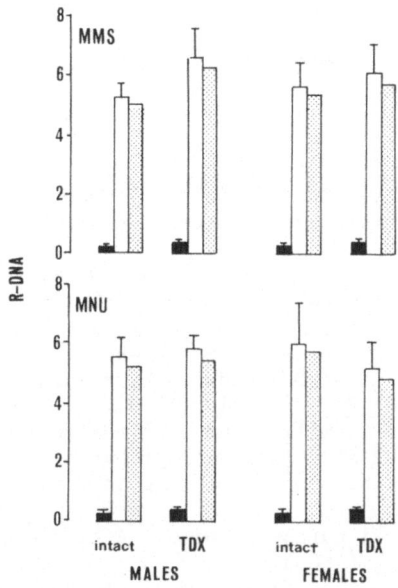

Fig. 3. Effect of thyroidectomy and T_3-treatment on the DNA
damaging capacity of alkylating agents in rat liver. For
the details of surgery, T_3-treatment and carcinogen administration see Materials and Methods. DNA damage was evaluated by alkaline sucrose gradient analysis and expressed
as R-DNA ratio. Each bar represents the mean ± SEM of 5-16
animals. TDX: thyroidectomized rats; TDX + T_3: thyroidectomized T_3-treated rats. ■ : carcinogen-untreated rats; □ :
carcinogen-treated rats; ▨ : R-DNA of carcinogen-treated
minus R-DNA of carcinogen-untreated rats.

sex differences in their metabolism[6,18,25]. Thyroid hormones are known to
modulate the activity of the liver microsomal mixed function oxidase system[11]. It is therefore possible that the inhibitory effect exerted by
thyroidectomy on the DNA damaging capacity of DL-ZAMI 1305, 2-AAF and DMNA
depends on a decrease of their metabolic activation in the liver of TDX animals. This is supported by the lack of effect of thyroidectomy on the genotoxic activity of the alkylating agents MMS and MNU. These molecules, which
show the same genotoxicity in the two sexes, do not require, in fact, metabolic activation to interact with cellular DNA[13]. It is also worth noting
that, after T_3 administration to TDX animals, both 2-AAF and DMNA induce a
DNA damage of higher amount than that observed in intact animals and that
DL-ZAMI 1305 shows a genotoxic activity in the liver of the T_3-treated TDX
male rats but not of the intact males. These findings further support the
role of thyroid hormones on the modulation of the genotoxic capacity of liver
hepatocarcinogens.

ACKNOWLEDGEMENTS

This work was supported in part by grants from C.N.R. (grant no. 85.
00011.04 and grant no. 85.02321.44 Progetto Finalizzato Oncologia) and from
the Ministero della Pubblica Istruzione (M.P.I. 40%, National Project of
Experimental Oncology). We wish to thank Miss Mirella Belleri for skilled
technical assistance and Miss Mariangela Errera for accurate typing of the
manuscript.

REFERENCES

1. F. Bielschowslky and W. H. Hall, Carcinogenesis in the thyroidectomized rat, Br. J. Cancer 7:358 (1953).
2. F. Bielschowslky, Carcinogenesis in the thyroidectomized rat: the effect of injected growth hormone, Br. J. Cancer 12:231 (1958).
3. F. Bielschowsky, M. Bielschowsky and E. K. Fletcher, Investigation on the role of thyroxine in the development of hepatomas in the hypophysectomized rats and pituitary dwarf mice, Br. J. Cancer 16:267 (1962).
4. R. Cox, I. Damjanov, S. E. Abanobi and D. S. R. Sarma, A method for measuring DNA damage and repair in the liver in vivo, Cancer Res. 47:2327 (1973).
5. I. Damjanov, R. Cox, D. S. R. Sarma and E. Farber, Patterns of damage and repair of liver DNA induced by carcinogenic methylating agents in vivo, Cancer Res. 33:2214 (1973).
6. J. R. De Baun, E. C. Miller and J. A. Miller, N-hydroxy-2-acetylaminofluorene sulfotransferase: its probable role in carcinogenesis and in protein-(methion-S-yl) binding in rat liver, Cancer Res. 30:577 (1970).
7. E. Farber, Chemical carcinogenesis. A biologic perspective, Am. J. Pathol. 106:271 (1982).
8. C. M. Goodall, Hepatic carcinogenesis in thyroidectomized rats: apparent blockade at the stage of initiation, Cancer Res. 26:1880 (1966).
9. D. L. Guernsey, A. Ong and C. Borek, Thyroid hormone modulation of X-ray-induced in vitro neoplastic transformation, Nature 288:591 (1980).
10. D. L. Guernsey, C. Borek and I. S. Edelman, Crucial role of thyroid hormone in X-ray-induced neoplastic transformation in cell culture, Proc. Natl. Acad. Sci. USA 78:5708 (1981).
11. J. E. A. Leakey, H. Mukhtar, J. R. Foust and J. R. Bend, Thyroid hormone-induced changes in hepatic monooxigenase system, heme oxygenase activity and epoxide hydrolase activity in adult male, female and immature rats, Chem. Biol. Interactions 40:257 (1982).
12. J. H. Leathem and H. B. Barken, Relationship between thyroid activity and liver tumor induction with 2-acetylaminofluorene, Cancer Res. 10:231 (1950).
13. E. C. Miller and J. A. Miller, Searches for ultimate chemical carcinogens and their reactions with cellular macromolecules, Cancer 47: 2327 (1981).
14. K. E. Paschkis, A. Cantarow and J. Stasney, Influence of thiouracil on carcinoma induced by 2-acetaminofluorene, Cancer Res. 8:257 (1948).
15. M. Presta, C. Mazzocchi, S. Ziliani and G. Ragnotti, In vitro and in vivo DNA damage of male and female rat liver nuclei by oncogenic and nononcogenic beta blockers, J. Natl. Cancer Inst. 70:747 (1983).
16. G. Ragnotti, M. Presta, L. Riboni and T. Zavanella, Liver tumors induced by a new beta-adrenoreceptor blocking agent in female rats, J. Natl. Cancer Inst. 68:669 (1982).
17. G. Ragnotti, M. Presta, J. A. M. Maier, M. Rusnati, G. Mazzoleni, F. Legati, R. Chiesa, M. Braga and D. Calovini, Critical role of gonadal hormones on the genotoxic activity of hepatocarcinogen DL-ZAMI 1305, Cancer Letters 36:253 (1987).
18. K. V. N. Rao and S. D. Vesselinovitch, Age and sex-associated diethylnitrosamine dialkylation activity of the mouse liver and hepatocarcinogenesis, Cancer Res. 33:1625 (1973).
19. M. D. Reuber, The thyroid gland and N-2-fluorenyldiacetamine carcinogenesis and cirrhosis of the liver in Wistar male rats, J. Natl. Cancer Inst. 35:959 (1965).
20. J. Segal, B. R. Troen and S. H. Ingbar, Influence of age and sex on the concentrations of thyroid hormone in serum in the rat, J. Endocr. 93:177 (1982).

21. H. Sidransky, B. P. Wagner and H. P. Morris, Sex difference in liver tumorogenesis in rats ingesting N-2-fluorenylacetamide, _J. Natl. Cancer Inst._ 26:151 (1961).

22. J. Short, R. F. Brown, A. Husakova, J. R. Gilbertson, R. Zemel and I. Lieberman, Induction of deoxyribonucleic acid synthesis in the liver of the intact animal, _J. Biol. Chem._ 247:1757 (1972).

23. J. Short, K. Klein, L. Kibert and P. Ove, Involvement of the iodo-thyronines in liver and hepatoma cell proliferation in the rat, _Cancer Res._ 40:2417 (1980).

24. S. D. Vesselinovitch, The sex-dependent difference in the development of liver tumors in mice administered dimethylnitrosamine, _Cancer Res._ 29:1024 (1969).

25. E. K. Weisburger, P. H. Grantham and J. H. Weisburger, Differences in the metabolism of N-hydroxy-N-2-fluorenylacetamide in male and female rats, _Biochemistry_ 3:808 (1964).

26. J. Zubroff and D. S. R. Sarma, A nonradioactive method for measuring DNA damage and its repair in nonproliferating tissues, _Anal. Biochem._ 70:387 (1976).

ACTIVATION OF URINARY BLADDER CARCINOGENS WITHIN THE TARGET ORGAN

Luisa Airoldi, Marina Bonfanti, Cinzia Magagnotti and
Roberto Fanelli

Laboratory of Environmental Pharmacology and Toxicology
Istituto di Ricerche Farmacologiche Mario Negri
Via Eritrea 62, 20157 Milan, Italy

The majority of human cancers in the most commonly involved organs are induced by environmental factors. Of the known exogenous carcinogenic agents, chemicals play an important role in tumor induction. Humans are exposed to chemical carcinogens through diet, tobacco, occupational and environmental settings. The list of chemicals known to be carcinogenic to man includes a broad range of structures such as aromatic amines, alkylating agents, vinyl chloride. Also, highly suspected of having carcinogenic activity in humans are aflatoxin B_1, polycyclic aromatic hydrocarbons and N-nitroso compounds. In many cases, epidemiological evidence has proved sufficient to identify certain chemicals or chemical mixtures as carcinogens.

The bladder is one of the cancer sites related to occupational factors. The first association between bladder cancer in man and occupational exposure to chemicals dates back to 1895 when Rehn reported a high incidence of bladder tumors among workers at a German dye manufacturing plant exposed to aromatic amines[1]. Subsequently, besides hazardous occupational exposure, bladder cancer has been associated with cigarette smoking, coffee and artificial sweetener consumption[2].

The understanding of the processes by which carcinogens induce neoplasia in their target tissues is largely derived from experimental findings in animal models. The development of animal models has greatly helped clarify the stages of carcinogenesis and is an essential step in the investigation of the effects of carcinogens in their target tissues, besides being a powerful tool in defining their metabolic pathways.

Several chemical agents are commonly employed to generate urinary bladder tumors in animals; among these, N-nitrosobutyl(4-hydroxybutyl)amine (BBN), N-[4-(5-nitro-2-furyl)-2-thiazolyl]formamide, 2-acetylaminofluorene and N-methylnitrosourea are widely utilized[3]. The histological characteristics of the tumors induced in animals by these chemicals are very similar to those reported for humans with bladder neoplasia[4].

Most of these compounds require metabolic activation to exert their toxic effect. The bulk of knowledge about the mechanism of action of bladder carcinogens derives from studies on the metabolism of arylamines. These are N-oxidised in the hepatic endoplasmic reticulum and the hydroxy derivatives are conjugated with glucuronic acid. The glucuronides then enter the circulation and are excreted into the urine. During their stay in the bladder in the slightly acidic urine, they are hydrolyzed to electrophilic species that reportedly bind to DNA[5,6].

In this model the bladder plays no role in the activation of N-aryl compounds. However, several studies show that in vitro preparations such as urothelial cell microsomal fractions from different animal species, human and rat cultured bladder epithelial cells or organ cultures, metabolize a variety of different substrates including polycyclic aromatic hydrocarbons, aromatic amines or nitrosamines to species that bind to DNA[7,10].

Thus it would appear that bladder epithelial cells can activate urinary bladder carcinogens.

We therefore decided to develop an in vitro model for studying the biotransformation of these chemicals.

The model has been recently described and consists of an isolated rat urinary bladder filled with urine containing the substrate to be metabolized and no cofactors. The bladder is immersed in 0.05 M phosphate buffer pH 7.4 and incubated at 37°C for different times[11].

The model is simple and handy; it reproduces closely the in vivo situation and the carcinogen introduced can be retained in contact with the urothelial cells for a reasonable period of time.

The rat urinary bladder carcinogen BBN was the compound chosen to test the bladder's capacity to metabolize the substrate used[12]. BBN's carcinogenicity is thought to depend on oxidation of the alcoholic group by the alcohol/aldehyde dehydrogenase enzymatic system to the acidic metabolite BCPN, the liver being the main site of metabolism[13].

We have observed that after an i.v. dose of 1 mg/kg BBN, the unchanged compound, its metabolites BCPN and the glucuronide of BBN (BBN-G) are found in the rat urine, the percentage of the administered dose excreted being 0.26 ± 0.16, 36 ± 2.5 and 11.7 ± 2.4 respectively (mean ± SE; manuscript submitted for publication). Thus, these products may be further transformed by the urothelium while in the bladder.

In order to test this hypothesis BBN (17 nmol) dissolved in 600 µl of urine previously sterilized by filtration was introduced into the isolated rat urinary bladder and incubated at 37°C for different times. The bladder wall is reportedly permeable to various nitrosamines[14], so substrate disappearance and metabolite formation were always measured inside the bladder and in the buffer where the bladder was immersed.

Disappearance of the substrate from the incubation system is reported in Fig. 1. BCPN formation was readily detectable 15 minutes after the start of the reaction and linear up to 120 minutes (Fig. 2). In order to test whether part of the substrate or metabolite formed was retained in bladder cells, some bladders were analyzed for BBN and BCPN content. The 60-minute urinary bladder homogenates contained 1.5 ± 0.5% BBN and 13.7 ± 0.3% BCPN. The sum of BCPN formed and unchanged BBN was about 88% of the substrate added, suggesting that BCPN represents the major metabolic pathway for BBN. Thus the urinary bladder appears to have the potential for metabolic activation of BBN to the compound considered responsible for tumor induction in the rodent urinary bladder and the bladder cells appear to contain the enzymatic system alcohol/aldehyde dehydrogenase thought to be involved in the transformation of BBN to BCPN.

BBN is the result of the ω-oxidation of N-nitrosodibutylamine (NDBA)

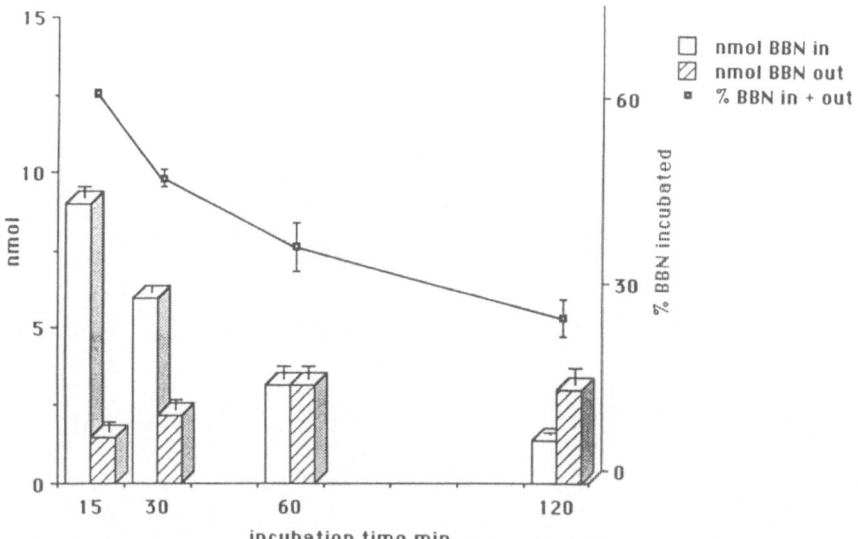

Fig. 1. Time-related disappearance of unchanged BBN from the isolated rat
urinary bladder. Columns indicate the nmol (mean ± SE) of substrate
measured inside (□) and outside (▨) the bladder. The line (⊢□⊣)
shows the percentage (mean ± SE) of unchanged BBN inside + outside.

by the P-450 dependent monooxygenase system. When NDBA was introduced into
the isolated bladder and incubated for 120 minutes at 37°C, BBN and BCPN
accounted for respectively 0.13 ± 0.02% and 0.06 ± 0.01% of the substrate
added, indicating that more than 30% of BBN formed was further oxidised to
BCPN.

Volatile nitrosamines are poorly excreted into the urine as such, but
they can be synthesized in the rat bladder[12,15,16]. Also, bacterial infection
of the urinary tract in man can promote the reduction of nitrate to nitrite[17].
Because of the higher nitrite availability, nitrosamine formation may occur.
The results suggest that whenever nitrosamines are formed, they can be activated within the bladder.

ß-Glucuronidase activity has been reported both in urine and bladder
cells[18]. We checked whether our in vitro system was suitable for studying
glucuronide hydrolysis within the bladder. BBN-G (17 nmol/600 μl) was introduced into the isolated bladder as a 0.05 M phosphate buffer solution, pH 7.4
(in order to measure urothelial ß-glucuronidase activity only) and as a urine
solution (in order to measure both urine and urothelial ß-glucuronidase activity). Fig. 3 (b and c) shows that in both systems the amount of glucuronide hydrolyzed reached a plateau at 4 h, accounting for about 1% of the incubated substrate. Part of BBN released was oxidised to BCPN. BCPN formation
was not detectable until 2 h from the start of the reaction and increased
slowly to 0.2% at 6 h.

BBN-G hydrolysis due to urinary ß-glucuronidase only is also shown in
Fig. 3 (a). BBN was released linearly, reaching 1.23 ± 0.02% of the glucuronide incubated at 6 h. No BCPN was detected at any time in this system and
no hydrolysis occurred when BBN-G was incubated in phosphate buffer.

Contrary to our expectations, the amount of BBN released in the presence
of urinary and urothelial ß-glucuronidase was no different from that released

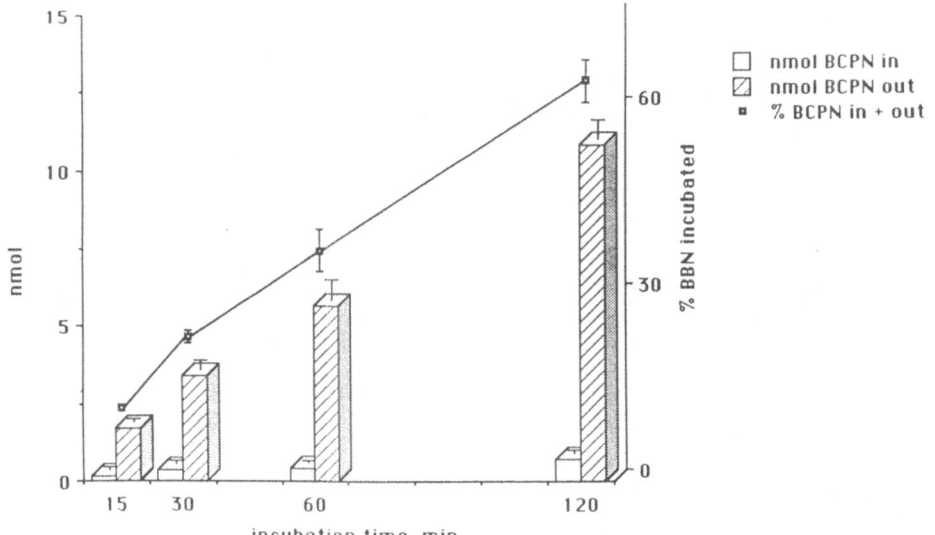

Fig. 2. Time-related formation of BCPN in the isolated rat urinary bladder. Columns indicate the nmol (mean ± SE) of metabolite measured inside (☐) and outside (▨) the bladder. The line (–☐–) shows the total percentage (mean ± SE) of BCPN formed inside + outside.

by urinary ß-glucuronidase alone. This might be the result of an inhibitory effect of urinary components exerted only on cellular ß-glucuronidase.

The results indicate that whenever an O-glucuronide is excreted in the urine it can be hydrolyzed by ß-glucuronidase, no matter whether cellular or urinary.

So far, we have considered only the metabolites of BBN retaining the nitroso moiety, but the most widely accepted metabolic pathway leading to the formation of electrophilic intermediates involves enzymatic hydroxylation at the carbon atom α to the nitroso group[19]. All subsequent steps are non-enzymatic reactions and the final result is the formation of a carbocation and molecular nitrogen in stoichiometric quantity. Thus determination of molecular nitrogen evolved can be used as an indicator of α-hydroxylation.

We have synthesized the doubly labeled ^{15}N-BBN (manuscript in preparation), incubated this substrate in the isolated rat bladder and measured the formation of $^{15}N_2$ by gas chromatography-mass spectrometry. Preliminary results indicate that the bladder metabolize BBN through the α-hydroxylation pathway.

The overall results lead to the conclusion that the urinary bladder's role in the activation of bladder carcinogens has long been underestimated. It is clear that the liver's potential for metabolic activation of chemicals far exceeds that of the bladder. However, activation of bladder carcinogens by urothelial cells results in the production of reactive species within the target organ and increases the probability of a toxic effect.

The model developed appears to offer an easy tool for studying in situ synthesis and metabolism of bladder carcinogens, and for identifying factors that can modify chemical metabolism and consequently their carcinogenic activity.

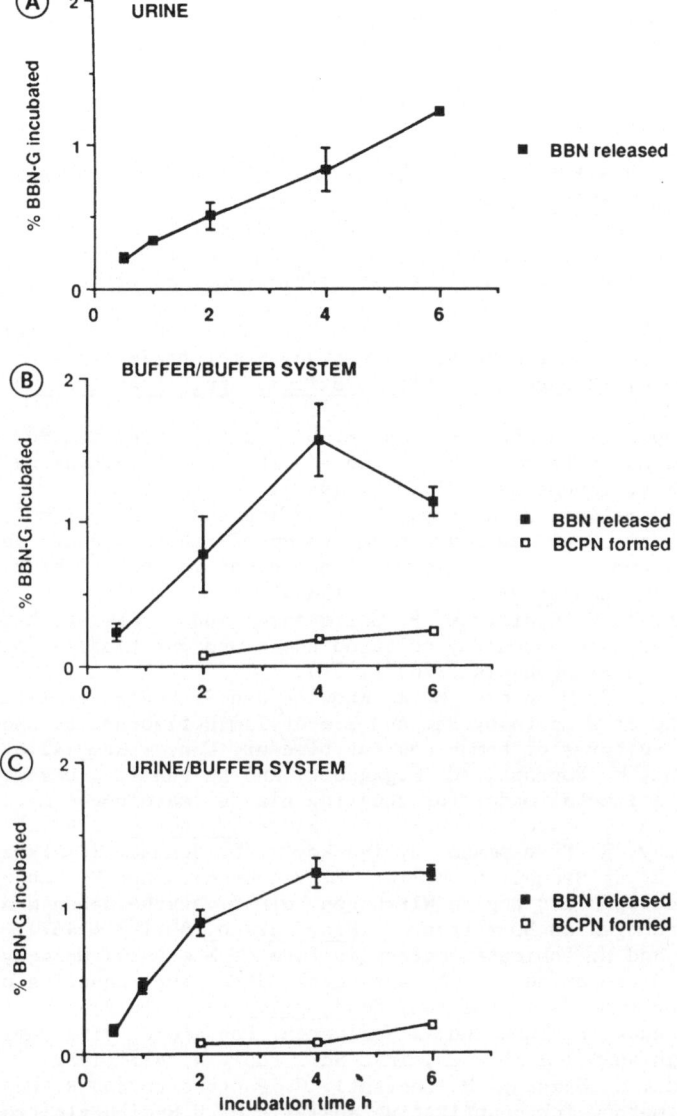

Fig. 3. Percentages of BBN release (mean ± SE) from BBN-G incubated in urine
         alone (a); BBN release and BCPN formation from BBN-G introduced in
         the isolated rat urinary bladder as a buffer solution (b); or as a
         urine solution (c).

ACKNOWLEDGEMENTS

    This work was supported by the Italian National Research Council, Spe-
cial Project "Oncology", Contract No. 86.02607.44. We thank J. Baggot and
the staff of the G. A. Pfeiffer Memorial Library who helped prepare the
manuscript.

REFERENCES

   1. L. Rehn, Blasengeschwultse bei Fuschin-Arbeitern. <u>Arch. Klin. Chir. L.</u>
       50:588 (1895).

2. G. R. Howe, J. D. Burch, A. B. Miller, G. M. Cook, J. Esteve, B. Morrison, P. Gordon, L. W. Chambers, G. Fodor and G. M. Winsor, Tobacco use, occupation coffee, various nutrients, and bladder cancer, J. Natl. Cancer Inst. 64:701 (1980).

3. S. M. Cohen, R. Hasegawa, R. E. Greenfield and L. B. Ellwein, Urinary bladder carcinogenesis, IARC Sci. Publ. 56:93 (1984).

4. N. Ito, S. Fukushima, T. Shirai, K. Nakanishi, R. Hasegawa and K. Imaida, Modifying factors in urinary bladder carcinogenesis, Environ. Health Perspect. 49:217 (1983).

5. F. F. Kadlubar, J. A. Miller and E. C. Miller, Hepatic microsomal N-glucuronidation and nucleic acid binding of N-hydroxy arylamines in relation to urinary bladder carcinogenesis, Cancer Res. 37:805 (1977).

6. J. M. Poupko, W. L. Hearn and J. L. Radomski, N-Glucuronidation of N-hydroxy aromatic amines: A mechanism for their transport and bladder-specific carcinogenicity, Toxicol. Appl. Pharmacol. 50:479 (1979).

7. J. M. Poupko, J. L. Radomski and W. L. Hearn, Bovine bladder mucosa microsomal cytochrome P-450 and 4-aminobiphenyl N-hydroxylase activity, Cancer Res. 41:1306 (1981).

8. R. R. Vanderslice, J. A. Boyd, T. E. Eling and R. M. Philpot, The cytochrome P-450 monooxygenase system of rabbit bladder mucosa: Enzyme components and isozyme 5-dependent metabolism of 2-aminofluorene, Cancer Res. 45:5851 (1985).

9. H. Autrup, R. C. Grafstrom, B. Christensen and J. Kieler, Metabolism of chemical carcinogens by cultured human and rat bladder epithelial cells, Carcinogenesis 2:763 (1981).

10. B. P. Moore, R. M. Hicks, M. A. Knowles and S. Redgrave, Metabolism and binding of benzo(a)pyrene and 2-acetylaminofluorene by short-term organ cultures of human and rat bladder, Cancer Res. 42:642 (1982).

11. L. Airoldi, M. Bonfanti, C. Magagnotti and R. Fanelli, Development of an experimental model for studying bladder carcinogen metabolism using the isolated rat urinary bladder, Cancer Res. 47:3697 (1987).

12. H. Druckrey, R. Preussmann, S. Ivankovic, D. Schmal, J. Afkham, G. Blum, H. D. Mennel, M. Muller, P. Petropoulos and H. Schneider, Organotrope carcinogene Wirkungen bei 65 verscheidenen N-nitrosoverbindungen an BD-ratten, Z. Krebsforsch. 69:103 (1967).

13. M. Okada and M. Ishidate, Metabolic fate of N-n-butyl-N-(4-hydroxybutyl)nitrosamine and its analogues, Selective induction of urinary bladder tumours in the rat, Xenobiotica 7:11 (1977).

14. J. S. Wishnok, K. Snow and V. Woolworth, Passage of nitrosamines through animal membranes, IARC Sci. Publ. 41:435 (1982).

15. L. Airoldi, C. Spagone, M. Bonfanti, C. Pantarotto and R. Fanelli, Rapid method for quantitative analysis of N,N-dibutylnitrosamine, N-butyl-N-(4-hydroxybutyl)nitrosamine and N-butyl-N-(3-carboxypropyl)nitrosamine in rat urine by gas chromatography-thermal energy analysis, J. Chromatogr. 276:402 (1983).

16. G. Hawksworth and M. J. Hill, The in vivo formation of N-nitrosamines in the rat bladder and their subsequent absorption, Br. J. Cancer 29:353 (1974).

17. R. M. Hicks, Nitrosamines as possible etiological agents in Bilharzial bladder cancer, in: "Banbury Report. Nitrosamines and Human Cancer", No. 12, P. N. Magee, ed., Cold Spring Harbor Laboratory, Cold Spring Harbor (1982).

18. E. Boyland, D. M. Wallace and D. C. Williams, The activity of the enzymes sulphatase and ß-gucuronidase in the urine, serum and bladder tissue, Br. J. Cancer 9:62 (1955).

19. R. Montesano and H. Bartsch, Mutagenic and carcinogenic N-nitroso compounds: Possible environmental hazards, Mutat. Res. 32:179 (1976).

LIPID PEROXIDATION AND BIOACTIVATION OF HALOGENATED HYDROCARBONS

IN RAT LIVER MITOCHONDRIA DURING EXPERIMENTAL SIDEROSIS

Aldo Tomasi, Emanuele Albano[1], Barbara Botti, Francesco
P. Corongiu[2], M. Assunta Dessì[2], Anna Iannone, Valeria
Franceschi, Vanio Vannini and Alberto Masini

Istituto di Patologia Generale, Via Campi 287
41100 Modena, [1]Dipartimento di Medicina a Oncologia
Sperimentale,Corso Raffaello 30, 10125 Torino,
[2]Dipartimento di Biologia Sperimentale, Sezione di
Patologia Sperimentale, Via Ospedale 76, 09100 Cagliari

INTRODUCTION

It is firmly extablished that increased amount of iron accumulated in
hepatic parenchymal cells is associated with tissue injury, fibrosis and ultimately, cirrhosis[1]. However, the pathogenetic mechanism of iron in determining the liver injury has not been experimentally proven[2]. Currently two
hypothesis have been put forward in order to explain the hepatocellular injury in chronic iron overload. The first implies that the excess iron largely occurring in lysosomes, physically disrupts these organelles with the release of cell damaging hydrolytic enzymes[3]. The second one presupposes that
the pathological accumulation of iron elicits membrane lipid peroxidation in
cellular organelles resulting in structural and functional alterations of
cell integrity[4]. Indeed, iron could promote the production of reactive oxygen species such as superoxide anion (O_2^-), hydroxy radicals (OH_4^-), and hydrogen peroxide (H_2O_2)[5]. The experimental evidence, gathered up to now, indicates that chronic iron overload may induce in vivo lipid peroxidation of
mitochondrial membranes[6-8]. Furthermore, the in vivo occurrence of lipid
peroxidation in the mitochondrial membranes has been suggested to be responsible for some anomalies in liver mitochondria isolated from rats made siderotic either by dietary iron[9,10] or by intraperitoneal injection of iron-
(III)- gluconate complex[11-13].

Along with these investigations, and taking into account[14] that mitochondria are able to metabolize the well known hepatotoxic and carcinogenic
agent, carbon tetrachloride (CCl_4) to form the trichloromethyl free redical,
we have used the spin trapping technique coupled to Electron Spin Resonance
spectroscopy (ESR) as a probe of a possible impairment of the mitochondrial
metabolic activity in the presence of excess iron.

In the present study the changes in iron concentration of the hepatic
tissue and of the mitochondrial fraction, isolated from either female or male
rats fed on a diet supplemented with carbonyl iron for a period of 60 days,
are presented. The susceptibility of the mitochondrial membrane to lipid
peroxidation, the mitochondrial functional integrity, the mitochondria Ca^{2+}

transport capability, and the ability in bioactivating CCl_4 have been investigated.

MATERIALS AND METHODS

Animals and Diet

Female and male albino rats (100-120 g) were purchased from Nossan (Corezzana, Milano, Italy). Rats were made siderotic by feeding a standard diet purchased from Dr. Piccioni (Brescia, Italy) supplemented with 25% (wt/wt) carbonyl iron.

Chemicals

Carbonyl iron was purchased from Fluka (Fluka AG, Buchs, CH-9470, Switzerland). Phenyl-t-butyl nitrone (PBN) was obtained from Aldrich Chemical Co (Milwaukee, WI, USA); all other chemicals and reagents were of the highest available purity.

Preparation of the Mitochondrial Fraction

Animals were killed by decapitation after an overnight starvation period. Liver mitochondria were prepared in 0.25 M sucrose according to a standard procedure[15]. The protein content of the final mitochondrial suspension was determined by the biuret method with bovine serum albumin as the standard.

Mitochondrial and Tissue Iron Determination

Mitochondrial and hepatic iron concentration was determined by an atomic absorption Perkin Elmer spectrophotometer (mod. 306) as follows: aliquots of 2 ml of either homogenate or mitochondrial fraction were digested by 2 ml HNO_3 and 1 ml H_2SO_4. The dried residue was added with 1 ml $HClO_4$ to obtain a clear solution which was then diluted with distilled water. Solutions of $(FeNO_3)_3$ in 0.5 N HNO_3 were used as standards. All the data are expressed as nmol iron per mg protein.

Lipid Peroxidation

Determination of the conjugated diene signal was performed according to Corongiu et al.[16] The lipid chloroform extracts were placed in glass test tubes with glass stoppers. The solvent was removed under vacuum at 40°C. The lipid extract was dissolved in cyclohexane in order to obtain a lipid concentration of 100 µg/ml and scanned from 300 to 220 nm and the absorbance and the second derivative spectrum were recorded. The presence of conjugated dienes was detected directly in the second derivative spectrum by the presence of minimum peaks that absorb at 233 and 242 nm. These absorbances have been elsewhere characterized as trans,trans and cis,trans hydroperoxydienes[17]. Before the scanning operation a background correction-memorized scan between Suprasil cells containing cyclohexane was performed in order to avoid spectral differences between sample and reference cell. Because artifactual peroxidation may occur during tissue sampling and processing, experiments were considered acceptable only when lipid extracts, prepared from control animals (at least 4 in number) and studied at the same time as the treated group, did not show any significant second derivative signal attributable to lipoperoxidative damage. In the Figure based on second derivative spectroscopy, an absorption peak in the conventional sense appears as an absorption minimum.

Mitochondrial Oxidative Metabolism

The metabolic parameters were assayed with a polarographic Clark type

oxygen electrode at 25°C in final volume of 3 ml. The concentration of mitochondria was 3 mg of protein per ml. The phosphorylative capacity was measured from the polarographic traces according to Chance and Williams[18]. The
respiratory states studied were those defined by Chance and Williams[18]. The
incubation standard medium was as follows: 100 mM NaCl; 10 mM TRIS-HCl
(pH 7.4); 10 mM Na,K-phosphate (pH 7.4) and 1.6 mM Na-pyruvate plus 0.4 mM
L-malate or 2 mM Na-succinate as respiratory substrates.

Mitochondrial Transmembrane Electrical Potential

The transmembrane potential ($\Delta\Psi$) was measured at 25°C in a final volume
of 1.5 ml by monitoring with a tetraphenylphosphonium (TPP$^+$) selective electrode, the movements of tetraphenylphosphonium across the mitochondrial membrane as in ref. 19. An inner mitochondrial volume of 1.1 µl/mg protein was
assumed. The metabolic medium for assaying the electrochemical parameters
had the following composition: 210 mM mannitol; 70 mM sucrose; 5 mM Hepes
pH 7.4 (MSH buffer); 5µM rotenone and 20 µM TPP$^+$. Mitochondria (3 mg pro-
tein/ml) were incubated at 25°C and then energized with 2.5 mM Na-succinate.

Spin Trapping

Spin trapping experiments were performed on freshly prepared and resuspended mitochondria (2 ml final volume, 10 mg of protein/ml) placed in a
50 ml Erlenmeyer flask fitted with a central well. CCl$_4$ (10 µl) was added to
the central well and the flasks were tightly closed. CCl$_4$ concentration
reached approx 0.15 mM into the incubation mixture[20]. The mitochondria were
supplemented with succinate (5 mM) and ADP (0.5 mM) along with the spin trapping agent PBN (25 mM). The experiments were performed under hypoxic conditions obtained by flushing for 10 min with moist O$_2$-free N$_2$. Incubations
were carried out at 37°C for 30 min; the suspensions were then extracted with
1 ml chloroform methanol (2:1, v/v) mixture, and the chloroform phase separated by centrifugation was used for ESR analysis. The presence of the
PBN-CCl$_3$ spin adduct was revealed by a Bruker 200D-SCR ESR spectrometer as
previously described[21].

RESULTS

The iron content of the liver tissue of female rats fed on a diet supplemented with 2.5% (wt/wt) carbonyl iron progressively increases from 16.2 ±
1.6 nmol/mg protein at zero time up to 156.1 ± 4.7. As shown in Table I the
Fe concentration of the hepatic tissue reaches a steady state level after 40
days of treatment. A very similar pattern is presented by the Fe content of
the liver tissue of male rats, but the maximum extent of Fe loading is much
lower. Fe concentration in males increases from 12.2 ± 3.6 nmol/mg protein
up to 50.2 ± 4.6 nmol/mg protein, which represents a 4 fold increase against
a 12 fold increase in the case of female rats. The changes in Fe content of
the mitochondrial fraction isolated from Fe treated rats present a similar
kinetic. It has to be stressed that there is a 12 fold increase in the Fe
content of mitochondria isolated from female rats against a 4 fold of the
male rats.

The induction of lipid peroxidation in vivo in mitochondrial membranes
appears to be associated with the attainment of a threshold value in the concentration of iron. Indeed, the second derivative spectrum of mitochondrial
lipids which exhibits two modifications at 233 nm and 242 nm characteristic
of trans,trans and cis,trans conjugated dienes is evident only in female
rats, after 40 days of treatment (not shown). It is noteworthy that no second derivative spectrum of conjugated dienes was observed during the treatment period in male rats (not shown).

The respiratory control index and the ADP/O ratio of mitochondria from both male and female treated rats, obtained either with a NAD-linked substrate, such as pyruvate plus malate, or with succinate, did not differ from controls at any time of treatment tested (not shown).

The determination of the membrane potential on the basis of movements of the lipid soluble cation tetraphenylphosphonium has proven useful to measure the energy state of mitochondria and to assess their intactness. Fig. 1 shows that mitochondria isolated from female rats after 60 days of treatment on addition of substrate immediately develop a normal $\Delta\Psi$ of about 180 mV. Similar values of membrane potential were found with mitochondria from male rats (not shown).

These data are therefore consistent with the above conclusion on the normal oxidative metabolism of iron treated mitochondria. However, the membrane potential pattern during the accumulation of a pulse of Ca^{2+} (i.e. 50 nmol/mg protein) revealed a marked difference between control mitochondria (Fig. 1A) and mitochondria from iron loaded female rats (Fig. 1B). In the latter case, the membrane potential trace after the drop, which corresponds to the energy utilized for the Ca^{2+} accumulation, does not return to the pre-Ca^{2+} level, as in the control, but suddenly decreases. When $\Delta\Psi$ has reached the lowest value, addition of the Ca^{2+} chelating agent EGTA promptly reverses the membrane potential to the normal value. This observation indicates that the inner membrane is not irreversibly depolarized. Thus suggesting that the membrane potential decrease is not due to the damage of the mitochondrion but rather to a continuous energy draining "Ca^{2+} cycling" process, which dissipates energy while reaccumulating the released Ca^{2+} into mitochondria. Similarly, when Ca^{2+} has been accumulated in iron loaded mitochondria, the addition of Ruthenium red, a specific inhibitor of the electrogenic Ca^{2+} uptake induces a constant increase of the membrane potential. On the other hand the addition of either BHT, an antioxidant, Desferal, a specific chelator of ferric iron, or BSA, a chelator of fatty acid, does not modify the $\Delta\Psi$ trace.

Table 1. Increase in iron concentration during dietary iron overload

Treatment time (days)	Fe concentration (nmol/mg protein)			
	Hepatic tissue		Mitochondrial fraction	
	male	female	male	female
0	12.2 ± 3.6	16.2 ± 1.6	4.8 ± 1.2	3.6 ± 0.7
20	39.6 ± 5.4	62.8 ± 2.1	11.5 ± 1.8	12.0 ± 2.0
40	48.4 ± 6.1	159.1 ± 6.5	19.1 ± 3.6	37.9 ± 4.5
60	50.2 ± 4.6	156.1 ± 4.7	18.3 ± 2.8	40.9 ± 8.7

Mean values for 3 to 5 different experiments are given ± S.D.
Iron was determined by atomic absorption spectroscopy.

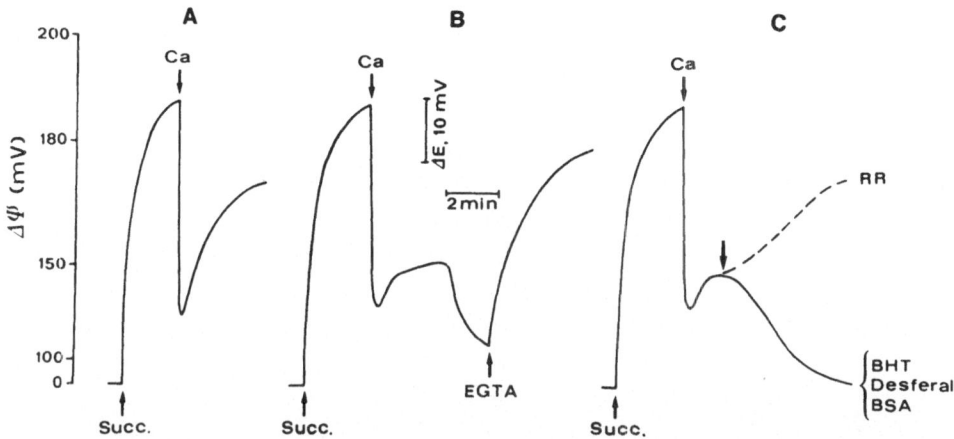

Fig. 1. Membrane potential of liver mitochondria isolated from rats treated
with carbonyl iron. Mitochondria in standard incubation medium, as
described in the Methods, were energized by the addition of 2.5 mM
succinate (Succ.). The arrows indicate the following additions: 150
μM $CaCl_2$ (Ca); 0.5 mM EGTA; 2 uM ruthenium red (RR); 30 μM butylated
hydroxy toluene (BHT); 2 mg/mg protein desferrioxamine (Desferal);
0.1% bovine serum albumin (BSA). 1.A: Control mitochondria; 1.B-C:
mitochondria from 60 days iron treated female rats. The mitochondrial potential ($\Delta\Psi$) was measured by tetraphenylphosphonium selective electrode, as described in the Methods. ΔE, electrode potential. The traces presented in the figures are representative of at
least three different experiments performed on a pool of 3 animals.

It has to be noted that mitochondria isolated either from female rats
after 20 days of treatment or from male rats at all treatment times will not
exhibit such anomalies in Ca^{2+} transport (not shown).

In parallel to this set of experiments, a different approach was used on
the same mitochondrial preparations with the aim of testing mitochondria
ability in bioactivating CCl_4. The mitochondria were incubated, as described
in Materials and Methods, in the presence of CCl_4 and the spin trapping agent
PBN, which is an efficient trap of the trichloromethyl free radical (Fig. 2).
Mitochondria from control and iron loaded rats were tested in such a way at
2, 6, 20, 40 and 60 days of treatment time. At any time of treatment no differences were observed between controls and treated rats.

DISCUSSION

The present results show that feeding rats on a diet supplemented with
carbonyl iron results in the accumulation of iron in the hepatic tissue which
is more pronounced in the case of female than male rats. The accumulation of
iron is progressive with the time up to nearly 40 days, when it reaches a
steady state value of approximately 160 nmol/mg protein for female rats and
of approx. 50 nmol/mg protein for male rats. These data confirm those of
Bonkowsky et al.[22] in rats treated parenterally with iron dextran. The
accumulation of iron in the hepatic tissue is paralleled by an increasing
concentration of iron in the mitochondrial fraction which reaches a steady
state value of about 40 nmol/mg protein within 40 days of treatment in the
case of female rats and of about 18 nmol/mg protein in the case of male rats.
The reason for the difference in the extent of iron accumulation between male
and female rats remains unexplained. The increased concentration of iron in

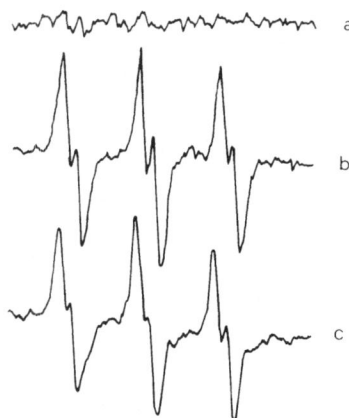

Fig. 2. ESR spectra of the $CCl_3^{·}$-PBN free radical adduct produced by isolated mitochondria exposed to CCl_4. Mitochondrial suspensions were incubated 30 min at 37°C under hypoxic conditions with 0.5 mM-ADP, 5
mM succinate, 25 mM PBN and in the presence of CCl_4 (10 µl added to
the center well of the incubation flasks). The traces represent: a)
controls incubated without CCl_4; b) mitochondria from normal rats
receiving CCl_4; c) mitochondria from iron overloaded rats receiving
CCl_4.

the hepatic tissue and in the mitochondrial fraction is associated with an
enhancement of in vivo lipid peroxidation of mitochondria membranes. The
process appears to be dependent on the attainment of a threshold level in
iron concentration, since it is observed in female rats when iron concentration exceeds 90 nmol/mg protein of liver tissue or about 30 nmol/mg mitochondrial protein. This finding agrees with the results of Bacon et al.[9,10], who
reports the occurrence of lipid peroxidation when the hepatic iron content
reached the above mentioned threshold value. Lipid peroxidation was not observed in microsomal membrane lipids at any treatment time as long as the
iron concentration was below the critical level, in accordance with previous
reports[9].

Iron in vivo may initiate free radical reactions, though the biochemical
mechanisms involved are not known. It has been suggested that in the presence of excess iron, the capability of hepatocytes to maintain iron in a protein bound non-reactive ferric iron may be exceeded, resulting in the release
of a low amount of reactive iron (ferrous iron or chelated iron). Reactive
iron can initiate membrane peroxidation either directly through a Fenton like
reaction, or by forming perferryl iron (FeO^{2+}), or a ternary complex between
arachidonic acid-ferrous iron and oxygen[23]. The in vivo lipoperoxidative reactions in the mitochondrial membranes, proved by the detection of conjugated
dienes, seems to be associated with a derangement in mitochondrial Ca^{2+}
transport. These anomalies have not been observed in mitochondria isolated
either from 20 days iron treated female rats or from male rats at all times of
treatment, where no lipid peroxidation was detected.

Lipid peroxidation and the ensuing membrane derangement will cause a
likely alteration in the flow of electrons through the respiratory chain. It
has been demonstrated that the mechanism responsible for the activation of
CCl_4 share some analogies with the univalent reduction of oxygen[14], hence the
spin trapping technique, which revealing the presence of the free radical
intermediate becomes, at the same time, a probe of the mitochondrial electron
flow. In our hands, trichloromethyl radical production did not show any significant difference in mitochondria from control or iron loaded rats at any

time of the treatment. Thus, it is plausible to maintain that under the experimental conditions, no major disruption has occurred in the mitochondrial electron flow.

The drop in membrane potential observed in mitochondria from 40 or 60 days iron treated female rats after they had accumulated a pulse of Ca^{2+} is apparently related to the energy dissipated to reaccumulate the released Ca^{2+}, a process which is known as "Ca^{2+} cycling". Indeed when the process of the energy dependent Ca^{2+} uptake is blocked by the addition of Ca^{2+} chelating agent EGTA or it is inhibited by the presence of the specific inhibitor Ruthenium red, the membrane potential returns to normal values. Therefore a net release of Ca^{2+} appears to be associated with the induction of lipid peroxidation. However this Ca^{2+} release does not result from gross alterations in the functional integrity of mitochondria. Indeed, the phosphorylative efficiency, as well as the respiratory control index of the mitochondria are normal. A further experimental support to this suggestion is that they present a normal membrane potential as well as a normal capability for Ca^{2+} uptake [24]. Some defects in the mitochondrial electron transport chain [10] as well as anomalies in the membrane potential[11-13] have been previously reported to occur in experimental siderosis. However those results were obtained in the presence of a concentration of iron much higher than that reported here.

Relevant to the present finding are the results obtained by the in vitro addition iron (III)-gluconate complex to liver mitochondria. An activation of a specific Ca^{2+} release route from mitochondria was observed to be brought about by the induction of lipid peroxidation by the ferric complex in mitochondrial membranes[25-28]. In the light of the present results we suggest that iron can be safely stored by mitochondria up to a fairly high concentration. Above a critical threshold value we have observed a derangement in the Ca^{2+} transport with a continuous energy drain which will eventually lead to a perturbation of the cellular energy state, a determining factor in the reversibility of the lesion. This event, given the role played by mitochondria in cell metabolism may then constitute the causal event of cell damage observed in the experimental hepatic siderosis.

ACKNOWLEDGEMENTS

This work has been supported in part by a grant of the Consiglio Nazionale delle Ricerche (Progetto Oncologia, contract No. 84.00837.44). A.I., F.P.C. and A.T. are grateful to the Association for International Cancer Research (London, U.K.) for financial support. F.P.C. wish to thank the Associazione Italiana per la Ricerca sul Cancro (Milano, Italy) for financial support.

REFERENCES

1. L. W. Powell, Hemochromatosis and related iron storage disease in: "Liver and Biliary Disease", R. Wright, G. H. Millward-Sandler, K. G. M. M. Alberti and S. Korran, eds., Bailliere Tindall Saunders Company, London (1985).
2. A. Jacobs, The pathology of iron overload, in: "Iron in Biochemistry and Medicine", A. Jacobs and M. Woorwood, eds., Academic Press, London (1980).
3. T. J. Peters and C. A. Seymour, Acid hydroxylase activities and lysosomal integrity in liver biopsies from patients with iron overload, Clin. Sci. Mol. Med. 50:75 (1976).
4. B. Halliwell and J. M. C. Gutteridge, The importance of free radicals and catalytic metal ions in human disease, Molec. Aspects Med. 8:89 (1985).

5. B. Halliwell and J. M. C. Gutteridge, Oxygen toxicity, oxygen radicals, transition metals and disease, Biochem. J. 219:1 (1984).

6. G. Hanstein, D. T. Heitmann, A. Sandy, H. L. Biesterfeld, H. H. Liem and U. Muller-Eberhard, Effects of hexachlorobenzene and iron loading on rat liver mitochondria, Biochim. Biophys. Acta 678:293 (1981).

7. B. R. Bacon, G. M. Brittenham, A. S. Tavill, C. E. McLaren, H. Park and R. O. Recknagel, Hepatic lipid peroxidation in vivo in rats with chronic dietary iron overload is dependent on hepatic iron concentration, Trans. Assoc. Amer. Phys. 96:146 (1983).

8. B. R. Bacon, A. S. Tavill, G. M. Brittenham, C. H. Park and R. O. Recknagel, Hepatotoxicity of chronic iron overload: Role of lipid peroxidation, in: "Free Radicals in Liver Injury", G. Poli, K. H. Cheeseman, M. U. Dianzani and T. F. Slater, eds., IRL Press, Oxford (1985).

9. B. R. Bacon, A. S. Tavill, G. M. Brittenham, C. H. Park and R. O. Recknagel, Hepatic lipid peroxidation in vivo in rats with chronic iron overload, J, Clin. Invest. 71:429 (1983).

10. B. R. Bacon, C. H. Park, G. M. Brittenham, R. O'Neil and A. S. Tavill, Hepatic mitochondrial oxidative metabolism in rats with chronic dietary iron overload, Hepatology 5:789 (1985).

11. T. Trenti, A. Masini, D. Ceccarelli-Stanzani, E. Rocchi and E. Ventura, Functional efficiency of mitochondrial membrane of rats with hepatic iron overload, Hepatology 4:777 (1984).

12. A. Masini, T. Trenti, E. Ventura, D. Ceccarelli-Stanzani and U. Muscatello, Functional efficiency of mitochondrial membrane of rats with hepatic chronic iron overload, Biochem. Biophys. Res. Commun. 124:462 (1984).

13. A. Masini, D. Ceccarelli-Stanzani, T. Trenti and E. Ventura, Transmembrane potential of liver mitochondria from hexachlorobenzene and iron treated rats, Biochim. Biophys. Acta 802:253 (1984).

14. A. Tomasi, E. Albano, S. Banni, B. Botti, F. Corongiu, M. A. Dessi, A. Iannone, V. Vannini and M. U. Dianzani, Free-radical metabolism of carbon tetrachloride in rat liver mitochondria, Biochem. J. 246:313 (1987).

15. A. Masini, D. Ceccarelli-Stanzani and U. Muscatello, The effect of oligomycin in rat liver mitochondria respiring in state 4, FEBS Lett. 160:137 (1984).

16. F. P. Corongiu, M. Lai and A. Milia, Carbon tetrachloride, bromotrichloromethane and ethanol acute intoxication, Biochem. J. 212:625 (1983).

17. F. P. Corongiu, M. A. Dessi, S. Vargiolu, G. Poli, K. H. Cheeseman, M. U. Dianzani and T. F. Slater, Antioxidant activity of α-tocopherol against lipid peroxidation in rat liver microsomes, in: "Free Radicals in Liver Injury", G. Poli, K. H. Cheeseman, M. U. Dianzani and T. F. Slater, eds., IRL Press, Oxford (1985).

18. B. Chance and G. R. Williams, The respiratory chain and oxidative phosphorylation, Adv. Enzymol. 17:65 (1956).

19. H. R. Lotscher, K. H. Winterhalter, E. Carafoli and C. Richter, The energy state of mitochondria during the transport of Ca , Eur. J. Biochem. 110:211 (1980).

20. G. Poli, E. Gravela, E. Albano and M. U. Dianzani, Studies on fatty liver with isolated hepatocytes, Exp. Mol. Pathol. 30:116 (1979).

21. E. Albano, K. A. K. Lott, T. F. Slater, A. Stier, M. C. R. Symons and A. Tomasi, Spin trapping studies on the free radical products formed by metabolic activation of carbon tetrachloride in rat liver microsomal fractions isolated hepatocytes and in vivo in the rat, Biochem. J. 204:593 (1982).

22. H. L. Bonkowsky, J. F. Healey, P. R. Sinclair, J. F. Sinclair and J. S. Pomeroy, Iron and the liver. Acute and long term effect of iron loading on hepatic heme metabolism, Biochem. J. 196:57 (1981).

23. R. T. Parmley, M. E. May, S. S. Spier, M. G. Buse and C. J. Alvarez, Ultrastructural distribution of inorganic iron in normal and iron loaded hepatic cell, Lab. Invest. 44:475 (1981).

24. A. Masini, T. Trenti, D. Ceccarelli and U. Muscatello, Mitochondrial involvement in causing cell injury in experimental hepatic iron overload, Ann. NY Acad. Sci. 488:517 (1986).

25. A. Masini, T. Trenti, E. Ventura, D. Ceccarelli and U. Muscatello, The effect of ferric iron complex on Ca^{2+} transport in isolated rat liver mitochondria, Biochem. Biophys. Res. Commun. 130:207 (1985).

26. A. Masini, T. Trenti, D. Ceccarelli-Stanzani and E Ventura, The effect of ferric iron complex on isolated rat liver mitochondria. I. Respiratory and electrochemical responses, Biochim. Biophys. Acta 810:20 (1985).

27. A. Masini T. Trenti, D. Ceccarelli-Stanzani and E. Ventura, The effect of ferric iron complex on isolated rat liver mitochondria. II. Ion movements, Biochim. Biophys. Acta 810:27 (1985).

28. A. Masini, T. Trenti, D. Ceccarelli and U. Muscatello, The effect of ferric iron complex on isolated rat liver mitochondria. III. Mechanistic aspects of iron induced calcium efflux, Biochim. Biophys. Acta 891:150 (1987).

LIPID PEROXIDATION, PROTEIN THIOLS, CALCIUM HOMEOSTASIS AND IMBALANCE OF

ANTIOXIDANT SYSTEMS IN BROMOBENZENE INDUCED LIVER DAMAGE

Alessandro F. Casini, Emilia Maellaro, Alfonso Pompella, Marco
Ferrali and Mario Comporti

Istituto di Patologia Generale, Università di Siena
Via del Laterino 8, 53100 Siena, Italy

INTRODUCTION

A line of research from our laboratory has been concerned, for several years, with the pathogenetic mechanisms of the liver damage produced by glutathione (GSH)-depleting agents. The present report deals with the main results obtained in in vivo studies with the use of the prototype aryl halide, bromobenzene.

The major route of bromobenzene metabolism is the formation of 3,4-bromobenzene epoxide[1,2]. The latter can covalently bind to cellular macromolecules, but can also be converted to chemically inert metabolites. The epoxide in fact (I) can be rearranged to form p-bromophenol[3], (II) can be converted to a dihydrodiol (3,4-bromophenyldihydrodiol) by an epoxide hydrase[2] or (III) can react with GSH to give a glutathionyl conjugate[4]. The latter reaction, which can occur even spontaneously in the cell environment, is largely accelerated by the catalysis of glutathione-transferases. Therefore, the conjugation with GSH represents the main pathway of bromobenzene metabolism and the conjugate is ultimately excreted in urine as a mercapturic acid[2,4].

Liver Glutathione Depletion Induced by Bromobenzene and Its Relation to Lipid Peroxidation and Necrosis

After bromobenzene administration to mice starved overnight (or fed a liquid glucose diet for two days in advance; such dietary regimens decrease the hepatic GSH stores), we have observed[5-7] the following situation: the hepatic GSH level is dramatically decreased (down to 30-15% of control values) at 3 h already and it decreases further at later times. Between 9 and 14 h liver necrosis (as assessed by the serum transaminase levels) appears in about 35% of the animals. The frequency of liver necrosis increases up to about 60% of the intoxicated animals at 18-20 h. Lipid peroxidation, as assessed by various procedures, shows a quite similar behavior. As shown in Fig. 1, when the individual values obtained at 14 h for serum transaminases or for lipid peroxidation (measured as the amount of carbonyl functions in liver phospholipids[8] in Fig. 1) are plotted against the corresponding hepatic GSH levels, it appears evident that both liver necrosis and lipid peroxidation occur only when the hepatic GSH depletion has reached critical values.

The treatment of the animals, after the intoxication, with the effective

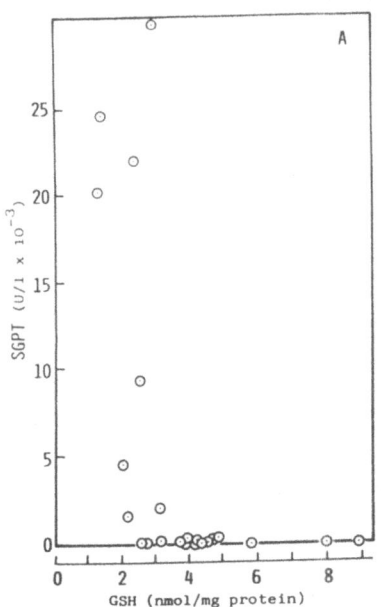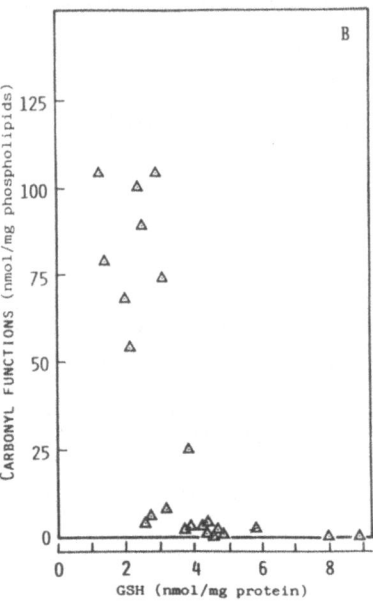

Fig. 1. A: Plot of serum transaminase (serum glutamate-pyruvate transaminase) levels against the corresponding hepatic glutathione (GSH)
contents in mice intoxicated with bromobenzene (13 mmoles/Kg body
wt. by mouth). B: Plot of lipid peroxidation values (amount of
carbonyl functions in liver phospholipids) against the corresponding
hepatic GSH contents in mice intoxicated with bromobenzene. The
values were obtained 14 h after poisoning. (Modified from Casini et
al.[5], with permission).

antioxidant Trolox C, a lower homolog of vitamin E, completely prevents both
lipid peroxidation and liver necrosis, while not changing at all the extent
of the covalent binding of bromobenzene metabolites to liver protein[5]. This
result therefore suggests that lipid peroxidation, rather than the covalent
binding, is the main pathogenetic mechansim responsible for the bromobenzene-
induced liver necrosis.

Relationships Between Lipid Peroxidation, Protein Thiols and Calcium Homeo-
stasis in Bromobenzene-induced Liver Damage

In recent years the concept has developed according to which a perturbation of calcium homeostasis is the crucial and irreversible event leading to
cell death in the liver damage produced by a number of toxins. As far as the
GSH-depleting agents are concerned, a great deal of experimental work[9-13] carried out in in vitro systems in which an oxidative stress (for instance that
caused by the redox cycling of menadione) is imposed to isolated hepatocytes,
suggests that the perturbation of calcium homeostasis is caused by alterations of protein sulphydryl groups. The latter alteration would be secondary
to GSH depletion[9-13]. We have therefore studied the relationships occurring
among GSH depletion, lipid peroxidation, loss of protein thiols, alteration
of calcium homeostasis and liver necrosis during the course of the in vivo
intoxication with bromobenzene.

In the experiment reported in Table 1 liver necrosis appears in some
animals at 9-12 h after the intoxication and increases subsequently in frequency and severity. Lipid peroxidation is already detectable, in some ani-

154

Table 1. Time-course of hepatic glutathione (GSH) depletion, liver damage (SGPT), lipid peroxidation (hepatic content of malonic dialdehyde, MDA) and decrease in protein thiol groups (protein -SH) after bromobenzene intoxication

	Time after intoxication				
	0 time	6 h	9 h	12 h	18 h
GSH (nmol/mg protein)	30.8 ± 2.1 (13)	2.5 ± 0.4 (4)	1.4 ± 0.2 (4)	1.2 ± 0.2 (7)	2.2 ± 0.3 (24)
SGPT (U/l)	38 ± 9 (14)	22 ± 4 (4)	151 ± 46 (4)	210 ± 93 (7)	2997 ± 879 (24)
MDA (pmol/mg protein)		8 ± 3 (4)	33 ± 6.6 (4)	82 ± 35 (7)	918 ± 282 (24)
Protein -SH (nmol/mg protein)	119.8 ± 3.9 (14)	123.5 ± 7.1 (4)	108.4 ± 3.5 (4)	103.7 ± 3.3 (7)	89.0 ± 3.3* (24)

Bromobenzene was given by gastric intubation at the dose of 13 mmol/kg body wt. Result are given as means ± S.E.M. *Significantly different from the 0 time value: P < 0.01. The number of the animals is reported in brackets. From Casini et al.[7], with permission.

mals, at 6-9 h and markedly increases afterwards showing in every instance a good statistical correlation with necrosis (r = 0.837; P < 0.001). Lipid peroxidation is given here as the malonic dialdehyde (MDA) content of the liver after it was checked that this assay correlates with other more sophisticated analyses to detect lipid peroxidation in vivo. Although the hepatic GSH depletion is nearly maximal at 6 h already, a significant decrease in protein sulphydryls can be seen at 12-18 h only. Such decrease in protein thiols seen at 12-18 h is correlated (r = -0.875; P < 0.001) with lipid peroxidation.

Results similar to those observed with the whole liver were also obtained with the isolated microsomal and mitochondrial fractions. As shown in Fig. 2, the decrease in protein thiols is correlated with lipid peroxidation in both microsomes and mitochondria.

In order to evaluate cellular calcium homeostasis, we measured the calcium sequestration activity of both the microsomal and the mitochondrial fraction, that is the capacity of liver mitochondria and microsomes to regulate the level of cytosolic calcium. As shown in Table 2, this activity is impaired in both fractions after bromobenzene intoxication. However, while the microsomal calcium pump decreases slowly after poisoning, the mitochondrial pump is significantly inhibited at 6 h already and the inhibition increases subsequently. The microsomal calcium pump is impaired in the animals in which extensive lipid peroxidation developed, but not in those in which lipid peroxidation was virtually absent. The plot (Fig. 3) of the individual values of the microsomal and mitochondrial calcium sequestration activities against the corresponding lipid peroxidation values shows that the inhibition of the calcium pumps is correlated with lipid peroxidation in both microsomes and mitochondria. A significant correlation was also found between the impairment of calcium pumps of microsomes and mitochondria and the corresponding losses of protein thiols (r = 0.894, P < 0.001 for microsomes; r = 0.861, P < 0.001 for mitochondria).

It seems therefore from the above results that lipid peroxidation is correlated with the decrease in protein thiols, the impairment of the micro-

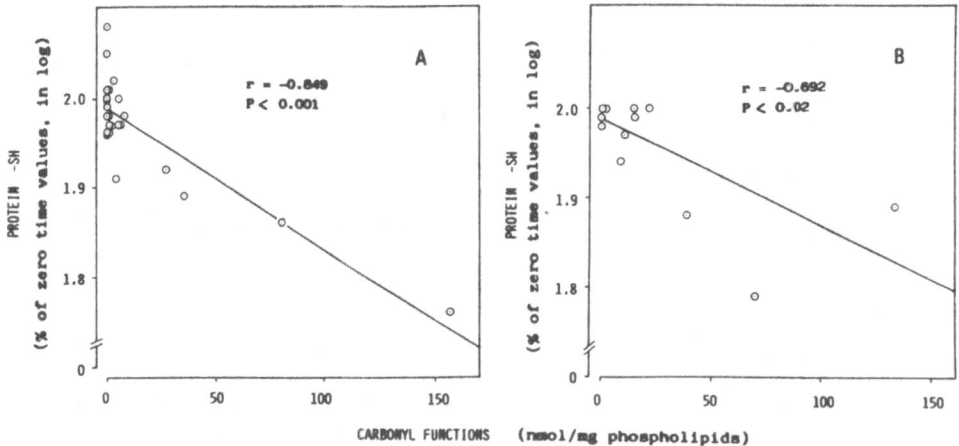

Fig. 2. Plot of the protein -SH groups against the corresponding lipid peroxidation values (amount of carbonyl functions in phospholipids) in both microsomes (panel A) and mitochondria (panel B) from liver of bromobenzene intoxicated mice. The values obtained 6-18 h after poisoning were used.

Table 2. Time-course of lipid peroxidation (carbonyl functions in phospholipids and calcium sequestration activity in liver microsomes and mitochondria after bromobenzene poisoning

Time after intoxication

		0 time	6 h	9 h	12 h	18 h
Microsomes	Carbonyl functions (nmol/mg phospholipids)		1.5 ± 1.5 [a] (4)	1.0 ± 0.9 [a] (4)	2.9 ± 1.4 (6)	31.4 ± 16.1 (10)
	Calcium uptake (nmol/mg protein/30 min)	122.5 ± 9.2 (11)		119.3 ± 7.7 (4)	123.8 ± 8.8 (6)	85.6 ± 23.4 (12)
Mitochondria	Carbonyl functions (nmol/mg phospholipids)		2.7 ± 2.6 [b] (4)			38.1 ± 15.6 (8)
	Calcium uptake (nmol/mg protein/5 min)	70.8 ± 0.6 (6)	38.7 ± 7.0* (4)			25.8 ± 7.8* (8)

Bromobenzene was given by gastric intubation at the dose of 13 mmol/Kg body wt. Results are given as means ± S.E.M. The number of the animals is reported in brackets. *Significantly different from 0 time value: $P < 0.05$. [a] Of 4 animals, only 1 (6 h) or 2 (9 h) showed detectable carbonyl functions (5.9 at 6 h; 0.5 and 3.6 nmol/mg phospholipids, at 9 h). [b] Of 4 animals, only 2 showed detectable carbonyl functions (0.5 and 10.4 nmol/mg phospholipids). From Casini et al.[7], with permission.

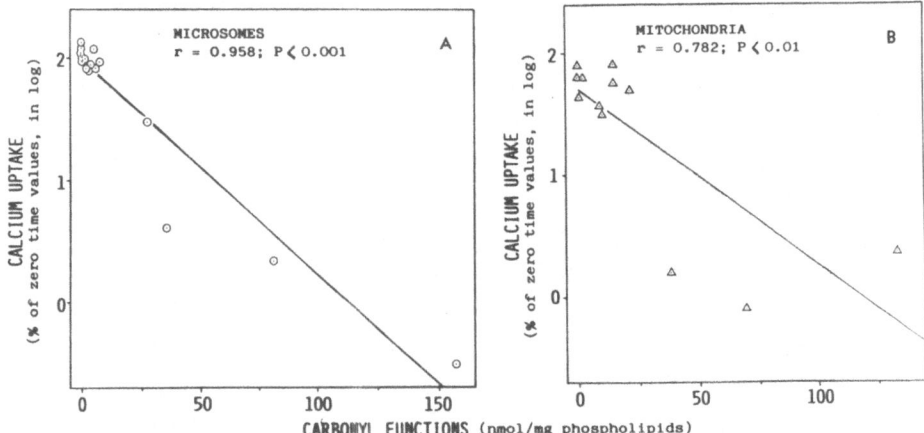

Fig. 3. Plot of the values for calcium uptake against the corresponding li-
         pid peroxidation values (amount of carbonyl functions in phospho-
         lipids) in both microsomes (panel A) and mitochondria (panel B)
         from livers of bromobenzene intoxicated mice. The values obtained
         6-18 h after poisoning were used. Calcium uptake was given as per
         cent values with respect to the zero time, i.e.; control values.
         The latter values are reported in Table 2. Bromobenzene intoxica-
         tion was performed as reported in Table 1. (From Casini et al.[7],
         with permission).

somal and mitochondrial calcium sequestration activity as well as with liver
cell death. In order to further evaluate the role of lipid peroxidation in
the bromobenzene-induced liver damage, experiments were carried out in which
desferrioxamine, a powerful iron chelator and inhibitor of lipid peroxidation, was administered to the animals after the intoxication. As shown in
Fig. 4, the treatment with desferrioxamine of the intoxicated animals prevents almost completely lipid peroxidation, liver necrosis and decrease in
protein thiols, while the GSH depletion was only minimally affected or not
affected at all, as in the experiment in which the animals were sacrificed
at 24 h (see below). In the experiment reported in Fig. 4 the animals were
killed 15 h after bromobenzene intoxication. Similar results were obtained
when the animals were killed 24 h after the intoxication, even if in some
occasional case high transaminase levels in the virtual absence of lipid
peroxidation were found. In these animals protein thiols were not decreased
as compared to controls. Despite the occurrence of such discrepancies, the
overall analysis of the data still showed a correlation (r = 0.490, P < 0.01)
between lipid peroxidation values and serum glutamate-pyruvate transaminase
(SGPT) levels.

On the basis of the present results we can conclude that in bromobenzene-induced liver injury a close correlation exists among lipid peroxidation, loss of protein -SH groups, impairment of the mechanism of regulation
of cellular calcium by mitochondria and microsomes and liver cell death. The
loss of protein -SH groups may well be an important factor in the bromobenzene-induced impairment of calcium sequestration activities of liver mitochondria and microsomes as suggested by Orrenius and Bellomo [9-13,15,16]. However, the decrease in protein thiols does not appear to be secondary to
hepatic GSH depletion, since first, the two phenomena show a completely different time course after the intoxication; and second, the desferrioxamine-
treatment of the intoxicated animals almost completely prevents the loss in
protein thiols, while not affecting the hepatic GSH depletion.

158

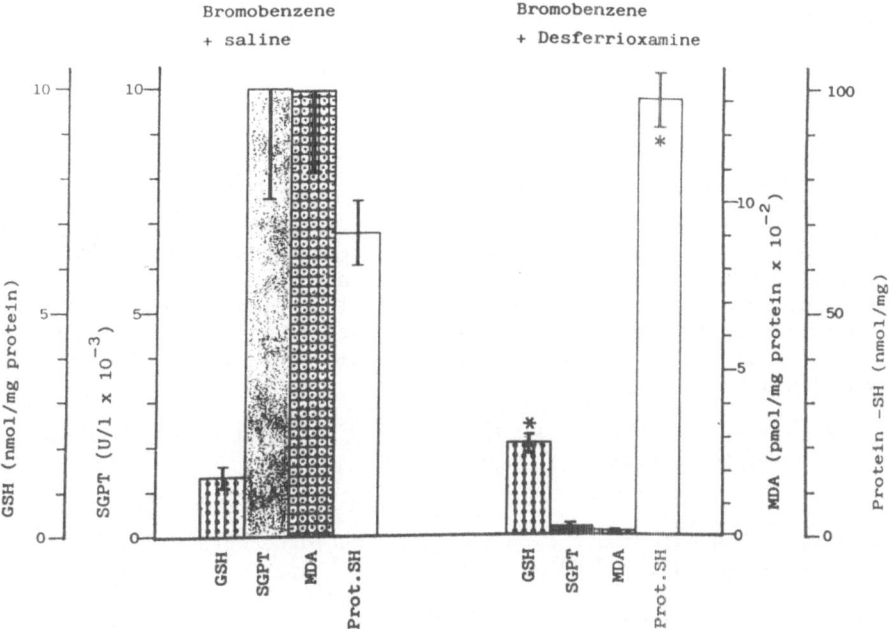

Fig. 4. Hepatic glutathione (GSH) depletion, liver damage (serum glutamate-
pyruvate transaminase, SGPT), lipid peroxidation (hepatic content
of malonic dialdehyde, MDA) and decrease in protein thiol groups
(protein -SH) in mice intoxicated with bromobenzene and given
either desferrioxamine or saline. The animals were sacrificed 15 h
after poisoning. Bromobenzene was given as reported in Table 1.
Desferrioxamine was given intraperitoneally, 7 and 13 h after the
intoxication, at the dose of 250 mg/Kg body wt. Results are given
as means ± S.E.M. *Significantly different from bromobenzene +
saline treated mice: P < 0.01. (From Casini et al.[7], with
permission).

Relationships Among the Various Antioxidant Systems of the Liver Cell in Bromobenzene Intoxication

In a recent study from our laboratory the relationships among the various antioxidant systems of the liver cell (namely vitamin E, glutathione and
ascorbic acid) were investigated under conditions of profound GSH depletion,
such as those produced by bromobenzene intoxication. It can be easily conceived that the abrupt loss of the antioxidant system represented by GSH affects the other defense systems against oxidative stress.

Vitamin E is the major antioxidant in the lipid environment of cellular
membranes. Because of the low ratio of tocopherol to polyunsaturated fatty
acids, a recycling system capable of reducing the oxidized forms of tocopherol has been postulated:

$$\left.\begin{array}{c} R^{\cdot} \\ RH \end{array}\right\} \quad \left\{\begin{array}{c} \alpha\text{-TOH} \\ \alpha\text{-TO}^{\cdot} \end{array}\right\} \quad \left\{\begin{array}{c} X^{\cdot} \\ XH \end{array}\right.$$

$$\left.\begin{array}{c} R^{\cdot} \\ RH \end{array}\right\} \quad \left\{\begin{array}{c} \alpha\text{-TOH} \\ \alpha\text{-TO}^{\cdot} \end{array}\right\} \quad \left\{\begin{array}{c} GS^{\cdot} \\ GSH \end{array}\right. \tag{1}$$

159

$$\begin{array}{cccc}
R^{\cdot} \downarrow & \left\{\begin{array}{c}\alpha\text{-TOH}\uparrow\\ \downarrow\alpha\text{-TO}^{\cdot}\end{array}\right\} & \left\{\begin{array}{c}\uparrow AA^{\cdot}\\ AA\downarrow\end{array}\right. & \left\{\begin{array}{c}AA^{\cdot}\\ \downarrow DHAA\end{array}\right. \\
RH \downarrow & & &
\end{array} \qquad (2)$$

AA = ascorbic acid
AA = semidehydroascorbic acid radical
DHAA = dehydroascorbic acid

According to Reddy et al.[17] and McCay et al.[18] this reducing system consists of a specific membrane enzyme, namely tocopheroxy radical reductase,
which utilizes GSH as a source of reducing equivalents (reaction 1). On the
other hand, according to Packer et al.[19], Willson[20] and Scarpa et al.[21], the
tocopherol-regenerating system consists of ascorbic acid that is converted
in the reaction to the semidehydroascorbic acid radical and then to dehydroascorbic acid (reaction 2).

The oxidized forms of ascorbic acid are converted back to the reduced
form by enzymatic systems such as the:

NADH-semidehydroascorbate reductase (semidehydroascorbic acid +
  NADH $\leftrightarrow$ ascorbate + NAD)

Dehydroascorbate reductase (dehydroascorbic acid + 2 GSH $\leftrightarrow$ ascorbic acid
  + GSSG)

Alternatively, GSH depletion may affect the other antioxidant systems,
if a lowered disposition of hydrogen peroxide by GSH peroxidase occurs.
Vitamin E may be consumed as a result of an increased formation of lipid per-
oxides in the membrane, and ascorbate may be involved by a direct interaction
with oxy radicals, especially with hydroxyl radicals.

The model of bromobenzene intoxication in vivo is characterized by a
long lag phase between the depletion of GSH and the onset of lipid peroxidation. In this situation the levels of the antioxidants can be measured well
before the appearance of lipid peroxidation.

Table 3 shows again an experiment in which the time-course of GSH depletion, liver necrosis and lipid peroxidation is followed after bromobenzene
intoxication. These data are similar to those already reported. The hepatic
GSH depletion is nearly maximal by 3 h already. The content of vitamin E remains unchanged during the first hours of the intoxication. A significant
decrease was observed at 9-12 h, before the development of lipid peroxidation. A further and now larger decrease occurs as lipid peroxidation markedly increases. Ascorbic acid is significantly increased during the first
hours after the intoxication, while a decrease can be seen at the latest time
(Table 3). The increase observed during the early times may be due to an increased synthesis as an adaptive response to increased consumption.

An interesting feature of this study is the variation of the level of
dehydroascorbic acid. Dehydroascorbic acid, in fact, markedly increases
during the first hours of the intoxication when a severe GSH depletion occurs
and no lipid peroxidation is seen (Table 3). The level of dehydroascorbic
acid then decreases even though remaining at higher values as compared to the
controls. The ascorbic/dehydroascorbic acid ratio shows the prevalence of
the oxidized over the reduced form throughout the intoxication period.

Two possible explanations can be proposed for the present results. On
the one hand, GSH may be involved in regenerating vitamin E through the
ascorbate-dehydroascorbate system. Alternatively, the GSH-dependent removal

Table 3. Hepatic glutathione (GSH) depletion, liver damage (SGPT), lipid peroxidation (hepatic content of malonic dialdehyde, MDA) and liver content of vitamin E (Vit. E), ascorbic acid and dehydroascorbic acid after bromobenzene intoxication

	Time after intoxication					
	0 time	3 h	9 h	12 h	15 h	18 h
GSH (nmol/mg protein)	25.1 ± 0.8 (47)	3.7 ± 0.2 (7)	2.1 ± 0.1 (6)	2.2 ± 0.2 (15)	2.4 ± 0.3 (14)	1.9 ± 0.3 (28)
SGPT (U/1)	58 ± 7 (49)	48 ± 30 (3)	35 ± 6 (6)	63 ± 15 (12)	2578 ± 1389 (14)	4669 ± 1545 (28)
MDA (pmol/mg protein)		0 (3)	0 (6)	3 ± 2 (15)	189 ± 85 (14)	1097 ± 406 (11)
Vit. E (pmol/mg protein)	124 ± 8 (40)	149 ± 35 (7)	108 ± 2 (6)	86 ± 22* (11)	67 ± 16* (14)	45 ± 16* (11)
Ascorbic acid (nmol/mg protein)	7.8 ± 0.3 (25)	11.7 ± 0.2* (7)		8.5 ± 0.7 (11)	6.6 ± 1.2 (14)	4.0 ± 0.8* (16)
Dehydroascorbic acid (nmol/mg protein)	0.37 ± 0.07 (19)	1.02 ± 0.21* (3)		1.55 ± 0.24* (10)	1.32 ± 0.16* (14)	0.59 ± 0.38 (5)
Ascorbic acid / Dehydroascorbic acid	21.0	11.5		5.5	5.0	6.8

Bromobenzene was given as reported in Table 1. Results are given as means ± S.E.M. *Significantly different from the 0 time value: $P < 0.05$, $P < 0.005$, $P < 0.001$. The number of the animals is reported in brackets.

of hydrogen peroxide, produced under normal metabolic conditions, may prevent oxy radical-induced changes in the balance between the reduced and oxidized forms of ascorbic acid and vitamin E. Thus, the increase in the oxidized form of ascorbic acid, that is dehydroascorbic acid, can be regarded as an index of oxidative stress occurring with GSH depletion prior to the development of lipid peroxidation.

ACKNOWLEDGMENTS

This work was supported by a grant from the Association for International Cancer Research (Great Britain). Additional funds were derived by a grant (no. 86.00370.44) from CNR (Italy) Special Project "Oncology", and from CNR Group of Gastroenterology.

REFERENCES

1. A. W. Azouz, D. V Parke and R. T. Williams, Studies in detoxification. 51. The determination of catechols in urine, and the formation of catechols in rabbits receiving halogenibenzenes and other compounds. Dihydroxylation in vivo, Biochem. J. 55:146 (1953).
2. J. W. Daly, D. M. Jerina and B. Witkop, Arene oxides and the NIH shift. The metabolism, toxicity and carcinogenicity of aromatic compounds, Experientia 278:1129 (1972).
3. D. Jerina, J. W. Daly, B. Witkop, P. Zaltzman-Nirenberg and S. Udenfriend, The role of the arene-oxepin systems in the metabolism of aromatic substances. III. Formation of 1,2-naphthalene oxide from naphthalene by liver microsomes, J. Am. Chem. Soc. 90:6525 (1968).
4. E. Boyland and L. F. Chasseaud, The role of glutathione and glutathione S-transferases in mercapturic acid biosynthesis, Adv. Enzymol. 32:173 (1969).
5. A. F. Casini, A. Pompella and M. Comporti, Liver glutathione depletion induced by bromobenzene, iodobenzene, and diethylmaleate poisoning and its relation to lipid peroxidation and necrosis, Am. J. Pathol. 118:225 (1985).
6. A. F. Casini, M. Ferrali, A. Pompella, E. Maellaro and M. Comporti, Lipid peroxidation and cellular damage in extrahepatic tissues of bromobenzene-intoxicated mice, Am. J. Pathol. 123:520 (1986).
7. A. F. Casini, E. Maellaro, A. Pompella, M. Ferrali and M. Comporti, Lipid peroxidation, protein thiols and calcium homeostasis in bromo-benzene-induced liver damage, Biochem. Pharmacol. 36:3689 (1987).
8. A. Benedetti, R. Fulceri, M. Ferrali, L. Ciccoli, H. Esterbauer and M. Comporti, Detection of carbonyl functions in phospholipids of liver microsomes in CCl_4- and $BrCCl_3$-poisoned rats, Biochim. Biophys. Acta 712:628 (1982).
9. G. Bellomo and S. Orrenius, Altered thiol and calcium homeostasis in oxidative hepatocellular injury, Hepatology 5:876 (1985).
10. D. Di Monte, D. Ross, G. Bellomo, L. Eklow and S. Orrenius, Alterations in intracellular thiol homeostasis during the metabolism of menadione by isolated rat hepatocytes, Arch. Biochem. Biophys. 235:334 (1984).
11. D. Di Monte, G. Bellomo, H. Thor, P. Nicotera and S. Orrenius, Menadione-induced cytotoxicity is associated with protein thiol oxidation and alteration in intracellular Ca^{2+} homeostasis, Arch. Biochem. Biophys. 235:343 (1984).
12. M. Moore, H. Thor, G. Moore, S. Nelson, P. Moldéus and S. Orrenius, The toxicity of acetaminophen and N-acetyl-p-benzoquinone imine in iso-lated hepatocytes is associated with thiol depletion and increased cytosolic Ca^{2+}, J. Biol. Chem. 260:13035 (1985).

13. S. Orrenius and G. Bellomo, Toxicological implications of perturbation of Ca^{2+} homeostasis in hepatocytes, in: "Calcium and Cell Function", vol. VI, W. Y. Cheung, ed., Academic Press, New York (1986).
14. A. Pompella, E. Maellaro, A. F. Casini, M. Ferrali, L. Ciccoli and M. Comporti, Measurement of lipid peroxidation in vivo: a comparison of different procedures, Lipids 22:206 (1987).
15. S. A. Jewell, G. Bellomo, H. Thor, S. Orrenius and M. T. Smith, Bleb formation in hepatocytes during drug metabolism is caused by disturbances in thiol and calcium ion homeostasis, Science 217:1257 (1982).
16. G. Bellomo, S. A. Jewell, M. T. Smith, H. Thore and S. Orrenius, Perturbation of Ca^{2+} homeostasis during hepatocyte injury, in: "Mechanisms of Hepatocyte Injury and Death", D. Keppler, H. Popper, L. Bianchi and W. Reutter, eds., MTP Press, Lancester, U.K. (1984).
17. C. C. Reddy, R. W. Scholz, C. E. Thomas and E. J. Massaro, Evidence for a possible protein-dependent regeneration of vitamin E in rat liver microsomes, Ann. N.Y. Acad. Sci. 393:193 (1982).
18. P. B. McCay, E. K. Lai, S. R. Powell and G. Breuggemann, Vitamin E functions as an electron shuttle for glutathione-dependent "free radical reductase" activity in biological membranes, Fed. Proc. 45:451 (1986).
19. J. E. Packer, T. F. Slater and R. L. Willson, Direct observation of a free radical interaction between vitamin E and vitamin C, Nature (Lond.) 278:737 (1979).
20. R. L. Willson, Free radical protection: why vitamin E, not vitamin C, ß-carotene or glutathione?, in Ciba Foundation Symposium 101: Biology of Vitamin E, Pitman Press, London (1983).
21. M. Scarpa, A. Rigo, M. Maiorino, F. Ursini and C. Gregolin, Formation of α-tocopherol radical and recycling of α-tocopherol by ascorbate during peroxidation of phosphatidylcholine liposomes. An electron paramagnetic resonance study, Biochim. Biophys. Acta. 801:215 (1984).

SECTION II

DEVELOPMENTAL STAGES OF CARCINOGENESIS

CELL PROLIFERATION AND CELL LOSS IN PROGRESSION IN LIVER CARCINOGENESIS:

A NEW HYPOTHESIS

Emmanuel Farber, Joel Rotstein, Leonard Harris, George Lee, and Zhi-Ying Chen

Departments of Pathology and of Biochemistry, Medical Sciences Building, University of Toronto, Toronto, Ontario Canada, M5S 1A8

INTRODUCTION

It is becoming well established that the development of hepatocellular carcinoma in the liver of the rat occurs through a number of steps (8 to 10 or more), some of which involve changes in a rare cell and others the progressive modulation of the phenotype in new hepatocyte populations[1]. Two major sequences appear in the majority of the models of liver carcinogenesis[1,2]. Sequence A, consisting of initiation and promotion, generates benign focal proliferations of hepatocytes ("clonal nodules"), a very small proportion (1-5%) of which persist[2]. Sequence B, consisting of progression, is the multistep process whereby the first precancerous lesion, the persistent hepatocyte nodule, slowly undergoes a series of changes leading to cancer. This sequence also includes the changes that occur in a bona fide cancer leading to increasing growth, invasion and metastasis[2]. Clearly, sequence B is the most relevant and most critical in the development of cancer. It can occur "spontaneously" and needs no external manipulation and is often not influenced in a major way by promoting environments.

The phenotype of the persistent nodule, the first true precancerous lesion, is very similar to that of the early nodules, the majority of which remodel. This similarity pertains to biochemical and enzyme patterns, architecture, cytological appearance and blood supply[3]. However, three phenotypic differences have been found so far[4-8] and these seem to be of some importance. These are: (a) cell proliferation without any apparent exogenous stimulus, involving at any moment in time from 4 to 8% of the cells, (b) cell loss, presumably by single cell necrosis with phagocytosis (so-called apoptosis), increasing as the level of cell proliferation increases, and (c) ability to grow as a nodule on transplantation into the spleen, with the ultimate development of hepatocellular carcinoma[6]. Nodules do not grow when transplanted to the liver, subcutaneous tissue or several other sites in syngeneic animals, other than the spleen. The cell proliferation, cell loss and transplantability to the spleen are shown by the cells of the persistent nodules but not by the cells of the liver surrounding the nodules.

The Balance between Cell Proliferation and Cell Loss

The persistent nodule is considered to be the first neoplastic or precancerous step, since seemingly autonomous growth is first seen in this cell population.

167

This cell population is interesting in that it retains some important aspects of normal control of cell proliferation and cell balance in the liver. It retains a clear-cut diurnal rhythm, responds to the strong mitogenic stimulus of partial hepatectomy as does the liver surrounding the nodules and the liver of normal control rats and maintains a balance between cell proliferation and cell loss. This balance results in only a very slow rate of growth of the nodules, even though the labeling index and the growth fraction may be as high as 8 to 10%[5]. This balance is maintained during the process of progression until cancer appears at which time the growth of the malignant neoplastic nodule population is much increased. Therefore, it is possible that a major factor in the evolution to cancer is the loss of this normal balance.

Thus, even though a major phenotypic property of cancer, seemingly autonomous growth, appears in the first persistent hepatocyte nodule, the balance between cell proliferation and cell loss is maintained until cancer appears much later. What is the basis for this balance?

Normal or Physiological Control of Liver Cell Mass or Number

The normal liver has a physiological control system whereby the cell number in the liver is kept more or less constant. When liver tissue is lost by operative removal or by cell death, the liver shows a regenerative response that compensates for the loss until the liver is returned pretty much to its original size (e.g. 9, 10). When liver responds to a mitogen by hyperplasia, again it compensates by losing cells so as to return to its original size[9,16]. Thus, the rat and presumably all other mammals and other animals have homeostatic mechanisms that control for the size of the liver or for the number of hepatocytes and other cells in the liver. The bases for these regulatory mechanisms are quite unknown.

Since this liver cellular homeostasis is active under a variety of experimental conditions including acute liver cell injury, is it operative in chronic liver disease such as liver carcinogenesis, and if so, when? Does the liver during hepatocarcinogenesis show compensatory regeneration following loss of liver and compensatory loss following primary hyperplasia of liver or liver cells?

New Hypothesis during Liver Carcinogenesis

As already indicated, the persistent nodule, at least in the resistant hepatocyte (RH) model, shows a new property - cell loss[5]. In this study, the degree of cell loss was measured and found to be slightly less than that of cell proliferation. Thus, the persistent nodule displays a balance between the level of cell proliferation and cell loss. For example, at 2 months post-initiation, the persistent nodules showed a growth fraction of about 4% and a necrotic index of about 3%; at 6 months, the corresponding values were 8% and 7% respectively.

This model for liver carcinogenesis, the RH model, is special in that the various cellular and tissue steps occur synchronously. Thus, it is theoretically possible to analyze the sequence of changes and to construct rational hypotheses.

In the RH model, the first clear-cut induction of a balance between cell proliferation and cell death is seen in the first persistent hepatocyte nodule at 2 months post-initiation[5]. No significant and reproducible level of cell death (so-called "apoptosis") was found during the earlier microscopic phase of nodule formation, i.e. in foci or islands[5]. Also, no significant level of cell death was observed in the majority of early nodules, those that remodel by differentiation[7,8].

The finding of a significant level of cell death in the first persistent nodule and not prior to this in either nodules or foci[5] seems to be in sharp contradiction to the quantitative observations of Schulte-Hermann and co-workers[15] and those of Columbano and colleagues with orotic acid[17]. In these two studies, single cell death with phagocytosis by liver cells (apoptosis) was found in microscopic foci of altered hepatocytes in rats initiated with nitrosomorpholine[15] or 1,2-dimethylhydrazine[17] and promoted with cyproterone acetate or phenobarbital[15] or orotic acid[17]. The degree of synchronization of lesion-development in these models is not known. With phenobarbital, the foci develop and expand asynchronously. The only measure of possible cell loss in these two studies was the number of apoptotic bodies per 100 hepatocytes. If it is assumed that the number of apoptotic bodies is a measure of or at least proportional to the degree of cell death, why are results with foci so different in the different models and why do persistent nodules in the RH model show a balanced degree of cell proliferation - cell death while precursor foci in the same model do not?

It is proposed that the differences between the cell dynamics in these different lesions is due to the manner in which they are derived. In the RH model, the stimulus for proliferation of the initiated hepatocytes is compensatory to cell loss. Initiation of liver carcinogenesis with many different carcinogens is associated with the appearance of a "resistance phenotype" in a rare hepatocyte[18,20]. The ultimate cancers arise from one or a few of such resistant hepatocytes[21]. Clonal expansion, promotion, is accomplished by providing a stimulus for cell proliferation by tissue removal (partial hepatectomy or cell death with CCl_4) while inhibiting the vast majority of the cells, the "uninitiated hepatocytes", from proliferating by brief exposure to 2-acetylaminofluorene. The rare (1 per 10^5-10^6) initiated hepatocyte, bearing the "resistance phenotype", can respond and does so by vigorous cell proliferation. Grossly visible hepatocyte nodules are readily generated within 10 days to 2 weeks. Growth of the nodules ceases when the liver has regained its former size.

The cessation of growth of the majority of nodules is rapidly followed by remodeling by redifferentiation to normal-looking adult liver[7,8]. These show an obviously preprogrammed remodeling to normal-looking liver. This normal-looking liver constitutes the major portion of the liver surrounding the later persistent nodules. In a few nodules, the remodeling by differentiation is either blocked or delayed and "autonomous" hepatocyte proliferation is somehow turned on[5].

Therefore, in the RH model, the clonal expansion to form nodules is a form of compensatory cell proliferation[1]. The few persistent hepatocyte nodules show a primary hyperplastic response above the normal control cell mass. Assuming that the basic homeostatic control for maintenance of liver cell mass is operating during hepatocarcinogenesis, it would be anticipated that the persistent hepatocyte nodule, being a primary hyperplastic lesion and not a compensatory one, should show cell loss as a response to cell proliferation. Thus, according to this hypothesis, cell loss would be an expected consequence of cell proliferation in the population of persistent hepatocyte nodules.

All of the focal lesions, precursor (foci or islands and early nodules) to the persistent hepatocyte nodule, being compensatory and not primary, would not be expected to show any significant level of cell loss as a balancing phenomenon for the compensatory cell proliferation.

Therefore, in the RH model, the absence of cell loss in foci and nodules and its presence in the persistent nodules are to be expected according to the hypothesis proposed.

These results are in striking contrast to those with some other models such as those using cyproterone acetate, phenobarbital or orotic acid as promoters. In these models, the slow asynchronous expansion by cell proliferation to form foci and ultimately hepatocyte nodules is due to a primary hyperplasia of unknown mechanism and not to compensatory cell proliferation, significant liver cell loss or cell death does not seem to be present during the long period of promotion. Regardless of mechanism, the presence of extra hepatocytes in the foci derived by a primary hyperplasia, should set in motion a compensatory cell loss that might well balance out or slow down the expansile growth of the foci [23]. Thus, the known very slow growth of foci (and nodules) in the models of liver carcinogenesis that utilize primary hyperplasia as a promoting environment would be quite predictable according to this unifying hypothesis.

New Hypothesis in Perspective

Viewed from this perspective, the persistent nodules retain the normal physiological homeostatic balance between cell proliferation and cell loss until, for reasons unknown, a disturbance in this balance supervenes with the first step in malignant behavior of the nodules. This could very well be closely related to a new hepatocyte population, creating a "nodule in a nodule" [24] in which the new few or rare cells that constitute the nodule in the persistent nodule have lost the ability to exercise the homeostatic balance between cell proliferation and cell loss.

Also, if the persistent hepatocyte nodule retains this basic physiological adaptive mechanism, it is obviously confined to the nodule and does not involve the whole liver. The liver cells surrounding the nodule do not show any large increase in cell proliferation or in cell loss, even though the few nodules within the same liver do. Thus, whatever mechanism is operating is doing so at a restricted geographic locus, the nodule. Conceivably, this could be a reflection of a fundamental control at the cell-to-cell level of organization.

Thus, if this hypothesis is shown to be viable and not destroyed by future critical experiments, it generates new perspectives on the carcinogenic process in the liver. From this viewpoint, at least two steps in the carcinogenic process become crucial and critical: (a) the acquisition of seemingly "autonomous" cell proliferation by hepatocytes in the persistent nodule, and (b) the nature of the balance, if any, between liver cell proliferation and liver cell loss in the earliest steps after a nodule becomes frankly malignant. Fundamental, then to (a) is whether the trigger for cell proliferation is a mechanism secondary to differentiation or vice versa. Is the lack or inhibition of differentiation of the persistent nodule merely a reflection of some as yet unknown trigger for cell proliferation? Fundamental to (b) becomes the elucidation of the physiological basis, followed by the biochemical and molecular bases for cell proliferation - cell loss homeostatic balance. This is a challenge that seems to be a high priority and urgent need at this time if we are to understand the fundamental mechanisms operating during the development of liver cancer with chemicals.

AKNOWLEDGMENTS

This work was supported by grants from the Medical Research Council of Canada (MT-5994), the National Cancer Institute of Canada and Health and Welfare, Canada.

REFERENCES

1. E. Farber and D. S. R. Sarma, Biology of Disease: Hepatocarcinogenesis: a dynamic cellular perspective, Lab. Invest. 56:4 (1987).

2. E. Farber, Some emerging general principles in the pathogenesis of hepatocellular carcinoma, Cancer Surveys 5:695 (1986).

3. E. Farber, Cellular biochemistry of the stepwise development of cancer with chemicals: G.H.A. Clowes Memorial Lecture, Cancer Res. 44:5463 (1984).

4. J. Rotstein, P. D. M. Macdonald, H. M. Rabes and E. Farber, Cell cycle kinetics of rat hepatocytes in early putative preneoplastic lesions in hepatocarcinogenesis, Cancer Res. 44:2913 (1984).

5. J. Rotstein, D. S. R. Sarma and E. Farber, Sequential alterations in growth control and cell dynamics of rat hepatocytes in early pre-cancerous steps in hepatocarcinogenesis, Cancer Res. 46:2377 (1986).

6. M. Tatematsu, G. Lee, M. A. Hayes and E. Farber, Progression of hepa-tocarcinogenesis: differences in growth and behaviour of transplants of early and late hepatocyte nodules in the spleen, Cancer Res. 47: 4699 (1987).

7. K. Enomoto and E. Farber, Kinetics of phenotypic maturation of re-modeling of hyperplastic nodules during liver carcinogenesis, Cancer Res. 42: 2330 (1982).

8. M. Tatematsu, Y. Nagamine and E. Farber, Redifferentiation as a basis for remodeling of carcinogen-induced hepatocyte nodules to normal-appearing liver, Cancer Res. 43:5049 (1983).

9. R. Schulte-Hermann, Adaptive liver growth induced by xenobiotic com-pounds and other stimuli, Crit. Rev. Toxicol. 3:97 (1974).

10. R. Schulte-Hermann, Reactions of the liver to injury: adaptation, in: "Toxic Injury of the Liver", E. Farber and M. M. Fisher, eds., Marcel Dekker, New York (1979).

11. N. Bohm and B. Moser, Reversible hyperplasia and hypertrophy of the mouse liver induced by a functional charge with phenobarbital, Beitr. Pathol. 157:283 (1976).

12. W. G. Levine, M. G. Ord and L. A. Stocken, Some biochemical changes associated with nafenopin-induced liver growth in the rat, Biochem. Pharmacol. 26:939 (1977).

13. R. Schulte-Hermann, V. Hoffman, W. Parzefall, M. Kallenbach, A. Gerhard and J. Schuppler, Adaptive responses of rat liver to the gestagen and anti-androgen cyproterone acetate and other inducers, II, Induc-tion of growth, Chem. Biol. Interact. 31:287 (1980).

14. A. Columbano, G. M. Ledda, P. Sirigu, T. Perra and P. Pani, Liver cell proliferation induced by a single dose of lead nitrate, Am. J. Pathol. 110:83 (1983).

15. W. Bursch, B. Lauer, I. Timmermann-Trosiener, G. Barthel, J. Schuppler and R. Schulte-Hermann, Controlled death (apoptosis) of normal and putative preneoplastic cells in rat liver following withdrawal of tumor promoters, Carcinogenesis 5:453 (1984).

16. A. Columbano, G. M. Ledda-Columbano, P. P. Coni, G. Faa, C. Liguori, G. Santa Cruz and P. Pani, Occurrence of cell death (apoptosis) during the involution of liver hyperplasia, Lab. Invest. 52:670 (1985).

17. A. Columbano, G. M. Ledda-Columbano, P. M. Rao, S. Rajalakshmi and D. S. R. Sarma, Occurrence of cell death (apoptosis) in preneo-plastic and neoplastic liver cells, Am. J. Pathol. 116:441 (1984).

18. D. Solt and E. Farber, New principle for the analysis of chemical car-cinogenesis, Nature (Lond) 268:702 (1976).

19. H. Tsuda, G. Lee and E. Farber, Induction of resistant hepatocytes as a new principle for the possible short-term in vivo test for car-cinogens, Cancer Res. 40:1157 (1980).

20. E. Farber, Resistance phenotype in the initiation and promotion of chemical hepatocarcinogenesis, Chemica Scripta 27A:131 (1987).

21. D. B. Solt, A. Medline, and E. Farber, Rapid emergence of carcinogen-induced hyperplastic lesions in a new model for the sequential ana-lysis of liver carcinogenesis, Am. J. Pathol. 88:595 (1977).

22. A. Columbano, G. M. Ledda-Columbano, G. Lee, S. Rajalakshmi, and D. S. R. Sarma, Inability of mitogen-induced liver hyperplasia to support

the induction of enzyme-altered islands induced by liver carcinogens <u>Cancer</u> <u>Res.</u> 47:5557 (1987).

23. R. Schulte-Hermann, I. Timmermann-Trosiener, and J. Schuppler, Response of liver foci in rats to hepatic tumor promoters, <u>Toxicol.</u> <u>Pathol.</u> 10:63 (1982).

24. H. Popper, S. S. Sternberg, B. C. Oser, and M. Oser, The carcinogenic effect of aramite in rats. A study of hepatic nodules, <u>Cancer(Phila)</u> 13:1035 (1960).

HUMAN PAPILLOMAVIRUS 16 DNA AND EPITHELIAL CELL IMMORTALIZATION

Joseph A. DiPaolo

Laboratory of Biology, Division of Cancer Etiology
National Cancer Institute, National Institutes of Health
Bethesda, Maryland 20892, USA

INTRODUCTION

Papillomaviruses have had a unique position in carcinogenesis studies. As early as 1907 Ciuffo[1] demonstrated that the infectious agent for the common wart persisted in filtered homogenates and thus was not due to a bacterial or protozean etiology. These findings raised the possibility that warts were caused by newly recognized submicroscopic entities that are now referred to as viruses. Shope and Hung[2] later isolated and characterized the first oncogenic DNA virus, referred to as cottontail rabbit papilloma. Subsequently, Rous and associates[3,4] demonstrated that when Shope papilloma virus was injected in rabbits, progression from papillomas to carcinomas would be accelerated by the application of either coal tar or a well characterized carcinogenic polycyclic hydrocarbon, 3-methylcholanthrene. This rabbit-virus model also responded to X-irradiation[5] which was shown to contribute to progression of papillomas to carcinomas. A decade later, it was realized that X-irradiation of juvenile laryngeal papilloma (HPV 6/11) resulted in carcinomas[6]. Today, treatment is by surgery or laser. Human papilloma viruses (HPV) are strongly associated with anogenital cancer, including cervical cancer. HPV's have been associated worldwide with 15% of the human cancers, most notably cancer of the uterine cervix, penis, vulva and anal region. HPV type 16, 18, 31, 33 and 39 have been identified in cervical cancer; over 90% of these neoplasias can be classified as positive. Current epidemiological evidence suggests that incidence of cervical cancer in women is greater in smokers than in non-smokers[7].

Originally, the papilloma viruses, along with polyoma, Simian vaccuolating virus 40, human BK and JC, were classified as papova viruses because all possess a closed, circular, double-stranded DNA genome complexed with histones and they are incapsulated in an isohedral virion[8]. Today this taxonomic association is considered incorrect. Papilloma viruses constitute a distinct group of viruses. Whereas the SV40 and polyoma virus group have capsids of 45 nm, those of the papilloma viruses have a diameter of 55 nm. This makes it possible to accommodate a 50% larger DNA, 7900 bp compared to 5250 bp for papillomas and polyoma, respectively. Analysis of DNA demonstrates that the sequences of DNA in their genetic organization of the open reading frame differ and that the patterns of RNA synthesis also differ. For example, SV40 and polyoma read from both DNA strands. In papilloma viruses, all the open reading frames are encoded and transcribed along only one of two DNA strands and the messages for early and late functioning require interspersed transcriptions and processing signals.

Most importantly, whereas SV40 and polyoma naturally infect a number of different organs and neural tissues and can be grown in vitro, most papillomas have a single host and multiply in mucosal or differentiated cutaneous epithelium at specific anatomical sites. The papilloma viruses have not been grown in vitro. They appear to require differentiated keratinocytes of epithelial for vegetative replication and remain undetectable in the basal layer. It is presumed but not proven that the replicative cycle of the papilloma virus is synchronized to epithelial cell differentiation. As the cell proceeds to differentiate from the basal to the spinous and granular stages, episomal DNA replication occurs in the cell nucleus. Only differentiated cells contain complete virus with protein coat and thus the complete virus is found in the stratum corneum.

Therefore, it is appropriate and relevant to transfect HPV DNA into human keratinocytes to determine their role in carcinogenesis. Recombinant HPV 16 DNA can immortalize human epithelial cells[9-11]. However, whereas transfection of 3T3 cells produce tumors in nude mice[12], no tumor has resulted from injection of immortalized keratinocytes[11].

MATERIALS AND METHODS

The epithelial cells used were derived from neonatal foreskins and cultured in a serum-free medium which favors the growth of keratinocytes and suppresses that of fibroblasts. The foreskin was obtained from healthy individuals as a result of circumcision. The epithelial cells were isolated from the foreskin by a collagenase float technique. Cells from individual foreskins were cultured in MCDB153-LB (unpublished results) which is a modification of the medium MCDB153 originally described by Boyce and Ham[13]. The new medium consists of a basal medium enriched by additional amino acids supplemented by hormones and a reduction of osmolarity. The primary cell divide approximately every 24 h and the untreated cells will multiply 40-50 times before undergoing senescence. Repeated utilization of this procedure alone has not resulted in either an extended lifespan or spontaneous transformation. Furthermore, the cells maintained their diploid karyotype.

Because papilloma viruses have not been grown in vitro, a recombinant plasmid (pMHPV16d) which has the neomycin antibiotic resistant gene for selection in mammalian cells and two head to tail HPV 16 copies cloned at the Bam H1 site was used (Fig. 1). The double HPV 16 construction insured that the L1 gene would be intact. This recombinant HPV 16 was transfected into epithelial cells using a modified calcium phosphate technique. The latter was necessary because the low calcium concentration and high phosphate content of the cell culture medium rendered it unsuitable for transfection of DNA. Therefore, for the transfection step, Dulbecco's medium was diluted 1 to 1 with water and used during the 4 h incubation with the plasmid. Neomycin-resistant colonies were selected with 100 micrograms of G418 per ml of medium. Southern and Northern analyses were performed using standard procedures[14]. All hybridizations were performed under stringent conditions using the intact 7.9-kpb BamH1 digested HPV 16 as a probe.

RESULTS AND DISCUSSION

Thus far 14 human keratinocyte lines have been established in serum-free medium after transfection of normal human keratinocytes strains with the recombinant HPV 16 dimer containing a neomycin-resistant gene for selection of colonies with HPV 16. Each line is originated from a different individual;

174

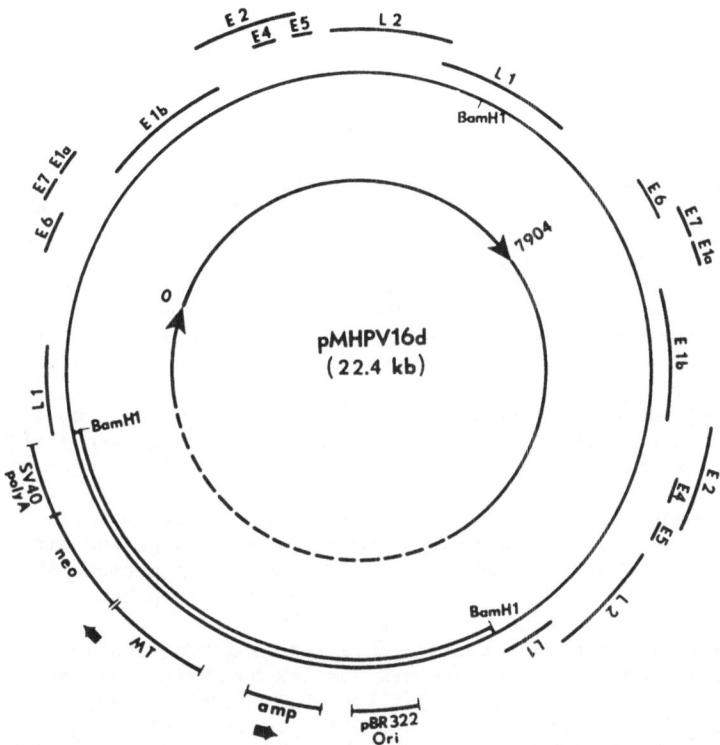

Fig. 1.   Structure and functional organization of recombinant HPV 16 DNA
pMHPV 16 dimer. The HPV 16 DNA was provided by L. Gissmann and H.
zur Hausen. Two, head-to-tail HPV 16 DNAs were inserted at the Bam
H1 site of pdMMTneo. O-7904 units indicate complete HPV 16 DNA.
Transcriptional orientation of mouse metallothionein (MT) promoter
and ampillicin (amp) gene shown by arrows. SV40, Simian virus 40;
neo, neomyein.

ordinarily between 10 and 20 G418-resistant colonies per microgram of plasmid
DNA were obtained. Non-transfected and vector transfected cells survived approximately 40-50 population doublings. The controls and vector treated
cells that undergo terminal differentiation are characterized by the formation of squames which tend to float away from the dish. The transfected HPV
16 keratinocytes maintain a morphology similar to that of undifferentiated
normal keratinocytes. These cell lines, which for practical purposes are
considered immortal, maintain a morphology similar to that of undifferentiated normal cells and continue to multiply at a rate similar to that of non-
treated controls and do not exhibit any growth crisis.

The immortality property of these cells can be uncoupled from differentiation. The addition of calcium greater than 0.3 mM or serum that is
calcium-free to the medium will induce differentiation of the cells. However
a subpopulation of transfected cells will continue to grow in the presence of
calcium or serum. Evidence for the origin and maintenance of epithelial properties of these cells is demonstrated by keratin analysis of control, transfected and serum selected cells. The non-treated controls exhibit a number
of keratins that are found in non-keratinizing squamous stratified epithelia.
With culturing and transfection, there is an increase of several keratins associated with culturing. Furthermore, after transfection as well as growth
in serum, there is an abundant amount of keratin 19 which is a major component of simple epithelia and is found not only in cultured keratinocytes but

also in squamous cell carcinomas. Thus, ample evidence indicates that the recipient cells as well as the transfected cells which become immortalized must be keratinocytes.

Chromosome analysis verified the human origin of these cells and provided evidence that the transfected keratinocytes undergo drastic chromosomal alterations. The non-treated cells are diploid and normal by G banding. After transfection and examination at the time after the controls have terminally differentiated, the chromosome pattern of the cells are anuploid, being either diploid, near diploid, or near triploid in chromosome number. The cell lines had a number of interesting characteristics often associated with gene amplification and/or cancers. The presence of homogenous staining regions and double minutes are definitive diagnosis for gene amplification. Each line had structural alterations characteristic for that specific line. The alterations might involve either translocation or deletion with some of the translocations being rather complex. The acquisition of continuous growth may be attribut- able to gene amplification and presence of HPV DNA in non-stained regions.

The presence of HPV 16 DNA in the transfected cells was verified by Southern blot analysis of DNA from transfected cells digested with Bam H1 and probed with the HPV 16 sequences. After Bam H1 digestion, the majority of the HPV 16 DNA in the cells was detected as a 7.9 Kbp band, indicating that most of the HPV 16 genome was intact. The pattern of additional bands, particularly at low populations, suggests rearrangement and integration. Furthermore, initially the pattern was complex but with additional culturing became simpler and stabilized, suggestive of a polyclonal population at low population doublings that became clonal with further culturing. Subsequent digestion with Eco RV, a non-cut enzyme for the plasmid, reduced the size of the initial bands thus providing additional evidence for integration. The DNA of a number of different cell lines was cut with either of two different non-cut enzymes for HPV 16d, Eco RV, or Xho 1 and probed with HPV 16 DNA. The HPV 16 DNA persisted as large molecular weight species associated with genomic DNA and not as monomeric episonal molecules. The different patterns produced by the non-cut enzymes again suggest integration of the HPV into the genomic DNA. HPV 16 messenger RNA expression in cell lines that have HPV 16 DNA integrated into the host genome was studied. Poly (A) + RNA extracted and purified from the cell lines was subjected to Northern blot analysis with the complete HPV 16 genome as a probe. All the cell lines had the same mRNA species, indicating that a number of messages were being produced.

Repeated tumorigenicity tests with cells grown in serum-free medium or serum-resistant cells that mimic carcinomas in morphology in culture have been negative in athymic nude mice. Immortality of human keratinocytes is associated with the integration of the HPV 16 genome. Integration and immortality represent important early events in the process leading to neoplastic transformation. Integration of human papilloma is associated primarily with carcinomas and occurs only rarely in the case of benign tumors.

Integration is also associated with such well known carcinomas as HeLa and SW756. The first demonstration of HPV 18 DNA integration near cellular proto-oncogenes and chromosomal fragile sites in HeLa cells and SW756 cells provides new insights of the role of viral integration in the development of cervical neoplasia [15-16]. Evidence is accumulating that in cervical cancer the critical integration site of human papilloma virus on human chromosomes is a non-random event because it is associated with fragile sites that may be either constitutive or heritable and also may be near the loci of known oncogenes. SW756 is a cervical carcinoma cell line which shares certain features with HeLa cells. Both cell lines have an abnormal chromosome constitution with complex rearrangements but a similar chromosome number; they contain 10 to 50 copies of HPV 18 DNA and have a similar HPV 18 DNA transcription pattern. A single HPV 18 integration site in SW756 cells was identified on chromosome 12 at band q13, integrated near Int-1 proto-oncogene in contrast to HeLa cells that have multiple copies of HPV 18 dispersed in four chromosomes. Three integration sites in HeLa cells on chromosomes 8, 9 and 22 correspond to the location of c-myc, c-abl and c-sis proto-oncogenes, respectively. The region 12q 13-22, which includes the HPV 18 integration site on SW756 cells, may carry important genes which are duplicated in certain lymphoproliferation disorders, particularly in chronic lymphocytic leukemia with trisomy 12.

When coupled with the correlation of the presence of certain papilloma viruses with genital dysplasias and with cancer plus the ability of HPV 16 to result in transformation of NIH 3T3 cells that express mRNAs associated with HPV (16) and which produce tumors in nude mice, the present results suggest that certain papilloma viruses do play a part in cancer formation. However, papilloma viruses are not sufficient by themselves to produce cancer. It is known that the presence of human papilloma 16 or 18 in women is not sufficient to assure the development of a high grade lesion into a carcinoma, because only a small percentage of such cases progress to cancer. This emphasizes that other etiological agents or cofactors must be involved to obtain uncontrolled growth and possible malignancy. Thus, the papilloma virus can be considered a necessary factor but is not by itself a sufficient cause of cervical cancer.

REFERENCES

1. G. Ciuffo, Innesto positivo con filtrato di verruca volgare, G. Ital. Mal. Venereol., 48:12 (1907).
2. R. E. Shope and E. W. Hurst, Infectious papillomatosis of rabbits; with a note on the histopathology, J. Exp. Med. 58:607 (1933).
3. P. Rous and W. F. Friedewald, The effect of chemical carcinogens on virus-induced rabbit carcinomas, J. Exp. Med. 79:511 (1944).
4. P. Rous and J. G. Kiss, The carcinogenic effect of a virus upon tarred skin, Science 83:468 (1936).
5. J. T. Syverton, R. A. Harvey, G. P. Berry and S. L. Warren, The Roentgen radiation of papilloma virus (Shope). I. The effect of x-rays upon papillomas on domestic rabbits, J. Exp. Med. 73:243 (1941).
6. P. H. Holinger and W. F. Rabbett, Late development of laryngeal and pharyngeal carcinoma in previously irradiated areas, Laryngoscope 63:105 (1953).
7. I. M. Sasson, N. J. Haley, E. L. Hoffman, D. Helljberg and S. Nilsson, Cigarette smoking and neoplasia of the uterine cervix: Smoke constituents in cervical mucus, N. Engl. J. Med. 312:315 (1985).
8. J. L. Melnick, Papova virus group, Science 135:1228 (1962).
9. J. A. DiPaolo, L. Pirisi, N. C. Popescu, S. Yasumoto and J. Doniger, Progressive changes induced in human and mouse cells by human papilloma-virus type 16 DNA, in: "Cancer Cells: Papillomaviruses", B. M. Steinberg, J. L. Brandsma, and L. B. Taichman, eds., Cold Spring Harbor Laboratory, New York (1987).
10. M. Durst, R. T. Dzarlieva-Petrusevska, P. Boukamp, N. E. Rusenig and L. Gissmann, Molecular and cytogenetic analysis of immortalized human primary keratinocytes obtained after transfection with human papilloma-virus type 16 DNA, Oncogene 1:251 (1987).
11. L. Pirisi, S. Yasumoto, J. Feller, J. Doniger and J. A. DiPaolo, Transformation of human fibroblasts and keratinocytes with human papilloma-virus type 16 DNA, J. Virol. 61:1061 (1987).
12. S. Yasumoto, J. Doniger and J. A. DiPaolo, Differential early viral gene expression in two stages of human papillomavirus type 16 DNA-induced malignant transformation, Mol. Cell. Biol. 7:2165 (1987).

13. S. T. Boyce and R. G. Ham, Calcium-regulated differentiation of normal human epidermal keratinocytes in chemically defined clonal culture and serum-free serial culture, J. Invest. Dermatol. 81 (suppl.):33 (1983).

14. T. Maniatis, E. F. Fritsch and J. Sambrook, "Molecular cloning: A laboratory manual", Cold Spring Harbor Laboratory, Cold Spring Harbor, New York (1982).

15. N. C. Popescu, J. A. DiPaolo and S. C. Amsbaugh, Integration sites of human papillomavirus 18 DNA sequences on HeLa cell chromosomes, Cytogenet. Cell Genet. 44:58 (1987).

16. N. E. Popescu, S. C. Amsbaugh and J. A. DiPaolo, Human papillomavirus 18 DNA is integrated at a single chromosome site in cervical carcinoma cell line SW756, J. Virol. 51:1682 (1987).

CELLULAR AND MOLECULAR STUDIES OF GROWTH, DIFFERENTIATION AND NEOPLASTIC

TRANSFORMATION OF HUMAN BRONCHIAL EPITHELIAL CELLS IN VITRO

Curtis C. Harris, Roger R. Reddel, Yang Ke, Andrea Pfeiffer,
George Mark, Tohru Masui, George Yoakum, Brenda I. Gerwin,
Paul Amstad and John F. Lechner

Laboratory of Human Carcinogenesis, Division of Cancer Etiology, National Cancer Institute, Bethesda, Maryland 20892

INTRODUCTION

Normal human cells in vitro appear to retain many normal phenotypic
properties, remain diploid, eventually undergo senescence and rarely, if ever
"spontaneously" transform to malignant cells. Retained properties may include synthesis of classes of proteins associated with specific cell types
such as collagens, keratins, or melanin; responsiveness to hormones; and
antigenic specificity. In addition, human cells with abnormal phenotypes
such as either enzymatic deficits or malignant properties frequently maintain
these phenotype in vitro. Human cells cultured in vitro have thus proven
to be extremely useful to scientists studying the molecular and biochemical
aspects of human carcinogenesis. Such studies have been facilitated by the
recent development of improved methods for culturing normal human epithelial
tissues and cells[1]. Chemically defined media have been developed for culturing many of these tissues and cells from normal organs, including those with
a high rate of cancer in humans. Serum-free media have several advantages in
studies of cultured human cells, including: (a) less experimental variability
compared to serum-containing media; (b) selective growth conditions for
normal cells of different types (e.g. epithelial versus fibroblastic) or for
normal versus malignant cells; (c) ease of identification of growth factors,
inhibitors of growth, and inducers of differentiation; and (d) ease of isolating and analyzing secreted cellular products. Advances in cell biology,
including the delineation of biochemical and morphological markers of specific cell types, have also facilitated the identification of cells in vitro
(including keratins as markers for epithelial cells and collagen types I and
III for identifying fibroblasts). These advances have created experimental
approaches to answering critical questions in human cell carcinogenesis[1,2].

This brief review will describe our recent studies concerning the molecular mechanisms controlling growth and squamous differentiation of normal
human bronchial epithelial (NHBE) cells and the dysregulation of these controls during the multistage process of carcinogenesis.

GROWTH AND DIFFERENTIATION

The balance between growth and terminal differentiation is strictly controlled in normal bronchial epithelial cells. Furthermore, carcinogenesis studies using murine epidermal cells suggest that defects in differentiation occur during tumor initiation and that selective clonal expansion of these initiated cells occurs during tumor promotion[3]. Studies using human bronchial epithelial cells are producing results supporting this hypothetical sequence of aberrations in control of growth and differentiation.

The concept of autocrine production of growth factors has been proposed to explain the uncontrolled growth of some neoplastic cells[4]. "Ectopic" hormones produced by carcinomas are candidates for autocrine growth factors. For example, gastrin-releasing peptide (the mammalian equivalent of bombesin) is secreted by most small cell carcinomas of the lung[5], and intracellular human chorionic gonadotropin is detected in many non-small cell carcinomas of the lung[6]. A monoclonal antibody to bombesin blocks the binding of the hormone to cellular receptors and inhibits clonal growth of small cell carcinomas in vitro and their growth as xenografts in vivo[7]. Both of these hormones enhance the growth of normal bronchial epithelial cells in vitro by binding to specific membrane receptors[8,9]. A second example is that 12-O-tetradecanoylphorbol-13-acetate (TPA) inhibits the growth of normal human colonic epithelial cells and is mitogenic in cultures of epithelial cells from adenomatous polyps[10]. Therefore, an imbalance between the pathways of growth and differentiation could provide a selective clonal expansion advantage for preneoplastic and neoplastic human cells, in the presence of an agent to which normal human epithelial cells respond by terminally differentiating[1].

Over the past few years, there has been considerable interest in the role that aberrant differentiation plays in carcinogenesis. It has been shown previously that NHBE cells, but not lung carcinoma cells, can be induced to undergo squamous differentiation by exposing them to serum, TPA or transforming growth factor beta (TGF-ß)[9].

Because of their short culture passage life span (4-5 subculturings) it is often difficult to do repeated experiments with the same culture of normal cells. To overcome this problem we recently transformed NHBE cells by infection with adenovirus 12-SV40 (Ad12-SV40) SV40 virus, or by transfection of normal cells with a plasmid containing SV40 large T antigen gene[11]. Ten different cultures of bronchial epithelial cells thus transformed have been studied for their ability to undergo squamous differentiation when exposed to TPA, TGF-ß or serum (Ke et al., unpublished results) All ten T-antigen positive cell cultures were significantly inhibited and differentiated when exposed to serum or TGF-ß. However, none differentiated when exposed to TPA. Fromm one such cell line, two subclones have been isolated one of which is induced by serum to stop dividing and to differentiate, and a second that not only fails to undergo squamous differentiation but is mitogenically stimulated when exposed to serum. These phenotypically very different subclones provide a new in vitro model for delineating the mechanism(s) of human bronchial epithelial cell squamous differentiation.

NEOPLASTIC TRANSFORMATION

Human cells have been transformed to malignant cells by oncogenic viruses[12] or transfected genetic elements of oncogenic DNA and RNA viruses[13,14]. For example, non-tumorigenic human skin keratinocyte cell lines became malignant if transfected with Ha- or Ki-ras oncogene or if treated with 4-nitroquinoline-1-oxide[15,16,17] and EBV "immortalized" human lymphocytes became malignant if treated with N-acetyoxy-2-acetylaminofluorene[18]. In these cases, the transformed cells were apparently immortal, and aneuploid, and produced progressively growing carcinomas in the athymic nude mouse assay. In the case of Ha-ras transfected bronchial and epidermal epithelial cells, the malignant epithelial cells continued to synthesize keratin.

A single transfected viral oncogene, v-Ha-ras, can cause a cascade of
events leading to neoplastic transformation of human bronchial epithelial
cells[14]. This cascade may have been due to enhanced genetic instability mediated by the transfected ras and is consistent with the hypothesis of clonal
evolution in neoplastic populations recently discussed by Nowell[19]. These
results are also consistent with the in vitro transformation of fibroblastic
cells from various types of rodent or human fibroblasts by the transfected
ras oncogene[20-22]. This may also be reflected in the more frequent finding
of carcinomas in humans than sarcomas, and with the relative ease of transformation of epithelial cells compared to other types of human cells[1].

The human bronchial epithelial cells transformed by v-Ha-ras were highly
invasive and metastatic from the primary subcutaneous injection site to
multiple organs, including liver, spleen, kidney and lung[14] (Yoakum et al.,
unpublished results). The transformed cells were also relatively resistant
in vitro to inducers of terminal squamous differentiation, such as TPA, and
produced an ectopic hormone and growth factor (human chorionic gonadotropin);
these findings are consistent with the hypothesis that neoplastic cells have
an imbalance in the control of their growth and differentiation pathways.

Human bronchial epithelial cells growing in continuous culture due to
the effects of an integrated SV40 T antigen gene are useful for studies of
multistage bronchial epithelial carcinogenesis. A summary of the effects of
oncogenes transfected into normal or SV40 T antigen "immortalized" human
bronchial epithelial cells is shown in Table 1. SV40 T antigen gene or ras
caused an enhanced lifespan of the normal bronchial epithelial cells but
rarely leads to "immortalized" cell lines. Although myc does not enhance
cellular lifespan, it appears to reduce responsiveness of the bronchial epithelial cell to serum-induced squamous differentiation (Amstad et al., unpublished results). Transfected ras has similar effects on the pathway of
squamous differentiation. As discussed above, v-Ha-ras rarely causes neoplastic transformation of normal human bronchial epithelial cells. However,
an Ad12-SV40 "immortalized" cell line BEAS-2B, is readily transformed by

Table 1. Effects of activated proto-oncogenes and SV40 large T antigen
gene on growth, differentiation and neoplastic transformation of
human bronchial epithelial cells in vitro

Transferred gene(s)	Population Doublings	Responsiveness to serum-induced squamous differentiation	Tumorigenicity[a]
vHa-ras	I[b]	D	+[c]
SV40 T	I	D or NC	-[d]
c-myc	NC	D	-
c-raf	NC	NC	-
SV40 T + vHa-ras	I	D	++++
SV40 T + vKi-ras	I	?	+++
SV40 T + c-myc	I	D	-
SV40 T + c-raf	I	?	-
SV40 T + c-myc + c-raf	I	D	++

[a]Progressively growing tumors in athymic nude mice. [b]I, increase; D, decrease; NC, no change; -, non-tumorigenic; +, tumorigenic. [c]In one of
seven cases. [d]< 30 passages.

v-Ha-<u>ras</u> or v-Ki-<u>ras</u> (Reddel et al., unpublished results). The combination of <u>myc</u> and <u>raf</u> can also cause neoplastic transformation of the BEAS-2B cell line.

REFERENCES

1. C. C. Harris, Human tissues, and cells in carcinogenesis research, <u>Cancer</u> <u>Res.</u> 47:1 (1987).
2. D. E. Brash, G. E. Mark, M. P. Farrell and C. C. Harris, Overview of human cells in genetic research: Altered phenotypes in human cells caused by transferred genes, <u>Somatic</u> <u>Cell</u> and <u>Mol.</u> <u>Genetics</u> 13:429 (1987).
3. S. H. Yuspa, Alterations in epidermal differentiation in skin carcinogenesis, <u>in</u>: "Interrelationship among Aging, Cancer and Differentiation", B. Pullman, ed., D. Reidel Publishing, New York (1985).
4. M. B. Sporn and A. B. Roberts, Autocrine growth factors and cancer, <u>Nature</u> <u>(Lond)</u> 313:747 (1985).
5. T. W. Moody, C. B. Pert, A. F. Gadzar, D. N. Carney and J. D. Minna, High levels of intracellular bombesin characterize human small-cell carcinoma, <u>Science</u> <u>(Wash.</u> <u>D.C.)</u> 214:1246 (1981).
6. B. F. Trump, T. Wilson and C. C. Harris, Recent progress in the pathology of lung neoplasms, <u>in</u>: "Lung Cancer", S. Ishikawa, Y. Hayata, and K. Suematsu, eds., Excerpta, Amsterdam (1982).
7. F. Cuttitta, D. N. Carney, J. Mulshine, T. W. Moody, J. Fedorko, A. Fischler and J. D. Minna, Bombesin-like peptides can function as autocrine growth factors in human small-cell lung cancer, <u>Nature</u> <u>(Lond)</u> 316:823 (1985).
8. J. C. Willey, J. F. Lechner and C. C. Harris, Bombesin and the C-terminal tetradecapeptide of gastrin-releasing peptide are growth factors for normal human bronchial epithelial cells, <u>Exp.</u> <u>Cell.</u> <u>Res.</u> 153:245 (1984).
9. C. C. Harris, G. H. Yoakum, J. F. Lechner, J. C. Willey, T. Masui, B. Gerwin, S. Schlegel and G. Mark, Growth, differentiation and neoplastic transformation of human bronchial epithelial cells, <u>in</u>: "Biochemical and Molecular Epidemiology of Cancer", C. C. Harris, ed., Alan R. Liss, New York (1986).
10. E. A. Friedman, Differential response of premalignant epithelial cell classes to phorbol ester tumor promoters and to deoxycholic acid, <u>Cancer</u> <u>Res.</u> 41:4588 (1981).
11. R. Reddel, K. Yang, B. I. Gerwin, M. McMenamin, J. F. Lechner, R. Hsu, D. E. Brash, J. B. Park, J. S. Rhim and C. C. Harris, Transformation of human bronchial epithelial cells by infection with SV40 or adenovirus-12 SV40 hybrid virus, or transfection via strontium phosphate coprecipitation with a plasmid containing SV40 early region genes, <u>Cancer</u> <u>Res.</u> (in press).
12. J. S. Rhim, G. Jay, P. Arnstein, F. M. Price, K. K. Sanford and S. A. Aaronson, Neoplastic transformation of human epidermal keratinocytes by AD12-SV-40 and Kirsten sarcoma viruses, <u>Science</u> <u>(Wash.</u> <u>D.C.)</u> 227:1250 (1985).
13. F. L. Graham, J. Smiley, W. C. Russell and R. Nairn, Characteristics of a human cell line trasformed by DNA from human adenovirus type 5, <u>J.</u> <u>Gen.</u> <u>Virol.</u> 36:59 (1977).
14. G. H. Yoakum, J. F. Lechner, E. Gabrielson, B. E. Korba, L. Malan-Shibley, J. C. Willey, M. G. Valerio, A. K. M. Shamsuddin, B. F. Trump and C. C. Harris, Transformation of human bronchial epithelial cells transfected by Harvey <u>ras</u> oncogene, <u>Science</u> 227:1174 (1985).
15. P. Boukamp, E. J. Stanbridge, P. A. Cerutti and N. E. Fusenig, Malignant transformation of 2 human-skin keratinocyte lines by Harvey-<u>ras</u> oncogene, <u>J.</u> <u>Invest.</u> <u>Dermatol.</u> 87:131 (1986).
16. N. E. Fusenig, P. Boukamp, D. Breitkreutz, S. Karjetta and R.

Dzarlieva-Petrusevska, Oncogenes and malignant transformation of human keratinocytes, _Abstract_ _of_ _Second_ _International_ _Conf._ _on_ _Anti-carcinogenesis_ _and_ _Radiation_ _Protection_, Gaithersburg, Maryland (1987).

17. J. S. Rhim, J. Fujita, P. Arnstein and A. A. Aaronson, Neoplastic conversion of human epidermal keratinocytes by adenovirus 12-SV-40 virus and chemical carcinogens, _Science_ _(Wash._ _D.C.)_ 232:385 (1986).

18. D. J. Kessler, C. A. Heilman, J. Cossman, R. T. Maguire and S. S. Thorgeirsson, Transformation of Epstein-Barr virus immortalized human B-cells by chemical carcinogens, _Cancer_ _Res._ 47:527 (1987).

19. P. C. Nowell, Mechanisms of tumor progression, _Cancer_ _Res._ 46:2203 (1986).

20. D. A. Spandidos and N. M. Wilkie, Malignant transformation of early passage rodent cells by a single mutated human oncogene, _Nature_ _(Lond.)_ 310:469 (1984).

21. R. Sager, K. Tanaka, C. C. Lau, Y. Ebina and A. Anisowicz, Resistance of human cells to tumorigenesis induced by cloned transforming genes, _Proc._ _Natl._ _Acad._ _Sci._ _USA_ 80:7601 (1983).

22. B. R. Franza, K. Maruyama, J. I. Garrels and H. E. Ruley, _In_ _vitro_ establishment is not sufficient prerequisite for transformation by activated _ras_ oncogenes, _Cell_ 44:409 (1986).

NEW INSIGHTS INTO TUMOR PROMOTION FROM MOLECULAR

STUDIES OF PROTEIN KINASE C

I. Bernard Weinstein, Gerard M. Housey, Mark D. Johnson,
Paul Kirschmeier, Catherine A. O'Brian, Wendy Hsiao
and Ling-Ling Hsieh

Comprehensive Cancer Center and Institute
of Cancer Research, Columbia University
College of Physicians and Surgeons
New York, New York 10032

INTRODUCTION: PATHWAYS OF SIGNAL TRANSDUCTION

Research on growth factors, growth factor receptors and signal transduction pathways have provided an exciting conceptual framework for understanding multistage carcinogenesis[1]. Fig. 1 displays in schematic form how
certain extracellular growth factors are perceived by cellular receptors,
which are often located at the cell surface, and how the occupancy of these
receptors leads to a cascade of signal transduction, through the cytoplasm
and eventually into the nucleus, thus altering patterns of gene expression.
This figure also emphasizes the central role that a series of protein kinase
enzymes plays in several pathways of signal transduction. A general theme
that has emerged is that the proto-oncogenes represent a subset of genes that
normally code for components in these pathways of signal transduction. Alterations in the structure and function of these proto-oncogenes can convert
them to "activated" oncogenes, which cause aberrations in signal transduction
and thus disrupt normal growth, differentiation and inter-cellular coordina-
tion.

A major pathway of signal transduction involves the turnover of phosphatidylinositol and the activation of a phospholipid- and Ca^{2+}-dependent serine
and threonine protein kinase, designated protein kinase C (PKC) (for review
see 2). It would appear that PKC plays a central role in a variety of mem-
brane-related signal transduction events. This is because several agonists
lead to the activation of a phospholipase C activity that hydrolyzes phospha-
tidylinositol-4,5-diphosphate to diacylglycerol (DAG) and inositol-1,4,5-tri-
phosphate (IP_3)[3]. DAG then activates PKC[2], and IP_3 binds to receptors present on the endoplasmic reticulum (ER), causing the release of Ca^{2+} from storage sites in the ER[3]. The resulting increase in cytoplasmic Ca^{2+} then activates several calmodulin-dependent enzymes (protein kinases, phosphatases,
phosphodiesterases), and also produces effects on the cytoskeleton. The fact
that the tumor promoter 12-0-tetradecanoyl phorbol-13-acetate (TPA), and
related tumor promoters, apparently act in place of DAG, and thus usurp the
function of PKC[1,2], provides a satisfying unity between the action of tumor
promoters and the current conceptual framework of growth control.

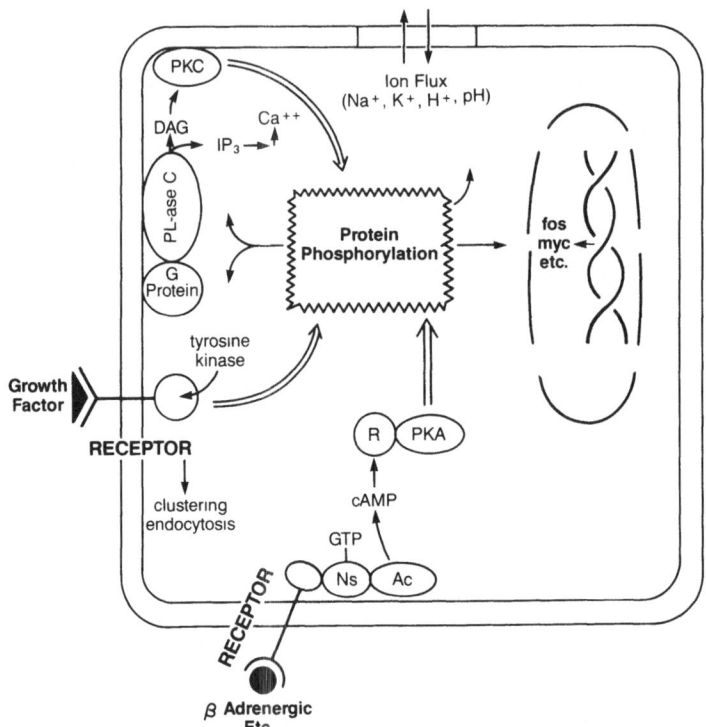

Fig. 1. A schematic diagram of a cell showing various pathways of signal transduction mediated by membrane-associated receptors. The growth factor
pathway applies to EGF, PDGF and insulin. The beta-adrenergic pathway
involves the coupling of a specific receptor via a G regulatory protein
(Ns) to adenylate cyclase (AC), and cyclic AMP (cAMP) binding to the
regulatory subunit (R) of protein kinase A (PKA). Various agonists
can activate phospholipase C (PL-ase C), presumably via a G protein,
leading to the release of diacylglycerol (DAG) and inositol-1,4,5-tri-
phosphate (IP_3). DAG activates protein kinase C (PKC) and IP_3 causes the
release of Ca^{2+} from the endoplasmic reticulum. This leads to a cascade
of events that alters the functions of membrane associated receptors,
ion channels and cytoplasmic proteins. Signals (undefined) also enter
the nucleus to induce the expression of various genes including c-<u>fos</u>
and c-<u>myc</u>. For details see text and ref. 1.

CLONING OF DNA SEQUENCES THAT ENCODE PKC

Because of the central role of PKC in signal transduction, growth control
and tumor promotion our laboratory has done extensive studies on the function
of PKC, and also cloned DNA sequences that encode this enzyme.

The enzyme has two domains, a catalytic domain, containing an ATP binding
site and the region to which protein substrates bind, and a regulatory domain
which is controlled by allosteric cofactors, i.e. lipid, Ca^{2+} and diacylglycerol
(DAG); or lipid and TPA. We hypothesize that the usual function of the regulatory domain is to inactivate the enzyme, by "closing" the catalytic site, and
that the binding of cofactors to the regulatory domain induces a conformational
change that opens the catalytic site and thus activates enzyme function[1]. Consistent with this scheme is evidence that limited proteolysis of the enzyme
yields a fragment that is about 66 Kd which is active in the absence of lipid and
other allosteric cofactors[2]. In addition, there exist inhibitors of PKC that

appear to act preferentially on the regulatory domain or the catalytic domain of the enzyme[2,4]. A recent paper by House and Kemp[5] is also consistent with this model. Indeed, they provide evidence that the regulatory domain contains a peptide sequence that acts as a pseudosubstrate inhibitor of the catalytic domain when the enzyme is in the inactive configuration.

Our laboratory[6,7] and other laboratories[8-11] have recently reported the isolation of cDNA clones that encode PKC. The results indicate that PKC belongs to a multigene family, with at least 4 separate genes that encode distinct isozymes of PKC. The deduced amino acid sequences of the isolated cDNA clones indicate remarkable homologies between the individual forms of PKC and also appreciable homologies with other protein kinases, especially the cyclic AMP-dependent protein kinase. The carboxyl-terminal portion comprises the catalytic domain, whereas the amino-terminal portion is the regulatory domain. The latter region contains two repeats of a cysteine-rich consensus sequence found in metal binding proteins and in several DNA binding proteins. We think that this may induce a specific conformation that is required for the interaction of this region of the enzyme with phospholipid and DAG, or TPA. The existence of multiple genes for protein kinase C, the preservation of these multiple forms during evolution, and the accumulating evidence that they are differentially expressed in different tissues [6,10,11,12], suggests that the individual forms may have subtle functional differences, but this requires further study.

OVEREXPRESSION OF PKC ALTERS GROWTH CONTROL

To better define the function of a specific form of PKC we have inserted the cDNA sequence for the β1 form of PKC into a retrovirus-derived vector developed in our laboratory (pMV7)[13], encapsidated the corresponding RNA into defective murine leukemia virus particles using the ψ2 system[14], and then infected the rat 6 fibroblast cell line with these viral particles. This yielded several cell lines that stably overexpress 20 to 53 times the normal level of PKC enzyme activity[7]. These cell lines also have an increase in high affinity phorbol ester receptors, when compared to control cells. Phosphorylation studies indicate that they display a marked increase in a 76 kilodalton ^{32}P-labelled protein, reflecting autophosphorylation of the overproduced PKC, as well as an increase in other phosphorylated proteins. The cell lines that overproduce PKC exhibit a dramatic change in morphology when exposed to TPA. Unlike control cells, which become refractory to the effects of TPA due to down-regulation of PKC, the overproducer cell lines continue to respond to repeated TPA treatments. In monolayer culture, these cell lines have a shorter exponential doubling time and grow to a higher saturation density and, when maintained at post-confluence, develop small, dense foci. In contrast to the control cells, which display complete anchorage dependence in the absence or presence of TPA, the cell lines that overproduce PKC form small colonies in soft agar in the absence of TPA, and larger and more frequent colonies in the presence of TPA. Thus, overproduction of a single form of PKC is sufficient to confer anchorage independent growth and other growth abnormalities in rat fibroblasts.

Taken together, the above results provide direct evidence that PKC plays a critical role in normal cellular growth control and that it mediates several of the cellular effects of the phorbol ester tumor promoters. These results also provide evidence that disturbances in the activation or down-regulation of PKC may be critical events in tumor promotion and multistage carcinogenesis. The exaggerated phosphorylation of specific cellular proteins in the cell lines that overproduce PKC may provide a useful tool for identifying these target proteins and their role in signal transduction. In recent studies we have found that these cells are also much more susceptible to transformation by an activated c-H-<u>ras</u> oncogene than control cells (unpublished data). Studies are in progress to determine whether they are also susceptible to malignant transformation by chemical carcinogens.

ISOLATION OF A cDNA WHOSE EXPRESSION IS INDUCED BY PKC ACTIVATION

TPA induces the expression of a number of cellular genes. We are also interested in the role of PKC activation in mediating these effects and the underlying molecular mechanisms. To identify specific genes that are regulated by TPA, we prepared a cDNA library using poly A$^+$ RNA obtained from quiescent C3H 10T 1/2 murine fibroblasts 4 h after treatment with TPA, and then screened this library for induced sequences, by the technique of differential hybridization[15]. We have isolated and characterized a cDNA clone (TPA-S1) whose corresponding mRNA is induced up to 20-fold in response to the treatment of cells with TPA, platelet derived growth factor (PDGF), epidermal growth factor (EGF), serum or diacylglycerol[15]. This effect is apparently at the level of transcription, and does not require de novo protein synthesis. We have designated the gene corresponding to TPA S-1 "phorbin" (phorbol ester induced). The role of PKC in the induction of phorbin is supported by the following lines of evidence: 1) agents that activate PKC, such as TPA, mezerein, PDGF, serum and 1-oleoyl-2-acetyl-glycerol (OAG), also increase phorbin mRNA levels; 2) protein kinase inhibitors with differential effects on PKC activity demonstrate the same relative inhibitory effects on the induction of phorbin by TPA; 3) down-regulation of PKC activity, by treatment of 10T 1/2 cells with TPA for 24 h, results in a loss of responsiveness to phorbin induction by subsequent TPA treatment; and 4) the above-described cell lines that overexpress PKC are extremely sensitive to TPA induction of phorbin RNA, and also yield an exaggerated and prolonged response (15 and unpublished studies).

Complete sequence analysis of the phorbin cDNA (730 bp) predicts a cysteine-rich, secreted protein with a molecular weight of 22.6 kd. The sequence of phorbin exhibits homology with two previously isolated cDNA sequences, designated "ERP" (erythroid-potentiating activity) and "TIMP" (tissue inhibitor of metalloproteinase). The significance of these homologies and the possible role of phorbin in cell proliferation and tumor promotion is under investigation.

RELEVANCE OF THE ABOVE STUDIES TO TUMOR PROMOTION IN VARIOUS TISSUES

Although the studies described above are concerned with the molecular mechanisms of action of TPA, a potent tumor promoter on mouse skin, we think that they have relevance to other types of tumor promoters and to tumor promotion in other tissues and species, including the process of multistage carcinogenesis in humans. For example, we have found that certain bile acids that are implicated in colon carcinogenesis can activate PKC and also induce translocation of PKC to the membrane fraction of cells[16,17]. Although it is unlikely that all classes of tumor promoters function by directly activating PKC, it is possible that they activate PKC indirectly by enhancing DAG production, or by distorting related pathways of signal transduction (Fig. 1). This merits further study, particularly with certain halogenated organic compounds that have tumor promoting activity in rodent liver. In recent studies on patterns of gene expression during rat liver regeneration and carcinogenesis[18,19], we have discovered that in both systems there is an early decrease in the abundance of EGF-receptor mRNA. This may be an important clue to an underlying change in signal transduction.

Elsewhere, we have discussed the desirability and feasibility of developing specific inhibitors of PKC and related pathways of signal transduction[1,4]. We have also reported promising early results in the development of inhibitors of PKC[1,4]. We are hopeful that further research in this field may provide novel strategies of cancer prevention and treatment.

ACKNOWLEDGEMENTS

We acknowledge the excellent secretarial assistance of Mrs. Nancy Mojica and Ms. Lintonia Sheppard. These studies were supported by NIH grant CA 02656 and funds from the Alma Toorock Memorial for Cancer Research. G. M. Housey is supported by the Medical Scientist Training Program. We thank Cheryl Fitzer, James Murphy and Myung-Hi Sonsorei Lee for valuable technical assistance.

REFERENCES

1. I. B. Weinstein, Growth factors, oncogenes and multistage carcinogenesis, J. Cell Biochem. 33:213 (1987).
2. Y. Nishizuka, Perspectives on the roles of protein kinase C in stimulus-response coupling, J. Natl. Cancer Inst. 76:363 (1986).
3. M. J. Berridge, Inositol triphosphate and diacylglycerol: Two interacting second messengers, Ann. Rev. of Biochem. 56:159 (1987).
4. C. A. O'Brian, R. M. Liskamp, D. H. Solomon and I. B. Weinstein, Triphenylethylenes: A new class of protein kinase C inhibitors, J. Natl. Cancer Inst. 76:1243 (1986).
5. C. House and B. E. Kemp, Protein kinase C contains a pseudosubstrate prototype in its regulatory domain, Science 238:1726 (1987).
6. G. M. Housey, C. A. O'Brian, M. D. Johnson, P. T. Kirschmeier and I. B. Weinstein, Isolation of cDNA clones encoding protein kinase C: Evidence for a novel protein kinase C-related gene family, Proc. Natl. Acad. Sci. USA 84:1065 (1987).
7. G. M. Housey, M. D. Johnson, W.-L. Hsiao, C. A. O'Brian, J. P. Murphy, P. Kirschmeier and I. B. Weinstein, Overproduction of protein kinase C causes disordered growth control in rat fibroblasts, Cell 52:343 (1988).
8. L. Coussens, P. J. Parker, L. Rhee, T. L. Yang-Feng, E. Chen, M. D. Waterfield, U. Francke and A. Ullrich, Multiple distinct forms of bovine and human protein kinase C suggest diversity in cellular signaling pathways, Science 233:859 (1986).
9. J. L. Knopf, M-H. Lee, L. A. Sultzman, R. W. Kriz, C. R. Loomis, R. M. Hewick and R. M. Bell, Cloning and expression of multiple protein kinase C cDNAs, Cell 46:491 (1986).
10. U. Kikkawa, K. Ogita, Y. Ono, Y. Asaoka, M. S. Shearman, F. Tomoko, K. Ase, K. Sekiguchi, K. Igarashi and Y. Nishizuka, The common structure and activities of four subspecies of rat brain protein kinase C family, FEBS Letters 223:212 (1987).
11. S. Ohno, H. Kawasaki, S. Imajoh, K. Suzuki, M. Inagaki, H. Yokohura, T. Sakoh and H. Hidaka, Tissue-specific expression of three distinct types of rabbit protein kinase C, Nature 325:161 (1987).
12. F. L. Huang, Y. Yoshida, H. Nakabayashi, J. L. Knopf, W. S. Young and K.-P. Huang, Immunochemical identification of protein kinase C isozymes as products of discrete genes, Biochem. Biophys. Res. Commun. 149:946 (1987).
13. P. T. Kirschmeier, G. M. Housey, M. D. Johson, A. S. Perkins and I. B. Weinstein, Construction and characterization of a retroviral vector demonstrating efficient expression of cloned cDNA sequences, DNA in press (1988).
14. R. Mann, R. C. Mulligan and D. Baltimore, Construction of a retrovirus packaging mutant and its use to produce helper-free defective retrovirus, Cell 33:153 (1983).
15. M. D. Johnson, G. M. Housey, P. Kirschmeier and I. B. Weinstein,

Molecular cloning of gene sequences regulated by tumor promoters and mitogens through protein kinase C, <u>Mol</u>. <u>Cell</u> <u>Biol</u>. 7:2821 (1987).

16. C. J. Fitzer, C. A. O'Brian, J. G. Guillem and I. B. Weinstein, The regulation of protein kinase C by chenodeoxycholate, deoxycholate and several related bile acids, <u>Carcinogenesis</u> 8:217 (1987).

17. J. G. Guillem, C. A. O'Brian, C. J. Fitzer, M. D. Johnson, K. A. Forde, P. LoGerfo and I. B. Weinstein, Studies on protein kinase C and colon carcinogenesis, <u>Arch</u>. <u>Sur</u>. 122:1475 (1987).

18. L. L. Hsieh, W.-L. Hsiao, C. Peraino, R. R. Maronpot and I. B. Weinstein, Expression of retroviral sequences and oncogenes in rat liver tumors induced by diethylnitrosamine, <u>Cancer</u> <u>Research</u> 47:3421 (1987).

19. L. L. Hsieh, C. Peraino and I. B. Weinstein, Expression of endogenous retrovirus-like sequences and cellular oncogenes during phenobarbital treatment and regeneration in rat liver, <u>Cancer</u> <u>Research</u> 48:265 (1988).

MODEL STUDIES ON THE IN VITRO ACTIVATION OF c-Ha-ras

PROTOONCOGENE BY DIETARY COMPONENTS

Christine M. Ireland, Eric Hebert[1], Colin S. Cooper
and David H. Phillips

Institute of Cancer Research, Chester Beatty Laboratories
Fulham Road, London SW3 6JB, UK and [1]Centre de Biophysique
Moleculaire, CNRS, 1A Avenue de la Recherche Scientifique
45071 Orleans, Cedex 2, France

INTRODUCTION

Evidence for the activation of ras protooncogenes by chemical carcinogens is provided both by the reproducible activation observed in chemically-
induced animal tumors (reviewed in 1) and by our demonstration that in vitro
modification of ras protooncogenes with ultimate carcinogens results in
transformed foci on transfection into NIH3T3 cells[2,3]. In the present study,
transforming oncogenes have been generated by the in vitro modification of
plasmid containing human c-Ha-ras with the model ultimate carcinogens
3-N,N-acetoxyacetylamino-4,6-dimethyl-dipyrido[1,2-a:3',2'-d]imidazole
(N-AcO-AGlu-P-3) and 1'-acetoxysafrole (AcO-safrole). The mutations produced
have been analyzed by selective amplification of the regions surrounding
codons 12 and 61 and subsequent oligonucleotide hybridization.

EXPERIMENTAL

The experimental procedure followed is illustrated in Fig. 1. Following in vitro modification, plasmid containing the human c-Ha-ras proto-
oncogene (pEC) is transfected into NIH3T3 cells. DNA from primary foci is
probed for the presence of human ras sequences and used in a second round of
transfection in order to reduce the copy number of ras genes present in the
secondary transformants. Secondary DNAs are subjected to polymerase chain
reaction (PCR)[4], which amplifies the ras sequences responsible for transforming activity. The procedure used for amplification is illustrated in
Fig. 2. Amplified DNAs are then probed with synthetic oligomers under
hybridization and washing conditions which allow detection of a single base
pair mismatch.

RESULTS

Fig. 3 shows the chemical structures of the two compounds used in this
study, together with the major sites of DNA modification for both compounds, and the results of primary and secondary transfection experiments.
Analysis of the mutations generated by both compounds is detailed in Table 1.
Two of the three foci induced by N-AcO-AGlu-P-3 treatment of pEC contain

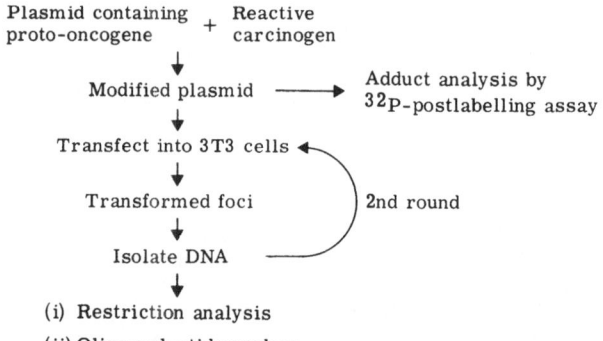

Plasmid containing proto-oncogene + Reactive carcinogen

↓

Modified plasmid ——→ Adduct analysis by ^{32}P-postlabelling assay

↓

Transfect into 3T3 cells ◄

↓   2nd round

Transformed foci

↓

Isolate DNA ——

↓

(i) Restriction analysis

(ii) Oligonucleotide probes

Fig. 1. Outline of the procedures for activating cellular protooncogenes by *in* *vitro* modification with DNA-damaging agents.

GC -> TA transversions at codon 61 of the <u>ras</u> gene, one each at position 1 and 3 of this codon. Two of the three AcOsafrole-induced foci also contain this type of mutation, one at base 1 of codon 61 and one at base 1 of codon 12 (data subsequently obtained). The remaining foci induced by each compound appear to contain mutations at neither of the regions examined.

DISCUSSION

The mutations induced by both compounds are to be expected given that

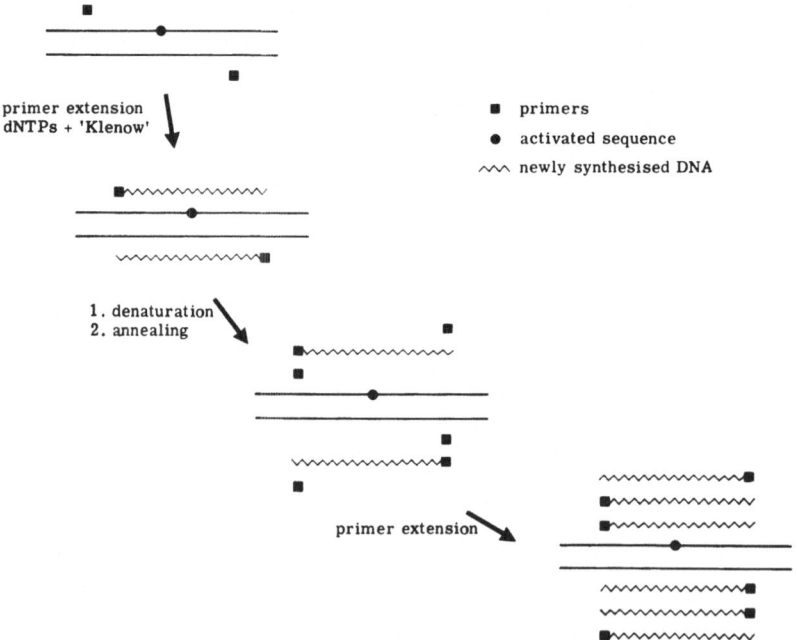

primer extension
dNTPs + 'Klenow'

■ primers
● activated sequence
∿∿ newly synthesised DNA

1. denaturation
2. annealing

primer extension

Fig. 2. Diagram of procedure used for amplification of activated <u>ras</u> sequences using the polymerase chain reaction. Cycle repeated x 25. Amplified sequence probed with synthetic 20-mers under conditions detecting single base pair mis match.

STRUCTURE	SITE OF DNA MODIFICATION	PRIMARY FOCI no.foci/no.dishes	SECONDARY FOCI no.foci/no.dishes	no.foci/μg
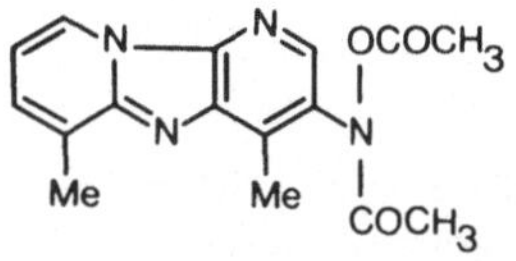 N-AcO-AGlu-P-3	C^8G	7/27	Ba 5/8 Bb 4/3 Bc 45/5 Be big 20/7	0.031 0.067 0.450 0.143
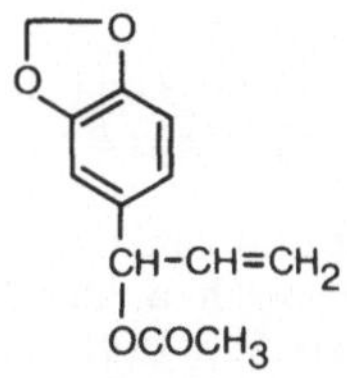 AcO-safrole	N^2G (major) N^6A (minor) + C^8G, N^7G	4/20	568 14/6 569 7/6 570 14/6	0.117 0.058 0.117

Fig. 3. Chemical structure, sites of DNA modification and focus-forming activity of N-AcO-AGlu-P-3 and AcO-safrole.

Table 1. Analysis of activating point mutations by oligonucleotide hybridization experiments

RESULTS OF DOT BLOT HYBRIDIZATION EXPERIMENTS

1. N-AcO-AGlu-P-3

focus	wild type sequence		mutated sequences codon 61							mutation
	codon 12 GGC	codon 61 CAG	pos^n 1 mix	pos^n 2 mix	pos^n 3 mix	base 1 G→T	base 1 G→C	base 3 G→T	base 3 G→C	
Ba										
Bb	+	+	–	–	–	–	–	–	–	not 12 or 61
Bc	+	–	–	–	+	–	–	+	–	61/3 G→T
Be big	+	–	+	–	–	+	–	–	–	61/1 G→T

2. a) AcO-safrole

focus	wild type sequence		mutated sequences codon 12			mutation
	codon 12 GGC	codon 61 CAG	pos^n 1 mix	pos^n 2 mix	base 1 G→A	
568	+	+	–	–	–	not 12
569	–	+	+	–	–	12/1 G→T or C
570	+	–	–	–	–	not 12

b)

focus	mutated sequences codon 61					mutation
	pos^n 1 mix	pos^n 2 mix	pos^n 3 mix	base 1 G→T	base 1 G→C	
568	–	–	–	–	–	not 12 or 61
569	–	–	–	–	–	as above
570	+	–	–	+	–	61/1 G→T

+ positive hybridization
− no signal observed

the primary site of DNA modification in each case is at guanine residues. N-AcO-AGlu-P-3, which shows some similarity in structural features and site of DNA modification with AAAF, produces mutations similar to those generated both in vitro[3] and in vivo[5] with the latter compound. Mutations produced in vivo by safrole[6], however, occur exclusively at adenine residues, suggesting that adenine adducts may present more genotoxic hazard in vivo than those occurring at guanine.

In conclusion, we believe that this model provides a useful method for detecting point mutations induced by chemical carcinogens in a gene relevant to carcinogenesesis and in a mammalian cell system.

REFERENCES

1. I. Guerrero and A. Pellicer, Mutational activation of oncogenes in animal model systems of carcinogenesis, Mutation Res. 185:293 (1987).
2. C. J. Marshall, K. H. Vousden and D. H. Phillips, Activation of c-Ha-ras-1 proto-oncogene by in vitro modification with a chemical carcinogen, benzo[a]pyrene diol epoxide, Nature 310:586 (1984).
3. K. H. Vousden, J. L. Bos, C. J. Marshall and D. H. Phillips, Mutations activating human c-Ha-ras 1 protooncogene (HRAS1) induced by chemical carcinogens and depurination, Proc. Natl. Acad. Sci. 83:1222 (1986).
4. R. K. Saiki, S. Scharf, F. Faloona, K. B. Mullis, G. T. Horn, H. A. Erlich and N. Arnheim, Enzymatic amplification of ß-globin genomic sequences and restriction site analysis for diagnosis of sickle cell anaemia, Science 230:1350 (1985).
5. R. W. Wiseman, S. J. Stowers, E. C. Miller, M. W. Anderson and J. A. Miller, Activating mutations of the c-Ha-ras protooncogene in chemically induced hepatomas of the male B6C3F1 mouse, Proc. Natl. Acad. Sci. 83:5825 (1986).
6. R. W. Wiseman, B. C. Stewart, D. Grenier, E. C. Miller and J. A. Miller, Characterization of c-Ha-ras proto-oncogene mutations in chemically induced hepatomas of B6C3F1 mouse, Proc. Amer. Assoc. Canc. Res. 28:147 (1987).

DNA FRAGMENTATION OR CHANGES IN CHROMATIN CONFORMATION. RESULTS WITH TWO
MODEL SYSTEMS: PROMOTION IN RAT LIVER CARCINOGENESIS AND PROLIFERATION IN
MASTOCYTES.

Silvio Parodi, Maurizio Taningher, Patrizia Russo, Paolo
Lusuriello, Michael Minks, Rossella Bordone, Franca
Marchesini, Viviana Pisano and Cecilia Balbi

Department of Clinical and Experimental Oncology, University
of Genoa/Centro Interuniversitario per la Ricerca sul Cancro
and Istituto Nazionale per la Ricerca sul Cancro
I-16132 Genoa, Italy

INTRODUCTION

We have recently observed that different promoting treatments in rat liver carcinogenesis induce an apparent DNA fragmentation, as monitored by the
alkaline elution technique (DNA elution through calibrated pores of filters
after lysis of nuclei with 2M NaCl) and sedimentation in alkaline sucrose
gradients [1,2]. A summary of the results obtained with 1% orotic acid in the
diet of rats for five weeks is shown in Table 1, and a summary of the results
obtained with a choline deficient diet for five days is shown in Table 2.
Birnboim has reported the induction of DNA fragmentation in leucocytes after
treatment with TPA [3].

It could be suspected that promoting treatments in many cases are accompanied by DNA damage and that they are not pure promoting treatments, or,
alternatively, some sort of DNA damage is involved also during promotion.
Indeed, the probability that promoting treatments in rat liver carcinogenesis
also induce significant amounts of DNA damage and repair has been shown by
Brambilla et al. [4], using the treatment of Solt and Farber (diethylnitrosamine
200 mg/kg i.p.; 2 weeks of basal diet; one week of basal diet + 0.02% acetylaminofluorene; partial hepatectomy; one week of basal diet + 0.02% acetylaminofluorene; ref. 5). The situation is further complicated by the results reported in this paper.

RESULTS OBTAINED WITH 1% OROTIC ACID IN THE DIET (RAT LIVER MODEL OF CARCINO-
GENESIS)

As we have indicated in the introduction, when analysed by alkaline elution DNA shows increased elution rates, after a treatment of promotion with
1% orotic acid in the diet for 5 weeks.

· We will briefly summarize here the principles of the alkaline elution
technique. Partially purified nuclei are prepared from liver cells with a
mild procedure, in the presence of EDTA, in order to avoid spurious DNA fragmentation. These nuclei are deposited on filters with pore diameters of
2-5 μm. The nuclei are lysed on the filters with different types of buffer.

195

Table 1. Effect of orotic acid diet (OA) followed by semisynthetic basal diet on alkaline elution rate of rat liver DNA

Treatment [a]	No. of animals	No. of elutions	Initial elution rate(K x 10^2)[b] $\bar{X} \pm SE$	Breaks/10^{10} daltons over controls
Basal diet (BD)	7	11	1.43 ± 0.14	
OA + 0 week BD	5	10	4.35 ± 0.39	7.9
OA + 1 week BD	5	12	7.59 ± 0.93	16.6
OA + 2 weeks BD	5	9	11.06 ± 1.47	26.0
OA + 3 weeks BD	3	6	7.88 ± 0.18	17.4

[a] The rats were fed on a basal diet containing 1% orotic acid for 5 weeks. Afterwards they were transferred to the basal diet. At different time periods thereafter as shown in table the rats were killed. [b] Elution rate measured at the third fraction. (Reported from Table 1 of ref. 1).

We, for instance, used a solution with 2 M NaCl + 0.2% sodium lauroylsarcosinate or a 2% sodium dodecylsulphate solution. As we will see later on, it is not irrelevant which buffer is used for lysing the nuclei. After lysis, DNA is eluted at pH 12 (alkaline elution) or at pH 10 (neutral elution) with a low ionic strength solution of tetraethylammonium hydroxide, in the presence of EDTA, at 25°C. The elution rate is regulated mainly by molecular weight, but DNA conformation can also play a role[6].

In addition to alkaline elution, in order to evaluate the effect of 1% orotic acid in the diet we have used a DNA unwinding method for measuring DNA fragmentation. DNA unwinding can be evaluated by a method proposed by

Table 2. Effect of choline deficient diet (CD) on alkaline elution rate of rat liver DNA

Treatment [a]	No. of animals	No. of elutions	Initial elution rate(K x 10^2)[b] $\bar{X} \pm SE$	Breaks/10^{10} daltons over control
Choline supplemented diet (CS)	14	14	1.48 ± 0.06	
CD for 3 days	9	16	2.89 ± 0.27 [c]	4.02
CD for 5 days	8	11	10.21 ± 1.23 [c]	23.78
CD for 5 days + CS for 7 days	6	12	2.42 ± 0.20 [c]	2.75
DMN (10 mg/kg)	4	8	13.92 ± 2.19 [c]	33.80

[a] The rats were maintained on CD diet for the number of days indicated. After exposure to the CD diet for 5 days, some of the rats were transferred to the choline supplemented diet for 7 days. [b] Elution rate measured at the third fraction. For DMN (10 mg/kg i.p., killed 2h after treatment) the average elution rate is given. [c] All results from the four experimental groups are statistically significant with respect to controls; $P < 0.001$ according to the Student's "t" test[20]. (Reported from Table 1 of ref. 2).

Birnboim and Jevcak[7]. Lysed nuclei that are incubated at pH 12 to induce DNA unwinding are brought back to pH 10 (at 25°C DNA is double stranded at pH 10) at different times after the beginning of the unwinding process. Non unwound DNA becomes renatured; unwound DNA becomes denatured. The quantum yield of intercalated ethidium bromide is very different for renatured and denatured DNA. As a consequence, the fluorescent probe allows a measurement of the degree of DNA unwinding. The unwinding rate is proportional to the degree of DNA fragmentation because each break is a starting point for DNA unwinding. The results obtained by evaluating DNA damage after five weeks of 1% orotic acid in the diet are shown in Table 3. Evaluated with this method the promoting treatment was completely negative in causing DNA breaks. The sensitivity of the method is at least equal to the sensitivity of alkaline elution.

Ourselves and others have shown that DNA viscosity in alkali is correlated with the degree of unwinding of the two strands[4,8-11]. In turn the DNA unwinding rate is correlated with DNA fragmentation. The method is more sensitive than the alkaline elution technique and the unwinding method devised by Birnboim and Jevcak[7].

If we compare the sensitivity of the two unwinding methods using as a reference the effect of the same chemical on the same target (dimethylnitrosamine (DMN) on the rat liver DNA), in our hands the viscosimetric method is about ten times more sensitive than the fluorometric unwinding method[8,12]. In spite of this fact we have failed to detect DNA fragmentation in the livers of rats fed orotic acid in the diet. A summary of the results obtained is shown in Table 4.

Recently we have found that the results obtained with the alkaline elution technique can be modified by the procedure of lysis of the nuclei before elution. Using 2 M NaCl + 0.2% sodium lauroylsarcosinate for the lysis of nuclei before elution we can observe an increased elution rate in liver DNA of rats fed with 1% orotic acid in the diet. By contrast, DNA elutes as slowly as control DNA if nuclei are lysed with 2% sodium dodecylsulphate (SDS).

A summary of the results obtained is shown in Table 5. It is worth noting that proteins of the nuclear matrix are present in significant amounts bound to DNA after the first type of lysis but they are completely absent

Table 3. Effect of orotic acid diet (OA) on rat liver DNA, as evaluated with the fluorometric assay of DNA unwinding

Treatment	No of animals	No of expt.[a]	% D.S. DNA (mean ± SE)[b]	Statistical significance[c]	Ratio over controls
Basal diet(BD)	18	18	26.90 ± 1.68	A, C, D	
OA + 2 wk BD	4	8	24.50 ± 3.18	A, B, E	0.910
DMN (0.3 mg/kg)[d]	5	5	17.00 ± 2.70	B, C, F	0.630
DMN (10 mg/kg)[d]	12	12	2.50 ± 0.99	D, E, F	0.093

[a] Each experiment was performed in triplicate. [b] The results are expressed as percentage of double stranded DNA (% D.S. DNA) remaining 1 h after beginning of the DNA unwinding process. [c] Data with the same letter are statistically different from each other at: $P < 0.1$ (A), $P < 0.05$ (B), $P < 0.01$ (C) and $P < 0.001$ (D), (E), (F), one tailed, according to the Student's "t" test[20]. [d] Injected i.p. 2 h before sacrifice.

Table 4. Effect of orotic acid diet (OA) followed by semisynthetic basal
         diet on viscosimetric behavior of liver DNA

Treatment	No of animals	No of expt.	Unwinding rate ζ + SE	Statistical significance[a]
Basal diet (BD)	5	5	0.53 + 0.07	
OA + 0 wk BD	3	3	0.67 + 0.20	N.S.
OA + 1 wk BD	2	2	0.52 + 0.03	N.S.
OA + 2 wk BD	3	3	0.59 + 0.24	N.S.
OA + 3 wk BD	2	2	0.34 + 0.06	N.S.
DMN (0.2 mg/kg)[b]	3	3	1.28 + 0.12	P < 0.01
DMN (0.6 mg/kg)[b]	3	3	>4[c]	P < 0.01

[a]According to the non-parametric Mann-Whitney U test[21] and to the parametric Student's "t" test[20]. [b]Injected i.p. 14 h before sacrifice. [c]Com-
pletely unwound from the beginning of measurement.

after lysis with 2% SDS (data not reported).

RESULTS OBTAINED BY STIMULATING MASTOCYTES TO PROLIFERATE WITH INTERLEUKINE 3

The murine mastocyte line PB-3[13] is dependent on interleukin-3 (IL-3) for
maintenance and growth. At limiting concentrations of IL-3 the cells enter a
quiescent phase in which there is minimal DNA synthesis and cell replication
ceases. The cells can be roused from quiescence by exposure to saturating
concentrations of IL-3, which in time leads to resumption of cell proliferation.

Populations of quiescent and exponentially growing cells have been used
to determine the extent of DNA fragmentation or chromatin conformational
changes during hormone induced stimulation to proliferation. As seen in Table
6 there is a considerable difference in the apparent elution rate of DNA from
growing and resting cells, which is suggestive of DNA fragmentation in the
growing cells. However, this difference between the cell states is reduced
following proteinase K treatment of chromatin (partial proteolysis). When 2%
SDS is used, and all proteins of the nuclear matrix are removed from DNA, the
elution rate for growing and non-growing cells is similar if not identical.

It is well known that topoisomerase II activity is very low in quiescent
cells and much higher in proliferating cells. Accordingly, breaks induced by
drugs that inhibit topoisomerase II, like ellipticine, are detectable only in
proliferating cells[14,15]. A similar result was observed with our cells. This
type of break really exists in DNA, and it has been documented that these
breaks are detectable even after lysis of nuclei with 2% SDS[16]. Even in this
respect our results are very similar to the ones already reported: ellipticine
was active only in proliferating cells, and this was the case after lysis with
either detergent (Table 6).

DISCUSSION

In this work we have reported the results obtained with two different
experimental models. In one case we have investigated liver cells in control
animals and liver cells in animal submitted to a treatment of promotion: 1%
orotic acid in the diet for five weeks. From the point of view of experimen-

Table 5. Effect of orotic acid diet (OA) on alkaline elution rate on liver DNA

Treatment[a]	No. of animals	No. of elutions	Fraction of DNA remaining on filter after 20 ml of eluting solution[b]	
			Nuclei lysed with solution A[c]	Nuclei lysed with solution B[d]
Basal diet (BD)	6	13	76.90 ± 1.31	73.52 ± 1.80
OA + 0 wk BD	5	13	17.70 ± 1.40	72.42 ± 2.20
OA + 2 wk BD	4	12	6.06 ± 0.19	69.73 ± 0.94
10 mg/kg DMN[e]	4	10	8.00 ± 0.52	9.94 ± 0.29

[a]The rats were fed on a basal diet containing 1% OA for 5 weeks. Afterwards they were transferred to the basal diet. At different time periods thereafter as shown in the table the rats were killed. [b]Values represent the mean ± SE. [c]Solution A: 2 M NaCl; 0.2% sodium lauroysarcosinate; 20 mM Na_2 EDTA (pH 10). [d]Solution B: 2% sodium dodecylsulphate; 20 mM Na_2 EDTA (pH 10). [e]Injected i.p. 4 h before sacrifice.

tal pathology the above treatment is a treatment of promotion and not of initiation. An apparent DNA fragmentation was however reported with a classical version of the alkaline elution technique[1]. While we have always been capable of confirming these results, we have found that no effect is observed when lysing the nuclei with 2% SDS instead of 2 M NaCl + 0.2% sodium lauroylsarcosinate. Both at neutral and alkaline pH significant amounts of protein of the nuclear matrix remain bound to DNA after the second type of lysis, but not after the first one (data not reported).

In addition, we have found that two different methods for measuring DNA breaks, in terms of DNA unwinding rates, gave completely negative results after treatment with 1% orotic acid in the diet. In the case of the Birnboim and Jevcak method 4.5 M urea was the main component of the lysing solution. In the case of the viscosimetric method 0.9 M NaCl + 0.9% SDS was the main component of the lysing solution.

In our opinion there are two possible explanations of the results obtained: 1) 2 M NaCl + 0.2% sodium lauroylsarcosinate lysis generates nucleoid-like structures[17] that is loops of DNA anchored to proteins of the nuclear matrix. These structures are sensitive to chromatin conformation and are eliminated by the lysis of nuclei with 2% SDS; 2) alternatively, endonucleases can be selectively activated only by lysis with 2 M NaCl + 0.2% sodium lauroylsarcosinate, and not by other types of lysis. In the second case it could be interesting to investigate the possible relation between the activity of such endonucleases and the accessibility of chromatin sites.

We have investigated the effects of interleukin-3 on the mastocyte line PB-3 as a second experimental model. The cell line PB-3 was maintained in the presence of a low level of interleukin-3, so that cells could survive at least two days. Interleukin-3 was added to surviving cells so that they started to proliferate after about 12 h.

Quiescent cells and proliferating cells are different in terms of levels of topoisomerase II. Topoisomerase II, an enzyme typical of proliferating cells appears only after the addition of interleukin-3. Accordingly, only

Table 6. DNA neutral elution rate (pH 9.6) in quiescent and interleukine-3 stimulated mastocytes (PB-3 line)

Elution rate (K x 10^2: mean ± SE)

	Cells without any treatment			Treatment with ellipticine 10 µM	
	Lysing buffer: 2 M NaCl + 0.2% SLS[a]		Lysing buffer: 2% SDS[b]	Lysing buffer: 2M NaCl + 0.2% SLS	Lysing buffer: 2% SDS
	- Proteinase K	+ Proteinase K	+ Proteinase K	+ Proteinase K	+ Proteinase K
Quiescent cells	2.95 ± 0.35 (9)[c]	1.96 ± 0.61 (8)	1.85 ± 0.42 (9)	3.84 ± 1.60 (2)	1.57 ± 0.99 (2)
Exponentially growing cells	10.52 ± 0.84 (9)	7.13 ± 0.79 (8)	1.41 ± 0.23 (9)	18.98 ± 5.35 (2)	18.36 ± 4.46 (2)

[a]SLS = sodium lauroylsarcosinate; [b]SDS = sodium dodecylsulphate; [c] in parentheses is reported the number of experiments.

proliferating cells are responsive to ellipticine, a drug typically dependent on topoisomerase II for its activity[18].

The DNA of proliferating cells eluted at an increased elution rate when the cells were lysed in the presence of 2 M NaCl, but not when they were lysed with 2% SDS. Even in this experimental model we have found a condition (cell proliferation) in which, alternatively: 1) chromatin conformation is changed; 2) or endonucleases are activated.

We have recently reported[19] that the incubation of L1210 nuclei with lithocholic acid induces similar effects: increased DNA elution rates after lysis with 2 M NaCl, but no effect either after lysis with 2% SDS, or when utilizing an unwinding method.

It seems that DNA elution (both at neutral or alkaline pH) detects not only DNA fragmentation, but also something else.

Whether this is direct chromatin conformation or accessibility of chromatin to endonucleases remains to be seen.

AKNOWLEDGEMENTS

Supported by CNR "Oncologia-SP 1", grant No. 86.00697.44 and "Medicina Preventiva e Riabilitativa-SP 5", grant No. 86.02062.56 and by AIRC (Associazione Italiana per la Ricerca sul Cancro).

REFERENCES

1. P. M. Rao, S. Rajalakshmi, A. Alam, D. S. R. Sarma, M. Pala and S. Parodi, Orotic acid, a promoter of liver carcinogenesis induces DNA damage in rat liver, Carcinogenesis 6:765 (1985).
2. T. H. Rushmore, E. Farber, A. K. Ghoshal, S. Parodi, M. Pala and M. Taningher, A choline-devoid diet, carcinogenic in the rat, induces DNA damage and repair, Carcinogenesis 7:1677 (1986).
3. H. C. Birnboim, DNA strand breakage in human leukocytes exposed to a tumor promoter, phorbol myristate acetate, Science 215:1247 (1982).
4. G. Brambilla, A. Martelli, A. Pino and L. Robbiano, Sequential analysis of DNA damage and repair during the development of carcinogen-induced rat liver hyperplastic lesions, Cancer Res. 46:3476 (1986).
5. D. Solt and E. Farber, New principle for the analysis of chemical carcinogenesis, Nature 263:701 (1976).
6. C. Balbi, M. Pala, S. Parodi, G. Figari, B. Cavazza, V. Trefiletti and E. Patrone, A simple model for DNA elution from filters, J. Theor. Biol. 118:183 (1986).
7. H. C. Birnboim and J. J. Jevcak, Fluorimetric method for rapid detection of DNA strand breaks in human white blood cells produced by low doses of radiation, Cancer Res. 41:1889 (1981).
8. S. Parodi, P. Carlo, A. Martelli, M. Taningher, R. Finollo, M. Pala and W. Giaretti, A circular channel crucible oscillating viscometer. Detection of DNA damage induced in vivo by exceedingly small doses of dimethylnitrosamine, J. Mol. Biol. 147:501 (1981).
9. S. Parodi, M. Pala, P. Russo, C. Balbi, M. L. Abelmoschi, M. Taningher, A. Zunino, L. Ottaggio, M. De Ferrari, A. Carbone and L. Santi, Alkaline DNA fragmentation, DNA disentanglement evaluated viscosimetrically and sister chromatid exchanges, after treatment in vivo with nitrofurantoin, Chem. Biol. Interact. 45:77 (1983).
10. G. Brambilla, P. Carlo, R. Finollo, F. A. Bignone, A. Ledda and E. Cajelli, Viscosimetric detection of liver DNA fragmentation in rats treated with minimal doses of chemical carcinogens, Cancer Res.

43:202 (1983).

11. P. Carlo, R. Finollo, A. Ledda, F. A. Bignone and G. Brambilla, Visco-simetric analysis of DNA damage in kidney and lung following expo-sure of rats to small doses of chemical carcinogens, Carcinogenesis 4:137 (1983).

12. M. Taningher, R. Bordone, P. Russo, S. Grilli, L. Santi and S. Parodi, Major discrepancies between results obtained with two different methods for evaluating DNA damage: alkaline elution and alkaline unwinding, Possible explanations, Anticancer Res. 7:669 (1987).

13. P. E. Ball, M. C. Controy, C. H. Heusser, J. M. Davis and J. F. Conscience, Spontaneous, in vitro, malignant transformation of a basophil/mast cell line, Differentiation 24:74 (1983).

14. M. M. S. Heck and W. Earnshaw, Topoisomerase II: a specific marker for cell proliferation, J. Cell. Biol. 103:2569 (1986).

15. D. M. Sullivan, B. S. Glisson, P. K. Hodges, S. Smallwood-Kentro and W. E. Ross, Proliferation dependence of Topoisomerase II mediated drug action, Biochemistry 25:2248 (1986).

16. W. E. Ross and M. O. Bradley, DNA double-strand breaks in mammalian cells after exposure to intercalating agents, Biochim. Biophys. Acta 654:129 (1981).

17. P. R. Cook and I. A. Brazell, Supercoils in human DNA, J. Cell Sci. 19:261 (1975).

18. F. Filipski and K. W. Kohn, Ellipticine-induced protein-associated DNA breaks in isolated L1210 nuclei, Biochim. Biophys. Acta 698:280 (1982).

19. P. Russo, M. Taningher, M. Pala, V. Pisano, P. Pedemonte, M. T. De Angeli, S. Carlone, L. Santi and S. Parodi, Characterization of the effects induced on DNA in mouse and hamster cells by lithocholic acid Cancer Res. 47:2866 (1987).

20. N. T. J. Bailey, "Statistical methods in biology", Unibooks, English Universities Press Ltd., London (1959).

21. S. Siegel, "Nonparametric statistics for behavioral sciences", McGraw Hill Book Company Inc., New York (1956).

CELLULAR AND MOLECULAR CHANGES IN EARLY STAGES OF HEPATOCARCINOGENESIS

Snorri S. Thorgeirsson, Peter Nagy and Ritva P. Evarts

Laboratory of Experimental Carcinogenesis
National Cancer Institute
Bethesda, Maryland 20892, USA

INTRODUCTION

The role of oval cells in liver regeneration and carcinogenesis remains controversial[1-7]. After partial hepatectomy no significant proliferation of oval cells is observed even though the liver undergoes a rapid restoration of its mass. In contrast, a prominent oval cell proliferation is frequently observed after the administration of carcinogens, especially azo-dyes and 2-AAF[1-6]. Oval cell proliferation is also observed when mature hepatocytes are unable to proliferate, due to nutritional and toxic effects of chemicals[7]. The fate of these oval cells particularly as it relates to questions regarding the precursor-product relationship to normal hepatocytes and histogenesis of hepatocellular carcinoma is a matter of considerable controversy (for review, see ref. 8). In the Solt-Farber protocol for production of hepatocellular carcinomas, administration of 2-AAF is used for selective growth inhibition of non-initiated hepatocytes, whereas partial hepatectomy provides a potent growth stimulus for resistant "initiated" hepatocytes[9]. In our earlier studies using in situ hybridization technique we observed a peculiar distribution of albumin mRNA in the liver two weeks after partial hepatectomy in Solt-Farber rats[10]. The albumin mRNA was present almost exclusively in rounded basophilic areas, some of which were gamma-glutamyl transpeptidase (GGT) positive, but many of them were negative for GGT[10]. Thus a severe "inhibition" of albumin expression was observed in the majority of the "old" hepatocytes, whereas the "new" basophilic hepatocytes were efficiently expressing the albumin gene. We therefore wanted to reexamine the effect of 2-AAF administration combined with partial hepatectomy, but without any initiating agent, on the expression of albumin mRNA using in situ hybridization. In addition, the role of oval cells as possible precursors of the basophilic albumin positive hepatocytes was addressed by studying the transfer of radiolabeled thymidine from oval cells to basophilic hepatocytes[11].

MATERIALS AND METHODS

Fischer male rats (150 g) were used in this experiment. Administration of 2-AAF was by gavage (1 mg/ml corn oil/day) 5 times during the first week at the end of which partial hepatectomy was performed. After one day recovery, 2-AAF administration was continued for 4 days. Total dose of 2-AAF was 9 mg/rat. Methyl-^{3}H-thymidine (80.9 Ci/mmol, New England Nuclear, Boston, MA), 1 μCi/g body weight, was administered i.p. to a group of animals starting at day 6 after partial hepatectomy at 2 PM, 8 PM, 2 AM and 8 AM. The last dose of 2-AAF was administered at 9 AM on the sixth day after partial hepatectomy. Animals were sacrificed at the time of partial hepatectomy and then at 3, 7, 9, 11 and 13 days after partial hepatectomy. Frozen serial sections were used throughout the experiment. For autoradiography the sections were fixed in paraformaldehyde for 20 min, washed in phosphate buffered saline and dehydrated in ethanols. Slides were dipped in Kodak NTB-2 emulsion diluted 1:1. Incubation time was 3 weeks. The slides were developed in Kodak D-19 developer (diluted 1:1) for 4 min, rinsed for 15 seconds in water and fixed for 4 min in Kodak fixer. After washing in water the slides were stained in hematoxylin and eosin. Histology, preparation of albumin riboprobes, in situ hybridization and autoradiography were performed as described earlier [10].

RESULTS AND DISCUSSION

At the time of partial hepatectomy extremely few oval cells were found in livers of 2-AAF treated rats. However, the normal distribution of albumin mRNA in the liver acinus, as described earlier [10], was not observed; i.e. decreasing density of silver grains around the central vein was absent. At day 3 after partial hepatectomy oval cell proliferation was evident around the portal tracts. Albumin mRNA was present in the oval cells showing a higher density of silver grains than in the surrounding hepatocytes. Animals sacrificed on day 7 after partial hepatectomy showed a prominent oval cell proliferation. These cells radiated from the portal tracts penetrating between acidophilic hepatocytes and reaching into zone 2 of the hepatic acinus. Albumin expression was still low in hepatocytes, and the oval cells continued to show a higher level of albumin expression than the acidophilic hepatocytes.

The nuclei of all oval cells became heavily labeled 6 h after the last injection of thymidine (Fig. 1, Panel 1, A,B,C). Fig. 2 shows the number of silver grains per nucleus of thymidine labeled cells at different time intervals after partial hepatectomy. High percentage (89%) of the oval cells became heavily labeled at day 7 (>150 grains per nuclei). At later time points, when basophilic small hepatocytes were present, it was impossible to obtain accurate number of labeled oval cells. Only 0.9% of acidophilic (old) hepatocytes were labeled at day 7. At day 9 after the partial hepatectomy a number of small basophilic hepatocytes were present in the periportal areas of the liver acinus. Their size was about 1/4 of the size of the acidophilic hepatocytes. A number of these acidophilic hepatocytes became trapped between the small basophilic cells. Oval cells and basophilic hepatocytes were always located in the same areas of the liver acinus. Albumin was predominantly expressed in these basophilic areas, whereas acidophilic hepatocytes showed only a low level of albumin expression. Thymidine labeling of the cells was almost exclusively observed in the areas where oval cells and basophilic cells were located (Fig. 1, Panel 2, A,B,C). On the 9th day after partial hepatectomy 91% of the small basophilic hepatocytes were labeled and their average grain count per nuclei was 75 ± 15 (mean S.D., 200 nuclei were counted). In contrast, only 0.7% of the acidophilic old hepatocytes were labeled at this time. At day 11 after partial hepatectomy the areas occupied by basophilic hepatocytes, which showed a disorganized growth pattern, was further increased. Oval cells were now more frequently observed at the periphery of the basophilic areas. Albumin continued to be expressed predominantly in the basophilic hepatocytes. At day 11 after partial hepatectomy 88% of the basophilic and 0.3% of the acidophilic hepatocytes were labeled with thymidine. The average number of silver grains per basophilic hepatocyte was further decreased to 47 ± 10/nucleus (Fig. 1, Panel 3, A,B,C). At day 13 the area occupied by oval cells was decreased, and the basophilic hepatocytes were arranged in trabecular patterns separated by sinusoids.

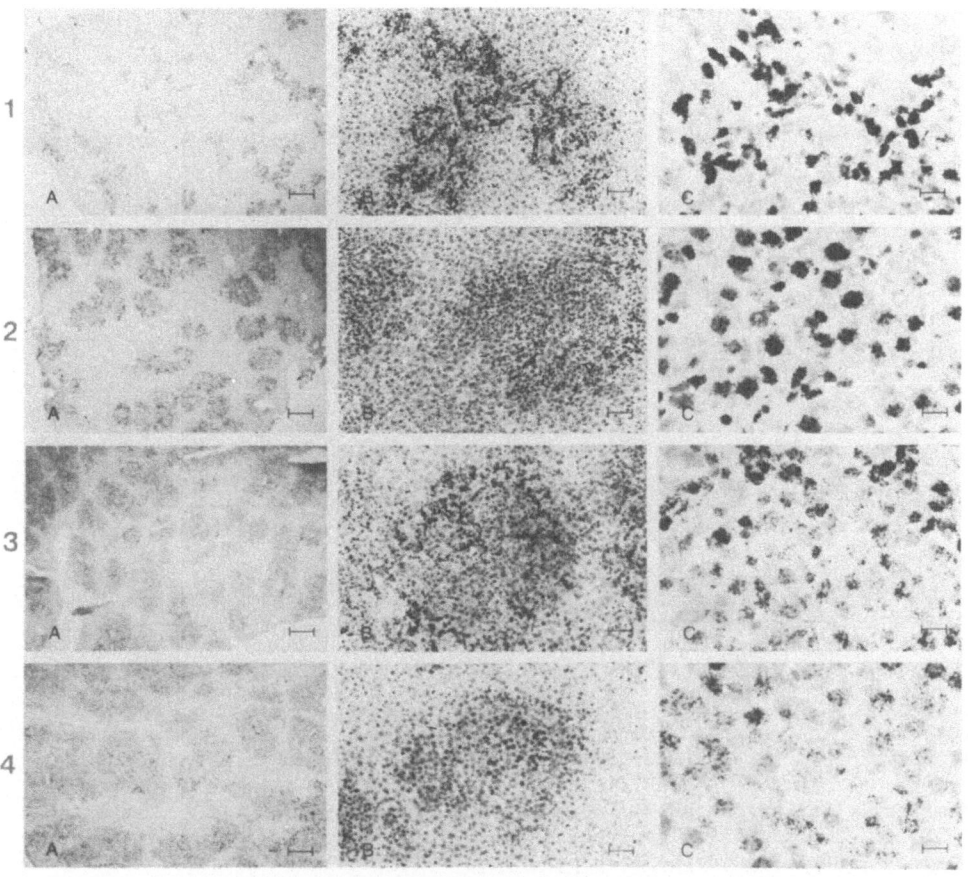

Fig. 1. Thymidine Labeling of Oval Cells and Basophilic Hepatocytes.
Panel 1. Thymidine labeling 7 days after PH. A: ^{3}H-labeled thymidine
was administered to the animals at 2 PM, 8 PM, 2 AM and 8 AM on the
6th day after PH. On the 7th day after PH animal was killed. Oval
cells became heavily labeled with thymidine. x20. Bar = 500 µm.
B: Higher magnification showing the distribution of labeled oval
cells. x100. Bar = 100 µm. C: Labeling of oval cell nuclei with
^{3}H-thymidine. x600. Bar = 17 µm. Panel 2. Thymidine labeling of
oval cells and basophilic hepatocytes 9 days after PH. A: x20.
Bar = 500 µm. B: Higher magnification showing the distribution of
labeled oval cells and basophilic hepatocytes. x100. Bar = 100 µm.
C: Thymidine labeling of the nuclei in the basophilic areas. x600.
Bar = 17 µm. Panel 3. Thymidine labeling of oval cells and basophilic hepatocytes 11 days after PH. A: x20. Bar = 500 µm. B: Higher
magnification showing the distribution of labeled oval cells and basophilic hepatocytes. x100. Bar = 100 µm. C: Thymidine labeling of
nuclei in the basophilic areas. x600. Bar = 17 µm. Panel 4. Thymidine labeling of oval cells and basophilic hepatocytes 13 days after
PH. A: x20. Bar = 500 µm. B: Higher magnification showing the distribution of labeled oval cells and basophilic hepatocytes. x100.
Bar = 100 µm. C: Thymidine labeling of nuclei in the basophilic
areas. x600. Bar = 17 µm.

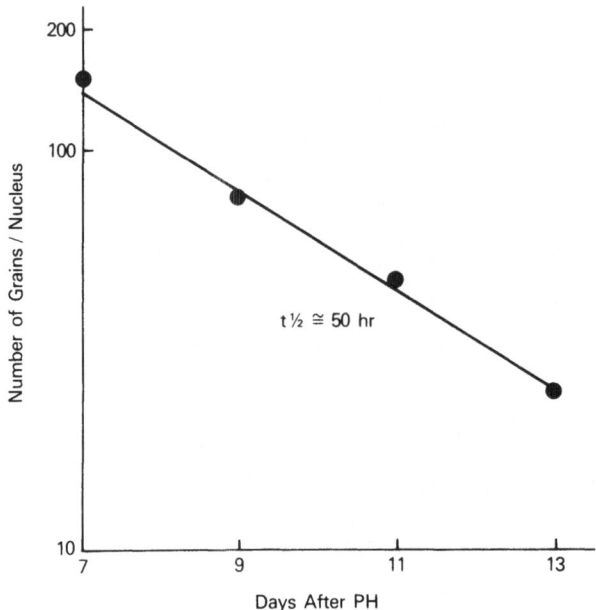

Fig. 2. ³H-Thymidine label of oval cells and basophilic hepatocytes
following partial hepatectomy (PH) and 2-AAF treatment.

Basophilia was less prominent and mitotic figures were more common in these
areas than observed earlier. A number of acidophilic hepatocytes were still
trapped between the basophilic cells. Albumin mRNA was again located predominantly in the basophilic areas of the liver acinus. A high percentage
(75%) of the basophilic hepatocytes were still labeled with thymidine, but
only 0.1% of the acidophilic hepatocytes were labeled. However, the number
of silver grains/hepatocyte was further decreased to 25 ± 7 silver grains per
nucleus (Fig. 1, Panel 4, A,B,C).

Tatematsu et al.[12] had earlier demonstrated that hepatocyte proliferation
following partial hepatectomy was virtually completely inhibited in the cau-
date lobe of the rat liver by dietary 2-AAF. Furthermore, Tatematsu et al.
showed that oval cells were heavily labeled with thymidine one day after
stopping the 2-AAF administration. The present study corroborates this observation. However, in our study virtually all basophilic hepatocytes became
labeled with thymidine 9, 11 and 13 days after partial hepatectomy. This is
in contrast to the findings in earlier studies aimed at establishing by
labeling experiments a precursor-product relationship between oval cells and
hepatocytes in vivo[11-13]. An important difference between our experimental
protocol and that of others appears to be the interval between thymidine
administration and the time at which the animals were sacrificed. Whereas
the remaining oval cells stayed heavily labeled during the experiment, the
average number of silver grains in the basophilic hepatocytes decreased from
74 to 25 in four days. This dilution of label implies that basophilic
hepatocytes have been cycling consecutively. However, whether or not replication of oval cells is required prior to differentiation into basophilic
hepatocytes cannot be resolved at this point.

In these experiments we have demonstrated a transfer of radiolabeled
thymidine from oval cells to the basophilic hepatocytes, and therefore oval
cells can be regarded as progenitors of these hepatocytes. Furthermore, the
presence of albumin mRNA both in oval cells and in the newly formed basophil-

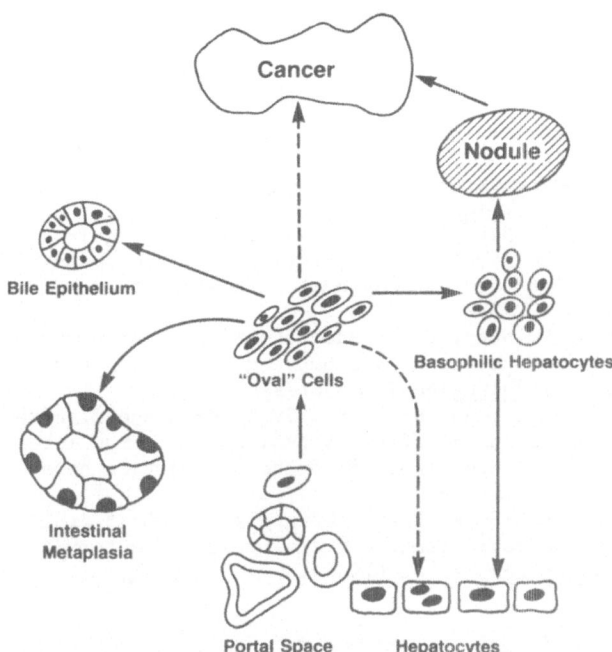

Fig. 3. Evolution of oval cells in chemical hepatocarcinogenesis.

ic hepatocytes also supports the precursor-product relationship between oval cells and basophilic hepatocytes. Finally, our results together with the already established progenitor role of oval cells for hepatocytes in azo-dye carcinogenesis [1-6], support the hypothesis (Fig. 3) that oval cells may play an important role in the histogenesis of hepatocellular carcinoma.

REFERENCES

1. J. M. Price, J. W. Harman, E. C. Miller and J. A. Miller, Progressive microscopic alterations in the livers of rats fed hepatic carcinogens 3'methyl-4-dimethylaminoazobenzene and 4'-fluoro-4-dimethylaminoazobenzene, Cancer Res. 12:192 (1952).
2. E. Farber, Similarities in the sequence of early histological changes induced in the liver of the rat by ethionine, 2-acetylaminofluorene, and 3'-methyl-4-dimethylaminoazobenzene, Cancer Res. 16:142 (1956).
3. Y. Ianaoka, Significance of the so-called oval cell proliferation during azo-dye hepatocarcinogenesis, Gann 58:355 (1967).
4. K. Ogawa, T. Minase and T. Onore, Demonstration of glucose 6-phosphatase activity in the oval cells of the rat liver and the significance of the oval cells in azo-dye carcinogenesis, Cancer Res. 34:3379 (1974).
5. H. Onda, Immunohistochemical studies on alpha-fetoprotein and alpha$_1$-acid glycoprotein during azodye hepatocarcinogenesis in rats, Gann 67:253 (1976).
6. K. Dempo, N. Chisaka, Y. Yoshida, A. Kaneko and T. Onre, Immunofluorescent study on alpha-fetoprotein-producing cells in the early stage of 3'methyl-4-dimethylaminoazobenzene carcinogenesis, Cancer Res. 35:1282 (1975).
7. J. W. Wilson and E. H. Leduc, Role of cholangioles in restoration of the liver of the mouse after dietary injury, J. Pathol. Bacteriol.

76:441 (1958).

8. N. Fausto, N. L. Thompson and L. Braun, Purification and culture of oval cells from rat liver, in: "Cell Separation: Methods and Selected Applications", Vol. 4, Academic Press, Inc., New York (1987).

9. D. B. Solt and E. Farber, New principle for the analysis of chemical carcinogenesis, Nature 263:702 (1976).

10. R. P. Evarts, P. Nagy, E. Marsden and S. S. Thorgeirsson, In situ hybridization studies on expression of albumin and alpha-fetoprotein during early stage of neoplastic transformation in rat liver, Cancer Res. 47:5469 (1987).

11. R. P. Evarts, P. Nagy, E. Marsden and S. S. Thorgeirsson, A precursor-product relationship exists between oval cells and hepatocytes in rat liver, Carcinogenesis in press (1987).

12. M. Tatematsu, R. H. Ho, K. Tohru, J. K. Ekem and E. Farber, Studies on the proliferation and fate of oval cells in the liver of rats treated with 2-acetylaminofluorene and partial hepatectomy, Am. J. Pathol. 114:418 (1984).

13. J. W. Grisham and E. A. Porta, Origin and fate of proliferated ductal cells in the rat: electron microscopic and autoradiographic studies, Exp. Mol. Pathol. 3:242 (1964).

SEQUENTIAL PHENOTYPIC CONVERSION OF RENAL EPITHELIAL CELLS DURING

NEOPLASTIC DEVELOPMENT

Peter Bannasch, Enrique Nogueira, Heide Zerban, Karin Beck
and Doris Mayer

Institut für Experimentelle Pathologie
Deutsches Krebsforschungszentrum
6900 Heidelberg, Im Neuenheimer Feld 280, FRG

INTRODUCTION

The discussion on the histogenesis of epithelial kidney tumors started
in human pathology about a century ago and remained controversial up to the
present. In experimental animals, especially in small rodents, renal tumors
have been produced by a variety of chemicals[1]. Systematic studies of the
sequence of cellular changes preceding the appearance of these tumors resulted in a number of interesting observations which did not only elucidate
specific problems of renal carcinogenesis but might also contribute important
new aspects to the unravelling of neoplastic development in general[2].

Our own experience in this field is based on a series of experiments in
rats which were treated orally with different nitrosamines (N-nitrosomorpholine, N-nitrosodiethylamine, N-nitrosodiethanolamine, N-nitrosoethyl-
ethanolamine), particularly N-nitrosomorpholine (NNM), or received a single
subcutaneous injection of 65 mg/kg body weight streptozotocin[3-12]. Using NNM,
which like many other renal carcinogens is at the same time a potent hepatocarcinogen in the rat, we were able to produce various epithelial kidney
tumors (epitheliomas) as defined cytologically (Fig. 1) such as chromophobic
and basophilic cell tumors, oncocytomas and clear or acidophilic (granular)
cell tumors[13,14]. Most of these tumor types have been known for a long time
from human pathology but in the case of the chromophobic tumors the experimental findings prompted Thoenes and colleagues[15] only recently to separate
the chromophobic renal cell carcinoma as a pathomorphologic entity from other
forms of renal cell carcinoma in man.

The renal tubular system is a complex structure composed of a number of
physiologically well defined segments each of which is lined by specifically
differentiated cells[16]. The various types of renal epitheliomas apparently
originate from different segments of this system such as the proximal tubule
and the collecting duct system indicating that differentiation of the target
cells of the carcinogen plays an important role in their commitment to certain types of tumor cells. In line with this concept the various tumor types
frequently develop concomitantly in the same kidney.

The appearance of all types of renal epitheliomas induced by the nitrosamines or streptozotocin was preceded by specific tubular alterations. These
preneoplastic tubular lesions emerged after long lag periods; they were

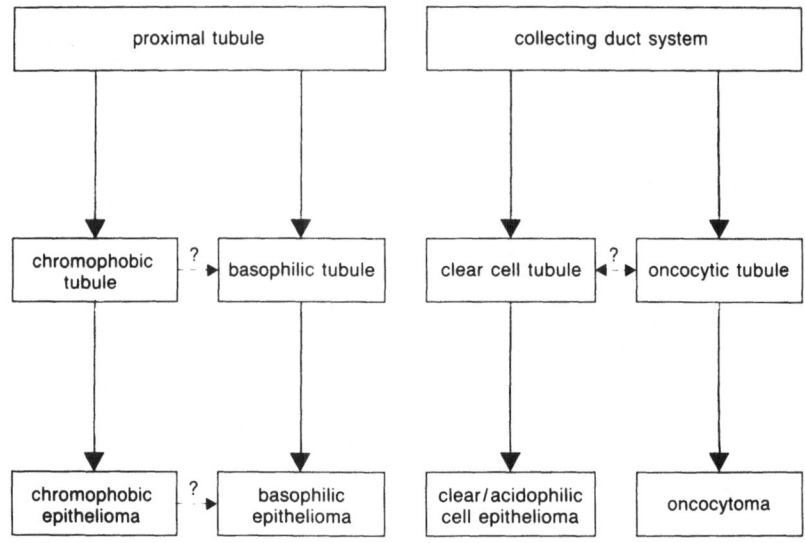

Fig. 1. Sequential cellular changes during renal carcinogenesis.

characterized by a phenotypic conversion of the normal renal epithelial cells. In a limited exposure stop protocol with NNM, the first phenotypic cellular changes were found weeks or even months after withdrawal of the carcinogen. There was no indication for an important role of cellular necrosis and hyperplasia in the development of the preneoplastic tubular lesions under our experimental conditions. In autoradiographic studies of similar stop experiments with NNM, no increase in cell proliferation was found in the rat kidney during the period of carcinogen administration [17]. The transformation of the preneoplastic tubules into tumors was followed by means of conventional light microscopy, enzyme histochemistry and electron microscopy step by step up to two years after cessation of carcinogenic treatment. A characteristic pattern of metabolic and morphologic changes was evident in both the preneoplastic tubules and the related tumors [8].

CHROMOPHOBIC AND BASOPHILIC CELL TUBULES AND TUMORS

The chromophobic and the basophilic lesions derived from the proximal nephron (Fig. 1). Chromophobic cell tubules and tumors were observed after treatment of rats with NNM [3,6,7] and N-nitrosodiethylamine [18]. The chromophobic tubular segments consisted of unusual, large, polyhedral cells which sometimes showed an apical brush border like normal cells of the proximal tubule. The finely reticular cytoplasm of the chromophobic cells was always PAS-negative but frequently rich in substances which gave a positive iron binding reaction according to Hale or stained with alcian blue. These histochemical reactions suggest an excessive storage of mucopolysaccharides (glycosaminoglycans, proteoglycans) which are probably enclosed in cytoplasmic vacuoles as seen at the ultrastructural level. However, the histochemical reactions were not always positive. This may be due either to insufficient preservation of readily soluble polysaccharides or to the storage of other substances, such as lipids which were eluted during tissue preparation.

The chromophobic tubules gave rise to chromophobic epitheliomas which usually also stored glycosaminoglycans as demonstrated histochemically. In addition, there was circumstantial evidence that the chromophobic tubules may transform into basophilic tubules and tumors. We draw this conclusion from

210

the frequent appearance of intermediate stages between Hale-positive and basophilic tubules and the intimate topographical association of many chromophobic tubules with basophilic tumors. However, basophilic tubules storing no or only low amounts of glycosaminoglycans may also develop from proximal tubules without a chromophobic intermediate stage. Transitions between such basophilic tubules and basophilic renal cell tumors have been documented by many authors.

The definitive basophilic epitheliomas are always poor in, or free from, histochemically demonstrable mucopolysaccharides but rich in free or membrane-bound ribosomes which are the ultrastructural equivalent of the pronounced cytoplasmic basophilia seen under the light microscope. The presence of a prominent brush border and the frequent occurrence of peroxisomes indicate the origin of the basophilic epitheliomas from the proximal nephron even at this late stage of tumor development[7]. Similar electron microscopical observations have been reported for basophilic epithelial tumors induced in rat kidney with various other chemicals [19-26].

Recently we investigated basophilic tubules and tumors produced in rat kidney with nitrosamines[10] or streptozotocin[8] by enzyme histochemical methods. Both the tubular lesions and the tumors were consistently characterized by the same enzyme histochemical pattern. The activities of glucose-6-phosphatase, succinate dehydrogenase, γ-glutamyltranspeptidase, alkaline and acid phosphatase were decreased, while those of the glyceraldehyde-3-phosphate dehydrogenase and glucose-6-phosphate dehydrogenase were increased. These observations support the postulated histogenetic connection between basophilic tubules and tumors. They are also in line with histochemical findings in basophilic renal epitheliomas induced by other chemical carcinogens[19,27-29]. The results of the histochemical studies suggest that a marked shift in energy metabolism from oxidative to glycolytic production of ATP with a corresponding reduction in mitochondrial respiration is characteristic of this tumor type. Since the pentose phosphate shunt, among its other functions, also provides pentoses for RNA and DNA synthesis, activation of this pathway is probably closely related both to the increase in ribosomes (basophilia) and to the enhanced cell proliferation in the basophilic lesions. A remarkable observation in the streptozotocin-induced, predominantly basophilic renal epitheliomas was the additional appearance of some clear cells storing glycogen in excess and a strong activity of glycogen phosphorylase throughout the tumor tissue. This unusual metabolic pattern of basophilic renal cell tumors may be due to the high levels of blood glucose in these animals which suffer from a streptozotocin-induced diabetes.

ONCOCYTIC TUBULES AND ONCOCYTOMAS

A third type of renal epithelioma which often develops in rats after treatment with chemical carcinogens is the oncocytoma[3,5,12,25], a tumor well known from human pathology [30]. The renal oncocytoma is a slowly growing tumor considered to be benign[13,31-34]. Light microscopically, this tumor is characterized by large polygonal cells with an intensely acidophilic and finely granular cytoplasm. Prestages of these tumors emerge as oncocytic tubules after long lag periods (Fig. 1). The electron microscope shows the typical fine structure of oncocytes: the cytoplasm is crowded with atypical mitochondria which exhibit abundant unusually long and densely packed cristae[9]. Sometimes, the severely altered mitochondria form whorl-like structures or they display an intra-mitochondrial accumulation of glycogen.

In contrast to the chromophobic and basophilic renal lesions the oncocytes derive from distal parts of the renal tubular system[5], precisely from the cortical collecting duct system consisting of the cortical collecting duct and the connecting tubule, and from the outer medullary collecting duct.

211

This has recently been shown at the light and electron microscopical level in both rats [11,18] and man [34].

Enzyme histochemical studies revealed that the oncocytic tubules and tumors induced in rats with NNM always exhibited activities of glycogen synthase, glycogen phosphorylase, glucose-6-phosphatase and glyceraldehyde-3-phosphate dehydrogenase which were comparable to those of the normal distal tubules [10]. As a rule, however, the activity of the mitochondrial enzyme succinate dehydrogenase was increased (corresponding to the abundance of mitochondria seen under the electron microscope), while that of the glucose-6-phosphate dehydrogenase was decreased throughout or in parts of the oncocytic lesions. This might explain the rarity of ribosomes and the slow growth of the oncocytic lesions.

CLEAR AND ACIDOPHILIC CELL TUBULES AND TUMORS

The clear and acidophilic (granular) rat renal cell tumors (Fig. 1) are very similar to the predominant human renal cell carcinoma. From a series of stop experiments with NNM, we have inferred that the clear cell tumors originate from glycogen-storing clear cell tubules in the rat [4]. In stop experiments, the tubular glycogenosis developed only some weeks or months after withdrawal of the carcinogen. All intermediate stages leading from the tubules storing glycogen in excess through small cystic or solid glycogenotic lesions to larger tumors composed of both clear glycogen storage cells and acidophilic cells poor in glycogen were observed.

Electron microscopically, the clear cells showed a considerable amount of monoparticulate glycogen but only a few mitochondria or profiles of the rough endoplasmic reticulum. In many cells, substantial parts of the stored glycogen were enclosed in autophagic vacuoles - a phenomenon well known from human glycogen storage disease type II. Of particular interest is the additional occurrence of lipid-storing cells in the clear cell tumors. The fine structure of the lipid bodies is very similar to structures described in the nervous system, kidney and other tissues in association with certain human storage diseases, such as mucopolysaccharidosis and gangliosidosis (cf. O'Brien, 35; Sandhoff and Harzer, 36). The acidophilic cells which in many tumors are mixed with the clear cells, exhibit a variable ultrastructure [4]. Their cytoplasm is either poor in both glycogen and organelles or it contains abundant mitochondria. However, unlike the mitochondria in oncocytes, these organelles frequently show a considerable rarefaction of their cristae and a finely granular dense matrix. The increase in mitochondria is most probably the reason for the granular appearance of many acidophilic cells under the light microscope.

The precise site of origin of the rat renal clear cell tubules and tumors within the tubular system remained obscure for a long time. In the last few months, however, we were able to show in serial sections of 36 small clear cell or mixed clear cell and acidophilic cell lesions that they were all connected with the collecting duct system [11]. This finding indicates a close histogenetic relationship between the oncocytic and the clear acidophilic cell tumors in the rat which is also suggested by the occasional common occurrence of oncocytes and clear glycogen storage cells in the same renal tubule [5].

CLEAR AND CHROMOPHOBIC RENAL CELL CARCINOMAS IN MAN

Some years ago we started studying human renal cell carcinomas by electron microscopy and cytochemistry [8,37]. Most of the tumors investigated consisted predominantly of clear cells and contained only small acidophilic (granular) cell components. Some of the tumors presented a chromophobic phenotype. In accordance with a number of earlier reports[38-40] but in contrast to our observation regarding the clear and acidophilic rat renal epitheliomas, our ultra-structural findings pointed to the proximal nephron as the site of origin of the human renal cell carcinoma. This intriguing discrepancy would need further investigation.

As to the results of the enzyme histochemical studies changes in carbohydrate metabolism are of particular interest: glycogen synthase and glycogen phosphorylase exhibited a high activity in all cells rich in glycogen. The activity of glucose-6-phosphatase was nearly totally lacking, while that of the glycolytic enzyme glyceraldehyde-3-phosphate dehydrogenase was usually increased in the tumors as compared to the proximal tubules of the remaining kidney. The activity of glucose-6-phosphate dehydrogenase was variable, in some tumors clearly elevated above the normal level, in others normal or even decreased. These histochemical results agree largely with the biochemical data reported by Weber and colleagues[41,42]. They mainly indicate a high glycolysis in the human renal clear and chromophobic cell carcinoma.

The apparently simultaneous activation of glycogen synthesis, glycolysis and sometimes also the pentose phosphate pathway in the clear cells is difficult to understand. However, recent biochemical studies in 11 human clear cell carcinomas did not only confirm the histochemical observations of increased activities of glycogen synthase and glycogen phosphorylase but also revealed that the concentration of glucose-6-phosphate is markedly increased[43]. This result suggests that the storage of various polysaccharides or lipids in preneoplastic and neoplastic renal lesions might be the consequence of an increased intracellular level of this central metabolite. The transformation of the preneoplastic into neoplastic lesions appears to be closely related to a basic shift of the carbohydrate metabolism due to an adaptive activation of alternative metabolic pathways, such as the pentose phosphate pathway or glycolysis. Similar metabolic aberrations have been observed during neoplastic development in other tissues, especially in the liver of various species[2]. Conventionally, the postulated changes in the flux of glucose would be explained by alterations in the requirement for glucose in the respective metabolic pathways. However, alternatively, it is conceivable that an excess supply with certain metabolites, such as glucose-6-phosphate, might trigger a metabolic cascade eventually resulting in the metabolism characteristic of tumor cells.

CONCLUSIONS

Systematic studies of the sequence of cellular changes during renal carcinogenesis induced in rats with different nitrosamines or streptozotocin and of human renal cell carcinomas led to the following main results and conclusions:

1. All types of epithelial kidney tumors known from human pathology, namely chromophobic and basophilic, oncocytic and clear or acidophilic (granular) cell tumors, may be induced by the chemicals.
2. Phenotypically altered epithelial cells resembling those in the tumors appear in single or multiple tubules long before unequivocal tumors develop. The progression from the preneoplastic tubular lesions to the tumors is an autogenous process independent of the further action of carcinogen.
3. Four different types of tubular lesions may be distinguished: (a) chromophobic and (b) basophilic tubules frequently accumulating acid mucopolysaccharides (glycosaminoglycans, proteoglycans); (c) oncocytic tubules exhibiting abundant atypical mitochondria, and (d) clear cell tubules storing glycogen in excess. Whereas the chromophobic and baso-

philic tubules derive from the proximal nephron, the oncocytic and clear or acidophilic cell tubules and tumors originate from the collecting duct system under the experimental conditions studied.

4. Each type of tubular lesion is apparently the precursor of a cytologically specific tumor type.

5. The well known aberration in carbohydrate metabolism in renal tumors might occur in response to a carcinogen-induced metabolic derangement which is frequently associated with excessive storage of polysaccharides or lipids. These storage phenomena are, perhaps, the consequence of a marked increase in glucose-6-phosphate as measured in human renal clear cell carcinomas.

6. An adaptation of cellular enzymes gradually activating alternative metabolic pathways, might be responsible for the ultimate neoplastic conversion of renal epithelial cells.

REFERENCES

1. Y. Hiasa and N. Ito, Experimental induction of renal tumors, Crit. Rev. Toxicol. 17:279 (1987).

2. P. Bannasch, Phenotypic cellular changes as indicators of stages during neoplastic development, in: "Theories of Carcinogenesis", O. H. Iversen, ed., Hemisphere Publishing Corporation, Washington (1988).

3. P. Bannasch, U. Schacht and E. Storch, Morphogenese und mikromorphologie epithelialer nierentumoren bei nitrosomorpholin-vergifteten ratten. I. Induktion und histologie der tumoren, Z. Krebsforsch. 81:311 (1974).

4. P. Bannasch, R. Krech and H. Zerban, Morphogenese und mikromorphologie epithelialer nierentumoren bei nitrosomorpholin-vergifteten ratten. II. Tubuläre glykogenose und die genese von klar-oder acidophil-zelligen tumoren, Z. Krebsforsch. 92:63 (1978a).

5. P. Bannasch, R. Krech and H. Zerban, Morphogenese und mikromorphologie epithelialer nierentumoren bei nitrosomorpholin-vergifteten ratten, III. Onkocytentubuli und Onkocytome, Z. Krebsforsch. 92:87 (1978b).

6. P. Bannasch, D. Mayer and R. Krech, Neoplastische und präneoplastische veränderungen bei ratten nach einmaliger oraler applikation von N-nitrosomorpholin, J. Cancer Res. Clin. Oncol. 94:233 (1979).

7. P. Bannasch, R. Krech and H. Zerban, Morphogenese und mikromorphologie epithelialer nierentumoren bei nitrosomorpholin-vergifteten ratten. IV. Tubuläre läsionen und basophile tumoren, J. Cancer Res. Clin. Oncol. 98:243 (1980).

8. P. Bannasch, H. J. Hacker, H. Tsuda and H. Zerban, Aberrant carbohydrate metabolism and metamorphosis during renal carcinogenesis, Advan. Enzyme Regul. 25:279 (1986a).

9. R. Krech, H. Zerban and P. Bannasch, Mitochondrial anomalies in renal oncocytes induced in rat by N-nitrosomorpholine, Eur. J. Cell Biol. 25:331 (1981).

10. H. Tsuda, H. J. Hacker, H. Katayama, T. Masui, N. Ito and P. Bannasch, Correlative histochemical studies on preneoplastic and neoplastic lesions in the kidney of rats treated with nitrosamines, Virchows Arch. B (Cell Path.) 51:385 (1986).

11. E. Nogueira and P. Bannasch, Histogenese von onkocytären und klar-zelligen nierentumoren der ratte, Verh. Dtsch. Ges. Path. 71:562 (1987a).

12. E. Nogueira and P. Bannasch, Oncocytic transformation of the rat renal collecting duct epithelium induced by various carcinogens, J. Cancer Res. Clin. Oncol. 113:S19 (1987b).

13. P. Bannasch, H. Zerban and H. J. Hacker, Oncocytoma, kidney, rat, in: "Monographs on Pathology of Laboratory Animals, Urinary System", T. C. Jones, U. Mohr, and R. D. Hunt, eds., Springer-Verlag, Berlin-Heidelberg-New York (1986b).

14. P. Bannasch and H. Zerban, Renal cell adenoma and carcinoma, in: "Monographs on Pathology of Laboratory Animal, Urinary System", T. C. Jones, U. Mohr and R. D. Hunt, eds., Springer-Verlag, Berlin-Heidelberg-New York (1986b).

15. W. Thoenes, S. Störkel and H.-J. Rumpelt, Human chromophobe cell renal carcinoma, Virchows Arch. B (Cell Pathol.) 48:207 (1985).

16. S. Bachmann, T. Sakai and W. Kriz, Nephron and collecting duct structure in the kidney, rat, in: "Monographs on Pathology of Laboratory Animals, Urinary System", T. C. Jones, U. Mohr and R. D. Hunt, eds., Springer-Verlag, Berlin-Heidelberg-New York-Tokyo (1986).

17. S. Hagen, Zur zellproliferation der rattenniere unter N-nitrosomorpholin in verschiedenen konzentrationen. Autoradiographische untersuchungen mit ^{3}H-thymidin, Inaug. Diss., Würzburg (1977).

18. E. Nogueira, Rat renal carcinogenesis after chronic simultaneous exposure to lead acetate and N-nitrosodiethylamine, Virchows Arch. B (Cell Path.) 53:365 (1987).

19. P. Mao and J. J. Molnar, The fine structure and histochemistry of lead induced renal tumors in rats, Am. J. Pathol. 50:571 (1967).

20. G. C. Hard and W. H. Butler, Ultrastructural aspects of renal adenocarcinoma induced in the rat by dimethylnitrosamine, Cancer Res. 31:366 (1971a).

21. G. C. Hard and W. H. Butler, Morphogenesis of epithelial neoplasms induced in the rat kidney by dimethylnitrosamine, Cancer Res. 31:1496 (1971b).

22. A. R. McGiven and H. J. Ireton, Renal epithelial dysplasia and neoplasia in rats given dimethylnitrosamine, J. Pathol. 108:187 (1972).

23. Z. Hruban, Y. Mochizuki, H. P. Morris and A. Slesers, Ultrastructure of Morris renal tumors, J. Natl. Cancer Inst. 50:1487 (1973).

24. L. P. Merkow, S. M. Epstein, M. Slifkin and M. Pardo, The ultrastructure of renal neoplasms induced by aflatoxin Bl, Cancer Res. 33:1608 (1973).

25. W. Gusek, Die ultrastruktur cycasin-induzierter nierenadenome, Virchows Arch. A (Path. Anat.) 365:221 (1975).

26. J. H. Dees, M. D. Reuber and B. F. Trump, Adenocarcinoma of the kidney. I. Ultrastructure of renal adenocarcinomas induced in rats by N-(4'-fluoro-4-biphenylyl)-acetamide, J. Natl. Cancer Inst. 57:779 (1976).

27. N. Ito, J. Johno, M. Marugami, Y. Konishi and Y. Hiasa, Histopathological and autoradiographic studies on kidney tumors induced by N-nitroso-dimethylamine in rat, Gann 57:595 (1966).

28. G. Jasmin and J. L. Riopelle, Transplantation de trois tumeurs rénales induites chez le rat par la diméthylnitrosamine, Int. J. Cancer 4:299 (1969).

29. B. M. Heatfield, D. E. Hinton and B. F. Trump, Adenocarcinoma of the kidney, II. Enzyme histochemistry of renal adenocarcinomas induced in rats by N-(4'-fluoro-4-biphenyl)acetamide, J. Natl. Cancer Inst. 57:795 (1976).

30. H. Hamperl, Onkocyten und Onkocytome, Virchows Arch. A (Path. Anat.) 335:452 (1962).

31. J. D. Van der Walt, H. A. S. Reid, R. A. Risdon and H. F. Shaw, Renal oncocytoma. A review of the literature and report of an unusual multicentric case, Virchows Arch. A (Path. Anat.) 398:291 (1983).

32. J. N. Eble and M. T. Hull, Morphologic features of renal oncocytoma: A light and electron microscopic study, Hum. Pathol. 15:1054 (1984).

33. M. M. Lieber, Renal oncocytoma, in: "Cancer of the Kidney", N. Javadpour, ed., Georg Thieme Verlag, New York (1984).

34. H. Zerban, E. Nogueira, G. Riedasch and P. Bannasch, Renal oncocytoma: origin from the collecting duct, Virchows Arch. B (Cell Path.) 52:375 (1987).

35. J. S. O'Brien, Tay-Sachs' disease and juvenile GM2-gangliosidosis, in: "Lysosomes and Storage Diseases", H. G. Hers and F. van Hoof, eds.,

Academic Press, New York-London (1973).

36. U. Sandhoff and K. Harzer, Total hexosaminidase deficiency in Tay-Sachs' disease (variant O), in: "Lysosomes and Storage Diseases" H. G. Hers and F. van Hoof, eds., Academic Press, New York-London (1973).

37. H. Zerban, H. J. Hacker, H. Palmtag and P. Bannasch, Cytomorphological and cytochemical analysis of renal cell carcinoma, Path. Res. Pract. 178:174 (1983).

38. C. Oberling, M. Rivière and F. Haguenau, Ultrastructure des épithéliomas à cellules claires du rein (Hypernéphromes ou tumeurs de Grawitz) et son implication pour l'histogénèse de ces tumeurs, Bull. Assn. Franc. Cancer 46:356 (1959).

39. R. Seljelid and J. L. E. Ericsson, An electron microscopic study of mitochondria in renal clear cell carcinoma, J. Microscopie 4:759 (1965).

40. M. Tannenbaum, Ultrastructural pathology of human renal cell tumors, in: "Pathology Annual", vol. 6, S. C. Sommers, ed., Butterworth, London (1971).

41. G. Weber, F. J. Goulding, R. C. Jackson and J. N. Eble, Biochemistry of human renal cell carcinoma, in: "Characterization and Treatment of Human Tumours", W. Davies and K. R. Harrap, eds., Exerpta Medica (1978).

42. G. Weber, Enzymic programs of human renal adenocarcinoma, in: "Renal Adenocarcinoma", G. Sufrin and S. A. Beckley, eds., UICC Publications, Geneva (1980).

43. D. Mayer and P. Bannasch, Activity of glycogen synthase and phosphorylase and glucose 6-phosphate content in renal clear cell carcinomas, J. Cancer Res. Clin. Oncol. in press (1988).

THE WOUND RESPONSE AS A KEY ELEMENT FOR AN UNDERSTANDING OF MULTISTAGE CARCINOGENESIS IN SKIN

Friedrich Marks[1], Gerhard Fürstenberger[1], Michael Gschwendt[1], Michael Rogers[1], Bärbel Schurich[1], Bernd Kaina[2] and Georg Bauer[3]

[1] German Cancer Research Center, Institute of Biochemistry, Heidelberg. [2] Kernforschungszentrum (Nuclear Research Center), Karlsruhe. [3] Institute of Medical Microbiology and Hygiene, University of Freiburg Federal Republic of Germany

INTRODUCTION

The general appearance of skin carcinoma is that of a steadily growing wound. Thus, Haddow's famous affirmation "the wound may be regarded as a tumor which heals itself"[1] may be also read the other way around, i.e. that a tumor may be regarded as a wound which does not heal. A huge amount of literature dealing with the assumed relationship between wound repair and carcinogenesis has indeed been accumulated (see ref. 1, 6). Only very recently, however, the methods of cell biology, biochemistry and molecular biology have reached a level where they enable the investigator to proof this relationship in clear-cut experimental approaches aiming at an understanding of the molecular mechanisms involved in both wound repair and carcinogenesis. One of the most exciting results of these novel approaches is the discovery that the majority of proto-oncogenes code for components of cellular pathways which are required for the transduction of growth-stimulating signals, especially of those provided by the peptide growth factors[2]. While the physiological role of such growth factors is still not entirely understood, there is accumulating evidence indicating that at least some of them (such as EGF, TGFα, TGFß, PDGF) may be involved in tissue repair and regeneration rather than in the control of everyday tissue growth[3].

For the investigation of both wound repair and carcinogenesis, mouse skin represents one of the most advanced model systems. The reason for this is not only that skin is easy to manipulate and to observe, but also that the skin model offers the invaluable possibility of subdividing the process of tumor development into several defined stages. Under proper experimental conditions these stages can be induced by distinct manipulations or agents.

Presently at least 4 stages of skin carcinogenesis can be distinguished, i.e. initiation, conversion, promotion and malignant progression (Fig. 1). Although these stages are first of all operationally defined by special experimental set ups, recent research has opened several avenues leading to a better understanding of the underlying biological and molecular mechanisms. It is hoped that these investigations will result in the evaluation of new preventive and therapeutic measures.

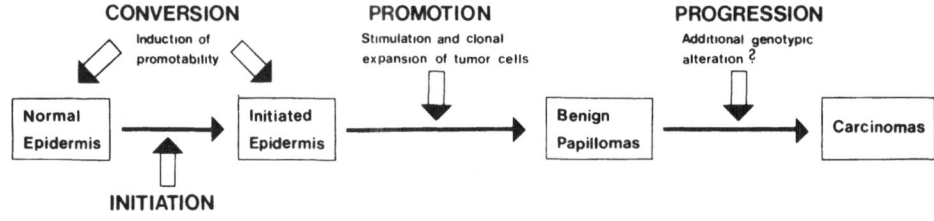

CONVERSION
Induction of
promotability

PROMOTION
Stimulation and clonal
expansion of tumor cells

PROGRESSION
Additional genotypic
alteration?

Normal
Epidermis → Initiated
Epidermis → Benign
Papillomas → Carcinomas

INITIATION
Genotypic alteration

Fig. 1. The stages of experimental carcinogenesis in mouse skin. Following
initiation, papilloma development is induced by continuous growth
stimulation resulting in clonal expansion of tumor cells (promotion). Papilloma growth can be promoted only if conversion has occurred within a certain period of time either prior to or after initiation. Using NMRI mice, initiation is carried out by "subthreshold" carcinogen (DMBA) treatment, conversion by limited applications
(1-4x) of the phorbol ester TPA or by single wounding and promotion
by continuous treatment with the phorbol ester RPA (once a week over
a period of 16-20 weeks) or by repeated wounding. Under these conditions progression to malignancy occurs spontaneously in 10% of the
papillomas within 8-10 months[11]. The rate of progression is increased
by carcinogen treatment of papilloma-bearing skin[17]. From Marks and
Fürstenberger[8].

Skin was the first and, until recently, the only tissue where a tumor-
inducing effect of mechanical wounding could be unequivocally demonstrated[4-7].
Recently, this phenomenon has been shown to be due to convertogenic and tumor-promoting effects of wounding.

THE STAGES OF EXPERIMENTAL SKIN CARCINOGENESIS (for a review see refs. 8-10)

Whether or not a distinct stage of skin carcinogenesis can be observed
depends entirely on the experimental conditions employed, i.e. animal species
and strain, tissue or cell type, agents used, regimen of treatment, time-
point of observation etc.. Thus the experimental approach of multistage carcinogenesis appears to be highly artificial. This does not devaluate, however, its heuristic value as far as an understanding of carcinogenesis in
general is concerned. To deal with misunderstandings which have been coming
up again and again in the literature, it has to be emphasized that the multistage approach represents a model situation created to investigate mechanisms
of tumor development in a reductionistic way rather than mimicking "spontaneous" carcinogenesis, i.e. the human situation.

Generally a multistage carcinogenesis experiment starts with initiation.
This is achieved by (local or systemic) treatment of the animal with a carcinogenic agent, such as a polycyclic aromatic hydrocarbon, UV-light etc.. If
a very low ("subthreshold") dose of a carcinogen is used, the animals will
not develop tumors. It is thought that under such conditions latent (or
"dormant") tumor cells are generated in the epidermis, but that latency period of tumor development starting from such cells surpasses the lifespan of
the animal. The existence of latent tumor cells can be experimentally demonstrated by subsequent tumor induction occurring in the stages called conversion and promotion. The latent tumor cells generated during initiation survive over the whole lifespan of the animal. This indicates that they are
either not subject to normal terminal differentiation or that initiation has
occurred in a self-regenerating stem cell population of epidermis. Dormancy
and persistence of initiated cells means that they have no proliferative advantage over the normal neighbour cells. When, in addition, their program of

differentiation does not greatly differ from that of normal epidermal cells, visible tumors will not develop spontaneously from such cells. It has indeed been shown that in mouse skin the great majority of tumors generated by subsequent promotion of such cells are non-autonomous, well-differentiated papillomas the growth and existence of which depends on continuous growth stimulation by an exogenous agent, i.e. the promoter[11]. In other words: for the promotion of tumor development in initiated skin nothing but a chronic hyperproliferative response seems to be required. This describes exactly the classical 2-stage carcinogenesis experiment where initiation is followed by repeated application of an irritant skin mitogen such as the phorbol ester TPA or by repeated wounding until papillomas become visible. It has been shown that this treatment has to be carried out in such a way that sustained epidermal hyperplasia develops[12]. Otherwise no tumor promoting effect is seen.

The conclusion that the induction of tumor development in initiated skin is solely equivalent with the induction of clonal expansion of initiated cells does not tell us the whole truth. There are, indeed, numerous compounds, which although being strong hyperplasiogenic agents, induce tumor growth in initiated skin only very poorly. This fact led to the discovery of an additional stage of skin carcinogenesis, i.e. conversion. The basic observation was that the tumor-inducing effect of limited, i.e. insufficient TPA-treatment on initiated skin could be "completed" by chronic application of skin irritants which per se exhibited only a very poor effect on tumor development[13-15]. The latter include turpentine[13], mezerein[14] and the phorbol ester retinoylphorbolacetate (RPA, ref. 15). Initially, this approach was called "two-stage tumor promotion", implying TPA to be a "complete" and the other agents to be "incomplete" tumor promoters[14,15]. Since the "first stage" (limited TPA-treatment) can, however, also be carried out prior to initiation[16] it is certainly not a stage of promotion, because under promotion the induction of tumor growth after initiation is understood. Therefore, the term conversion[13] has been introduced and the term promotion has been restricted to the chronic hyperproliferative process induced subsequently to initiation and conversion by agents such as RPA etc.[16].

Since without convertogenic treatment, initiated cells do not respond to the promoting agent by clonal expansion, conversion may be operationally defined as induction of promotability. The mechanisms hidden behind this definition are still not very well understood.

As already mentioned, in the course of the standardized initiation-conversion-promotion experiment primarily papillomas arise. In comparison, carcinoma development is a delayed and low yield event[11]. Whether malignant tumors develop from papillomas (malignant progression) or along an independent route is still a matter of dispute. Carcinoma development depends on conversion and promotion only in that it occurs after papilloma development. As soon as the maximum of papilloma development is reached, the rate of carcinoma development is independent of further promoter application[17]. This observation strongly indicates a progression from the benign to the malignant phenotype. New data on the clonality of skin tumors are, on the other hand, apparently not consistent with such a mechanism[18,19].

MECHANISTIC ASPECTS OF MULTISTAGE CARCINOGENESIS IN SKIN

Initiation and Malignant Progression: Genotoxic Events

Initiation is generally believed to be due to a genotoxic event which results in somatic mutation, oncogene activation etc.. Such a hypothesis most easily explains the irreversibility of the initiated state and is strongly supported by a steadily increasing body of evidence showing the

219

ability of initiating carcinogens to interact with the cellular genome. Moreover, the Harvey-ras proto-oncogene has been shown to be constantly activated and point-mutated in skin papillomas and carcinomas generated along the multistage route with 7,12-dimethylbenz(a)anthracene as an initiating agent[20]. This does not mean, however, that H-ras activation is the only genetic key event of initiation in skin since with other initiating agents no such clearcut correlation between H-ras mutation and carcinogenesis has been observed[20,21]. Since treatment of papillomatous mouse skin with initiating agents leads to a higher carcinoma incidence, malignant progression is thought to result from additional genotoxic effects[17]. This concept is also supported by the fact that the papilloma/carcinoma spectrum in skin depends entirely on the type and the dose of the initiator.

Promotion: Clonal Expansion of Tumor Cells by Continuous Growth Stimulation

As already mentioned, promotion may be understood as clonal expansion of papilloma cells in the course of a chronic hyperproliferative process induced in epidermis. Thus, the mechanisms underlying promotion should be the same as those underlying the induction of epidermal hyperproliferation.

Depending on the stimulus two types of hyperproliferative response can be experimentally induced in mouse skin. They differ from each other in that either epidermal hyperplasia develops or not (unbalanced versus balanced hyperproliferation, see ref. 22). Especially the hyperplastic reaction results in profound morphological and biochemical changes and has, therefore, been called "hyperplastic transformation" of skin (Fig. 2 and ref. 23). Hyperplastic transformation seems to be the general response of skin to injury such as chemical irritation, irradiation or mechanical wounding. The ability of the skin to respond to injury by hyperplastic transformation is subject to ontogenetic development: in the NMRI mouse the response cannot be elicited prior to the second week after birth[22]. All skin tumor promoters induce hyperplastic transformation, i.e. are irritant hyperplasiogenic agents which somehow evoke a response similar to that seen after wounding. The wound response proper is thought to be controlled by endogenous factors ("wound hormones") of systemic or local origin. Recently, the transforming growth factors TGFα[24,25] and TGFß[26] have been shown to be involved in skin wound healing. TGFα has been found to be released by keratinocytes and stimulate re-epithelization of skin wounds probably via an autocrine mechanism[27]. Hyperplastic transformation is mediated by eicosanoids, such as prostaglandins, being released from epidermal cells upon stimulation. In mouse skin eicosanoids seem to fulfil at least two purposes, i.e. to induce epidermal hyperproliferation in synergism with other stimuli and to mediate the inflammatory response which consistently accompanies hyperplastic development in epidermis[28]. Thus, skin inflammation may be an important condition for tumor promotion. Consequently, proinflammatory mediators such as active oxygen species, proteases, eicosanoids and others have been repeatedly shown or at least proposed as playing a key role in tumor promotion (for a review see ref. 29, 30). Further characteristic biochemical features are the induction of the enzyme ornithine decarboxylase (which results in polyamine formation) and a desensitization of epidermal cells for systemic and local growth-inhibitory signals such as catecholamines and epidermal chalone. A central role in hyperplastic transformation of epidermis and thus in skin tumor promotion seems to be played by the IP_3-DAG signal transduction system. This is indicated by the fact that phorbol esters, which are among the most active inducers of epidermal hyperplasia and tumor promotion, specifically activate PKC by mimicking the effect of DAG[31]. Actually, within the phorbol ester series a strong correlation exists between hyperplasiogenic efficacy and PKC-stimulatory activity. In addition, the strong promoters are potent inducers of the arachidonic cascade. Whether the latter response is linked to PKC activation or occurs independently, remains to be established. Although the cellular substrate proteins of epidermal PKC are still not known, there is

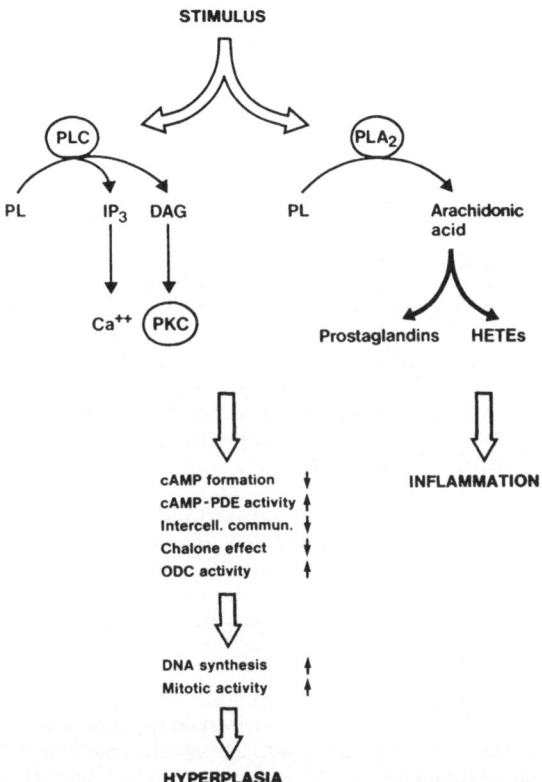

Fig. 2. Mechanistic aspects of hyperplastic transformation in mouse skin. It
is assumed that an important event in the induction of hyperplastic
transformation consists in an activation of two pathways of signal
transduction, i.e. the IP_3/DAG- and the arachidonic acid cascade,
via receptor-mediated stimulation by an endogeneous stimulus ("wound
hormone") of phosphatidyl-inositolphosphate-specific phospholipase C
(PLC) and activation of phospholipase A_2 (PLA_2; whether PLA_2 activation occurs independently of or as a consequence of PLC activation
remains to be established). The mediators released, i.e. inositol-
trisphosphate (IP_3), diacylglycerol (DAG), prostaglandins and hy-
droxylated eicosatetraenoic acids (HETEs, ref. 88,89) give rise to
the responses listed in the lower part of the diagram, whereby
prostaglandins mediate the hyperplastic and HETEs probably the inflammatory response. Strong tumor promoters such as the phorbol
esters act by mimicking the stimulatory effect of DAG on protein
kinase C (PKC) and inducing the arachidonic acid cascade.

indirect evidence that several events occurring in the course of hyperplastic
transformation, such as ODC-induction[32], inhibition of intracellular communication[33] and others, are linked to PKC-catalyzed protein phosphorylation.
Under physiological conditions the IP_3-DAG cascade is activated by exogeneous
signals such as hormones via interaction with a membrane receptor[34,35]. The
endogeneous activator of the cascade in epidermis has not yet been identified. While the concept of DAG as playing a key role in hyperplastic transformation is thus generally accepted, the role of IP_3 is less clear. The main
function of IP_3 is that of an intracellular Ca^{2+}-mobilizing signal. Thus,
Ca^{2+}-dependent reactions occurring in the course of hyperplastic transformation may be controlled via IP_3 release. Among those reactions, the phosphorylation of ribosomal elongation factor 2 by a Ca^{2+}/calmodulin-dependent

proteinkinase might be related to hyperplastic transformation and tumor promotion in that an inhibition by cyclosporine A of both processes correlated with an inhibition of EF-2 phosphorylation [36]. Another Ca^{2+}-dependent reaction essential for hyperplastic transformation is the activation of phospholipase A_2, i.e. the rate-limiting step of arachidonic acid metabolism (see above).

Conversion: Stem Cell Activation by Chromosomal Damage?

As compared with initiation and promotion, the mechanisms of which are thought to be understood, at least in principle, as somatic mutation and chronic induction of hyperplastic transformation, the mechanism of conversion is still widely unknown. Obviously conversion somehow sensitizes initiated cells for promotion. In other words: an initiated cell appears to be unable to respond to a hyperplasiogenic stimulus by papilloma formation, unless initiation has occurred in skin pretreated with a convertogenic agent or convertogenic treatment follows initiation. Theoretically two possibilities exist to explain the insensibility of non-converted initiated cells to a promoter:

1. Like a normal epidermal cell the initiated cell is stimulated by the promoter to proliferate but does not express its neoplastic phenotype, i.e. its behavior including its proliferative capacity is indistinguishable from that of a normal epidermal cell.
2. The initiated cell is unable to respond to the promoter by hyperproliferation.

In the first case the initiated cell is looked upon as a "recessive mutant" which has to be phenotypically activated by the convertogenic agent [13]. Such a mechanism appears to be incompatible with the "inverted experiment" where conversion is induced prior to initiation [16]. On the other hand, conversion may create a long-lasting situation in the tissue which allows a spontaneous phenotypic expression as soon as initiation occurs. An attractive new concept - admittedly purely speculative as far as mouse skin is concerned - says that the neoplastic phenotype is under the control of so-called tumor suppressor genes which have to be eliminated or inactivated before neoplastic development can occur [37,38]. In the framework of our first concept this would mean that conversion is due to transient suppressor gene inactivation.

In the second case it is assumed that in the target cells of initiation the molecular machinery required for the hyperplastic response is not fully developed but has to be induced by the convertogenic agent. Such a hypothesis would be compatible with the concept of initiation occurring in a primitive stem cell compartment of epidermis and with the ontogeny of the hyperplastic response, as mentioned above. Conversion would then be due to "stem cell activation" whatever this means. It has been shown that in mouse epidermis a slowly proliferating cell population proposed to consist of stem cells ("label-retaining cells") is stimulated to proliferate by the convertogenic promoter TPA but not by the non-convertogenic promoter RPA [39].

As can be seen, both concepts imply that the convertogenic agent has an effect on gene expression. There are indeed numerous reports on genetic effects of phorbol ester tumor promoters, including the induction of sister chromatid exchanges in fibroblasts [40], of DNA single strand breaks in leukocytes [41] and keratinocytes [42,43], of chromosomal aberrations in yeast [44] and leukocytes [45,46] and of the replication of integrated viral genomes [47-50]. In addition, the expression of distinct genes such as the ODC-gene, c-fos, c-myc [51-56] and others occurs upon phorbol ester treatment of different cell types. However, most of these effects are shared by both convertogenic and non-convertogenic phorbol esters [53] and, therefore, are apparently useless as far as the elucidation of the convertogenic process is concerned. What has to be looked are conversion-specific events. One approach is to compare the effects of a convertogenic promoter such as TPA with those of a non-convertogenic substance such as RPA on the target tissue of multistage carcinogenesis, i.e. mouse epidermis. This has been done with the result that both agents behave practically identical, for example as far as the reactions summarized in Fig. 2 are concerned (except prostaglandin $F_{2\alpha}$ formation, see ref. 28). Recently, however, a clear-cut difference was found in that the convertogenic tumor promoter TPA induces chromosomal aberrations in NMRI mouse epidermis in vivo (unpublished results) and in primary mouse keratinocytes in vitro[57,58], whereas the non-convertogenic tumor promoter RPA does not (Fig. 3).

The conclusion that conversion may be somehow related to chromosomal damage (clastogenicity) has been strongly supported by the finding that methylmethane sulfonate (MMS) exhibits both convertogenic (Fig. 4) as well as chromosome-damaging efficacy in the mouse skin system (unpublished results). MMS is known for its strong clastogenic effect, whereas its gene-mutagenic potency is rater low in animal cells[59-63]. Actually, in mouse skin, MMS is almost inactive as an initiator (Fig. 4) indicating again that there is a relationship between initiation and mutation rather than between initiation and clastogenicity. As shown in Fig. 4 the convertogenic effect of MMS is augmented by cotreatment with RPA. The possible reason for this synergistic effect is explained below. An event which might be related to clastogenicity is the modification of DNA bases in phorbol-ester-treated polymorphonuclear leukocytes which is mediated by intracellular formation of hydrogen peroxide[64]. TPA has been found as being much more efficient in inducing this reaction than RPA.

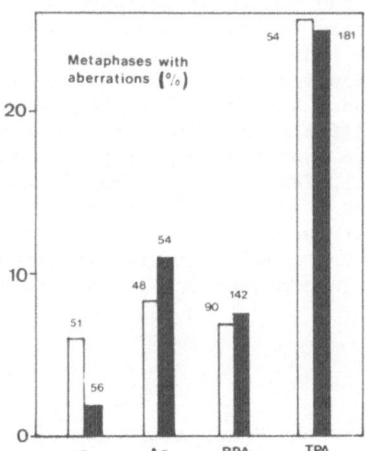 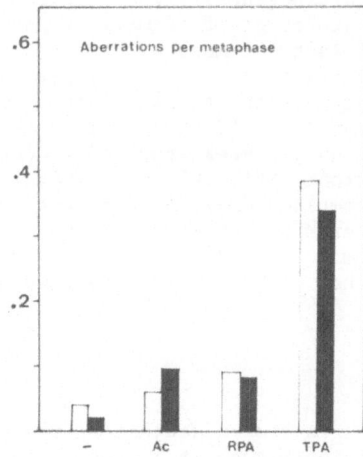

Fig. 3.   Clastogenic effect of the convertogenic tumor promoter TPA in mouse epidermis in vivo. The left diagram shows the percentage of metaphases with chromosomal aberrations (mainly breaks, gaps and fragments), the right diagram the number of chromosomal aberrations per metaphase in basal epidermal keratinocytes obtained from mice locally treated once with acetone (Ac, 0.1 ml), or phorbol esters RPA and TPA (10 nmol, 0.1 ml acetone). 48 h after treatment the animals were killed and basal cell fractions 3 and 4 were obtained from skin preparations according to ref. 90. After in vitro cultivation of these cells for 2 or 3 days[90] metaphases were obtained and analyzed according to ref. 91. Black columns: basal keratinocyte-fraction 3; empty columns: basal keratinocyte-fraction 4; labelling of columns: numbers of metaphases analyzed.

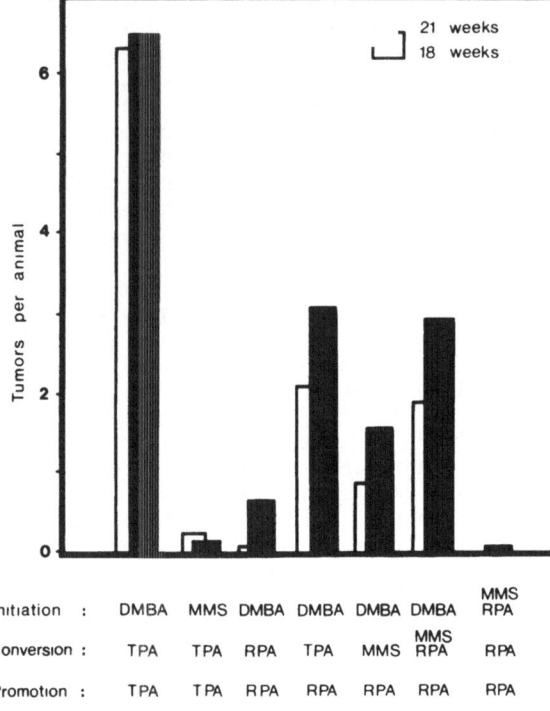

Fig. 4. Methylmethane sulfonate (MMS) as a convertogenic agent in mouse skin
carcinogenesis. The diagram shows the number of papillomas per mice
(groups of 16 animals) found 18 weeks (empty columns) and 21 weeks
(black columns) after treatment as indicated below the ordinate. For
initiation (DMBA) a single local application at the beginning of the
experiment was made, for conversion two applications (MMS or TPA at
day 7 and 10 after initiation) and for promotion (RPA) one application per week starting 1 week after conversion. Dose per application: DMBA, 100 nmol, MMS, 100 µmol, TPA, 20 nmol, RPA, 10 nmol. For
combined MMS/RPA-treatment (columns 6 and 7), RPA was given 6 h after MMS. Note that MMS while exhibiting no initiating efficacy (columns 2 and 7) shows a distinct convertogenic effect which equals
that of TPA (column 4) when given in combination with RPA (column 6)

These results indicate that the genetic events assumed to be essential
for conversion may occur in the course of a clastogenic reaction rather than
being due to gene mutation. The concept of chromosomal damage playing an
important role in tumor development has been recently put forward by other
authors[65,66]. However, no distinction was made between conversion and promotion. Free radicals and oxygen species have been postulated as clastogenic
factors. Recently, arachidonic acid metabolites generated along the lipoxygenase pathway, such as hydroperoxy eicosatetraenoic acids, were identified
as clastogenic factors in leukocytes[67]. Similar arachidonic acid metabolites
are also probably involved in the clastogenic effect of TPA on NMRI mouse
epidermis as indicated by the inhibitory effect of ETYA (eicosatetraynoic
acid) on both the induction of chromosomal aberrations (unpublished results)
and conversion[68]. In the dose applied in these experiments, ETYA inhibits
the lipoxygenase pathway of arachidonic acid metabolism much stronger than
the cyclooxygenase pathway.

Conversion by "Wound Hormones"

As yet no decision can be made as to whether the chromosomal aberrations observed after TPA treatment indicate more specific events such as gene amplification and translocation or are symptoms of cell damage and cell-death. In this connection, the observation that wounding exerts both a strong convertogenic (Fig. 5) and promoting effect in initiated skin may provide a further clue for a deeper understanding of conversion, since it shows that conversion can be entirely induced by endogeneous factors, i.e. somehow resembles a physiological process.

Looking for such factors with the qualities of a "wound hormone" (i.e. being locally released upon wounding and acting on cells of the wounded tissue), we have recently found a platelet-derived polypeptide that exerts convertogenic efficacy in the initiated mouse skin in vivo (Fig. 6). Because of a special in vitro activity this peptide has been called Epstein-Barr-Virus-inducing factor (EIF, ref. 69). EIF was considered as a candidate for a convertogenic agent since it exhibits a strong synergism with TPA in several in vitro systems including a "tumor-promoting" effect in $C_3H10T1/2$ cells[70,71]. According to its biological properties, EIF is a member of the TGFß-family of growth factors[72]. Recently, TGFß has indeed been shown to play a role in skin wound healing[26]. TGFß/EIF stimulates fibroblast proliferation and inhibits epidermal cell growth. The latter may be the reason why the convertogenic effect of TGFß/EIF is seen only when the intracutaneous injection of

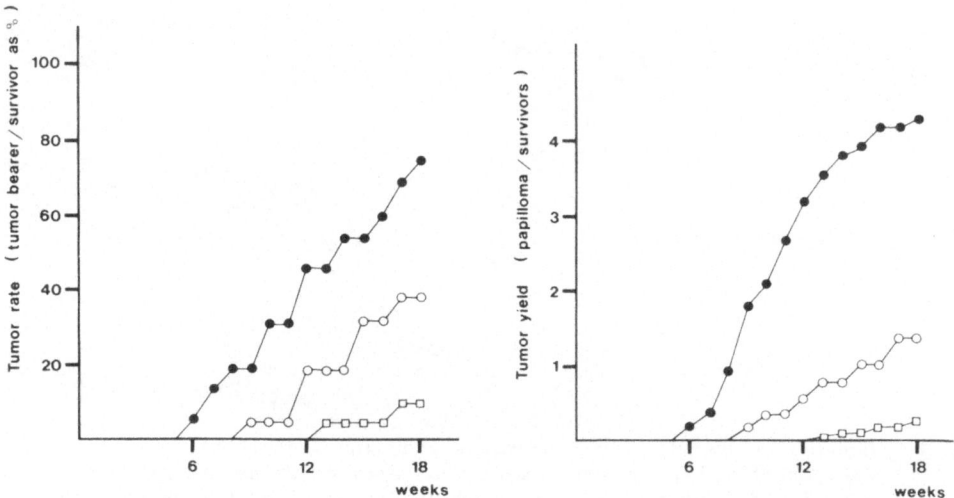

Fig. 5. Skin wounding as convertogenic treatment. The figure shows the result of a three-stage carcinogenesis experiment carried out with groups of 16 NMRI mice each. Initiation was carried out by topical application of 100 nmol 7,12-dimethylbenz(a)anthracene (in 0.1 ml acetone) at zero time. One week later, the animals were treated once with 10 nmol RPA (□), or 10 nmol TPA (●) or wounding (six cuts of 1 cm length into the initiated back skin). Two weeks later, promotion was started by local application of 10 nmol RPA and continued over the whole period of the experiment (2 applications per week). Note that the rate of tumor development (left diagram) is approximately identical in wounded and TPA-treated animals, whereas the tumor yield (right diagram) is lower in wounded as compared with TPA-treated mice, probably because the "converted" skin area is much smaller upon skin cutting than upon painting with a TPA solution.

the factor is combined with a mitogenic stimulus provided by simultaneous application of the non-convertogenic phorbol ester RPA (Fig. 6). It has indeed been shown that the induction of DNA replication is a necessary but not sufficient condition for conversion[73]. Thus EIF seems to "complete" the tumor-inducing effect of RPA for the convertogenic component and may thus be regarded as a "pure" convertogenic agent (whereas TPA is a promoter/mitogen with additional convertogenic potency!). It is hoped that the investigation of the mechanism of action of EIF alias TGF in epidermis will finally result in an understanding of the mechanism of conversion.

In a skin wound the mitogenic stimulus provided in our experiment by RPA (Fig. 6) must come from additional endogenous factors. TGFα (and EGF) is a good candidate since it has been shown to stimulate the re-epithelization of wounds probably via an autocrine mechanism[24,25]. Actually, TGFα can entirely replace RPA in an experiment such as shown in Fig. 6 (unpublished). We postulate, therefore, that the convertogenic and tumor-promoting effect of skin-wounding in initiated skin is due to a combined action of endogeneous mitogens such as TGFα with factors of the TGFß-family and eicosanoids as well as other "mediators of inflammation".

Is Conversion Due to "Programmed Cell Death"?

Recently the concept has been put forward that DNA strand breaks, such as induced by clastogenic factors may cause cell death along a specific mole-

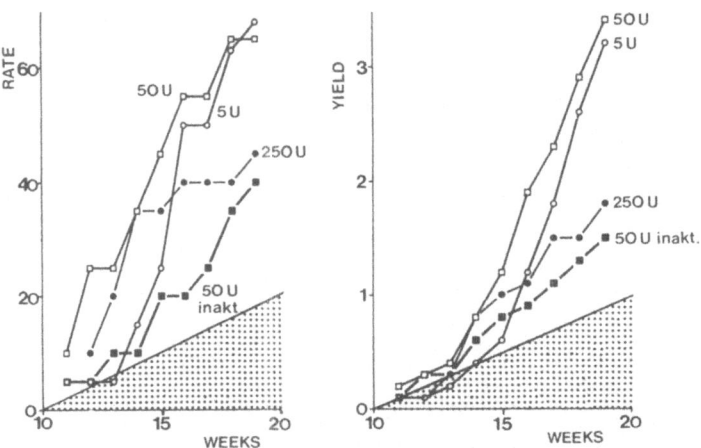

Fig. 6. Conversion by "Epstein-Barr-Virus inducing factor, EIF", a member of the TGFß-family. The experiment was carried out as described for Fig. 4 and 5 except that wounding or TPA application was replaced by a 4-point intracutaneous injection of EIF (O, 5 units;□, 50 units;●, 250 units;■, 50 units heat-inactivated) combined with a local application of 10 nmol phorbol ester RPA in 0.1 ml acetone. The hatched zones represent the results of control experiments where saline was injected instead of EIF. The low convertogenic effect of this control treatment is probably due to wounding by injection. When EIF was injected but the accompanying RPA application was omitted the resulting values did not differ from the controls (not shown). Left diagram: tumor rate (% papilloma-bearing animals per group). Right diagram: tumor yield (number of papillomas per animal). Note that the EIF effect is highest at a dose of 50 units, declining at higher doses. 1 Unit is the dose of EIF which induces half maximal Epstein-Barr early antigen production in Raji cells when given together with the synergism producing agent iodo-deoxyuridine.

cular mechanism. This "cellular euthanasia"[74] is thought to be mediated by the enzyme ADP-ribosyltransferase which is activated upon DNA-damage. By catalyzing ADP-ribosylation of various proteins, especially in the cell nucleus, ADP-ribosyltransferase causes the depletion of the cell for NAD^+ and, in turn, for ATP. These events may ultimately result in cell death. Cell death is probably an important event in the wound response[75] since it guarantees a) a cleansing of the wounded area from damaged and postmitotic cells which would otherwise hinder a re-epithelization of the wound and b) an activation of stem cell proliferation and maturation (provided that the stem cells are not subject to killing). Under physiological conditions cell death may be subject to hormonal regulation rather than being induced by exogeneous clastogenic agents. Such "programmed cell death" or apoptosis is best exemplified by the killing effect of glucocorticoids on lymphocytes[76]. It has been proposed that the formation of DNA single strand breaks initiating apoptosis is catalyzed by a nuclear Ca^{2+}/Mg^{2+}-dependent DNase which is probably activated upon Ca^{2+}-translocation into the cell nucleus[77]. These Ca^{2+}-ions may be released from mitochondria upon stimulation by lipid radicals such as those generated along the pathways of arachidonic acid metabolism[78]. Convertogenic agents such as TPA have indeed been shown to induce a pronounced translocation of Ca^{2+}-ions from the cytoplasm into the nucleus[79]. If one expects the hypothesis of induced cell death as being an important condition for epidermal stem cell activation and, thus, for tissue regeneration and conversion, one is left with the question, how endogeneous factors such as TGFß/EIF fit such a concept. An answer must remain speculative as long as the mechanism of action of TGFß/EIF in epidermis is not understood. As already mentioned above, TGFß/EIF inhibits epidermal cell proliferation. Recently it has been proposed that this inhibition results from a stimulation of terminal differentiation[80]. When measuring the activity of epidermal transglutaminase as a biochemical parameter of keratinocyte differentiation we were, however, unable to find any evidence for such a mechanism in the mouse skin system[72]. Perhaps the growth inhibitory effect of TGFß/EIF indicates an apoptogenic mechanism leading to programmed cell death in wounded skin?

THE POTENTIAL IMPORTANCE OF THE MULTISTAGE MODEL FOR A PREVENTION OF HUMAN CANCER

The multistage model of skin carcinogenesis is certainly a highly artificial approach which allows the investigator to dissect the complex pattern of tumor development so that the process can be analyzed stage by stage. The argument that such a situation never occurs in daily life and makes the model irrelevant for the human situation is frequently raised, but nevertheless trivial, since one should not demand more from an animal model than it can provide, i.e. an insight into biological pathways along which tumors may develop. The most important results which have emerged from the mouse skin model are:

1. Tumor development can be induced by a combination of treatments each of which appears to be harmless when carried out alone, but which potentiate each other in a strongly synergistic mode.
2. The initiating effect of a carcinogenic agent on the tissue is irreversible but remains latent provided the dose of the agent is low.
3. The post-initiation stages of skin carcinogenesis, i.e. conversion and promotion, proceed along physiological pathways normally involved in defense against harmful external influences, wound response and tissue regeneration. They facilitate the formation of papillomas from initiated cells. In contrast to initiation, the converted and promoted state is reversible so long as autonomous papillomas have not developed.
4. The progression from the benign to the malignant state probably requir-

es additional effects on the cellular genome; the preceeding clonal expansion of papilloma cells due to promotion greatly increases the probability of such genetic alterations.

At least in skin, initiation is probably an everyday-event, thought to be mostly due to external influences such as UV-light. The close relationship of conversion and promotion to physiological processes allows the proposal that the development of tumors from initiated cells could be induced not only by exogenous (environmental) but, perhaps with a lower frequency, also by endogenous factors. This may perhaps help to explain the well-known effect of certain hormones on human cancer development.

The problem of whether and how the mouse skin model is of relevance for an understanding of human cancer development has not yet been settled. Since the pro's and contra's have been recently discussed in depth[10,81,82] it is unnecessary to repeat this discussion here. Epidemiological studies indicate that promoting influences may be important, or even critical, in the development of certain human cancers. The most striking evidence for this is the rather long latency period of most malignant diseases. In addition, risk reversibility and the synergistic effect of many carcinogenic agents show a striking parallelism to the situation in the skin model. For example, the carcinogenic effects of tobacco smoke and asbestos or ionizing radiation potentiate each other, as do the carcinogenic effects of initiators and promoters in the animal model. The risk of getting lung cancer decreases once smoking is stopped showing that the process has a reversible component (promotion?). In fact, this is, again, exactly what one would predict from the skin model, where the risk of getting skin carcinomas is drastically reduced when promoter treatment is stopped before the endpoint of papilloma development is reached.

This brings us to another practical impact of the multistage model, i.e. that it may help to develop measures of cancer prevention. A patient could indeed live with initiated or even fully transformed cells provided one would be able to inhibit the post-initiation stages of carcinogenesis. This requires, of course, a timely diagnosis of pre-malignant states.

Numerous agents have been described as being able to inhibit promotion and conversion and thus bring tumor development to a halt in the mouse skin model[83,84]. Generally spoken, all compounds which are able to inhibit the hyperplastic transformation of skin have to be regarded as potential anti-promoters. Such an effect has been shown for inhibitors of DNA replication, RNA- and protein biosynthesis, for glucocorticosteroids, non-steroidal antiphlogistica, antioxidants, vitamin A acid and its derivatives, inhibitors of ornithine decarboxylase and protease inhibitors. Epidermal G_1-chalone, a local endogenous inhibitor of epidermal cell proliferation, does not inhibit tumor promotion because upon hyperplastic transformation epidermis becomes desensitized for this factor [23].

The effect of non-steroidal antiphlogistica such as indomethacin, aspirin or ETYA is most probably due to an inhibition of arachidonic acid metabolism. It has indeed been shown that an early release of prostaglandin E_2 in mouse skin is obligatory for the triggering of the hyperplastic response while a delayed formation of prostaglandin $F_{2\alpha}$ as well as of arachidonic acid metabolites generated along the lipoxygenase pathway is essential for tumor development, i.e. promotion and conversion[28,68,89]. Whether the antiproliferative and anti-promoting effect of glucocorticosteroids is also related to arachidonic acid metabolism (inhibition of phospholipase A_2) is not yet clear. Recently drugs known as immunosuppressants have been found to be especially powerful inhibitors of hyperplastic transformation including skin inflammation and tumor promotion. These compounds include the antibiotics cyclosporine[85] and didemnin[86]. When the effects of different cyclosporine

derivatives were compared a correlation between anti-promoting/anti-prolife-
rative potency and immunosuppressive efficacy of these compounds was not observed [87]. Instead the inhibition of hyperproliferation seems to correlate
with the inhibition of phosphorylation of an epidermal 100 kD protein identified as ribosomal elongation factor EF-2 [36]. EF-2-phosphorylation is apparently catalyzed by a Ca^{2+}/calmodulin-dependent proteinkinase, and the cyclosporine effect is due to an interaction of the antibiotic with calmodulin [36].

Thus, at least in the mouse skin model, several biochemical pathways exist the blockade of which may result in an inhibition of tumor development,
i.e. promotion/conversion. It must be emphasized, however, that in other organs multistage carcinogenesis may proceed along other cellular pathways and
may, therefore, be sensitive to other types of inhibitors. It may thus be
impossible to design a generalized scheme for the prevention of the post-initiation stages of tumor development by extrapolating the results obtained by
studying experimental skin carcinogenesis. Rather preventive measures have
to be determined specifically for every tissue and tumor type.

The observation that wounding exerts a strong convertogenic/promoting
effect in the mouse skin model appears to be especially harassing as far as
the human situation is concerned. One may, of course, argue that the wound
effect is restricted to animals and to skin, but one has to be aware of the
fact that such an argument actually lacks any substantiation. If we accept
the model character of animal experiments, the purpose of which is to open
our eyes for biological pathways along which human diseases could develop as
well, then we feel that the skin model of carcinogenesis has created an urgent demand for a re-evaluation and indepth investigation of the effects on
wounding (including the taking of biopsies!) and chronic inflammation on the
development and progression of neoplastic diseases in man.

REFERENCES

1. A. Haddow, Molecular repair, wound healing and carcinogenesis: tumor
   production a possible overhealing? Adv. Cancer Res. 16:181 (1972).
2. F. Marks, What's new in oncogenes and growth factors? Pathology Res.
   Pract. 182:831 (1988).
3. M. B. Sporn and A. B. Roberts, Peptide growth factors and inflammation,
   tissue repair and cancer, J. Clin. Invest. 78:329 (1986).
4. H. Hennings and R. K. Boutwell, Studies on the mechanism of skin tumor
   promotion, Cancer Res. 30:312 (1970).
5. I. Clark-Lewis and A. W. Murray, Tumor promotion and the induction of
   epidermal ornithine decarboxylase activity in mechanically stimulated mouse skin, Cancer Res. 38:494 (1978).
6. T. S. Argyris, Regeneration and the mechanism of epidermal tumor promotion, CRC Crit. Rev. Toxicol. 14:211 (1986).
7. G. Fürstenberger and F. Marks, Growth stimulation and tumor promotion in skin, J. Invest. Dermatol. 81:157s (1983).
8. F. Marks and G. Fürstenberger, Multistage carcinogenesis: The mouse
   skin model, in: "Accomplishments in Oncology", Vol. 2/1, H. zur
   Hausen and J. R. Schlehofer, eds., Lippincott, Philadelphia (1987).
9. F. Marks and G. Fürstenberger, From the normal cell to cancer: The
   multistep process of experimental skin carcinogenesis, in: "Concepts
   and Theories in Carcinogenesis", A. P. Maskens, P. Ebbesen and A.
   Burny, eds., Excerpta Medica, Amsterdam (1987).
10. F. Marks and G. Fürstenberger, Multistage carcinogenesis in animal
    skin: The reductionist's approach in cancer research, in: "Theories
    of Carcinogenesis", O. H. Iversen, ed., Hemisphere, Washington D.C.
    (1988).
11. F. J. Burns, M. Vanderlaan, E. Snyder and R. E. Albert, Induction and
    progression kinetics of mouse skin papillomas, in: "Mechanisms of

Tumor Promotion and Cocarcinogenesis", T. J. Slaga, A. Sivak and R. K. Boutwell, eds., Raven Press, New York (1978).

12. E. E. Sisskin, T. Gray and J. C. Barrett, Correlation between sensitivity to tumor promotion and sustained epidermal hyperplasia of mice and rats treated with TPA, Carcinogenesis 3:403 (1982).

13. R. K. Boutwell, Some biological aspects of skin carcinogenesis, Progr. Exp. Tumor Res. 4:207 (1964).

14. T. J. Slaga, S. M. Fischer, K. Nelson and G. L. Gleason, Studies on the mechanism of skin tumor promotion: Evidence for several stages of promotion, Proc. Natl. Acad. Sci. USA 77:3659 (1980).

15. G. Fürstenberger, D. L. Berry, B. Sorg and F. Marks, Skin tumor promotion by phorbol esters is a two-stage process, Proc. Natl. Acad. Sci. USA 78:7722 (1981).

16. G. Fürstenberger, V. Kinzel, M. Schwarz and F. Marks, Partial inversion of the initiation-promotion sequence of multistage tumorigenesis in the skin of NMRI mice, Science 230:76 (1985).

17. H. Hennings, R. Shores, M. L. Wenk, E. F. Spangler, R. Tarone and S. H. Yuspa, Malignant conversion of mouse skin tumors is increased by tumor initiators and unaffected by tumor promoters, Nature 304:67 (1983).

18. A. L. Reddy, M. Caldwell and P. J. Fialkow, Studies of skin tumorigenesis in PGK mosaic mice: Many promoter-independent papillomas and carcinomas do not develop from pre-existing promoter-dependent papillomas, Int. J. Cancer 39:261 (1987).

19. A. L. Reddy, M. Caldwell and P. J. Fialkow, Sequential studies of skin tumorigenesis in phosphoglycerate kinase mosaic mice: Effect of resumption of promotion on regressed papillomas, Cancer Res. 47:1947 (1987).

20. M. Quintanilla, K. Brown, M. Ramsden and A. Balmain, Carcinogen-specific mutation and amplification of Ha-ras during mouse skin carcinogenesis, Nature 322:78 (1986).

21. A. Balmain and K. Brown, Oncogene activation in chemical carcinogenesis, Adv. Cancer Res. in press (1987).

22. F. Marks, S. Bertsch, G. Fürstenberger and H. Richter, Growth control in mouse epidermis: Facts and speculations, in: "Psoriasis: Cell Proliferation", N. A. Wright and R. S. Camplejohn, eds., Churchill Livingstone, Edinburgh (1983).

23. F. Marks, G. Fürstenberger, M. Ganss, H. Richter and D. Seemann, Hyperplastic transformation: The response of mouse skin to injury, Brit. J. Dermatol. 109, Suppl. 25:18 (1983).

24. G. S. Schultz, M. White and R. Mitchell, Epithelial wound healing enhanced by transforming growth factor-α and by Vaccinia growth factor, Science 235:350 (1987).

25. A. B. Schreiber, M. E. Winkler and R. Derynck, Transforming growth factor-α: A more potent angiogenic mediator than epidermal growth factor, Science 232:1250 (1986).

26. A. B. Roberts, M. B. Sporn, R. K. Assoian, J. M. Smith, N. S. Roche, L. M. Wakefield, U. I. Heine, L. A. Liotta, V. Falanger, J. H. Keteri and A. S. Fanci, Transforming growth factor type ß: Rapid induction of fibrosis and angiogenesis in vivo and stimulation of collagen formation in vitro, Proc. Natl. Acad. Sci. USA 83:4167 (1986).

27. R. J. Coffey, R. Derynck, J. N. Wilcox, T. S. Bringman, A. S. Goustin, H. L. Moses and M. R. Pittelkow, Production and auto-induction of transforming growth factor-α in human keratinocytes, Nature 328:817 (1987).

28. G. Fürstenberger and F. Marks, Prostaglandins, epidermal hyperplasia and skin tumor promotion, in: "Arachidonic Acid Metabolism and Tumor Promotion", S. M. Fischer and T. J. Slaga, eds., Nijhoff, Boston (1985).

29. F. Marks and G. Fürstenberger, Tumor promotion in skin: Are active oxygen species involved? in: "Oxidative Stress", H. Sies, ed.,

Academic Press, New York (1985).

30. W. Troll and R. Wiesner, The role of oxygen radicals as possible mechanism of tumor promotion, Ann. Rev. Pharmacol. Toxicol. 25:509 (1985).

31. M. Castagna, Y. Takai, K. Kaibuchi, K. Sano, V. Kikkawa and Y. Nishizuka, Direct activation of calcium-activated, phospholipid-dependent protein-kinase by tumor-promoting phorbol esters, J. Biol. Chem. 257:7847 (1982).

32. R. C. Smart, M. Huang and A. H. Conney, sn-1,2-Diacylglycerols mimic the effects of TPA in vivo by inducing biochemical changes associated with tumor promotion in mouse epidermis, Carcinogenesis 7:1865 (1986).

33. H. Yamasaki, Aberrant control of intercellular communication and cell differentiation during carcinogenesis, in: "Concepts and Theories in Carcinogenesis", A. P. Maskens, P. Ebbesen and A. Burny, eds., Excerpta Medica, Amsterdam (1987).

34. A. A. Abdel-Latif, Calcium-mobilizing receptors, polyphosphoinositides, and the generation of second messengers, Pharmacol. Rev. 38:228 (1986).

35. M. J. Berridge, Inositol lipids and cell proliferation, Biochim. Biophys. Acta 907:33 (1987).

36. M. Gschwendt, W. Kittstein and F. Marks, Cyclosporin A inhibits phorbol ester-induced hyperplastic transformation and tumor promotion in mouse skin probably by suppression of Ca /calmodulin-dependent processes such as phosphorylation of elongation factor 2, Skin Pharmacol. in press (1988).

37. R. Sager, Genetic suppression of tumor formation, Adv. Cancer Res. 44:43 (1985).

38. A. C. Knudson, Hereditary cancer, oncogenes and anti-oncogenes, Cancer Res. 45:1437 (1985).

39. J. A. McCutcheon, J. C. Bickenbach and J. C. Mackenzie, Effect on label-retaining cells of tumor promoters and differing levels of hyperplasia, J. Dental Res. 64:298 (1985).

40. A. R. Kinsella and M. Radman, Tumor promoter induces sister chromatid exchanges: Relevance to mechanisms of carcinogenesis, Proc. Natl. Acad. Sci. USA 75:6149 (1978).

41. H. C. Birnboim, DNA strand breakage in human leukocytes exposed to a tumor promoter, phorbol myristate acetate, Science 211:1247 (1982).

42. D. R. Dutton and G. T. Bowden, Indirect induction of clastogenic effect in epidermal cells by a tumor promoter, Carcinogensis 6:1279 (1985).

43. J. A. Hartley, N. W. Gibson, L. A. Zwelling and S. H. Yuspa, The association of DNA strand breaks and terminal differentiation in mouse epidermal cells exposed to tumor promoters, Cancer Res. 45:4864 (1985).

44. J. M. Parry, E. M. Parry and J. C. Barrett, Tumor promoters induce mitotic aneuploidy in yeast, Nature 294:263 (1981).

45. D. F. Callen and J. H. Ford, Chromosome abnormalities in chronic lymphocytic leukemia revealed by TPA as mitogen, Cancer Genet. Cytogenet. 10:87 (1983).

46. I. Emerit and P. A. Cerutti, Tumor promoter phorbol-12-myristate-13-acetate induces chromosomal damage via indirect action, Nature 293:144 (1981).

47. R. J. Imbra and M. Karin, Phorbol ester induces the transcriptional stimulatory activity of the SV40 enhancer, Nature 323:555 (1986).

48. P. B. Fisher, I. B. Weinstein, D. Eisenberg and H. S. Ginsberg, Interactions between adenovirus, a tumor promoter, and chemical carcinogens in transformation of rat embryo cell cultures, Proc. Natl. Acad. Sci. USA 75:2311 (1978).

49. H. zur Hausen, F. J. O'Neill and U. K. Freese, Persisting oncogenic herpes-virus induced by the tumor promoter TPA, Nature 272:373 (1978).

50. H. zur Hausen, G. W. Bornkamm, R. Schmidt and E. Hecker, Tumor initiators and promoters in the induction of Epstein-Barr virus, Proc. Natl. Acad. Sci. USA 76:782 (1979).

51. A. K. Verma, D. Erickson and B. J. Dolnick, Increased mouse epidermal ornithine decarboxylase activity by the tumor promoter TPA involves increased amounts of both enzyme protein and messenger RNA, Biochem. J. 237:297 (1986).

52. S. K. Gilmour, A. K. Verma, Th. Madara and T. G. O'Brien, Regulation of ornithine decarboxylase gene expression in mouse epidermis and in epidermal tumors during two-stage tumorigenesis, Cancer Res. 47:1221 (1987).

53. S. Rose-John, G. Fürstenberger, P. Krieg, E. Besemfelder, G. Rincke and F. Marks, Differential effects of phorbol esters on c-fos and c-myc and ornithine decarboxylase gene expression in mouse skin in vivo, Carcinogenesis: in press (1988).

54. M. E. Greenberg and E. B. Ziff, Stimulation of 3T3 cells induces transcription of the c-fos protooncogene, Nature 311:433 (1984).

55. R. Mueller, R. Bravo, J. Burckhardt and T. Curran, Induction of c-fos-gene and protein by growth factors precedes activation of c-myc, Nature 312:716 (1984).

56. G. P. Dotto, M. Z. Gilman, M. Maruyama and R. A. Weinberg, c-myc and c-fos expression in differentiating mouse primary keratinocytes, Embo. J. 5:2853 (1986).

57. R. T. Petrusevska, G. Fürstenberger, F. Marks and N. E. Fusenig, Cytogenetic effects caused by phorbol ester tumor promoters in primary mouse keratinocyte cultures: Correlation with the convertogenic activity of TPA in multistage skin carcinogenesis, Carcinogenesis, in press (1988).

58. R. T. Petrusevska, N. Pohlmann and N. E. Fusenig, Induction of chromosomal abberations in primary mouse keratinocyte cultures by tumor-promoting phorbol esters and inhibition of cytogenetic effects by antipromoters, in: "Accomplishments in Oncology", Vol. 2/1, H. zur Hausen and J. R. Schlehofer, eds., Lippincott, Philadelphia (1987).

59. E. Vogel and A. T. Natarajan, The relation between reaction kinetics and mutagenic actions of monofunctional alkylating agents in higher eukaryotic systems. II. Total and partial sex-chromosome loss in Drosophila, Mutation Res. 62:101 (1979).

60. E. Vogel and A. T. Natarajan, The relation between reaction kinetics and mutagenic action of monofunctional alkylating agents in higher eukaryotic systems: Interspecies comparisons, in: "Chemical Mutagens", Vol. 7, F. J. de Serres and A. Hollaender, eds., Plenum Press, New York (1982).

61. D. B. Couch, N. L. Forbes and A. W. Hsie, Comparative mutagenicity of alkyl sulfate and alkane sulfonate derivatives in Chinese hamster ovary cells, Mutation Res. 57:217 (1978).

62. S. M. Morris, R. H. Heflich, D. T. Beranek and R. L. Kodell, Alkylation-induced sister-chromatid exchanges correlate with reduced cell survival, not mutations, Mutation Res. 105:163 (1982).

63. A. T. Natarajan, J. W. I. M. Simons, E. W. Vogel and A. A. van Zeeland, Relationship between cell killing, chromosomal aberrations, sister-chromatid exchanges and point mutations induced by monofunctional alkylating agents in Chinese hamster cells. A correlation with different ethylation products in DNA, Mutation Res. 128:31 (1984).

64. K. Frenkel and K. Chrzan, Hydrogen peroxide formation and DNA base modification by tumor promoter-activated polymorphonuclear leukocytes, Carcinogenesis 8:455 (1987).

65. P. A. Cerutti, Pro-oxidant states and tumor promotion, Science 227:375 (1985).

66. J. Emerit and P. A. Cerutti, Eicosanoids and chromosomal damage, in: "Icosanoids and Cancer", H. Thaler-Dao, A. Crastes de Paulet and R. Paoletti, eds., Raven Press, New York (1984).

67. T. Ochi and P. Cerutti, Clastogenic action of hydroperoxy-5,8,11,13-icosatetraenoic acids on mouse embryo fibroblasts C3H/10T1/2, Proc. Natl. Acad. Sci. USA 84:990 (1987).

68. G. Fürstenberger, M. Gschwendt, H. Hagedorn and F. Marks, Modulation of the conversion stage of multistep carcinogenesis in mouse skin by eicosanoids, in: "Prostaglandins in Cancer Research", E. Garaci, R. Paoletti and M. G. Santoro, eds., Springer, Berlin-Heidelberg (1987).

69. G. Bauer, P. Höfler and M. Simon, Epstein-Barr-Virus induction by a serum factor: Purification of a high molecular weight protein that is responsible for induction, J. Biol. Chem. 257:11405 (1982).

70. G. Bauer, P. Höfler and M. Simon, Epstein-Barr-Virus induction by a serum factor: Characterization of the purified factor and the mechanism of its activation, J. Biol. Chem. 257:11411 (1982).

71. G. Bauer, Epstein-Barr-Virus inducing factor: A growth factor with tumor promoting activity, J. Cancer Res. Clin. Oncol. 109:A48 (1985).

72. M. Rogers, G. Fürstenberger, G. Bauer, P. Höfler and F. Marks, EIF (Epstein-Barr-Virus inducing factor), a peptide factor with TFGß-activity, is a convertogenic agent ("stage I tumor promoter") in mouse skin in vivo. Proc. 3rd Int. Congress Hormones and Cancer, Raven Press, New York (1988) (in press).

73. V. Kinzel, H. Loehrke, K. Goerttler, G. Fürstenberger and F. Marks, Suppression of the first stage of TPA-effected tumor promotion in mouse skin by non-toxic inhibition of DNA synthesis, Proc. Natl. Acad. Sci. USA 81:5858 (1984).

74. J. C. Gaal, K. R. Smith and C. K. Pearson, Cellular euthanasia mediated by a nuclear enzyme: a central role for nuclear ADP-ribosylation in cellular metabolism, Trends Bioch. Sci. 12:129 (1987).

75. C. B. Croft and D. Tarin, Ultrastructural studies of wound healing in mouse skin. I. Epithelial behavior, J. Anat. 105:63 (1970).

76. J. J. Cohen and R. C. Duke, Glucocorticoid activation of a Ca^{2+}-dependent endonuclease in lymphocyte nuclei leads to cell death, J. Immunol. 132:38 (1984).

77. E. Duvall and A. H. Wyllie, Death and the cell, Immunology Today 7:115 (1986).

78. C. Richter, B. Frei and P. A. Cerutti, Mobilization of mitochondrial Ca^{2+} by hydroperoxyeicosatetraenoid acid, Biochim. Biophys. Res. Commun. 143:609 (1987).

79. P. Csermely, R. Fodor and J. Somogyi, The tumor promoter TPA elecits redistribution of heavy metals in subcellular fractions of rabbit thymocytes as measured by plasma emission spectroscopy, Carcinogenesis 8:1663 (1987).

80. T. Masui, L. M. Wakefield, J. F. Lechner, M. A. Laveck, M. B. Sporn and C. C. Harris, Type ß transforming growth factor is the primary differentiation-inducing serum factor for normal human bronchial cells, Proc. Natl. Acad. Sci. USA 83:2438 (1986).

81. O. H. Iversen, ed., "Theories of Carcinogenesis", Hemisphere, Washington (1988).

82. A. P. Maskens, P. Ebbesen and A. Burny, eds., "Concepts and Theories in Carcinogenesis", Excerpta Medica, Amsterdam (1987).

83. J. J. Slaga, Can tumor promotion be effectively inhibited? in: "Models, Mechanisms and Etiology of Tumor Promotion", M. Börszönyi, K. Lapis, N. E. Day and H. Yamasaki, eds., Int. Agency for Res. on Cancer, IARC, Lyon (1984).

84. T. J. Slaga, Multistage tumor promotion and specificity of inhibition, in: "Mechanisms of Tumor Promotion", Vol II, T. J. Slaga, ed., CRC Press Boca Raton, Florida (1984).

85. M. Gschwendt, W. Kittstein and F. Marks, Cyclosporin A inhibits phorbol ester-induced cellular proliferation and tumor promotion as well as phosphorylation of a 100-kD protein in mouse epidermis,

Carcinogenesis 8:203 (1987).

86. M. Gschwendt, W. Kittstein and F. Marks, Didemnin B inhibits biological effects of tumor promoting phorbol esters on mouse skin, as well as phosphorylation of a 100 kD protein in mouse epidermis cytosol, Cancer Letters 34:187 (1987).

87. M. Gschwendt, W. Kittstein and F. Marks, The weak immunosuppressant cyclosporine D as well as the immunologically inactive cyclosporine H are potent inhibitors in vivo of phorbol ester TPA-induced biological effects in mouse skin and of Ca^{2+}/calmodulin-dependent EF-2 phosphorylation in vitro, Biochem. Biophys. Res. Commun. 150:545 (1988). press (1988).

88. M. Gschwendt, G. Fürstenberger, W. Kittstein, E. Besemfelder, W. E. Hull, H. Hagedorn, H. J. Opferkuch and F. Marks, Generation of the arachidonic acid metabolite 8-HETE by extracts of mouse skin treated with phorbol ester in vivo; identification by ^{1}H-n.m.r. and GC-MS spectroscopy, Carcinogenesis 7:449 (1986).

89. S. M. Fischer, G. Fürstenberger, F. Marks and T. J. Slaga, Events associated with mouse skin tumor promotion with aspect to arachidonic acid metabolism: A comparison between SENCAR and NMRI mice, Cancer Res. 47:3174 (1987).

90. M. Gross, G. Fürstenberger and F. Marks, Isolation, characterization, and in vitro cultivation of keratinocyte subfractions from adult NMRI mouse epidermis: epidermal target cells for phorbol esters, Exp. Cell Res. 171:460 (1987).

91. R. T. Dzarlieva-Petrusevska, N. E. Fusenig, Tumor promoter 12-0-tetradecanoylphorbol-13-acetate (TPA)-induced chromosome aberrations in mouse keratinocyte cell lines: a possible genetic mechanism of tumor promotion, Carcinogenesis 6:1447 (1985).

POSSIBLE MECHANISM OF OROTIC ACID INDUCED LIVER TUMOR PROMOTION IN THE RAT

Ezio Laconi, Pier P. Coni, Giuseppina Pichiri, Prema M. Rao,
Srinivasan Rajalakshmi and Dittakavi S. R. Sarma

Department of Pathology, Medical Sciences Building
University of Toronto, Toronto, Ontario, Canada M5S 1A8

INTRODUCTION

One of the important advances in the study of experimental liver carcinogenesis is the development of model systems which permit an analysis of the
initiation, promotion and progression phases of the carcinogenic process[1-6].
We have recently demonstrated that orotic acid (OA), a precursor for pyrimidine nucleotide biosynthesis is an effective promoter of liver[6-9] and duo-
denal[10] cancer in the rat. In view of the fact that OA is a normal cellular
constituent whose levels can be perturbed in a variety of ways[11] and that it
had the potential to be a multiorgan tumor promoter, it became important to
understand the mechanisms by which OA exerted this effect.

Operationally, promotion results in focal proliferation of initiated
hepatocytes to form foci of enzymes altered hepatocytes and hepatic nodules.
Therefore, any mechanism proposed for promotion should be able to explain how
focal proliferation is induced during promotion. One of the hypotheses considered is that promoters create and environment to which the initiated and
non initiated hepatocytes respond differently. Such a differential can be
generated in many ways[12,13]. For example (see Fig. 1), the promoter can
selectively stimulate growth of the initiated hepatocytes to form nodules.
Alternatively, it can exert a mitoinhibitory effect on the noninitiated hepatocytes while permitting the initiated hepatocytes to develop into foci and
nodules. In the latter case, the initiated hepatocytes will have to be resistant to the effects of the promoter. Yet another reason for focal proliferation could reside in the property of the initiated hepatocytes themselves, independent of the promoter, in that they have a lower threshold for
growth stimuli and therefore respond to them better. The promoting effect of
OA has been examined against this background.

EFFECT OF OA ON DNA SYNTHESIS AND LIVER CELL PROLIFERATION

Feeding OA for 4 to 5 weeks to either initiated or uninitiated rats did
not induce liver cell proliferation, monitored either as incorporation of
tritiated thymidine into hepatic DNA or as the cumulative labeling index following implantation of osmotic minipumps containing tritiated thymidine[7].
Furthermore, OA did not induce ornithine decarboxylase, and enzyme that is
associated with liver cell proliferation (S. Vasudevan and D. S. R. Sarma,
unpublished observations). Experiments were also carried out to see if OA is

235

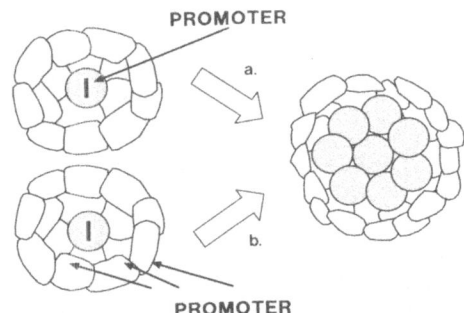

PROMOTER

PROMOTER

Fig. 1.   Schematic representation showing that promoters expand initiated
hepatocyte population either by selectively stimulating the initiated hepatocytes (a) or by selectively exerting a mitoinhibitory
effect on the surrounding cell population while permitting the initiated hepatocytes to respond to growth stimuli (b). I, initiated
hepatocyte.

mitogenic to hepatocytes in the γ-glutamyltransferase (γ-GT) positive foci.
The results indicate that the hepatocytes in the γ-GT positive foci, which
develop following initiation with diethylnitrosamine and promotion with
choline-deficient diet, also did not exhibit an increased labeling index upon
exposure to OA[14]. This finding is in contrast to the results reported with
other liver tumor promoters like phenobarbital (PB), α-hexachlorocyclohexane
and cyproterone acetate[15]. All these different types of evidence seem to
justify the conclusion that OA is not an inducer of liver cell proliferation.

The next question asked was whether OA is an inhibitor of normal hepatocyte proliferation. Feeding OA for 6 to 10 weeks did not block liver cell
DNA synthesis following 2/3 partial hepatectomy, although it did delay the
peak of DNA synthesis[16]. This result suggests that OA is not an inhibitor of
hepatic DNA synthesis.

The above observations indicate that OA is neither an inducer of liver
cell proliferation nor an inhibitor of DNA synthesis in rat liver. However,
the possibility exists that OA might inhibit the response of hepatocytes to
some growth stimuli other than those induced by partial hepatectomy. This
aspect is being studied in detail now and some of the results are presented
here.

IN VITRO RESPONSE OF HEPATOCYTES TO EPIDERMAL GROWTH FACTOR (EGF) AND ITS
MODULATION BY OA

Male Fischer 344 rats were exposed to a semisynthetic basal diet containing 1% OA for 6 weeks coupled with 2/3 partial hepatectomy at the end of
the second week. This protocol has a promoting effect in carcinogen initiated rats. The results presented in Fig. 2 indicate that hepatocytes from
the OA fed rats have a lower labeling index compared to the controls. In
addition, hepatocytes isolated from the OA fed rats were also less responsive
to stimulation by EGF as indicated by the lower labeling index (Fig. 2).
Since in these experiments the rats were on OA diet at the time the hepatocytes were isolated, the possibility had to be considered that OA or its
metabolites might interfere with the response of hepatocytes to EGF. In the
next experiment therefore, the same protocol was employed except that the
rats were transferred from OA to basal diet and maintained on a basal diet
for 4 weeks before sacrifice. Under these conditions the increased uridine
nucleotides in the liver resulting from the OA diet return to normal.

236

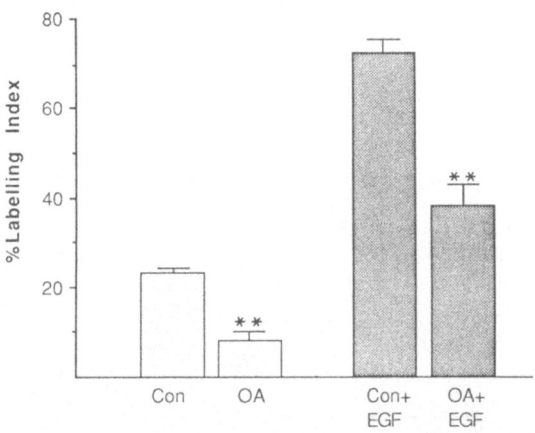

Fig. 2. Effect of orotic acid feeding on the EGF-induced labeling index in
isolated hepatocytes. Male Fischer 344 rats weighing 150 g were
exposed either to a semisynthetic basal diet (diet 101 from Dyets
Inc. Bethlehem, PA) or to the basal diet containing 1% orotic acid.
Two weeks later they were subjected to 2/3 partial hepatectomy and
they were continued on their respective diets for another four weeks
prior to sacrifice. Routinely, 2.0×10^5 viable hepatocytes were
cultured on 35 mm dishes coated with collagen (vitrogen, 60 μg per
dish) (Collagen Corp., Palo Alto, CA) in modified William E medium
containing fetal bovine serum (10% v:v), insulin (20 U/I), L-gluta-
mine (2 mM), HEPES (10 mM), penicillin (100 U/ml) and streptomycin
(100 μg/ml)[26]. After an attachment period of 3 h at 37°C in
air-CO_2 (95:5) the medium and non-attached cells were removed. At
this time, medium was changed to serum-free modified William E
medium supplemented with L-proline (2 mM), and sodium pyruvate (10
mM). Appropriate dishes also contained EGF (20 ng/ml)[27], and
[^{3}H]thymidine (5 μCi/dish; sp. activity 80.9 Ci/mmol). After 48 h
of incubation, the cells were washed in cold PBS, fixed in 10%
buffered formalin, and processed for autoradiography. The labeling index was determined on approximately 800 hepatocytes per dish.
Values are the average of 3 dishes ± SE. The experiment was repeated twice with similar patterns of results. ** P < 0.01.

Hepatocytes isolated from such rats also had a lower labeling index both in
the presence and absence of EGF (data not presented). This observation is
very provocative in that it suggests that the promoting regimen confers on
the hepatocytes a property which persists for a long time and exerts its
effect even in the absence of the promoter. It will be of great interest to
determine whether fixation of some lesion is involved in the promotion phase,
albeit in the uninitiated surrounding liver, a phenomenon analogous to the
fixation of a carcinogen induced lesion during the initiation phase.

In the next series of experiments the effect of orotic acid added in
vitro on the response of hepatocytes to EGF was investigated. As can be seen
from Fig. 3 addition of orotic acid inhibited, in a dose dependent fashion,
the growth response of isolated hepatocytes to EGF. It is too early to speculate whether the in vivo and the in vitro effects of orotic acid on EGF-
induced labeling index of hepatocytes are the same.

The observations that orotic acid did not inhibit liver DNA synthesis in
vivo following 2/3 partial hepatectomy but did inhibit the labeling index of
hepatocytes in response to EGF suggest that orotic acid is not a general
inhibitor of DNA synthesis but exerts it mitoinhibitory effect by interfering

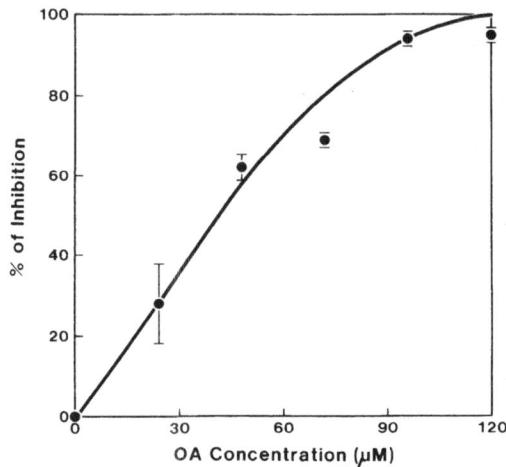

Fig. 3. Effect of in vitro addition of orotic acid on EGF-induced labeling index in isolated hepatocytes. Experimental details are the same as detailed in the legend to Fig. 2. After an attachment period of 3 h, the medium was changed to serum-free modified William E medium supplemented with L-proline (2 mM), and sodium pyruvate (10 mM). EGF, orotic acid (as orotic acid methyl ester, Sigma Chemical Co., St. Louis, Missouri), and tritiated thymidine were added and the experiment was terminated 48 h later. Values are the mean ± SE of 3 dishes. The experiment was repeated more than 4 times with similar pattern of results.

with the response of hepatocytes to certain growth stimuli such as insulin and EGF. Based on these findings it is tempting to speculate that the promoting effect of orotic acid may be in its ability to inhibit the noninitiated hepatocytes from responding to certain endogenous growth stimuli, while permitting the initiated ones to respond. Obviously, the endogenous growth stimuli relevant for orotic acid mediated liver tumor promotion need to be determined.

The next question we have asked was what is the differential between initiated and noninitiated hepatocytes which makes the latter less responsive than the initiated ones to growth stimuli in the presence of OA. As pointed out earlier, promotion mediated through the mito-inhibitory effect of the promoter on the surrounding implies that the initiated cells are resistant to some effects of the promoter. In fact, we have found that the hepatic nodules are resistant to OA in that, upon exposure to OA, they do not show an increase in uridine nucleotides to the same extent as the surrounding liver[17,18]. If we extrapolate these findings to initiated cells, then being resistant to OA they would respond to growth factors. The surrounding liver cells, on the other hand, being susceptible to OA effects would be less or unresponsive to growth stimuli. It is significant that similar observations have been made with the promoter PB. Thus, hepatocytes from rats fed PB for months showed a lower labeling index in response to EGF in vitro[19] and partial hepatectomy in vivo[20,21] (E. Laconi and E. Farber, unpublished observations). In addition, PB at high concentrations inhibited the response of hepatocytes to EGF in vitro[19] although it had a stimulatory effect at lower concentrations[22,23]. It is interesting to note that Betschart et al.[24,25] have reported a decrease in insulin receptors in hepatocytes from rats fed PB or a choline deficient diet.

The concept of differential mitoinhibition was first formulated and

utilized in the development of the resistant hepatocyte model for the generation of hepatic nodules[2]. In this protocol, the initiated rats are subjected to a brief exposure to 2-acetylaminofluorene (2-AAF) coupled with partial
hepatectomy or necrogenic dose of CCl_4. The 2-AAF exerts a mito-inhibitory
effect on the surrounding liver but permits the initiated cells to respond to
the cell proliferative stimulus generated by partial hepatectomy or CCl_4.
However, in this case, the 2-AAF apparently acts by inhibiting DNA synthesis
in the surrounding liver cells. The resistant hepatocyte model and perhaps
OA and PB all seem to act by exerting a mito-inhibitory effect on the surrounding. However, the mechanism of creating this inhibitory environment is
different: in the resistant hepatocyte model it may be by inhibiting DNA synthesis while with OA and perhaps PB it may be by inhibiting the response of
hepatocytes to growth stimuli (Figs. 4 and 5).

A third mechanism postulated for focal proliferation was that initiated
cells may have a lower threshold for growth stimuli by virtue of which they
respond better and grow. In experiments designed to examine this aspect,
rats were initiated with diethylnitrosamine (200 mg/kg i.p.) and subsequently
exposed to either a basal diet or one containing 1% OA for 15 weeks. Following a 2/3 partial hepatectomy the kinetics of labeling indices in the hepatocytes in the γ-GT positive foci as well as surrounding liver was determined. The results show that 16 h post partial hepatectomy, the labeling
index of the hepatocytes in the foci was 12-16% compared to less than 2% in
the surrounding liver[16]. These results suggest that hepatocytes in the foci
are more responsive to proliferative stimuli than the surrounding perhaps
because of the acquisition of a new state of differentiation i.e. they do not
go back to "GO"stage at the end of the cell cycle as normal hepatocytes do,
but are somewhere in between "GO" and G1 and therefore enter the cell cycle
earlier when triggered by a stimulus. Such a differential response to growth
stimulus can become very significant to account for the observed selective
growth of focal lesions in the absence of an exogenously applied promoting
treatment.

Thus, there are perhaps at least two types of differentials by virtue of
which initiated hepatocytes grow into nodules. One is intrinsic to the
initiated hepatocytes and independent of the promoter while the other is
created by the promoting regimen acting on the surrounding noninitiated hepatocyte population. In concert, they give an advantage to the initiated
hepatocytes such that they respond better to endogenous growth stimuli and
grow into nodules.

ACKNOWLEDGEMENTS

We wish to thank Lori Cutler for her excellent secretarial assistance.
The study was supported in part by U.S. P.H.S. Grant Ca 37077 from the

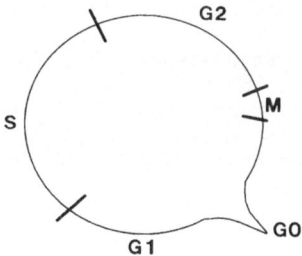

Fig. 4. Schematic representation showing that mito-inhibition
can be achieved by interferring with the cell cycle.

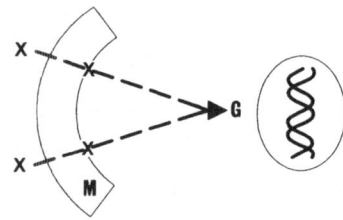

Fig. 5. Schematic representation showing that mito-inhibition can be achieved by interferring with the response of the cell to growth factors. G, growth stimulus. M, membrane.

National Cancer Institute, by grants from National Cancer Institute, Canada and from the Ministry of Labour, Occupational Health, Ontario, Canada. E. L. and P. P. C. were supported by Associazione Italiana per la Ricerca sul Cancro, Italy.

REFERENCES

1. C. Peraino, R. J. M. Fry and E. F. Staffeldt, Reduction and enhancement by phenobarbital of hepatocarcinogenesis induced in rat by 2-acetyl-aminofluorene, Cancer Res. 48:1506 (1971).
2. D. B. Solt and E. Farber, New principle for the analysis of chemical carcinogenesis, Nature (Lond) 263:702 (1976).
3. H. C. Pitot, L. Barsness, T. Goldsworthy and T. Kitagawa, Biochemical characterization of stages of hepatocarcinogenesis after a single dose of diethylnitrosamine, Nature (Lond) 271:456 (1978).
4. E. Farber and D. S. R. Sarma, Biology of disease: Hepatocarcinogenesis: A dynamic cellular perspective, Lab. Invest. 56:4 (1987).
5. E. Scherer and P. Emmelot, Initiation-promotion-initiation: induction of secondary foci and micro-carcinoma within islands of precancerous liver cells in the rat, J. Cancer Res. Clin. Oncol. 105:A-25 (1983).
6. A. Columbano, G. M. Ledda-Columbano, P. M. Rao, S. Rajalakshmi and D. S. R. Sarma, Dietary orotic acid, a new selective growth stimulus for carcinogen altered hepatocytes in rat, Cancer Letters 16:191 (1982).
7. P. M. Rao, Y. Nagamine, R-K. Ho, M. W. Roomi, C. Laurier, S. Rajalakshmi and D. S. R. Sarma, Dietary orotic acid enhances the incidence of γ-glutamyltransferase positive foci in rat liver induced by chemical carcinogens, Carcinogenesis 4:1541 (1983).
8. C. Laurier, M. Tatematsu, P. M. Rao, S. Rajalakshmi and D. S. R. Sarma Promotion by orotic acid of liver carcinogenesis in rats initiated by 1,2-dimethylhydrazine, Cancer Res. 44:2186 (1984).
9. P. M. Rao, Y. Nagamine, M. W. Roomi, S. Rajalakshmi and D. S. R. Sarma Orotic acid, a new promoter for liver carcinogenesis, Toxicol. Pathol. 12:173 (1984).
10. P. M. Rao, E. Laconi, S. Rajalakshmi and D. S. R. Sarma, Orotic acid, a liver tumor promoter, also promotes carcinogenesis of the intestine, Proc. Am. Assoc. Cancer Res. 27:142 (1986).
11. S. Vasudevan, E. Laconi, S. E. Abanobi, P. M. Rao, S. Rajalakshmi and D. S. R. Sarma, Effect of glycine on the induction of orotic aciduria and urinary bladder tumorigenesis in the rat, Toxicologic Pathol. 15:194 (1987).
12. E. Farber and R. G. Cameron, The sequential analysis of cancer development, Adv. Cancer Res. 35:125 (1980).
13. D. S. R. Sarma, P. M. Rao and S. Rajalakshmi, Liver tumor promotion by chemicals. Models and Mechanisms, Cancer Surveys 5:781 (1987).

14. E. Laconi, S. Vasudevan, P. M. Rao, S. Rajalakshmi and D. S. R. Sarma, "Orotic acid, a novel tumor promoter: studies on the mechanism of its action", Banbury Report, Cold Spring Harbor, New York 25:85 (1987).

15. R. Schulte-Hermann, G. Ohde, J. Schuppler and I. Timmermann-Trosiener, Enhanced proliferation of putative preneoplastic cells in rat liver following treatment with the tumor promoters phenobarbital, hexachlorocyclohexane, steroid compounds and nafenopin, Cancer Res. 41: 2556 (1981).

16. E. Laconi, S. Vasudevan, P. M. Rao, S. Rajalakshmi and D. S. R. Sarma, Enhanced response of initiated cells to partial hepatectomy during rat liver carcinogenesis, Fed. Proc. 46:745 (1987).

17. M. A. Lea, V. Oliphant, A. Luke and J. V. Tesoriero, Uptake and metabolism of orotate in normal and neoplastic tissues, Proc. Am. Assoc. Cancer Res. 27:18 (1986).

18. E. Laconi, S. Vasudevan, P. M. Rao, S. Rajalakshmi and D. S. R. Sarma, Hepatic nodules have a characteristic pattern of nucleotide pools distinct from that of the surrounding liver, Proc. Am. Assoc. Cancer Res. 28:169 (1987).

19. P. M. Eckl, W. R. Whitcomb, S. A. Meyer and R. L. Jirtle, Effects of phenobarbital and calcium on the growth of normal and preneoplastic hepatocytes, Proc. Am. Assoc. Cancer Res. 28:171 (1987).

20. H. Barbason, C. Rassenfosse and E. H. Betz, Promotion mechanism of phenobarbital and partial hepatectomy in DENA hepatocarcinogenesis-cell kinetics effect, Br. J. Cancer, 47:517 (1983).

21. C. Peraino, W. L. Richards and F. J. Stevens, Multistage hepatocarcinogenesis, in: "Mechanisms of Tumor Promotion: Tumor Promotion in Internal Organs", T. J. Slaga, ed., CRC Press Inc., Boca Raton (1983).

22. U. Armato, P. G. Andreis and F. Romano, Exogenous Cu, Zn-superoxide dismutase suppresses the stimulation of neonatal rat hepatocytes growth by tumor promoters, Carcinogenesis 5:1547 (1984).

23. A. M. Edwards and C. M. Lucas, Phenobarbital and some other liver tumor promoters stimulate DNA synthesis in cultured rat hepatocytes, Biochem. Biophys. Res. Commun. 131:103 (1985).

24. J. M. Betschart, C. Gupta, H. Shinozuka and M. A. Virji, Modulation of hepatocyte insulin and glucagon receptors in rats fed phenobarbital, a tumor promoter, Proc. Am. Assoc. Cancer Res. 28:168 (1987).

25. J. Betschart, M. A. Virji, M. I. Perera and H. Shinozuka, Alterations in hepatocyte insulin receptors in rats fed a choline-deficient diet Cancer Res. 46:4425 (1986).

26. M. A. Hayes, E. Roberts, M. W. Roomi, S. H. Safe, E. Farber and R. G. Cameron, Comparative influences of different PB-type and 3-MC-type polychorinated biphenyl-induced phenotypes on cytocidal hepatoxicity of bromobenzene and acetaminophen, Toxicol. Appl. Pharmacol 76:118 (1984).

27. E. Roberts, E. Farber and M. A. Hayes, Effects of epidermal growth factor on labelling index of hepatocytes from normal liver, preneoplastic nodules and hepatocellular carcinoma, Proc. Am. Assoc. Cancer Res. 27:212 (1986).

THE ROLE OF INITIATING AND PROMOTING PROPERTIES IN AROMATIC

AMINE CARCINOGENESIS

Hans-Günter Neumann

Institute of Pharmacology and Toxicology of the
University of Wurzburg, FRG

INTRODUCTION

Investigations as to the metabolism of carcinogenic aromatic amines have been fundamental in establishing the concept that ultimate reactive electrophiles are responsible for the damage of critical cellular macromolecules and that the DNA-lesions thus produced represent a key event in the generation of tumors[1]. Putative promutagenic DNA-adducts have been identified in many cases and attempts have been made to correlate the extent of formation and the persistence of such adducts with the carcinogenic potency of the parent amine. Although the basic principle has been confirmed in many instances, the correlations of genotoxic effects of aromatic amines with tumor formation, particularly with tissue-specific tumors were not satisfactory[2]. For these poor correlations many reasons can be envisaged, one of which is related to the multistage concept of carcinogenesis. According to this concept, irreversible genotoxic effects may well correlate with initiation but not necessarily with the end point tumor which is reached only via a multistep process in which some of the steps could be independent of the early DNA-damaging events.

In this context, it appears to be necessary to corroborate the proposed role of identifiable DNA-adducts for tumor initiation and to look for additional properties of aromatic amines which could be responsible for effects in later stages which make the amines complete carcinogens.

We have selected a set of well known aromatic amine model compounds for comparison of their properties in the rat at different investigational levels. The compounds are: 2-acetylaminofluorene (AAF), 2-acetylaminophenanthrene (AAP), and trans-4-acetylaminostilbene (AAS). The comparison involves (1) the biological end point tumor and preneoplastic lesions, (2) the primary reactions as measured by macromolecular binding and adduct formation, as well as by the interaction with cellular receptors, and (3) consequences of the primary reactions at the molecular and the biochemical level.

INITIATION VERSUS TUMOR FORMATION

AAF is a well established liver carcinogen in rats. With moderate doses (0.02% in the diet for 18 days) Peraino et al.[3] obtained a 30% yield of liver tumors. With a similar dose of AAP (0.02% in the diet for 21 days) Scribner and Mottet[4] observed tumors of the mammary gland (55%) and the Zymbal's gland (10%), but none in the liver. When a diet containing 0.005% AAS was fed for 13 weeks, tumors of the Zymbal's gland developed in 80% of the rats and neither liver nor mammary tumors occurred[5]. Thus, by definition, all three compounds are complete carcinogens but have different target tissues. All three aromatic amides are metabolically activated in the rat, and the tissue dose of reactive metabolites has been assessed by measuring macromolecular binding. In rat liver, which is a target tissue only for AAF, the tissue dose of reactive metabolites is comparable for all three compounds. Scribner and Koponen[6] compared total DNA-binding in liver after the administration of AAF and AAP. DNA-binding indices of 95 and 110, respectively, can be calculated from their data. Similarly, we have compared DNA-binding of AAF and AAS and find DNA-binding indices of 95 and 225, respectively (Ruthsatz and Neumann, unpublished) which indicates that AAS actually binds more efficiently to liver DNA than does AAF, although it produces no tumors in this tissue.

There are, of course, many possible explanations as to why DNA-binding does not correlate with carcinogenicity in liver. Different DNA-adducts with different biological potential may be involved or minor adducts, not yet identified, could be responsible for the biological effect. In addition, repair systems could respond differently and the persistence of DNA-lesions may not have been accounted for sufficiently. Finally, and most difficult to assess, the DNA-adducts measured could be generated to a greater extent in non-target cells within the tissue or in non-critical regions of the target DNA.

The major DNA adducts of AAF and AAP have been known for some time[7,8], those of AAS were identified only very recently (Fig. 1; 9, 10; Franz and Neumann, unpublished). The structures of the adducts of all three amides are consistent with promutagenic properties, and this is underlined by the mutagenicity in bacterial systems of all the amides when metabolically activated[11]. The DNA-damage generated by AAS metabolites appears to be particularly efficient in several test systems[12]. This could be due to the formation of cyclic adducts and also to the generation of DNA-DNA interstrand cross-links which we have observed recently (Ruthsatz and Neumann, unpublished). Neither type of DNA-damage is known for AAF or AAP.

All the arguments questioning the relevance for tumor formation of the levels and the kind of DNA damage observed in rat liver with AAP and AAS could be overcome if these effects could be related to tumor initiation. With closely similar protocols it is indeed possible to demonstrate that all three aromatic amides produce critical, i.e. initiating lesions in liver cells which can give rise to formation of liver tumors when the initiating treatment is followed by an appropriate promoting regimen (Table 1). This does not prove that the DNA-adducts under discussion are responsible for the initiating lesions, but considering the available information they seem to be the most likely candidates for a cause-effect relationship. On the basis of this initiation-promotion model in rat liver, AAF can be defined as a complete carcinogen, AAP and AAS as initiators for this tissue.

This raises the question whether DNA adducts in other tissues would also indicate the presence of initiated cells. Next to liver, rat kidney accumulates most DNA adducts with AAS[13] but tumors have never been observed in this tissue. We have recently stimulated cell proliferation in one of the rat kidneys by unilateral nephrectomy, or imposed additional strain by ß-cyclodextrin, a nephrotoxic antibiotic, after an initiating treatment with AAS. As a result kidney tumors developed in some of the animals, indicating an initiating effect of AAS in this tissue (Table 2, Hoffmann, Romen and Neumann, unpublished).

In order to avoid the necessity of exogenous stimulation of growth, we have also administered the three aromatic amides to newborn rats. Five

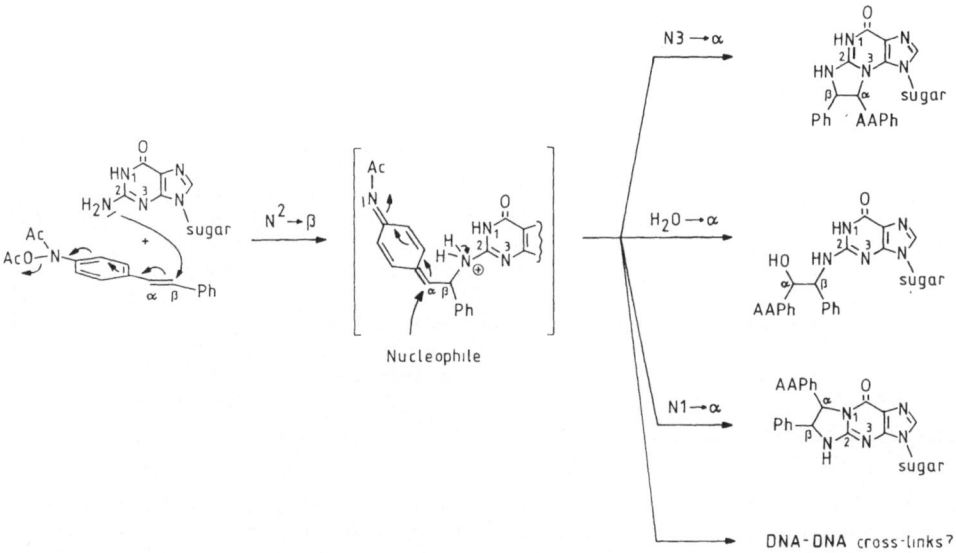

Fig. 1. The major DNA adducts of AAS

doses, totalling 110 μmol/kg, were applied between day 5 and 13 after birth
to male and female Wistar rats, and livers were screened for gamma-GT positive foci, glucose-6-phosphatase-negative foci, and glycogen storage islands
after 15, 30 and 50 weeks. Livers from AAS treated animals gave clearly
positive results at all time points. The number of foci appears to remain
rather constant with time, but the average size increased by 50 weeks. Tumors did not develop during this time (Horter, Kress, Kirchner and Neumann,
unpublished). The applied dose was not sufficient to produce any detectable
effects with AAF. A borderline effect was seen with AAP at the last time
point. These results underline the strong initiating activity of AAS in rat
liver. It is therefore the more surprising that this compound is an incomplete liver carcinogen in adult rats. Evidently, it lacks some activity necessary to promote initiated liver cells.

WHAT MAKES AAF A COMPLETE HEPATOCARCINOGEN?

Cell proliferation is necessary to complete the process of initiation,
i.e. to establish a permanent lesion from the initial DNA damage[14]. In this
sense an incomplete carcinogen could be just an incomplete initiator, and any

Table 1. Carcinogenicity of aromatic amines in rat liver in initiation-
        promotion experiments

Compound	Treatment	Time	Second treatment	Liver tumors
AAF [a]	0.02% diet	18 d	DDT 0.05% diet	77/103
AAP [b]	0.02% diet	21 d	PH, DDT 0.05% diet	30/30
AAS [c]	8 x 6 mg/kg po	20 d	DDT 0.05% diet	6/10

[a] Peraino et al.[3]; [b] Scribner, N. K., personal communication; [c] Hilpert et
al.[34].

245

Table 2. The initiating activity of AAS in rats

Compound	Dose (μmol/kg)	Time	Second treatment	Tumors	Incid.
AAS	8 x 25	20 d		Zymbal's gl.	5/9
AAS	8 x 25	20 d	PH, phenobarbital	Liver	8/11
				Sebaceous gl.	5/11
AAS	4 x 20	9 d	ß-Cyclodextrin	Kidney	3/10
				Sebaceous gl.	1/10
AAS	4 x 20	9 d	Unil. nephrectomy + gentamycin	Kidney	3/10
				Sebaceous gl.	3/10
AAS	5 x 22	8 d	Newborn animals	Liver	12/12
				Preneopl. l.	

property which stimulates cell turnover would supplement it. Cytotoxic effects are usually quoted as fulfilling this purpose but they are also made responsible for late effects during chronic administration, providing selective pressure for initiated cells[15].

AAF is said to be hepatotoxic. AAS is, according to our experience, not hepatotoxic[16]. This difference is difficult to comprehend if reactive metabolites are considered responsible for toxicity since AAS metabolites appear to modify proteins, RNA and DNA more efficiently than those of AAF. However, covalent binding is possibly not representative of all the primary reactions taking place. We have studied the possibility that radicals are generated, or that reactive oxygen is formed by redox cycling of metabolites[17] and that these processes contribute differently to overall cellular strain. We have measured the efflux of oxidized glutathione into the bile after i.p. injection of the test compounds [18]. This is considered to be a sensitive method of analyzing oxidative stress which overwhelms the cellular defense mechanisms [19]. Single injection of AAF and AAS into rats at a dose (1 mmol/kg) considerably higher than that applied in chronic feeding experiments did not increase the efflux of oxidized glutathione. The same dose of the parent amines AF and AS was toxic for the animals but this could not be attributed to liver toxicity. The efflux of oxidized glutathione was increased neither at the high dose, which due to toxicity was not measured under circumstances directly comparable with those in the above experiment, nor with a non-toxic dose (0.1 mmol/kg). A negative result was also obtained with a massive dose (2.65 mmol/kg) of paracetamol, which is in line with the observations of Smith and Mitchell[20]. Menadione was used as a positive control in our experiments and, much to our surprise, doses above 0.4 mmol/kg of this established and direct redox-cycler were necessary to elevate the efflux of glutathione. This demonstrates the high capacity of rat liver to cope with oxidative stress, and we consider it highly unlikely that aromatic amines cause cytotoxic effects through the generation of radicals or reactive oxygen in liver under almost any conditions relevant in vivo.

Reactive oxygen produced in vivo by menadione is able to produce DNA-strand breaks. This can be demonstrated by using the alkaline elution assay. The effect can be increased by lowering the glutathione levels in liver. Compared with the effects of X-rays and methyl-methanesulfonate, however, menadione is only very weakly active. With aromatic amines no effects could be detected (Hillesheim and Neumann, unpublished; c.f. also Ruthsatz et al., 7). It is therefore unlikely that oxidative stress contributes significantly not only to cytotoxicity but also to the genotoxic effects of aromatic amines in rat liver.

These results raise questions as to the acute toxicity of AAF in rat liver. We have measured the following parameters after i.p. injection of 1 mmol/kg AAF to female Wistar rats: serum transaminases (GOT, GPT), bile flow, oxidized glutathione in bile, and thiobarbituric acid-reactive material in liver. We could not find any deviations from normal values. In addition, we analyzed the perfusate of isolated perfused livers and saw no changes in LDH activity, total glutathione nor the lactate/pyruvate ratio. Bile flow and total glutathione in bile were also unchanged in these preparations [18]. So far, we have been unable to detect any biochemical signs of acute liver toxicity with AAF exposures exceeding those used in carcinogenicity tests. This is in line with previous observations indicating that liver damage and necrosis of liver cells does not occur during the early phase of chronic AAF administration [21-23].

These results are at variance with the idea that macromolecular damage caused by reactive metabolites is responsible for the promoting properties of AAF and it seems to be reasonable to look for alternative mechanisms [2]. Recently, we observed that AAF and AAS act synergistically in initiating rat liver cells [24]. The initiated animals were subjected to partial hepatectomy, phenobarbital was administered as a promoter and preneoplastic lesions were analyzed. The biological effects were more than additive when the two compounds were combined, which is presently poorly understood, but, in addition, the biological effect depended on the sequence in which the initiators were applied. Enzyme altered foci developed much faster when AAF was given after AAS. Since the pattern of DNA adducts and the extent of DNA binding were independent of the administration protocol, we concluded that AAF affects the development of lesions by effects unrelated to adduct formation [7,24].

In a first attempt to demonstrate at the molecular level the possibility that aromatic amines may be biologically active without being metabolized to reactive electrophiles, Cikryt and his coworkers studied the affinity of several aromatic amides to the rat hepatic aromatic hydrocarbons (Ah) receptor. This approach was triggered by the well known fact that AAF is able to induce drug metabolizing enzymes [25], whereas the N-dimethyl-derivative of trans-4-aminostilbene, which is metabolized to AAS, is inactive in this respect [26]. AAF and AAP decrease the binding of 2,3,7,8-tetrachlorodibenzo-p-dioxin (TCDD) in a concentration-dependent manner. The 50% inhibition concentrations are 5 and 13 µM, respectively, which for AAF is in the range of concentrations reached in rat liver cytosol in vivo after a moderate 100 µmol/kg dose. AAS and the non-carcinogenic 4-acetylaminofluorene do not compete with TCDD [27].

These authors also studied the inducibility of the Ah-receptor in rat liver [28]. AAF is inactive in this respect, but AAS, although without measurable affinity to the receptor, is able to double the receptor concentration in vivo. Assuming that Ah-receptor mediated effects are related to tumor promotion [29], a hypothesis can be formulated to explain the strong synergistic effects of the combination AAS-AAF in the above mentioned experiments. In addition to the genotoxic effects of the two compounds, AAS induced the Ah-receptor concentration and AAF administered afterwards produced a receptor-mediated promoting response which became effective in AAS-initiated cells.

The induction of ornithine decarboxylase (ODC) has also been suggested to correlate with tumor promotion [30] and to be Ah-receptor-mediated [31]. AAF, AAS, 4-acetylaminofluorene, and methylcholanthrene all increase ODC activity in rat liver [32]. The time courses of the induction are somewhat different and parallel the concentrations of the inducers in liver cytosol. According to these experiments, ODC-induction does not correlate with the apparent promoting properties or the Ah-receptor affinity of the tested compounds in rat liver.

More work is certainly necessary to investigate other possibilities of inter-ference with receptor-mediated processes, particularly those encountered in the transfer of growth signals from the cellular membrane to the nucleus. Gene products of cellular proto-oncogenes are interesting candidates for such studies [33].

CONCLUSIONS

There is increasing evidence that the three model compounds, AAF, AAP and AAS, produce critical DNA lesions in rat liver which may be relevant for tumor initiation, but are not sufficient to trigger the whole process of tumor formation. Additional growth stimuli are necessary and, in the case of the complete liver carcinogen AAF, may be produced by specific, receptor-mediated processes rather than by acute toxicity. It is interesting to note that the growth of the target tissues for AAP and AAS, the mammary gland and the Zymbal's gland are sex hormone-dependent. Initiated cells may therefore be subjected to endogenous promotion.

ACKNOWLEDGEMENTS

Work carried out in this laboratory was supported by the Deutsche Forschungsgemeinschaft (SFB 172).

REFERENCES

1. J. A. Miller and E. C. Miller, The metabolic activation of carcinogenic aromatic amines and amides, Progr. Exp. Tumor Res. 11:273 (1969).
2. H.-G. Neumann, Role of extent and persistence of DNA modifications in chemical carcinogenesis of aromatic amines, Recent Results in Cancer Research 84:77 (1983).
3. C. Peraino, R. J. M. Fry, E. Staffeldt and J. P. Christopher, Comparative enhancing effects of phenobarbital, amobarbital, diphenyl-hydantoin, and dichlordiphenyltrichloroethane on 2-acetylamino-fluorene-induced hepatic tumorigenesis in the rat, Cancer Res. 35: 2884 (1975).
4. J. D. Scribner and N. K. Mottet, DDT acceleration of mammary gland tumors induced in the male Sprague-Dawley rat by 2-acetamidophenan-anthrene, Carcinogenesis 2:1235 (1981).
5. R. A. Andersen, M. Enomoto, E. C. Miller and J. A. Miller, Carcino-genesis and inhibition of the Walker 256 tumor in the rat by trans-4-acetylaminostilbene, its N-hydroxy metabolite, and related com-pounds, Cancer Res. 24:128 (1964).
6. J. D. Scribner and G. Koponen, Binding of the carcinogen 2-acetamido-phenanthrene to rat liver nucleic acids: lack of correlation with carcinogenic activity, and failure of the hydroxamic acid ester model for in vivo activation, Chem-Biol. Interactions 28:201 (1979).
7. M. Ruthsatz, R. Franz and H.-G. Neumann, DNA-damage, initiation and promotion by aromatic amines, in: "Primary Changes and Control Factors in Carcinogenesis", T. Friedberg and F. Oesch, eds., Deutscher Fachschriften-Verlag, Wiesbaden (1986).
8. H.-G. Neumann, The role of DNA-damage in chemical carcinogenesis of aromatic amines, J. Cancer Res. Clin. Oncol. 112:100 (1986).
9. R. Franz, H.-R. Schulten and H.-G. Neumann, Identification of nucleic acid adducts from trans-4-acetylaminostilbene derivatives, Chem-Biol Interactions 59:281 (1986).
10. R. Franz and H.-G. Neumann, Reaction of trans-4-N-acetoxy-N-acetyl-aminostilbene with guanosine, deoxyguanosine, RNA and DNA in vitro: predominant product is a cyclic N^2,N3-guanine adduct, Chem-Biol.

Interactions 62:143 (1987).

11. J. D. Scribner, S. R. Fist and N. K. Scribner, Mechanism of action of carcinogenic aromatic amines: an investigation using mutagenesis in in bacteria, Chem-Biol. Interactions 26:11 (1979).

12. V. M. Maher, R. H. Heflich and J. J. McCormick, Repair of DNA damage induced in human fibroblasts by N-substituted aryl compounds, Natl. Cancer Inst. Monographs 58:217 (1981).

13. D. Hilpert and H.-G. Neumann, Accumulation and elimination of macro-molecular lesions in susceptible and non-susceptible rat tissues after repeated administration of trans-4-acetylaminostilbene, Chem-Biol. Interactions 54:85 (1985).

14. T. Kitagawa, H. C. Pitot, E. C. Miller and J. A. Miller, Promotion by dietary phenobarbital of hepatocarcinogenesis by 2-methyl-N,N-dimethyl-4-aminoazobenzene in the rat, Cancer Res. 39:112 (1979).

15. E. Farber and R. Cameron, The sequential analysis of cancer development Adv. Cancer Res. 31:125 (1980).

16. P. Marquardt, W. Romen and H.-G. Neumann, Tissue specific acute toxic effects of the carcinogen trans-4-dimethylaminostilbene, Arch. Toxicol. 56:151 (1985).

17. A. Stier, R. Clauss, A. Lücke and I. Reitz, Radicals in carcinogenesis by aromatic amines, in: "Free Radicals, Lipid Peroxidation and Cancer", D. C. H. McBrien and T. F. Slater, eds., Academic Press, London, New York (1982).

18. W. Hillesheim and H.-G. Neumann, Aromatic amines do not produce oxidative stress in vivo, Biol. Chem. Hoppe-Seyler 368:1057 (1987).

19. H. Sies, R. Brigelius, H. Wefers, A. Müller and E. Cadenas, Cellular redox changes and responses to drugs and toxic agents, Fund. Appl. Toxicol. 3:200 (1983).

20. C. V. Smith and J. R. Mitchell, Acetaminophen hepatotoxicity in vivo is not accompanied by oxidative stress, Biochem. Biophys. Res. Commun. 133:329 (1985).

21. A. K. Laird and A. D. Barton, Cell growth and the development of tumors Nature (London) 183:1655 (1959).

22. B. Flaks, Changes in the fine structure of rat hepatocytes during the early phases of chronic 2-acetylaminofluorene intoxication, Chem-Biol. Interactions 2:129 (1970).

23. P. O. Seglen, G. Saeter and P. E. Schwartze, Nuclear alterations during hepatocarcinogenesis: promotion by 2-acetylaminofluorene, Abstracts European Meeting on Experimental hepatocarcinogenesis, Spa, Belgium (1987).

24. J. Kuchlbauer, W. Romen and H.-G. Neumann, Syncarcinogenic effects on the initiation of rat liver tumors by trans-4-acetylaminostilbene and 2-acetylaminofluorene, Carcinogenesis 6:1337 (1985).

25. A. Aström and J. W. DePierre, Characterization of the induction of drug metabolizing enzymes by 2-acetylaminofluorene, Biochim. Biophys. Acta 673:225 (1981).

26. H.-G. Neumann, The metabolism of repeatedly administered trans-4-dimethylaminostilbene and 4-dimethylaminobibenzyl, Z. Krebsforsch. 79:60 (1973).

27. P. Cikryt and M. Göttlicher, The affinity of carcinogenic aromatic amines to the rat hepatic aromatic hydrocarbon receptor (Ah) in vitro, J. Cancer Res. Clin. Oncol. 113:S13 (1987).

28. M. Göttlicher and P. Cikryt, Induction of the aromatic hydrocarbon (Ah) receptor by trans-4-acetylaminostilbene in rat liver. Comparison with other aromatic amines, Carcinogenesis 8:1021 (1987).

29. V. Ivanovic and I. B. Weinstein, Benzo(a)pyrene and other inducers of cytochrome P_1-450 inhibit binding of epidermal growth factor to cell surface receptors, Carcinogenesis 3:505 (1982).

30. A. K. Verma and R. K. Boutwell, Effects of dose and duration of treat-ment with the tumor promoting agent, 12-0-tetradecanoylphorbol-13-acetate, on mouse skin carcinogenesis, Carcinogenesis 1:271 (1980).

31. D. W. Nebert, N. M. Jensen, J. W. Perry and T. Oka, Association between ornithine decarboxylase induction and the Ah locus in mice treated with polycyclic aromatic compounds, J. Biol. Chem. 255:6836 (1980).

32. M. Göttlicher and P. Cikryt, Induction of ornithine decarboxylase by aromatic amines in rat liver, Cancer Lett. 35:65 (1987).

33. D. L. Hwang, A. Roitman, B. I. Carr, G. Barseghian and A. Lev-Ran, Insulin and epidermal growth factor receptors in rat liver after administration of the hepatocarcinogen 2-acetylaminofluorene: ligand binding and autophosphorylation, Cancer Res. 46:1955 (1987).

34. D. Hilpert, W. Romen and H.-G. Neumann, The role of partial hepatectomy and of promoters in the formation of tumors in non-target tissues of trans-4-acetylaminostilbene in rats, Carcinogenesis 4:1519 (1983).

LONG-TERM FATE OF THE HEPATOCYTE HYPERPLASTIC FOCI INDUCED BY A TWO STAGE

CARCINOGENETIC PROTOCOL

Gino Malvaldi, Elisabetta Chieli, Adonella Marradi and
Michela Saviozzi

Istituto di Patologia Generale
Scuola Medica, Università di Pisa
Via Roma 55, 56100 Pisa, Italy

INTRODUCTION

The appearance of a liver hepatocellular neoplasm is always preceded by
the development of histochemically detectable hepatocyte focal lesions (HFL)
which are believed to be related to the development of hepatoma (for recent
reviews: Bannasch[1], Moore and Kitagawa[2], Farber and Sarma[3]). A large discrepancy, however, usually exists between the number of HFL and the number of
tumors per liver. Since initiation is thought as an irreversible step in the
natural history of neoplasia and HFL are considered a cell population derived
from initiated hepatocytes[4], HFL as such should be stable in time, i.e. they
should not disappear after withdrawal of the inducing carcinogen(s). This
has been actually observed in some models of liver carcinogenesis[5-7]. In
others, however, many of the HFL present at the end of the carcinogen administration cycle seem to disappear with time[8-10].

We have reported previously a two stage model of liver carcinogenesis in
which many HFL are obtained when a single diethylnitrosamine (DEN) injection
is followed by administration of thiobenzamide (TB), a thiono containing compound endowed with dose dependent hepatotoxic properties[11,12]. This report
deals with the results obtained by investigating the natural history of the
HFL induced either by applying the basic protocol (DEN followed by 5 weeks of
TB feeding) or by varying the length and/or timing of TB administration
cycle.

MATERIALS AND METHODS

Male Sprague-Dawley rats bred in our colony under standard conditions
were used; they were housed in steel-bottomed polycarbonate cages and had
free access to food and water.

In a first experiment, 3-week-old rats were given a single injection of
DEN as initiator (150 mg/kg b.wt., i.p.) and weaned; 2 weeks later the animals were fed standard diet added with TB (1 g/kg of diet) for 5 weeks; the
animals were then returned to normal diet for 7 months. Two control groups
were used: the one was given DEN only, the other was fed TB-containing diet
without prior DEN initiation. Animals from all groups were killed both at
the end of the promotion cycle and at monthly intervals. Three month after

251

withdrawal of TB from the diet, a subgroup of rats from the main group was fed again TB-added diet. Half of the animals were killed after 2 weeks TB refeeding, the others 4 weeks later.

In a second experiment, young adult rats were initiated as before and divided into 3 groups, along with adequate controls; these animals were fed TB-added diet for 1, 2 or 3 months, respectively, and killed both at the end of each TB cycle and after a 4-month recovery period.

Before killing each animal was overloaded with iron dextran[13] and starved for 24 h, to deplete glycogen stores. The liver was excised and weighed; fragments from each lobe were fixed in cold 80% ethanol and quickly embedded in paraffin; serial sections were stained with hematoxylin and eosin as well as reacted for glycogen (PAS), iron (Perls) and gamma glutamyltranspeptidase activity (GGT)[14].

The stained sections were photographed; the slides were projected and the sections as well as the HFL inside them were outlined with different color pencils. This composite drawing was used to collect (by planimeter) the primary data concerning the number and the total area of HFL per section. When necessary, raw data were corrected according to the stereological procedures of Campbell et al.[15] In selected instances, each HFL was classified as positive to one, two, three or all the four markers here employed. Mitotic and apoptotic indices were identified according to established criteria on liver sections stained with hematoxylin and eosin; at least 10,000 cells/animal were counted both within and outside HFL. Statistical analysis was performed by means of the Student's t and chi-square tests; percentage confidence limits were calculated according to Swinscow[16].

RESULTS

Effects of TB Withdrawal on the Number, Volume and Phenotypic Complexity of HF

After 5 weeks of continuous TB administration the liver of all DEN-initiated animals showed a number and a total volume of HFL far greater than that of the controls (DEN-only or TB-only treated animals) (Figs. 1 and 2).

All the observed HFL were round or oval in shape; furthermore, most of them showed an homogenous staining. The mitotic index as well as the apoptotic index of HFL were higher than those of the surrounding liver parenchyma (Figs. 3a and 3b).

If the HFL are grouped according to being positive to one or more of the markers used to identify them, it can be observed that 1) the GGT positivity was the most represented marker, followed by the iron exclusion, starvation resistant glycogen storage and appearance in H & E (which was considered as one marker only, without further specifications such as clear, eosinophilic or hyperbasophilic cells) (results not shown); 2) the % incidence of the HFL displaying one, two and three markers was greater than that observed for HFL displaying four markers (Table 1); and 3) the mean volume of the HFL displaying 4 markers was significantly higher than that of the HF displaying 1, 2 or 3 markers (Table 1).

Upon withdrawal of TB from the diet, the number of HFL per liver was roughly unchanged for one month, then underwent a clear-cut decrease; a significant increase was again observed during the last months of observation (Fig. 1). Also the total HFL volume per liver showed a similar trend, but here the values observed after 7 months of recovery were far higher than those observed at the end of TB cycle (Fig. 2).

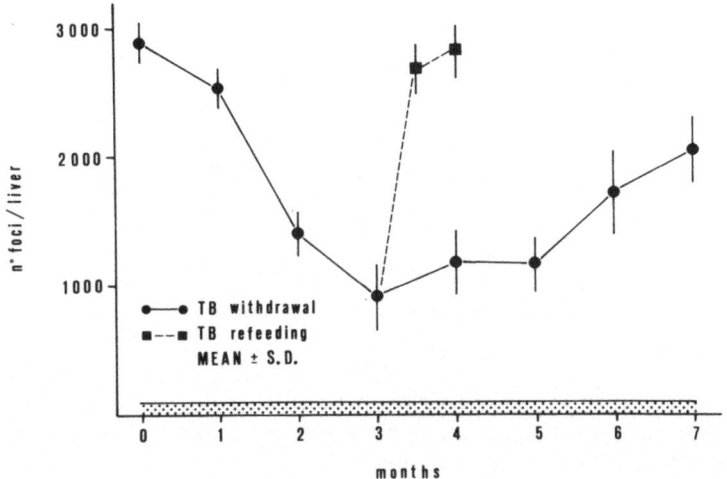

Fig. 1. Changes with time of the number of HFL per liver in DEN initiated rats after TB withdrawal and refeeding. The dashed area indicates the control values ± SD.

During the first 4 months of recovery, a large majority of HFL showed an irregular outline and a patchy response to the reagents used; subsequently, nearly all the HFL resumed an homogenous appearance. Also mitotic and apoptotic indexes within HFL showed fluctuations, even if they were always higher than those observed in the surrounding liver parenchima (Figs. 3a and 3b).

Finally, no changes were seen, as far as the phenotypic complexity of HFL is concerned, with respect to the pattern observed at the end of TB administration period. Four and seven months after TB withdrawal from the diet the % incidence of the foci displaying one or more markers was roughly the same as before. At these times also the mean volume of HFL displaying 4 markers was greater than those showing a simpler phenotype (Table 1).

Effect of Refeeding TB to DEN-initiated, TB-promoted Rats

To verify if the HFL which had disappeared after TB withdrawal from the diet were actually lost or were no longer visible by disappearance of the phenotypic markers used to identify them, TB-added diet was again administered to DEN-initiated, TB-promoted rats 3 months after withdrawal of TB from the diet. After 2 weeks of refeeding an abrupt increase in the number as well as in the size of HFL was observed (Figs. 1 and 2). No further changes were seen after 4 weeks refeeding. At this time the mitotic activity within HF was greater than that observed in the surrounding parenchima (Fig. 3a) which, in turn, was higher than that of the controls; it was also slightly greater than that observed within HFL from the rats of the main group. Also apoptosis increased again in the HFL cells as well as in the surrounding liver parenchima both 2 and 4 weeks after TB readministration.

Effect of the Length of the Promotion Cycle upon Disappearance of HFL after TB Withdrawal

To verify if a partial disappearance of HFL always follows the withdrawal of the promoter or some relationship exist with the length of the promotion cycle, DEN-initiated rats were given TB for 1, 2 or 3 months, then returned to a normal diet for 4 months.

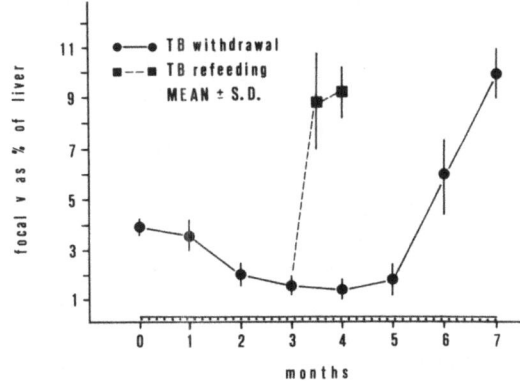

Fig. 2. Changes with time of the HFL total volume (as % of the liver volume)
in DEN initiated rats after TB withdrawal and refeeding. The dashed
area indicates control values ± SD.

The obtained results are summarized in Table 2; a good portion of HFL
induced by 1 month feeding disappeared (as observed in the first experiment),
whereas this did not occur when HFL were induced by a more prolonged TB promotion cycle. A similar trend was observed also as far of the total HFL volume was concerned. In considering these results it should be recalled that
TB administration for 3 months resulted in a liver cirrhosis both in uninitiated and DEN-initiated rats.

DISCUSSION

The reported results indicate that many of the HFL elicited by a short-
term administration of TB to DEN-initiated rats slowly disappeared after
withdrawal of the promoter, but appeared again 1) several months later, without further treatments, and 2) shortly during a second TB cycle. On the
contrary, if the promotion length was adequate, HFL did not disappear upon TB
withdrawal from the diet.

The decrease in the number and total volume of HFL during recovery is
similar to that observed in other models of liver carcinogenesis[2,17].

The mechanism involved in the HFL disappearance closely recalls redifferentiation or "remodeling" as suggested by Tatematsu et al.[18]; even if, for
the sake of enumeration, the actual HFL outline can be always traced with
reasonable accuracy by superimposing serial sections stained for the 4
markers here employed, most HFL become patchy with time. This suggests that
many of the HFL are phenotypically unstable and can revert to a normal looking phenotype upon withdrawal of the promoter. During this time, however,
the apoptotic index within HFL was always greater than that of the surrounding liver tissue. This finding is similar to that reported in other models
of liver carcinogenesis[19,20]. Since apoptosis is believed to mark programmed
cell death[21], some cell loss within HFL cannot be excluded. On the other
hand, also mitotic index of HFL is higher than that of the perifocal tissue.
Collectively, these findings suggest an increased cell turnover rather than a
relevant decrease in the HFL cell number. The hypothesis that many of the
HFL-forming cells actually remain, even if in a phenotypically hidden state,
is further supported by the results obtained by refeeding TB during recovery.
Two weeks of further promotion, carried out at a time in which both number
and size of HFL were at their lowest, restored the previous values; as far as
size is concerned, the increase in HFL total volume is far higher than that

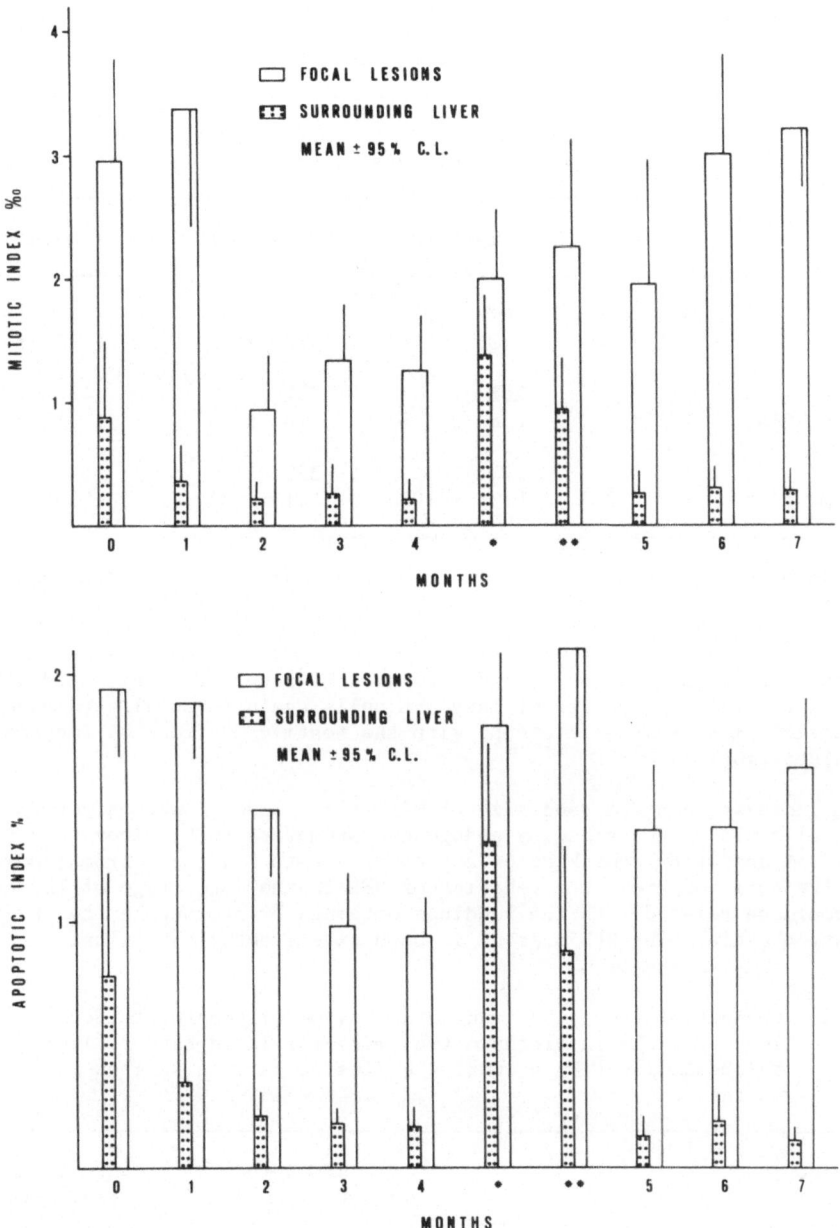

Fig. 3. Mitotic (a) and apoptotic (b) index of hepatocytes within and outside HFL in the months following TB administration for 5 weeks to
DEN-initiated rats.

observed after 2 weeks of TB administration to freshly initiated rats[12]; two
additional weeks of TB did not induce further increase. At these times the
mitotic index within HFL did not show obvious changes with respect to controls (i.e. rats from the main group undergoing recovery), the apoptotic
index being somewhat increased. None of these findings is by itself conclusive; however, if they are considered together, it seems unlikely that reappearance and increase in size of HFL after a very short TB administration

Table 1. Changes of the phenotypic complexity of the hepatocyte focal lesions in diethylnitrosamine-initiated rats after 5 weeks of thiobenzamide feeding and during recovery

Treatment	% of HFL displaying one to four markers (mean volume of each type of HFL in μ^3)			
	1 marker	2 markers	3 markers	4 markers
5 W TB	27% (83 ± 8)	29% (136 ± 16)	32% (164 ± 68)	12% (284 ± 95)
5 W TB + 4 M norm. diet	28% (27 ± 7)	29% (82 ± 5)	29% (105 ± 29)	14% (214 ± 89)
5 W TB + 7 M norm. diet	26% (53 ± 27)	27% (136 ± 108)	32% (619 ± 87)	15% (2430 ± 820)

Values in brackets are mean ± SE. Phenotypic analysis of 100 HFL at each time point.

cycle could depend solely on multiple division of a very few initiated cells. It seems more likely, a switch of many HFL cells again to an altered phenotype, possibly more suitable to cope with the hostile environment imposed by TB administration.

The increase in number and size of HFL as seen many months after TB withdrawal could be sustained by endogenous promoters such as hormones or bile acids[22] and/or by some kind of autocrine growth factors. From a more speculative point of view, the behavior of HFL in the late stage of TB recovery could be related with the findings obtained by prolonging the TB administration cycle. The HFL present in both experimental situations are dif-

Table 2. Changes of the number and total volume of hepatocyte focal lesions (HFL) in diethylnitrosamine-initiated rats following thiobenzamide (TB) promotion cycles of various lengths and 4 months recovery

Treatment	HFL number[a] (no./cm^3)	HFL volume[a] (cu.mm/cm^3)
1 month TB	357 ± 35	31 ± 8
1 month TB and 4 months recovery	126 ± 40	13 ± 3
2 months TB	569 ± 60	174 ± 72
2 months TB and 4 months recovery	533 ± 85	161 ± 68
3 months TB	517 ± 35	306 ± 96
3 months TB and 4 months recovery[b]	412 ± 102	324 ± 62

[a]Means ± SE; n = 5. [b]In this group 2 rats out of 5 harbored hepatoma; the volume calculations were corrected accordingly.

ferent from those appearing after a short TB promotion in that they do not regress. Such biological difference, which is brought about either by a time factor or by a prolonged promotion by xenobiotics, suggests that some changes (which here lack a morphohistochemical correspondence) had occurred in the HFL cell populations. Possibly, the term "progression" should be used to indicate these late changes in the HFL behavior.

In this study, the characterization of the phenotypic complexity of the HFL by the use of 4 markers showed no predictive value, neither to indicate the likelihood of the focal lesions to disappear nor to identify those HFL more prone than others to develop ultimate neoplastic changes. In fact, the incidence of HFL displaying one or more markers was roughly the same throughout the entire length of the study, i.e. HFL did not lose or gain markers during their natural history. These results are in line with those reported by Peraino et al.[23] and by Goldsworthy and Pitot[24]. The use of more than one marker, however, while confirming GGT as the most representative one, revealed also a positive correlation between the HFL size and the number of markers per focus. Both findings are in agreement with those of Goldsworthy and Pitot[24] and Estadella et al.[25], even if they used different markers and different experimental models.

ACKNOWLEDGEMENT

This work was supported by the C.N.R., P.F. "Oncologia".

REFERENCES

1. P. Bannasch, Preneoplastic lesions as end points in carcinogenicity testing, 1, Hepatic preneoplasia, Carcinogenesis 7:698 (1986).
2. M. A. Moore and T. Kitagawa, Hepatocarcinogenesis in the rat: The effect of promoters and carcinogens in vivo and in vitro, Int. Rev. Cytol. 101:125 (1986).
3. E. Farber and D. S. R. Sarma, Biology of disease, Hepatocarcinogenesis: a dynamic cellular perspective, Lab. Invest. 56:4 (1987).
4. D. B. Solt, E. Cayama, D. S. R. Sarma and E. Farber, Persistence of resistant putative preneoplastic hepatocytes induced by N-Nitrosodiethylamine or N-methyl-N-nitrosourea, Cancer Res. 40:1112 (1980).
5. E. Scherer, M. Hoffmann, P. Emmelot and M. Friedrich-Freksa, Quantitative study on foci of altered liver cells induced in the rat by a single dose of diethylnitrosamine and partial hepatectomy, J. Natl. Cancer Inst. 49:93 (1972).
6. A. M. Moore, D. Mayer and P. Bannasch, The dose dependence and sequential appearance of putative preneoplastic populations induced in the rat liver by stop experiments with N-nitrosomorpholine, Carcinogenesis 3:1429 (1982).
7. T. Goldsworthy, M. A. Campbell and M. C. Pitot, The natural history and dose response characteristics of enzyme altered foci in rat liver following phenobarbital and diethylnitrosamine administration, Carcinogenesis 5:67 (1984).
8. G. M. Williams and K. Watanabe, Quantitative kinetics of development of N-2 fluorenylecetamide-induced altered hyperplastic hepatocellular foci resistant to iron accumulation and of the reversion or persistence following removal of carcinogen, J. Natl. Cancer Inst. 61:113 (1978).
9. M. Tatematsu, Y. Nagamine and E. Farber, Redifferentiation as a basis for remodeling of carcinogen induced hepatocytes nodules to normal appearing liver, Cancer Res. 43:5049 (1983).
10. K. Enomoto and E. Farber, Kinetics of phenotypic maturation or remodeling of hyperplastic nodules during liver carcinogenesis, Cancer

Res. 42:2330 (1982).

11. G. Malvaldi, E. Chieli and M. Saviozzi, Promotive effects of thiobenz-amide on liver carcinogenesis, Gann 74:469 (1983).

12. G. Malvaldi, E. Chieli and M. Saviozzi, Characterization of the promoting activity of Thiobenzamide on liver carcinogenesis, Toxicol. Pathol. 14:370 (1986).

13. N. Hirota and G. M. Williams, The sensitivity and heterogeneity of histochemical markers for altered foci involved in liver carcinogenesis, Am. J. Pathol. 95:317 (1979).

14. A. M. Rutenburg, M. Kim, J. W. Fischbein, J. S. Hanker, H. L. Wasserkrug and A. M. Seligman, Histochemical and ultrastructural demonstration of gamma glutamyltranspeptidase activity, J. Histochem. Cytochem. 17:517 (1969).

15. M. A. Campbell, H. C. Pitot, V. R. Potter and B. A. Laishes, Application of quantitative stereology to the evaluation of enzyme altered foci in rat liver, Cancer Res. 41:465 (1982).

16. T. D. V. Swinscow, "Statistics at square one", British Medical Association, London (1978).

17. M. A. Moore, M. J. Hacker and P. Bannasch, Phenotypic instability in focal and nodular lesions induced in a short-term system in the rat liver, Carcinogenesis 4:595 (1983).

18. M. Tatematsu, T. Takano, R. Hasegawa, K. Imaida, J. Nakanowatari and N. Ito, A sequential quantitative study of the reversibility or irreversibility of liver hyperplastic nodules in rats exposed to hepatocarcinogens, Gann 71:483 (1980).

19. A. Columbano, G. M. Ledda-Columbano, P. M. Rao, S. Rajalakshmi and D. S. R. Sarma, Occurrence of cell death (apoptosis) in preneoplastic and neoplastic liver cells, Am. J. Pathol. 116:441 (1984).

20. W. Bursch, B. Lauer, I. Timmermann-Troisiener, G. Barthel, J. Schuppler and R. Schulte-Hermann, Controlled death (apoptosis) of normal and putative preneoplastic cells in rat liver following withdrawal of tumor promoters, Carcinogenesis 5:453 (1984).

21. A. H. Wyllie, J. F. R. Kerr and A. R. Currie, Cell death: the significance of apoptosis, Int. Rev. Cytol. 68:251 (1980).

22. R. G. Cameron, K. Imaida, H. Tsuda and H. Ito, Promotive effects of steroids and bile acids on hepatocarcinogenesis initiated by diethylnitrosamine, Cancer Res. 42:2426 (1982).

23. C. Peraino, E. F. Staffeldt, B. A. Carnes, V. A. Ludeman, J. A. Blomquist and S. V. Vesselinovitch, Characterization of histochemically detectable altered hepatocyte foci and their relationship to hepatic tumorigenesis in rats treated once with diethylnitrosamine or Benzo(a)-pyrene withine one day after birth, Cancer Res. 44:3340 (1984).

24. T. L. Goldsworthy and H. C. Pitot, The quantitative analysis and stability of histochemical markers of altered hepatic foci in rat liver following initiation by diethylnitrosamine administration and promotion with phenobarbital, Carcinogenesis 6:1261 (1985).

25. M. D. Estadella, M. J. Pujol and J. Domingo, Enzyme pattern and growth rate of liver preneoplastic clones during carcinogenesis by diethylnitrosamine, Oncology 41:276 (1984).

EFFECTS OF CHRONIC THIOACETAMIDE ADMINISTRATION ON LIVER DRUG METABOLIZING

SYSTEM AND ON THE DEVELOPMENT OF HEPATOCYTE FOCI, NODULES AND TUMORS

Pier G. Gervasi, Vincenzo Longo, Mino Marzano, Michela Saviozzi[1] and Gino Malvaldi[1]

Istituto di Mutagenesi e Differenziamento, CNR
Via Svezia 10, 56100 Pisa, Italy. [1]Istituto di Patologia
Generale dell'Università, Via Roma, 56100 Pisa, Italy

Preneoplastic hepatocyte nodules (HN) generated in the rat liver by different experimental protocols display a common pattern in the xenobiotic metabolism, namely a decrease of the phase I components and an increase of the phase II ones[1]. If initiated hepatocytes share with HN these metabolic properties (which explain the observed increase of HN resistance to many liver toxins), then any mild but prolonged liver injury by chemicals needing bioactivation should cause, in addition to liver cirrhosis, also the appearance of HN and tumors, providing the liver has been exposed to initiating stimuli.

To verify the hypothesis, diethylnitrosoamine (DEN)-initiated as well as uninitiated rats (male Sprague Dawley) were administered thioacetamide (TAA) at low dose (2 mg/day/100 g b.w.) for 6 months. TAA is a compound devoid of mutagenic activity[2] but it is well known to depress cytochrome P-450-dependent drug metabolism when acutely administered in vivo[3]. During 6 months both HN incidence and changes in the drug metabolizing system (DMS) were followed at monthly intervals.

In the uninitiated rats a regenerative hyperplastic liver cirrhosis slowly developed upon TAA chronic administration. A few gamma-glutamyl-transpeptidase-positive, PAS-positive hepatocyte focal lesions were seen from the 3rd month onward, their cumulative volume never exceeding 0.1% of the liver volume. By contrast in the DEN-initiated TAA-treated rats the liver was macronodular because of the appearance and growth of many HN (Fig. 1). At the end of the TAA cycle over 40% of the liver parenchyma was constituted by HN and enzyme-altered hepatocyte foci. 3 rats out of 7 harboured an hepatoma.

During TAA administration both uninitiated and DEN-initiated rats underwent a progressive decrease of the cytochrome P-450 liver content as well as of the activity of aminopyrine N-demethylase, ethoxycoumarin O-deethylase and ethoxyresorufin O-deethylase. Six months later the values of these parameters in the liver of rats from both groups were I) decreased (up to 10-30% with respect to those found in normal, age-matched controls) II) superimposable to those observed in HN which in turn were similar to those reported in HN generated by other carcinogenetic protocols. On the contrary, the components of the phase II of DMS were markedly enhanced. In particular, the activities of microsomal epoxide hydrase and UDP-glucuronyl transferase as well

259

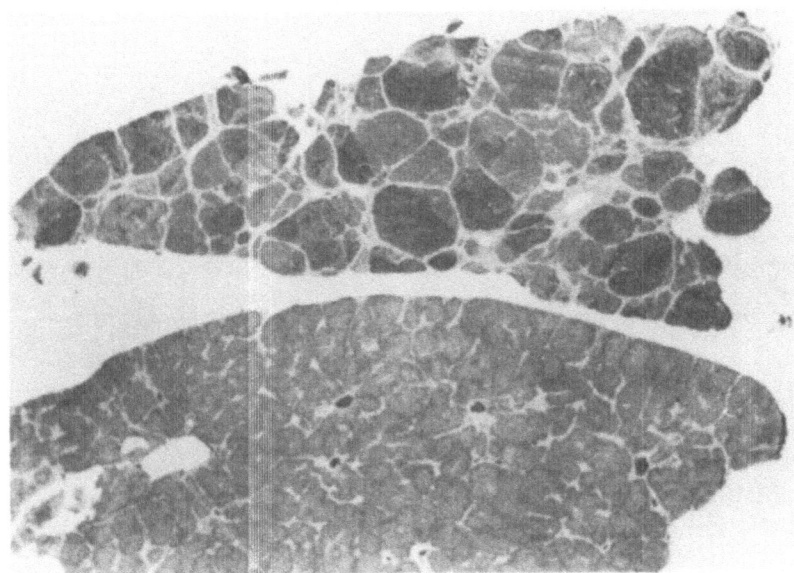

Fig. 1. Appearance of liver from DEN-initiated (upper half) and uninitiated
(lower half) rats treated with TAA for 6 months. A striking difference in the extent of liver cirrhosis as well as in the number of
GGT-positive (dark) nodules is seen.

as cytosolic glutathione-S-transferase were increased over the controls in
both experimental groups and NH (about 2-3 times), whereas the increase of
benzaldehyde dehydrogenase activity (about ten times compared to the control
value) was evident only in the liver of DEN-initiated/TAA treated rats and in
HN.

To assess the initiating activity of TAA the compound was given in a
single high dose (100-250 mg/Kg b.w.) to 1-week old, 4-week old as well as to
young partially hepatectomized rats of both sexes, since high mitotic activity of the hepatocytes enhances the initiation process. This TAA treatment
was followed by a standard promoting protocol (2-acetylaminofluorene in the
diet plus a single carbon tetrachloride dose) (2-AAF/CCl$_4$).

Also in these experiments the number and the area of the hepatocyte foci
displaying gamma-glutamyltranspeptidase activity and starvation resistant
glycogen storage were used as end-point of the assay.

The overall findings indicated that the initiating activity of TAA was
quite low, if any, because no differences were seen in the number of hepatocyte foci between TAA initiated, 2-AAF/CCl$_4$ promoted animals and their controls in the 3 experimental situations investigated.

In conclusion, chronic administration of TAA at low dose showed a weak,
if any initial ability on liver carcinogenesis, but provided strong promoting stimuli for already initiated hepatocytes. In rather sharp contrast
with other liver promoters and inducers of cytochrome P-450 (phenobarbital
et al.[4]), the DMS pattern of the liver parenchyma of TAA treated rats was
similar to that displayed by HN.

ACKNOWLEDGEMENTS

Supported by CNR, PF "Oncologia" and M.P.I. 40% and 60% funds.

REFERENCES

1. M. W. Roomi, R. K. Ho, D. S. R. Sarma and E. Farber, A common biochem-
    ical pattern in preneoplastic hepatocyte nodules in four differ-
    ent models in the rat, Cancer Res. 45:564 (1985).
2. J. McCann, E. Choi, E. Yamasaki and B. N. Ames, Detection of carcino-
    gens as mutagens in the Salmonella/microsome test; assay of 300
    chemicals, Proc. Nat. Acad. Sci. 72:5135 (1975).
3. A. L. Hunter and R. A. Neal, Inhibition of hepatic mixed function
    oxidase activity in vitro and in vivo by various thiono-sulfur
    containing compounds, Biochem. Pharmacol. 24:2199 (1975).
4. R. Schulte-Hermann, Tumor promotion in the liver, Arch. Toxicol. 57:
    147 (1985).

CELL DEATH BY APOPTOSIS IN NORMAL, PRENEOPLASTIC AND NEOPLASTIC TISSUE

Rolf Schulte-Hermann, Wilfried Bursch, Lazlo Fesus[1], and
Bettina Kraupp

Institute of Tumorbiology-Cancer Research, University of
Vienna, 1090 Vienna, Borschkegasse 8a, Austria
[1] Department of Biochemistry, University Medical School of
Debrecen, Debrecen, Hungary

INTRODUCTION

Cells of hiher organisms can die through different causes and under different circumstances which can be classified broadly into at least 3 different categories:

1. Massive tissue injury such as after hypoxia or after CCl_4 in the liver.
   Some authors restrict use of the term necrosis to cell death occurring
   after massive tissue damage[1,2].
2. Terminal differentiation of tissues such as skin, intestine or blood
   cells.
3. Regression of organs, elimination of excessive cells, cell turnover in
   healthy tissues. During development certain organs regress such as the
   Mullerian duct in the male embryo. In the adult organism atrophy or
   removal of hyperplasia may occur in hormone-dependent organs. The type
   of cell death involved has recently been designated "apoptosis"[1,3-8].
   It is conceived as a genetically encoded cellular suicide program that
   can be activated in situations where cell elimination appears physiologically advantageous or even necessary. Examples of physiological
   states in which apoptosis is believed to occur are given in Table 1.
   It appears that apoptosis is a widespread phenomenon which has been
   observed in a variety of species throughout the animal kingdom and in
   various organs during developmental stages and adulthood.

So far the process of apoptosis is not very well characterized, and the
distinction from damage induced cell death is not very clear (Table 2).
Thus, some authors consider death induced by certain types of damage, such as
councilman bodies in human hepatitis or radiation induced cell death as belonging to the apoptosis category[1,2]. Indeed, due to the lack of specific
markers of apoptosis and unequivocal discrimination from damage induced death
is very difficult. Apoptosis was postulated to exhibit specific morphological features; these differed from those seen after certain types of damage
induced cell death (e.g. CCl_4 in the liver). Histologically visible signs of
cell death by apoptosis as found in the liver are schematically depicted in
Fig. 1; the sequence of events shown was first suggested by Kerr, Wyllie and
Currie[1,3]. However, since it is difficult to use morphology as a sole criterion for a certain biological process we have so far mainly relied upon a

Table 1. Occurrence of apoptosis

Hydra in fasting state[30]
Nematodes: maturation[11]
Insects during metamorphosis[1]
Amphibia: loss of tadpole tail[31]
Mammalia:
  Embryonic development (e.g. elimination of interdigital webs)[1]
  Hormone-dependent organs: during atrophy[4-8]
   Liver: elimination of induced hyperplasia[18,20]
   Liver: atrophy after starvation
   Liver: preneoplastic foci[27,28,32]
   Lymphatic cells: glucocorticoid-induced death[9,10,12]
   T-Lymphocytes: IL-2 withdrawal
              → termination of immunresponse (?)[22]
  Tumors: regression of estrogen-dependent kidney tumors

functional criterion of apoptosis. This is based on the fact that apoptosis - in contrast with other types of cell death - can be inhibited by growth stimuli (Table 2). The usefulness of this criterion will be demonstrated below. In addition, it can be hoped that specific biochemical markers of apoptosis gradually become available (Table 2).

Evidence suggesting that apoptosis is a genetically determined, endogenous suicide program includes the following findings:

1. Inhibitors of RNA and protein synthesis prevent initiation of cell death in cultured lymphocytes by glucocorticoids[9,10].
2. Recently mutants of the nematode <u>Caenorabditis elegans</u> have been found which are deficient in certain steps of cell death (deathless mutants)[11].
3. Likewise, a mouse thymoma cell mutant has been found that is resistent to induction of cell death by glucocorticoids and by natural killer cells, but not to cytolysis by complement which does not act via an apoptotic process[12]. It may be of interest to add that apoptosis appears to be receptor mediated since certain leukemia cells without glucocorticoid receptors do not undergo apoptosis in response to this hormone[9].

An <u>in vitro</u> correlate of apoptosis may be the type of cell death occurring in growth factor dependent cells after complete removal of the growth factor. This phenomenon has been known as long as the growth factors themselves[13]. It is in fact surprising that enormous efforts have been made to

Table 2. Characteristics of apoptosis

Occurrence in specific physiological situations[1-3]
Toxic damage not necessary[2]
Morphology[1,3,19]
Induction by withdrawal of growth stimuli[18]
Inhibition by growth stimuli[18]

DNA fragmentation to nucleosomes[10,21-23,33]
Expression of transglutaminase (?)[25]

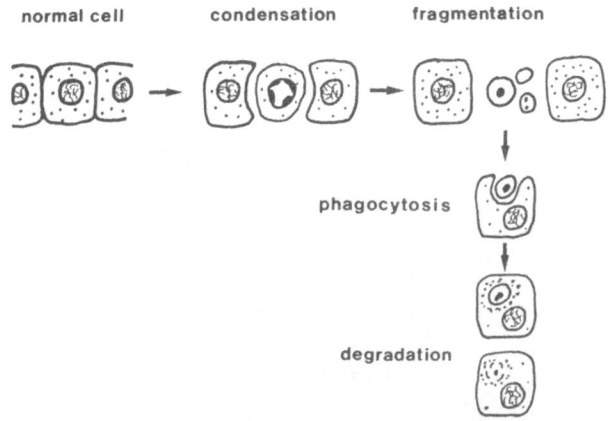

normal cell   condensation   fragmentation

phagocytosis

degradation

Fig. 1. Hypothetical diagram of the process of apoptosis during regression
of liver hyperplasia.

understand the effect of growth factors on cell <u>proliferation</u> while almost no
attention was paid until very recently to the mechanism of cell <u>death</u> after
growth factor depletion.

RESULTS AND DISCUSSION

Apoptosis in Hyperplastic Rat Liver After Treatment with Hepatomitogens

Many compounds of xenobiotic origin such as the environmental pollutant
alpha-hexachlorocyclohexane (α-HCH), the synthetic sex steroid cyproterone
acetate (CPA), phenobarbital (PB), the peroxisome proliferator nafenopin and
many others have been found to induce liver growth in rodents which includes
cell multiplication (hyperplasia)[14-17]. This liver hyperplasia can be reversible depending on the biological halflife of the inducing chemical[18]. We have
found that during the period of regression of DNA no evidence of damage induced necrosis occurred. Instead histological signs of apoptosis were detected[19]. Counting the number of such apoptotic cells and apoptotic bodies
revealed a clear increase during the period of DNA elimination (Fig.2).
Similar findings were made by Columbano in another model in which liver
hyperplasia was induced by treatment with lead nitrate. Again a massive up-
surge of apoptotic figures occurred during the period of DNA elimination[20].

To obtain further proof that truly apoptosis is involved in the observed
regression of liver mass we adopted the functional criterion mentioned above
(Table 2). As seen in Fig. 2 the concentration of the hepatomitogenic drug
CPA falls rapidly and is very low 48 h after the last treatment (day 8).
Therefore we wondered whether the drop of the hepatomitogen levels in the
liver might provide the signal for initiation of cell death. To check this
hypothesis we retreated animals with CPA on day 8, i.e. at the maximum of
apoptotic activity. As may be seen in Fig. 3 retreatment indeed dramatically
decreased the incidence of apoptosis four hours later and not only CPA was
active; also other hepatomitogens such as PB, α-HCH and nafenopin were
equally effective in inhibiting cell death. These findings strongly suggest
that cell death does not occur in response to tissue damage that might have
been produced by CPA but rather is a regulatory phenomenon serving to remove
the excessive hyperplasia and can be inhibited by a variety of chemically unrelated mitogenic signals.

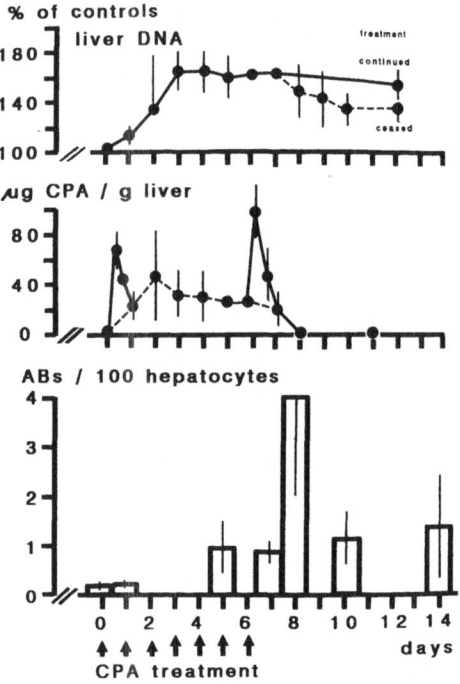

Fig. 2. Effect of CPA treatment and CPA withdrawal on DNA content, CPA concentration and on incidence of apoptoses in rat liver. Before com-
mencing treatment, the rats (Wistar, female) were adapted for 2-3
weeks to an inverted light-dark rhythm (L:D 12:12h) with lights off
from 9 a.m. to 9 p.m. and lights on from 9 p.m. to 9 a.m.; food was
available for 5 h from 9 a.m. to 2 p.m. (see Fig. 5). The animals
were treated with cyproterone acetate (CPA) as indicated at the
bottom of the figure; small arrows: 100 mg/kg/day, large arrows:
130 mg/kg/day. Liver DNA (mg DNA/100g rat) is expressed as percentage of solvent-treated controls. Solid line: rats killed 24 h
after the last treatment; treatment continued: initial CPA treatment as indicated below the abscissa followed by further treatment
with 130 mg CPA/kg/day until day 11; treatment ceased (dashed
line): rats killed between 2 and 5 days after the last (7th) treatment. CPA concentration (µg/g liver) was determined 24 h after
each dose (dashed line) and at closer intervals after the first and
seventh dose (solid line). The CPA determinations were performed
by B. Düsterberg, Schering, Berlin, FRG. Apoptoses (ABs/100 hepatocytes): the number of apoptotic bodies (ABs, see ref. 19 for
morphological description) found in histological sections is expressed as percentage of intact hepatocytes.

Apoptosis in Rat Liver After Starvation and Feeding

As a further test of this concept we studied another situation where
liver DNA is removed, namely liver atrophy occurring in response to starvation. As seen in Fig. 4 during 8 days of starvation liver mass decreased by
approx 60%, liver DNA by aprox. 25%. Concomitantly we found considerable
numbers of apoptotic figures at all times investigated; the highest levels
were found on days 7 and 8 (Fig. 4).

The fasting/feeding state of the animals may also have profound influence on the height of the apoptotic peak after CPA withdrawal as shown above.

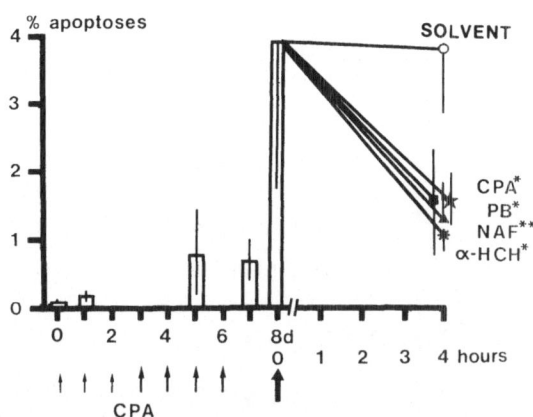

Fig. 3. Effect of CPA and other liver mitogens on the incidence of
apoptoses in rat liver. Treatment: day 0-6: see Fig. 2; day 7:
no treatment; day 8: rats were treated with a single oral dose
(130 mg/kg) of either CPA, nafenopin (NAF), phenobarbital (PB) or
alpha-hexachlorocyclohexane (α-HCH). The compounds were dissolved
in an aequeous solution of 0.09 % Myrj (Serva, Heidelberg). Solvent: rats treated with 0.09 % Myrj alone. Means (± SD) of 3-5
rats are given. *P < 0.05 **P < 0.01 CPA, PB, NAF or α-HCH treated
rats vs. rats treated with vehicle.

In the experiments presented in Fig. 2 the animals were subjected to a feeding rhythm which provided access to food for only 5 hours per day, and were
always sacrificed before the daily feeding period. As shown in Fig. 5 consumption of the scheduled meal on day 8 rapidly decreased the incidence of
apoptosis; if the meal was denied, the number of apoptotic bodies remained
high. Thus not only the hepatomitogens but also feeding apparently can inhibit the initiation of apoptosis. Further studies have indicated that the
main constituents of the diet i.e. carbohydrates, fat and protein if given
alone were all able to suppress apoptosis (data not shown). Because of the
cyclic eating behaviour of normal rats another hypothesis emerging from these
data is that apoptotic activity exhibits a diurnal rhythm in the liver. To
our knowledge this possibility has not yet been investigated but it should be
taken into account in the design of experiments.

Phagocytosis of Apoptotic Fragments by Intact Hepatocytes

The sequence of histologically discernible stages of apoptosis predicts
phagocytosis of apoptotic fragments by intact hepatocytes. This remained a
point of concern because hepatocytes are not known to be capable of phagocytosis. However, it was found that the extracellular apoptotic bodies did not
exhibit signs of autodigestion. Only the intra-hepatocytic bodies did so to
a variable extent[19]. Furthermore, during a period of CPA treatment we have
labeled all proliferating hepatocytes by continuous infusion of ^{3}H-thymidine.
After withdrawal of CPA we checked whether apoptotic bodies in labeled hepatocytes were also labeled. This was rarely the case; rather in the majority
of labeled cells the apoptotic bodies were unlabeled (Fig. 6). This result
provides a clear proof that the chromatin in the apoptotic bodies was not
derived from chromatin of the host cell, i.e. was from an extrahepatocellular
source. In addition, this experiment suggests that non-labeled cells (which
did not participate in proliferation during CPA treatment) were preferred for
death by apoptosis. The basis of this selective effect is unknown.

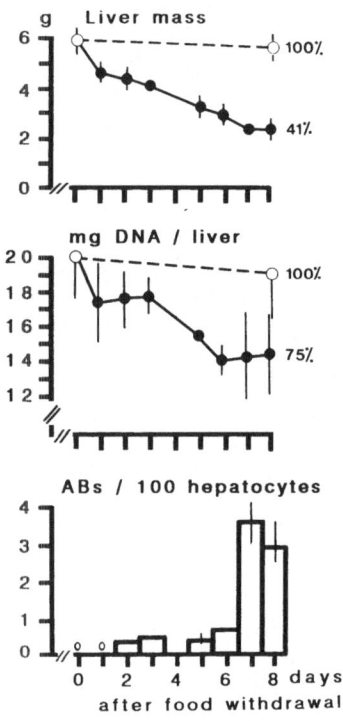

Fig. 4. Effect of food withdrawal on liver mass, liver DNA content and incidence of apoptoses in rat liver. Animals: female Wistar rats, 150-160 g at food withdrawal. Rats were killed between 1 and 8 days after food withdrawal. (a) liver mass (g liver/rat), (b) mg DNA/liver; a and b:○fed rats, ● starved rats, (c) number of apoptotic bodies (ABs) per 100 intact hepatocytes. Means of 3-5 animals are given. Vertical bars: in (a) and (b) SD (were not given smaller than symbols), in (c) 95% confidence limits.

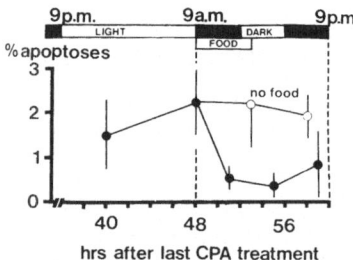

Fig. 5. Effect of food uptake on the incidence of apoptoses in rat liver. Feeding and CPA treatment (3 x 100 plus 4 x 130 mg/kg/d): see Fig. 2. Rats were killed between 40 and 59 h after the last CPA treatment (corresponds to day 8, see Fig. 2,3). (O) no food at 48 h after the last CPA treatment (●) rats on sheduled feeding.

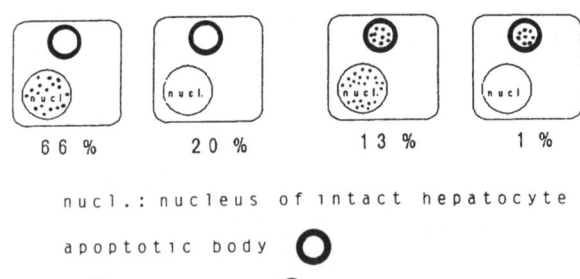

nucl.: nucleus of intact hepatocyte

apoptotic body ◯

⊚ labelled; ◯ unlabelled

Fig. 6. Occurrence of labelled DNA in nuclei and ABs of the same hepatocytes. Continuous infusion of [³H]-methyl-thymidine ([³H]-TdR) was
used to label all hepatocytes which proliferate during a CPA treatment for 7 days (see fig. 2). Rats were killed two days after cessation of CPA, (at the peak incidence of apoptosis in the liver)
and [³H]-TdR labeling of chromatin in hepatocytic nuclei and in ABs
was determined by autoradiography of histological sections of the
liver. The total number of intra-hepatocellular ABs with chromatin
was taken as 100%; the data shown indicate the relative frequencies
of [³H]-labeling of ABs as well as their occurrence in hepatocytes
with or without [³H]-labeled nuclei (details: see ref. 19).

Duration of Apoptosis

For some time we were concerned about the apparent discrepancy between
the amount of DNA elimination which in our CPA model was as much as 25% in a
matter of a few days, and the low levels of apoptotic cells which were never
more than a few percent per liver (see Fig. 2 and 5). The inhibition of
apoptosis by CPA and/or feeding provided the opportunity to measure the duration of the entire apoptotic process and of the individual stages. As
shown in Fig. 7a there was no change in the incidence of apoptotic bodies for
about 1 h after treatment. Then a rapid decrease occurred which reached a
minimum at 4 h. Thus approx. 1 h seems to be required between administration of the inhibitory regimen (CPA + feeding) and the first visible manifestation of cell death. A similar lag phase has been observed in lymphocytes in vitro[21]. Extra- and intra-hepatocytic apoptotic bodies (AB) tended
to disappear with linear kinetics in a semilogarithmic plot (Fig. 7b). Remarkable, extrahepatocellular AB were not only less frequent but also disappeared more rapidly than the intrahepatocellular AB (half-life 30 min versus
70 min). This consists completely with the sequence of apoptotic events
shown in Fig. 1. We conclude from this experiment that apoptosis is a rapid
phenomenon. The histologically visible parts of it may last on average
approximately 2.5 h. This can explain why at any given timepoint only relatively few apoptotic cells are found even in situations where considerable
amounts of DNA disappear from the liver.

Biochemistry of Apoptosis

Little is known on the complex and coordinated events that must occur at
the beginning and in the further course of apoptosis. So far the earliest
discernable metabolic change known is an activation of an endonuclease which
degrades chromatin into nucleosomes[9,10,21-23]. We have recently detected that
apoptotic cells possess high amounts of the enzyme transglutaminase, an
enzyme involved in crosslinking of proteins via glutamine and lysine residues[24,25]. As shown immunocytochemically the enzyme is present within the
liver tissue in the endothelial cell compartment. Among hepatocytes, however, only apoptotic bodies show a reaction (Fig. 8). Further support for a

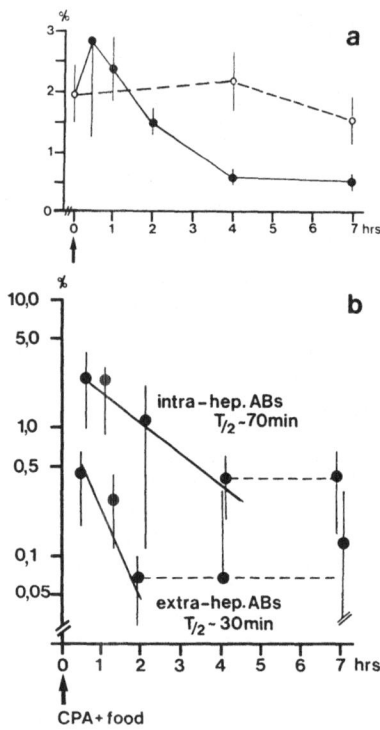

Fig. 7. Time course of decrease of AB incidence after CPA and food administration. Rats received CPA for 7 days as described in figure 2. Two
days after the last CPA dosing (i.e. at the peak incidence of ABs),
rats were retreated with a single dose of CPA (130 mg/kg). Additionally, rats received food according to the feeding schedule (see
Fig. 5). In a) the incidence of the total number of ABs (intra-,
extra- hepatocellular ABs, ABs with and without chromatin) is given,
in b) we differentiated between extra- or intra- hepatocellular
localization of the ABS. ●——● : CPA and food at "0"; ○——○ : no food
and no CPA at "0". 7000 - 8000 hepatocytes were scored per liver for
determination of the AB incidence. Each point represents the mean
(±) SD of 5-6 animals.

correlation between transglutaminase and apoptosis is provided by a roughly
parallel increase and decrease of enzyme activity and apoptosis during the
course of the experiment (Fig. 9). The function of transglutaminase during
apoptosis is unclear. The enzyme may be involved in the incapacitation of
essential proteins, in membrane alteration in the separation of the cell from
its neighbours, in the fragmentation process etc. Perhaps the most important
aspect is that it may provide the first marker to label individual apoptotic
cells.

In conclusion the findings presented so far point to an important role
for apoptosis as a regulator of cell number in adult liver and other tissues.
This suggests the question whether or not a disturbance of controls of apopto-
sis.might be involved in the imbalance of growth during tumorigenesis.

Apoptosis in Tumorigenesis and Tumor Growth

We have previously studied the effect of liver tumor promoters on the
growth of putative preneoplastic foci in rat liver. An early discovery was

270

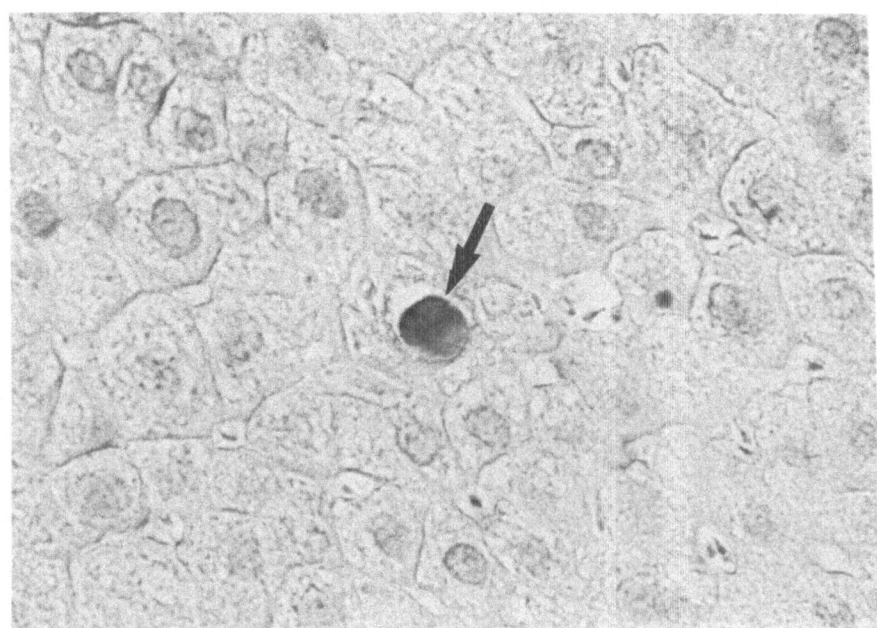

Fig. 8.   Immunohistochemical staining for transglutaminase in rat liver after
CPA withdrawal. Acetone fixed liver specimens were embedded in paraplast and 5 μ thick sections were exposed to antihuman transglutaminase immunoglobulin. The reaction was accomplished by the unlabeled
antibody peroxidase-antiperoxidase technique [34]. The sections were
counter-stained with Meyer's hemalaun. (↑) transglutaminase positive
apoptotic body; about 200x.

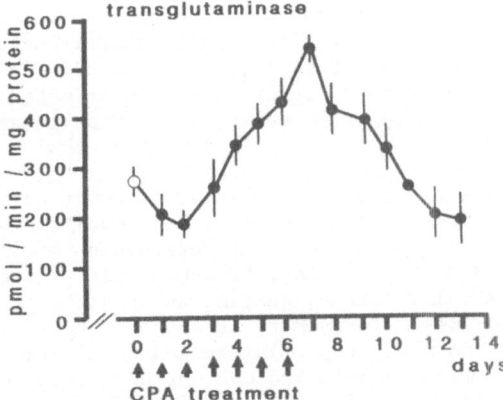

Fig. 9.   Effect of CPA treatment and CPA withdrawal on transglutaminase
activity in rat liver. CPA treatment: see Fig. 2. The transglutaminase activity was determined as described elsewhere [35]. (O) controls;
(●) CPA treated rats. Means (± SD) of 3-5 rats are given.

271

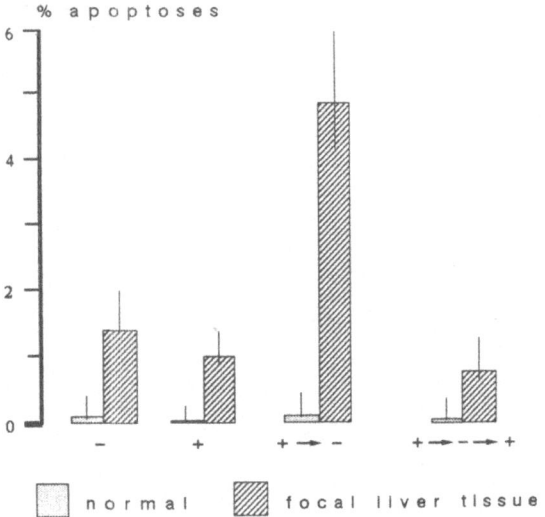

Fig. 10. Effect of phenobarbital (PB) treatment and PB withdrawal on the
incidence of ABs in normal liver and in liver foci. Rats were
treated with a single dose of NNM (250 mg/kg). One subgroup received no further treatment (-). The other NNM treated rats received PB (50 mg/kg/d via the food) for 7 months and then were
subjected to one of the following protocols: (1) continuous
treatment until necropsy (+); (2) PB withdrawal until necropsy
(+ → -); (3) PB withdrawal for 39 days, followed by PB treatment
(50 mg/kg/d via the food) for 3 days (+ → - → +). The total
number of ABs found in normal or in focal liver tissue is given,
vertical bars indicate the 95% confidence limits.

that these foci exhibit relatively high rates of cell proliferation but grow
only slowly at least at early stages[26]. With phenobarbital, foci growth could
be dramatically enhanced without a persistent significant enhancement of cell
proliferation in foci[26]. This perplexing situation was resolved when we found
that in preneoplastic foci the incidence of apoptosis is much higher than in
normal liver and that the tumor promoter phenobarbital inhibits apoptosis in
foci as it does in normal liver (Fig. 10, c.f. Fig. 3). This result suggests
that at least in this model tumor promotion may be effected largely through
inhibition of death of preneoplastic cells, i.e. by prolongation of their
life-time[27,28,29].

To study the potential role of apoptosis for tumor growth we selected
an estrogen dependent hamster kidney tumor. The results show that in this
malignant tumor cell death is an important determinant of the growth rate.
As in the experiments described so far, treatment with the mitogen (Diethylstilbestrol) depresses cell death suggesting an apoptotic type of death (see
Bursch et al., this volume). In summary it appears that apoptosis is important for the growth of normal, preneoplastic and neoplastic tissues. Closer
studies into this phenomenon should broaden our understanding of tumor growth
and help develop new strategies for therapy of tumors.

ACKNOWLEDGEMENTS

The excellent technical assistance of G. Barthel is gratefully acknow-
ledged.

REFERENCES

1. A. H. Wyllie, J. F. R. Kerr and A. R. Currie, Cell death: the signifi-
    cance of apoptosis. Int. Rev. Cytol. 68:251 (1980).
2. J. Searle, B. V. Harmon, C. J. Bishop and J. F. R. Kerr, The signifi-
    cance of cell death by apoptosis in hepatobiliary disease. J.
    Gastroent. and Hepatol. 2:77 (1987).
3. J. F. R. Kerr, A. H. Wyllie and A. R. Currie, Apoptosis: a basic bio-
    logical phenomen with wide-ranging implications in tissue kinetics,
    Br. J. Cancer 26:239 (1972).
4. D. Hopwood and D. A. Levison, Atrophy and apoptosis in the cyclical
    human endometrium, J. Pathology 119:159 (1976).
5. B. A. Sandow, N. B. West, R. L. Norman and R. M. Brenner, Hormonal con-
    trol of apoptosis in hamster uterine luminal epithelium, American
    J. Anatomy 156:15 (1979).
6. D. J. P. Ferguson and T. J. Anderson, Morphological evaluation of cell
    turnover in relation to the menstrual cycle in the "resting" human
    breast, Br. J. Cancer 44:177 (1981).
7. R. Stiens and B. Helpap, Die Regression der Rattenprostata nach
    Kastration, Histologische, morphometrische und zellkinetische
    Untersuchungen unter Berücksichtigung der Apoptose, Pathol. Res.
    Pract. 172:73 (1981).
8. J. D. O'Shea and P. J. Wright, Involution and Regeneration of the endo-
    metrium following paturition in the ewe, Cell Tissue Research
    236:477 (1984).
9. A. H. Wyllie, R. G. Morris, A. L. Smith and D. Dunlop, Chromatin
    cleavage in apoptosis: Association with condensed chromatin mor-
    phology and dependence on macromolecular synthesis, J. Pathology
    142:67 (1984).
10. J. J. Cohen and R. C. Duke, Glucocorticoid activation of a calcium-
    dependent endonuclease in thymocyte nuclei leads to cell death,
    J. Immunology 132:38 (1984).
11. H. M. Ellis and H. R. Horvitz, Genetic control of programmed cell death
    in the C. elegans, Cell 44:817 (1986).
12. D. S. Ucker, Cytotoxic T lymphocytes and glucocorticoids activate an
    endogenous suicide process in target cells, Nature 327:62 (1987).
13. R. Levi-Montalcini, The nerve growth factor thirty-five years later,
    In Vitro Cellular and Developmental Biology 23:227 (1987).
14. T. S. Argyris and D. R. Magnus, The stimulation of liver growth and de-
    methylase activity following phenobarbital treatment, Dev. Biol.
    17:187 (1968).
15. R. Schulte-Hermann, Induction of liver growth by xenobiotic compounds
    and other stimuli, Crit. Rev. Toxicol. 3:97 (1974).
16. W. G. Levine, M. G. Ord and L. A. Stocken, Some biochemical changes
    associated with nafenopin induced liver growth in the rat, Biochem.
    Pharmacol. 26:939 (1977).
17. R. Schulte-Hermann, V. Hoffmann, W. Parzefall, M. Kallenbach, A.
    Gerhardt and J. Schuppler, Adaptive responses of rat liver to the
    gestagen and anti-androgen cyproterone acetate and other inducers.
    II. Induction of growth, Chem. Biol. Interact 31:287 (1980).
18. W. Bursch, B. Düsterberg and R. Schulte-Hermann, Growth, regression and
    cell death in rat liver as related to tissue levels of the hepato-
    mitogen cyproterone acetate, Arch. Toxicol. 59:221 (1986).
19. W. Bursch, H. S. Taper, B. Lauer and R. Schulte-Hermann, Quantitative
    histological and histochemical studies on the occurrence and stages
    of controlled cell death (apoptosis) during regression of rat liver
    hyperplasia, Virch. Arch. Cell Pathol 50:153 (1985).
20. A. Columbano, G. M. Ledda-Columbano, G. Coni, G. Faa, C. Ligouri, G.
    Santa-Cruz and P. Pani, Occurrence of cell death (apoptosis) during
    the involution of liver hyperplasia, Lab. Invest. 52:670 (1985).
21. A. H. Wyllie, Glucocorticoid-induced thymocyte apoptosis is associated

with endogenous endonuclease activation, Nature 284:555 (1980).

22. R. C. Duke and J. J. Cohen, Il-2 Addiction: Withdrawal of growth factor activates a suicide program in dependent T-cells, Lymphokine Research 5:289 (1986).

23. V. N. Afanas'ev, B. A. Korol, Y. A. Mantsygin, P. A. Nelipovich, V. A. Pechatnikov and S. R. Umansky, Flow cytometry and biochemical analysis of DNA degradation characteristic of two types of cell death, FEBS 194:347 (1986).

24. L. Fesus, E. F. Szucs, K. E. Barrett, D. D. Metcalfe and J. E. Folk, Activation of transglutaminase and production of protein-bound gamma-glutamylhistamine in stimulated mouse mast cells, J. Biol. Chem. 260:13771 (1985).

25. L. Fesus, V. Thomazy and A. Falus, Induction and activation of tissue transglutaminase during programmed cell death, FEBS Letters 224:104 (1987).

26. R. Schulte-Hermann, J. Schuppler, I. Timmermann-Trosiener, G. Ohde, W. Bursch and H. Berger, The role of growth of normal and preneoplastic cell populations for tumor promotion in rat liver, Environmental Health Perspect. Vol 50 (1983).

27. W. Bursch, B. Lauer, T. Timmermann-Trosiener, G. Barthel, J. Schuppler and R. Schulte-Hermann, Controlled cell death (apoptosis) of normal and putative preneoplastic cells in rat liver following withdrawal of tumor promoters, Carcinogenesis 5:453 (1984).

28. F. Feo, R. Garcea, L. Daino and R. Pascale, Mechanism of the inhibition of liver hepatocarcinogenesis promotion by S-adenosyl-L-methionine. Proceedings of the European Meeting on Experimental Hepatocarcinogenesis (Spa, Belgium, May 1987) Plenum Press, in press (1987).

29. R. Schulte-Hermann, W. Parzefall and W. Bursch, Role of stimulation of liver growth by chemicals in hepatocarcinogenesis. Banbury Report 25: Nongenotoxic Mechanisms in Carcinogenesis, Cold Spring Harbor Laboratory, (1986).

30. T. C. G. Bosch and C. N. David, Growth regulator in Hydra: Relationship between epithelial cell cycle, length and growth rate, Dev. Biol. 104:161 (1984).

31. J. F. R. Kerr, B. Harmon and J. Searle, An electronmicroscope study of cell deletion in the anurian tadpole tail during spontaneous metamorphosis with special reference of apoptosis of striated muscle fibres, J. Cell Sci. 14:571 (1974).

32. A. Columbano, G. M. Ledda-Columbano, P. M. Rao, S. Rajalakshmi and D.S.R. Sarma, Occurrence of cell death (apoptosis) in preneoplastic and neoplastic liver cells, Am. J. Pathol 116:441 (1984).

33. D. S. Schmid, J. P. Tite and N. H. Ruddle, DNA fragmentation: Manifestation of target cell destruction mediated by cytotoxic T-cell lines, lymphotoxin-secreting helper T-cell clones, and cell-free lymphotoxin-containing supernatant, Proc. Natl. Acad. Sci. USA 83:1881 (1986).

34. L. A. Sternberger, P. H. Hendy, I. I. Cuculies and H. G. Meyer, The unlabeled antibody enzyme method of immunohistochemistry: preparation and properties of soluble antigen-antibody complex (horseradish peroxidase-antiperoxidase) and its use in identification of spirochaetes, J. Histochem. Cytochem. 18:315 (1970).

35. L. Fesus and G. Arato, Quantitation of tissue transglutaminase by a sandwich ELISA system, J. Immunol. Methods 94:131 (1986).

ROLE OF CELL DEATH FOR GROWTH AND REGRESSION OF HORMONE-DEPENDENT H-301

HAMSTER KIDNEY TUMORS

Wilfried Bursch[1], Joachim Liehr[2], David Sirbasku[3], and Rolf Schulte-Hermann[1]

[1]Institut für Tumorbiologie-Krebsforschung, Universitat Wien
[2]Department of Pharmacology and Toxicology, University of Texas Medical Branch, Galveston, and [3]Department of Biochemistry and Molecular Biology, University of Texas Medical School, Houston

Cell death by apoptosis is conceived as an important regulatory process for the control of cell number in normal and preneoplastic tissue[1-5]. In the present study, the potential role of apoptosis for tumor growth was investigated. As a model system we used an estrogen dependent hamster kidney tumor cell line, which was designated H-301 line by Sirbasku and Kirkland[6]. H-301 cells were previously found to form tumors upon inoculation into hamsters treated with estrogen or diethylstilbestrol (DES[7]). A typical growth pattern of such tumors is shown in Fig. 1. Within 18 days after inoculation of a H-301 cell suspension tumors of about 1 g had developed. Continuation of DES treatment caused about a doubling of tumor weight within 8 days. However, when DES treatment was stopped tumor weight decreased by about 80% within 4 days after DES withdrawal. In Fig. 1 it is also shown that tumor growth could be reinitiated when DES treatment was resumed after an interruption for 4 days. Within 2 days after resumption of DES administration the tumors reached the weight they had before the interruption of DES treatment.

The morphological appearance of the tumors is shown in Fig. 2. The H-301 cells formed solid, vascularized tumors (Fig. 2a). Mitoses could be seen in all stages (not shown). Furthermore, various morphological signs of cell death could be observed in growing as well as involuting tumors:

1. Small or large, coherent areas of cell residues, which were designated "necrotic area" (Fig. 2b).
2. Small, condensed cell residues, which usually occurred isolated and irregularly distributed in the solid parts of the tumors and which were designated "single cell death" (Fig. 2c).

In an attempt to estimate the role of mitoses, single cell death and necrosis for the actual growth rate of the H-301 tumors, their incidence was determined in H&E stained histological sections obtained from tumors at the various stages of growth and regression. The following results were obtained.

DES Treatment (Fig. 3, left panel)

Mitotic activity (about 2%) appears to be sufficient to prevail cell loss by single cell death (about 8%) or necrosis since tumors grow steadily.

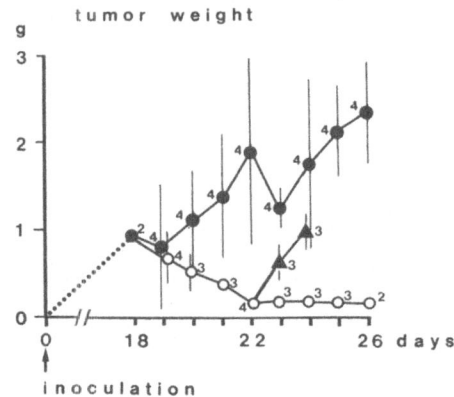

Fig. 1. Effect of DES treatment, DES withdrawal and DES retreatment on the weight of H-301 tumors. A 39 mg pellet consisting of 90% DES and 10% cholesterol was implanted subcutaneously under the skin of the back of male syrian hamsters. The animals absorb 200-300 µg of DES/ hamster/day from these pellets. Day 0: 1 day after implantation of the DES pellet 3,6 x 10^6 H-301 cells were inoculated subcutaneously into the area of the left posterior thorax. Day 0 until day 18: all hamsters were treated with DES and thereafter subjected to one of the following protocols: (●) continued DES treatment until necropsy, (○) DES withdrawal by removal of the pellet, (▲) DES withdrawal for 4 days, followed by readministration of DES: one intraperitoneal injection of 25 µg of DES/hamster and infusion of 0,21 µg DES/hamster/min by the means of Alzet osmotic minipumps, implanted subcutaneously. Means of one representative experiment are given, the number of animals is indicated at the symbols. Vertical bars indicate standard deviation and where not shown it is smaller than symbols.

DES Withdrawal (Fig. 3, right panel)
_____

Tumor size was reduced by about 80% within 3-4 days. Coincidentally, mitotic activity decreased to about 1/10 and the rate of single cell death increased (2-fold at its maximum). The incidence of necrotic areas appears not to be affected by DES withdrawal.

DES Retreatment (Fig. 4)
_____

The decrease of mitotic activity and the temporary increase of single cell death after removal of the DES pellet suggests that mitosis and single cell death may depend on presence or absence of DES. Consequently we checked whether or not retreatment with DES would stimulate mitosis and inhibit single cell death. After a period of 4 days DES withdrawal, the hosts were retreated with DES and it was found that (a) mitotic activity increased between 8 and 24 h post readministration of DES to about 2%, i.e. the rate of mitosis found in tumors exposed continuously to DES. The time course of increase of mitototic activity indicates that the H-301 cells passed a prereplicative period. (b) The incidence of single cell death was reduced by about 80% within 8 h after DES administration, suggesting an immediate inhibition of single cell death. Lastly, (c) the incidence of necrosis appears not to be affected by DES retreatment.

In conclusion, the results indicate that DES exerts two effects on H-301

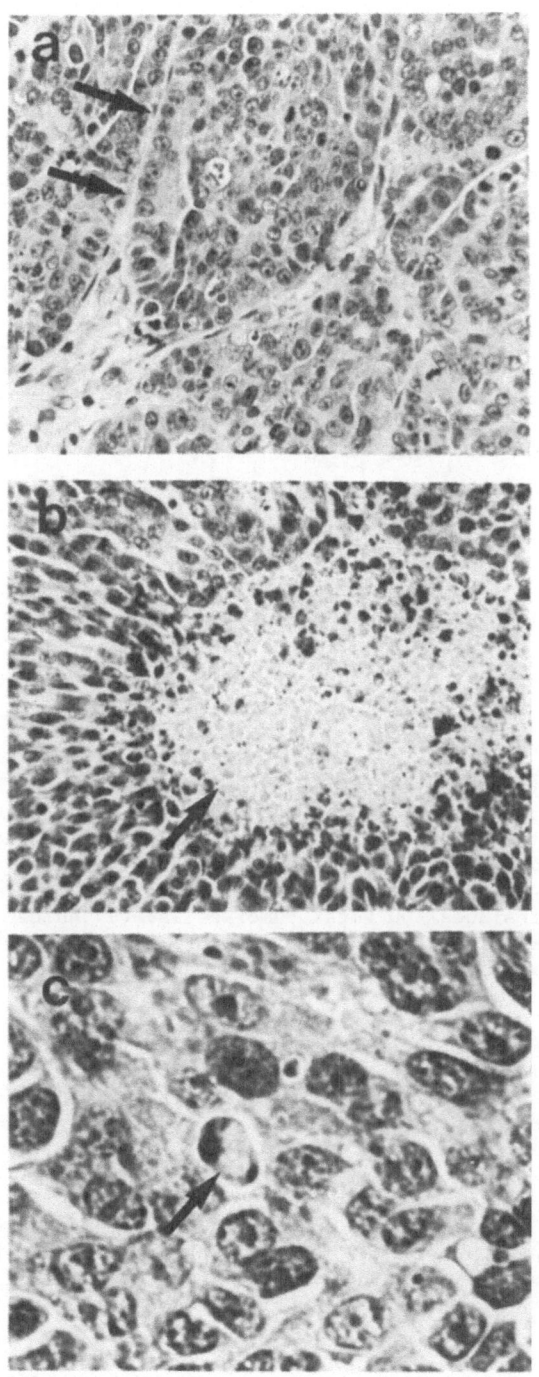

Fig. 2.   Tumor at 18 days after inoculation of a H-301 cell
          suspension and continuous DES treatment. a: Small
          layer of connective tissue surrounding solid cord
          of tumor cells (↑); about 250x. b: Necrotic area,
          about 250x; H&E. c: Single cell death, about 1000x.

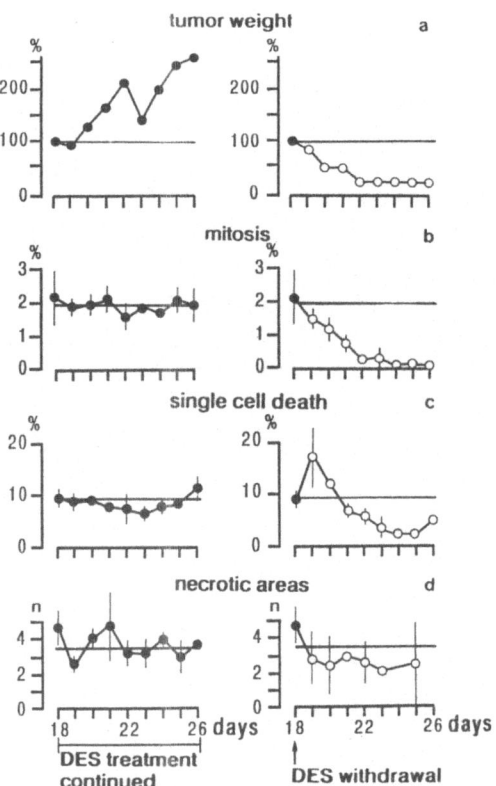

Fig. 3. Effect of DES treatment and DES withdrawal on the incidence of mitoses, single cell death and necrotic areas in H-301 tumors. Treatment: see Fig. 1. (a) Tumor weight, tumor weights of the animals killed between day 19 and day 26 are expressed as percentage of the tumor weight of day 18. (b) Mitoses/100 tumor cells. (c) Single cell residues/100 vital tumor cells. (d) number of necrotic areas/microscopic field. Left panel: DES treatment; right panel: DES withdrawal. Each point is the mean (±SD) of 3-4 animals.

tumor cells. Firstly, stimulation of cell proliferation and, secondly, inhibition of single cell death. On the other hand, the occurrence of necrotic areas appears not to be affected by DES. These findings suggest that the actual growth rate of the H-301 tumors predominantly depends on the ratio between cell proliferation and cell loss by single cell death. Furthermore, the inhibitory effect of DES on the occurrence of single cell death suggests that this type of cell death reflects "apoptosis" as described in other hormone dependent organs (mamma, prostrate) or during regression of hormonally induced liver hyperplasia[2,5,8-10] rather than focal or diffuse necrosis as may occur in tumors consequent to lack of oxygen or nutrient supply.

REFERENCES

1. J. F. R. Kerr, A. H. Wyllie and A. R. Currie, Apoptosis: a basic biological phenomenon with wide-ranging implications in tissue kinetics Br. J. Cancer 26:239 (1972).
2. A. H. Wyllie, J. F. R. Kerr and A. R. Currie, Cell death: the significance of apoptosis, Int. Rev. Cytol. 68:251 (1980).
3. W. Bursch, B. Düsterberg and R. Schulte-Hermann, Growth, regression

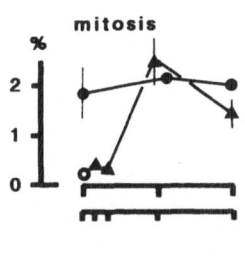

mitosis

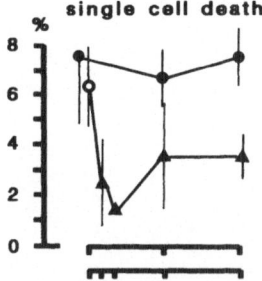

single cell death

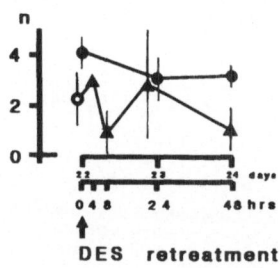

necrotic areas / microscopic field

DES retreatment

Fig. 4. Effect of retreatment with DES on the incidence of mitoses, single cell death and necrotic areas in involuting H-301 tumors. Treatment: see Fig. 1. (a) Mitoses/100 tumor cells; (b) single cell residues/100 vital tumor cells; (c) number of necrotic areas/ microscopic field. (○) DES withdrawal for 4 days; (▲) DES retreatment; (●) continuous DES treatment. Each point is the mean (± SD) of 3-4 animals. Note the different scales of the abscissa.

and cell death in rat liver as related to tissue levels of the hepatomitogen cyproterone acetate, Arch. Toxicol. 59:221 (1986).

4. A. Columbano, G. M. Ledda-Columbano, P. M. Rao, S. Rajalakshmi and D. S. R. Sarma, Occurrence of cell death (apoptosis) in preneoplastic and neoplastic liver cells, Am. J. Pathol. 116:441 (1984).

5. R. Schulte-Hermann, W. Bursch, L. Fesus and B. Kraupp, Cell death by apoptosis in normal, preneoplastic and neoplastic tissue (this volume).

6. D. A. Sirbasku and W. L. Kirkland, Control of cell growth. IV. Growth properties of a new cell line established from an estrogen-dependent kidney tumor of the Syrian hamster, Endocrinology 98:1260 (1976).

7. J. G. Liehr and D. A. Sirbasku, Estrogen-dependent kidney tumors, in: "Tissue Culture of Epithelial cells". M. Taub, ed., Plenum Publishing Corporation, New York (1985).

8. D. J. P. Ferguson and T. J. Anderson, Morphological evaluation of cell turnover in relation to the menstrual cycle in the "resting" human breast, Br. J. Cancer 44:177 (1981).

9. R. Stiens and B. Helpap, Die Regression der Rattenprostata nach Kastration. Histologische, morphometrische und zellkinetische

Untersuchungen unter Berucksichtigung der Apoptose, <u>Pathol.</u> <u>Res.</u> <u>Pract.</u> 172:73 (1981).

10. W. Bursch, B. Lauer, T. Timmermann-Trosiener, G. Barthel, J. Schuppler and R. Schulte-Hermann, Controlled cell death (apoptosis) of normal and putative preneoplastic cells in rat liver following withdrawal of tumor promoters, <u>Carcinogenesis</u> 5:453 (1984).

CAN APOPTOSIS INFLUENCE INITIATION OF CHEMICAL HEPATOCARCINOGENESIS?

Amedeo Columbano, Giovanna M. Ledda-Columbano, Pierpaolo
Coni, Maria G. Ennas, Dittakavi S. R. Sarma[1] and Paolo Pani

Istituto di Farmacologia e Patologia Biochimica, Università
di Cagliari, Via Porcell, 4, 09100 Cagliari, Italy
[1]Department of Pathology, Medical Sciences Building, Toronto
M5S 1A8, Ontario, Canada

CELL PROLIFERATION AND INITIATION OF LIVER CARCINOGENESIS

Although the mechanism of action of cell proliferation in the carcinogenic process has not been clearly defined so far, it appears that cell proliferation is essential for the initiation phase, and it also characterizes
the promotion phase, as well as the progression from hepatocyte nodules to
cancer[1-7]. Undoubtedly, most of the studies aimed to determine the exact role
of cell proliferation in the development of liver cancer have been focused on
the initiation step. The evidence that stems out from these studies is that
at least one round of cell proliferation is necessary for the accomplishment
of the initiation process; replication of DNA containing carcinogen-induced
miscoding lesions such as 0^6-alkylguanine[8], 0^4-alkylthymine[9], or non-coding
lesions such as apurinic sites[10], or certain types of gap filling mechanisms[11],
have been implicated as the key event.

Regardless of the mechanism(s) by which cell proliferation acts as a
necessary component in the initiation phase, its need in the process is made
evident by several series of experimental evidences; I) carcinogens which do
not induce liver cancer when administered to adult rats, whose liver has a
very poor cell turnover, do induce cancer development when given to newborn
rats whose liver is actively dividing[12]; II) when single non-necrogenic doses
of methylnitrosourea (MNU), or 1,2-dimethylhydrazine (DMH), themselves not
resulting in the formation of enzyme altered foci, are given shortly before
or after partial hepatectomy (PH), a high number of initiated cells can be
monitored as gamma-glutamyltransferase (GGT) positive foci in the liver of
rats, irrespective of the promoting procedure employed[2]; III) inhibition of
dimethylnitrosamine (DMNA) or diethylnitrosamine (DENA)-induced liver necrosis, and therefore of the compensatory proliferation, results in a drastic
reduction in the number of putative preneoplastic lesions induced by these
carcinogens[13].

COMPENSATORY CELL PROLIFERATION VERSUS MITOGEN-INDUCED HYPERPLASIA

Interestingly, in all these studies the proliferative stimulus for cell
proliferation was of compensatory type following PH or necrosis by hepatoxins
or by the carcinogens themselves.

On the contrary, although it is known that liver cells may be stimulated to proliferate also by a variety of non-necrogenic compounds[14], the effect of primary hyperplasia induced by mitogens on the initiation of hepatocarcinogenesis has not been studied.

These two types of cell proliferation have at least two major differences: while in the compensatory type, cell loss is the primary event, and regeneration is triggered in order to replace the cells that have been lost, in the direct hyperplasia the mitogenic effect is the primary event, and it results in an excess of cells which are removed within days or a few weeks as a result of a "controlled" type of cell death, namely apoptosis[15,16]. In addition, unlike the compensatory cell proliferative stimulus, mitogen-induced cell proliferative stimulus has to overcome the growth control mechanism normally operating in the organ; it is likely that the mechanism by which cells respond to these two types of stimuli might also be different.

The present article summarizes the results of a number of our studies designed to determine whether these two types of cell proliferation could also exert a different effect on the initiation phase of liver carcinogenesis. In addition, the possibility of a different effect of these two types of proliferative stimuli on the promotion phase, is also briefly discussed.

MITOGEN-INDUCED LIVER HYPERPLASIA AND INITIATION

In order to determine whether cell proliferation induced by mitogens could be different from compensatory cell proliferation, as far as the initiation step is concerned, we selected as a liver mitogen, lead nitrate[17,18]; this compound , when administered in a single dose to Wistar rats, is able to induce a dramatic increase in liver weight and DNA content, with a maximum between 3 and 4 days, a time point when both liver and DNA are almost doubled when compared to controls. The massive response of the hepatocytes to lead nitrate is not likely to be the consequence of cell killing by this agent, since only a slight increase in serum transaminase activity was observed with the higher dose, and no increase at all was seen when a lower dose (5 μmoles/100g) was used.

Regardless of the mechanism(s) by which lead nitrate stimulates liver cell proliferation, starting from the third day after treatment, a rapid involution of the excess liver mass, accompanied by an elimination of the excess DNA takes place, so that within 15 days the liver regains its original size. During the regression of liver hyperplasia, which appears to be a common phenomenon following withdrawal of several mitogens, the occurrence of a particular mode of cell death, apoptosis, considered to be a "programmed" or "controlled" type of cell death can be observed[15,16,19]. Thus, in about 12 days 50% of liver cells (excess cells) are presumably eliminated. On the basis of these findings a critical question arises: how initiated cells will respond in a situation wherein a massive cell elimination is taking place? Are they resistant to this particular mode of cell death, or will they be eliminated together with the normal liver cells? It is obvious that if a consistent number of carcinogen-altered cells, including the initiated ones, is removed, then only a low number of preneoplastic lesions should develop.

To test this question, we used an experimental protocol where cell proliferation was induced by lead nitrate; during "S" phase, a single non-necrogenic dose of MNU, a direct methylating carcinogen, was given. Following a recovery period of 15 days, the animals were subjected to a promoting regimen according to the protocol described by Cayama et al.[1]. The results presented in Fig. 1, show that no GGT-positive foci developed in the liver of rats treated with the carcinogen during mitogen-induced hyperplasia; on the contrary, several GGT-positive foci were observed in liver of rats given the

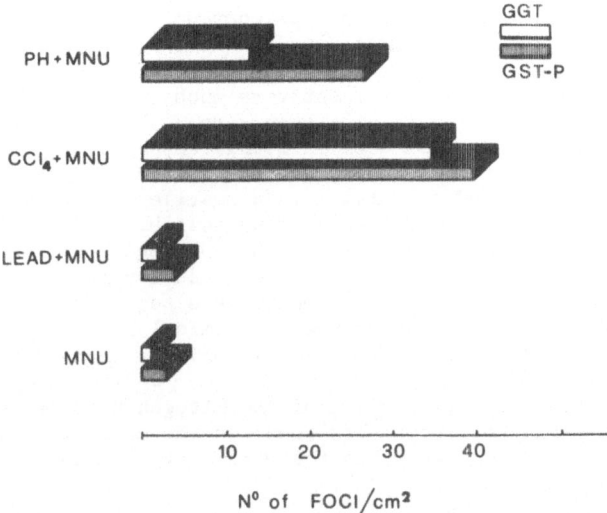

Fig. 1. Effect of different proliferative stimuli on the induction of enzyme
altered foci by MNU. MNU was injected at a dose of 60 mg/kg to male
Wistar rats, during cell proliferation induced by 100 μmoles/kg of
lead nitrate or 2 ml/kg of CCl₄, or PH. After a two week recovery
period, the initiated hepatocytes were assayed as GGT or GST-P positive foci using the resistant hepatocyte model[1].

carcinogen during compensatory cell proliferation. Similar results were
obtained when phenotypic markers such as the placental form of glutathione
S-transferase (GST-P), and adenosin triphosphatase (ATPase) were used to
identify putative preneoplastic populations. It is important to note that
under our experimental conditions no inhibition of MNU-induced DNA alkylation
was observed in lead-treated rat liver, and that the number of proliferating
cells at the time of carcinogen administration in lead-treated rat liver was
the same as that of regenerating liver following CCl₄ [20].

The failure to support initiation by MNU was made evident not only by
the absence of putative preneoplastic foci at 1.5 months after carcinogen
administration, as shown in Fig. 1, but also by an almost complete lack of
enzyme altered foci or hepatic nodules even at 10 months after carcinogen
treatment, a time point when MNU coupled with PH already resulted in the appearance, in some animals, of hepatocellular carcinoma (data not shown).

Thus, we face a condition fulfilling the two prerequisites necessary
for the accomplishment of the initiation step; a) a vigorous proliferative
process, plus, b) a DNA containing carcinogen-induced lesions; nevertheless,
we failed to achieve preneoplastic foci or nodules.

Is that due to the fact that, I) despite a _permissive_ condition (cell
proliferation + DNA damage) hepatocytes have not been initiated, or alternatively, II) initiated cells have been formed, but they have been lost during
the regression of liver hyperplasia?

As far as the former possibility is concerned, it should be considered
the possibility that pretreatment with lead nitrate could somehow modify some
biochemical properties of the hepatocyte, thus, making them "resistant" to
the carcinogenic attack. As a matter of fact, lead nitrate induces in rat
liver, several biochemical changes that modify the phenotypic pattern of the

hepatocyte[21,22]; acute treatment with this compound inhibits the activity of Phase I enzymes, it induces the activity of Phase II enzymes, such as that of GGT and DT-diaphorase, stimulates the appearance of GST-P, an enzyme which is almost undetectable in normal liver but very much increased in preneoplastic lesions[23], inhibits the activity of ATPase, etc. The occurrence of this special biochemical pattern not seen in fetal, neonatal or regenerating liver suggests that lead nitrate may induce in the hepatocyte a new state of differentiation better adapted to survive in a hostile environment, similar to that exhibited by hepatocyte nodules[24]. The possibility that the "resistant" pattern induced by lead nitrate could somehow protect the liver cells from some critical damage induced by the carcinogen during the initiation process was therefore considered; thus, experiments were performed using three other hepatic mitogens that do not induce the biochemical pattern induced by lead nitrate; cyproterone acetate[25], ethylene dibromide[26] and nafenopin[27].

As shown in Fig. 2, irrespective of the mitogen used no formation of foci of enzyme altered hepatocyte was seen; on the contrary, several putative preneoplastic lesions were seen when the carcinogen was administered during liver regeneration[28].

This experiment suggests that the lack of foci formation is not likely due to some specific effect of lead nitrate; however, it does not rule out the possibility that pretreatment with mitogens could modify the response of the liver to carcinogens.

In order to eliminate the latter possibility, the experimental protocol was changed in such way that the carcinogen was given to a normal rat liver, and was immediately followed by proliferative stimuli (see Fig. 3). When this protocol was adopted, preneoplastic lesions were seen only in animals whose liver was induced to proliferate by a regenerative stimulus, but not in rats post-treated with mitogens (Fig. 4). Thus, these results suggest that the lack of appearance of enzyme altered foci when the proliferative stimulus is induced by mitogens cannot be due to a peculiar effect of the inducers of liver growth.

Evidence, albeit indirect, thus, is accumulating suggesting that the absence of preneoplastic lesions might not be due to a lack of formation of initiated cells; the possibility that it may depend upon the elimination of initiated cells during the regression of liver hyperplasia should then be considered (see Fig. 5).

CELL LOSS (APOPTOSIS) AND INITIATION

As mentioned earlier, during the involution of liver hyperplasia there is a massive deletion of cells by apoptosis; this process stops once the liver has reached its original size[16]. Is apoptosis a random phenomenon or is it selective for some cell populations? And if the second case is correct, does apoptosis preferentially occurs in the newly made cells, or alternatively, in those cells that had lost their replicative capacity?

Preliminary experiments carried out to answer this critical question and based on the labeling of liver cells after lead nitrate treatment, revealed that the process of cell removal seems to be quite a random phenomenon. Indeed, autoradiographic analysis showed that the percentage of labeled apoptotic bodies (representing the remnants of cells that have divided following lead nitrate administration) was 54-58% of the total number of apoptotic bodies. Although these findings need to be confirmed, nevertheless they are very intriguing; if apoptosis is randomly distributed, why virtually no initiated cells are seen when carcinogen administration is coupled with a mitogenic stimulus? Does it mean that initiated cells are more prone to

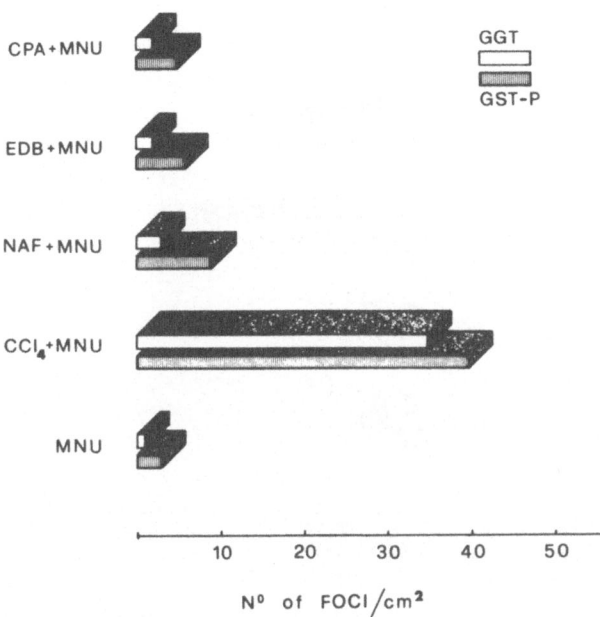

Fig. 2. Male Wistar rats were given CCl₄ (2 ml/kg, i.g.), ethylene dibromide
(100 mg/kg, i.g.), cyproterone acetate (100 mg/kg, i.g.), or
nafenopin (200 mg/kg, i.g.). In some rats the proliferative stimulus
was achieved through PH. During DNA synthesis a non-necrogenic dose
of MNU (60 mg/kg, i.p.) was administered at 20 h after PH, 24 h
after ethylene dibromide or cyproterone acetate, 30 h after
nafenopin, and 42 h after CCl₄. The initiated hepatocytes were
assayed as GGT or GST-P positive foci as described in Fig. 1.

apoptosis than normal hepatocytes? Obviously, this is a critical question
and its understanding will be of significance in the knowledge of the biochemical mechanisms operating in the process of apoptosis, and possibly, of
the nature of initiated cells.

Regardless of the fact that initiated cells may be killed in a selective
or random matter, it is quite conceivable that by inhibiting the apoptotic
process it should be possible to "save" the initiated cells, and by consequence to induce the occurrence of enzyme altered foci, following an oppor-
tune promotion procedure.

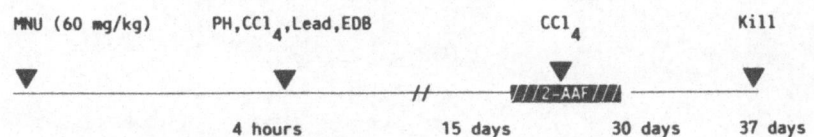

Fig. 3. Experimental protocol to study the effect of treatment with various
proliferative stimuli _following_ carcinogen administration on the
induction of enzyme altered foci. Male Wistar rats were given a
single dose of MNU (60 mg/kg) and 4 h later CCl₄, lead nitrate, or
ethylene dibromide were given at the doses described in the previous
figures. One group of animals was also subjected to PH.

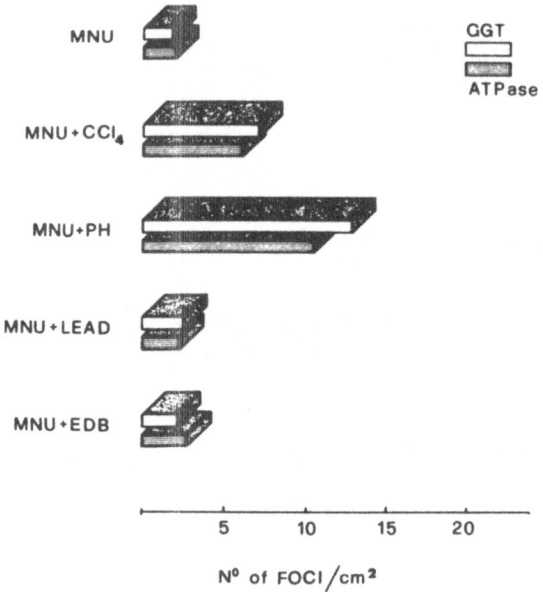

Fig. 4. Quantitation of GGT and ATPase foci. For experimental protocol see Fig. 4.

In an attempt to approach this problem we reasoned that if apoptosis is triggered in order to eliminate the excess tissue, then, by reducing the hyperplastic liver mass (and the number of excess cells) by PH, after mitogen

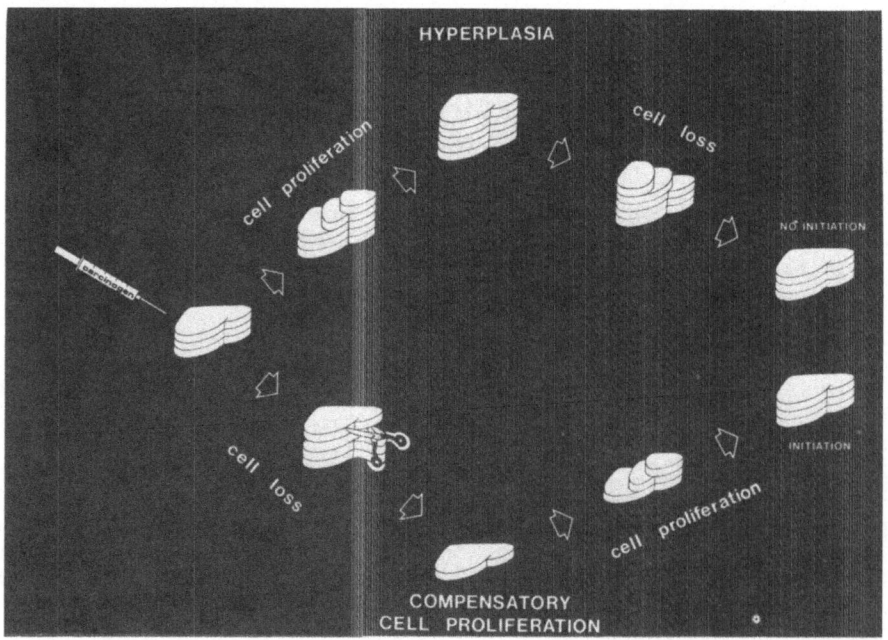

Fig. 5. Schematic representation of the possible different fate of initiated cells depending upon the nature of the proliferative stimulus.

treatment, we would theroretically be able to inhibit the apoptotic process and consequently to save initiated cells (see Fig. 6).

Preliminary results indicate that the administration of a non-necrogenic dose of DENA followed by mitogen treatment (24 h later), and then by PH (48 h later), results in the appearance of enzyme altered foci. These results need to be confirmed; nevertheless they suggest that inhibition of apoptosis may be possible by making useless its occurrence (in this case by removing the excess cells by surgery).

The results of our studies on the role of cell proliferation on the initiation phase of chemical hepatocarcinogenesis suggest the following: a) mitogen-induced liver hyperplasia coupled with carcinogen treatment unlike regenerative cell proliferation does not induce the development of putative preneoplastic lesions, and, b) the lack of formation of preneoplastic lesions may not depend upon the inability of this type of proliferative process to originate initiated cells, but rather to an elimination of initiated cells shortly after their formation.

LIVER REGENERATION AND PROMOTION

Sustained or repeated damage to the liver and compensatory regeneration appears to have promoting effect on the development of preneoplastic and neoplastic lesions. Pound and McGuire[5,29] have shown that repeated PH or 7 necrogenic doses of CCl_4 subsequent to carcinogen administration, enhance the number of preneoplastic and neoplastic nodules. An increase in the yields of hepatocellular tumors due to ionizing irradiation has been reported by the action of a hepatonecrogenic dose of CCl_4[30], and by PH after a dose of X-irradiation[31]. More recently, it has been shown that feeding rats a choline deficient diet which induces a severe liver necrosis and compensatory cell proliferation[32,33], also acts as a liver tumor promoter[34]. This relationship between cell proliferation and promotion is also observed in many other organs and tissues such as skin, buccal mucosa, intestine (for a review see 35). Thus, this albeit indirect evidence indicates that regenerative cell proliferation may provide a promoting stimulus for development of preneoplastic and neoplastic lesions. The evidence, however, is far less clear on the effect of liver mitogens on preneoplasia or neoplasia.

MITOGEN-INDUCED LIVER HYPERPLASIA AND PROMOTION

Several compounds including sex steroids, phenobarbital, chlorinated hydrocarbons and hypolipidemic drugs that stimulate the growth of normal liver, have also been shown to possess tumor promoting activity[14,36]; however, recently several reports have cast doubts on the ability of hypolipidemic agents to promote the development of enzyme altered foci[37,38]; other studies even reported an accelerated regression of preneoplastic lesions by hypolipidemic drugs[39]; more recently another inducer of liver growth, di-(2-ethylhexyl)phtalate (DEHP) which is a liver tumor inducer after chronic treatment[40], has been reported to inhibit the emergence of enzyme altered foci in the short term assay[41]. Thus, the assumption that mitogen-induced cell proliferation must be considered a promoting factor is questionable, at least as far as inducers of liver growth such as BR931, DEHP and nafenopin are concerned. Phenobarbital is perhaps the liver tumor promoter best studied; although its mechanism of action is by no means clear, its capacity to induce hepatic DNA synthesis[42], has been considered as a major component of its promoting ability. However, the facts that most of the liver growth induced by this compound is due to hypertrophy rather than to hyperplasia, together with the finding that the chronic treatment with phenobarbital induces only a small initial rise in DNA synthesis of normal as well as preneoplastic hepa-

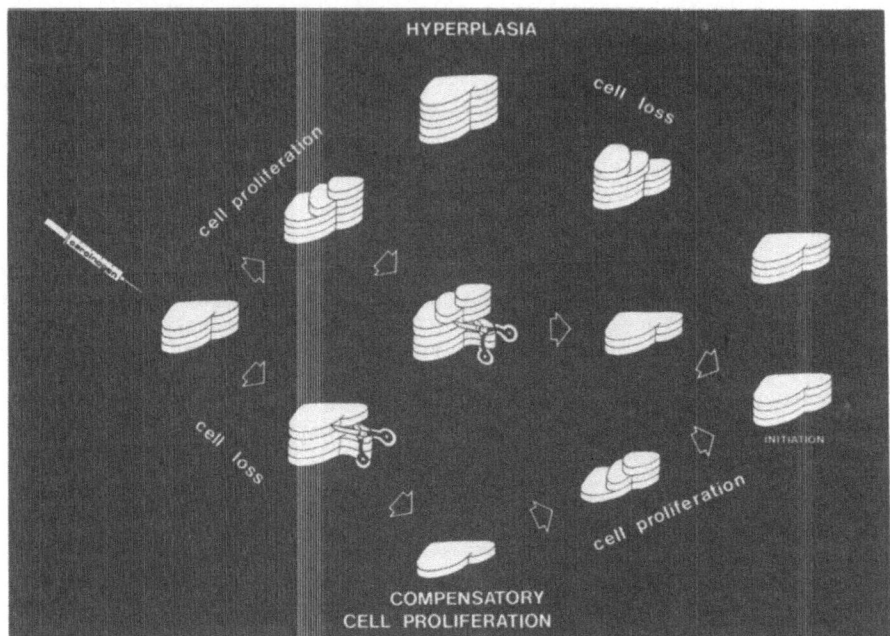

Fig. 6. Representation of the possible inhibitory effect of PH on the process of cell removal which follows the initial hyperplasia induced
by mitogens, and of its effect in the development of enzyme altered
foci.

tocytes[43] suggest that mechanisms other than cell proliferation should be
considered as responsible for the promoting effect of this compound. More
recently, inhibition of apoptosis, a process occurring at a very high rate in
hepatic foci and nodules[15,44], and of phenotypic remodelling of foci and nodular cells[36], have been suggested as the factors responsible for the enlargement of preneoplastic lesions.

Thus, not much evidence is available in the liver system to really prove
that the capacity to induce primary hyperplasia is per se what makes a chemical a tumor promoter.

One of the conditions that should be considered if one wants to determine whether a compound is a promoter because of its mitogenic capacity, is
that the chemical should induce DNA synthesis not only in the very first initial period, but throughout the experimental period; continuous feeding,
thus is not a good system because the liver after the first 2-3 days does not
respond any more to the mitogenic stimulus. A better system should be to
inject the mitogen at various time intervals in such a way that liver cells
may respond any time the proliferative stimulus is applied.

We have recently designed an experiment aimed to determine whether repeated injections with the mitogen lead nitrate could exert a promoting effect on the development of GGT-positive foci induced by an initiating dose of
DENA. Surprisingly, the results indicated that as many as 20 injections of
lead nitrate did not have any promoting effect on the growth of GGT-positive
foci; the lack of promoting activity is shown by the fact that no increase in
the size or in the number of GGT-positive foci was seen in DENA + lead nitrate treated rat liver when compared to the group treated with DENA alone
(Table 1). It is important to note that lead nitrate maintained some capacity to stimulate DNA synthesis even after the 20th injection. These

Table 1. Effect of repeated injections with lead nitrate on GGT foci in rat liver after DENA administration

Treatment [a]	Body weight	Liver weight(%)	GGT$^+$ Foci	
			No./cm^2	Diameter (μm)
DENA + Lead nitrate	549 ± 31	3.7 ± 0.1	8.8 ± 2.9	166 ± 17
DENA + H$_2$O	557 ± 27	2.9 ± 0.1	19.7 ± 3.6	174 ± 17
Saline + Lead nitrate	535 ± 19	3.7 ± 0.1	0.9 ± 0.2	N. D.
Saline + H$_2$O	586 ± 40	2.9 ± 0.2	1.6 ± 0.8	N. D.

Values are the average of 5 rats ± SE. Lead nitrate was injected at a dose of 5 μmoles/100g of body weight, i.v., once every 15-20 days.
[a] Time of sacrifice 11 months.

results are quite surprising since it was expected that the administration of a powerful mitogen such as lead nitrate should promote the growth of preneoplastic lesions. The fact that lead nitrate despite its mitogenic activity[17,18], and its capacity to induce the expression of an altered phenotype similar to that of preneoplastic hepatocytes[21,22], fails to promote the growth of enzyme altered cells may suggest that the waves of apoptosis that follow the mitogenic stimuli, may eliminate the newly made islands cells, thus limiting the expansion of preneoplastic lesions. If this were to be the case, the process of apoptosis that follows the initial hyperplasia would compete with the proliferative process, thus, playing an opposite role to mitosis; in this respect, apoptosis might be defined as a compensatory cell death.

An in depth study of the balance between cell proliferation and cell loss might provide very useful informations on the dynamics of the carcinogenic process. The detection of compounds that may alter this balance by favoring the loss of focal or nodular cells through apoptosis, would be of particular help for the understanding of this process.

ACKNOWLEDGEMENTS

Supported by funds from C.N.R. (Medicina Preventiva e Riabilitativa, No. 86.01727.56), Italy, and NATO (0184/86).

REFERENCES

1. E. Cayama, H. Tsuda, D. S. R. Sarma and E. Farber, Initiation of chemical carcinogenesis requires cell proliferation, Nature 275:60 (1978).
2. A. Columbano, P. M. Rao, S. Rajalakshmi and D. S. R. Sarma, Requirement of cell proliferation for the initiation of liver carcinogenesis as assayed by three different procedures, Cancer Res. 41:2079 (1981).
3. J. O. Laws, Tissue regeneration and tumor development, Br. J. Cancer 13:669 (1959).
4. A. W. Pound, Carcinogenesis and cell proliferation, N. Z. Med. J. 67:88 (1968).
5. A. W. Pound and I. J. McGuire, Repeated partial hepatectomies as a pro-

moting stimulus for carcinogenic response of liver to nitrosamines in rats, Br. J. Cancer 37:585 (1978).

6. I. Foulds, "Neoplastic development", Vol. 1, Academic Press, New York (1969).

7. E. Farber, Sequential events in chemical carcinogenesis, in: "Cancer I: A Comprehensive Treatise", F. Becker, ed., Plenum Publ. Corp., New York (1982).

8. A. Loveless, Possible relevance of O^6-alkylation of deoxyguanosine to the mutagenicity and carcinogenicity of nitrosamines and nitrosamides, Nature 223:206 (1969).

9. J. A. Swendberg, M. C. Dryoff, M. A. Bedell, J. A. Popp, N. Huh, U. Kirstein and M. F. Rajewsky, O^4-ethyldeoxythimidine, but not O^6-ethyldeoxyguanosine, accumulates in hepatocyte DNA of rats exposed continuously to diethylnitrosamine, Proc. Natl. Acad. Sci. USA 81:1692 (1984).

10. S. Rajalakshmi and D. S. R. Sarma, Replication of hepatic DNA in rats treated with dimethylnitrosamine, Chem-Biol. Interact. 11:245 (1975).

11. A. Columbano, G. M. Ledda-Columbano, P. M. Rao, S. Rajalakshmi and D. S. R. Sarma, In vivo replication of carcinogen-modified DNA: Presence of dimethylnitrosamine-induced N-7-methylguanine and O^6-methylguanine in the parental and daughter strands of the in vivo replicated hybrid DNA, Biochem. Arch. 1:121 (1985).

12. V. M. Craddock, Cell proliferation and experimental liver cancer, in: "Liver Cell Cancer", H. M. Cameron, D. A. Linsell and G. P. Warwick, eds., Elsevier/North Holland Biomedical Press, Amsterdam (1976).

13. T. S. Ying, D. S. R. Sarma and E. Farber, The sequential analysis of liver cell necrosis, Am. J. Pathol. 99:159 (1980).

14. R. Schulte-Hermann, Reactions of the liver to injury: adaptation, in: "Toxic Injury of the Liver", E. Farber, M. M. Fisher, eds., Marcel Dekker, New York, (1979).

15. W. Bursch, B. Lauer, I. Timmermann-Trosiener, G. Barthel, J. Schuppler, and R. Schulte-Hermann, Controlled cell death (apoptosis) of normal and putative preneoplastic cells in rat liver following withdrawal of tumor promoters, Carcinogenesis 5:453 (1984).

16. A. Columbano, G. M. Ledda-Columbano, P. Coni, G. Faa, C. Liguori, G. Santacruz and P. Pani, Occurrence of cell death (apoptosis) during the involution of liver hyperplasia, Lab. Invest. 52:670 (1985).

17. A. Columbano, G. M. Ledda, P. Sirigu, T. Perra and P. Pani, Liver cell proliferation induced by a single dose of lead nitrate, Am. J. Pathol. 110:83 (1983).

18. G. M. Ledda-Columbano, A. Columbano, G. Faa and P. Pani, Lead and liver cell proliferation: effect of repeated administrations, Am. J. Pathol. 113:315 (1983).

19. A. H. Wyllie, J. F. R. Kerr and A. R. Currie, Cell death: the significance of apoptosis, Int. Rev. Cytol. 68:251 (1980).

20. A. Columbano, G. M. Ledda-Columbano, P. Coni and P. Pani, Failure of mitogen-induced cell proliferation to achieve initiation of rat liver carcinogenesis, Carcinogenesis 8:345 (1987).

21. M. W. Roomi, A. Columbano, G. M. Ledda-Columbano and D. S. R. Sarma, Lead nitrate induces biochemical properties characteristic of hepatocyte nodules, Carcinogenesis 7:1643 (1986).

22. G. M. Ledda-Columbano, A. Columbano, M. G. Ennas, M. Curto, M. G. DeMontis, P. Pani, M. W. Roomi and D. S. R. Sarma, Induction by lead nitrate of a differentiated hepatic phenotype similar to that of hepatocyte nodules in their cytoprotective machinery, Proc. Fourth Sardinian International Meeting, Alghero, Italy (1987).

23. K. Sato, A. Kitahara, K. Satoh, T. Ishikawa, M. Tatematsu and N. Ito, The placental form of glutathione S-transferase as a new marker protein for preneoplasia in rat chemical hepatocarcinogenesis, Gann, 75:199 (1984).

24. M. W. Roomi, R. K. Ho, D. S. R. Sarma and E. Farber, A common biochemical pattern in preneoplastic hepatocyte nodules generated in four different models in the rat, Cancer Res. 45:564 (1985).

25. R. Schulte-Hermann, V. Hoffman, W. Parzefall, M. Kallenbach, A. Gerhardt and J. Schuppler, Adaptive response of rat liver to the gestagen and anti-androgen cyproterone acetate and other inducers, II. Induction of growth, Chem-Biol. Interact. 31:287 (1980).

26. E. Nachtomi and E. Farber, Ethylene dibromide as a mitogen for liver, Lab. Invest. 38:279 (1978).

27. W. G. Levine, M. G. Ord and L. A. Stocken, Some biochemical changes associated with nafenopin-induced liver growth in the rat, Biochem. Pharmacol. 26:939 (1977).

28. A. Columbano, G. M. Ledda-Columbano, G. Lee, S. Rajalakshmi and D. S. R. Sarma, Inability of mitogen-induced liver hyperplasia to support the induction of enzyme-altered islands induced by liver carcinogens, Cancer Res. 47:5557 (1987).

29. A. W. Pound and L. J. McGuire, Influence of repeated liver regeneration on hepatic carcinogenesis by diethylnitrosamine in mice, Br. J. Cancer 37:595 (1978).

30. L. J. Cole and P. C. Nowell, Radiation carcinogenesis: the sequence of events, Science 150:1782 (1965).

31. N. Haran-Ghera, N. Trainin, L. Fiore-Donati and I. Beremblum, A possible two stage mechanism in rhabdomyosarcoma induction in rats, Br. J. Cancer 16:653 (1962).

32. L. I. Giambarresi, S. L. Katyal and B. Lombardi, Promotion of liver carcinogenesis in the rat by a choline-devoid diet: role of liver cell necrosis and regeneration, Br. J. Cancer 46:825 (1982).

33. A. K. Ghoshal, M. Ahluwalia and E. Farber, The rapid induction of liver cell death in rats fed a choline-deficient methionine-low diet, Am. J. Pathol. 113:309 (1983).

34. M. A. Sells, S. L. Katyal, S. Sell, H. Shinozuka and B. Lombardi, Induction of foci of altered γ-glutamyltranspeptidase positive hepatocytes in carcinogen-treated rats fed a choline deficient diet, Br. J. Cancer 40:274 (1979).

35. T. S. Argyris, Regeneration and the mechanism of epidermal tumor promotion, CRC Crit. Rev. Toxicol., 14:211 (1985).

36. R. Schulte-Hermann, Tumor promotion in the liver, Arch. Toxicol. 57:147 (1985).

37. W. Staubli, P. Bentley, F. Bieri, E. Frohlich and F. Waechter, Inhibitory effect of nafenopin upon the development of diethylnitrosamine-induced enzyme-altered foci within the rat liver, Carcinogenesis 5:41 (1984).

38. A. B. DeAngelo and C. T. Garret, Inhibition of development of preneoplastic lesions in the livers of rats fed a weakly carcinogenic environmental contaminant, Cancer Lett. 20:199 (1983).

39. M. I. R. Perera and H. Shinozuka, Accelerated regression of carcinogen-induced preneoplastic hepatocyte foci by peroxisome proliferators BR931, 4-chloro-6-(2,3-xylidino)-2-pyrimidynilthio(N-B-hydroxyethyl) acetamide, and di-(2-ethylhexyl)phtalate, Carcinogenesis 5:1193 (1984).

40. J. M. Ward, J. M. Rice, D. Creasia, P. Lynch and C. Riggs, Dissimilar pattern of promotion by di-(2-ethylhexyl)phtalate and phenobarbital of hepatocellular neoplasia initiated by diethylnitrosamine in B6C3F1 mice, Carcinogenesis 4:1021 (1983).

41. A. B. DeAngelo, C. T. Garrett and A. E. Queral, Inhibition of phenobarbital and dietary choline deficiency promoted preneoplastic lesions in rat liver by environmental contaminant di(2-ethylhexyl)-phtalate, Cancer Lett. 23:323 (1984).

42. C. Peraino, R. J. M. Fry, E. Staffeldt and J. P. Cristopher, Comparative enhancing effects of phenobarbital, amobarbital, diphenyl-hydantoin, and dichlorodiphenyltrichloroethane on 2-acetylaminoflu-

orene-induced hepatic tumorigenesis in the rat, <u>Cancer</u> <u>Res.</u> 35:2884 (1975).

43. R. Schulte-Hermann, J. Schuppler, I. Timmermann-Trosiener, G. Ohde, W. Bursch and H. Berger, The role of growth of normal and preneoplastic cell populations for tumor promotion in rat liver, <u>Environ.</u> <u>Health</u> <u>Perspect.</u> 50:185 (1983).

44. A. Columbano, G. M. Ledda-Columbano, P. M. Rao, S. Rajalakshmi and D. S. R. Sarma, Occurrence of cell death (apoptosis) in preneoplastic and neoplastic liver cells: A sequential study, <u>Am.</u> <u>J.</u> <u>Pathol.</u> 116:441 (1984).

ANALYSIS OF THE BIOLOGY OF INITIATION USING THE RESISTANT HEPATOCYTE MODEL

Ross G. Cameron

Department of Pathology, Medical Sciences Building
University of Toronto
Toronto, Ontario, Canada M5S 1A8

INTRODUCTION

Critical to the design of models for the analysis of the hepatocellular lineage of carcinogenesis was the demonstration in vivo that key populations of hepatocytes arising during hepatocarcinogenesis showed a relative resistance to a variety of hepatotoxins[1-3]. It was then possible to show that in a cytotoxic environment, these resistant (R^+) hepatocytes could be very selectively stimulated to grow while the surrounding hepatocytes did not[4-11]. A more direct visualization of R^+ hepatocytes prior to such clonal expansion was not possible until a marker for R^+ hepatocytes in nodules was identified, namely, the glutathione S-transferase P-form or GST-P[12-15]. GST-P positive (GST-P) single hepatocytes were first shown by Moore et al.[14] in livers with nodules. We found singles, doublets and groups of GST-P$^+$ hepatocytes in vivo within days to weeks after initiation with diethylnitrosamine (DEN) alone, prior to selection or promotion[15]. Using the resistant hepatocyte model[4-11], we found GST-P$^+$ nodules of significant size within several days after the start of resistance-selection suggesting that the nodules did not arise de novo from single cells and that GST-P$^+$ groups of hepatocytes are precursors of nodules[15]. A design for more direct visualization of hepatocellular lineage was based on studies which quite conclusively showed that for hepatocytes to become initiated they must undergo cell proliferation[16-20]. We then demonstrated that hepatocytes proliferating in response to DEN-induced necrosis could incorporate bromodeoxyuridine (BUDR) and be seen in tissue sections using antibodies to BUDR and immunohistochemistry[15]. This paper reveals results of studies to look for putative initiated hepatocytes, namely, resistant (R^+), GST-P$^+$ and proliferating (BUDR$^+$).

METHODS

Initiation and BUDR Incorporation

A series of experiments were carried out to establish doses of DEN which induced a significant number of GST-P$^+$ hepatocytes and at the same dose a minimal number of proliferating hepatocytes (BUDR$^+$) arising in response to DEN-induced necrosis so as not to obscure the field with BUDR$^+$ cells. Fischer 344 adult male rats of 140 to 160 g b. wt. (Charles River Breeding Laboratories, Wilmington, MA) were given DEN as a single dose i.p. in saline of either 1, 5, 10, 15, 25, 50 or 200 mg/kg b. wt. and then 3 to 5 rats per group sacrificed at 3, 7 or 14 days after DEN. BUDR at 100 mg/kg b. wt. was given starting at 12 h after DEN at 25 mg/kg b. wt. and a total of 6 doses, one each 6 h was given. Resistance-selection was as before[11] with 2-acetyl-aminofluorene and partial hepatectomy. Representative standard midline sections of all lobes were fixed in formalin or Carnoy's solution[6], and also in cold acetone for immunohistochemistry as described previously by Kaku et al.[21]. Quantitation of GST-P$^+$, BUDR$^+$ hepatocytes was done over entire liver sections under the light microscope by two observers. BUDR was purchased from Sigma Chemical Co., St. Louis, MO. Antibodies to BUDR were monoclonal from Becton Dickenson, and antibodies to nodule GST-P and placental GST-P were generous gifts of Dr. T. H. Rushmore of Toronto and Dr. Kiyomi Sato at Hirosaki, Japan, respectively.

RESULTS

DEN-induced necrosis and hepatocyte proliferation was seen at 25 mg/kg b. wt. (Fig. 1). At a dose of 200 mg/kg DEN, there was a very wide zone 3 necrosis and striking hepatocyte proliferation. GST-P$^+$ singles, doublets and groups were found at all but the lowest dose of DEN used (Figs. 2, 3). With increasing dosage of DEN there was a proportionate increase in GST-P$^+$ hepatocytes as singles, doublets or groups (Table 1, Fig. 4). Based on these results the dose of 25 mg/kg b. wt. of DEN was chosen to initiate, in a search for visible R$^+$, BUDR$^+$, GST-P$^+$ hepatocytes. When DEN exposure was followed by resistance-selection with 2-acetylaminofluorene and partial hepatectomy, large GST-P$^+$ nodules appeared within 3 days (Fig. 5) suggesting that GST-P$^+$ groups were precursors for the nodules.

At three days and 7 days after DEN exposure BUDR$^+$ (Fig. 6) or GST-P$^+$ (Fig. 2) hepatocytes were found (Table 2). Hepatocytes showing both BUDR$^+$ and GST-P$^+$ using serial contiguous sections were not evident at 3 and 7 days after DEN.

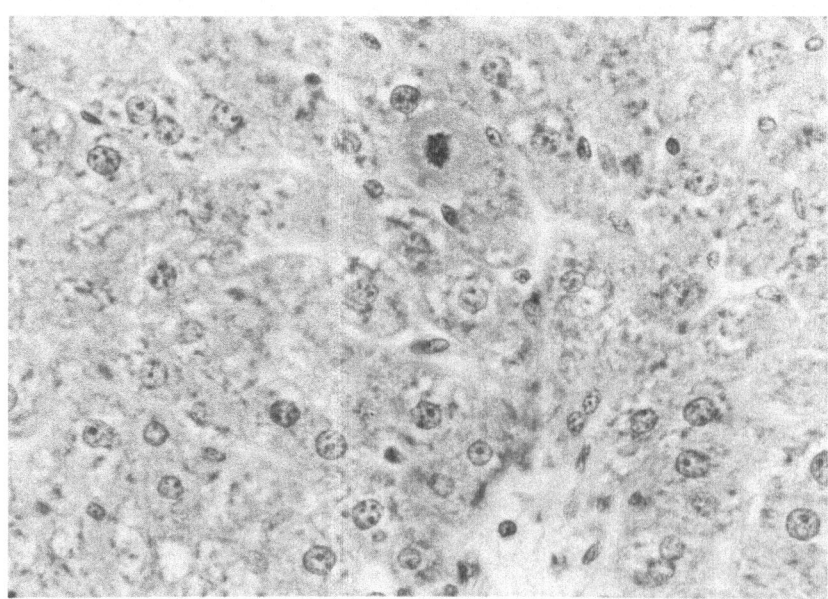

Fig. 1. Representative section of liver of rat at 3 days after exposure to a single dose of 25 mg/kg b. wt. i.p. of DEN. Mitoses such as the one evident were found in most lobules (haematoxylin and eosin stain, x 400).

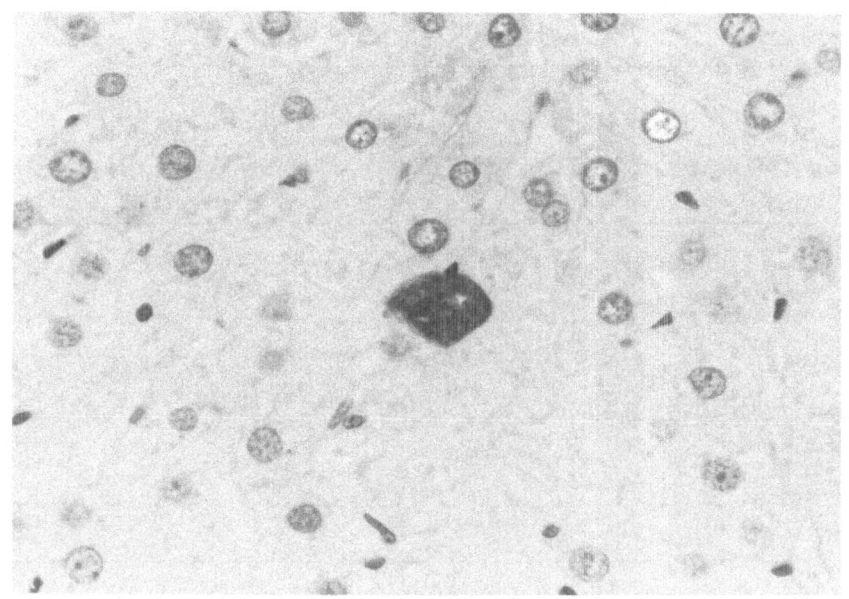

Fig. 2. Section of liver of rat at 7 days after exposure to 25 mg/kg of DEN. A single GST-P[+] hepatocyte is evident. Similar GST-P[+] hepatocytes were found through all zones of the liver (see Table 2). Stained with PAP method[21] using antibodies to nodule GST-P. x 400.

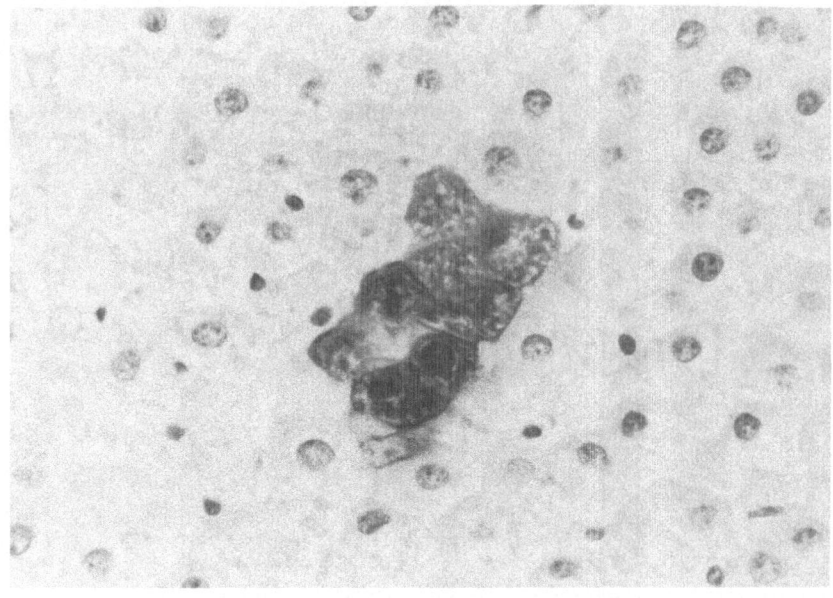

Fig. 3. Section of liver at 14 days after 50 mg/kg dose of DEN. A group (connected) of GST-P[+] hepatocytes is evident with a maximum cell "diameter" of 4 positive cells. (PAP method using antibodies to nodule GST-P, x 400).

Table 1. Numbers of GST-P[+] singles, doublets and groups of hepatocytes _in vivo_ in response to a single dose exposure to diethylnitrosamine (DEN)

Number[a] of GST-P[+] hepatocytes/cm^2	DEN dose (mg/kg b. wt.)					
	1	5	10	25	50[b]	200
1 week after DEN						
Singles [c]				38[a]	46,118[b]	147
Doublets				6	2,12	28
Groups				4	0,1	23
Total				48	48,131	198
2 weeks after DEN						
Singles	3	3	3		17,53	42
Doublets	1	2	1		8,9	9
Groups	0	1	1		2,10	21
Total	4	6	5		27,72	72

[a]Numbers are means per cm^2 liver section of total counts from livers of 3 to 5 rats. GST-P immunohistochemistry and quantitation is given in the methods section of the text. [b]Numbers are values of totals from each individual liver. Values from two rats are shown. [c]Singles, doublets and groups of hepatocytes are defined in the text and in the figure legends to Figs. 2 and 3.

DISCUSSION

In this paper we document the appearance in response to initiation of BUDR[+] cells or of GST-P[+] hepatocytes which respond to resistance-selection (R[+]) to form nodules. Quantitation of GST-P[+], BUDR[+] singles, doublets and groups and of R[+] nodules (results not shown, see ref. 15), points to a population of hepatocytes which has at least two of 3 properties proposed for

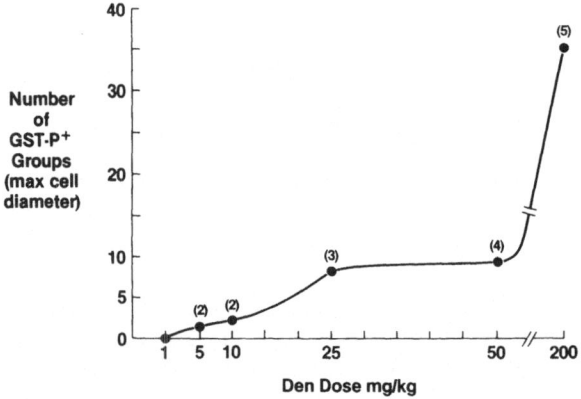

Fig. 4. Graph showing quantitation of the DEN dose response of induction of GST-P[+] groups of hepatocytes in rat liver (see also Table 1).

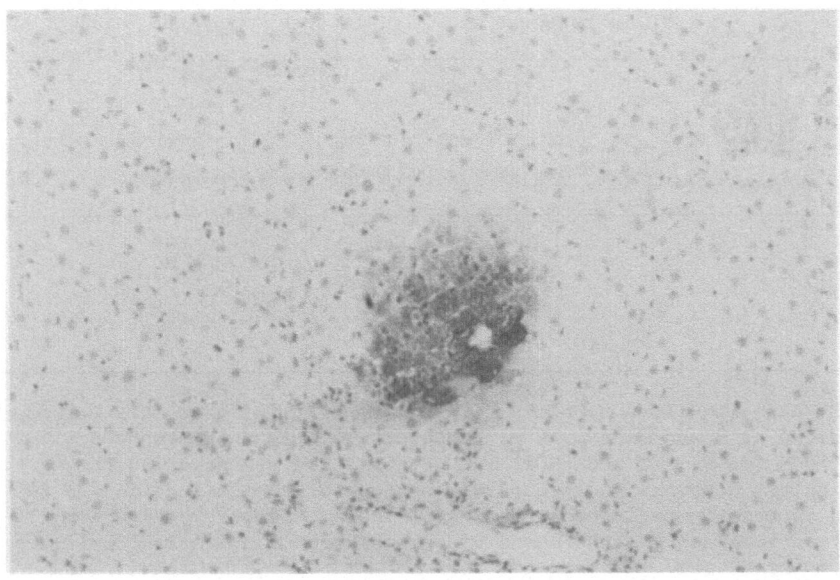

Fig. 5. Section of liver at 72 h after the partial hepatectomy part of resistance-selection (see methods) in a rat given 25 mg/kg of DEN. GST-P$^+$ hepatocytes are evident occupying most of the field of view. PAP method, x 400.

putative initiated hepatocytes (Fig. 7). In addition, the results suggest that many hepatocytes have one or two properties but not all three, see also ref. 15. The difficulty in our studies to date to demonstrate BUDR$^+$ and GST-P$^+$ in the same hepatocytes could be related to two factors, namely, to the interference of the BUDR with expression of GST-P or the generation of GST-P$^+$ hepatocytes in response to DEN only after 3 days. We are now planning to incorporate BUDR over a much longer period than 3 days after DEN.

The dose-related response of numbers of GST-P$^+$ hepatocytes shown here which appears independent of the cell proliferative response to DEN-induced necrosis, is intriguing since initiation of R$^+$, GST-P$^+$ nodules is a discontinuous response starting above doses of DEN 20-25 mg/kg [4-16]. Whether the dose dependency of induction of singles, doublets or groups of GST-P$^+$ hepatocytes induced by DEN has a close association with biochemical responses is an open question [22].

The ability to identify and to isolate hepatocytes which are precursors for cancer development continues to be a critical endeavour. Molecular studies of persistent nodules compared to surrounding tissue and cancers have revealed specific alterations in gene expression of hepatocytes [23-27]. It is hypothesized that these persistent genetic changes are built in at the time of initiation [15]. This model for identification of initiated hepatocytes would seem to be a very promising beginning in the pursuit of understanding the molecular and cellular bases of the cancer process.

ACKNOWLEDGEMENTS

I wish to thank Professor Emmanuel Farber for continuing discussions and encouragement. I am most grateful for the excellent technical work of

Table 2. Numbers of GST-P$^+$ and of proliferating (BUDR$^+$) hepatocytes in vivo in response to a single dose exposure to DEN [a]

Number[b] of GST-P$^+$, BUDR$^+$ hepatocytes/cm^2	3 days		7 days	
	GST-P$^+$	BUDR$^+$	GST-P$^+$	BUDR$^+$
Singles[c]	19[d]	134	57	73
Doublets	1	36	7	26
Groups	0	0	5	1
Total	20	170	69	100

[a]DEN was given as a single dose of 25 mg/kg b. wt. and rats were sacrificed at 3 and 7 days after DEN exposure. [b]Immunohistochemical methods for GST-P$^+$ and BUDR$^+$ and quantitation are given in the text. [c]Singles, doublets and groups of hepatocytes are defined in the text and in the figure legends to Figs. 2 and 3. [d]Numbers are per cm^2 liver section and are the means of values from quantitation of livers from three rats.

Mrs. Dianna Armstrong and Mr. Ken Ekem, and the invaluable assistance of Lori Cutler with the manuscript. We thank Dr. Tom H. Rushmore and Dr. Kiyomi Sato for antibodies to nodule GST-P and placental GST-P respectively. This work was supported by a grant from the Medical Research Council of Canada, MA 9187.

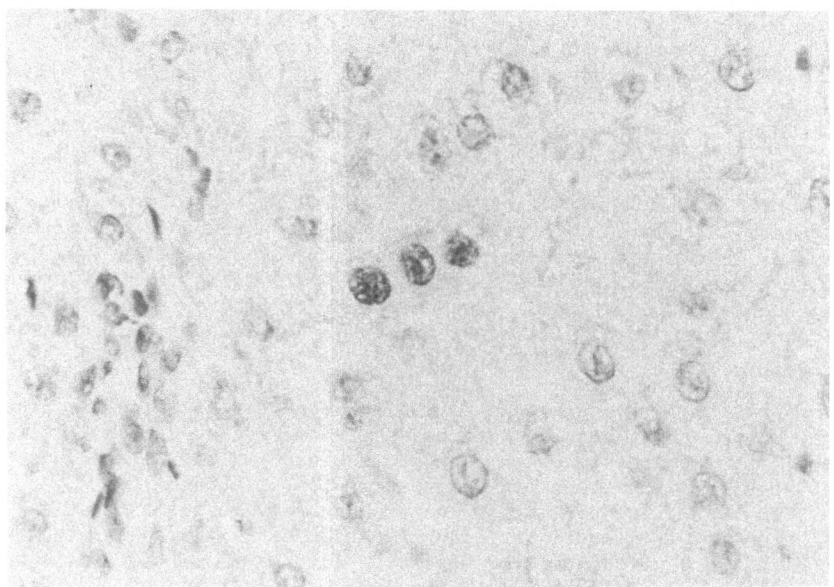

Fig. 6. Section of liver at 7 days after 25 mg/kg of DEN as in Fig. 2 of a rat given BUDR. Three BUDR$^+$ hepatocytes in a row, staining with anti-BUDR in their nuclei are evident. Similar BUDR$^+$ hepatocytes were found throughout the liver tissue (see Table 2) and in other BUDR-loaded rats when examined at 3 and 7 days after DEN. (PAP method using monoclonal antibodies to BUDR, x 400).

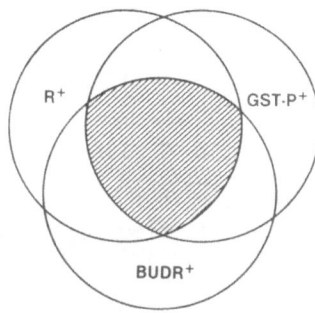

Fig. 7. Venn diagram with three circles each representing a specific
population of hepatocytes. One circle represents hepatocytes
showing resistance or R⁺, a second group is proliferating and
is BUDR⁺ and the third circle includes all GST-P⁺ hepatocytes.
Initiated hepatocytes are proposed to be that population of
hepatocytes which has all three phenotypic properties.

REFERENCES

1. E. Farber, S. Parker and M. Gruenstein, The resistance of putative pre-
    malignant liver cell populations, hyperplastic nodules to the acute
    cytotoxic effects of some hepatocarcinogens, Cancer Res. 36:3879
    (1976).
2. R. Cameron, G. D. Sweeney, K. Jones, G. Lee and E. Farber, A relative
    deficiency of cytochrome P-450 and aryl hydrocarbon (benzo[a]pyrene)
    hydroxylase in hyperplastic nodules induced by 2-acetylaminofluorene
    in rat liver, Cancer Res. 36:3888 (1976).
3. D. J. Judah, R. F. Legg and G. E. Neal, Development of resistance to
    cytotoxicity during aflatoxin carcinogenesis, Nature (Lond) 265:343
    (1977).
4. D. B. Solt and E. Farber, New principle for the analysis of chemical
    carcinogenesis, Nature 262:701 (1976).
5. E. Farber, D. Solt, R. Cameron, B. Laishes, K. Ogawa and A. Medline,
    Newer insights into the pathogenesis of liver carcinogenesis, Am. J.
    Pathol. 89:477 (1977).
6. D. B. Solt, A. Medline and E. Farber, Rapid emergence of carcinogen-
    induced hyperplastic lesions in a new model for the sequential
    analysis of liver carcinogenesis, Am. J. Pathol. 88:595 (1977).
7. E. Farber and R. Cameron, The sequential analysis of cancer
    development, Adv. Cancer Res. 31:125 (1980).
8. H. Tsuda, G. Lee and E. Farber, The induction of resistant hepatocytes
    as a new principle for a possible short term in vivo test for car-
    cinogens, Cancer Res. 40:1157 (1980).
9. M. Tatematsu, Y. Nagamine and E. Farber, Redifferentiation as a basis
    for remodelling of carcinogen-induced hepatocyte nodules to normal
    appearing liver, Cancer Res. 43:5049 (1983).
10. M. A. Hayes, E. Roberts and E. Farber, Initiation and selection of re-
    sistant hepatocyte nodules in rat given the pyrrolizidine alkaloids
    lasiocarpine and senecionine, Cancer Res. 45:3726 (1985).
11. M. A. Hayes, E. Roberts, S. H. Safe, E. Farber and R. G. Cameron,
    Influence of different polychlorinated biphenyls on cytocidal, mito-
    inhibitory, and nodule-selecting activities of N-2-fluorenylacetamide
    in rat liver, J. Nat. Cancer Inst. 76:683 (1986).
12. K. Sato, A. Kitihara, K. Satoh, T. Ishikawa, M. Tatematsu and M. Ito,
    The placental form of glutathione S-transferase as a new marker
    protein for preneoplasia in rat chemical hepatocarcinogenesis, Gann
    75:199 (1984).

13. M. Tatematsu, Y. Mera, N. Ito, K. Satoh and K. Sato, Relative merits of immunohistochemical demonstration of placental A, B, and C forms of glutathione S-transferase and histochemical demonstration of gamma glutamyl transpeptidase as markers of altered foci during liver carcinogenesis in rats, Carcinogenesis 6:1621 (1985).

14. M. A. Moore, K. Nakagawa, K. Satoh, T. Ishikawa and K. Sato, Single GST-P positive liver cells - putative initiated hepatocytes, Carcinogenesis 8:483 (1987).

15. R. G. Cameron, Comparison of GST-P versus GGT as markers of hepatocellular lineage during analyses of initiation of carcinogenesis, Cancer Invest. in press (1988).

16. T. S. Ying, K. Enomoto, D. S. R. Sarma and E. Farber, Effects of delays in the cell cycle on the induction of preneoplastic and neoplastic lesions in rat liver by 1,2-dimethylhydrazine, Cancer Res. 42:876 (1982).

17. E. Cayama, H. Tsuda, D. S. R. Sarma and E. Farber, Initiation of chemical carcinogenesis requires cell proliferation, Nature (Lond) 275:60 (1978).

18. D. B. Solt, E. Cayama, H. Tsuda, K. Enomoto, G. Lee and E. Farber, Promotion of liver cancer development by brief exposure to dietary 2-acetylaminofluorene plus partial hepatectomy or carbon tetrachloride, Cancer Res. 43:188 (1983).

19. A. Columbano, S. Rajalakshmi and D. S. R. Sarma, Requirement of cell proliferation for the initiation of liver carcinogenesis as assayed by three different procedures, Cancer Res. 41:2079 (1981).

20. W. K. Kaufmann, D. G. Kaufman, J. M., Rice and M. L. Wenk, Reversible inhibition of rat hepatocyte proliferation by hydrocortisone and its effect on cell cycle-dependent hepatocarcinogenesis by N-methyl-N-nitrosourea, Cancer Res. 41:4653 (1981).

21. T. Kaku, J. K. Ekem, C. Lindayen, D. J. Bailey, A. W. P. Van Nostrand and E. Farber, Comparison of formalin and acetone-fixation for immunohistochemical detection of carcinoembryonic antigen (CEA) and keratin, Am. J. Clin. Pathol. 80:806 (1983).

22. E. Scherer and P. Emmelot, Precancerous transformation in rat liver by diethylnitrosamine in relation to repair of alkylated site in DNA, Mutat. Res. 46:155 (1977).

23. C. B. Pickett, J. B. Williams, A. Y. H. Lu and R. G. Cameron, Regulation of glutathione S-transferase and DT-diaphorase mRNA in persistent hepatocyte nodules during chemical hepatocarcinogenesis, Proc. Natl. Acad. Sci. USA 81:5091 (1984).

24. J. B. Williams, A. Y. H. Lu, R. G. Cameron and C. B. Pickett, Rat liver DT-diaphorase: construction of a DT-diaphorase cDNA clone and regulation of DT-diaphorase mRNA by 3-methylcholanthrene and in persistent hepatocyte nodules induced by chemical carcinogens, J. Biol. Chem. 261:5524 (1986).

25. K. Satoh, A. Katahara, Y. Soma, Y. Inaba, I. Hayami and K. Sato, Purification, induction and distribution of placental glutathione transferase: a new marker enzyme for preneoplastic cells in the rat chemical carcinogenesis, Proc. Natl. Acad. Sci. USA 82:3964 (1985).

26. A. Okuda, M. Sakai and M. Muramatsu, The structure of the rat glutathione S-transferase P gene and related pseudogenes, J. Biol. Chem. 262:3858 (1987).

27. T. H. Rushmore, R. N. S. Sharma, M. W. Roomi, L. Harris, K. Satoh, K. Sato, R. K. Murray and E. Farber, Identification of a characteristic cytosolic polypeptide of rat preneoplastic hepatocyte nodules as placental glutathione S-transferase, Biochem. Biophys. Res. Commun. 143:98 (1987).

THEORETICAL CALCULATIONS OF THE ELECTROSTATIC POTENTIAL OF SOME

TUMOR PROMOTERS

Alastair F. Cuthbertson, Derek Higgins and
Colin Thomson

NFCR Project, Dept. of Chemistry
University of St. Andrews
St. Andrews, Fife KY16 9ST, Scotland

INTRODUCTION

An understanding of the molecular basis of tumor promotion is essential
if we are to fully understand those molecular events which occur between the
original initiating step, and the subsequent development of a malignant tumor[1,2]. It is very likely that the long time period observed for the development of cancer in humans reflects the consequences of exposure to tumor promoters over several years.

Unfortunately, although our knowledge of tumor promotion in mouse skin
is extensive[3-8], and a large variety of tumor promoters have been used in
these studies, it is unlikely that these compounds play a role in most human
cancer, since they are rather exotic chemicals which are not of widespread
occurence in the environment.

Nevertheless, an understanding of the molecular basis of their effects
is of great importance, because it should help us in the search for promoters
which are involved in human cancer.

In mouse skin, numerous studies have been carried out with the 12,13 diesters of the tetracyclic diterpene alcohol, phorbol[9,10], and more recently, a
number of totally different chemicals have been found which have very similar
biological effects to the phorbol esters. Among these compounds, called TPA
like promoters[9,10], are the teleocidins, which are based on the indole alkaloid structure[7,11], aplysiatoxins, which are polyacetates[12] and ingenols[7,13].

The wealth of structure-activity data on these promoters has continued
to increase, especially since the discovery that these four classes of tumor
promoter all bind with high affinity to a receptor protein which has been
shown to be Protein Kinase-C (PKC), a Ca^{2+}/phospholipid dependent kinase
which plays a pivotal role in the important pathway involved in phospholipid
metabolism[14,15]. The natural substrates for PKC are diacylglycerols (DAG),
and it is believed that TPA type promoters usurp the role of the DAG[15].

In order to understand the structure activity data at the molecular level, we need more information on the molecular structure of the various promoters, and their interactions with PKC. Unfortunately, although we know
something of the active site amino acid sequence in PKC, we do not yet have

any structural data on this enzyme[16]. Despite this limitation, we have used the methods of theoretical chemistry in an attempt to rationalise the experimental data, and describe the methodology and our preliminary results in this paper. Our conclusions are that more detailed studies on these molecules using theoretical techniques will ultimately enable us to more fully understand the molecular basis of the tumor promotion step in Chemical Carcinogenesis.

THEORETICAL CHEMISTRY BACKGROUND

During the last ten years, applications of the methods of theoretical chemistry to a large variety of biological problems have become feasible, due to very important advances in the methodology and computational resources. Of particular importance is our ability to calculate a theoretical structure of molecules containing up to $\sim$100 atoms, using gradient techniques, which enable one to locate energy minima and transition states on the potential surface. Structures can also be computed for larger molecules, such as polypeptides or polynucleotides at a lower level of theory.

There are two principle methods which we use. In the first, the method of Molecular Mechanics[17], the total energy is calculated from an empirical force field which has been extensively tested on small molecules. This method is very fast, but since it is non-quantum mechanical, it only yields structural information. However, the results are in very good agreement with experiment if the relevant force field parameters are known, and the method can be used for very large systems, such as large polypeptides[18].

The second method is the semi-empirical MNDO method, which is a self consistent field quantum chemical technique, extensively developed and tested by Dewar[19,20]. The approximations and parametrisation of the method are well documented, and the results of the calculations give the molecular wave function, the energy minima (or transition state structures), and a variety of molecular properties, including $\Delta H_f^{\ominus}$ (298K) and the vibrational frequencies. The computed minimum energy structure is described in terms of the usual 3N-6 internal coordinates i.e. bond lengths, interbond angles and torsion angles. Agreement with experimental geometries and properties for a large variety of molecules is very good, and we therefore have confidence in the predictions of the method for other molecules.

It is perhaps less widely appreciated amongst biochemists that the behaviour of even the largest molecules is determined by the electrons in the molecule, and the electronic charge distribution $\rho(r)$ obtained from the wave function $\Psi(r)$, MUST be considered if we are to fully understand the molecular basis of biochemical processes. The information available from calculations of Ψ and related properties is much greater than information available from just the spatial positions of the atoms in the molecule. This is especially true of molecules with a complicated 3-dimensional arrangement of the atoms in space. Stereo-electronic effects can rarely be predicted just on the basis of the atomic positions.

The ability to calculate the wave function Ψ for the molecules involved in various aspects of chemical carcinogenesis in principle should enable us to explore in much greater detail the reasons for the observed structure-activity data on these molecules.

THE MOLECULAR ELECTROSTATIC POTENTIAL AND COMPUTER GRAPHICS

One molecular property which is very easily computed from the wave function is the molecular electrostatic potential (MEP), which we can define as

302

$$V(r) = - \int \frac{\rho(r')}{|r' - r|} \, dr' + \sum_m \frac{Z_m}{|R_m - r|}$$

where $\rho(r')$ is the electronic charge density at r', and Z_m and R_m are the charge and postions of the m^{+n} nucleus $V(r)$ can be regarded as the potential of the molecule at r due to the electronic charge distribution of the molecule[21]. It is clear from numerous previous applications that this quantity is of considerable use in exploring structure activity relationships in pharmacology, and we believe will also be important in the applications to tumor promotion.

For a complicated molecule, the most effective way of displaying the potential in three dimensions is not obvious. The availability of colour graphics displays has, however, made the pictorial representation of $V(r)$ much easier, and we have developed programs to do this based on the following methodology.

$V(r)$ is computed from the wave function at a series of points on a Van der Waals surface surrounding each atom in the molecule, using the algorithm due to Conolly[22]. The resulting potentials are then displayed on a color graphics terminal, with a range of colors. Typically the most negative potentials are displayed in red, zero potential in yellow, and positive potentials in blue, with intermediate potentials in the appropriate intermediate colors. The resulting displays show quite clearly the important variations in potential over the whole molecular framework, the kinds of potential generated by particular functional groups in the molecule, and, most important, the changes in the appearance of the MEP maps which occur when different substituents are introduced into the molecule. We believe that a detailed analysis of these maps will provide much more detailed information related to structure-activity relationships than has been obtained up to now by an analysis based solely on the spatial positions of key functional groups. This is because the intermolecular interactions, and therefore the binding, are determined by $\rho(r)$, and hence are reflected in $V(r)$.

PREVIOUS WORK AND STRUCTURAL DATA

Fortunately, it is possible to confirm the reliability of the computations because crystallographic data is available for phorbol[23], teleocidin[24], and ingenol[25]. The basic ring structures of these compounds was used as a starting structure for the molecular modelling studies on model compounds, whilst the completely optimized structures of phorbol, teleocidin and ingenol were compared in detail with the X-ray structures. There has been one previous study of phorbol, using an older method MINDO/3 (a precurser to MNDO)[26], but in this study and some more recent modelling studies[27,28], there was no discussion of the wave function or qualities related to it.

In the earlier attempts to understand why these chemically quite different molecules interact with the same receptor, Weinstein and coworkers, suggested, on the basis of the available structure-activity data, that there are key functional groups in the different molecules which have to be in a particular spatial arrangement for binding[28,29]. Relatively crude comparisons were made of these molecules in this light, and some molecular mechanics calculations confirmed that these conformations correspond to low energy structures.

COMPUTATIONAL STRATEGY

Our approach has been to calculate the molecular wave function for these molecules, and for model compounds related to them, and to obtain as a result, more detailed information than is possible using a non-quantum mechanical method such as molecular mechanics.

Since these molecules are quite large in terms of quantum mechanical calculations, we have chosen first to investigate model compounds containing the key functional groups which are necessary in the phorbol and teleocidin promoters. In these model compounds, we do not include the long chain ester groupings (in phorbols), nor the terpene ring in teleocidins. It is of course true that phorbol itself is not a promoter, but the ester group is probably necessary for lipid solubility rather than binding.

In our first attempts, we in fact carried out calculations in which an even simpler model for the phorbol system was used, essentially based on a 2 membered ring system[30]. However, to obtain the correct conformation of the second seven-membered ring, the full structure of the phorbol ring system needs to be included. We report in this paper, our results on the more realistic models for phorbol and teleocidin type promoters. This work is incomplete, since the studies of the structure activity relationships will demand similar computations for a large number of the substituted derivatives, and this work is in progress and will be reported in detail elsewhere.

As mentioned earlier, we are now able to calculate fully optimized geometries for these molecules using current methodologies. To reduce the computer time required, we use the following strategy.

Starting from the crystal structure coordinates of the basic ring system, we fully optimize the structure using the AMBER molecular mechanics program[31-33].

We use the all atom force field[32], the calculations are very fast, and the resulting structure provides a good starting geometry for a subsequent calculation, again with full optimization, carried out with the MNDO SCF method[19,20]. The MNDO optimized structure can be compared with X-ray structure for the complete molecules.

The MNDO wave function is then used to calculate the Molecular Electrostatic Potential (MEP) in 3-dimensions.

Finally, for the smaller model compounds, we have calculated an ab-initio SCF wave function using an STO-3G basis set at the MNDO optimized geometry[34]. The results of this calculation are displayed in the same way, and in general do not differ significantly from the MNDO results. Fortunately, the main features of the MEP can be reliably calculated without doing an ab-initio calculation, which is much more expensive.

The calculations described in the present paper were carried out either on the VAX 11/785 computer at the University of St. Andrews, or the MICROVAX-2/GPX work station in the authors laboratory. The MEP were displayed both on a Tektronix 4109 color terminal or on the VT290 color terminal of the workstation, using either the display facilities of the molecular modelling program CHEMX[35], or our own in-house graphics programs[36,37], which are under continuous development.

RESULTS

Comparison of the Atom Positions in the Crystal Structures

Fig. 1 gives the numbering scheme and chemical structures for TPA, Teleocidin and Ingenol. A fitting algorithum in CHEMX was used to fit

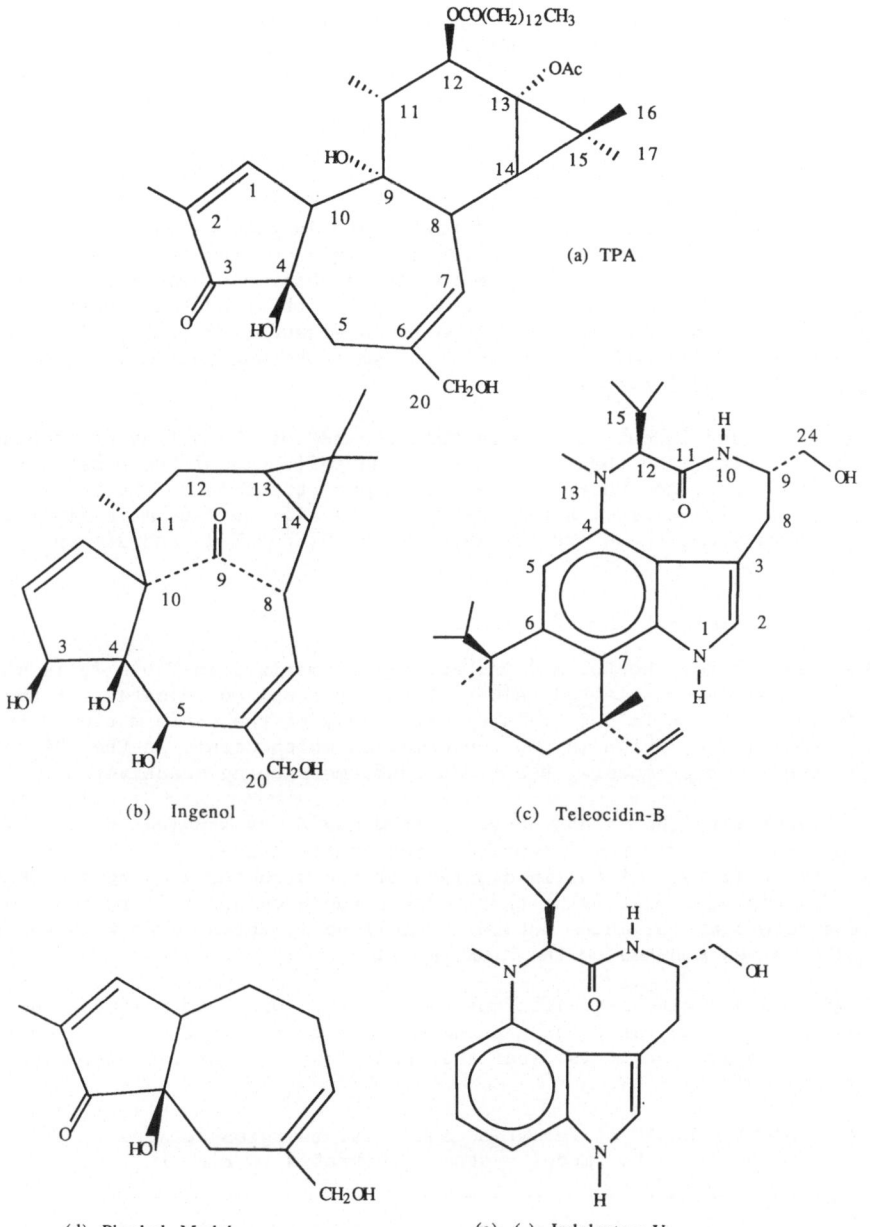

Fig. 1. Chemical formulae and numbering scheme for (a) TPA (b) Ingenol
(c) Teleocidin (d) Phorbol Model Structure (e) (-) Indolactam-V.

atoms O3, O4, O9, O20 in phorbol and ingenol with O11, N13, N1 and O24 in teleocidin. The best fit was obtained with TPA-ingenol, with a mean deviation
of 0.71A. Details of this are given in our earlier papers on the close correspondence of the position of these key groups in the different molecules.

Calculated Structures and Potential of Models for Phorbol: 2 Ring System

The compound illustrated in Fig. 1 was chosen as a model, and several

305

different substituents attached to the rings were investigated (Table 1). These different substituents result in different tumor promoting activities in the real phorbol esters, and we hoped to see differences in the MEP which would correlate with these differences. The geometries were completely optimized at the MNDO level, and the MEP computed from a single point ab-initio SCF calculation. Table 1 lists the heats of formation and the total energies computed at the STO-3G level.

The calculated bond lengths and angles are in good agreement with the crystal structure data on the relevant rings[23,25], but the dihedral angles are not; the calculations giving a more stable chair conformation of the 7-membered cycloheptene ring, in contrast to the observed boat conformation. Details of these calculations are to be found elsewhere[30], but it is clear that a more realistic model is essential, namely to use the full phorbol molecule as the model system.

We also found in our early work that the MEP of the different compounds are rather similar, and that in this case the influence of the missing ring might be important, and therefore we turned our attention to phorbol itself. Since this is a lot bigger, substituent effects will be much more time consuming to study, but these are feasible on our MICROVAX-2, and are underway.

Computed Structures for Phorbol and Ingenol

Our results for phorbol and ingenol are summarized in Table 2, in which we list $\Delta H_f^{\ominus}$ and some <u>selected</u> values of the internal coordinates for the two molecules. These preliminary results refer only to the conformation with ßOH group at position 4, which is the conformation which occurs in the TPA and ingenol type tumor promoters, the 4-OHα conformers being inactive.

Agreement with the values obtained from the X-ray studies is generally very good, particularly the 7 membered ring conformation. The close agreement is particularly striking in displays of the structures using the CHEMX program. It is also gratifying that these structures are well reproduced using the molecular mechanics method. A full optimization with this technique takes a few minutes on the MICROVAX-2!

Figs. 2 and 3 give six different views of the MEP maps computed for these molecules. The variations in the MEP can only be properly appreciated in a color display, but it is clear that analysis of these for different

Table 1.   Heats of formation and total energies for the phorbol model system illustrated in Fig. 1

Substituent	$\Delta H_f^{\ominus}$ (MNDO) kcal.mol^{-1}	$\Delta H_f^{\ominus}$ (STO-3G/MNDO) atomic units
4-OHß	-103.7	-679.652
4-OHα	-108.1	-679.660
4-Hß	- 68.1	-605.831
4-Hα	- 69.9	-605.834
4-OHß, 9-OH	-142.7	-753.477
4-OHß, 6-Me	- 61.6	-605.833
4-OHß, 9-OH 6-7 epoxide	-162.2	-827.329
4-OMeß, 9-OH	-132.2	-792.051

Table 2. Comparison of selected internal coordinates for the calculated and X-ray structures of phorbol and ingenol

	Phorbol		Ingenol	
	Calc	X-ray	Calc	X-ray
Bond Lengths				
C1-C2	1.358	1.341	1.350	1.320
C2-C3	1.499	1.460	1.532	1.485
C3-O3	1.220	1.220	1.397	1.471
C4-O4	1.400	1.427	1.387	1.414
C6-C7	1.356	1.327	1.360	1.338
C8-C9	1.599	1.546	1.548	1.527
C9-O9	1.410	1.461	1.223	1.207
Interbond Angles				
C1-C2-C3	109.29	108.5	110.2	109.4
O3-C3-C2	127.44	126.7	113.8	109.4
C6-C7-C8	128.10	127.5	120.4	119.6
Torsion Angles				
C1-C2-C3-C4	358.5	355.1	-16.1	-18.9
C5-C4-C3-C2	224.2	230.7	265.5	271.7
C7-C8-C9-C10	46.5	43.2		
O9-C9-C10-C1			15.3	21.9
$\Delta H_f^{\ominus}$ (298K)	-178.6275		-121.244	

substituted phorbols and ingenols should be most helpful in interpreting the structure activity data.

Computed Structure for a Teleocidin Model

Table 3 gives the computed structure for one possible configuration of a model for teleocidin: namely, the Indolelactom -V reported by Irie et al.[38]. Fig. 4 gives six different views of the MEP of this molecule. The structural parameters are again reproduced quite well, although the X-ray value for R(C8-C9) seems long: our value of 1.563A looks more reasonable. The lactam ring conformation is quite flexible, and we are currently exploring the PE surface for this molecule.

DISCUSSION

The calculation of the MEP on Van der Waals surfaces, as depicted in Figs. 2, 3, and 4, should furnish very detailed information which is relevant to the interaction of these molecules with their receptor. Analysis of these maps, together with the computation of similar MEP for other conformers, and for different substituents in these molecules, should enable us to shed more light on the structural and electronic features of these molecules which are important in determining the observed differences in the biological activities in these series[7,10,38]. Such studies are underway and the results will be reported elsewhere[39]. It must be emphasised that it is essential to use color displays of the MEP to obtain the maximum of information, and therefore Figs. 2-4 can only give a crude impression of the potential of these methods,

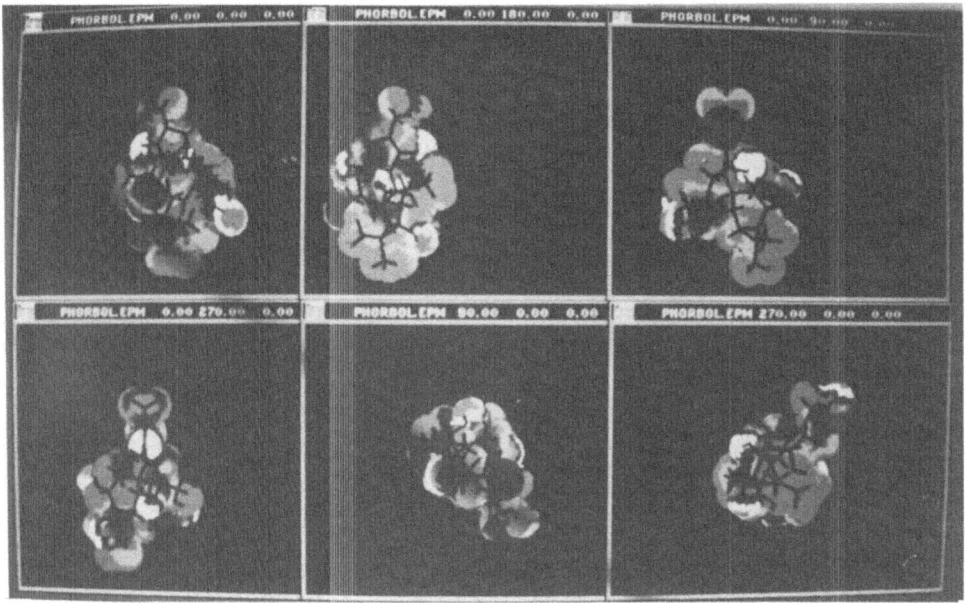

Fig. 2. Six different views of the MEP of the Phorbol optimized MNDO
structure.

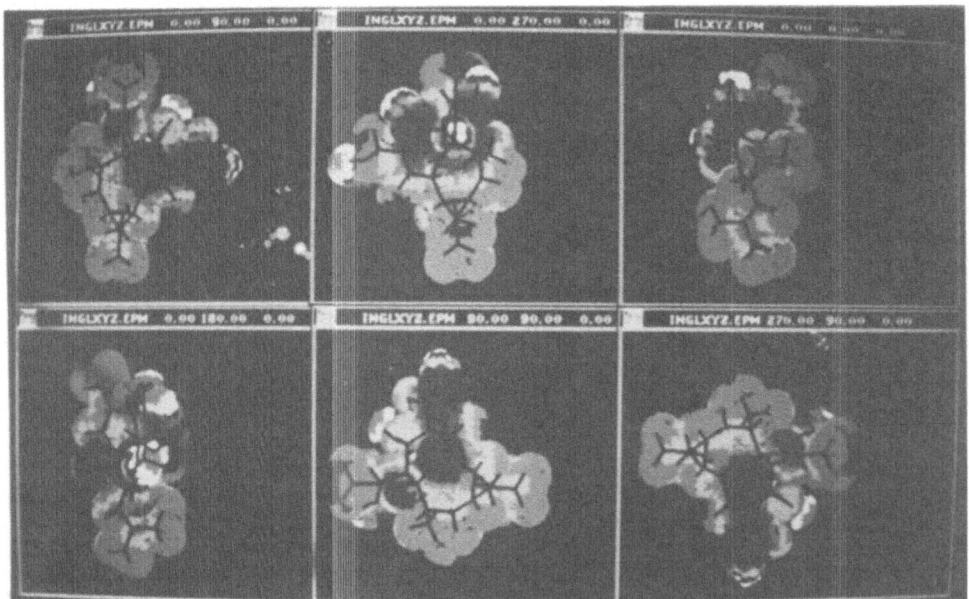

Fig. 3. Six different views of the MEP of the Ingenol optimized MNDO structure.

Table 3. Comparison of selected internal coordinates with X-ray
values for the Teleocidin model (-)-Indolactam-V

	Calc	X-ray
Bond lengths		
N1-H1	0.993	1.000
C2-C3	1.396	1.407
C4-N13	1.430	1.416
C12-N13	1.474	1.450
C11-O11	1.231	1.230
C12-O15	1.579	1.537
C3-C8	1.502	1.450
C8-C9	1.563	1.606
Interbond Angles		
C5-C4-N13	119.6	120.1
N13-C12-C15	116.8	116.0
C3-C8-C9	116.7	112.3
Torsion Angles		
C12-N13-C4-C5	236.1	238.7
C11-C12-N13-C4	222.3	213.8
C24-C9-N10-C11	202.4	180.4
O11-C11-C12-N13	254.1	251.9

$\Delta H_f^{\ominus}$ (298K) = -12.423

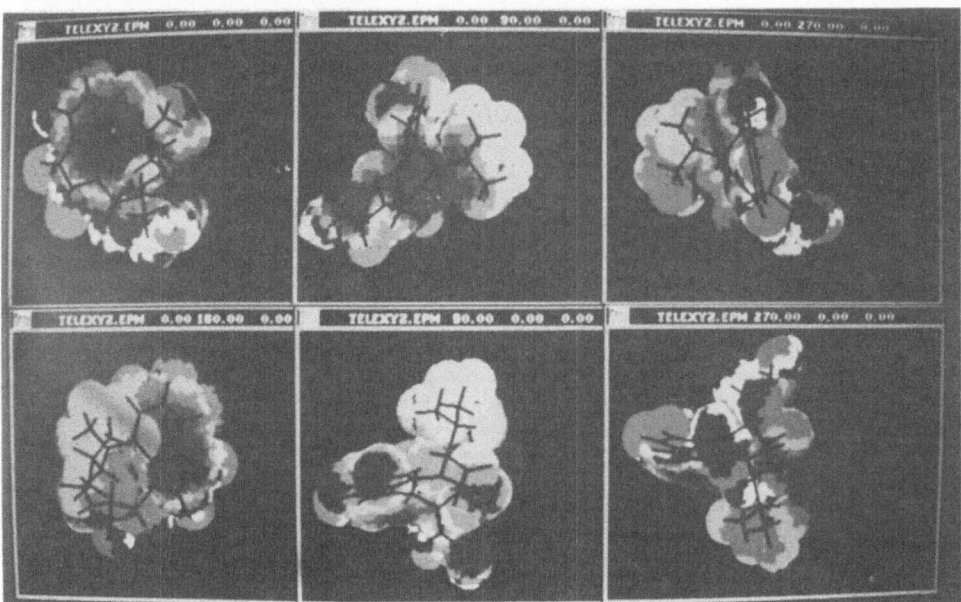

Fig. 4.   Six different views of the MEP of the (-)-Indolactam-V optimized
MNDO structure.

which we believe will be invaluable in the study of these and other tumor promoters and inhibitors of tumor promotion[40].

ACKNOWLEDGEMENTS

We are indebted to DEC for the donation of the MICROVAX-2/GPX work station, and to the National Foundation for Cancer Research for continued financial support.

REFERENCES

1. Chemical carcinogens, 2nd Ed. A. C. S. Monograph No. 182, C. E. Searle, ed., American Chemical Society, Washington (1984).
2. Cancer, a comprehensive review, F. F. Becker, ed., Plenum Press, New York, vol. 1-5 (1975-80).
3. Mechanisms of tumor promotion and cocarcinogenesis, vol. 2, Carcinogenesis: a comprehensive treatise, T. J. Slaga, A. Sivak and R. K. Boutwell, eds., Raven Press, New York (1978).
4. Models, mechanisms and etiology of tumor promotion, IARC Monograph No. 56, IARC, Lyon (1984).
5. Co-carcinogenesis and biological effects of tumor promotors, vol. 7, Carcinogenesis: a comprehensive treatise, E. Hecker, N. F. Fusenig, W. Kunz, F. Marks and H. W. Theilmann, eds., Raven Press, New York (1982).
6. Mechanisms of tumor promotion, vols. I, II, III, T. J. Slaga, ed., CRC Press, Boca Raton, Fla (1984).
7. Cellular interactions by environmental tumor promotors, H. Fujiki and T. Sugimura, eds., Scientific Soc. Press, Tokyo (1984).
8. Tumor promotion and human cancer, T. J. Slaga and R. Montesano, eds., Cancer Survey vol.3 no.4 (1983).
9. Irritant diterpene ester promoters of mouse skin: contributions to etiologies of environmental cancer and to biochemical mechanisms, E. Hecker in ref. 7.
10. E. Hecker, Cell membrane associated protein kinase C as receptor of diterpene esters: co-carcinogens of the tumor promoter type and the phenotypic expression of tumors, Arzneim-Forsch/Drug. Res 35 1980 (1985).
11. H. Fujiki, M. Mori, M. Nakayasu, M. Terada, T. Sugimura and R. E. Moore, Indole alkaloids: Dihydroteleocidin B, teleocidin and lyngbyatoxin A as members of a new class of tumor promoters, Proc. Natl. Acad. USA 78:3872 (1981).
12. H. Fujiki, M. Suganuma, M. Nakayasu, H. Hoshino, R. E. Moore and T. Sugimura, A third class of new tumor promoters, polyacetates (debromoaplysiatoxin and aplysiatoxin), can differentiate biological actions relevant to tumour promoters, Gann 73:495 (1982).
13. H. J. Opferkuch and E. Hecker, On the active principles of the spurge family (Euphorbiaceae)IV Skin irritant and tumour promoting diterpene esters from Euphorbia ingens, E. Mey. J. Cancer Res. Clin. Oncol. 103:255 (1982).
14. K. B. Delclos, D. S. Nagle and P. M. Blumberg, Specific binding of phorbol ester tumour promoters to mouse skin, Cell. 19:1025 (1980).
15. C. Ashendel, The phorbol ester receptor: a phospholipid regulated protein kinase, Biochem. Biophys. Acta 822:219 (1986).
16. P. J. Parker, L. Coussens, N. Totty, L. Rhee, S. Young, E. Chen, S. Stabel, M. D. Waterfield and A. Ullrich, The complete primary structure of protein kinase-C - the major phorbol ester receptor, Science 233:853 (1986).
17. U. Burkert and N. L. Allinger, "Molecular Mechanics", ACS Monograph, n. 177, ACS, Washington (1982).

18. B. Venkataraghvan and R. J. Feldman, "Macromolecular Structure and Specificity: Computer Assisted Modelling and Applications", Ann. New York Acad. Sci. vol. 239 (1985).

19. M. J. S. Dewar and W. Thiel, Ground states of molecules 38. The MNDO method. Approximations and parameters, J. Amer. Chem. Soc. 99:4899 (1977).

20. M. J. S. Dewar, Computing calculated reactions, Chemistry in Britain 11:97 (1975).

21. P. Politzer and D. G. Truhlar, eds., "Chemical Applications of Atomic and Molecular Electrostatic Potentials", Plenum Press, New York (1981).

22. M. L. Conolly, Molecular surface program, QCPE n. 429, Quantum Chemistry Program Echange, Indiana, USA, Science 221:709 (1983).

23. V. F. Brandl, M. Röhrl, K. Zechmeister and W. Hoppe, Rontgenstrukturanalysen von neophorbol C31 H35 09 BR und phorbol C20 H28 06, Acta Cryst. B27:171 (1971).

24. H. Harada, N. Sakabe, Y. Hirata, Y. Tomiie and I. Nitta, The X-ray structure determination of dihydroteleocidin-B monobromoacetate, Bull. Chem. Soc. Japan 39:1773 (1966).

25. K. Zechmeister, F. Brandl, W. Hoppe, E. Hecker, H. J. Opferkuch and W. Adolf, Structure determination of the new tetracyclic diterpene ingenol-triacetate with triple product methods, Tet. Letts 4075 (1970).

26. G. R. Pack, The molecular structure and charge distribution of phorbol: parent compound of the tumour promoting phorbol diesters, Cancer Biochem. Biophys. 5:183 (1981).

27. A. M. Jeffrey and R. M. Liskamp, Computer-assisted molecular modelling of tumour promoters:rationale for the activity of phorbol ester, teleocidin B and aplysiatoxin, Proc. Natl. Acad. Sci. 83:241 (1986).

28. P. A. Wender, K. F. Koehler, N. A. Sharkey, M. L. Dell'Aquila and P. M. Blumberg, Analysis of the phorbol ester pharmacophore on protein kinase C as a guide to the rational design of new classes of analogs, Proc. Natl. Acad. Sci. USA 83:4214 (1986).

29. A. Horowitz, H. Fujiki, I. B. Weinstein, A. Jeffrey, E. Okin, R. E. Moore and T. Sugimura, Comparative effects of aplysiatoxin, debromoaplysiatoxin and teleocidin on receptor binding and phospholipid metabolism, Cancer Res. 43:1529 (1983).

30. A. F. Cuthbertson and C. Thomson, Electrostatic potentials of some tumour promoters, J. Mol. Graph. 5:92 (1987).

31. P. K. Weiner and P. A. Kollman, AMBER: assisted model building with energy refinement, a general program for modelling molecules and their interactions, J. Comp. Chem. 2:287 (1981).

32. S. J. Weiner, P. A. Kollman, N. T. Nguyen and D. A. Case, An all atom force field for simulations of protein and nucleic acids, J. Comp. Chem. 7:230 (1986).

33. P. A. Kollman, P. K. Weiner, J. Coldwell and V. C. Singh, Amber version 3.0, U.C.S.F., San Francisco (1987).

34. W. J. Henre, L. Radom, P. von R. Schleyer and J. A. Pople, in: "Ab-Initio Molecular Orbital Theory", J. Wiley and Sons, eds., New York (1980).

35. "Chem X", Chemical Design, Oxford Jan (1987).

36. C. Edge, Ph. D. Thesis, University of St. Andrews (1987).

37. D. Higgins, unpublished work.

38. K. Irie, N. Hagiwara, H. Tokuda and K. Koshimizu, Structure-activity studies of the indole alkaloid tumour promoters, Carcinogenesis 8:547 (1987).

39. C. Thomson, D. Higgins and J. Wilkie, work in progress.

40. J. Barker and C. Thomson, work in progress.

A CELL CULTURE MODEL FOR STUDY OF CHEMICALLY-DERIVED AND SPONTANEOUS MOUSE

LUNG ALVEOLOGENIC CARCINOMA

Garry J. Smith[1], Jacqueline M. Bentel[1], Tarja A. Savolainen[1],
John G. Steel[2] and Christine K. Loo[1]

[1]Carcinogenesis Research Unit, School of Pathology, University
of New South Wales, P. O. Box 1, Kensington 2033, Australia
and [2]C.S.I.R.O. Division of Molecular Biology, P. O. Box 184
North Ryde, 2113, Australia

INTRODUCTION

Lung adenomas are among the most common tumors of mice and may occur spontaneously or be induced by carcinogens such as urethane[1]. A proportion of the benign adenomas reportedly progress to adenocarcinomas[1]. The alveolar adenoma is believed to derive from an epithelial cell of lung alveolus, the type II pneumocyte.

RESULTS AND DISCUSSION

An epithelial cell type related to the type II pneumocyte has been established in culture (Fig. 1). This cell strain, designated NAL1A, from normal adult mouse lung, is type II pneumocyte-related based upon ultrastructural observation of lamellar structures, analysis of phospholipid profiles, and identification with a type II pneumocyte-specific polyclonal antiserum[2]. In-

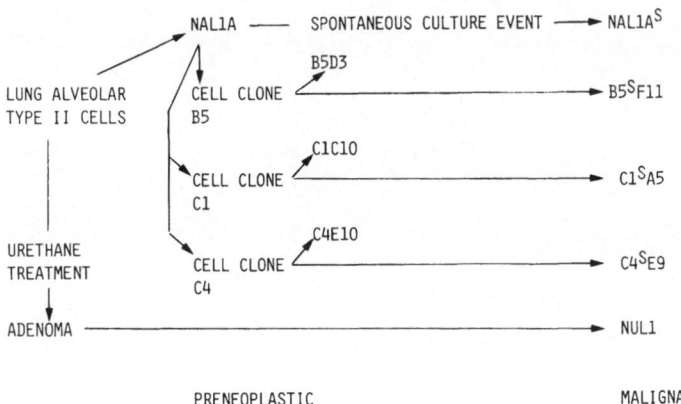

Fig. 1. Derivation of type II pneumocyte-related and urethane-induced adenoma cell strains.

dependent cultures of a malignant cell strain have been derived from urethane-derived mouse lung adenomas (Fig. 1) which have been established in culture[3]. The adenoma-derived strain is malignant by the criteria of anchorage-independent growth[4], and subcutaneous formation of invasive poorly differentiated carcinomas when implanted in mice (Fig. 2).

Cloned cell lines of NAL1A, namely clones B5, C1 and C4, spontaneously generated morphological variant cells similar to NUL1. A secondary round of cloning enabled rescue of "sibling" pairs of non-malignant and morphological variant cell lines from these clones. These non-malignant and variant clones were designated, respectively, B5D3 and B5^{S}F11, C1C10 and C1^{S}A5, C4E10 and C4^{S}E9 (Fig. 1). The spontaneous morphological variant cell clones have exhibited malignant characteristics similar to the chemical tumor-derived cell strain NUL1.

The spontaneously malignant and tumor-derived cell strains have exhibited a number of changes in cell membrane-associated features which correlate with increased malignant properties. These include a reduction in EGF receptor activity[4] and decreased pericellular fibronectin (Fig. 3). The synthesis and secretion of fibronectin, determined by [^{35}S] Methionine incorporation and precipitation with a fibronectin-specific polyclonal antiserum is absent in both malignant cell types (Steele et al., submitted for publication). A tight correlation has been observed between onset of malignant phenotype and loss of fibronectin protein synthesis and pericellular secretion.

An analysis of the molecular basis for the loss of fibronectin expression in these minimal deviation variants may help to elucidate the molecular events underlying onset of the malignant phenotype. Comparison of such molecular events in the chemically derived versus spontaneously generated malignant cells described above may provide information on the mechanism of chemical carcinogenesis in the mouse lung adenoma system. The similarities between the cells generated by the chemical and spontaneous mechanisms sug-

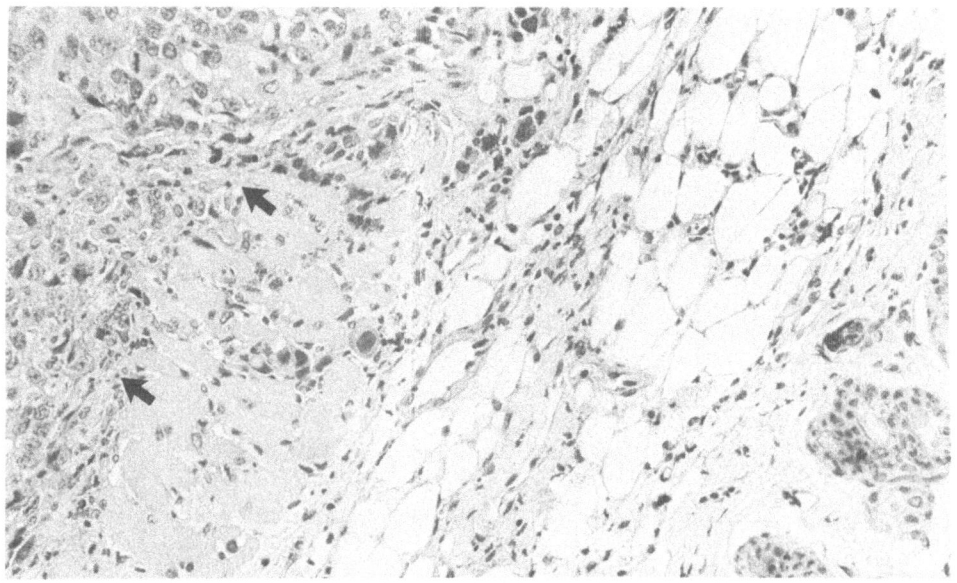

Fig. 2. An invasive poorly differentiated carcinoma (arrowed) formed subcutaneously in a BALB/c mouse 6 weeks after injection of 10^6 cells of B5^{S}F11.

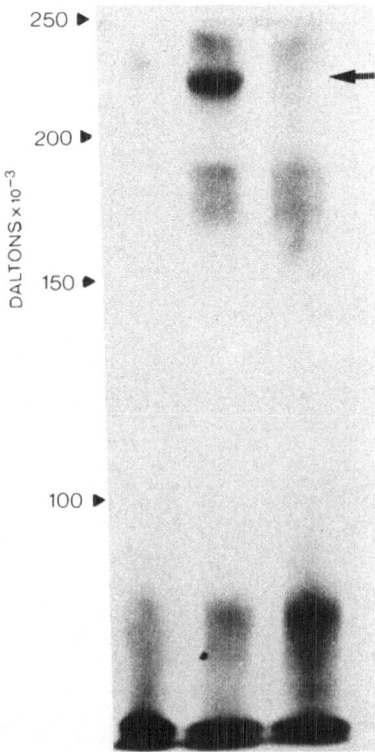

Fig. 3. [^{125}I] labeled cell surface fibronectin (arrowed) is present in non-malignant (1) and absent in malignant (2) cell strains as observed by PAGE.

gests that similar molecular events may underly the two phenomena of neoplastic transformation.

ACKNOWLEDGEMENTS

This work was supported by grants from The New South Wales State Cancer Council, National Health and Medical Research Council and University of New South Wales - C.S.I.R.O.

REFERENCES

1. M. B. Shimkin and G. D. Stoner, Lung tumours in mice: Application to carcinogenesis bioassay, Adv. Cancer Res. 21:1 (1975).
2. G. J. Smith, R. K. Kumar, C. P. Hristoforidis and A. W. J. Lykke, Expression of a type 2 pneumocyte-specific antigen by a cell strain from normal adult mouse lung, Cell Biol. Int. Reports 9:1115 (1985).
3. G. J. Smith and J. M. Bentel, Malignant epithelial cell strains cultured from BALB/c mouse lung adenoma, Cell Biol. Int. Reports 11:111 (1987).
4. G. J. Smith, F. A. Bennet, J. G. Steele and J. M. Bentel, Onset of neoplastic phenotype in an epithelial cell strain from adult BALB/c mouse lung alveolus, J. Natl. Cancer Inst. 76:73 (1986).

STUDIES ON METABOLIC PERTURBATIONS AND TUMOR PROMOTION: GLYCINE INDUCES THE SYNTHESIS OF OROTIC ACID, A TUMOR PROMOTER AND THE EXPRESSION OF CERTAIN CELL CYCLE DEPENDENT GENES

Shanti Vasudevan, George Lee, Ijaz A. Qureshi[1], Prema M. Rao,
Srinivasan Rajalakshmi and Dittakavi S. R. Sarma

Department of Pathology, Medical Sciences Building
University of Toronto, Toronto, Ontario, Canada M5S 1A8
and [1]Department of Pediatrics, Hospital Sainte-Justine
University of Montreal, 3175 Chemin Cote Sainte-Catherine
Montreal, Quebec, Canada H3T 1C5

Recently we have demonstrated that orotic acid (OA), a normal cellular constituent is an efficient promoter of liver carcinogenesis in the rat[1-4]. Being a precursor of pyrimidine nucleotides, feeding OA results in an increased synthesis of uridine nucleotides in the liver. This increase in uridine nucleotides is associated with a decrease in adenine nucleotides thereby creating an imbalance in the nucleotide pool. It was hypothesized that creation of such an imbalance in the nucleotide pool is an important factor in the mechanism by which OA exerts its promoting effect[4,5]. One of the significant aspects of this hypothesis is that since each cell has its own nucleotide pool pattern geared to its needs, disruption of this pattern could be one mechanism to achieve promotion in a variety of organs. Indeed OA has been shown to promote carcinogenesis in the duodenum initiated by azoxymethane[6].

If the creation of an imbalance in nucleotide pool by increased OA synthesis is a key factor in promotion, then other conditions which induce synthesis of OA need to be explored for their promoting ability. Among these will be included (I) dietary manipulations like feeding excess amino acids or proteins or feeding diets deficient in arginine, (II) metabolic and genetic disorders, particularly of the urea cycle, e.g. deficiency of ornithine transcarbamylase (OTC) and arginase, defects in ornithine transport into mitochondria, etc. We have found that feeding rats a diet deficient in arginine results not only in an increased synthesis of OA but also a nucleotide pool pattern in the liver similar to that found in rats fed OA, i.e., increased uridine nucleotides and decreased adenine nucleotides[4,7]. Furthermore, such a dietary regimen also had a promoting effect in carcinogen initiated rats[4]. It is of interest to note that patients with hypertyrosinemia exhibit a high incidence of hepatocellular carcinoma[8]. Whether orotic acid or altered nucleotide pools play any role in the pathogenesis of cancer development in these patients is yet to be determined. Similarly the sparse fur (spf) mutant male mouse which is characterized by an over 95% deficiency in OTC, synthesizes and excretes large amounts of OA[9]. Here again we have found that there is an increased synthesis of uridine nucleotides and an imbalance in hepatic nucleotide pool, a pattern similar to that encountered in rats fed orotic acid (Fig. 1). Whether these conditions have a promoting effect in carcinogen initiated mice is currently under investigation. The

317

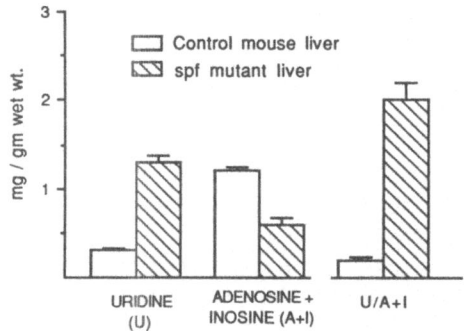

Fig. 1. Pyrimidine and purine ribonucleosides/ribonucleotides in sparse fur (spf) mutant male mouse liver. Immediately after sacrifice the liver tissue was quickly frozen in liquid nitrogen. A 0.6 M perchloric acid soluble nucleoside/nucleotide pool was prepared by the method of Brown et al.[20] and converted to nucleosides by phosphodiesterase and alkaline phosphatase[21]. The ribonucleosides were separated from deoxyribonucleosides using a boronate affigel 601 column[22]. The ribonucleosides of cytosine, uracil (U), adenine (A), inosine (I) and guanine were separated using high pressure liquid chromatography on a μbondapak C_{18} column (Waters) using a 0.01 M ammonium phosphate buffer pH 5.1. For simplicity sake only the values for U, A and I are presented in the figure. The values are the average of 7 spf mice and 4 control mice $\pm$ S.E.

foregoing emphasizes the need to examine the nucleotide pool patterns generated in other genetic disorders leading to orotic aciduria and determine whether these genetic disorders increase the risk of tumor promotion. Also of importance would be other metabolic disorders like folic acid deficiency wherein high levels of dUMP are found as a consequence of the decreased activity of dihydrofolate reductase[10]. Although no data on promotion are available, evidence has been presented which suggests that folic acid deficiency is associated with cell transformation in vitro[10].

MECHANISM OF INDUCTION OF OA SYNTHESIS BY AMINO ACIDS

The finding that a simple dietary manipulation like feeding excess amino acids can endogenously generate a potential multiorgan tumor promoter like OA prompted our investigations into the mechanisms involved in the induction of OA synthesis. It has been hypothesized earlier that the increased ammonia formed from excess amino acids (or proteins) causes an increase in the mitochondrial carbomoylphosphate (CP) levels which surpasses the capacity of the urea cycles system. The excess CP then leaks out into the cytosol where it is converted into OA (Fig. 2). Our experiments to elucidate the mechanisms involved in the stimulation of OA synthesis by amino acids have revealed that, in both rats and mice, the induction is a transcription and translation dependent event[11]. Thus, administration of excess glycine to rats caused a stimulation of OA synthesis which was inhibited by both actinomycin D and cycloheximide[11]. The inhibitor sensitive step appears to be the induction of ornithine decarboxylase (ODC), an enzyme which metabolizes ornithine and thereby depletes its reserves[12]. Since ornithine plays a pivotal role in the utilization of mitochondrial CP for urea synthesis, a decrease in its levels limits the availability of this key intermediate for urea synthesis. Although administration of glycine increases the levels of both CP and ornithine, the simultaneous induction of ODC alters the ratio of CP to ornithine in favour of CP. Under these conditions the excess mitochondrial CP becomes

318

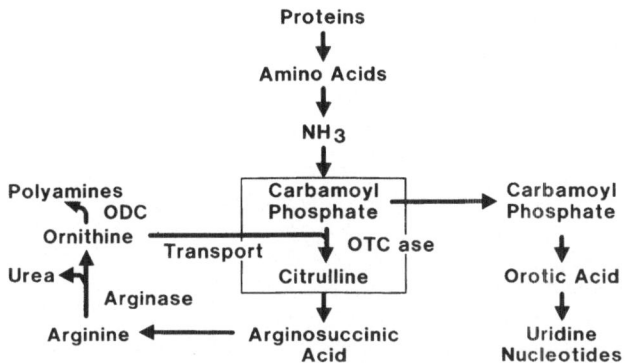

Fig. 2. Schematic representation of the synthesis of orotic acid from excess amino acid.

available for the synthesis of OA. Inhibition of induction of ODC by cyclo-heximide also shuts off the amino acid stimulated OA synthesis.

INDUCTION OF CELL CYCLE DEPENDENT GENES BY GLYCINE

The induction of ODC and OA by glycine brings into focus another aspect of the problem. ODC is an enzyme normally induced during cell proliferation and is involved in the biogenesis of polyamines which are important for DNA synthesis. The question therefore arises whether glycine also induces the expression of other genes that are normally activated during the cell cycle. Our results indicate that administration of glycine induces not only the synthesis of ODC but also the expression of some other cell cycle dependent genes viz., c-fos, c-myc and c-Ha-ras (Table 1). The kinetics of expression of all these genes is similar. Their mRNAs can be detected as early as 20 min after injection of glycine, peak at 1 h and decline by 2 h indicating the transient nature of the induction (data not shown). However, there was no induction of the expression of ribonucleotide reductase, another gene function associated with DNA synthesis (Table 1).

Despite induction of some cell cycle dependent genes, glycine did not induce DNA synthesis (data not presented). In this context, a point that merits consideration is the temporal sequence of expression of c-fos, c-myc, c-ras and ODC during cell cycle. Both in vivo and in vitro, cells respond to growth stimuli by expressing these genes in a temporal sequence[13-17]. After glycine administration however, there is no temporal sequence in their expression. One wonders whether sequential expression of these genes rather than expression per se is necessary for DNA synthesis.

This observation is provocative. For example proteins have been shown to induce DNA synthesis in hepatocytes and liver cell proliferation in the rat in vivo[18]. However, the role of individual amino acids in the induction of cell cycle is not known. Do they merely serve as building blocks for the increased demand in protein synthesis? Or do they in addition, participate in the process by inducing the expression of different cell cycle dependent genes? The observation that glycine induces the expression of c-fos, c-myc, c-ras and ODC but not ribonucleotide reductase and DNA synthesis lends support to the thesis that different amino acids can induce the expression of different genes that are implicated in cell cycle. This leads to the suggestion that amino acids either singly or in mixtures serve as transducer molecules for signals affecting cell proliferation. Induction of ODC by growth factors has already been shown to be mediated by amino acids[19].

319

Table 1. Induction of the expression of cell cycle dependent genes by glycine

Genes probed	Fold Induction
c-myc	44
c-fos	12
c-Ha-ras	2.6
Ornithine decarboxylase	3.0
Ribonucleotide reductase	1.0

Rats were injected with glycine (2.5 mmoles/100 g i.p.) or with an equivalent volume of 0.9% NaCl and killed 1 h after the injection. Livers were removed and total RNA was extracted[23] from pooled livers of 3 rats sacrificed per time point. Poly(A$^+$) RNA was isolated by two cycles of oligo (dT) cellulose[24]. Ten μg of poly(A$^+$) RNA from each time point were fractionated on 1% agarose gels[25], transferred to gene screen plus (New England Nuclear) and immobilized with UV[26]. Various cDNA probes were labelled with ^{32}P-dCTP by random priming[27]. Blots were hybridized at 65°C in the absence of formamide and washed in 2 x SSC at 65°C and the bands were revealed by autoradiography[28]. Probes for v-fos, c-myc and v-Ha-ras were purchased from Oncor Inc., Maryland. Ribonucleotide reductase probe was a generous gift from Dr. William Lewis and Mr. Enrico Wensing, Department of Microbiology, University of Toronto. Quantitations were done using videodensitometric scanning technique. The values represent fold induction of expression of specific polyA RNA compared to 0.9% NaCl injection.

The results presented here suggest that certain metabolic perturbations can induce not only the synthesis of orotic acid, a tumor promoter but also the expression of some cell cycle dependent genes. The significance of these observations in terms of tumor promotion needs further study.

ACKNOWLEDGEMENTS

We wish to thank Ayesha Alam for her excellent technical assistance and Lori Cutler for her able secretarial assistance. The study was supported in part by U.S. P.H.S. Grants CA37077 and CA 23958 from the National Cancer Institute, by grants from National Cancer Institute, Canada and by grants from Medical Research Council (MA9124 awarded to I.A.Q.). S.V. and G.L. are supported by Ontario Graduate Scholarships.

REFERENCES

1. A. Columbano, G. M. Ledda, P. M. Rao, S. Rajalakshmi and D. S. R. Sarma, Dietary orotic acid, a new selective growth stimulus for carcinogen altered hepatocytes in rat, Cancer Letters 16:191 (1982).
2. P. M. Rao, K. Nagamine, R. K. Ho, M. W. Roomi, C. Laurier, S. Rajalakshmi and D. S. R. Sarma, Dietary orotic acid enhances the incidence of γ-glutamyltransferase positive foci in rat liver induced by chemical carcinogens, Carcinogenesis 4:1541 (1983).
3. C. Laurier, M. Tatematsu, P. M. Rao, S. Rajalakshmi and D. S. R. Sarma, Promotion by orotic acid of liver carcinogenesis in rats initiated by 1,2-dimethylhydrazine, Cancer Res. 44:2186 (1984).
4. P. M. Rao, Y. Nagamine, M. W. Roomi, S. Rajalakshmi and D. S. R. Sarma,

Orotic acid, a new promoter for experimental liver carcinogenesis, Toxicol. Pathol. 12:173 (1984).

5. P. M. Rao, S. Vasudevan, E. Laconi, S. Rajalakshmi and D. S. R. Sarma, Imbalances of pyrimidine and purine nucleotides and their implications in liver carcinogenesis, in: "Nutritional Diseases: Research Directions in Comparative Pathobiology", D. Scarpeli and G. Migaki, eds., Alan R. Liss. Inc., New York (1986).

6. P. M. Rao, E. Laconi, S. Rajalakshmi and D. S. R. Sarma, Orotic acid, a liver tumor promoter also promotes carcinogenesis in the intestine, Proc. Am. Assoc. Cancer Res. 27:142 (1986).

7. J. A. Milner, Mechanism for fatty liver induction in rats fed arginine deficient diets, J. Nutr. 109:663 (1979).

8. A. G. Weinberg, C. E. Mize and H. G. Worthem, The occurrence of hepatoma in the chronic form of hereditary hypertyrosinemia, J. Pediatrics 88:434 (1976).

9. I. A. Qureshi, J. Letarte and R. Ouellet, Activities of orotate metabolizing enzyme complex and various urea cycle enzymes in mutant mice whith ornithine transcarbamylase deficiency, Experientia 38:308 (1982).

10. C. L. Krumdieck, Role of folate deficiency in carcinogenesis, in: "Nutrition Factors in the Induction and Maintenance of Malignancy", C. S. Butteworth and M. L. Hutchinson, eds., Bristol Myers Nutrition Symposium 2, Academic Press, New York (1983).

11. S. Vasudevan, E. Laconi, S. E. Abanobi, P. M. Rao, S. Rajalakshmi and D. S. R. Sarma, Regulation of synthesis of orotic acid (OA), a liver tumor promoter, induced by amino acids, Fed. Proc. 44:1339 (1985).

12. S. Vasudevan, E. Laconi, P. M. Rao, S. Rajalakshmi and D. S. R. Sarma, Interrelationship between ornithine decarboxylase (ODC) and orotic acid (OA) synthesis induced by amino acids, Fed. Proc. 46:753 (1987).

13. L. Kaczmarek, Biology of diseases: protooncogene expression during cell cycle, Lab. Invest. 54:365 (1986).

14. N. L. Thompson, J. Z. Mead, L. Braun, M. Goyette, P. R. Shank and N. Fausto, Sequential protooncogene expression during rat liver regeneration, Cancer Res. 46:3111 (1986).

15. H. M. Lachman, K. S. Matton, E. L. Skoultchi and C. L. Schildkraut, c-myc mRNA levels in the cell cycle change in mouse erythroleukemia cells following inducer treatment, Proc. Natl. Acad. Sci. USA 82:5323 (1985).

16. D. T. Denhardt, D. R. Edwards and C. L. J. Parfelt, Gene expression during mammalian cell cycle, Biochim. Biophys. Acta 865:83 (1986).

17. R. Baserga, "The Biology of Cell Reproduction", Harvard University Press, Cambridge, MA, (1985).

18. J. Short, N. B. Armstrong, M. A. Kolitsky, R. A. Mitchell, R. Zemel and I. Liberman, Amino acids and the control of nuclear DNA replication in the liver, in: "Control of Proliferation in Animal Cells", B. Clarkson and R. Baserga, eds., Cold Spring Harbor Laboratory, Cold Spring Harbor, N.Y. (1974).

19. C. A. Reinhart, Jr. and E. S. Canellakis, Induction of ornithine decarboxylase activity by insulin and growth factors is mediated by amino acids, Proc. Natl. Acad. Sci. USA 82:4365 (1985).

20. E. G. Brown, R. P. Newton and N. M. Shaw, Analysis of free nucleotide pools of mammalian tissues by high pressure liquid chromatography, Analyt. Biochem. 123:378 (1982).

21. D. H. Swenson and P. D. Lawley, Alkylation of deoxyribonucleic acid by carcinogens dimethylsulphate, ethylmethanesulphonate, N-ethyl-N-nitrosourea and N-methyl-N-nitrosourea. Relative reactivity of the phosphodiester site thymidyl (3'-5')thymidine, Biochem. J. 171:575 (1978).

22. S. M. Payne and B. N. Ames, A procedure for rapid extradtion and high pressure liquid chromatographic separation of the nucleotides and

other small molecules from bacterial cells, <u>Analyt. Biochem.</u> 123:151 (1982).

23. J. M. Chigwin, R. J. Przybyla, T. Macdonald and W. J. Rutter, Isolation of biologically active ribonucleic acid from sources enriched in ribonuclease, <u>Biochemistry</u> 18:5294 (1979).

24. H. Aviv and P. Leder, Purification of biologically active globin messenger RNA by chromatography on oligothymidylic acid cellulose, <u>Proc. Natl. Acad. Sci.</u> USA 69:1408 (1972).

25. T. Maniatis, E. F. Fritisch and J. Sambrook, "Molecular Cloning: A Laboratory Manual", Cold Spring Harbor Laboratory, Cold Spring Harbor, N.Y. (1982).

26. E. W. Khandjian, UV cross linking of RNA to nylon membrane enhances hybridization signals, <u>Mol. Biol. Reports</u> 11:107 (1986).

27. A. Feinberg and B. Vogelstein, A technique for radiolabilling DNA restriction endonuclease fragments to high specific activity, <u>Analyt. Biochem.</u> 132:6 (1984).

28. New England Nuclear, Boston. "Gene Screen Instruction Manual", Catalogue No. NEF-972.

COMPARATIVE STUDY ON THE EFFECT OF DIFFERENT TREATMENT SCHEDULES ON SOME CARBOHYDRATE METABOLIZING ENZYME ACTIVITIES IN RATS DURING HEPATOCARCINOGENESIS

Ulrich Gerbracht[1], Günter Weiße[2], Bernd Schlatterer[3], Manfred Reinacher[4], Rolf Schulte-Hermann[5] and Erich Eigenbrodt[1]

[1]Institut für Biochemie und Endokrinologie, Justus-Liebig-Universität Giessen, FRG. [2]Institut für Toxicologie, E. Merck Darmstadt, FRG. [3]Umweltbundesamt, Berlin, FRG. [4]Institut für Veterinar-Pathologie, Justus-Liebig-Universität Giessen, FRG. [5]Institut für Tumorbiologie-Krebsforschung, Universität Wien Austria

INTRODUCTION

Phenobarbital (PB) is a well known tumor promoter, which in combination with an initiating drug is able to accelerate the development of γ-glutamyl-transferase (γ-GT)-positive foci and to induce tumor formation after long term feeding[1-4]. PB is also found to stimulate liver growth[1,3,5]. The hypolipidaemic drugs nafenopin and clofibrate induce liver tumors when fed to rats or mice[6-9]. They do not, however, stimulate growth of γ-GT-positive foci[3,10,11]. Recently we have shown some alterations in carbohydrate metabolizing enzymes in hepatocarcinomas induced with NNM and promoted by long term feeding with phenobarbital or clofibrate[12]. Pyruvate Kinase (PK) and fructose-1,6-biphosphatase (FBPase) activities were reduced and malic enzyme activity was increased in these rat liver tumors with and without PB or clofibrate feeding. The activity of γ-GT, however, was dependent on the applicated drugs since only PB induced γ-GT-positive hepatocarcinomas. In this study activities of some enzymes were recorded after short term feeding of PB, clofibrate (Clof) or nafenopin (Naf) for 6 and 16 weeks. The results were compared with data obtained from long term application of these compounds for 64 weeks. Pyruvate kinase isoenzyme type L (LPK), malic enzyme (ME) and γ-glutamyltransferase were also investigated by immunohistological and histochemical methods.

MATERIALS AND METHODS

Female Wistar rats, 4 or 16 weeks of age and weighing 50 or 160 g, respectively, were obtained from Zentralinstitut für Tierzucht, Hannover, FRG. They were kept in air conditioned rooms and received food and water ad libitum. In experiment 1 the rats weighing 50 g were treated with a single dose of methylnitrosourea (MNU) dissolved in water and applied by stomach tube in a concentration of 60 mg/kg body weight. The day before administration of MNU rats were hepatectomized (2/3) under ether narcosis. Beginning one day after MNU application animals received PB (50 mg/kg b. wt./day),

---

*This paper is a part of the doctoral thesis of U. Gerbracht.

Naf (50 mg/kg b. wt./day) or Clof (250 mg/kg b. wt./day) with the food. Control rats were either untreated or pretreated only with the initiator MNU. The animals were killed after 16 weeks.

In experiment 2 rats were treated with a single dose of aflatoxin B_1 (Afl), dissolved in olive oil and administered by stomach tube, in a concentration of 5 mg/kg b. wt.. The control group received olive oil without aflatoxin B_1. Beginning 3 weeks later the rats received PB (50 mg/kg b. wt./day) or Naf (100 mg/kg b. wt./day) with the food. The animals were killed 6, 16 or 64 weeks after initiation.

For histology liver slices of approximately 5 mm thickness were fixed in ice cold acetone. Following embedding in Paraplast[R] histological sections of 5 µm were cut. The γ-GT reaction was demonstrated by the method of Rutenburg et al.[13]. Some liver slices of approximately 3 mm thickness were frozen directly at -150°C in isopentane and stored at -70°C. Cryostat sections were then cut at a thickness of about 5 µm and stained consecutively for ME and LPK. ME was demonstrated according to Lojda et al.[14]. The demonstration of LPK type L with the peroxidase antiperoxidase (PAP) method was done as described earlier[15].

γ-GT-positive liver foci were counted and the areas of these foci were measured by a Kontron-MOP-Videoplan. The area of the liver section was determined by using a planimeter.

For biochemical investigations residual liver slices were stored at -70°C. The livers were homogenized with a teflon pestle in a buffer containing 50 mM Tris/Cl (pH 7.2), 10 mM mercaptoethanol, 2 mM 5-aminohexanid (PMSF) and 1 mM EDTA. The pH was adjusted to 7.2 with HCl and the homogenate kept at 4°C. The homogenate was centrifuged at 50,000 g for 20 minutes and the supernatant was used for enzyme assays. In the case of the two membrane bound enzymes γ-GT and alkaline phosphatase the homogenates were centrifuged in a minifuge after addition of 1% Triton X-100. All enzyme activities were measured at 37°C in an automatical photometer (Eppendorf ACP 5040).

The activity of PK (EC 2.7.1.40) was measured by the method of Eigenbrodt and Schoner[16], FBPase (EC 3.1.3.11) according to McPherson[17], malic enzyme (EC 1.1.1.40) in accordance with Zelewski[18], glucose-6-phosphate dehydrogenase (G6PDH) (EC 1.1.1.49) as described by Lohr and Waller[19], enolase (EC 4.2.1.11) was measured by the method of Bergmeyer[20], and lactate dehydrogenase (LDH) (EC 1.1.1.27) according to Bergmeyer[21]. The activity of γ-glutamyltransferase (EC 2.3.2.2) was determined according to Persijn[22] and of alkaline phosphatase (AP) (EC 3.1.3.1) following the method of the "Optimierte Standardmethode" (DGKCH) Test combination of Boehringer, Mannheim, FRG[23].

Statistical Methods

All values in the tables are mean ± standard deviation. Those significantly different from the appropriate control group by Student's t-test are marked as follows: (*) = P < 0.05; (**) = P < 0.01; (***) = P < 0.001.

RESULTS

Animal Weights and Liver Growth

No difference in body weights between all animal groups were found (Table 1). The same Table also illustrates the change in the ratio of liver weight to body weight. This ratio showed a progressive increase in Naf and PB treated rats. Liver growth was only slightly induced with Clof (Table 1).

<u>γ-Glutamyltransferase Histochemistry</u>

Table 1 shows the appearance and growth of γ-GT-expressing liver foci after treatment with PB, Naf and Clof. PB given after MNU accelerated the occurrence of γ-GT foci and led to an induction of foci growth. Naf had no effect on the appearance of γ-GT-positive foci. A slight increase in foci area was found when Clof was fed subsequently to MNU. Both hypolipidaemic drugs did not increase the number of γ-GT-positive liver foci (Table 1).

<u>Pyruvate Kinase Isoenzyme Type L Immunohistochemistry</u>

Immunohistological investigation revealed that Naf leads to a strong diffuse decrease in LPK staining (Fig. 1B). In contrast to the strong decrease with Naf, PB reduced LPK mainly in liver foci (Fig. 1C).

<u>Malic Enzyme</u>

Naf caused an increased histochemical staining for ME activity in the liver sections mainly the periveneous zone. The expression of this enzyme was less intense when PB was applied and was even lower in the group treated only with MNU (Fig. 2B). Preneoplastic liver foci occurring in the PB group were excessively stained by ME reaction (Fig. 2C).

<u>Biochemical Determination of Enzyme Activities</u>

A strong decrease in the activity of PK was recorded in the Naf treated groups. The two other compounds, PB and Clof also decreased PK activity significantly but to a lesser degree than did Naf. The strongest inhibitory effect on FBPase activity was found with Naf but was also significant after feeding with PB. Clof application resulted only in a slight reduction in FBPase activity (Table 2).

PB, Naf and Clof all induced the activity of ME. The most effective substance was Naf which increased the activity to about 800%. Both PB and Clof raised ME activity to about 150% (Table 2).

The activity of G6PDH was significantly elevated in rats fed Naf but was not altered in other groups (Table 2).

LDH was induced after exposure to the hyplipidaemic drugs and was strongly decreased after administration of PB.

While PB increased γ-GT and decreased AP activity, Naf strongly induced AP but reduced γ-GT activity. The inhibitory effect upon γ-GT was also present after feeding of Clof. Naf given without prior initiation caused quantitatively the same alterations as did Naf when applied in combination with MNU. Similar data were recorded for Naf treatment after administration of Afl (Table 3).

In experiment 2 (time course of alterations after initiation by Afl) rats were kept for up to 64 weeks to study the effect of prolonged treatment of these drugs. Table 3 summarizes the results. Afl initiation led to a permanent increase in ME (Fig. 3), G6PDH, γ-GT and AP. PB and Naf decreased PK, enolase and FBPase but increased the activity of ME to a higher degree than Afl alone did. In addition, a strong induction of G6PDH, LDH and AP was observed in Naf exposed rats, whereas PB reduced activity of LDH. As to be expected, PB increased γ-GT activity. Reduction of PK and LDH and stimulation of G6PDH and γ-GT activities were more pronounced after longer application of PB, resulting in significant differences of these activities between the animals investigated after 6 weeks and those sacrificed after 64 weeks. (Fig. 3, Table 3).

Table 1. Body weight, liver weight, liver weight in % of body weight, number and area of γ-GT-positive foci after different treatment schedules

Treatment schedule (sacrificed after 16 weeks)	Body weight (g)	Liver weight (g)	Liver weight % of b. wt.	Number of γ-GT foci/cm²	Area of γ-GT positive foci mm²/cm²
PH/O	2.37 ± 7.72	6.58 ± 0.50	2.74 ± 0.37	0.48 ± 0.67	0.014 ± 0.019
PH/Naf	231.1 ± 8.62	11.76 ± 1.77[a]***	5.02 ± 0.48[a]***	1.34 ± 1.71	0.0086 ± 0.0013
PH/MNU/O	240.1 ± 21.5	7.05 ± 0.71	2.94 ± 0.31	6.05 ± 2.07	0.0686 ± 0.0206
PH/MNU/PB	234.0 ± 18.7	8.17 ± 0.72[b]**	3.50 ± 0.25[b]***	28.70 ± 17.68	1.091 ± 1.397
PH/MNU/Naf	231.5 ± 25.2	10.84 ± 1.46[b]***	4.68 ± 0.35[b]***	6.41 ± 1.75	0.0655 ± 0.0365
PH/MNU/Clof	228.7 ± 17.9	7.24 ± 0.45	3.17 ± 0.18[b]*	7.76 ± 8.59	0.203 ± 0.328

Statistics: [a] significantly different from control group (PH/O); [b] significantly different from group pretreated with MNU (PH/MNU/O).

Table 2. Alterations of enzyme activities after 16 weeks dependent on different treatment schedules

Treatment schedule	PK U/g	FBPase U/g	ME U/g	G6PDH U/g	LDH U/g	Alk. Ph. mU/g	GGT mU/g
PH/O (5)	67.20 ± 16.66	10.11 ± 0.66	2.73 ± 0.54	4.53 ± 1.96	710 ± 81	430 ± 47	188.1 ± 224.4
PH/Naf (4)	35.71[a]*** ± 4.21	6.56[a]*** ± 0.45	24.96[a]*** ± 1.46	12.48[a]*** ± 2.59	914[a]** ± 127	2689[a]*** ± 875	34.84 ± 3.74
PH/MNU/O (16)	67.32 ± 12.02	9.24 ± 1.31	3.15 ± 1.19	5.02 ± 1.94	656 ± 121	603 ± 455	245.3 ± 153.4
PH/MNU/PB (7)	55.97[b]* ± 10.54	7.98[b]* ± 0.74	4.59[b]** ± 0.91	3.76 ± 0.61	518 ± 69	428 ± 101	424.5 ± 250.7
PH/MNU/Clof (5)	51.72[b]** ± 6.27	8.29 ± 1.85	4.83[b]** ± 1.47	4.65 ± 1.85	802 ± 50	639 ± 85	97.8 ± 80.0
PH/MNU/Naf (12)	34.44[b]*** ± 6.38	6.24[b]*** ± 0.98	25.50[b]*** ± 3.26	12.67[b]*** ± 3.98	905[b]** ± 101	2112[b]*** ± 511	55.73 ± 32.58

Statistics: [a] significantly different from the control group (PH/O); [b] significantly different from the group pretreated with MNU (PH/MNU/O).

Table 3. Time course of alterations of enzyme activities induced by different treatment schedules

		FBPase U/g	Enolase U/g	Alk. Ph. mU/g	LDH U/g	G6PDH U/g
0/0 (4)	6 W.	13.41 ± 1.46	35.07 ± 3.38	515.7 ± 40.0	823.7 ± 168.6	8.43 ± 4.23
Afla/O (4)		11.26[a]* ± 1.15	26.78[a]* ± 4.34	1456.0 ± 800.0	677.7[a]* ± 82.4	13.62 ± 5.24
Afla/PB (6)		8.87[b]** ± 1.15	28.05 ± 1.83	740.3 ± 136.1	608.2 ± 109.3	7.98 ± 1.96
Afla/Naf (6)		7.05[b]*** ± 1.29	22.92[b]*** ± 1.27	2457.0 ± 734.0	1122.9[b]*** ± 168.4	32.81[b]*** ± 7.52
0/0 (4)	16 W.	14.50 ± 1.16	40.31 ± 3.30	610.8 ± 72.2	947.6 ± 221.7	7.85 ± 4.29
Afla/O (4)		12.69[a]* ± 0.82	37.98 ± 2.45	685.0 ± 150.2	816.4 ± 91.1	11.92 ± 7.51
Afla/PB (10)		9.08[b]*** ± 1.01	30.58[b]*** ± 3.73	548.6 ± 160.3	472.9[b]*** ± 106.8	8.32 ± 4.42
Afla/Naf (10)		8.42[b]*** ± 0.66	25.66[b]*** ± 1.31	2931.0[b]*** ± 613.0	1204.2[b]*** ± 106.9	26.83[b]*** ± 7.01
0/0 (2)	64 W.	12.76	31.86	501.6	661.7	8.31
Afla/O (6)		12.66 ± 0.72	36.70 ± 5.03	959.6 ± 459.9	776.8 ± 190.0	12.67 ± 5.67
Afla/PB (4)		8.88[b]*** ± 0.88	31.14[b]* ± 2.96	895.6 ± 331.6	440.8[b]*** ± 52.3[c]*	12.58 ± 6.44
Afla/Naf (4)		8.96[b]*** ± 0.51	25.78[b]*** ± 2.74	3376.0[b]*** ± 413.0[c]*	1218.4[b]*** ± 71.5	25.94[b]*** ± 4.54

Statistics: [a] significantly different from the control group (0/0); [b] significantly different from the group pretreated with aflatoxin B (Afla/O); [c] significant difference between the groups receiving the respective drugs for 6 and 64 weeks.

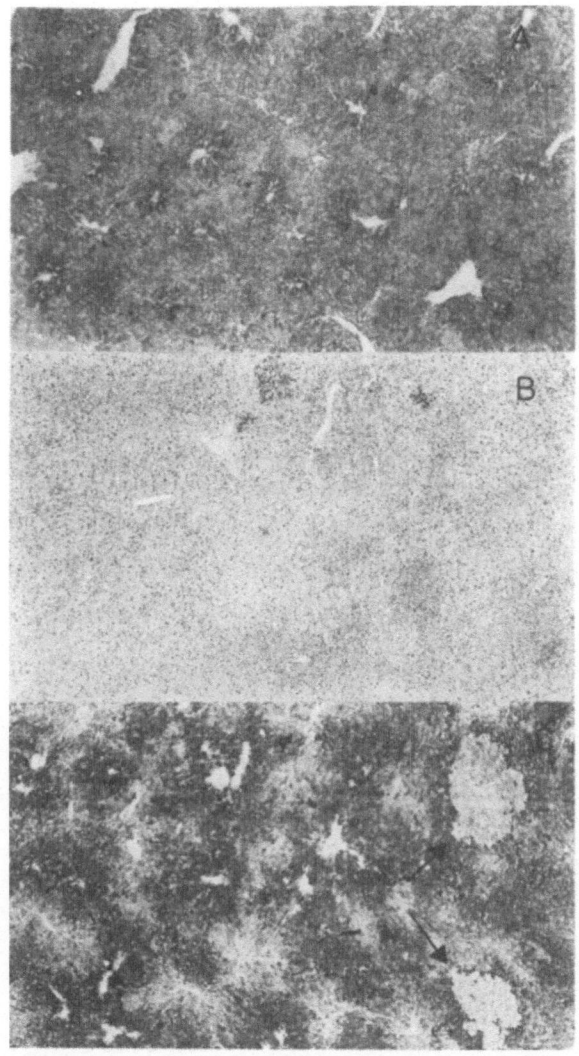

Fig. 1. Immunohistological demonstration of LPK in liver. LPK staining is strongly reduced in the whole liver after Naf feeding, whereas the treatment with MNU in combination with PB leads to LPK deficient foci. A) PH/MNU/O; B) PH/MNU/Naf; C) PH/MNU/PB.

DISCUSSION

PB, Naf and Clof are non-genotoxic compounds leading to an acceleration of tumor development in liver of rats when given in combination with a classical carcinogen[2,4,9]. Application of these non-genotoxic substances also causes liver growth (Table 1)[1,3,5,6]. Hepatomegaly induced with Naf and Clof is known to be due to both hyperplasia and hypertrophy[6]. This growth response is apparently different from normal growth or from liver growth occurring during regeneration because it is associated with increases in specific hepatic functions such as drug metabolism and peroxisome proliferation[1].

Feeding PB resulted in an appearance of γ-GT-positive foci which are known to present before hepatocarcinomas developed (Table 1)[2,4]. The

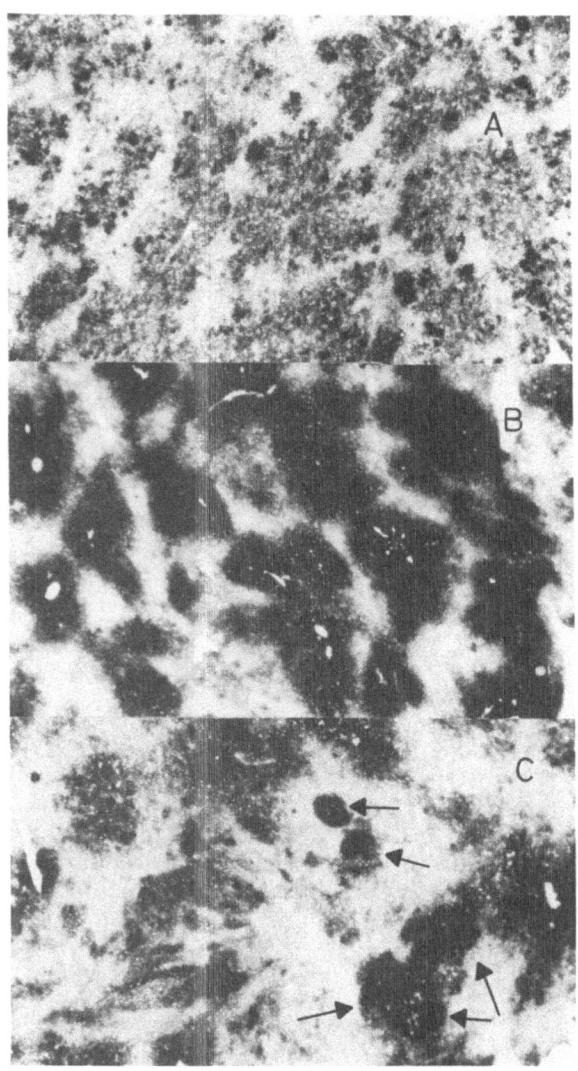

Fig. 2. Histochemical demonstration of NADP dependent ME in liver sections. Naf caused an increased histochemical staining for ME activity in the periveneous zone, whereas treatment with PB leads to ME foci (indicated by arrow).

inability of Naf to induce growth of γ-GT-positive liver foci was reported by other authors [10], too, and is in accordance with our own data [11,12]. Di(2-ethylhexyl)phthalate and WY 14,643, two other potent peroxisome proliferation inducing compounds, showed also carcinogenic effects [6,24,25] and led to a suppression of γ-GT in liver foci [26,27]. Clof was also found to have a suppressive effect on γ-GT activity in biochemical assay, although this compound is able to stimulate slightly γ-GT-positive foci (Tables 1-3). The hypolipidaemic effect of Naf is about 5 times as strong as that of Clof [28], suggesting that suppression of γ-GT activity in homogenates as well as in focal lesions might be related to the pharmacological potency of peroxisome proliferators.

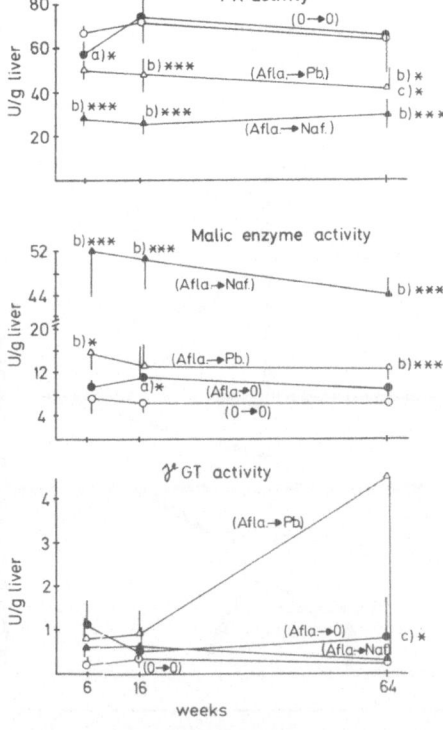

Fig. 3. Activity after 6, 16 and 64 weeks of treatment with Afla/0 (●—●),
Afla/PB (△—△) and Afla/Naf (▲—▲) of NADP dependent ME, PK and
γ-GT. a) Significantly different from control group (○—○). b) Significantly different from group pretreated with Afl. c) Significant
difference between the groups receiving the respective drugs for 6
and 64 weeks.

The inhibitory effect of Naf on γ-GT activity, in focal lesions as well
as in liver homogenate, was also observed in rats initiated by N-2-fluorenyl-
acetamide (AAF), whereas Clof decreased this activity only in liver homogenates[3]. In accordance with our data, this study describes also a strong
induction of AP after feeding of Naf but not after Clof administration[3].

The comparison of carbohydrate metabolizing enzyme activities in rats
treated with PB, Naf or Clof subsequently to an initiation with MNU or Afl
reveals that the application of each of these compounds resulted in a decrease of PK, FBPase and enolase activities but in an increase of ME activity. Recently we have shown that the same alterations also occur in hepatocarcinomas of rats induced with NNM and followed by PB or Clof treatment[12].
In this previous study an interesting finding was observed with PK : Clof
inhibited the activity of PK with and without initiation by NNM to nearly the
same degree. In contrast PB showed a strong suppression of PK activity only
when combined with NNM administration[12]. These data are in accordance with
those presented in this study. While Naf led to an immediate and permanent
decrease in PK, the inhibitory effect of PB was more pronounced with prolonged application (Fig. 3). The latter effect may be due to occurrence of
more and larger L-PK-negative foci[29] during the time course of PB administration. A reduction of PK in nodules induced by DEN and followed by a selection procedure with AAF and CCl₄ was also found[30] and confirm our data.

Preneoplastic Hepatocyte

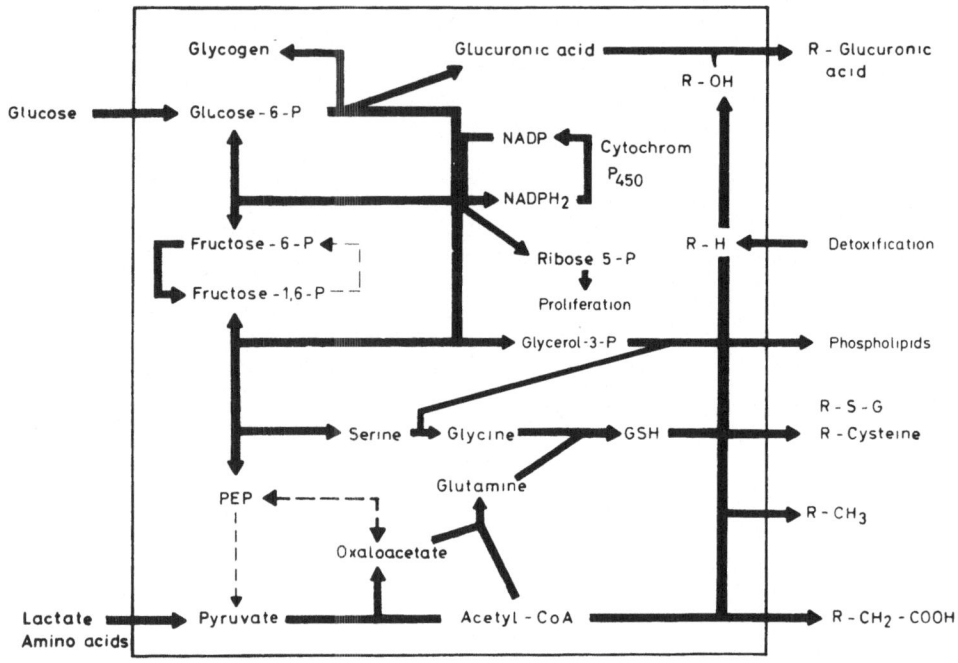

Hepatocytes

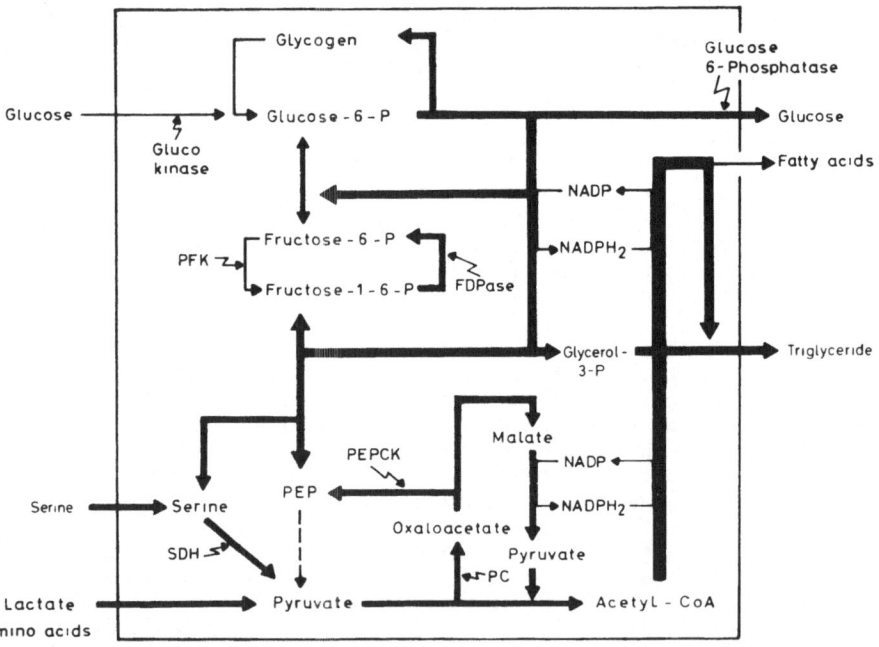

Fig. 4. Diagrammatic representation of carbohydrate enzyme activities and
fluxes of carbohydrate metabolites in normal and preneoplastic
hepatocytes.

In our study an increase of G6PDH activity was observed with prolonged PB feeding whereas Naf induced activity within 6 weeks and no further increase was reached after treatment for 64 weeks. This is in accordance with the observation of other authors[32] that G6PDH activity increases in late but not in early liver foci.

Previously it was shown by using histochemical methods that the ME is strongly induced in preneoplastic foci[31], a result confirmed by our results of PB administration. Furthermore, Naf induced ME in the periveneous zones of the liver but no foci occurred.

Normal liver produces triglycerides and cholesterol, from glucose or amino acids, in well fed animals, and glucose, from amino acids or glycerol, in starved animals (Fig. 4). The measured enzyme alterations in preneoplastic foci should lead to changes in metabolite fluxes as indicated in Fig. 4. Glucose is channeled into the pentose-phosphate pathway to provide reducing potency (NADPH), needed for detoxification, and ribose-5-phosphate, supporting proliferative processes. The block in glycolysis at the level of pyruvate kinase leads to an accumulation of carbohydrate metabolites between glucose 6-phosphate and phosphoenol pyruvate, thus channeling increased amounts of glycolytic metabolites to serine and glycine syntheses[12]. These data support that demonstration of alterations in ME, PK and FBPase activities may be helpful to develop a short term screening assay for hepatic carcinogenes which also responds to hypolipidaemic drugs, which are negative in all other short term tests.

ACKNOWLEDGEMENTS

Supported by the Umweltbundesamt (10603060, E.E.) and Bundesministerium für Forschung und Technologie (CMT 32 A, E.E., M.R.).

REFERENCES

1. R. Schulte-Hermann, Tumor promotion in the liver, Arch. Toxicol. 57:147 (1985).
2. M. A. Pereira, S. L. Herren-Freund, A. L. Britt and M. M. Khouny, Effect of coadministration of phenobarbital sodium on NNM-induced GGT-positive foci and hepatocellular carcinoma in rats, JNCI 72:741 (1983).
3. S. Numoto, K. Furukawa, K. Furuya and M. Williams, Effects of hepato-carcinogenic peroxisome-proliferating hypolipidaemic agents Clofibrate and Nafenopin on the rat liver cell membrane enzymes γ-GT and alkaline phosphatase and on the early stages of liver carcinogenesis, Carcinogenesis 5:1603 (1984).
4. T. Kitagawa and H. Sugano, Enhancing effect of phenobarbital on the development of enzyme-altered islands and hepatocellular carcinomas initiated by 3'-Methyl-4-(dimethylamino)azobenzene or diethylnitrosamine, Gann 69:679 (1978).
5. R. Schulte-Hermann, Induction of liver growth by xenobiotic compounds and other stimuli, Crit. Rev. Toxicol. 3:97 (1974).
6. J. K. Reddy and N. D. Lalwai, Carcinogenesis by hepatic peroxisome proliferators: Evaluation of the risk of hypolipidaemic drugs and industrial plasticizers to humans, Crit. Rev. Toxicol. 12:1 (1983).
7. J. M. Reddy and M. S. Rao, Malignant tumors in rats fed nafenopin, a hepatic peroxisome proliferator, J. Natl. Cancer Inst. 59:1645 (1977).
8. J. K. Reddy and S. A. Qureshi, Tumorigenicity of the hypolipidaemic peroxisome proliferator ethyl-α-p-chlorophenoxyisobutyrate (clofibrate) in rats, Br. J. Cancer 40:476 (1979).

9. Y. Mochizuki, K. Furukawa and N. Sawada, Effects of simultaneous administration of clofibrate with diethylnitrosamine on hepatic tumorigenesis in the rat, Cancer Letters 19:99 (1983).

10. W. Staubli, P. Bentley, F. Bieri, E. Frohlich and F. Waechter, Inhibitory effect of nafenopin upon the development of DENA-induced enzyme-altered foci within the rat liver, Carcinogenesis 5:41 (1984).

11. U. Gerbracht, I. Timmermann-Trosiener and R. Schulte-Hermann, Studies on regression of foci after withdrawal of tumor promoters, Fd. Chem. Toxic. 23:881 (1985).

12. U. Gerbracht, E. Roth, K. Becker, M. Reinacher and E. Eigenbrodt, A study of the activities of carbohydrate metabolizing enzymes and the levels of carbohydrate metabolites and amino acids in normal liver and in hepatocellular carcinoma, in: "Experimental Hepatocarcinogenesis", M. Roberfroid and V. Preat, eds., Plenum Press, New York, (1988).

13. A. M. Rutenburg, H. Kim, H. W. Fischbein, J. S. Hanker, H. L. Wasserkrug and A. M. Seligman, Histochemical and ultrastructural demonstration of γ-glutamyl transpeptidase activity, J. Histochem. Cytochem. 17:517 (1969).

14. Z. Lojda, R. Gossrau and T. H. Schiebler, "Enzyme Histochemistry: A Laboratory Method", Springer-Verlag, Berlin, Heidelberg, New York (1979).

15. G. Fischer, M. Domingo, D. Lodder, N. Katz, M. Reinacher and E. Eigenbrodt, Immunohistochemical demonstration of decreased L-pyruvate kinase in enzyme altered rat liver lesions produced by different carcinogens, Virchows Arch. B 53:359 (1987).

16. E. Eigenbrodt and W. Schoner, Purification and properties of the pyruvate kinase isoenzymes type L and M from chicken liver, Hoppe-Seyler's Z. Physiol. Chem. 358:1033 (1977).

17. A. McPherson, D. Burkey and P. Stankiewicz, Crystalline alkaline form fructose-1,6-dephosphatase, J. Biol. Chem., 252:7031 (1977).

18. M. Zelewski and J. Swierczynski, The effect of clofibrate feeding on the NADP-linked dehydrogenase activity in rat tissue, Biochim. Biophys. Acta 758:152 (1983).

19. G. W. Löhr and H. D. Waller, Glucose-6-phosphat-Dehydrogenase, in: "Methoden der enzymatischen Analyse", H. Bergmeyer, ed., Verlag Chemie, Weinheim (1974).

20. U. Bergmeyer, K. Grawehn and M. Graßl, Enolase, in: "Methoden der Enzymatischen Analyse", H. Bergmeyer, ed., Verlag Chemie, Weinheim, (1974).

21. U. Bergmeyer, K. Grawehn and M. Graßl, Lactat-Dehydrogenase, in: "Methoden der enzymatischen Analyse", H. Bergmeyer, ed., Verlag Chemie, Weinheim (1974).

22. J. P. Persijn and W. Van der Silk, L-γ-Glutamyltransferase, J. Clin. Chem. Clin. Biochem. 14:421 (1976).

23. German Society for Clinical Chemistry, Alkaline "Optimierte Standard-Gesellschaft für Klinische Chemie (1976).

24. J. K. Reddy, M. S. Rao, D. L. Azarnoff and S. Sell, Mitogenic and carcinogenic effects of a hypolipidaemic peroxisome proliferator (Wy-14,643) in rat and mouse liver, Cancer Res. 39:151 (1979).

25. M. S. Rao, N. D. Lalweni and J. K. Reddy, Sequential histologic study of rat liver during peroxisome proliferator [4-Chloro-6-(2,3-xylidino)-2-pyrimidinyl-thio]-acetic acid (Wy-14,643)-induced carcinogenesis, J. Natl. Cancer Inst. 73:983 (1984).

26. A. B. Deangelo and C. T. Garret, Inhibition of development of praeneoplastic lesions in the livers of rats fed a weakly carcinogenic environmental contaminant, Cancer Letters 20:199 (1983).

27. M. S. Rao, N. D. Lalwani, D. S. Scarpelli and J. K. Reddy, The absence of GGT activity in putative praeneoplastic lesions and in hepatocellular carcinomas induced in rats by the hypolipidaemic

methode" conforming to the recommendations of the Deutsche peroxisome proliferator Wy-14,643, <u>Carcinogenesis</u> 3:1231 (1982).

28. M. Best and C. Duncan, Lipid effects of a phenolic ether (Su-13437) in the rat. Comparison with CPIB, <u>Artherosclerosis</u> 12:185 (1970).

29. M. Reinacher, E. Eigenbrodt, U. Gerbracht, G. Zenk, I. Timmermann-Trosiener, P. Bentley, F. Waechter and R. Schulte-Hermann, Pyruvate kinase isozymes in altered foci and carcinoma of rat liver, <u>Carcinogenesis</u> 7:1351 (1986).

30. G. M. Ledda-Columbano, A. Columbano, S. Dessì, P. Coni, C. Chiodino and P. Pani, Enhancement of cholesterol synthesis and pentose phosphate pathway activity in proliferating hepatocyte nodules, <u>Carcinogenesis</u> 6:1371 (1985).

31. M. A. Moore, H. Tsuda and N. Ito, Dehydrogenase histochemistry of N-ethyl-N-hydroxyethylnitrosamine-induced focal liver lesions in the rat increase in NADPH-generating capacity, <u>Carcinogenesis</u> 7:339 (1986).

32. D. Mayer, M. Moore and P. Bannasch, Biochemical correlation of glycogen content and activity of some enzymes of carbohydrate metabolism in rat liver during early stages of carcinogenesis, <u>J. Cancer Res. Clin. Oncol.</u> 104:99 (1982).

EXPRESSION OF A NOVEL N-ACETYLGLUCOSAMINYLTRANSFERASE IN RAT HEPATIC NODULES

Rosa Pascale[1], Saroja Narasimhan[2] and Srinivasan Rajalakshmi[1]

[1]Department of Pathology, Medical Sciences Building
University of Toronto, Toronto, Ontario, Canada M5S 1A8
[2]Department of Biochemistry, The Hospital for Sick Children
Toronto, Ontario M5G 1X8

The hallmark of liver cancer research in recent years has been the development of experimental models by which initiation, promotion and progression phases can be studied[1-4]. Even though several agents act as promoters, yet many of them are organ specific. Nonetheless, promotion by orotic acid is elegant in that by selecting the initiating carcinogen both cancer in the liver[5,6] as well as intestine[7] can be achieved. Being an intermediate in the de novo biosynthesis of pyrimidine nucleotides, orotic acid is rapidly metabolized by the liver to uridine nucleotides, which on accumulation creates an imbalance in the pool sizes of nucleotides. Interestingly, promotion by orotic acid can be reversed either by blocking the conversion of orotic acid into uridine nucleotides or by trapping the accumulated uridine nucleotides[8]. These observations suggest that the pool sizes of nucleotides may have an important role in the promotion phase of the carcinogenic process. An understanding of the metabolic principles underlying orotic acid induced tumor promotion therefore requires a study on the effect of an imbalance of nucleotides on macromolecular biogenesis involving a template process such as nucleic acid synthesis or a non-template process like glycosylation. Glycosylation being a non-template process is regulated by a number of factors including the level and availability of nucleotide sugars.

It has become evident from a number of investigations that the biosynthesis of N-linked carbohydrate units is initiated in the rough endoplasmic reticulum (RER) by co-translational transfer of oligosaccharides assembled on lipid carrier to polypeptide chain. This event is followed by a series of processing reactions which occur as the protein migrates from RER through the cisternae of Golgi complex generating a variety of structures in N-linked saccharides. UDP-N-acetylglucosamine and UDP-galactose are involved as donors of sugar moieties in reactions involving the transfer of N-acetylglucosamine (Gn) and galactose (Gal) during the course of synthesis of complex-type structures termed bisected or non-bisected bi, tri, and tetra-antennary structures. A highly specific N-acetylglucosaminyl-transferase (GnTase) is required for each type of linkage[9,10]. Consistent and striking changes have been found in the carbohydrate moieties of glycoproteins derived from transformed cells regardless of the means by which they are transformed whether by viruses, mutagens, transfection with DNA obtained from neoplastic cells or activated H-ras[11,22]. Recently some of these changes were shown to be due to an increase in the amounts of tri- and tetra-antennary structures[13,14]. A second alteration in the oligosaccharide structure unique to γ-glutamyltrans-

337

peptidase (γ-GT) isolated from carcinoma of the liver and pancreas has been the detection of bisecting Gn residues [15,16]. The molecular basis for these structural alterations in the carbohydrate moieties of N-linked glycans is perhaps due to an alteration in the concentrations of nucleotides and the relative amounts and activities of the various glycosyltransferases such as GnTases I, II, III, IV and V.

In view of the foregoing considerations, we initiated studies on glycosylation of glycoproteins during hepatocarcinogenesis. In the present study the pattern of activities of GnTase I, II, III and UDP-galactose: N-acetyl-glucosamine ß1-4 galactosyltransferase (GalTase) has been examined in hepatic nodules. The activities of GnTase I, II, III and GalTase are determined under optimal conditions for each one of them using total microsomal membranes as the enzyme source. The tissues that served as controls have been tissues surrounding the nodules, regenerating liver 24 h after 2/3 partial hepatectomy and unoperated normal rat liver. The acceptors used in the assay are glycopeptides having defined structures; the donors of sugar moieties are radio-labeled UDP-N-acetylglucosamine and UDP-galactose. The products of the reactions are characterised after purification by column chromatography on Con A and 360 MHz high resolution proton NMR spectroscopy. The methods used for the preparation of acceptors, assay of enzyme activities, and characterization of products were those described by Narasimhan and colleagues [17-21].

The data presented in Table 1 demonstrate that in hepatic nodules promoted by OA there is a significant activity of GnTase III. The negligible activity seen in surrounding, regenerating and resting liver represent radioactivity, resulting from pooling of the values from several fractions (Fig. 1). The most striking features of the study are the results shown in Table 2. In order to ascertain whether the expression of GnTase III is unique to nodules promoted by orotic acid or a property of nodules, GnTase III activity was assayed in nodules produced after promotion with choline deficient diet [22] and by resistant hepatocyte model. The data show that the nodules promoted by all the 3 models exhibited the activity of GnTase III. The expression of GnTase III activity by nodules promoted with different promoters taken together with the failure to detect in the liver surrounding the nodules suggests that the expression of GnTase III is more associated with either promotion or evolution of nodules rather than an effect of promoters. A lack of GnTase III in regenerating liver is consistent with the explanation that expression of GnTase III is not relatable to an enhanced proliferative capacity normally seen in nodules. The induction of GnTase III in nodules is of interest since these are population of cells from which hepatocellular carcinoma can develop. The γ-glutamyltranspeptidase (γ-GT) from hepatoma and pancreatic carci-

Table 1. Activity of GnTase III in hepatic nodules promoted by orotic acid

Enzyme source	GnTase III Activity (nmoles/hr/mg protein)
Hepatic nodules [a]	1.82; 1.75; 2.10
Surrounding liver	0.02
Regenerating liver	0.02

The method used to assay the enzyme and characterize the product is described earlier [19]. [a]1-2 gms of nodules were pooled from 3-4 rats; from each rat 2-5 nodules were taken.

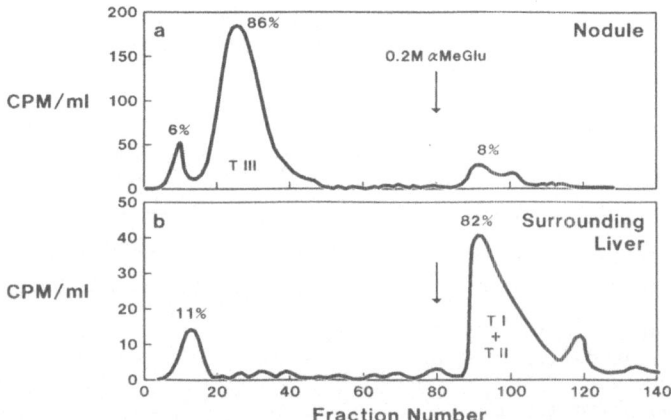

Fig. 1. a) Elution profiles on Con A-Sepharose(0.7 x 17cm). Total ^{14}C-pro-
ducts of hepatic nodules. Total radioactivity applied to the column
was 2,800 cpm. The arrows indicate the point of application of
buffer containing 0.2 M methyl α-D-glucoside. Numbers near the peaks
indicate the % radioactivity recovered in the peak relative to the
total radioactivity recovered from the column. Recoveries from the
column were greater than 87%. b) The same as a) except the total
[14]C-products were obtained using surrounding liver as the enzyme
source.

noma display bisected Gn residue, which is absent in γ-GT present in normal
liver or pancreas[15,16]. The demonstration of the induction of GnTase III in
preneoplastic cell population is consistent with the detection of bisected Gn
residues in the glycoprotein of neoplastic cells. It is worthy to note that
according to available literature, none of the proteins synthesised and secreted by liver contain bisected Gn residue in their carbohydrate moieties of
N-linked glycans[23]. The activity of GnTase III has been found in high levels
in hen oviduct membrane and correspondingly, avian glycoproteins originating
from this tissue contain bisected Gn residues[19]. An induced activity of
GnTase III was found in a lectin resistant cell line (Lec 10 CHO)[24]. It merits consideration whether the expression of GnTase III activity by hepatic
nodules is related to its resistant phenotype. Hepatic nodules promoted with
orotic acid acquire resistance to nucleotides imbalance inducable by orotic
acid[25], the nodules promoted by choline deficient diet are resistant to fat
accumulation normally induced by choline deficiency[22] and the nodules promoted
by resistant hepatocyte model are resistant to the mitoinhibitory effect of
2-acetylaminofluorene[2].

The bisecting Gn residue introduced by GnTase III, in N-linked complex
type carbohydrates seems to influence the intrinsic molecular and biological
properties of these molecules[26,27]. Bisecting Gn residue has been shown to
exert a profound influence on the activity of glycosidases, glycosyltransferases, and galactosyltransferases[19,20]. The decreased activities of GnTase I,
II, and GalTase seen in nodules (Table 3) is consistent with these observations. However, there may be several other explanations which cannot be
ruled out. Being glycoproteins, glycosyltransferases besides their role in
the synthesis of complex structures found in glycoproteins, may have other
biological roles such as receptors at cell surface or carriers of antigenic
determinants for various cell types. In consistence with these views it was
found that GalTase at the cell surface recognises lactosaminoglycan substrates on adjacent cell surface or in the extracellular matrix. Results
also suggested that GalTase participates during contact inhibition of growth

339

Table 2. Activity of GnTase III in hepatic nodules pro-
          moted by orotic acid, choline deficient diet
          and by resistant hepatocyte model

Promoting regimen	GnTase III activity (nmoles/h/mg protein)	
	Nodules	Surrounding liver
1% orotic acid [a]	2.0	0.02
choline deficient diet [b]	0.9	0.00
Resistant hepatocyte model [c]	1.0	N.D.

The method of assay and characterisation of product is
described earlier [19]. N.D. not done. [a] Rats were initiated
with 1,2-dimethylhydrazine(100 mg/kg i.p.) given 18 h
after 2/3 partial hepatectomy. A week later they were
promoted with 1% orotic acid. Rats were killed 32 wk
later. [b] Rats were initiated with diethylnitrosamine
(200 mg/kg i.p.) and promoted with a choline deficient
diet. Twenty weeks after initiation, they were subjected
to 2/3 partial hepatectomy and were killed 7 wk thereafter. [c] Rats were initiated with diethylnitrosamine
(200 mg/kg i.p.) and promoted with 3 daily injections of
20 mg/kg 2-acetylaminofluorene on 14,15,16th day following initiation. CCl_4 (2 ml/kg i.g.)was given on 16th day
and PH 4 wk later. Animals were killed 32 wk after promotion.

in fibroblasts and in the binding of cytosolic T lymphocytes to their target
cells [28].

    The biological behavior of cancer cells are quite distinct from normal
cells consisting of an impaired cell-cell communication, lack of contact

Table 3. Pattern of glycosyltransferases in hepatic
          nodules promoted by orotic acid

Enzyme source	GnTase I	GnTase II	GalTase
	(nmoles/hr/mg protein)		
Hepatic nodules	4.0	22.7	19.2
Surrounding liver	11.1	29.8	37.8

GnTase I, N-acetylglucosaminyltransferase I. GnTase II,
N-acetylglucosaminyltransferase II. GalTase UDP-galac-
tose:N-acetylglucosamine ß-4 galactosyl transferase. The
method of assay and characterization of products are
described earlier [19,20]. Other details are described in the
legend to Table 1.

inhibition of growth, capacity to metastasise and ability of anchorage independent growth. More importantly glycoproteins with receptor functions are involved in signal perception, an alteration in glycosylation pattern can modulate their functions. The results described in this communication are consistent with the foregoing views and further provide a molecular basis for changes observable in cellular behavior during hepatocarcinogenesis. However, an interesting aspect of the study has been the demonstration of both the induction of GnTase III (an enzyme not associated with liver) and alterations in other hepatic glycosyltransferases including galactosyltransferase in preneoplastic cell population. Since nodules can develop into malignant tumor, a detailed investigation on glycosyltransferases including cloning of these genes will be useful for understanding their role in hepatocarcinogenesis.

ACKNOWLEDGEMENTS

We wish to thank Drs. E. Laconi, D.S.R. Sarma, P.M. Rao and H. Schachter for helpful discussion and Lori Cutler for her excellent secretarial assistance. The study was supported in part by U.S. P.H.S. grants CA 45361 and CA 37077 from the National Cancer Institute, by grants from National Cancer Institute, Canada. R. P. was supported by Associazione Italiana per la Ricerca sul Cancro, Italy.

REFERENCES

1. C. Peraino, R. J. M. Fry and E. F. Staffeldt, Reduction and enhancement by phenobarbital of hepatocarcinogenesis induced in the rat by 2-acetylaminofluorene, Cancer Res. 48:1506 (1971).
2. D. B. Solt and E. Farber, New Principle for the analysis of chemical carcinogenesis, Nature (Lond) 263:702 (1976).
3. H. C. Pitot, L. Barsness, T. Goldsworthy and T. Kitagawa, Biochemical characterisation of stages of hepatocarcinogenesis after a single dose of diethylnitrosamine, Nature (Lond) 271:456 (1978).
4. D. S. R. Sarma, P. M. Rao and S. Rajalakshmi, Liver tumor promotion by chemicals: Models and mechanisms, Cancer Surveys 5:781 (1986).
5. C. Laurier, M. Tatematsu, P. M. Rao, S. Rajalakshmi and D. S. R. Sarma, Promotion by orotic acid of liver carcinogenesis in rats initiated by 1,2-dimethylhydrazine, Cancer Res. 44:2186 (1984).
6. P. M. Rao, Y. Nagamine, M. W. Roomi, S. Rajalakshmi and D. S. R. Sarma Orotic acid, a new promoter for liver carcinogenesis, Toxicologic. Path. 12:173 (1984).
7. P. M. Rao, E. Laconi, S. Rajalakshmi and D. S. R. Sarma, Orotic acid, a liver tumor promoter, also promotes carcinogenesis of the intestine, Proc. Amer. Assoc. Cancer Res. 27:142 (1986).
8. P. M. Rao, E. Laconi, S. Vasudevan, A. Denda, S. Rajagopal, S. Rajalakshmi and D. S. R. Sarma, Dietary and metabolic manipulations of the carcinogenic process: Role of nucleotide pool imbalances in carcinogenesis, Toxicologic. Path. 15:190 (1987).
9. R. Kornfeld and S. Kornfeld, Assembly of asparagine-linked oligosaccharides, Ann. Rev. Biochem. 54:631 (1985).
10. H. Schachter, Biosynthetic controls that determine the branching and microheterogeneity of protein-bound oligosaccharides, Biochem. Cell. Biol. 64:163 (1986).
11. J. C. Collard, P. Van Beek, J. W. G. Jansen and J. F. Schijven, Transfection by human oncogenes: concomitant induction of tumorigenicity and tumor associated membrane alterations, Int. J. Cancer 35:207 (1985).
12. U. V. Santer, F. Gilbert and M. C. Glick, Change in glycosylation of membrane glycoproteins after transfection on NIH 3T3 with human

341

tumor DNA, Cancer Res. 44:3730 (1984).

13. M. Pierce and J. Arango, Rous sarcoma virus-transformed baby hamster kidney cells expresses higher levels of asparagine-linked tri- and tetraantenary glycopeptides containing [GlcNAc-γ(1,6) Man-α(1,6)-Man] and poly-N-acetyllactosamine sequences than baby hamster kidney cells, J. Biol. Chem. 261:10771 (1986).

14. J. W. Dennis, S. Laferte, C. Waghorne, M. L. Breitman and R. S. Kerbel, ß1-6 branching of Asn-linked oligosaccharides is directly associated with metastasis, Science 236:582 (1987).

15. K. Yamashita, A. Hitoi, N. Taniguchi, N. Yokosawa, T. Yutaka and A. Kobata, Comparative study of the sugar chains of γ-glutamyltranspeptidases purified from rat liver and rat-AH-66 hepatoma cells, Cancer Res. 43:5059 (1983).

16. N. Yamaguchi, K. Kawai and T. Ashihara, Discrimination of γ-glutamyl-transpeptidase from normal and carcinomatous pancreas, Clinica Chimica Acta 154:133 (1986).

17. S. Narasimhan, N. Harpaz, G. Longmore, J. P. Carver, A. A. Grey and H. Schachter, Control of glycoprotein synthesis. The purification by preparative high voltage paper electrophoresis in borate of glyco-peptides containing high mannose and complex oligosaccharide chains linked to asparagine, J. Biol. Chem. 255:4876 (1980).

18. S. Narasimhan, J. R. Wilson, E. Martin and H. Schachter, A structural basis for four distinct elution profiles on con A-sepharose affinity chromatography of glycopeptidases, Can. J. Biochem. 57:83 (1979).

19. S. Narasimhan, Control of glycoprotein synthesis. UDP-GlcNAc: glyco-peptide ß4-N-acetylglucosaminyltransferase III. An enzyme in hen oviduct which adds GlcNAc in ß1-4 linkage to the ß linked mannose of trimannosyl core of N-glycosyl oligosaccharides, J. Biol. Chem. 257:10235 (1982).

20. S. Narasimhan, J. C. Freed and H. Schachter, Control of glycoprotein synthesis. Bovine milk UDP-galactose: N-acetylglucosamine ß-4 galactosyltransferase catalyses the preferential transfer of galactose to the GlcNAc ß1,2 Manα 1,3-branch of both bisected and non-bisected complex biantennary asparagine-linked oligosaccharides, Biochemistry 24:1694 (1985).

21. S. Narasimhan, J. C. Freed and H. Schachter, The effect of a "bisec-ting" N-acetylglucosaminyl group on the binding of biantennary, complex oligosaccharides to concanavalin A, Phaseolus vulgaris erythro-agglutinin [E-PHA] and Ricinus communis agglutinin (RCA-120) immobilised on agarose, Carbohydrate Res. 149:65 (1986).

22. M. A. Sells, S. L. Katyal, S. Sell, H. Shinozuka and B. Lombardi, Induction of foci of altered γ-glutamyltranspeptidase positive hepatocytes in carcinogen-treated rats fed a choline-deficient diet, Br. J. Cancer 20:274 (1979).

23. A. Kobata, The carbohydrates of glycoproteins, in: "Biology of Carbo-hydrates", V. Ginsberg and P. W. Robbins, eds., John Wiley and Sons, Vol. 2 (1984).

24. C. Campbell and P. Stanley, A dominant mutation to ricin resistance in chinese hamster ovary cells induces UDP-GlcNAc: glycopeptide ß-4N-acetylglucosaminyltransferase III activity, J. Biol. Chem. 259:13370 (1984).

25. E. Laconi, S. Vasudevan, P. M. Rao, S. Rajalakshmi and D. S. R. Sarma, Hepatic nodules have a characteristic pattern of nucleotide pools distinct from that of the surrounding liver, Proc. Amer. Assoc. Cancer Res. 28:169 (1987).

26. J. R. Brisson and J. P. Carver, Solution conformation of asparagine-linked oligosaccharides: α(1-2)-, α1-3-, ß(1-2)-, and ß1-4 linked units, Biochemistry 22:3671 (1983).

27. J. R. Brisson and J. P. Carver, Solution conformation of asparagine-linked oligosaccharides: (1-6)linked moiety, Biochemistry 22:3680 (1983).

28. B. D. Shur, E. M. Bayna, R. B. Runyan, J. S. Reichner, D. M. Scully and E. Kurt-Jones, The receptor function of cell surface glycosyl-trans-ferases during mammalian fertilisation, development and immune recognition, <u>in</u>: " Colloque INSERUM/CNRS, Cellular and Pathological Aspects of Glycoconjugate Metabolism", H. Dreyfus, R. Manarelli, L. Freysz, and G. Rebel, eds., Paris (1984).

CYTOSOLIC AND PLASMAMEMBRANES ENZYMATIC ACTIVITIES DURING DIETHYLNITROSAMINE-INDUCED CARCINOGENESIS IN RAT LIVER

Mario U. Dianzani, Rosa A. Canuto, Margherita Ferro[1], Anna M. Bassi[1], Luciana Paradisi, Maria E. Biocca and Giuliana Muzio

Dept. of Experimental Medicine and Oncology, University of Turin,Corso Raffaello 30, 10125, Turin, Italy
[1]Institute of General Pathology, University of Genoa
Via L. B. Alberti 2, 16132 Genoa, Italy

INTRODUCTION

It is well known that lipid peroxidation is strongly decreased in hepatomas with respect to normal liver and the extent of this decrease is proportional to that of dedifferentiation[1,2].

Hepatomas or cell lines with low dedifferentiation display sometimes normal peroxidation values, but these are not further stimulated by the addition of ADP-iron or ascorbate, as normal liver does. In the course of carcinogenesis, decrease of lipid peroxidation occurs as early as at the stage of reversible nodules, both in the ethionine[3] and in diethylnitrosamine-(DEN)-partial hepatectomy model[4].

The decline in lipid peroxidation has been shown not only by measuring the amounts of thiobarbituric (TBA)-reacting substances accumulating in the tissue during incubation at 37°C, but even by evaluating the accumulation in the incubation medium of the whole aldehyde pool, including aldehydes belonging to the 4-hydroxy-2,3-trans-unsaturated series[4].

4-hydroxy-nonenal (HNE), the most studied member of this toxic series, has been shown to display a lot of biological and toxic actions[5,6]. Some of them take place at extremely low concentration (from 10^{-6} to 10^{-8}-10^{-9} M), suggesting the possibility for a biological role. Among the actions displayed at low concentrations, the most important are: 1) the stimulation, followed by inhibition, of plasmamembrane adenylate cyclase[7]; 2) the chemotactic power towards polymorphonuclear leucocytes[8]; 3) the inhibition in the synthesis of mRNA for the oncogene c-myc, seen in a cultured erythroleukemic cell line[9]; 4) the re-expression by this cell line of genes for gamma-globin and, to a lesser extent, for beta globin[10]. As these results seem to speak in favor of a possible effect of this aldehyde on cell differentiation, the real meaning of lipid peroxidation taking place in "normal" tissues becomes questionable.

In the present paper, we studied the behavior of aldehyde-metabolism enzymes both in nodules and hepatomas at different stages of carcinogenesis, by using either the DEN-hepatectomy model previously used, as described by Solt et al.[11] or the DEN-CCl$_4$ model as described by Roomi et al.[12]. In fact, the decrease in TBA-test, as well as in other aldehydes accumulation, could also have been produced by increase in the activity of the aldehyde-removing enzymes.

Moreover, we studied the behavior of the same enzymes in HTC cells, a poorly differentiated line of rat hepatoma cells in culture. This line is compared to MH_1C_1, cell line displaying rather high differentiation.

The studied enzymes were the NAD- and NADP-dependent aldehyde dehydrogenase (E.C. 1.2.1.3 and E.C. 1.2.1.5 respectively, ALDH), the NADH-dependent alcohol dehydrogenase (E.C. 1.1.1.1 ADH), and the NADPH-dependent aldehyde reductase (E.C. 1.1.1.2 ALRed), the glutathione-S-transferase (E.C. 2.5.1.18, GST).

In rat liver the two reducing enzymes and GST are found mainly in the cytosolic fraction[13-15]; whereas multimolecular forms of ALDH are present in cytosolic, mitochondria and microsomes[16,17].

For these reasons GST, ADH and ALRed activities were studied in the cytosol, whereas ALDH was studied in the cytosol, as well as in mitochondria and in microsomes of normal liver, of nodules of different age and of hepatoma. In HTC cells, all the enzymatic assays were carried out only in the cytosolic fraction.

Moreover, we studied glucagon- and fluoride-stimulated adenylate cyclase as well as guanylate cyclase from plasmamembranes isolated from either normal liver or hepatomas obtained after DEN-CCl$_4$ treatment.

MATERIALS AND METHODS

Diethylnitrosamine Carcinogenesis

Male F-344 rats, weighing 120-150 g at the beginning, were used. The schematic representation of hepatocarcinogenic procedure according to Roomi et al.[12] is explained in Fig. 1. The rats of each group were killed by decapitation. Their livers were removed and rinsed in ice-cold isolation medium containing 70 mM sucrose, 220 mM mannitol, 20 mM Tris-HCl, 2 mM Tris-EGTA, 0.1% (w:v) bovine serum albumin (fraction 5, defatted). Preneoplastic nodules were carefully dissected from perinodular liver as previously described[13].

Histological specimens were prepared from each of used tissue, after fixation in a mixture containing 95% ethanol, 40% formaldehyde solution and

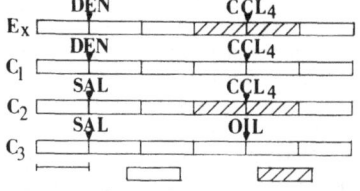

Fig. 1. Diagramatic representation of experimental (Ex) and control (C_1, C_2, C_3) regimens: DEN, diethylnitrosamine 200 mg per kg body weight, i.p.; SAL, 0.9% NaCl solution (equivalent volume i.p.); AAF, 2-acetylaminofluorene; CCl$_4$, 2 ml per kg body weight, diluted 1:1 with corn oil, i.g.; OIL, corn oil 2 ml per kg body weight, i.g.

acetic acid (6:3:1 by vol.). Paraffin sections were routinely stained with hematoxylin and eosin.

γ-Glutamyltranspeptidase was demonstrated histochemically according to the method of Rutenburg et al.[18]

Monolayer Culture of MH_1C_1 and HTC Cell Lines

Hepatoma cell lines were obtained from the American Type Culture Collection (Rockville, Md, USA).

MH_1C_1 and HTC cells are clones isolated from Morris rat hepatoma 7795[19] and 7288C[20], respectively. Cells were grown in plastic flasks at 37°C in 95% air and 5% CO_2 atmosphere. MH_1C_1 cell line was grown in the presence of Ham F10 medium supplemented with 10% horse serum and 7.5% new-born calf serum; HTC cell line was grown in the presence of MEM plus NEA medium supplemented with 5% fetal bovine serum.

Media were additioned with 100 U/ml penicillin and 0.1 g mg/ml streptomycin. Confluent monolayers of MH_1C_1 and HTC cell cultures were washed with cold HBSS and then exposed to trypsin/EDTA (0.05-0.02%) solution for cell harvest. Cell suspensions were washed three times with HBSS and sedimented by centrifugation at 600 x g for 10 min.

Isolation of Subcellular Fractions

Rat hepatocytes were isolated as described elsewhere by Poli et al.[21].

Homogenates from tissues and cell lines were prepared in Potter-Elvehjem homogenizer with 3 strokes of tightly fitting pestle in a volume of isolation medium corresponding to 2.5 times their weight and were diluted to 20% (w:v) with the same medium.

Diluted homogenates were centrifuged at 10,000 g_{min}. From the collected supernatants the mitochondrial fraction was isolated at 133,000 g_{min} and washed 3 times.

Microsomes were isolated from the post-mitochondrial supernatant by centrifuging at 6×10^6 g_{min}. Mitochondrial and microsomal fractions were finally resuspended in a volume of medium containing 250 mM sucrose, 20 mM Tris-HCl, 1 mM Tris-EGTA so as to have 1 g of tissue per ml.

Post-microsomal supernatants from 20% homogenates were used as cytosolic fractions. The purity of subcellular fractions was determined by specific enzyme markers.

Enzyme Assays

The activity of ALDH was determined as previously described[22]. NADH-dependent ADH and ALRed activities of the cytosolic fraction were measured in 200 mM potassium phosphate buffer (pH 7) according to Scrivastava et al.[23]. The activity of GST was determined according to Alin et al.[24]. All enzymatic assays were carried out at 30°C by using HNE and benzaldehyde (BA) as substrates.

Protein content was measured by a biuret procedure[25].

Preparation of Plasmamembranes

Liver and hepatoma were removed and homogenized in Hepes-glycylglycine buffer at pH 7.4. The homogenate was filtered through cheesecloth, diluted

in the homogenizing media and mixed with Percoll-2.5 M sucrose (9:1), which
formed a continuous gravity gradient and centrifuged with a vertical rotor at
17,000 g for 15 min. Only membranes floating at the top of the cellulose
nitrate tubes were removed, resuspended, washed in the Hepes-glycylglycine
buffer and stored in liquid nitrogen. The resulting preparation allowed a
recovery of 10% of the adenylate cyclase and guanylate cyclase activity with
respect to the homogenate[7]. The quality and the degree of purification of
the isolated plasmamembranes were assessed by measuring the activities of the
enzyme markers. The recovery was 60 to 80% of the original adenylate cyclase
and guanylate cyclase activities.

RESULTS

     Fig. 2 gives information on ALDH activities, as studied with 0.1 mM HNE
as substrate, either with NAD^+ or $NADP^+$ as cofactors. It shows that the
enzymatic activity increases with time in the cytosolic fraction of both
preneoplastic nodules and hepatoma, independently of the cofactor used,
whereas in the perinodular liver the enzyme activity returns soon to the
control values. No change is evident in cytosol of rats receiving control
treatments C_1 and C_2 in comparison with C_3 treatment. A similar behavior is
also observed in the homogenates of the various tissues so far examined (data
not shown).

     By using BA as substrate (Fig. 3), we got qualitatively similar results
The values of the enzymatic activity, however, are higher with this substrate
than with HNE.

     The extent of the increase in $NADP^+$-dependent activity of both preneo-
plastic nodules and hepatoma is higher than that of NAD^+-dependent activity;
in our hand this result was found to be independent of the substrate we used.

     No change is evident in the mitochondrial ALDH activity from nodules and
hepatoma with respect to the same fraction obtained from the control liver,

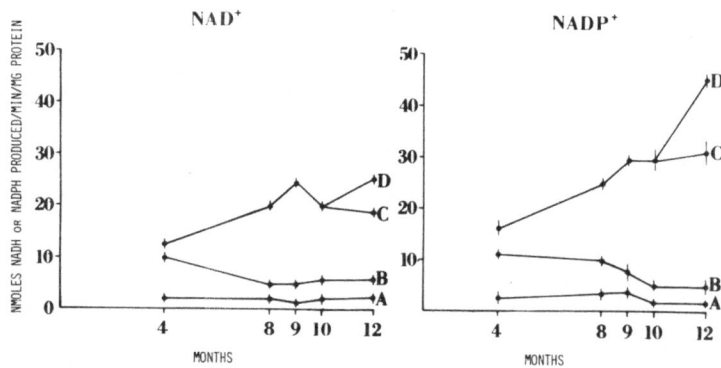

Fig. 2.   HNE-aldehyde dehydrogenase activity in cytosol from the liver of
          control rats and from hepatoma, preneoplastic nodules and perinodu-
          lar liver. a) Aldehyde dehydrogenase activity NAD^+-dependent, b) al-
          dehyde dehydrogenase activity $NADP^+$-dependent. The data are means ±
          standard deviation. The assay mixture (1 ml) contained 50 mM sodium
          pyrophosphate (pH 8.8), 1 mM pyrazole, 0.002 mM rotenone, 0.1 mM HNE
          and an appropriate amount of enzyme. The reaction was started by the
          addition of 1 mM NAD^+ or 2.5 mM $NADP^+$. No change from C_3 is evident
          in liver of rats when C_1 and C_2 treatments were used. A, control;
          B, perinodular liver; C, nodules; D, hepatoma.

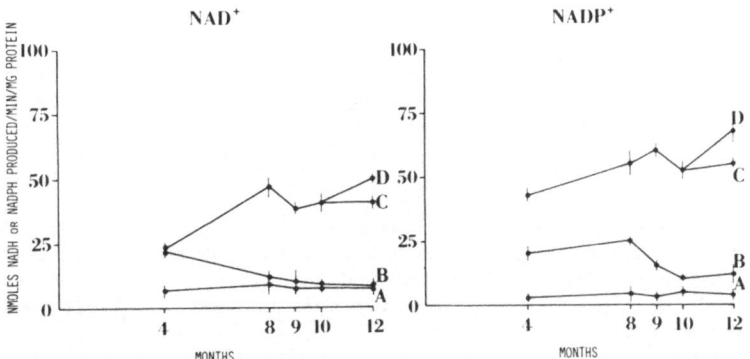

Fig. 3. BA-aldehyde dehydrogenase activity in cytosol from the liver of control rats and from hepatoma, preneoplastic nodules and perinodular
liver. a) Aldehyde dehydrogenase activity NAD^+-dependent, b) aldehyde dehydrogenase activity $NADP^+$-dependent. The data are means ±
standard deviation. The assay mixture (1 ml) contained 50 mM sodium
pyrophosphate (ph 8.8), 1 mM pyrazole, 0.002 mM rotenone, 0.1 mM BA
and an appropriate amount of enzyme. The reaction was started by the
addition of 1 mM NAD^+ or 2.5 mM $NADP^+$. No change from C_3 is evident
in liver of rats when C_1 and C_2 treatments were used. A, control;
B, perinodular liver; C, nodules; D, hepatoma.

Table 1. Aldehyde dehydrogenase activity in mitochondria and
microsomes from the liver of control rats and from
hepatoma, preneoplastic nodules and perinodular liver

		Mitochondria		Microsomes	
		HNE	BA	HNE	BA
Control	NAD	4.31	2.63	8.51	2.47
	NADP	1.91	1.24	4.17	1.86
Perinodular	NAD	4.40	2.21	9.29	2.11
liver	NADP	1.17	0.92	4.42	2.09
Nodules	NAD	4.25	3.49	8.54	7.55
	NADP	2.22	0.97	8.73	6.68
Hepatoma	NAD	2.68	4.07	9.89	8.12
	NADP	1.99	1.93	10.77	11.31

Data are expressed as nmoles of NADH or NADPH produced per min
per mg of protein. The values are the mean of 2 experiments.
Abbreviations : HNE, 4-hydroxy-2,3-nonenal; BA, benzaldehyde. No
change from C_3 is evident in liver of rats when C_1 and C_2 treatments were used.

using either HNE or BA as substrates (Table 1). Microsomes from both nodules and hepatoma show a little increase in NAD^+- and $NADP^+$-dependent ALDH activity with both HNE and BA (Table 1).

Table 2 shows the activities of NAD^+- and $NADP^+$-dependent ALDH in the cytosol obtained from normal rat hepatocytes, MH_1C_1 and HTC cells. If compared with hepatocytes, the two hepatoma cell lines have higher enzymatic activities; this is particularly evident in the case of HTC cells using BA as substrate.

Fig. 4 reports the behavior of NADPH-dependent ALRed in the cytosolic fractions. This activity increases 4-5 fold for HNE and 2 fold for BA both in preneoplastic nodules at 12 months and in hepatoma with respect to the control liver; no significant change for this enzymatic activity was found in the perinodular liver.

No changes in NADH-dependent ADH activity were observed (Tables 3, 4). Table 5 gives results obtained for the activities of ADH and ALRed in hepatoma cell lines. There is a decrease in the activity of ADH, more consistent in MH_1C_1 than in HTC, whereas there is a decrease of ALRed activity in MH_1C_1 and a significant increase of the same enzymatic activity in HTC. Fig. 5 refers to GST activity in cytosolic fractions of various tissues, by using 0.05 mM HNE as substrate. A significant increase was seen both in preneoplastic nodules and in hepatoma.

Table 6 shows a substantial decrease of GST in MH_1C_1 and HTC cells with respect to normal hepatocytes. This result is in contrast with those obtained for DEN-CCl_4 induced nodules and hepatomas.

Table 7 shows that basal membrane-bound adenylate cyclase of control rat liver plamamembranes is stimulated by glucagon, GTP + glucagon and fluoride as previously shown[7]. With DEN-CCl_4 induced hepatoma plasmamembranes, a large increase in basal adenylate cyclase activity was seen, whereas, hormonal stimulation, as exerted either by glucagon or by GTP + glucagon, is lower

Table 2. Aldehyde dehydrogenase activity in cytosol of hepatocytes, MH C cells and HTC

	NAD		NADP	
	HNE	BA	HNE	BA
Hepatocytes	7.67 ± 1.70	6.86 ± 1.32	1.64 ± 0.29	2.78 ± 0.51
MH_1C_1	10.12 ± 4.16	23.60 ± 5.81[b]	3.87 ± 1.43[a]	20.76 ± 9.21[b]
HTC	27.66 ± 4.09[b]	101.35 ± 17.99[b]	44.97 ± 14.91[b]	261.87 ± 42.16[b]

Data are expressed as nmoles of NADH or NADPH produced per min per mg of cytosol protein. The values are the mean ± SD of 5-6 experiments. Abbreviations: HNE, 4-hydroxy-2,3-nonenal; BA, benzaldehyde. P < 0.05; P < 0.001 (hepatoma cells vs. hepatocytes).

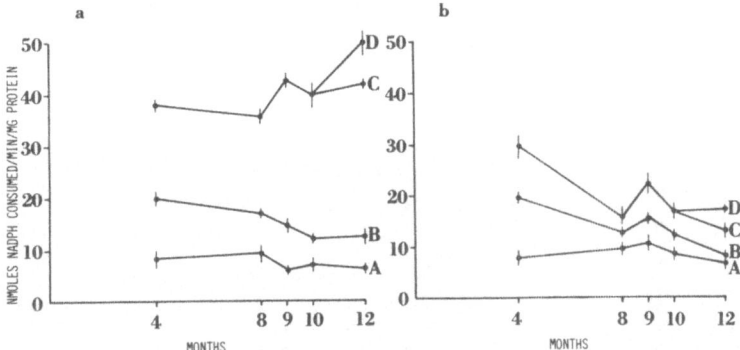

Fig. 4. Aldehyde reductase activity in cytosol from the liver of control rats and from hepatoma, preneoplastic nodules and perinodular liver. a) HNE-aldehyde reductase; b) BA-aldehyde reductase. The data are means ± standard deviation. The assay mixture (1 ml) contained 200 mM potassium phosphate buffer (pH 7.0), 0.1 mM HNE or BA and an appropriate amount of enzyme. The reaction was started by the addition of 0.1 mM NADPH. No change from C_3 is evident in liver of rats when C_1 and C_2 treatments were used. A, control; B, perinodular liver; C, nodules; D, hepatoma.

than that seen in control plasmamembranes. Fluoride-stimulation of adenylate cyclase activity is lacking in hepatoma plasmamembranes.

Fig. 6 shows that basal guanylate cyclase activity of hepatoma plasmamembranes is about 40% lower than that of the normal liver. The addition of 1 μM HNE resulted in an inhibition of the same enzymatic activity in both

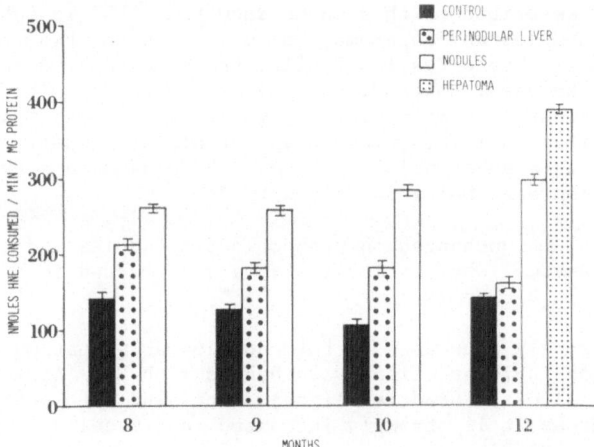

Fig. 5. Glutathione-S-transferase activity in cytosol from the liver of control rats and from hepatoma, preneoplastic nodules and perinodular liver. The data are the means ± standard deviation. The assay mixture (1 ml) contained 100 mM sodium phosphate buffer (pH 6.5), 0.5 mM reduced glutathione, 0.05 mM HNE and an appropriate amount of enzyme. No change from C_3 is evident in liver of rats when C_1 and C_2 treatments were used. A, control; B, perinodular liver; C, nodules; D, hepatoma.

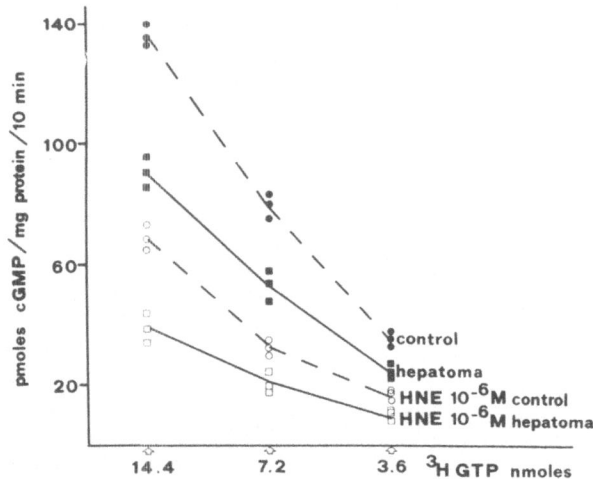

Fig. 6. Effect of HNE on guanylate cyclase activity in normal and hepatoma plasmamembranes. The values represent triple determinations of three experiments. Partially purified plasmamembranes (20-40 μg of protein) were incubated with 0.1 mM of GTP, [³H]-GTP (at the concentrations indicated in Fig.) and 1 μM HNE for 10 min at 37°C. The cGMP formed by guanylate cyclase was determined as described with some modifications [7].

control and hepatoma plasmamembranes (for GTP amounts ranging from 14.4 to 3.6 nmoles).

DISCUSSION

The results described in this paper show that ALDH is substantially increased both in nodules and hepatoma, as well as in the two studied hepatoma-derived cell lines. These results confirm those recently described by Lindahl et al.[26] who used, however, only BA as substrate. They explained their results suggesting a molecular change in the enzymes that would become more specific for aromatic aldehydes than for alifatic ones. Our results show, however, that a substantial increase (15 fold) occurs even with HNE as substrate, this increase being 34 fold with BA.

ADH, however, is unchanged in preneoplastic nodules and the corresponding hepatoma, whereas it is strongly decreased in MH_1C_1 cells and to a less extent in HTC cells.

ALRed is strongly increased in both nodules and hepatoma with HNE as substrate, whereas it is practically unchanged with BA. As far as regards the hepatoma cell lines, ALRed is slightly decreased with both HNE and BA in MH_1C_1 cells, whereas it is strongly increased in HTC cells with HNE and to a less extent also with BA.

GST is increased in both preneoplastic nodules and hepatoma, whereas it is significantly decreased in the hepatoma cell lines, and especially in HTC cells using HNE as substrate.

These results show that there are substantial differences in the behavior of aldehyde-removing enzymes in different types of cells and that there are also differences related to the two substrates.

Table 3. HNE alcohol dehydrogenase activity in cytosol from the liver of control rats and from hepatoma, preneoplastic nodules and perinodular liver

	4 months	8 months	9 months	10 months	12 months
Normal liver	42.49	39.31	38.35	37.89	33.64
Perinodular liver	46.45	39.23	39.88	35.85	35.04
Nodules	48.16	40.27	51.34	44.76	39.80
Hepatoma					44.55

Data are expressed as nmoles of NADH consumed per min per mg of cytosol protein. The values are the mean of 3-4 experiments and standard deviations were 15% or less of the mean. Abbreviations: HNE, 4-hydroxy-2,3-nonenal. No change from C_3 is evident in liver of rats when C_1 and C_2 treatments were used.

Table 4. Benzaldehyde alcohol dehydrogenase activity in cytosol from the liver of control rats and from hepatoma, preneoplastic nodules and perinodular liver

	4 months	8 months	9 months	10 months	12 months
Normal liver	84.58	79.11	70.70	77.89	77.69
Perinodular liver	73.12	73.50	66.73	69.71	71.28
Nodules	66.21	78.43	77.43	60.67	67.05
Hepatoma					77.35

Data are expressed as nmoles of NADH consumed per min per mg of cytosol protein. The values are the mean of 3-4 experiments and standard deviations were 15% or less of the mean. No change from C_3 is evident in liver of rats when C_1 and C_2 treatments were used.

354

Table 5. Activities of alcohol dehydrogenase NADH-dependent and of aldehyde reductase in cytosol of hepatocytes, MH$_1$C$_1$ cells and HTC cells

	Alcohol dehydrogenase NADH-dependent		Aldehyde reductase	
	HNE	BA	HNE	BA
Hepatocytes	58.33 ± 15.30	109.99 ± 16.80	7.62 ± 1.80	7.65 ± 1.67
MH$_1$C$_1$	1.68 ± 0.51^a	3.21 ± 1.37^a	5.79 ± 1.66	2.74 ± 1.29^a
HTC	28.48 ± 3.28^b	71.98 ± 11.12^b	26.24 ± 1.39^a	12.79 ± 2.52^b

Data are expressed as nmoles of NADH consumed per min per mg of cytosol protein for alcohol dehydrogenase and as nmoles of NADPH consumed per min per mg of cytosol protein for aldehyde reductase. The values are the mean ± SD of 5-6 experiments. Abbreviations: HNE, 4-hydroxy-2,3-nonenal; BA, benzaldehyde. [a] P < 0.001; [b] P < 0.05 (hepatoma cells vs. hepatocytes).

Table 6. Glutathione-S-transferase activity in cytosol of hepatocytes, MH_1C_1 cells and HTC cells

Substrate	Hepatocytes	MH_1C_1	HTC
CDNB	1721.3 ± 180.0	1052.7 ± 122.8 [a]	865.3 ± 53.4 [a]
HNE	280.5 ± 54.9	171.1 ± 32.5 [a]	30.1 ± 7.2 [a]

Data are expressed as nmoles of CDNB or HNE consumed per min per mg of cytosol protein. The values are the mean ± SD of 4 experiments. Abbreviations: CDNB, 1-chloro-2,4-dinitrobenzene; HNE, 4-hydroxy-2,3-nonenal. [a]$p < 0.001$ (hepatoma cells vs. hepatocytes).

Table 7. Effect of glucagon, GTP + glucagon and sodium fluoride on adenylate cyclase activity of plasmamembrane fractions from rat liver and hepatoma

| Treatment | cAMP (pmoles/mg protein/min) formed | | | |
	Basal	Glucagon 10^{-6} M	GTP + Glucagon	NaF 10 mM
Normal liver	18.7 ± 8.7	67.4 ± 8.7	72.6 ± 9.3	102.3 ± 15.4
Hepatoma	47.3 ± 13.2	94.6 ± 14.6	98.6 ± 18.5	34.7 ± 5.4

Results are given as means ± SD for three experiments.

It is remarkable, however, that the data found for preneoplastic nodules and hepatoma by using DEN-CCl$_4$ carcinogenesis model are practically identical to those seen with the DEN-hepatectomy model. This might suggest that the study of carcinogenesis model is more suitable for this type of investigation than that of the hepatoma cell lines cultivated in vitro, where the general feature may be also dependent upon different reasons related to the in vitro growth.

If it is true, the present experiments show a tendency towards a net increase of HNE removal, starting as early as at the stage of the preneoplastic nodules and continuing in hepatoma.

In any case, the present data show that one cannot judge the behavior of lipid peroxidation in tumors by the single measurement of MDA accumulation or also by HPLC-estimation of the whole accumulated aldehyde pool and that an evaluation of the removal rate is always needed.

As far as regards the behavior of plasmamembrane adenylate cyclase, the present data are still preliminary. It is, however, interesting to note that basal adenylate cyclase is increased in hepatoma plasmamembranes, whereas basal guanylate cyclase is decreased. Moreover, hormonal stimulation of adenylate cyclase by either glucagon or GTP + glucagon is decreased in hepatoma plasmamembranes. Fluoride stimulation is practically suppressed in hepatoma.

ACKNOWLEDGEMENTS

This work was supported by a grant from the Centro Nazionale delle Ricerche, Roma, Italy (Progetto Finalizzato Oncologia).

REFERENCES

1. M. U. Dianzani, R. A. Canuto, M. A. Rossi, G. Poli, R. Garcea, M. E. Biocca, G. Cecchini, F. Biasi, M. Ferro and A. M. Bassi, Further experiments on lipid peroxidation in transplanted and experimental hepatomas, Toxicol. Pathol. 12:189 (1984).
2. T. J. Player, Lipid peroxidation in rat liver, hepatomas and regenerating liver, in: "Free Radicals, Lipid Peroxidation and Cancer", D. C. H. McBrien and T. F. Slater, eds., Academic Press, London (1982).
3. E. Gravela, F. Feo, R. A. Canuto, R. Garcea and L. Gabriel, Functional and structural alteration of liver ergastoplasmic membranes during DL-ethionine hepatocarcinogenesis, Cancer Res. 35:3041 (1975).
4. M. U. Dianzani, G. Poli, R. A. Canuto, M. A. Rossi, M. E. Biocca, F. Biasi, G. Cecchini, G. Muzio, M. Ferro and H. Esterbauer, New data on kinetics of lipid peroxidation in experimental hepatomas and preneoplastic nodules, Toxicol. Pathol., 14:404 (1986).
5. S. Hauptlorenz, H. Esterbauer, W. Moll, R. Pumpel, E. Schauenstein and B. Puschendorf, Effects of the lipid peroxidation product 4-hydroxynonenal and related aldehydes on proliferation and viability of cultured Ehrlich ascites tumor cells, Biochem. Pharmacol. 34:3803 (1985).
6. M. U. Dianzani, Biochemical effects of saturated and unsaturated aldehyde, in: "Free Radicals, Lipid Peroxidation and Cancer, D. C. H. McBrien and T. F. Slater, eds., Academic Press, London (1982).
7. L. Paradisi, C. Panagini, M. Parola, G. Barrera and M. U. Dianzani, Effects of 4-hydroxynonenal on adenylate cyclase and 5'-nucleotidase activities in rat liver plasmamembranes, Chem. Biol. Interact. 53:209 (1985).

8. M. Curzio, H. Esterbauer, C. Di Mauro, G. Cecchini and M. U. Dianzani, Chemotactic activity of the lipid peroxidation product 4-hydroxynonenal and homologous hydroxyalkenals, *Biol. Chem. Hopper-Seyler* 367:321 (1986).

9. G. Barrera, S. Martinotti, V. Fazio, V. Manzari, L. Paradisi, M. Parola, L. Frati and M. U. Dianzani, Effect of 4-hydroxynonenal on c-*myc* expression, *Toxicol. Pathol.* (1987) in press.

10. G. Barrera, V. M. Fazio, S. Martinotti, V. Manzari, M. G. Farace, M. U. Dianzani and L. Frati, Differenziazione ed inibizione di c-*myc* in cellule K562 attraverso prodotti della lipoperossidazione cellulare, Atti V Congresso Nazionale ABCD, Sorrento (1986).

11. D. B. Solt, A. Medline and E. Farber, Rapid emergence of carcinogen-induced hyperplastic lesions in a new model for the sequential analysis of liver carcinogenesis, *Amer. J. Pathol.* 88:595 (1977).

12. M. W. Roomi, R. K. Ho, D. S. R. Sarma and E. Farber, A common biochemical pattern in preoplastic hepatocyte nodules generated in four different models in the rat, *Cancer Res.* 45:564 (1985).

13. T. G. Flynn, Aldehyde reductases: monomeric NADPH-dependent oxidoreductases with multifunctional potential, *Biochem. Pharmacol.* 31:2705 (1982).

14. H. Esterbauer, H. Zollner and J. Lang, Metabolism of the lipid peroxidation product 4-hydroxynonenal by isolated hepatocytes and by liver cytosolic fractions, *Biochem. J.* 228:363 (1985).

15. L. F. Chasseaud, The role of glutathione and glutathione-S-transferases in the metabolism of chemical carcinogens and other electophilic agents, *Adv. Cancer Res.* 29:175 (1979).

16. S. O. C. Tottmar, H. Petterson and K. H. Kiessling, The subcellular distribution and properties of aldehyde dehydrogenase in rat liver, *Biochem. J.* 135:577 (1973).

17. R. Lindahl, Subcellular distribution and properties of aldehyde dehydrogenase from 2-acetylaminofluorene-induced rat hepatomas, *Biochem. J.* 183:55 (1979).

18. A. M. Rutenburg, H. Kim, J. W. Fishbein, H. C. Wassemberg and A. M. Seligman, Histochemical and ultrastructural demonstration of γ-glutamyltranspeptidase activity, *J. Histochem. Cytochem.* 17:517 (1969).

19. A. H. Tashjian Jr., F. C. Bancroft, U. I. Richardson, M. B. Goldlust, F. A. Rommel and P. Ofner, Multiple differentiated functions in an unusual clonal strain of hepatoma cells, *In Vitro* 6:32 (1970).

20. E. B. Thompson, G. M. Tomkins and J. F. Curran, Induction of tyrosine transaminase by steroid hormones in a newly established tissue culture cell line, *Proc. Natl. Acad. Sci USA* 56:296 (1966).

21. G. Poli, E. Gravela, E. Albano and M. U. Dianzani, Studies on fatty liver with isolated hepatocytes. II. The action of carbon tetrachloride on lipid peroxidation, protein and triglyceride synthesis and secretion, *Exp. Mol. Pathol.* 30:116 (1979).

22. R. A. Canuto, R. Garcea, M. E. Biocca, R. Pascale, L. Pirisi and F. Feo, The subcellular distribution and properties of aldehyde dehydrogenase of hepatoma AH-130, *Eur. J. Cancer Clin. Oncol.* 19:389 (1983).

23. S. K. Srivastava, L. H. Ansari, G. A. Hair and B. Das, Aldose and aldehyde reductases in human tissues, *Biochim. Biophys. Acta* 800:220 (1984).

24. P. Alin, U. H. Danielson and B. Mannervik, 4-hydroxyalk-2-enals are substrates for glutathione transferase, *FEBS Lett.* 179:267 (1985).

25. A. G. Gornall, C. J. Bardawill and M. David, Determination of serum proteins by means of the biuret reaction, *J. Biol. Chem.* 177:751 (1949).

26. R. Lindahl and S. Evces, Changes in aldehyde dehydrogenase activity during diethylnitrosamine-initiated rat hepatocarcinogenesis, *Carcinogenesis* 8:785 (1987).

CHEMOPREVENTION OF EXPERIMENTAL TUMORIGENESIS BY

DEHYDROEPIANDROSTERONE AND STRUCTURAL ANALOGS

Arthur G. Schwartz[1], Laura L. Pashko[1], Laura A. Hastings[1],
Jeannette H. Whitcomb[1] and Marvin L. Lewbart[2]

[1]Fels Research Institute and Department of Microbiology
Temple University Medical School, Philadelphia, PA 19140 and
[2]Steroid Laboratory, Crozer-Chester Hospital, Chester PA 19013

INTRODUCTION

The human adrenal cortex secretes three classes of steroid hormone: glucocorticoid, mineralcorticoid and the so-called adrenal androgens, dehydroepiandrosterone (DHEA) and DHEA-sulfate. The term adrenal androgen is not
strictly accurate since DHEA per se is not androgenic and only through metabolism to steroids such as testosterone does it exert such action. DHEA-
sulfate is also a principal source of estrogen, through placental metabolism,
in the pregnant female[1,2]. However, in the normal male and female the proportion of sex steroids derived from adrenal DHEA is very small compared to the
gonadal contribution, and the significance of the adrenal secretion of DHEA
and DHEA-sulfate remains unclear[3].

DHEA AND BREAST CANCER

DHEA and DHEA-sulfate plasma levels undergo a marked and progressive
age-related decline beginning in the second decade and eventually fall to
10-20% of their maximal values in the seventh decade[4]. In contrast, during
the normal aging process basal plasma levels of cortisol and aldosterone show
little change[5,6]. The stimulation of cortisol and aldosterone levels following ACTH administration also shows no apparent change with age, whereas
stimulability of DHEA and DHEA-sulfate declines markedly[7].

In 1962 Bulbrook et al.[8] reported that women with primary operable
breast cancer excrete subnormal amounts of 11-deoxy-17-ketosteroids (derived
primarily from DHEA and DHEA-sulfate) prior to mastectomy and suggested that
this hormone abnormality might precede the onset of the disease. In order to
test this hypothesis, Bulbrook et al. undertook a prospective study in which
24-h urine specimens were collected from approximately 5000 women on the
island of Guernsey with no apparent breast cancer and who were followed for
nine years. At the end of that time 27 women had developed malignant breast
tumors. The excretion of various urinary steroids in the women who had developed breast cancer was compared with that of 187 carefully matched controls from the same population that had not developed a tumor, and it was
found that the excretion of androsterone and etiocholanolone (two principal
metabolites of DHEA) was lower in the women who had developed breast cancer[9].

CANCER PREVENTIVE ACTION

Over the past several years this laboratory and others have found that oral administration of DHEA to laboratory mice and rats produces a broad spectrum of cancer preventive action, inhibiting the development of spontaneous breast cancer[10] and chemically-induced tumors of the lung[11], colon[12] and thyroid[13], as well as pre-neoplastic foci in the liver[14]. Topical application of either DHEA or the synthetic steroid, 3 ß-methyl-5-androsten-17-one, to the skin of mice inhibits the development of 7,12-dimethylbenz(a)anthracene (DMBA)-initiated and tetradecanoylphorbol-13-acetate (TPA)-promoted papillomas at both the initiation and promotion stage[15] and also blocked the formation of skin papillomas and carcinomas produced by multiple topical application of DMBA[16].

DHEA is a potent uncompetitive inhibitor of mammalian glucose-6-phosphate dehydrogenase (G6PDH), but not of algal or yeast enzyme[17,18]. G6PDH is the first enzyme in the pentose-phosphate pathway, a major source of extramitochondrial NADPH. NADPH is an obligate co-factor for specific biosynthetic processes, including fatty acid and cholesterol biosynthesis and the synthesis of ribonucleotides and deoxyribonucleotides.

ANTI-INTITIATION BY DHEA

In 1975 our laboratory demonstrated that DHEA protected cultured rat liver epithelial-like cells and hamster embryonic fibroblasts against DMBA- and aflatoxin B_1-induced cytotoxicity and transformation, and inhibited the rate of metabolism of [^{3}H]DMBA to water-soluble products[19]. Carcinogens such as DMBA and aflatoxin B_1 require metabolic activation by mixed-function oxidases, NADPH-requiring enzymes, to reactive forms which are cytotoxic and carcinogenic[20]. Very probably DHEA protected cultured cells against these carcinogens by reducing NADPH production that, in turn, inhibited their metabolic activation.

Feo et al.[21] recently reported that cultured fibroblasts from individuals with the Mediterranean variant of G6PDH deficiency are less sensitive to the cytotoxic and transforming effects of benzo(a)pyrene (BP) and are less efficient in metabolizing [^{3}H]-BP into water-soluble products than are fibroblasts from normal individuals. Treatment of normal fibroblasts with DHEA mimicked the effect of G6PDH deficiency. The authors also reported a marked deficiency in pentose phosphate shunt activity and a lowering of the NADPH/NADP ratio in the G6PDH deficient fibroblasts. Similar results were obtained with lymphocytes from G6PDH deficient individuals[22].

In the mouse either topical application of DHEA to the skin at 100 μg or 400 μg[15] or oral administration in the diet at 0.6% for two weeks[23] inhibits the rate of binding of [^{3}H]DMBA to skin DNA. Again, inhibition of [^{3}H]DMBA-binding to skin DNA may result from depression of G6PDH activity and inhibition of carcinogen activation. Very probably the reduction in [^{3}H]DMBA-binding to skin DNA by DHEA accounts for the anti-initiating activity of the steroid against DMBA-initiated papillomas[15] as well as for the inhibition in development of DMBA-induced papillomas and carcinomas in the complete carcinogenesis model[16].

INHIBITION OF DNA SYNTHESIS

Topical DHEA treatment also inhibits papilloma formation when applied 1 h before TPA treatment, indicating that the steroid also blocks tumor promotion. The following experimental evidence suggests that DHEA inhibition of tumor promotion may also result from G6PDH inhibition by the steroid. Tumor promoters such as TPA produce diverse effects on mammalian cells, and the mechanism by which they enhance skin papilloma formation is not fully understood. TPA application to mouse skin stimulates epidermal DNA synthesis and hyperplasia [24,25]. Kinzel et al.[25] have provided strong experimental support for the notion that the induction of DNA synthesis by TPA is a necessary condition for its tumor-promoting action. The tumor-promotion phase can be subdivided into two stages. Stage I is brought about by a single application of TPA; it must be followed in stage II by repeated treatment with incomplete promoters such as 12-0-retinoylphorbol-13-acetate or mezerein. Kinzel et al.[25] produced tumors in mice by an initiating dose of DMBA followed by a single treatment with TPA (stage I) and then twice weekly applications of 12-0-retinoylphorbol-13-acetate (stage II). They found that a single intraperitoneal injection of mice with hydroxyurea, a reversible non-toxic inhibitor of DNA synthesis in mouse epidermis, given at different times before or after treatment with TPA interfered with tumor formation. Hydroxyurea treatment produced an almost complete inhibition of tumor formation if administered 18 h after TPA application, i.e., at the time of maximal DNA synthesis.

A single intraperitoneal injection of DHEA (10 mg/kg) into ICR mice immediately before TPA application also abolishes the TPA stimulation in epidermal [^{3}H]thymidine incorporation [26]. The synthetic steroid, 16α-Br-epiandrosterone (Epi-Br), a compound that is 30 to 50 times more potent as an inhibitor of mammalian G6PDH [27], is also much more active as an inhibitor of the TPA stimulation of [^{3}H]thymidine incorporation - an injected dose of 0.4 mg/kg of Epi-Br was more effective than 10 mg/kg of DHEA[26]. Of the various sex steroids and a glucocorticoid that were tested, all of which are inactive G6PDHinhibitors, only corticosterone inhibited the rate of [^{3}H]thymidine incorporation. Glucocorticoids are known to inhibit epidermal DNA synthesis [28].

Specific 3T3 fibroblast clones can, under carefully defined conditions, be stimulated to undergo highly efficient differentiation into adipocytes [29]. Adipocyte differentiation is a multistep process, the mechanism of which is unclear. Increases in thymidine incorporation, in cell number, and in translatable mRNAs precede the elevation of marker enzymes of lipid synthesis, and of increased incorporation of acetate into lipid [29]. Using the above assay of conversion of 3T3 preadipocytes to adipocytes, Gordon et al.[30] found that DHEA and Epi-Br blocked the conversion to adipocytes of the 3T3-L1 and 3T3-F442A mouse embryo fibroblast clones. They also observed that Epi-Br was much more potent than DHEA in blocking conversion and suggested that G6PDH inhibition by these steroids very likely accounted for this effect.

Two major functions usually ascribed to the oxidative branch of the pentose phosphate cycle are the generation of NADPH for reductive biosynthesis and other specific metabolic reactions as well as the formation of ribose-5-phosphate, which is utilized in nucleotide biosynthesis via the intermediate 5-phosporibosyl-1-pyrophosphate. Purine ribonucleotide and thymidylic acid biosynthesis are dependent upon tetrahydrofolic acid, which requires NADPH for its reductive synthesis from folic acid. Also the enzymatic formation of deoxyribonucleotide diphosphates from their corresponding ribonucleotide diphosphates by ribonucleotide reductase is NADPH dependent.

If DHEA and related steroids repress DNA synthesis in cells by reducing ribonucleotide and deoxyribonucleotide synthesis as a result of G6PDH inhibition, then provision of these nucleosides would be expected to reverse the DHEA-induced inhibition. This was indeed shown to be the case with cultured HeLa TCRC-2 cells. DHEA at a concentration of 10^{-5} M inhibited the growth rate of these cells in culture, and this growth inhibition was almost completely overcome by adding to the culture medium a mixture of the deoxynucleosides of adenine, guanine, thymine and cytosine [31]. Gordon et al.[32] have also found that the addition of the four ribonucleosides (uridine, cytidine, adenosine and guanosine) to cultured 3T3-L1 cells almost completely prevented the blocking action of Epi-Br on the differentiation of these cells to adipocytes.

Feo et al. have found that ribo- or deoxyribonucleosides are also effective in vivo in overcoming an anti-promotional effect of DHEA in rat liver. Preneoplastic liver foci, which are γ-glutamyltranspeptidase (GGT) positive, are produced in rats by an initiating dose of diethylnitrosamine followed, 15 days later, by a 15-day feeding of a diet containing 2-acetylaminofluorene with a partial hepatectomy at the midpoint of this feeding and then two weeks of feeding a phenobarbital-containing diet. If DHEA (0.6%) was administered in the diet during the period of phenobarbital treatment, there was a 50% reduction in the percent of liver occupied by GGT-positive foci and a 70% decrease in the labeling index of foci cells. These inhibitory effects of DHEA were prevented by the i.p. injection of a mixture of the ribo- or deoxyribonucleosides of adenine, guanine, cytosine and uracil/thymine [33].

INHIBITION OF SUPEROXIDE FORMATION

In addition to an NADPH requirement for mixed function oxidase activity and for specific enzymatic reactions in the biosynthesis of ribonucleotides and deoxyribonucleotides, NADPH is a coenzyme for a membrane-bound oxidase found in granulocytes and macrophages which generates superoxide anion (O_2^-)[34]. O_2^- and other forms of reactive oxygen can induce deletion mutations and chromosomal aberrations [35,36] and may exert critical effects in the tumor promotion process. X-irradiation of cultured hamster embryonic fibroblasts followed by TPA treatment induces malignant transformation, and this process can be blocked by treating with superoxide dismutase (SOD) along with TPA [37]. Kensler et al.[38] found that topical application of a low-molecular weight copper complex, bis[(3,5-diisopropyl-salicylato)(0,0)copper(II)] that has SOD-mimetic action, together with TPA in the two-stage skin tumor system markedly inhibited papilloma formation. TPA rapidly stimulates O_2^- formation by granulocytes in vitro [39]. DHEA inhibits this stimulation and the synthetic steroid, Epi-Br, is a more potent inhibitor, again suggesting that G6PDH inhibition and a reduction in NADPH pool size is the probable mechanism of inhibition [40].

Thus three processes which contribute to tumor development: 1) metabolic activation of a carcinogen through the action of mixed-function oxidases, 2) tumor promoter stimulation of cell proliferation, and 3) tumor promoter stimulation of O_2^- formation are all inhibited by DHEA, probably as a result of G6PDH inhibition and a lowering of the NADPH cellular pool. We have recently obtained evidence indicating that these first two processes may also be inhibited in mouse epidermis following a regimen of 40% food restriction for two weeks, suggesting a similarity in the mechanism by which underfeeding and DHEA inhibit tumorigenesis. Specifically, reducing the food intake of A/J mice (60% of ad libitum fed) for two weeks inhibits the rate of binding of [³H]DMBA to mouse skin DNA by 50% [41]. A similar reduction in [³H]DMBA binding to skin DNA is produced by topical application of DHEA immediately before DMBA treatment. The TPA stimulation in the rate of [³H]thymidine incorporation in mouse epidermis is also abolished in food restricted animals, an effect that is also seen with topical DHEA treatment [41]. We found that mouse epidermal G6PDH activity was depressed by 60% following two weeks of food restriction [41], and when considered with the evidence that the inhibition of [³H]DMBA binding to DNA and TPA stimulation in [³H]thymidine incorporation by DHEA very possibly results from G6PDH inhibition, it is reasonable to assume that the depression in epidermal G6PDH activity following underfeeding may exert a similar effect. Similarly, Dessi et al. found that lead nitrate stimulation of rat hepatic G6PDH activity, DNA synthesis rate, and cell proliferation were all depressed in food restricted rats [42]. Although further experimental evidence in other organs on G6PDH activity levels following food restriction is clearly needed, it is an intriguing possibility that under-feeding depresses G6PDH activity, with a consequent inhibition of initiation and promotion, and that this may contribute to the tumor preventive activity of this regimen.

INHIBITION OF TPA STIMULATION OF PROSTAGLANDIN E_2 LEVELS

A fourth process important in experimental tumorigenesis is also blocked by treatment of mice with DHEA and specific structural analogs. Preliminary evidence suggests that this effect of DHEA, again, may result from G6PDH inhibition.

Stimulated production of various eicosanoid compounds, such as leuko-trienes and prostaglandins, is believed to play an important role in the enhancement of tumorigenesis by tumor promoters, such as TPA in the skin[43] and bile salts in the colon [44]. Arachidonic acid is metabolized to different eicosanoids through both the 5-lipoxygenase pathway, which generates various leukotrienes, and the cyclooxygenase pathway, which is the source of specific prostaglandins, prostacyclins, and thromboxanes[45,46]. Inhibition of lipoxygen-ase blocks TPA-induced tumor promotion in mouse skin [43] and the work of Fischer and Adams[47] suggests that a major source of TPA stimulated reactive oxygen produced in isolated mouse epidermal cells is due to the metabolism of arachidonic acid via the lipoxygenase pathway. Fürstenberger and Marks[48] have found that TPA-induced increase in prostaglandin E_2 (PGE_2) content in mouse epidermis may be an obligatory event in the TPA enhancement of the DNA synthesis rate, which plays a central role in tumor promotion[25]. Indometha-cin treatment blocks both the TPA stimulation in epidermal PGE_2 content as well as the stimulation in [³H]thymidine incorporation.

Topical application of TPA to skin of CD-1 mice stimulates epidermal PGE_2 levels 9-fold 24 h later. We have found that treatment of mice for 10 days with a diet containing 0.2% of either 8354 or 8356, two synthetic DHEA analogs which are more potent G6PDH inhibitors than DHEA, prior to TPA appli-cation produces a 15-fold reduction in the TPA stimulation in PGE_2 levels. Treatment with 0.2% DHEA containing diet reduces the stimulated skin content by 3-fold, and 0.2% eticholanolone, a weak G6PDH inhibitor, reduces it only 1.3-fold [49]. The overall correlation with G6PDH inhibitory activity again suggests a possible role for G6PDH inhibition in the mechanism of action. One possible explanation is that these steroids reduce the rate of conversion of linoleic acid to arachidonic acid, a biosynthesis requiring NADPH [50]. How-ever, further work is needed to establish the mechanism of this biological action of the DHEA class of steroid.

ACKNOWLEDGEMENTS

This work was supported by Grant CA-38574 from the National Cancer Institute, Cancer Prevention Grant SIG-6 from the American Cancer Society, the Advanced Technology Center of Southeastern Pennsylvania and the Samuel S. Fels Fund.

We thank Denise Fairman and Tanya Howard for their excellent technical assistance.

REFERENCES

1. E. E. Baulieu and F. Dray, Conversion of [³H]dehydroisoandrosterone (3 ß-hydroxy-Δ^5-androsten-17-one) sulfate to [³H]estrogens in normal

pregnant women, J. Clin. Endocrinol. Metab. 23:1298 (1963).

2. P. K. Siiteri and P. C. MacDonald, The utilization of circulating dehydroisoandrosterone sulfate for estrogen synthesis during human pregnancy, Steroids 2:713 (1963).

3. S. Lieberman, Pictorial endocrinology and empirical hormonology, The Sir Henry Dale Lecture for 1986, J. Endocr. 111:519 (1986).

4. N. Orentriech, J. L. Brind, R. L. Rizer and J. H. Vogelman, Age changes and sex differences in serum dehydroepiandrosterone sulfate concentrations throughout adulthood, J. Clin. Endocrinol. Metab. 59:551 (1984).

5. C. Gherondache, L. Romanoff and G. Pincus, Steroid hormones in aging men, in: "Endocrines and Aging", L. Gitman, ed.,Charles Thomas, Springfield, IL (1967).

6. P. Weidman, S. deMyttenaere-Bursztein, M. Maxwell, and J. deLima, Effect of aging on plasma renin and aldosterone in normal man, Kidney Int. 8:325 (1975).

7. L. Parker, T. Gral, V. Perrigo and R. Shorosky, Decreased adrenal androgen sensitivity to ACTH during aging, Metabolism 30:601 (1981).

8. R. D. Bulbrook, J. L. Hayward, C. C. Spicer and B. S. Thomas, Abnormal excretion of urinary steroids by women with early breast cancer Lancet ii:1238 (1962).

9. R. D. Bulbrook, J. L. Hayward and C. C. Spicer, Relation between urinary androgen and corticoid excretion and subsequent breast cancer, Lancet ii:395 (1971).

10. A. G. Schwartz, Inhibition of spontaneous breast cancer formation in female C3H (A^{vy}/a) mice by long-term treatment with dehydroepiandrosterone, Cancer Res. 39:1129 1979)(.

11. A. G. Schwartz and R. H. Tannen, Inhibition of 7,12-dimethylbenz(a)anthracene- and urethan-induced lung tumor formation in A/J mice by long-term treatment with dehydroepiandrosterone, Carcinogenesis 2:1335 (1981).

12. J. W. Nyce, P. N. Magee, G. C. Hard and A. G. Schwartz, Inhibition of 1,2-dimethylhydrazine-induced colon tumorigenesis in Balb/c mice by dehydroepiandrosterone, Carcinogenesis 5:57 (1984).

13. M. A. Moore, W. Thamavit, A. Tsuda, K. Sato, A. Ichihara and N. Ito, Modifying influence of dehydroepiandrosterone on the development of dihydroxy-di-n-propylnitrosamine-initiated lesions in the thyroid, lung and liver of F344 rats, Carcinogenesis 7:311 (1986).

14. R. Garcea, L. Daino, R. Pascale, S. Frassetto, P. Cozzolino, M. E. Ruggiu and F. Feo, Inhibition by dehydroepiandrosterone of liver preneoplastic foci formation in rats after initiation-selection in experimental carcinogenesis, Toxicol. Path. 15:164 (1987).

15. L. L. Pashko, R. J. Rovito, J. R. Williams, E. L. Sobel and A. G. Schwartz, Dehydroepiandrosterone (DHEA) and 3 ß-methylandrost-5-en-17-one: inhibitors of 7,12-dimethylbenz(a)anthracene (DMBA)-initiated and 12-O-tetradecanoylphorbol-13-acetate (TPA)-promoted skin papilloma formation in mice, Carcinogenesis 5:463 (1984).

16. L. L. Pashko, G. C. Hard, R. J. Rovito, J. R. Williams, E. L. Sobel and A. G. Schwartz, Inhibition of 7,12-dimethylbenz(a)anthracene-induced skin papillomas and carcinomas by dehydroepiandrosterone and 3 ß-methylandrost-5-en-17-one in mice, Cancer Res. 45:164 (1985).

17. P. A. Marks and J. Banks, Inhibition of mammalian glucose-6-phosphate dehydrogenase by steroids, Proc. Natl. Acad. Sci. USA 46:447 (1960).

18. G. W. Oertel and P. Benes, The effects of steroids on glucose-6-phosphate dehydrogenase, J. Steroid Biochem. 3:493 (1972).

19. A. G. Schwartz and A. Perantoni, Protective effect of dehydroepiandrosterone against aflatoxin B$_1$ and 7,12-dimethylbenz(a)anthracene-induced cytotoxicity and transformation in cultured cells, Cancer Res. 35:2482 (1975).

20. J. A. Miller, Carcinogenesis by chemicals: an overview, Cancer Res. 30:559 (1970).

21. F. Feo, L. Pirisi, R. Pascale, L. Daino, S. Frassetto, R. Garcea and L. Gaspa, Modulatory effect of glucose-6-phosphate dehydrogenase deficiency on benzo(a)pyrene toxicity and transforming activity for in vitro-cultured human skin fibroblasts, Cancer Res. 44:3419 (1984).

22. F. Feo, L. Pirisi, R. Pascale, L. Daino, S. Frassetto, S. Zanetti and R. Garcea, R., 1984, Modulatory mechanisms of chemical carcinogenesis: the role of the NADPH pool in benzo(a)pyrene activation, Toxicol. Path. 11:261 (1984).

23. L. L.Pashko and A. G. Schwartz, Effect of food restriction, dehydroepiandrosterone, or obesity on the binding of [^{3}H]-7,12-dimethylbenz(a)anthracene to mouse skin DNA, J. Gerontol. 38:8 (1983).

24. T. S. Argyris, Nature of epidermal hyperplasia produced by mezerein, a weak tumor promoter, in initiated skin of mice, Cancer Res. 43:1768 (1983).

25. V. Kinzel, H. Lochrke, L. Goertler, G. Fürstenberger and F. Marks, Suppression of the first stage of phorbol 12-tetradecanoate-13-acetate-effected tumor promotion in mouse skin by non-toxic inhibition of DNA synthesis, Proc. Natl. Acad. Sci. USA 81:5858 (1984).

26. L. L. Pashko, A. G. Schwartz, M. Abou-Gharbia and D. Swern, Inhibition of DNA synthesis in mouse epidermis and breast epithelium by dehydroepiandrosterone and related steroids, Carcinogenesis 2:717 (1981).

27. R. Raineri and H. R. Levy, On the specificity of steroid interaction with mammary gland glucose-6-phosphate dehydrogenase, Biochemistry 9:2233 (1970).

28. C. W. Castor and B. L. Baker, The local action of adrenocortical steroids on epidermis and connective tissue of the skin, J. Endocrinol. 47:234 (1950).

29. H. Green and O. Kehinde, An established preadipose cell line and its differentiation in culture II. Factors affecting the adipose conversion, Cell 5:19 (1975).

30. G. B. Gordon, J. A. Newitt, L. M. Shantz, D. E. Weng and P. Talalay, Inhibition of the conversion of 3T3 fibroblast clones to adipocytes by dehydroepiandrosterone and related anticarcinogenic steroids, Cancer Res. 46:3389 (1986).

31. C. R. Dworkin, S. D. Gorman, L. L. Pashko, V. J. Cristofallo and A. G. Schwartz, Inhibition of growth of HeLa and WI-38 cells by dehydroepiandrosterone and its reversal by ribo- and deoxyribonucleosides, Life Sci. 38:1451 (1986).

32. G. B. Gordon, L. M. Shantz and P. Talalay, Modulation of growth, differentiation and carcinogenesis by dehydroepiandrosterone, Adv. Enzyme Regulation 26:355 (1987).

33. F. Feo, R. Garcea, L. Daino, S. Frassetto, P. Cozzolino, M. E. Ruggiu, M. G. Vannini, R. Pascale, L. Lenzerini, M. M. Simile and M. Puddu, Inhibition of hepatocarcinogenesis promotion by dehydroepiandrosterone and its reversal by ribo- and deoxyribonucleosides, Fourth Sardinian International Meeting: Models and Mechanisms in Chemical Carcinogenesis, abstract, (1987).

34. B. M. Babior, The enzymatic basis for O_2^- production by human neutrophils, Can. J. Physiol. Pharmacol. 60:1353 (1982).

35. I. Emerit and P. Cerutti, Clastogenic action of tumor promoter phorbol-12-myristate-13-acetate in mixed human leukocyte cultures, Carcinogenesis 4:1313 (1983).

36. A. W. Hsie, L. Recio, D. S. Katz, C. Q. Lee, M. Wagner and R. L. Schenley, Evidence for reactive oxygen species inducing mutations in mammalian cells, Proc. Natl. Acad. Sci. USA 83:9616 (1986).

37. C. Borek and W. Troll, Modifiers of free radicals inhibit in vitro the oncogenic actions of X-rays, bleomycin and the tumor promoter 12-0-tetradecanoylphorbol-13-acetate, Proc. Natl. Acad. Sci. USA 80:1304 (1983).

38. T. W. Kensler, D. M. Bush and W. J. Kozumbo, Inhibition of tumor promotion by a biomimetic superoxide dismutase, Science 221:75 (1983).

39. B. D. Goldstein, G. Witz, M. Amoruso, D. S. Stone, and W. Troll, Stimulation of human polymorphonuclear leukocyte superoxide anion radical production by tumor promoters, Cancer Lett. 11:257 (1981).

40. J. M. Whitcomb and A. G. Schwartz, Dehydroepiandrosterone and 16α-Brepiandrosterone inhibit 12-0-tetradecanoylphorbol-13-acetate stimulation of superoxide radical production by human polymorphonuclear leukocytes, Carcinogenesis 6:333 (1985).

41. A. G.Schwartz and L. L. Pashko, Food restriction inhibits [^{3}H] 7,12-dimethylbenz(a)anthracene binding to mouse skin DNA and tetradecanoylphorbol-13-acetate stimulation of epidermal [^{3}H]thymidine incorporation, Anticancer Res. 6:1279 (1986).

42. S. Dessi, B. Batetta, D. Pulixi, A. Carrucciu, M. Armeni, M. F. Mulas and P. Pani, Modifying influence of fasting on DNA synthesis, cholesterol metabolism and HMP shunt enzymes in liver hyperplasia induced by lead nitrate, Fourth Sardinian International Meeting: Models and Mechanisms in Chemical Carcinogenesis, abstract, (1987).

43. E. Aizu, T. Nakadate, S. Yamamoto and R. Kato, Inhibition of 12-0-tetradecanoylphorbol-13-acetate mediated epidermal ornithine decarboxylase induction and skin tumor promotion by new lipoxygenase inhibitors lacking protein kinase C inhibitory effects, Carcinogenesis 7:1809 (1986).

44. P. A. Craven, J. Pfanstiel and F. R. DeRubertis, Role of activation of protein kinase C in the stimulation of colonic epithelial proliferation and reactive oxygen formation by bile acids, J. Clin. Invest. 79:532 (1987).

45. Y. S. Bakhle, Synthesis and catabolism of cyclo-oxygenase product, Brit. Med. Bull. 39:214 (1983).

46. G. W. Taylor and H. R. Morris, Lipoxygenase pathways, Brit. Med. Bull. 39:219 (1983).

47. S. M. Fischer and L. M. Adams, Suppression of tumor promoter-induced chemiluminescence in mouse epidermal cells by several inhibitors of arachidonic acid metabolism, Cancer Res. 45:3130 (1985).

48. G. Fürstenberger and F. Marks, Indomethacin inhibition of cell proliferation induced by the phorbolester TPA is reversed by prostaglandin E$_2$ in mouse epidermis, Biochem. Biophys. Res. Commun. 84:1103 (1978).

49. L. A. Hastings, L. L. Pashko, M. L. Lewbart and A. G. Schwartz, Dehydroepiandrosterone and two structural analogs inhibit tetradecanoylphorbol-13-acetate stimulation of prostaglandin E$_2$ content in mouse skin, submitted.

50. M. Nagao, T. Ishibashi, T. Okayasu and Y. Imai, Possible involvement of NADPH-cytochrome P-450 reductase and cytochrome b$_5$ on ß-ketostearoyl-CoA reduction in microsomal fatty acid chain elongation supported by NADPH, FEBS Letters 155:11 (1983).

INHIBITION OF INITIATION AND PROMOTION STEPS OF CARCINOGENESIS

BY GLUCOSE-6-PHOSPHATE DEHYDROGENASE DEFICIENCY

Francesco Feo, Renato Garcea, Lucia Daino, Serenella Frassetto, Patrizia Cozzolino, Maria E. Ruggiu, Maria G. Vannini, Rosa Pascale, Luciano Lenzerini, Maria M. Simile and Marco Puddu

Istituto di Patologia generale, Università di Sassari
07100 Sassari, Italy

INTRODUCTION

A number of observations indicate that deficiency of glucose-6-phosphate dehydrogenase (G6PD), either genetically transmitted or caused by dehydroepiandrosterone (DHEA)[1], or some related steroids, is associated with an anti-tumor effect. Some epidemiologial evidence of a decreased tumor incidence in G6PD-deficient subjects has been reported [2-5]. Although these observations, do not definitively prove the existence of clear relationships between G6PD-deficiency and tumor incidence, they suggest a negative correlation. In accordance with this hypothesis retrospective studies have shown subnormal plasma levels of DHEA or DHEA-sulfate in women with breast cancer[6,7]. Moreover, in a prospective study a subnormal urinary excretion of the DHEA metabolites androsterone and etiocholanolone has been found to be associated with enhanced breast cancer risk [8]. DHEA mimics many of the effects of the genetically transmitted G6PD deficiency as concerns the resistance of in vitro cultured cells to the toxic and transforming effects of carcinogens. Following the first observation by Schwartz[9] that long-term per os treatment with DHEA inhibits the formation of spontaneous tumors in mice, different other studies have shown that this treatment also prevents the development of chemically induced tumors in the same animal[10-13], as well as in the rat [14]. DHEA is known to exhibit an anti-obesity effect (cf. ref. 15 for review). Even if in some rat strains DHEA causes a decrease in food intake, it has been suggested that the anti-obesity effect is linked to a decreased food utilization more than to a reduced intake[16]. Alternatively, an increased ß-oxidation of fatty acids in peroxisomes, could represent an "energy wasting" pathway in DHEA-treated animals [17]. Reduction in food intake is known to decrease the susceptibility of different tissues to cancer [18]. However, the possibility that the DHEA anti-tumor action is linked to food restriction, is ruled out by the observation that topic application of DHEA, while not reducing body weight, greatly inhibits development of skin tumor in mice [12].

CARCINOGEN METABOLISM IN G6PD-DEFICIENT CELLS

It has been hypothesized that the fall in NADPH pool consequent to G6PD-deficiency results in a decreased capacity of deficient cells to metabolize carcinogens by the NADPH-dependent mixed function oxygenases, this could lead to a decreased tumor initiation[19-21]. In order to test the above hypothesis,

369

different studies have analyzed the metabolism of various carcinogenic compounds in G6PD-deficient cells. The results may be summarized as follows: (a) rat liver epithelial like cells, cultured in vitro in the presence of DHEA[19], or in vitro cultured G6PD-deficient human lymphocytes and fibroblasts[20-22] are more resistant than controls to the toxic effect of various carcinogens. (b) Lymphocytes carrying the Mediterranean variant of G6PD exhibit a lower aryl hydrocarbon hydroxylase activity, when this enzymatic activity is tested in a reaction system containing exogenous G6P and NADP$^+$ and endogenous G6PD; the addition of exogenous G6PD restores the control values[22]. Moreover, NADPH cytochrome c reductase is lower in homogenates of G6PD-deficient cells when tested in the above conditions. (c) In vitro cultured epithelial like cells produce lower amounts of total water-soluble metabolites of 7,12-dimethyl-benzo(a)anthracene (DMBA) when incubated with DHEA or epiandrosterone[19]. A decreased synthesis of BaP metabolites, mutagenic for his$^-$ Samonella typhimurium, occurs in lymphocytes carrying the G6PD variant. This effect is enhanced by further reducing G6PD activity by addition of DHEA to the deficient cells. (d) Isolation of single organic-soluble or water-soluble BaP metabolites produced by G6PD-deficient lymphocytes and fibroblasts during incubation in vitro with BaP, has shown that the deficient cells do not undergo evident modifications of BaP metabolism, excepting for a great fall in the metabolite production[23,24]. G6PD-deficient lymphocytes show a decreased ability to form BaP-7,8-diol-9,10-epoxide and BaP-DNA adducts. Low DNA methylation has been found in the deficient fibroblasts after incubation with dimethylnitrosamine[20], while inhibition of [^{3}H]DMBA binding to skin DNA has been obtained by oral administration of diet containing 0.6% DHEA[25] or by topical application of DHEA to the skin[12].

Recent evidence seems to indicate that NADPH liver content is rate-limiting in BaP oxidation by the mixed function oxygenases[24]. This suggests that a fall in NADPH pool may explain the reduced ability of G6PD-deficient cells to metabolize carcinogens. The possibility of a reduced carcinogenesis initiation in G6PD-deficient cells, which could explain the anti-tumor effect of G6PD deficiency, may be thus envisaged. This hypothesis seems to be substantiated by Pashko et al.'s interesting observation of an anti-initiating effect of DHEA during DMBA-induced mouse skin carcinogenesis[12].

EFFECT OF G6PD-DEFICIENCY ON TUMOR PROMOTION

A clear anti-promoting effect of DHEA during the development DMBA-induced and 12-O-tetradecanoylphorbol-13-acetate (TPA)-promoted skin papillomas has also been shown[12]. This observation is particularly important in view of the fact that preneoplastic and neoplastic lesions of the colon and breast[26], skin[27], oral mucosa[28], and liver[29] exhibit an elevation in G6PD-activity. A high G6PD activity is crucial for rapid cell growth. Recent experiments, performed in the Ito's as well as in our laboratory, have extended the study of DHEA effect to liver carcinogenesis promotion[14,30-33]. Enzyme-altered foci (EAF), which develop in the liver after initiation with N-ethyl-N-hydroxy-ethylnitrosamine or after initiation with diethylnitrosamine and then selection according to the resistant hepatocyte (RH) model[34] of experimental hepatocarcinogenesis, exhibit a large increase in G6PD, malic enzyme (ME) and isocitric dehydrogenase (ICD) activities[29,33]. The administration of 0.6% DHEA with the diet, during the development of EAF, causes a great fall in liver and EAF G6PD activity as well as in the development of EAF[14,30-33], while it enhances the liver ME and ICD activities[33]. A high histochemical reaction for ME activity has also been found in EAF of DHEA-treated animals[33]. Thus, EAF show a high NADPH-generating potential which presumably is not modified by DHEA.

These observations indicate that, at least in the liver, the DHEA anti-promoting effect cannot be attributed to the production of NADPH amounts insufficient for NADPH-dependent functions, such as DNA and cholesterol synthesis which are particularly active in rapidly growing tissues. Alternative explanations deal on an inhibitory effect of DHEA on DNA synthesis and/or free radical production.

DHEA and some related steroids inhibit DNA synthesis of TPA-stimulated mouse skin cells [35]. The potency in overcoming TPA-stimulated DNA synthesis is correlated to the capacities of different compounds to inhibit G6PD. Administration of 0.6% DHEA, during the development of EAF, in rats subjected to the initiation/selection treatments, causes a decrease in DNA synthesis in the focal cells [32,33,36], associated with phenotypic reversion of these cells [33,36]. In addition, DHEA slows the growth of different cell lines in culture [37,38], and inhibits in vitro differentiation of 3T3 fibroblasts to adipocytes, a phenomenon closely linked to cell proliferation, lipid synthesis and morphological changes [38]. Interestingly, the addition of a mixture of four ribonucleosides of adenine, guanine, cytosine and uracil (RNs) to the reaction medium results in a partial reversal of DHEA inhibition of growth of cultured cells [37]. A complete reversal has been observed with a mixture of four deoxiribonucleosides of adenine, guanine, cytosine and thymine (DRNs) [37]. DRNs also prevent the inhibition by DHEA of in vitro differentiation of 3T3 cells [38].

These observations could indicate that ribose deficiency is crucial for growth inhibition by DHEA. In the attempts to evaluate if similar behavior occurred in vivo during hepatocarcinogenesis promotion, we have studied the effect of RNs and DRNs on the inhibition by DHEA of EAF development, after initiation/selection, according to the RH model. RNs or DRNs administered to rats together with DHEA, completely overcome growth inhibition of EAF by DHEA, while the DHEA-induced inhibition of G6PD and stimulation of ME activities are not affected [33]. On the basis of these findings it has been concluded that reduced availability of nucleosides may be responsible for growth inhibition by DHEA and the DHEA anti-promoting effect is largely dependent on inhibition of DNA synthesis. Even though RNA synthesis has not been studied, a similar conclusion may also apply to this synthesis in DHEA-treated rats. Interestingly, RNs and DRNs equally reverse DHEA inhibition [33]. This indicates that hepatocytes of DHEA-treated rats, produce enough NADPH to ensure the reduction of RNs to DRNs. It may thus be hypothesized that growth inhibition by DHEA is largely linked to reduced availability of pentose phosphates for DNA synthesis. Production of ribulose-5-phosphate is indeed decreased in G6PD-deficient hepatocytes [33].

Reactive species of oxygen have been implicated in the development of liver, skin and mammary gland carcinogenesis [39-41]. The formation of O_2^- and its dismutation product H_2O_2 seems to be critical for promotion by TPA [42]. This suggests that modulation of O_2^- production may influence promotion by TPA and, maybe, other promoters. G6PD-deficiency, either genetically transmitted or caused by DHEA or related hormones, markedly reduces the capacity of human polymorphonuclear leukocytes (PMNs) to produce O_2^- under TPA stimulation [43,44]. This phenomenon is apparently correlated with the degree of G6PD deficiency. Interestingly, activation by promoters of human PMNs to produce H_2O_2, and incubation of activated PMNs with DNA, causes the formation of modified thymine. The extent of this phenomenon is correlated with the promoting activitity of the various compounds tested [46]. In addition, in vitro exposure of mouse fibroblasts to human PMNs, stimulated to produce reactive oxygen species, causes the development of both malignant and benign tumors in nude mice injected with treated cells, but not in those receiving control cells [47]. These findings, considered together, seem to indicate that O_2^- production could be involved in different steps of carcinogenesis. Inhibition of the production of these intermediates by DHEA or related hormones, could be a mechanism through which these hormones prevent tumor development. Further work however is necessary in order to substantiate this hypothesis.

CONCLUSIONS

Initiation, promotion and progression are the main steps in cancer development. Available evidence indicates that DHEA interferes with the first two mechanisms. Experiments are presently in progress in our laboratory to analyze DHEA effect on the progression of stable preneoplastic nodules to cancer. The interference of G6PD deficiency with some steps of cancer development could explain its anti-tumor effect. Unfortunately, the absence of reliable epidemiological studies hinder any definitive conclusion on the susceptibility to cancer of subjects carrying the Mediterranean variant of G6PD. It should be noted that it is extremely difficult to assess this susceptibility, due to the relatively limited population carrying the variant enzyme and the difficult to control all the factors which modulate, positively or negatively, carcinogenesis in each individual, independently of G6PD activity. However, on the basis of available knowledge it may be envisaged that the use of hormones which inhibit G6PD activity could be a promising mean to inhibit development of most preneoplastic lesions.

Different factors may modulate the various steps of the carcinogenic process. The use of experimental models to study positive modulation could be important in cancer prevention, especially if the same or analogous modulatory mechanisms may be expected to occur in man, so representing risk factors. Negative modulators may be used to prevent cancer development and obtain the reversion of preneoplastic tissue to normally differentiated tissue. Among the various substances which control various stages of tumor development, DHEA is particularly interesting due to the fact that it is a naturally occurring and non toxic compound. Additional work however is needed to really assess the therapeutic potentiality of this compound.

ACKNOWLEDGEMENTS

Supported by grants from CNR (Prog. Fin. Ingegn. Genet.), MPI (program 60%) and Regione Autonoma Sardegna (Prog. Fin. Ass. Sanità).

REFERENCES

1. R. Ranieri and H. R. Levy, On the specificity of steroid interaction with mammalian glucose-6-phosphate dehydrogenase, Biochemistry 9: 2233 (1970).
2. P. Beaconfield, T. Rainsburg and G. Kalton, Glucose-6-phosphate dehydrogenase deficiency and the incidence of cancer, Oncologia 19:11 (1965).
3. S. N. Naik and D. E. Anderson, The association between glucose-6-phosphate dehydrogenase deficiency and cancer in American Negroes, Oncologia 25 356 (1971).
4. E. Sulis, Glucose-6-phosphate dehydrogenase and cancer, Lancet 2:1335 (1972).
5. M. Mbensa, C. Rwakunda and R. L. Verwilighen, Glucose-6-phosphate dehydrogenase deficiency and malignant hepatoma in Bantu population, E. Afr. med. J. 57:17 (1978).
6. D. Y. Wang, R. D. Bulbrok, M. Herian and J. L. Hayward, Studies on the sulphate esters of dehydroepiandrosterone and androsterone in the blood of women with breast cancer, Eur. J. Cancer 10:477 (1974).
7. B. Zumoff, J. Levin, R. S. Rosenfeld, M. Markham, G. W. Strain and D. K. Fukushima, Abnormal 24-hr mean plasma concentrations of dehydroisoandrosterone and dehydroisoandrosterone sulfate in women with primary operable breast cancer, Cancer Res. 41:3360 (1981).
8. R. D. Bulbrook, J. L. Hayward and C.C. Spicer, Relation between urinary androgen and corticoid excretion and subsequent breast cancer, Lancet

2:395 (1971).

9. A. G. Schwartz, Inhibition of spontaneous breast cancer formation in female C3H-A /A mice by long-term treatment with dehydroepiandrosterone, Cancer Res. 39:1129 (1979).

10. A. G. Schwartz and R. H. Tannen, Inhibition of 7,12-dimethylbenz(a)-anthracene- and uretan-induced lung tumor formation in A/J mice by long-term treatment with dehydroepiandrosterone, Carcinogenesis 2: 1335 (1981).

11. J. W. Nyce, P. N. Magee, G. C. Hard and A. G. Schwartz. Inhibition of 1,2-dimethylhydrazine-induced colon tumorigenesis in Balb/c mice by dehydroepiandrosterone, Carcinogenesis 5:57 (1984).

12. L. L. Pashko, R. J. Rovito, J. R. Williams, E. L. Sobel and A. G. Schwartz, Dehydroepiandrosterone (DHEA) and 3ß-methylandrost-5-en-17-one: inhibitors of 7,12-dimethylbenz(a)anthracene (DMBA)-initiated and 12-O-tetradecanoylphorbol-13-acetate (TPA)-promoted skin papilloma formation in mice, Carcinogenesis 5:463 (1984).

13. L. L. Pashko, G. C. Hard, R. J. Rovito, J. R. Williams, E. L. Sobel and A. G. Schwartz, Inhibition of 7,12-dimethylbenz(a)anthracene-induced skin papillomas and carcinomas by 3ß-methylandrost-5-en-17-one, Cancer Res. 45:164 (1985).

14. M. A. Moore, N. Thamavit, H. Tsuda, K. Sato, A. Ichihara and N. Ito, Modifying influence of dehydroepiandrosterone on the development of dihydroxy-di-n-propylnitrosamine-initiated lesions in the thyroid, lung and liver of F344 rats, Carcinogenesis 7:311 (1986).

15. A. G. Schwartz, L. Pashko and J. M. Whitcomb, Inhibition of tumor development by dehydroepiandrosterone and related steroids, Toxicol. Pathol. 14:357 (1986).

16. M. P. Cleary, N. Fox, B. Lazin and J. T. Billheimer, Comparison of the effects of dehydroepiandrosterone treatments to ad libitum and pair-feeding in the obese Zucker rat, Nutr. Res. 5:1247 (1985).

17. B. Leighton, A. R. Tagliaferro and E. A. Newsholme, The effect of de-hydroepiandrosterone acetate on liver peroxisomal enzyme activities of male and female rats. J. Nutr. 117:1287 (1987).

18. M. Tucker, The effect of long-term food restriction on tumors in rodents, Int. J. Cancer 23:803 (1979).

19. A. G. Schwartz and A. Perantoni, Protective effect of dehydroepiandrosterone against aflatoxin B1 and 7,12-dimethylbenzanthracene-induced cytotoxicity and transformation in cultured cells, Cancer Res. 35: 2482 (1975).

20. F. Feo, L. Pirisi, R. Pascale, S. Zanetti, L. Daino and V. Laspina, Modulatory effect of glucose-6-phosphate dehydrogenase deficiency on the carcinogen activation by in vitro growing human fibroblasts, in: "Membranes in Tumor Growth", T. Galeotti, A. Cittadini and S. Papa eds., Elsevier, Amsterdam (1982).

21. F. Feo, L. Pirisi, R. Pascale, L. Daino, S. Frassetto and R. Garcea, Modulatory effect of glucose-6-phosphate dehydrogenase deficiency on the benzo(a)pyrene toxicity and transforming activity for in vitro cultured human skin fibroblasts, Cancer Res. 328:3419 (1984).

22. F. Feo, L. Pirisi, R. Pascale, L. Daino, S. Frassetto, S. Zanetti and R. Garcea, Modulatory mechanisms of chemical carcinogenesis: the role of the NADPH pool in the benzo(a)pyrene activation, Toxicol. Pathol. 12:262 (1984).

23. F. Feo, M.E. Ruggiu, L. Lenzerini, R. Garcea, L. Daino, S. Frassetto, V. Addis, L. Gaspa and R. Pascale, Benzo(a)pyrene metabolism by lympho-cytes from normal individuals carrying the Mediterranean variant of glucose-6-phosphate dehydrogenase, Int. J. Cancer 39:560 (1987).

24. R. Pascale, L. Daino, M. E. Ruggiu, M. G. Vannini, R. Garcea, S. Frassetto, L. Lenzerini, L. Gaspa, M. M. Simile, M. Puddu and F. Feo, Relationships between NADPH content and benzo(a)pyrene metabolism in normal and glucose-6-phosphate dehydrogenase-deficient human fibro-blasts, this volume.

25. L. L. Pashko and A. G. Schwartz, Effect of food restriction, dehydroepiandrosterone or obesity on the binding of H-7,12-dimethylbenz(a) anthracene to mouse skin DNA, J. Gerontol. 38:8 (1983).

26. J. Koudstaal, B. Makkink and S. H. Overdiep, Enzyme histochemical patterns in human tumors.II. Oxidoreductases in carcinoma of the colon and breast, Eur. J. Cancer 11:111 (1975).

27. G. Heyden, Histochemical investigation of malignant cells, Histochemistry 39:327 (1974).

28. A. W. Evans, N. W. Johmsin and R. G. Butcher, A quantitative histochemical study of the glucose-6-phosphate dehydrogenase activity in premalignant and malignant lesions of human and oral mucosa, Histochem. J., 15:483 (1983).

29. M. A. Moore, A. Tsuda, and N. Ito, Dehydrogenase histochemistry of N-ethyl-N-hydroxyethylnitrosamine-induced focal liver lesions in the rat - Increase in NADPH-generating capacity. Carcinogenesis 7:339 (1986).

30. R. Garcea, L. Daino, R. Pascale, S. Frassetto, P. Cozzolino, M. E. Ruggiu and F. Feo, Inhibition by dehydroepiandrosterone of liver putative preneoplastic foci formation in rats subjected to in-initiation-selection process of experimental carcinogenesis, Proc. 3rd Sardinian International Meeting: " Agents and Processes in Chemical Carcinogenesis (Abstracts)", STEF, Cagliari (1985).

31. M. A. Moore, N. Thamavit, A. Ichihara, K. Sato and N. Ito, Influence of dehydroepiandrosterone and butylated hydroxyanisole treatment during the induction phase of at liver nodular lesions in short-term system, Carcinogenesis 7:1059 (1986).

32. R. Garcea, L. Daino, R. Pascale, S. Frassetto, P. Cozzolino, M. E. Ruggiu and F. Feo, Inhibition by dehydroepiandrosterone of liver preneoplastic foci formation in rats after initiation-selection of experimental carcinogenesis, Toxicol. Pathol. 15:164 (1987).

33. R. Garcea, L. Daino, S. Frassetto, P. Cozzolino, M. E. Ruggiu, M. G. Vannini. R. Pascale, L. Lenzerini, M. M. Simile, M. Puddu and F. Feo, Reversal by ribo- and deoxyribo-nucleosides of dehydroepiandrosterone-induced inhibition of enzyme-altered foci in the liver of rats subjected to the initiation-selection process of experimental carcinogenesis, Carcinogenesis, in press (1988).

34. D. B. Solt, A. Medline and E. Farber, Rapid emergence of carcinogen-induced hyperplastic lesions in a new model for sequential analysis of liver carcinogenesis, Am. J. Physiol. 88:595 (1977).

35. A. G. Schwartz, L. Pashko and J. M. Whitcomb, Inhibition of tumor development by dehydroepiandrosterone and related steroids, Toxicol. Pathol. 14:357 (1986).

36. M. A. Moore, T. Nakamura and N. Ito, Immunohistochemically demonstrable glucose-6-phosphate dehydrogenase, γ-glutamyl transpeptidase, ornithine decarboxylase and glutathione S-transferase enzymes: absence of direct correlation with cell proliferation in at liver putative preneoplastic lesions, Carcinogenesis 7:1419 (1986).

37. C. R. Dworkin, D. D. Gorman, L. L. Pashko, V. J. Cristofalo and A. G. Schwartz, Inhibition of growth of HeLa and WI-38 cells by dehydroepiandrosterone and its reversal by ribo- and deoxyribo-nucleosides, Life Sci. 38:145 (1986).

38. G. B. Gordon L. M. Shantz and P. Talalay, Modulation of growth, differentiation and carcinogenesis by dehydroepiandrosterone, Adv. Enz. Res. 26:355 (1987.

39. T. J. Slaga, A. J. P. Klein-Szanto, L. L. Triplett, L. P. Yotti and J. E. Trosko, Skin tumor promoting activity of benzoylperoxide, a widely used free radical generating compound, Science 213:1023 (1981).

40. P. M. Horvarth and C. Ip, Synergistic effect of vitamin E and selenium in the chemoprevention of mammary carcinogenesis in rats, Cancer Res. 43:5335 (1983).

41. S. K. Goel, N. D. Lalwani and J. K. Reddy, Peroxisome proliferation and

lipid peroxidation in rat liver, Cancer Res. 46:1324 (1986).

42. A. R. Kennedy, W. Troll and J. B. Little, Role of free radicals in the initiation and promotion of radiation transformation, Carcinogenesis 5:1213 (1984).

43. J. M. Whitcomb and A. G. Schwartz, Dehydroepiandrosterone and 16-Br-epi-androsterone inhibit 12-O-tetradecanoylphorbol-13-acetate stimulation of superoxide radical production by human polymorphonuclear leukocytes, Carcinogenesis 6:333 (1985).

44. R. Pascale, R. Garcea, M. E. Ruggiu, L. Daino, S. Frassetto, M. G. Vannini, P. Cozzolino, L. Lenzerini, F. Feo and A. G. Schwartz, Decreased stimulation by 12-O-tetradecanoylphorbol-13-acetate of superoxide radical production by polymorphonuclear leukocytes carrying the Mediterranean variant of glucose-6-phosphate dehydrogenase, Carcinogenesis 8:1567 (1987).

45. K. Frenkel and K. Chrzan, Hydrogen peroxide formation and DNA base modification by tumor promoter-activated polymorphonuclear leukocytes, Carcinogenesis 5:1547 (1987).

46. S. A. Weitzman, A. B. Weitberg, E. P. Clark and T. P. Stossel, Phagocytes as carcinogens: malignant transformation produced by human neutrophils, Science 227:1231 (1985).

ADDITIVE EFFECT OF LOW-DOSE DIETHYLNITROSAMINE ON THE INITIATION

OF LIVER CARCINOGENESIS IN RATS

Masahiro Tsutsumi[1], Seiichi Takahashi[1], Dai Nakae[1], Kazumi Shiraiwa[1], Tetsuo Kinugasa[1], Kenji Kamino[2], Ayumi Denda[1] and Yoichi Konishi[1]

[1]Department of Oncological Pathology, Cancer Center, Nara Medical College, 840 Shijo-cho, Kashihara, Nara 634, Japan and [2]Institute of Pathology, University of Düsseldorf, Moorenstr 5, 4000 Düsseldorf, German Federal Republic

INTRODUCTION

Simultaneous or exposure to low-dose carcinogens existing in the environment is an important causal factor in human carcinogenesis. It is gener- ally accepted that carcinogenesis is a multi-step process consisting of two qualitatively different stages, initiation and promotion[1]. Initiation is a transient "flash" phenomenon involving exposure to a carcinogen and may occur frequently in humans.

It has been reported that the initiation of carcinogenesis by alkylating agents results from the formation of pro-mutagenic alkylating nucleotides in the DNA of target cells[2], after which it is necessary for the resulting DNA lesion to be fixed by cell proliferation[3]. In rat liver carcinogenesis, an initiation assay has been established[3,4], in which phenotypically altered enzyme foci can be delineated as putative preneoplastic lesions which will finally evolve into hepatocellular carcinoma. Also in animal experiments, inhibitory and additive[5,6] effects on cancer induction of continuous exposure to two different carcinogens have been reported.

Previously, we described the persistent effect of low dose of preadministered diethylnitrosamine (DEN) on the induction of enzyme-altered foci in rat liver and stressed the important role of low-dose carcinogen exposure in the establishment of initiation[7]. In the present investigation, we studied the additive effects of pretreatment with various low doses of DEN and other carcinogens on a complete initiation consisting of the same various doses of DEN plus partial hepatectomy (PH), in order to assess the summation effect of low-dose carcinogen exposure in vivo.

MATERIALS AND METHODS

Animals

Fully grown male Fischer 344 rats (Shizuoka Laboratory Animal Center, Shizuoka, Japan), 18-20 weeks old, weighing approximately 270 g each were used. The rats were housed in wire cages in an air-conditioned room at 22°C and 60% humidity under a daily cycle of alternating 12-hour periods of light and dark. The rats were maintained on a commercial stock diet, Oriental MF (Oriental Yeast Co., Ltd., Tokyo, Japan), except when being subjected to the selection pressure described in the experimental regimen section, and given water ad libitum. All rats were killed under ether anesthesia when required.

Chemicals

DEN (Wako Chemical Co., Ltd., Osaka, Japan) was dissolved in 0.9% NaCl solution (saline, 5 mg/ml). 2-Acetylaminofluorene (AAF) was purchased from Tokyo Kasei Co., Ltd., Tokyo, Japan. Carbon tetrachloride (CCl_4, Nakarai Chemicals Co., Ltd., Kyoto, Japan) and was diluted 1:1 with corn oil. Reagents for histochemical examination were obtained from Nakarai Chemicals Co., Ltd.. DHPN (Nakarai Chemicals Co., Ltd., Kyoto, Japan) was dissolved in 0.9% NaCl solution (saline 5 mg/ml). B(a)P (Sigma Chemicals, Mo., USA) was dissolved in corn oil (30 mg/ml). MNU (Nakarai Chemicals Co., Ltd., Kyoto, Japan) was dissolved in 0.1 M sodium citrate buffer (pH 6.0, 10 mg/ml). DMH (Aldrich Chemicals, Wis., USA) was dissolved in 0.4 nM EDTA-saline (pH 6.6, 20 mg/ml). MNNG (Nakarai Chemical Co., Ltd., Kyoto, Japan) was dissolved in DMSO (5 mg/ml).

Experimental Regimen

This study consisted of 3 series of experiments. The concept and details of the methods used for the complete initiation and selection pressure have been described previously[7].

Dose-dependent Effect of DEN Pretreatment on a Complete Initiation. The experimental schedule used in this investigation (experiment 1) is schematically illustrated in Fig. 1. The dose of preadministered DEN was 0, 1, 2, 5 or 10 mg/kg body weight and that of postadministered DEN was 10 mg/kg body weight. Rats which had received intraperitoneal (i.p.) injection of the above doses of DEN were subjected to complete initiation consisting of i.p. injection of DEN postadministration and PH, followed by selection pressure with a diet containing 0.02% AAF plus CCl_4. Each complete initiation was performed 4 weeks after DEN preadministration. PH was performed using the method of Higgins and Anderson[8] 4 h after DEN postadministration, and selection pressure was started 2 weeks after complete initiation. All rats were killed 9 weeks after DEN preadministration and liver foci were analyzed histochemically.

Dose-dependent Effect of DEN Posttreatment on a Fixed DEN Pretreatment Dose. The experimental schedule used for this investigation (experiment 2) is schematically illustrated in Fig. 1. The dose of preadministered DEN was 10 mg/kg body weight and that of postadministered DEN was 0, 1, 2, 5 or 10 mg/kg body weight. Complete initiation followed by selection pressure were performed in the same manner as for experiment 1. All rats were killed 9 weeks after DEN preadministration and liver foci were analyzed histochemically.

Effect of Various Chemical Pretreatments on Complete Initiation. The experimental schedule used for this investigation (experiment 3) was essentially the same as that used in experiment 1 except that various chemicals were preadministered instead of DEN (Fig. 2). In various experimental groups, rats received 10 mg/kg b.w. DEN, 500 mg/kg b.w. DHPN, 200 mg/kg b.w. B(a)P, 60 mg/kg b.w. MNU, 100 mg/kg b.w. DMH or 80 mg/kg b.w. MNNG pretreatment prior to complete initiation consisting of single i.p. injection of 10 mg/kg b.w. DEN plus PH 4 weeks later, followed by selection pressure. In the control groups, rats received the same carcinogen as pretreatment prior to saline injection plus PH 4 weeks later, followed by selection pressure. Liver foci were analyzed histochemically.

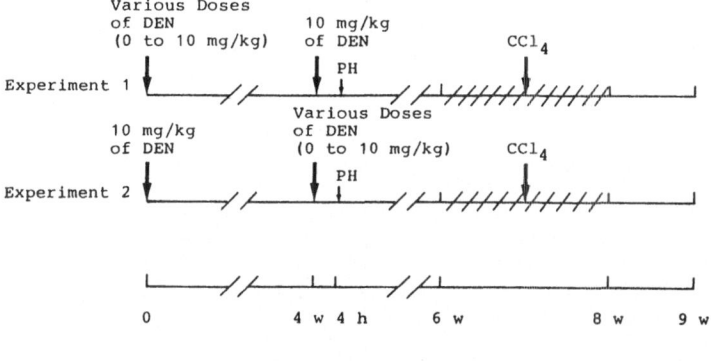

Fig. 1. Experimental schedule for effect of pre- and post-administered low
doses of DEN on the induction of γ-GTP-positive foci in rat liver.

Histological and Histochemical Studies

Immediately after the rats had been killed, the liver was removed and
weighed. Tissue slices taken from three lobes of the liver were fixed in
ethanol at 4°C for subsequent histochemical study and serial slices from the
three lobes were fixed in 10% buffered formalin for histological study. Histochemically, γ-glutamyltranspeptidase (γ-GTP) activity was studied using the
method of Rutenberg et al.[9]. The numbers and sizes of γ-GTP-positive foci
were analyzed with an image analyzer model HTB-c995 (Hamamatsu Television
Co., Ltd., Shizuoka, Japan) connected to a Desktop Computer System-45
(Hewlett-Packard Co., USA). The numbers of foci per unit volume of liver
were calculated using the procedure of Campbell et al.[10]. For histological
study, fixed liver tissues were sectioned and routinely stained with
hematoxylin and eosin.

RESULTS

Additive Effect of a Threshold Dose of Preadministered DEN and Complete Initiation in Experiment 1

The results of experiment 1 are shown in Table 1. Final body and liver
weights of rats which received 0, 1, 2, 5 or 10 mg/kg body weight DEN as a
pretreatment did not show any significant differences. No additive effect of
pre-administered DEN at doses of 1, 2 or 5 mg/kg body weight, followed by

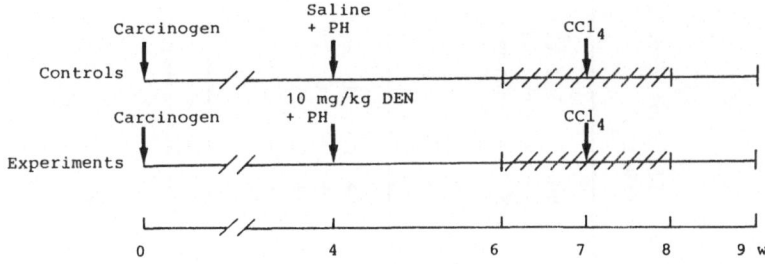

Fig. 2. Effect of various carcinogens preadministered 4 weeks before the
initiation by DEN plus partial hepatectomy on the induction of
γ-GTP-positive foci in rat liver.

Table 1. Effect of preadministered low-dose DEN on complete initiation[a] judged by γ-GTP-positive focus induction in rat liver

Group	Dose of DEN	Effective no. of animals	Body weight Initial	Body weight Final	Liver weight	No. of γ-GTP foci/cm²	Area of foci mm²/cm²	Mean diameter (μm)	No. of γ-GTP foci/cm³
1	0 + 10	10	265 ± 14[b]	262 ± 17	6.81 ± 0.51	13.3 ± 5.4	6.0 ± 3.0	740 ± 113	129 ± 40
2	1 + 10	10	274 ± 13	260 ± 26	6.50 ± 0.76	12.3 ± 5.4	4.9 ± 3.3	688 ± 103	137 ± 65
3	2 + 10	10	267 ± 10	251 ± 27	6.35 ± 0.76	14.4 ± 4.7	5.1 ± 2.4	650 ± 110	153 ± 49
4	5 + 10	10	266 ± 8	257 ± 7	6.10 ± 0.58	13.1 ± 7.0	3.8 ± 2.3	590 ± 78	153 ± 76
5	10 + 10	7	266 ± 11	252 ± 15	6.47 ± 0.46	23.5 ± 8.3[c]	12.2 ± 6.6[c]	789 ± 137	239 ± 85[c]

[a]Described in experimental regimen. [b]Values are mean ± SD. [c]Significantly different from group 1 (P < 0.05).

Table 2. Effect of postadministered low-dose DEN[a] on a fixed dose of preadministered DEN judged by γ-GTP-positive focus induction in rat liver

Group	Dose of DEN	Effective No. of animals	Body weight Initial	Body weight Final	Liver weight	No. of γ-GTP foci/cm²	Area of foci mm²/cm²	Mean diameter (μm)	No. of γ-GTP foci/cm³
1	10 + 10	7	266 ± 11[b]	252 ± 15	6.47 ± 0.46	23.6 ± 8.3[c]	12.2 ± 6.7[c]	789 ± 137	239 ± 85[b]
2	0 + 10	10	265 ± 14	262 ± 17	6.81 ± 0.51	13.3 ± 5.4[c]	6.0 ± 3.0[c]	740 ± 113	129 ± 40[c]
3	10 + 5	9	275 ± 14	261 ± 20	6.85 ± 0.83	9.6 ± 3.7	3.4 ± 2.1	645 ± 149	100 ± 36
4	0 + 5	8	305 ± 18	279 ± 15	6.48 ± 0.48	4.7 ± 3.6	1.8 ± 1.5	688 ± 76	56 ± 45
5	10 + 2	8	274 ± 10	236 ± 44	6.85 ± 0.63	4.1 ± 2.1	1.8 ± 1.5	773 ± 362	53 ± 46
6	10 + 1	8	271 ± 12	260 ± 18	7.09 ± 0.94	4.2 ± 2.3	1.0 ± 0.8	574 ± 164	53 ± 35
7	10 + 0	10	280 ± 7	258 ± 17	6.66 ± 0.77	2.9 ± 2.5	1.2 ± 1.3	649 ± 182	28 ± 15

[a]Given as the complete initiation described in experimental regimen. [b]Values are mean ± SD. [c]Significantly different from group 7 (P < 0.05).

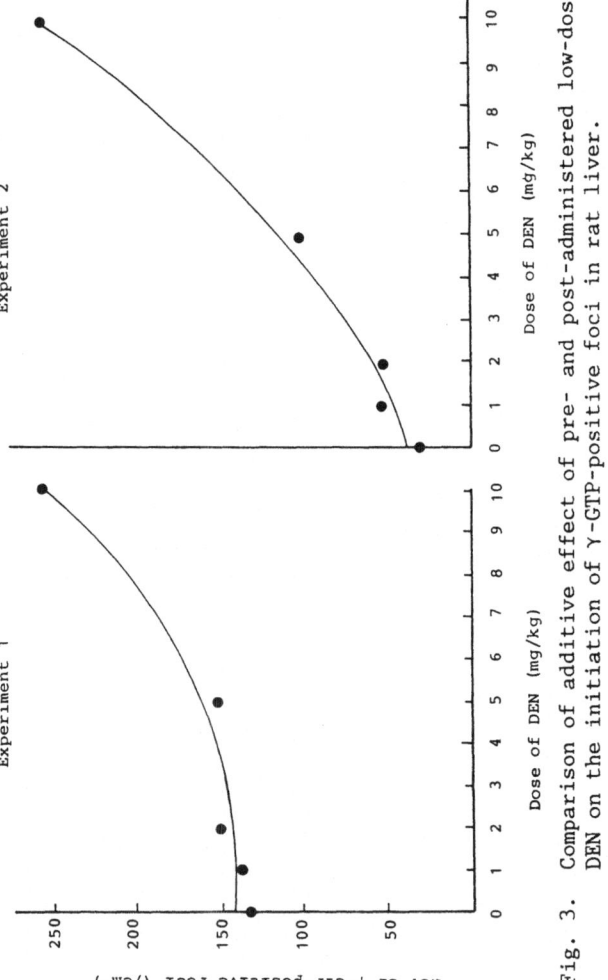

Fig. 3. Comparison of additive effect of pre- and post-administered low-dose DEN on the initiation of γ-GTP-positive foci in rat liver.

complete initiation, was observed. However, an additive effect of preadministered DEN at 10 mg/kg body weight and complete initiation was observed, and the numbers and areas of γ-GTP-positive foci in the liver of rats which received 10 mg/kg body weight of preadministered DEN were increased significantly in comparison with those of foci in the liver of rats which received preadministration with 0, 1, 2 or 5 mg/kg body weight DEN. DEN at 10 mg/kg body weight is thus probably the threshold pretreatment dose necessary to produce an additive effect on complete initiation under the conditions used in this experiment.

Additive Effect of a Threshold Dose of Postadministered DEN and a Fixed Dose of Preadministered DEN in Experiment 2

The results of experiment 2 are shown in Table 2. Final body and liver weights of rats which received 0, 1, 2, 5 or 10 mg/kg b.w. DEN as a posttreatment did not show any significant differences. No additive effect of 10 mg/kg b.w. DEN as a pretreatment and 1, 2 or 5 mg/kg b.w. DEN as a posttreatment was observed. However, an additive effect of 10 mg/kg b.w. DEN as a pretreatment and the same dose of DEN as a post-treatment was observed, and the numbers and areas of γ-GTP-positive foci were increased significantly in comparison with those of foci in the liver of rats which received 0, 1, 2 or 5 mg/kg b.w. DEN postadministration. DEN at 10 mg/kg b.w. is thus probably the threshold posttreatment dose necessary to produce an additive effect with 10 mg/kg b.w. DEN pre-treatment under the conditions used here.

Comparison of Additive Effect of Pre- and Post-administered DEN on the Induction of γ-GTP-positive Foci in Experiments 1 and 2

The results obtained with regard to the induction of γ-GTP-positive foci from the above two series of experiments are schematically illustrated in Fig. 3. The numbers of foci observed in the liver of rats which received 0, 1, 2, or 5 mg/kg b.w. in experiment 1 were greater than those in experiment 2, since complete initiation with 10 mg/kg b.w. DEN had been performed. It is clear from these two experiments that an additive effect for the induction of γ-GTP-positive foci was recognized when 10 mg/kg b.w. DEN was employed as a pretreatment and the same dose of DEN was used as a posttreatment. A total dose of 20 mg/kg b.w. DEN is thus sufficient to induce γ-GTP-positive foci in rat liver under these conditions.

Effect of Pretreatment with Various Chemicals on Complete Initiation in Experiment 3

The results of experiment 3 are shown in Table 3. Doses of 500 mg/kg b.w. DHPN, 200 mg/kg b.w. B(a)P, 60 mg/kg b.w. MNU, 100 mg/kg b.w. DMH and 80 mg/kg b.w. MNNG induced almost the same number of γ-GTP-positive foci without complete initiation, as was the case with 10 mg/kg b.w. DEN previously. An additive effect on complete initiation consisting of 10 mg/kg b.w. DEN and PH was shown for DEN, DHPN or B(a)P preadministration but not for MNU, DMH or MNNG preadministration.

DISCUSSION

The results of this study demonstrated that in order to produce and additive effect of pretreated DEN and complete initiation, definite doses of DEN are needed which are sufficient to induce γ-GTP-positive foci in rat liver. A DEN dose of 10 mg/kg b.w., as a pretreatment, and the same dose of DEN as a posttreatment were capable of inducing GTP-positive foci, but doses of less than 10 mg/kg b.w. DEN given as both pre- and post-treatment were not. Previously, we reported that DEN preadministered at 10 mg/kg b.w. had a persistent effect on long-delayed DEN treatment given as an initiator when PH

Table 3. Effect of pretreatment with various carcinogen 4 weeks before the initiation by DEN + partial hepatectomy on the induction of γ-GTP-positive foci in rat liver

Test compound	Initiation (DEN + PH)	Effective no. of animals	No. of γ-GTP foci/cm^2	No. of γ-GTP foci/cm^3
Saline	−	10	0.6 ± 0.5[a]	7 ± 9
	+	10	16.2 ± 6.3	140 ± 55
DEN (10 mg/kg)	−	5	3.2 ± 2.0[b]	36 ± 18[b]
	+	6	31.2 ± 5.8[b]	326 ± 62[b]
DHPN (500 mg/kg)	−	10	2.8 ± 2.2	28 ± 24[b]
	+	7	27.4 ± 5.0[b]	258 ± 52[b]
B(a)P (200 mg/kg)	−	9	2.8 ± 2.4	30 ± 26[b]
	+	10	29.9 ± 8.8[b]	290 ± 90[b]
MNU (60 mg/kg)	−	10	3.7 ± 1.2	31 ± 16
	+	10	18.7 ± 7.8	167 ± 82
DMH (100 mg/kg)	−	8	3.6 ± 2.4	32 ± 26
	+	10	17.7 ± 6.5	162 ± 68
MNNG (80 mg/kg)	−	6	2.1 ± 1.1	28 ± 18
	+	6	17.4 ± 6.3	170 ± 66

[a]Values are mean ± SD. [b]Significantly different from control group (saline pretreatment and DEN plus PH) ($P < 0.05$).

and selection pressure were applied[7]. The present results confirm those obtained previously that DEN given as a pre- and post-treatment at a total of 20 mg/kg b.w. produces and additive effect on the induction of γ-GTP-positive foci. It is known that alkylating agents cause alkylation of DNA under physiological conditions, and extensive studies on the formation of DNA adducts, and their removal have been conducted[12]. O^6-Alkylguanine is inferred to be a principle component pro-mutagenic adducts[13]. However, it has recently been suggested that O^2- and O^4-alkyldeoxythymidine might also be important in rat liver carcinogenesis since both species persist and accumulate in the lesions induced by DEN[14,15] in contrast with the relatively rapid removal of O^6-ethylguanine[16]. DNA synthesis increases the likelihood that a preformed repairable DNA defect with the potential for focus induction will become genetically fixed as an irreversible lesion. Either DNA synthesis occurring prior to repair, or error-prone repair during or following DNA synthesis could be instrumental in this fixation[17]. In the present study, DEN pretreatment might have induced insufficiently initiated cells which were unable to evolve into phenotypically altered enzyme foci through cell proliferation alone, but which would have been able to do so following DEN posttreatment and PH, i.e., the complete initiation.

The effect of other preadministered carcinogens on this complete initiation consisting of DEN and PH was examined. Additive effects of respectively preadministered DHPN and B(a)P were observed on the complete initiation but preadministered MNU, DMH or MNNG had no effect. DHPN is known to induce alkylating adducts[18], B(a)P induces bulky adducts of DNA[19], while MNU, DMH and MNNG induce methylating DNA adducts[20,21,22]. The present results therefore suggest that additive effects of carcinogens might depend on the pattern of DNA damage they induce. Furthermore, these results might indicate that non-specific DNA damage by environmental chemicals is not generally involved in the establishment of initiation and that an additive effect of DNA damage sufficient to promote initiation may require the formation of a specific type of adduct.

The precise mechanisms involved in the additive effects are unknown. However, it is reported[23] that transient estrogen stimulation is sufficient to induce long-lasting alterations in the ability of a tissue to respond to subsequent hormonal treatment and that these altered response characteristics persist for several months. For this reason, this effect has been called the estrogen "memory" effect. Carcinogenesis is a multi-step process initiated by DNA damage, gene mutation, gene rearrangement and gene translocation, which ends in phenotypic transformation to cancer cells[24]. It is possible that gene expression could be changed through heritable alterations in chromatin or DNA structure and that hormones could induce alterations in DNA and chromatin structure[23,25]. Although further detailed studies will be required in this context it seems clear that low-dose DEN is able to induce alteration of gene expression and produce the type of additive effect occurring in carcinogenesis.

ACKNOWLEDGEMENTS

This work was supported by Grant-in-Aid for Cancer Research from the Ministry of Education, Science and Culture, and by a Grant-in-Aid for Cancer Research from the Ministry of Health and Welfare for the Comprehensive 10-year Strategy for Cancer Control, Japan.

REFERENCES

1. E. Farber, The multistep nature of cancer development, Cancer Res. 44:4217 (1984).

2. M. C. Dyroff, F. C. Richardson, J. A. Popp, M. A. Bedell and J. A. Swenberg, Correlation of O^4-ethyldeoxythymidine accumulation with hepatic initiation and hepatocellular carcinoma in rats continuously administered diethylnitrosamine, Carcinogenesis 7:241 (1986).

3. E. Cayama, H. Tsuda, D. S. R. Sarma and E. Farber, Initiation of chemical carcinogenesis requires cell proliferation, Nature 275:60 (1978).

4. D. Solt and E. Farber, New principle for the analysis of chemical carcinogenesis, Nature 263:701 (1976).

5. M. Moore and T. Kitagawa, Hepatocarcinogenesis in the rat: The effect of promoters and carcinogenesis, in vivo and in vitro, in: "International Review of Cytology", Academic Press, New York (1986).

6. N. Ito, M. Tatematsu, K. Nakanishi, R. Hasegawa, T. Takano, K. Imaida and T. Ogiso, The effect of various chemicals on the development of hyperplastic liver nodules in hepatectomized rats treated with N-nitro-sodiethylamine or 2-fluorenylacetamide, Gann 71:832 (1980).

7. S. Takahashi, M. Tsutsumi, D. Nakae, A. Denda, T. Kinugasa and Y. Konishi, Persistent effect of low dose of preadministered diethylnitrosamine on induction of enzyme-altered foci in rat liver, Carcinogenesis 8:509 (1987).

8. G. M. Higgins and R. M. Anderson, Experimental pathology of the liver, 1. Restoration of the liver of the white rat following partial surgical removal, Arch. Pathol. Lab. Med. 12:1186 (1931).

9. A. M. Rutenberg, H. Kim, J. W. Fischbein, J. S. Hanker, H. K. Wasserkrug and A. J. Seligman, Histochemical and ultrastructural demonstration of gamma-glutamyltranspeptidase activity, J. Histochem. Cytochem. 17:517 (1969).

10. H. A. Campbell, H. C. Pitot, V. P. Potter and B. A. Laishes, Application of quantitative stereology to the evaluation of enzyme-altered foci in rat liver, Cancer Res. 42:465 (1982).

11. G. P. Margison and P. J. O'Connor, Nucleic acid modification by N-nitroso compounds, in: "Chemical Carcinogenesis and DNA", P. L. Grover, ed., CRC Press, Boca Raton, Fla. (1981).

12. A. E. Pegg, Formation and metabolism of alkylated nucleosides: possible role in carcinogenesis by nitroso compounds and alkylating agents, Adv. Cancer Res. 25:195 (1977).

13. R. Montesano, Alkylation of DNA and tissue specificities in nitrosamine carcinogenesis, J. Cell. Biochem. 17:259 (1981).

14. F. C. Richardson, M. C. Dyroff, J. A. Boucheron and J. A. Swenberg, Differential repair of O^4-alkylthymidine following exposure to methylating and ethylating hepatocarcinogenesis, Carcinogenesis 6:625 (1985).

15. E. Scherer, A. P. Timmer and P. Emmelot, Formation by diethylnitrosamine and persistence of O^4-ethylthymidine in rat liver in vivo, Cancer Lett. 10:1 (1980).

16. E. Scherer, A. P. Steward and P. Emmelot, Effects of formation of O^6-ethylguanine in, and its removal from, liver DNA of rats receiving diethylnitrosamine, Chem. Biol. Interactions 19:1 (1977).

17. S. M. D'Ambrosio and R. B. Setlow, Enhancement of postreplication repair in Chinese hamster cells, Proc. Natl. Acad. Sci. USA 73:2396 (1976).

18. T. A. Lawson, A. S. Helgeson, C. J. Grandjean, L. Wallcave and D. Nagel, The formation of N-nitrosomethyl(2-oxopropyl)amine from N-nitroso-bis(2-oxopropyl)amine in vivo, Carcinogenesis 2:845 (1981).

19. M. A. Pereira, Dose response for benzo(a)pyrene adducts in mouse epidermal DNA, Cancer Res. 39:2556 (1979).

20. D. T. Beranek, C. C. Weis and D. H. Swenson, A comprehensive quantitative analysis of methylated and ethylated DNA using high-pressure liquid chromatography, Carcinogenesis 1:595 (1980).

21. P. D. Lawley and C. J. Thatcher, Methylation of deoxyribonucleic acid

in cultured mammalian cells by N-methyl-N-nitroso-N-nitrosoguani-
dine, <u>Biochem. J.</u> 116:693 (1970).

22. P. J. O'Connor, M. J. Capps and A. W. Craig, Comparative studies of
hepatocarcinogen N.N-dimethylnitrosamine <u>in vivo</u>: reaction sites in
rat liver DNA and the significance of their relative stabilities,
<u>Brit. J. Cancer</u> 27:153 (1973).

23. S. P. Tam, R. J. G. Haché and R. G. Deeley, Estrogen memory effect in
human hepatocytes during repeated cell division without hormone,
<u>Science</u> 234:1234 (1986).

24. T. Sugimura and M. Miwa, Poly(ADP-ribose) and cancer research,
<u>Carcinogenesis</u> 4:1503 (1983).

25. H. Weintraub, Assembly and propagation of repressed and derepressed
chromosomal states, <u>Cell</u> 42:705 (1985).

MODULATION OF LIVER CARCINOGENESIS BY SURGERY

Véronique Préat[1,3], Jean-Claude Pector[2] and Marcel Roberfroid[1]

[1]Université Catholique de Louvain, 73.69 Unité de Biochimie
Toxicologique et Cancérologique, 1200 Brussels, Belgium
[2]Université Libre de Bruxelles, Institut Bordet, Service de
Chirurgie, 1000 Brussels, Belgium. [3]Chargé de recherches
F.N.R.S.

MODULATION OF RAT LIVER CARCINOGENESIS

Modulation of Carcinogenesis

Carcinogenesis is a complex progressive and multistep process in which
phases have been identified mainly based on operational rather than on biological evidences[1,2].

It is generally admitted that an "initiation" is necessary for the induction of cancer. However, it is not known exactly what initiation is and
what the relevant consequences of this initiation are. It is also well accepted that, if initiation is necessary, it can either be sufficient or insufficient to induce cancer, so that after initiation, the carcinogenic process can either evolve naturally per se up to malignancy or need to be modulated in order to induce the appearance of cancer. Indeed, it has been
shown in many cases that treatments, whatever their nature is, can influence
the carcinogenic process induced by an initiation leading to a modification
of cancer incidence, cancer yield and/or latency period.

We have recently reinforced the concept of modulation to account for
these modifications of the carcinogenic process by variable treatments.
Positive modulation is defined as a treatment that increases the incidence
and/or the yield of cancer and/or shortens the latency period preceding their
appearance whereas negative modulation is a treatment that decreases the incidence and/or the yield of cancer and/or enhances the latency period[2],
(Fig. 1).

Nature of the Modulating Factors of Liver Carcinogenesis

While reviewing the modulation of liver carcinogenesis, it appeared that
the nature of the modulating factors can be very different[3]. Indeed, the
modulating factors can be:
  1. Carcinogens
  2. Non genotoxic xenobiotics
  3. Endogenous compounds
  4. Dietary unbalance
  5. Surgery

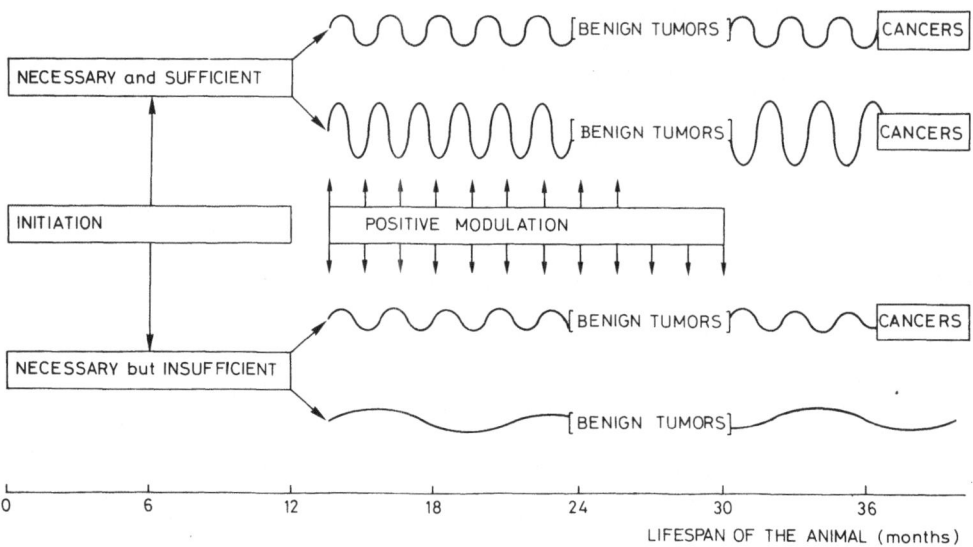

Fig. 1. Representation of the concept of modulation.

It is well known that prolonged exposure to carcinogen is more potent to induce tumor than a single or a short-term exposure[4-6]. The same observation applies to a treatment with a second carcinogen[7,8].

After an initiation, the chronic treatment with a non genotoxic can also lead to an increased incidence and/or yield of liver cancer and/or shortened latency period. Since the original observation by Peraino[9], that phenobarbital (PB) fed chronically to rats after a short 2-acetylaminofluorene treatment, can promote liver cancer development, many reports confirmed this finding and extended it to other drugs, pesticides or food additives[10-14]. The research on negative modulators or chemopreventive agents of liver carcinogenesis has not yet given very conclusive evidences.

Endogenous compounds, when administered exogenously can also act as positive modulators of liver carcinogenesis. Indeed, estrogens[15,16], bile acids (Delzenne, personal communication[17]) or orotic acid[18], have been shown to promote cancer development when administered after an initiation.

Dietary imbalances can also exert such a positive modulating effect. Choline methionine deficient diet[19] or high fat diet[20] indeed promote cancer development.

We reported previously that a surgical operation such as a portacaval shunt also enhances liver cancer when performed after initiation[21]. The aim of the present paper is to analyze the effect of surgery on rat liver carcinogenesis. It will focus on the effect of portal derivations and partial hepatectomy on the hepatocarcinogenic process induced by the Solt and Farber protocol[7].

Effect of the Modulating Factors on the Evolution of the Hepatocarcinogenic Process

After the administration of a hepatocarcinogen, so called preneoplastic and neoplastic lesions develop in the liver parenchyma before the appearance of liver cancer. These lesions, foci and nodules, might very likely be the

precursors of hepatocellular carcinomas [22]. The modulating factors can influence the growth of foci into nodules and their transformation in hepatocellular carcinomas. Indeed, it is generally admitted that positive modulators promote the development of the foci and nodules as well as the one of
hepatocellular carcinomas so that most of the researchers analyzed the effect
of the modulating factors in the early stages rather than at the end point,
cancer [10,17-19,23-24].

However, we have shown that the modulating factors can differently modulate the carcinogenic process [3]. While comparing the effect of different
xenobiotics, on the development of foci and nodules in early stages (1-3
months) and cancer stage (6-7 months), we found that compounds like PB or DDT
increase foci and nodules as well as cancer development whereas compounds
like nafenopin inhibit foci and nodules development but strongly enhance the
yield and incidence of cancer. Butylated hydroxytoluene promotes nodule development but has no effect on cancer formation[11,12]. These results clearly
indicate a lack of correlation between the number of premalignant lesions
(foci and nodules) and the number of cancers which develop later on.

Thus, the modulating factors can differently modulate the evolution of
an initiated liver to cancer, the ratio number of foci and nodules/liver as
well as the ratio number of foci and nodules in early stages/number of cancers in the end stage [3].

EFFECT OF PORTAL DERIVATIONS ON RAT LIVER CARCINOGENESIS

Since we reported earlier that portacaval shunt exerts a promoting effect on liver cancer development [21], the effect of two portal derivations,
portacaval shunt (PCS) and portacaval transposition (PCT) were compared in
order to better understand the mechanism of action of PCS.

As shown in Fig. 2, PCS is a surgical operation which consists in de-
rivating the portal blood supply into the vena cava and ligaturing the portal
vein. In PCT, the portal blood supply is also derivated into the vena cava
but the caval blood goes into the portal vein.

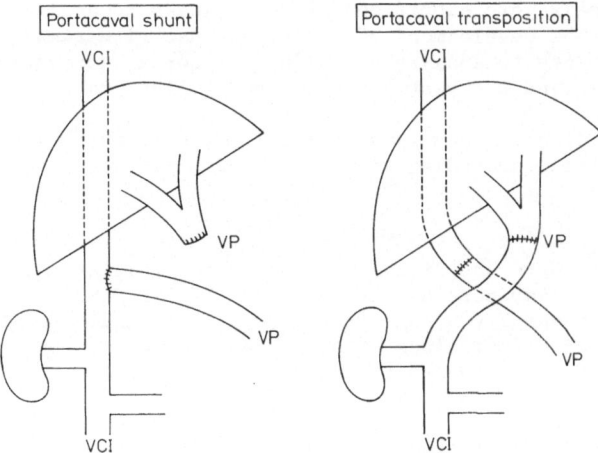

Fig. 2.   Schematic representation of the PCS and PCT.

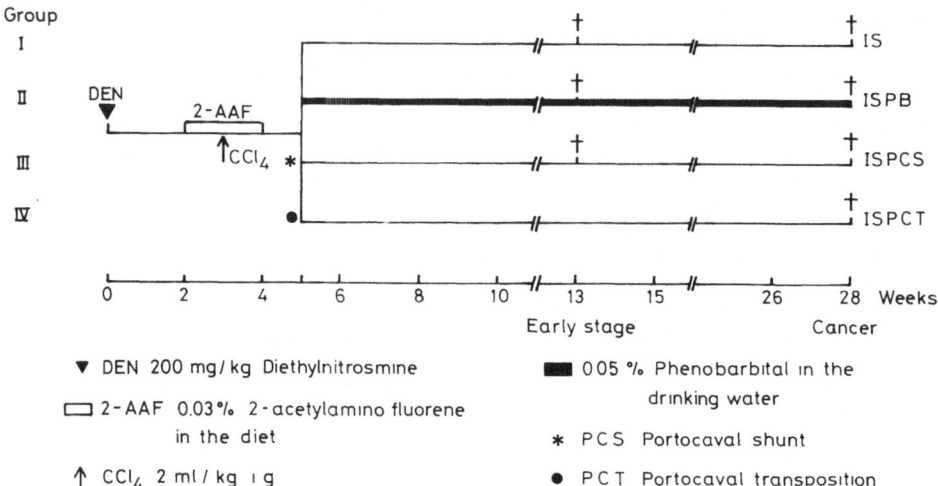

Fig. 3. Experimental protocol used to analyze the effect of portal derivations on rat liver carcinogenesis induced by the initiation/selec-
tion protocol.

Experimental Protocol

The experimental protocol used to analyze the effect of portal derivations on rat liver carcinogenesis is schematized in Fig. 3.

Male Wistar rats weighing 180 g (ICO: Wi IOPSAF/Han) were injected with
a single dose (i.p.) of 200 mg/kg of diethylnitrosamine (DEN) (Initiation I).
2 weeks later, they were submitted to a selection (S) procedure[7] consisting
in a 2-week-feeding with 0.03% of 2-acetylaminofluorene (2-AAF) and in the
middle of this treatment, a necrogenic dose of 2 ml/kg of CCl_4[25]. One week
after S, the animals were divided in 4 groups: group I received a basal diet
(IS); group II received 0.05% of phenobarbital as a sodium salt in their
drinking water (IS PB); group III was submitted to a PCS (IS PCS) whereas
group IV was submitted to a PCT (IS PCT).

Control groups were non-treated animals (group V), animals submitted to
PCS (group VI) or PCT (group VII).

8 to 10 animals were sacrificed after 13 weeks to analyze the effect of
the treatment on the development of early lesions. The remaining rats were
killed after 28 weeks to analyze the effect of portal derivations on the incidence, the yield and the histological type of liver cancer[11,12].

Routine histological analysis of the liver were performed. Hepatic lesions were classified according to Squirre and Levitt [26].

Results

When the rats were sacrificed 13 weeks after initiation, surprisingly
20% of the rats already bore cancer in group III (IS PCS) whereas no cancer
was observed in groups I (IS) and II (IS PB). This indicates that PCS has a
positive modulating effect since it shortened the latency period and increased cancer incidence as compared to the control group IS.

As described previously[11,25], the lesions observed in group I consisted
mainly in clear and tigroid foci. In group II, they were mainly eosinophilic
foci and mixed nodules. In group III, mixed nodules were predominantly seen.
The histochemical pattern of the lesions is presently under study.

When the animals of group I were sacrificed after 28 weeks, no liver cancer was observed (Table 1). Foci and nodules, mainly with a mixed cell population predominantly clear, were detected in the parenchyma.

In group II, the incidence of cancer was 63% since 12 rats out of 19 bore cancer with a yield of 2 tumors per rat bearing tumors. The cancers were mainly well differentiated hepatocellular carcinomas. Mixed nodules were observed in the parenchyma.

In group III, 11 rats out of 14 bore cancer (78% incidence) with a yield of 2.4 tumors. These cancers were well differentiated hepatocellular carcinomas or hepatocellular carcinomas with a glandular pattern.

In group IV, 3 rats out of 10 bore 1 tumor which were well differentiated hepatocellular carcinomas.

The statistical analysis (Table 1) indicates that after 28 weeks PB, PCS and PCT significantly increase the incidence of cancer as compared to the control group IS. It appeared also that PCS and PB are more potent positive modulators than PCT ($P < 0.05$).

The liver parenchyma of rats submitted only to PCS (group IV) showed no preneoplastic nor malignant lesions after 28 weeks. However its architecture and its histochemical pattern were modified. Zone I hepatocytes were smaller and contained glycogen whereas zone III hepatocytes were larger and had less if any glycogen. Some oval cells were seen in portal areas[21]. Very few alterations were observed in group VII (PCT).

Discussion

The experiment reported here confirms that surgery can positively modulate liver carcinogenesis. It clearly shows that portal derivations such as PCS or PCT exert a promoting effect or positive modulating effect since as compared to the control group (IS), they significantly enhance the incidence and the yield of liver cancers (Table 1). PCS also decreases the latency period preceding their appearance. If both portal derivations are positive modulators, it appears that PCS is more potent than PCT (Table 1).

Therefore, modifications of the venous blood supply to the liver can modulate liver cancer development. A suppression of this venous blood supply

Table 1. Yield, incidence and histological type of liver cancer observed in groups I (IS), II (IS PB), III (IS PCS) and IV (IS PCT) after 28 weeks

Treatment	Yield of liver cancer	Incidence of liver cancer	Hepatocellular carcinomas		Other liver cancer
			I^{26}	IV^{26}	
IS	0	0/20			
IS PB	2	12/19 (63%)[a]	12	2	1
IS PCS	2.4	11/14 (76%)[a]	10	4	
IS PCT	1	3/10 (30%)[a,b]	3		

[a] $P < 0.05$ versus IS; [b] $P < 0.05$ versus IS PCS.

393

is more efficient than a shift from portal to caval blood. The mechanism(s) by which changes in venous blood supply influence the process are unknown and difficult to determine because of the pleiotropic effects of PCS and PCT on the metabolism [27].

The modulating treatments can influence the carcinogenic pathway by modifying the evolution of foci and nodules to cancers. Like PB, PCS belongs to the class of modulating factors which increase foci and nodules as well as cancer development [3]. However, PCS seems to be more efficient than PB since already 13 weeks after initiation, 20% of the rats bore cancers.

EFFECT OF PARTIAL HEPATECTOMY (PH) ON LIVER CARCINOGENESIS

Effect of PH Performed 8 Weeks before Initiation

PH is known to enhance liver carcinogenesis when performed several hours before or after the administration of a carcinogen. This effect has been attributed to the cell proliferation induced by PH. This cell proliferation seems to be a necessary step for the fixation of DNA damage leading to the initiation of carcinogenesis [28].

Based on the observations of H. Bartsch (personal communication) indicating that a PH performed 10 weeks before chronic administration of DEN enhanced cancer development when compared to a control group, we investigated whether PH could also exert such an enhancing effect when performed several weeks before the initiation-selection protocol.

The experimental protocol designed to answer this question is schematized in Fig. 4. It consists in performing a 70% PH or a sham operation 8 weeks before submitting the rats to a IS PB protocol described above [11,25], namely an initiation with 200 mg/kg of DEN, a selection with 2-acetylaminofluorene and CCl_4 followed by chronic PB administration.

The results indicate that PH performed several weeks before initiation could enhance cancer development as compared to control rats since the incidence of liver cancer increased from 0% in sham operated rats to 70% in hepatectomized rats. It should be noticed that in the sham operated group the cancer incidence was lower than in former experiments [11-13,21]. This could be explained by an increased age and weight which might reduce the sensitivity to carcinogen.

This experiment indicates that liver carcinogenesis can be premodulated since a surgical operation performed several weeks before the initiation can influence the carcinogenic process. It suggests that PH may have a "memory" effect on liver. Similar results were found in skin carcinogenesis when initiation-promotion sequence was inverted [29] and when wounding before initiation increased cancer incidence (Marks, personal communication).

Effect of PH Performed after Initiation

PH had little if any effect on the development of hepatocellular carcinomas when performed after chronic administration of carcinogen [30,31]. However, in the view of the potential "memory" effects of PH on liver cells, we checked whether PH performed several weeks after initiation could exert a positive modulating effect on rat liver carcinogenesis. The experimental protocol used is schematized in Fig. 5. Again, the incidence of cancer was compared between a control group submitted to IS and a test group submitted to a 70% PH one week after the end of the selection (IS PH).

The results show that PH had a slight enhancing effect when performed

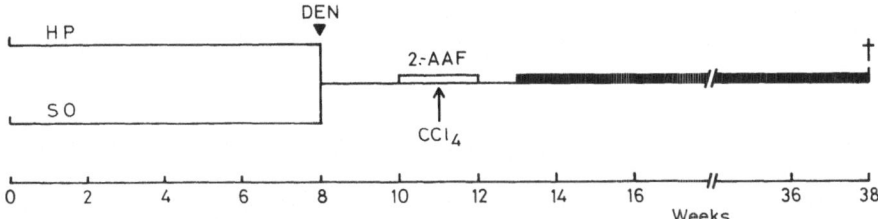

Fig. 4.   Experimental protocol used to analyze the effect of PH performed
         8 weeks before initiation/selection/promotion protocol.

after initiation-selection since the cancer incidence was 40% in the IS PH
group as compared to 10% in the IS group and 88% in the IS PB group.

      These two experiments reported here show that PH performed several weeks
before or after initiation can positively modulate liver carcinogenesis.
They suggest that cell proliferation could play a role in the modulation of
liver carcinogenesis.

GENERAL CONCLUSIONS

   1.   The nature of the modulating factors of rat hepatocarcinogenesis can be
        very different.
   2.   Surgery (portal derivation or partial hepatectomy) can positively modu-
        late liver carcinogenesis.
   3.   The paradigm of "promotion" (increase in cancer incidence by a chronic
        treatment with a xenobiotic administered after initiation) could be

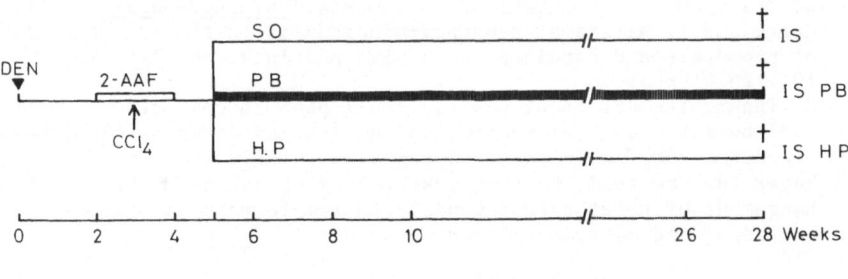

Fig. 5.   Experimental protocol used to analyze the effect of PH performed
         several weeks after initiation.

rethought since: (a) carcinogenesis can be modulated before initiation, and (b) a surgical operation having a relatively punctual effect can act as a positive modulator.

REFERENCES

1. E. Farber and D. S. R. Sarma, Biology of disease hepatocarcinogenesis, A dynamic perspective, Laboratory Investigation 56:4 (1987).
2. M. Roberfroid, From normal cell to cancer: an overview introducing the concept of modulation of carcinogenesis, in: "Concepts and Theories in Carcinogenesis, A. Maskens, ed., Excepta Medica Congress Series, Elsevier Scientific Publishers, Amsterdam (1987).
3. V. Préat and N. Delzenne, Modulating factors of hepatocarcinogenesis, in: "Experimental Hepatocarcinogenesis", M. Roberfroid and V. Préat, eds., Plenum Press, New York (1988).
4. P. Bannasch, The cytoplasm of hepatocytes during Carcinogenesis, Rec. Results Cancer Res. 19:1 (1968).
5. G. Teebor and F. Becker, Regression and persistance of hyperplastic nodules induced by N-2-fluorenylacetamide and their relationship to hepatocarcinogenesis, Cancer Res. 31:1 (1971).
6. H. Barbazon, V. Smoliar, A. Fridman-Manduzio and E. Betz, Effect of discontinuation of chronic feeding of diethylnitrosamine on the development of hepatomas in adult rats, Br. J. Cancer 40:260 (1979).
7. D. Solt and E. Farber, New principle for the analysis of chemical carcinogenesis, Nature 263:701 (1976).
8. Van Rensselaer Potter, Use of sequential applications of initiators in the production of hepatomas in the rat, Cancer Res. 44:2733 (1984).
9. C. Peraino, R. Fry and E. Staffeld, Reduction and enhancement by phenobarbital of hepatocarcinogenesis induced in rats by 2-acetylaminofluorene, Cancer Res. 31:1506 (1971).
10. H. Pitot and A. Sirica, The stages of initiation and promotion in hepatocarcinogenesis, Biochim. Biophys. Acta 605:191 (1980).
11. V. Préat, J. de Gerlache, M. Lans, H. S. Taper and M. Roberfroid, Comparative analysis of the promoting effect of phenobarbital, DDT, nafenopin and butylated hydroxytoluene in rat hepatocarcinogenesis, Carcinogenesis 7:1025 (1986).
12. V. Préat, M. Lans, J. de Gerlache, H. S. Taper and M. Roberfroid, Comparison of the biological effects of phenobarbital and nafenopin on rat hepatocarcinogenesis, Jp. J. Cancer Res. 77:629 (1986).
13. V. Préat, J. de Gerlache, M. Lans and M. Roberfroid, Promoting effect of oxazepam in rat hepatocarcinogenesis, Carcinogenesis 8:97 (1987).
14. M. Moore and T. Kitagawa, Hepatocarcinogenesis in the rat: the effect of promoters and carcinogens in vivo and in vitro, Int. Rev. Cyt. 101:125 (1986).
15. H. S. Taper, The effect of Oestradiol-17-phenylpropionate and Oestradiol benzoate on N-nitrosomorpholine induced liver carcinogenesis in ovarectomised female rats, Cancer 42:462 (1978).
16. J. Yager, H. Campbell, D. Longnecker, B. Roebuck and M. Benoit, Enhancement of hepatocarcinogenesis in female rats by ethinyl estrodiol and mestranol but not estradiol, Cancer Res. 44:3862 (1984).
17. R. Cameron, K. Imaido and N. Ito, Promotive effect of steroids and bile acids on hepatocarcinogenesis initiated by diethylnitrosamine, Cancer Res. 42:2426 (1982).
18. P. M. Rao, K. Wagamine, K. Ho, W. Roomi, C. Laurier, S. Rajalakshmi and D. S. R. Sarma, Dietary orotic acid enhances the incidence of gamma-glutamyl-transferase positive foci in rat liver induced by chemical carcinogens, Carcinogenesis 4:1541 (1983).
19. M. A. Sells, S. Katyal, S. Sell, H. Shinozuka and B. Lombardi, Induction of foci of altered gamma-glutamyl-transpeptidase positive

cytes in carcinogen treated rats fed a choline deficient diet, <u>Br. J. Cancer</u> 40:274 (1979).

20. J. de Gerlache, H. Taper, M. Lans, V. Préat and M. Roberfroid, Dietary Modulation of rat liver carcinogenesis, <u>Carcinogenesis</u> 8:337 (1987).

21. V. Préat, J. C. Pector, H. S. Taper, M. Lans, J. de Gerlache and M. Roberfroid, Promoting effect of portacaval anastomosis in hepato-hepatocarcinogenesis, <u>Carcinogenesis</u> 5:1151 (1984).

22. P. Bannasch, Preneoplastic lesions as end point in carcinogenesis testing, I. Hepatic Neoplasia, <u>Carcinogenesis</u> 7:689 (1986).

23. G. Williams, Phenotypic properties of preneoplastic liver lesions and application to detection of carcinogens and tumor promoters, <u>Toxicol Pathol.</u> 10:3 (1982).

24. N. Ito, H. Tsuda, R. Hasegawa and R. Imaida, Comparison of the promoting effects of various agents in induction of preneoplastic lesions in rat liver, <u>Envir. H. Per.</u> 50:131 (1983).

25. M. Lans, J. de Gerlache, V. Préat, H. S. Taper and M. Roberfroid, Phenobarbital as a promoter in the initiation/selection process of experimental rat hepatocarcinogenesis, <u>Carcinogenesis</u> 4:141 (1983).

26. R. Squirre and M. Levitt, Report of a workshop on classification of specific hepatocellular lesions in rats, <u>Cancer Res.</u> 35:3214 (1975).

27. J. C. Pector, Conséquences métaboliques des dérivations portales chez le rat: Thèse d'agrégation de l'enseignement supérieur, Université Libre de Bruxelles (1981).

28. E. Farber, Chemical carcinogenesis: a biological perspective, <u>Am. J. Pathol.</u> 108:271 (1982).

29. G. Fürstenberger, V. Kinzelt, M. Schwarz and F. Marks, Partial inversion of the initiation-promotion sequences of multistage tumorigenesis in the skin NMRI mice, <u>Science</u> 230:76 (1985).

30. M. F. Rajewski, W. Dauber and H. Frankenberg, Liver carcinogenesis by diethylnitrosamine in the rat, <u>Science</u> 152:83 (1966).

31. H. Barbason, C. Rassenfosse and E. Betz, Promotion mechanism of phenobarbital and partial hepatectomy in DENA hepatocarcinogenesis and cell kinetics effect, <u>Br. J. Cancer</u> 47:517 (1983).

ANALYSIS OF THE EFFECTS OF MODIFYING AGENTS ON SIX DIFFERENT PHENOTYPES
IN PRENEOPLASTIC FOCI IN THE LIVER IN MEDIUM-TERM BIOASSAY MODEL IN RATS

Hiroyuki Tsuda[1], Satoshi Uwagawa[1], Toyohiko Aoki[1], Shoji Fukushima[1], Katsumi Imaida[1], Nobuyuki Ito[1], Kiyomi Sato[2], Toshikazu Nakamura[3] and Franz Oesch[4]

[1] First Department of Pathology, Nagoya City University Medical School, 1 Kawasumi, Mizuho-cho, Mizuho-ku Nagoya 467, Japan
[2] Second Department of Biochemistry, Hirosaki School of Medicine, Zaifu-cho 5, Hirosaki 036, Japan
[3] Institute for Enzyme Research, University of Tokushima School of Medicine, Tokushima 770, Japan
[4] Institute of Toxicology, University of Mainz, Obere Zahlbacher Straße, D-6500, Mainz, FRG

INTRODUCTION

Recently a great deal of interest has been expressed in characterizing the altered enzyme phenotype of putative preneoplastic rat liver lesions. In particular, attention has been given to the changes in drug metabolizing potential, conferring physiological advantage to initiated cells, and their usefulness as marker lesions for the analysis of the development of neoplasia[1-2].

The present study reports the effects of promoting and inhibitory agents on expression of multiple enzyme phenotypes of putative preneoplastic cells and their surrounding hepatocytes which were induced in a medium-term bioassay system developed in this laboratory[3-6]. The levels of glutathione-S-transferase P-form (GST-P), glucose-6-phosphate dehydrogenase (G6PD), γ-glutamyl transpeptidase (γ-GT), adenosine triphosphatase (ATPase), cytochrome P-450PB$_{3a}$, and microsomal epoxide hydrolase with broad substrate specificity (mEHb) were compared along with cellular proliferation which was monitored by the incorporation of bromodeoxyuridine (BrdU) in the nuclei of hepatocytes (labeling index, L.I.).

MATERIALS AND METHODS

Six-week-old male F344 rats (Charles River Japan Inc., Atsugi, Japan) were initially given a single dose (200 mg/kg, i.p.) of diethylnitrosamine (DEN) (Tokyo Chemical Industry Co., Ltd., Tokyo, Japan). Two weeks after the initial dose the animals were treated with 4 different test compounds for 6 weeks, and were subjected to two-thirds partial hepatectomy (PH) in week 3. BrdU (Tokyo Chemical Industry Co., Ltd.) was injected continuously during the last week (week 8) using an osmotic minipump (Alza Corporation, Palo Alto, CA USA) implanted subcutaneously into the back of the rats (3 rats each). Control rats were given DEN (+ PH) alone. The compounds tested were known promoting agents: 2-amino-3-methylimidazo(4,5-f)quinoline (IQ), a carcinogenic pyrolysis product, a gift from Dr. M. Nagao (National Cancer Center Research Institute, Tokyo, Japan)[7], phenobarbital (PB) (Iwaki Seiyaku Co., Tokyo, Japan); and inhibitory agents: butylated hydroxyanisole (BHA)[8] (Wako Pure Chemical Industries Ltd., Osaka, Japan) and di(2-ethylhexyl)phtalate (DEHP) (a peroxisome proliferator reported as a hepatocarcinogen which apparently caused a decrease in the appearance of γ-GT positive (γ-GT$^+$) foci, Tokyo Chemical Industry Co., Ltd.)[9].

The rats were sacrificed at the end of week 8 and the livers were immediately frozen in isopentane cooled at -140°C, or fixed in ice-cold acetone. Complete serial 4-6 µm sections were cut on a cryostat for the histochemical demonstration of G6PD, ATPase and γ-GT and immunohistochemical demonstration of GST-P, P-450PB$_{3a}$, mEHb and incorporated BrdU (Fig. 1).

The antibodies, GST-P[12], P-450PB$_{3a}$, mEHb[14], G6PD[15] and anti-BrdU (Becton Dickinson, Mountain View, CA, USA) were applied immunohistochemically using the avidin-biotin-peroxidase complex method[11](Vectastain ABC kit, Vector Laboratories Inc., Burlingame, CA, USA).

The localization and conformity of enzyme-altered foci were measured on a photograph enlarged approximately 10 times.

RESULTS

Quantitative values (number/cm^2) of all different phenotypes, except γ-GT, were largest in IQ-treated rats, followed by that of GST-P, G6PD and ATPase in PB-treated group. BHA and DEHP clearly inhibited the induction of foci positive for GST-P, G6PD and γ-GT, or negative for ATPase, when compared to the control group given DEN alone. Such modifying effects were not clearly observed in assays by P-450PB$_{3a}$ and mEHb. Binding for P-450PB$_{3a}$ was mostly (> 90%) positive (increased) in all treatment groups.

Values in the number of BrdU positive (BrdU$^+$) foci constituted approximately 20 to 40% of the individual values for GST-P, G6PD, γ-GT and ATPase, all, except γ-GT, being proportional to that of respective phenotypes in PB-treated rats. The number of γ-GT$^+$ foci was greater in PB-treated rats than in IQ-treated rats; however, values of BrdU$^+$ foci within γ-GT$^+$ foci were larger

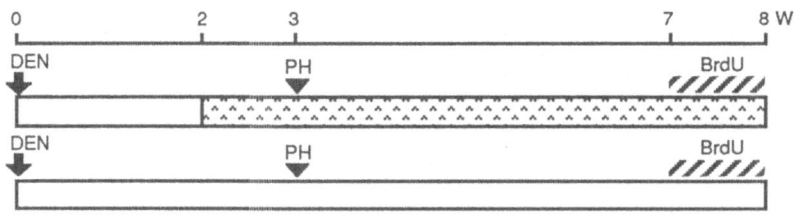

Animals ; 6-week-old F344 male rat
↓ : Diethylnitrosamine (DEN), 200 mg/kg, i.p.
▼ : Partial hepatectomy (PH)
[⌃⌃] : 2-Amino-3-methylimidazo(4,5-f)qunoline (IQ, 0.05%)
Phenobarbital (PB, 0.05%)
Butylated hydroxyanisole (BHA, 1%)
Di(2-ethylhexyl)phthalate (DEHP, 0.3%)
BrdU : s.c. injection of BrdU (150 mg/kg) by minipump

Fig. 1. Experimental protocol.

in IQ-treated than PB-treated rats. Values for BrdU$^+$ foci among P-450PB$_{3a}$ positive (P-450PB$_{3a}^+$) and mEHb positive (mEHb$^+$) foci ranged 34 to 64% (Figs. 2-4).

L.I. of BrdU$^+$ foci within each treatment group ranged from 41% (control) to 70% (BHA); however, that of surrounding hepatocytes did not show an obvious difference (13 to 19%).

The conformity class distribution clearly showed that total number of foci count decreased as number of conforming enzymes increased. Foci simultaneously expressing two enzymes occurred most frequently in all treated groups (Fig. 5). In contrast, the incidence of BrdU$^+$ foci conforming with the other six enzyme phenotypes clearly increased alone with the number of conforming phenotypes (Fig. 6).

Among the foci expressing 2 enzyme phenotype, combination of GST-P + G6PD gave the largest number followed by G6PD + ATPase and GST-P + ATPase. The number of foci observed in BHA or DEHP-treated groups was too few to count for analisys, due to their potent inhibition in the development of foci.

DISCUSSION

Our results clearly show that modifying agents assayed in the current protocol caused an increase or decrease in the number of enzyme-altered foci, without any preferential effects on specific phenotypes such as GST-P, G6PD, γ-GT and ATPase. However, modifying potential of the chemicals was not clearly shown, except in the IQ-treated group, when assayed by the expression of P-450PB$_{3a}$ or mEHb.

It should be noted that the number of foci positive for BrdU was proportional to the total number of foci. However, the ratio of BrdU$^+$ foci in individual phenotype was no more than 50% of their total number, indicating that all foci are not evenly undergoing proliferation (mitosis).

Quantitative analysis of simultaneous expression of 6 different pheno-

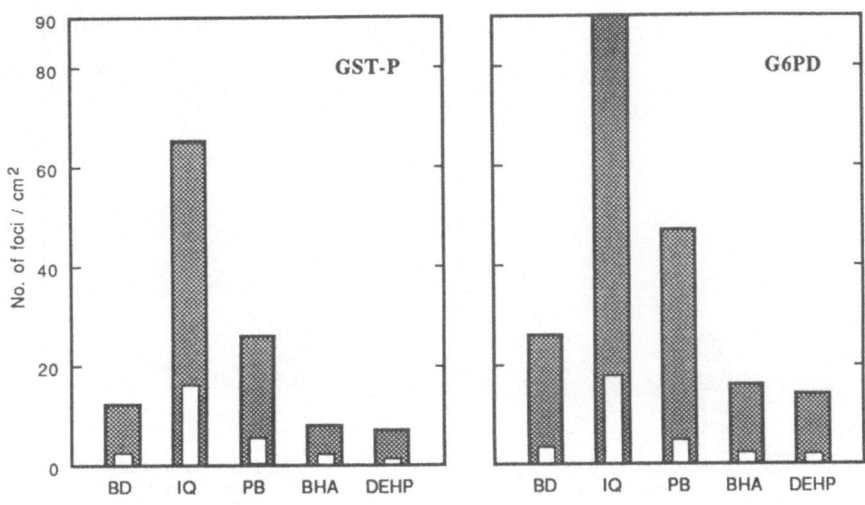

Fig. 2.   Number of GST-P$^+$ and G6PD$^+$ foci and their conformity to BrdU$^+$ foci (□).

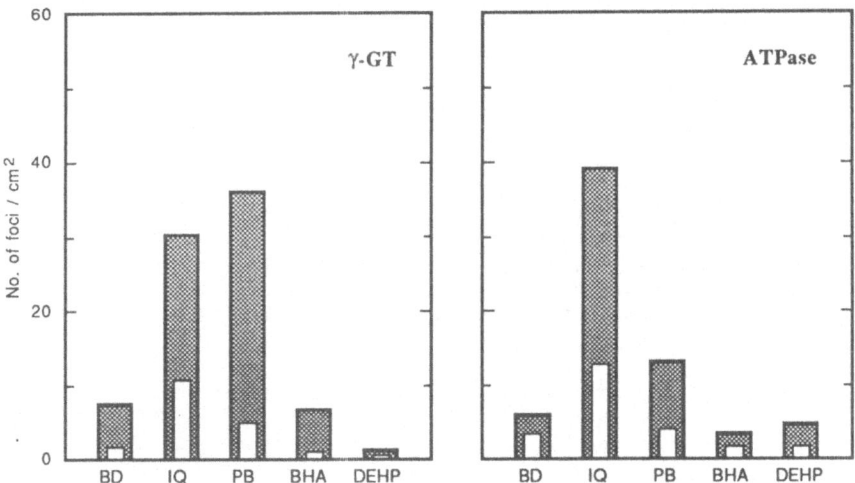

Fig. 3. Number of γ-GT⁺ and ATPase⁺ foci and their conformity to
BrdU foci (☐).

types showed that in animals given IQ or PB, foci expressing 2 enzymes gave
maximum number. The number of foci gradually decreased as the number of
conforming phenotypes increased. However this trend was not obvious in the
group given inhibitory agents, BHA or DEHP. This may indicate that phenotypic expression of foci induced by such inhibitory agents is variable.

On the other hand, obvious increases in the incidence of BrdU⁺ foci
among the higher range of conformity (4 to 6 phenotypes) strongly indicate
that multiple expression of different phenotypes within individual lesions is
closely related to the potential for neoplastic proliferation. When
analyzing number of conforming phenotypes with respect to foci count (sensitivity) and their positivity to BdrU (proliferative potential), the combina-

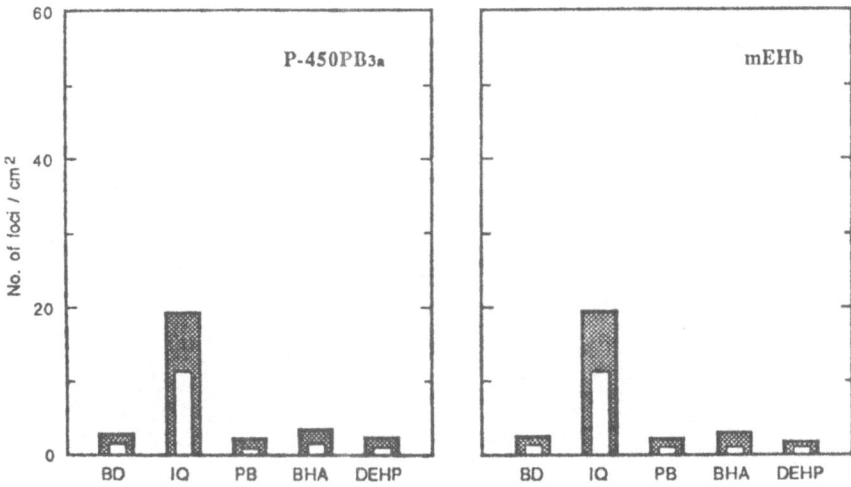

Fig. 4. Number of P-450PB$_{3a}$⁺ and mEHb⁺ foci and their conformity to
BrdU⁺ foci (☐).

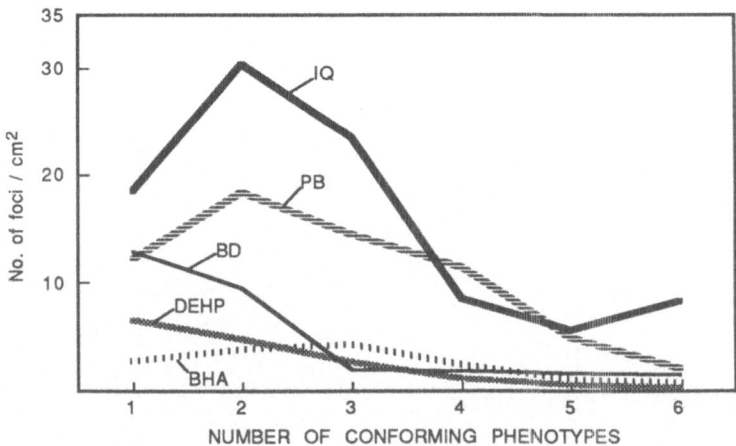

Fig. 5. Number of foci in different conformity classes.

tion of GST-P + G6PD or GST-P + G6PD + ATPase were found to be most suitable (Fig. 7).

We observed a large number of foci showing positive binding to P-450PB$_{3a}$ although binding of foci cells to cytochrome P-450 species was generally decreased[16,17]. This observation may correspond to that noted in earlier-stage lesions induced by DEN or N-nitrosomorpholine[18]. However, further investigation of the altered expression of drug metabolizing enzymes, including that of cytochrome P-450 species, under different experimental conditions appears warranted.

ACKNOWLEDGMENTS

Supported by Grants-in-aid for Cancer Research from the Ministry of Health and Welfare and from the Ministry of Education, Science and Culture, and for a Comprehensive 10 year Strategy for Cancer Control, and by a grant from the Deutsche Forschungsgemeinschaft.

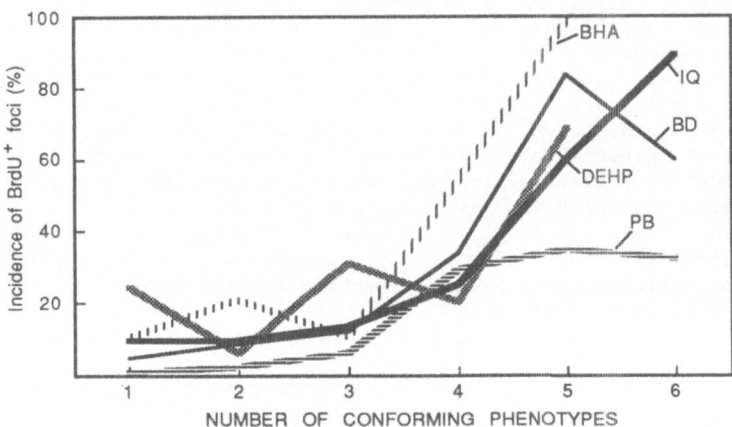

Fig. 6. Incidence of BrdU foci in different conformity classes.

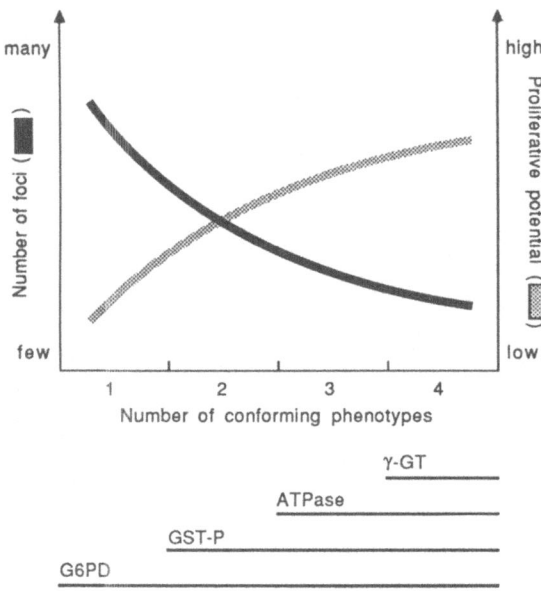

Fig. 7. Schematic correlation of 4 phenotypes to the number of foci and their proliferative potential.

REFERENCES

1. P. Bannasch, Commentary: Preneoplastic lesions as end points in carcinogenicity testing, I. Hepatic neoplasia, Carcinogenesis 7:698 (1986).

2. E. Farber and R. Cameron, The sequential analysis of cancer development Adv. Cancer Res. 31:125 (1980).

3. D. Solt and E. Farber, New principle for the analysis of chemical carcinogenesis, Nature (Lond) 263:701 (1976).

4. N. Ito, H. Tsuda, R. Hasegawa and K. Imaida, Sequential observation of pathomorphologic alterations in preneoplastic lesions during the promoting stage of hepatocarcinogenesis and the development of short-term test system for hepatopromoters and hepatocarcinogens, Toxicol. Pathol. 10:37 (1982).

5. M. Tatsematsu, R. Hasegawa, K. Imaida, H. Tsuda and N. Ito, Survey of various chemicals for initiating and promoting activities in a short-term in vivo system based on generation of hyperplastic liver nodules in rats, Carcinogenesis 4:381 (1983).

6. H. Tsuda, R. Hasegawa, K. Imaida, T. Masui, M. A. Moore and N. Ito, Modifying potential of thirty-one chemicals on the short-term development of γ-glutamyl transpeptidase-positive foci in diethylnitrosamine-initiated rat liver, Gann 75:876 (1984).

7. S. Takayama, Y. Nakatsuru, M. Masuda, H. Ohgaki, S. Sato and T. Sugimura, Demonstration of carcinogenicity in F344 rats of 2-amino-3-methylimidazo(4,5-f)quinoline from broiled sardine, fried beef and beef extract, Gann 75:467 (1984).

8. H. Tsuda, T. Sakata, T. Masui, K. Imaida and N. Ito, Modifying effects of butylated hydroxyanisole, ethoxyquin and acetaminophen on induction of neoplastic lesions in rat liver and kidney initiated by N-ethyl-N-hydroxyethylnitrosamine, Carcinogenesis 5:525 (1984).

9. W. M. Kluwe, J. K. Haseman, J. F. Douglas and J. E. Huff, The carcinogenicity of dietary di(2-ethylhexyl)phthalate (DEHP) in Fischer 344 rats and B6C3F mice, J. Toxicol. Environ. Health 10:797 (1982).

10. H. J. Hacker, M. A. Moore, D. Mayer and P. Bannasch, Correlative histochemistry of some enzymes of carbohydrate metabolism in preneoplastic and neoplastic lesions in rat liver, Carcinogenesis 3:1265 (1982).

11. S. M. Hsu, L. Raine and H. Fanger, Use of avidin-biotin-peroxidase complex (ABC) in immunoperoxidase techniques: A comparison between ABC and unlabelled antibody (PAP) procedures, J. Histochem. Cytochem. 29:577 (1981).

12. K. Satoh, A. Kitahara, Y. Soma, Y. Inaba, I. Hyama and K. Sato, Purification, induction and distribution of placental glutathione transferase: a new marker enzyme for preneoplastic cells in rat chemical carcinogenesis, Proc. Natl. Acad. Sci. USA 82:3964 (1985).

13. C. R. Wolf, E. Moll, T. Friedberg, F. Oesch, A. Buchmann, W. D. Kuhlmann and H. W. Kunz, Characterization, localization and regulation of a novel phenobarbital-inducible form of cytochrome P-450 compared with three further cyt. P-450 isoenzymes, NADPH P-450 reductase, glutathione transferases and microsomal epoxide hydrolase, Carcinogenesis (Lond) 5:993 (1984).

14. P. Bentley and F. Oesch, Purification of rat liver epoxide hydrolase to apparent homogeneity, FEBS Lett. 59:291 (1975).

15. T. Nakamura, K. Yoshimoto, K. Aoyama and A. Ichikawa, Hormonal regulation of glucose-6-phosphate dehydrogenase and lipogenesis in primary cultures of rat hepatocytes, J. Biochem. 91:681 (1982).

16. R. Cameron, G. D. Sweeney, K. Jones, G. Lee and E. Farber, A relative deficiency of cytochrome P-450 and aryl hydrocarbon (benzo(a)pyrene) hydroxylase in hyperplastic nodules induced by 2-acetylaminofluorene in rat liver, Cancer Res. 36:3888 (1976).

17. H. Tsuda, M. A. Moore, M. Asamoto, T. Inoue, S. Fukushima, N. Ito, K. Satoh, Z. Amelizad and F. Oesch, Immunohistochemically demonstrated altered expression of cytochrome P-450 molecular forms and epoxide hydrolase in N-ethyl-N-hydroxyethylnitrosamine-induced rat kidney and liver lesions, Carcinogenesis 8:711 (1987).

18. H. W. Kunz, A. Buchmann, M. Schwartz, R. Schmitt, W. D. Kuhlmann, C. R. Wolf and F. Oesch, Expression and inducibility of drug-metabolizing enzymes in preneoplastic lesions of rat liver during nitrosamine-induced hepatocarcinogenesis, Arch. Toxicol. 60:198 (1987).

S-ADENOSYLMETHIONINE ANTIPROMOTION AND ANTIPROGRESSION EFFECT IN HEPATOCARCINOGENESIS. ITS ASSOCIATION WITH INHIBITION OF DNA METHYLATION AND GENE EXPRESSION

Francesco Feo, Renato Garcea, Lucia Daino, Rosa Pascale, Serenella Frassetto, Patrizia Cozzolino, Maria G. Vannini, Maria E. Ruggiu, Marilena Simile and Marco Puddu

Istituto di Patologia Generale, Università di Sassari 07100 Sassari, Italy

INTRODUCTION

The identification of several steps in the carcinogenic process, has permitted a new approach in the prevention of cancer development. It is conceivable that interference with one or more of the recognizable steps leads to the breakdown of the carcinogenic process. In recent years some attempts to interfere with carcinogenic promotion, at different levels, have been made using retinoids[1,2], 1α,25-dihydroxyvitamin D_3, a hormonally active metabolite of vitamin D^3, antioxidants[4,5], α-difluoromethylornithine[6,7], putrescine[8], indomethacin[9], protease inhibitors[10], inhibitors of arachidonic acid metabolism[11], protein kinase C inhibitors[12], dehydroepiandrosterone and some related hormones[13-16], and S-adenosylmethionine[17-21] (SAM).

SAM is the most important lipotrope. Different reports have shown that it enters liver cells[22-27], and may be administered at relatively high doses without any toxic effect in rat (50% lethal dose, 1.5 g/Kg) and man[28]. Different observations correlate hepatocarcinogenesis promotion and polyamine synthesis to the liver lipotrope content. In fact, chronic administration of a lipotrope-deficient diet enhances rat liver carcinogenesis induced by various carcinogens[29,30]. Moreover, recent evidence indicates that this diet, without any added carcinogen and with barely detectable levels of some carcinogens, is associated with the development of liver tumors[31-34]. Treatment of rats with a lipotrope-deficient diet leads to an increase in ornithine decarboxylase (ODC) and SAM decarboxylase activities[35], two key enzymes of polyamine synthesis. This treatment leads to a fall in hepatic SAM content[36]. A fall in SAM level is also caused by long-term feeding of phenobarbital (PB)[37], a known promoting agent[38,39] which induces ODC activity[40]. Finally, recent studies in our laboratory have shown that in Wistar rats subjected to the initiation of hepatocarcinogenesis by diethylnitrosamine (DENA) followed by a selection treatment according to Solt and Farber[41], the appearance of enzyme-altered foci (EAF) is coupled with a great fall in liver SAM content[17-21]. This effect is enhanced by the administration of PB for 16 weeks after the end of the selection treatment. In more than 95% of rats persistent neoplastic nodules (NN) and hepatocellular carcinomas (HC) develop and exhibit a low SAM content[19-21]. A great increase in ODC activity and in polyamine synthesis occurs in the liver of carcinogen-treated rats. These changes are further enhanced by PB[18]. Evidence has been given indicating that the increase in polyamine synthesis is not a mere consequence of liver regeneration, in rats subjected to a partial hepatectomy during the selection treatment, but it parallels the development of EAF. In addition, a high ODC activity and a low lipotrope content have been demonstrated in both NN and HC[19-21].

THE SAM ANTIPROMOTING EFFECT

Growing tissues, either normal or preneoplastic, need relatively high amounts of SAM for nucleic acid, phospholipid and polyamine synthesis. In order to investigate the relationships between SAM liver content and carcinogenesis, we have attempted to reconstitute the liver SAM pool by administering exogenous SAM to rats subjected to the initiation/selection treatments in accordance with Solt and Farber[41], followed, or not followed, by the administration for 16 weeks of a diet supplied with 0.05% PB. We have studied, in these rats, the effects of SAM administration on the development of preneoplastic liver.

When high SAM doses (384 µmol/Kg,day) were given to carcinogen-treated rats, starting at the end of the selections treatment, there was a great fall in liver ODC activity and polyamine content[17-21], associated with a great inhibition of the development of EAF and a complete prevention of NN and HC development[19-21]. These changes were coupled with accumulation of 5'-methylthioadenosine (5'-MTA) in the liver[20,21].

5'-MTA is a SAM catabolite formed by direct cleavage of SAM, catalyzed by SAM hydrolase, or as an end product of polyamine synthesis[42]. 5'-MTA however, greatly reduces this synthesis since it is a strong inhibitor of spermine and spermidine synthetases[43,44], ODC[18-20] and SAM decarboxylase[45]. SAM does not inhibit ODC activity in vitro. Inhibition occurs after preincubation of liver homogenates with SAM, before preparing crude ODC (cytosolic fraction)[20]. This effect is enhanced by adenine, an inhibitor of 5'-MTA catabolism[41]. In contrast, 5'-MTA does not need preincubation to inhibit ODC activity in vitro[20]. It also inhibits in vitro growth of isolated hepatocytes[21]. In addition, a low 5'-MTA content has been found in rapidly growing tissues such as NN and HC, and in the liver of carcinogen-treated rats[20,21].

These observations point to an antipromoting effect of SAM during hepatocarcinogenesis presumably caused by 5'-MTA accumulation and growth inhibition during the treatment of rats with high SAM amounts.

Mechanisms of disappearance of preneoplastic cells, during hepatocarcinogenesis promotion include phenotypic reversion[46,47], and controlled cell death (apoptosis)[48]. Thus, it would be expected that these mechanisms are involved in the SAM antipromoting effect. Indeed we observed that SAM injection during the development of EAF causes a great increase in phenotypic reversion of γ-glutamyltranspeptidase (GGT)-positive foci (see also below) coupled with a significant fall in labeling index[21]. As known, PB inhibits remodeling[47]. This effect is completely prevented by SAM[21]. In view of the positive correlation between phenotypic complexity and growth capacity of EAF[49], we examined if decrease in growth and redifferentiation, in SAM-treated rats, are associated with reduction in phenotypic complexity of EAF. Three markers have been determined histochemically: GGT, glucose-6-phosphate dehydrogenase and glucose-6-phosphatase. Three weeks after starting 2-AAF feeding in the rats subjected to the initiation/selection treatment, GGT-positive foci, those expressing a high glucose-6-phosphate dehydrogenase activity, and those characterized by the absence of glucose-6-phosphatase represent, respectively, 15, 7, and 4 p. cent. At the 7th week these figures are 33, 22, and 15 p. cent. As shown in Fig. 1, EAF expressing respectively 1, 2 or 3 markers are 32, 46 and 22 p. cent at the 3rd week, and 41, 38

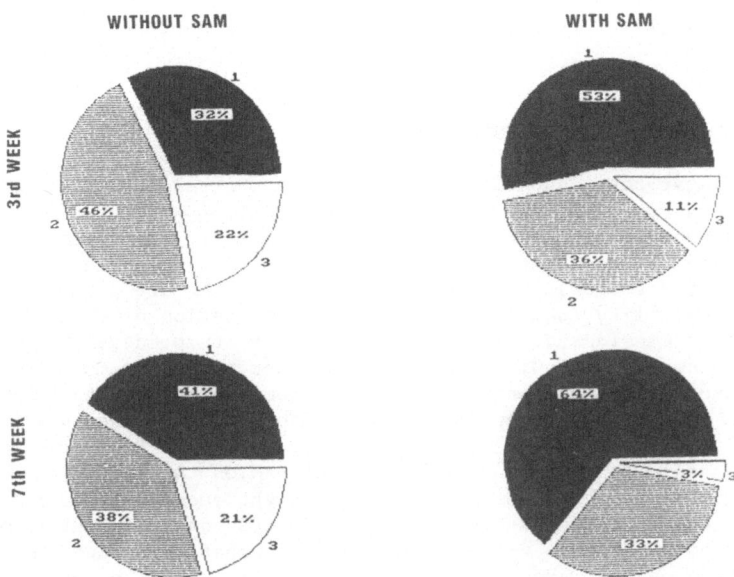

Fig. 1. Relative percentages of EAF showing 1, 2 or 3 markers. Male Wistar
rats received a single initiating DENA dose (150 mg/Kg, i.p.) and,
after 15 days, a selective regimen consisting of 0.03% 2-AAF in the
diet for 2 weeks, with a partial hepatectomy at the midpoint of
this treatment. They were then fed a diet supplemented with 0.05%
PB. When indicated 384 μmol/Kg,day of SAM were injected i.m. (6
daily doses), starting at the end of 2-AAF feeding. The rats were
sacrificed 3 and 7 weeks after starting 2-AAF. Enzyme markers were:
GGT, glucose-6-phosphate dehydrogenase and glucose-6-phosphatase.
Data are means of 5 different animals. Standard deviations were
lower than 10% of the means.

and 21 p. cent, at the 7th week. Clearly, SAM causes a great rise in the
number of foci exhibiting only one of the three markers used.

The possibility that controlled cell death plays a role in the disappearance of preneoplastic cells, in SAM-treated rats, has also been considered. The percentage of apoptotic bodies in preneoplastic liver is
relatively low during promoter treatment, and a sharp increase occurs after
arrest of this treatment[48]. SAM injection causes a significant increase in
apoptotic bodies in GGT-positive foci, either during or after PB administra-
tration[21]. In addition, SAM does not modify the relative percentages of
intracellular and extracellular apoptotic bodies, which should rule out the
possibility that SAM affects the endocytic phase[50] of the apoptotic process,
nor does SAM modify food consumption by rats, which is another factor which
regulates the apoptotic process[50]. However, controlled cell death is a rapid
phenomenon. The histologically visible parts of it may last approximately 2.5
h in the liver[50]. So, an eventual slowing or accelerating effect of SAM on
this process could result in, respectively, over- or under-evaluation of the
phenomenon.

On the basis of the above observations it has been concluded that SAM
causes a decrease in DNA synthesis, an increase in phenotypic reversion and,
apparently, in controlled cell death of putative-preneoplastic cells. Since
rapid growth, low remodeling and apoptosis characterize promotion[51], our re-

sults are indicative of an inhibitory effect of exogenous SAM on carcinogenesis promotion.

INHIBITION OF NODULE PROGRESSION BY SAM

Male Wistar rats, subjected to the initiation/selection treatment, followed by 0.05% PB in the diet for 16 weeks, develop, 4-6 weeks after arresting PB treatment, persistent neoplastic nodules[19,20]. Thirty-two weeks after the appearance of the first nodules, HC develop[20]. A good deal of evidence (cf. ref. 52 for review) indicates that persistent NN are precursors of HC. Thus, the possibility that SAM inhibits the progression of persistent nodules to HC has been considered. The chance to prevent tumor development by blocking the development of preneoplastic tissue by SAM appears to be of interest, due to the existence of different human preneoplastic lesions whose progression to malignant neoplasia is well known (see ref. 53 for review).

The data in Table 1 show the effect of SAM on the growth of NN. No differences in body weight and relative liver weight occur between rat groups. It appears that a prolonged treatment with SAM (56 days), started after the development of NN, causes 42-45% decreases in number and weight of visible nodules. These changes are paralleled by a 42% fall in labeling index. The data in Table 1 also show that there occurs a great decrease in SAM/S-adenosylhomocysteine (SAH) ratio in NN. This depends on a substantial decrease in SAM content, with respect to normal liver, while SAH, a SAM catabolite formed during the methylation reactions, does not undergo any change. SAM treatment, by increasing the nodule SAM content, without affecting SAH content (data not shown), induces a significant rise in SAM/SAH ratio.

Histochemical analysis has shown that under SAM treatment, NN undergo a decrease in size and progressively lose their uniformity for GGT reaction. Some nodules exhibit a very weak reaction for this enzymatic marker (data not shown).

Table 1. Effect of SAM on the growth of neoplastic nodules

Treatment	Body wt (g)	RLW	Nodules No/liver	g/100 g of liver	LI	SAM/SAH
DENA/2-AAF/PH/ PB	391 ± 8	5.98 ± 0.11	58 ± 8	33.1 ± 3.7	5.57 ± 0.81	0.95 ± 0.03
DENA/2-AAF/PH/ PB/SAM	362 ± 16	5.81 ± 0.28	31[a] ± 6	18.1[a] ± 1.5	3.23[a] ± 0.83	1.44[a] ± 0.09

The same conditions as in Fig. 1. SAM treatment (384 μmol/Kg,day; 6 daily doses, i.m.) was started 24 weeks after the beginning of the selection treatment and continued for 56 days. Then the rats were killed. To determine labeling index (LI) 45 μCi/day,100 g body wt. of [3H]thymidine (89 Ci/mmol) were injected continuously to rats during the last 7 days by osmotic minipumps implanted subcutaneously. LI was determined by counting 8,000 nodular cells per liver. SAM and SAH were determined by a hplc method as in 20. Data are means ± SD of 5 animals. RLW, relative liver weight (g per 100 g of body wt). [a]"t"-Test: with SAM vs without SAM, P < 0.001.

It thus appears that SAM exerts an antiprogression effect for liver NN. Nevertheless, further experiments are needed to assess if this effect results in the prevention of the development of HC.

COMPARATIVE EVALUATION OF SAM AND METHIONINE FOR ANTITUMOR EFFECT

According to recent observations methionine supply partially prevents the growth and progression of ethionine-induced liver carcinomas[54], and benzo(a)pyrene-induced skin carcinomas[55]. Interestingly, dietary supplementation of rats with 2% DL-methionine and 1% choline chloride prevents rapid increase in N-2-guanine tRNA methyltransferase II activity, otherwise seen in response to cancer promoting doses of 2-AAF[56]. In contrast with these findings, however, long-term administration of L-methionine to rats, during promotion of DENA-initiated hepatocarcinogenesis, was observed to not influence the yield of HC [57].

We directly compared the effect of equimolar amounts of L-methionine and SAM, in vivo, on the development of the preneoplastic tissue. Either 384 μmol/Kg,day of SAM or L-methionine were given i.m. to Wistar rats subjected to initiation/selection treatments, followed by PB. The lipotrope treatment was started at the end of the selection step and was continued for 5 weeks. As shown in Table 2, at the end of this treatment, 26% of liver is occupied by GGT-positive foci in the rats that do not receive lipotropes. There is a significant decrease in the foci size and number in SAM-treated rats which leads to a 61% fall in the percentage of liver occupied by puta-

Table 2. Effect of SAM and L-methionine on the development of GGT-positive foci, ODC activity and 5'-MTA accumulation

Treatments	Foci (No/mm^2)	Surface area (mm^2)	% of GGT-positive parenchyma	LI	ODC activity	5'-MTA (μg/g)
2-AAF/PH/PB				0.60[a] ± 0.01	39.41 ± 3.41	1.05 ± 0.02
DENA/2-AAF/PH/PB	0.62 ± 0.12	0.43 ± 0.10	26.06 ± 2.02	1.98[b] ± 0.04	79.62[b] ± 3.90	0.70[b] ± 0.05
DENA/2-AAF/PH/ PB/SAM	0.45[c] ± 0.09	0.21[c] ± 0.03	9.79[c] ± 2.48	0.78[c] ± 0.05	56.61[c] ± 0.10	1.54[c] ± 0.04
DENA/2-AAF/PH PB/Methionine	0.63 ± 0.08	0.46 ± 0.04	26.46 ± 1.27	1.94 ± 0.02	79.54 ± 1.41	0.89 ± 0.08

Male Wistar rats were subjected to the initiation/selection/PB treatments as shown in the legend of Fig. 1. When indicated the rats received 384 μmol/Kg, day of SAM or L-methionine (6 daily doses i.m.). Rats were sacrificed 7 weeks after starting selection treatment. Data are means ± SD of 4-5 animals. Labeling index (LI) was determined in parenchymal cells for uninitiated rats, and in GGT-positive foci for other rat groups, as described in the legend of Fig. 1. LI in the surrounding of uninitiated livers was 0.25 ± 0.01. ODC activity was determined as in 18 and is expressed as pmol of $^{14}CO_2$ released/h, mg protein from [$^{14}CO_2$]ornithine. 5'-MTA was determined by a hplc method as in 20. "t"-Test: [a]GGT-positive foci vs surrounding: P < 0.001. [b]Initiated vs uninitiated: P < 0.001. [c]With vs without SAM: at least P < 0.01.

tive preneoplastic cells. Table 2 also shows a great rise in ODC activity
and a fall in 5'-MTA content in the liver of initiated rats, compared to
controls. As expected (cf. 21), SAM causes a great drop in ODC activity,
accumulation of 5'-MTA, and a great fall in DNA synthesis in focal cells.
However, when equimolar amounts of L-methionine are substituted for SAM in
carcinogen-treated rats, no modification of any of the parameters studied
takes place. Similarly, no modification occurs after administration of
methionine plus adenosine (384 μmol/Kg,day of each; not shown). The lack of
L-methionine effect is also proved by the fact that SAM, but not methionine, partially reconstitutes the SAM liver content, which otherwise undergoes a large decrease in rats subjected to the initiation/selection/PB treatments (not shown). In keeping with these observations we found that DNA
synthesis of primary cultures of hepatocytes isolated from normal rats
undergo a 85% decrease in the presence of 1 mM SAM, but it is not affected by
the same amount of L-methionine[21].

Increase in hepatic SAM levels following methionine administration in
vivo [58], or after liver perfusion with methionine[59], has been reported. On
the other hand, conflicting data have been published as concerns SAM uptake
by liver cells. Hoffman et al.[59] showed that low SAM concentrations are not
utilized by in vitro perfused liver. However, other observations indicate
that SAM enters liver cells in vitro[22-25], and in vivo[22,26,27]. It should be
considered that SAM uptake by perfused liver may be largely influenced by the
relative rates of perfusion and SAM diffusion to extracellular and cellular
spaces.

Perhaps, conflicting results on SAM and L-methionine antitumor effects
may be explained on the basis of variations of the SAM intracellular pool.
In vivo studies have shown that SAM liver content undergoes a significant
rise only in rats fed a diet containing not less than 3% methionine[58]. The
aforementioned inhibition of liver tumor progression has been observed in
rats receiving, for 91 days, about 840 μmol/Kg,day of L-methionine[54], which
represents a much higher total dose than 384 μmol/Kg,day of SAM for 56 days
(Table 1). Need for high methionine amounts to obtain an antitumor effect,
is also proved by the results of recent work with NIH/3T3 cells, carrying the
human Ha-ras-1 proto-oncogene stably integrated in their DNA and displaying a
transformed phenotype[60]. These cells acquire a normal phenotype, are no
longer tumorigenic when injected into nude mice, and lose the Ha-ras-1 gene
when cultured with DL-methionine amounts as high as 25 mM. We could thus
conclude that SAM is more efficient than methionine (or methionine plus
adenosine) in contrasting tumor development. A higher SAM efficacy, as
compared to the methionine efficacy, has been found for reconstitution of the
liver SAM and GSH levels in ethanol intoxicated rats (Pascale et al., unpublished results). A corollary of these observations is that SAM does not
undergo any splitting into methionine and adenosine before entering liver
cells. Available data do not permit an explanation of the difference in
efficiency of SAM and methionine. Perhaps, less SAM is available to hepatocytes of rats treated with methionine than to those treated with SAM, since
methionine is partially used for protein synthesis (see also ref. 27). In
addition, we could hypothesize that product inhibition of SAM synthetase[61]
limits the SAM liver level. The liver has an additional SAM synthetase which
may not be product-inhibited[62]. However, the product-inhibited enzyme may be
induced by high methionine levels[61].

SAM TREATMENT, DNA METHYLATION AND GENE EXPRESSION

A high SAH content competitively inhibits methylation reactions[63-65].
The recovery of SAM/SAH ratio, consequent to SAM treatment of rats bearing
NN (Table 1), envisages the possibility that the reconstitution of SAM pool
is coupled with an increased DNA methylation.

Table 3. Effect of SAM on DNA methylation in carcinogen-treated rats

Treatments	Weeks after the start of 2-AAF feeding					
	3	5	7	16	24 Foci	24 Nodules
Untreated control	3.58 ± 0.09	3.68 ± 0.06	3.61 ± 0.06	3.59 ± 0.03	3.60 ± 0.05	
Uninitiated control	3.60 ± 0.02	3.61 ± 0.01	3.57 ± 0.04	3.61 ± 0.06	3.59 ± 0.05	
DENA/2-AAF/PH/PB	3.29 ± 0.01[a]	3.12 ± 0.03[a]	2.76 ± 0.05[a]	2.77 ± 0.02[a]	3.00 ± 0.04[a]	1.72 ± 0.09[a]
DENA/2-AAF/PH/PB + SAM (96)		3.28 ± 0.03[b]	3.27 ± 0.03[b]			
DENA/2-AAF/PH/PB + SAM (384)	3.60 ± 0.01[b]	3.62 ± 0.03[b]	3.48 ± 0.08[b]	3.55 ± 0.06[b]	3.07 ± 0.02[b]	

The same conditions as in Fig. 1. DNA was extracted from liver or NN and hydrolyzed with 80% (v:v) formic acid. 5-mC content was determined by hplc of hydrolyzates. 5-mC content is expressed as percent of total DNA cytosine content. Data are means ± SD of 3-5 animals. The daily doses of SAM in μmol/Kg are in parentheses 5-mC content of DNA from uninitiated controls treated with SAM (either at low or high dose) did not differ from uninitiated control and were not included in the Table. "t"-Test: [a]different from uninitiated control for at least P < 0.005; [b]different from DENA/2-AAF/PH/PB for at least P < 0.005.

413

Choline-deficient diet causes a drop in the liver SAM content as well as in SAM/SAH ratio [36]. This results in an undermethylated environment, and in a fall in methylation reactions. Undermethylation of DNA [66-68], RNA [69], and phospholipid [70,71] has been observed in liver during chronic treatment with a lipotrope-deficient diet. A large drop in the liver SAM content has also been documented during the development of EAF in the liver of rats subjected to the initiation/selection protocol, and in NN [17-21]. Thus, we considered the possibility that DNA from preneoplastic tissue is undermethylated and that SAM treatment could influence DNA methylation.

In the liver of rats subjected to the initiation/selection/PB treatments there is, between the 3rd and the 24th week after starting 2-AAF feeding, a significant decrease in the overall DNA methylation, with respect to uninitiated controls (Table 3). A greater fall occurs in NN. The maximum fall in liver DNA methylation was observed at the 7th and 16th week, which coincides with the maximun development of GGT-positive foci in Wistar rats, but not with the maximum value of DNA synthesis in GGT-positive cells [18-20]. SAM administration for 16 weeks after the end of the selection treatment causes, at all doses used, a complete or near complete reversion of DNA undermethylation. A 25% increase in 5-mC content has also been observed in DNA from NN of rats which received SAM for 56 days after the appearance of NN (2.17 ± 0.25 p. cent of total cytosine; SD, n = 3). About the same degree of liver DNA undermethylation as in PB-treated rats occurs in the rats subjected to initiation/selection treatments without PB, which is completely reversed by SAM (not shown). This indicates that PB does not interfere with the changes in DNA methylation observed during development of GGT-positive foci.

SAM antipromoting effect has been demonstrated using relatively high amounts of SAM which cause accumulation of 5'-MTA [20,21], a strong growth inhi-

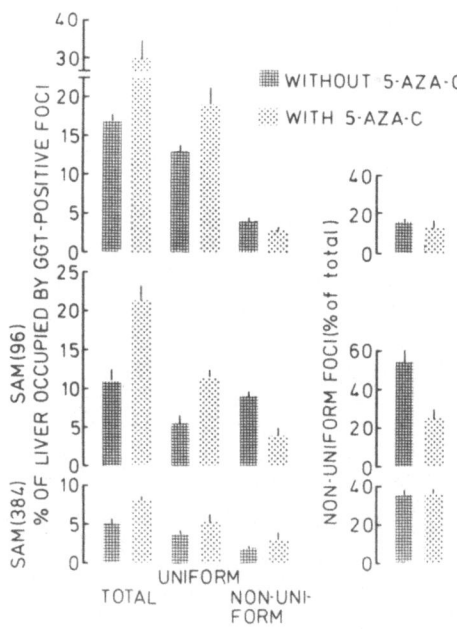

Fig. 2.   Effect of 5-AzaC on the development of GGT-positive foci. The rats were subjected to the initiation/selection treatments and, one week after the end of 2-AAF feeding, to 5-AzaC (2 μmol/Kg) and/or the indicated doses of SAM (μmol/kg,day) for 10 days.

bitor. This renders it difficult to assess if DNA methylation is primarily involved in the SAM antipromoting effect. Since DNA methylation has also been observed with a SAM dose as low as 96 µmol/Kg,day (Table 3), we evaluated the antipromoting effect of this SAM dose and the interference with this effect, of 5-azacytidine (5-AzaC), a known inhibitor of DNA methylation[72-74].

In the rats subjected to DENA/2-AAF/PH and then to the administration of low SAM doses (96 µmol/Kg,day) for 3 weeks, the liver occupied by GGT-positive foci is 35% lower than in controls (Fig. 2). This largely depends on a decrease in the size without any significant change in foci number (not shown), and is coupled with a large rise in the percentage of foci showing a non-uniform pattern. A 70% fall in GGT-positive liver occurs in the rats subjected to a high SAM dose (384 µmol/Kg,day). 5-AzaC, given to rats for 10 days, starting one week after the end of 2-AAF feeding, causes a great increase in GGT-positive liver, largely linked to a rise in DNA synthesis of GGT-positive hepatocytes (not shown). The enhancing effect of 5-AzaC is surprising. 5-AzaC, at the dose used in this study, does not induce any change in body and liver weights. This dose does not appear to be toxic: all rats survived a 10-day treatment with 5-AzaC. 5-AzaC does not exhibit any selecting effect if given in place of 2-AAF in the initiation/selection protocol. Interestingly, non-toxic doses of 5-AzaC enhance mitotic activity and DNA synthesis in 24 h regenerating liver when given before partial hepatectomy[74]. In addition, a prolonged treatment of Balb/c mice with 2 mg/Kg of 5-AzaC has recently been shown to induce a significant rise in the incidence of lymphomas, lung, and skin tumors, mammary carcinomas and a variety of other tumors[75]. The data in Fig. 2 also show that 5-AzaC does not interfere with the antipromoting effect of high SAM doses, while it largely prevents the decrease in GGT-positive liver and the increase in non-uniform foci produced by low SAM doses. Biochemical analysis demonstrated that 5-AzaC does not modify SAM, SAH and 5'-MTA liver contents and liver ODC activity; nor does it interfere with the effect of SAM, at all doses used, on these parameters (not shown). However, 5-AzaC causes, as expected, a great fall in 5-mC content of liver DNA (1.37 ± 0.09 p. cent of total cytosine; SD, n = 4) which is not modified by SAM treatments.

These results indicate that the administration of SAM doses that do not cause 5'-MTA accumulation and do not inhibit ODC activity, is followed by an increase in SAM/SAH ratio, growth inhibition and a great increase in phenotypic reversion. All of these effects are partially reversed by a hypomethylating agent, which could indicate that DNA methylation may be involved in the antipromoting effect of low SAM doses.

The relationships between DNA methylation and gene expression have been widely investigated (see ref. 76 for review). Even if various exceptions do exist, a good deal of data indicate that undermethylation of some specific sites of DNA is a pre-condition necessary, even if not sufficient, for expression of some genes[76-78]. Even if the role of this phenomenon in the promotion process is not yet fully clear, it could be of interest to assess if variations in the availability of methyl donor could influence proto-oncogene expression.

We have studied the SAM effect on the expression of c-Ha-ras, c-Ki-ras and c-myc in NN induced by DENA following the Solt and Farber protocol, followed by PB. The results of dot blot analysis (Fig. 3) clearly show that a 2-day treatment with 384 µmol/Kg,day of SAM causes a small but significant decrease in the expression of c-Ha-ras, c-Ki-ras and c-myc proto-oncogenes. A greater decrease in the expression of all three genes occurs after a 7-day treatment with SAM, and a much greater fall after a 60-day treatment. These results have been confirmed for c-Ha-ras by Northern analysis (Fig. 4), which shows a large decrease in the gene expression as a consequence of a 60-day treatment with SAM.

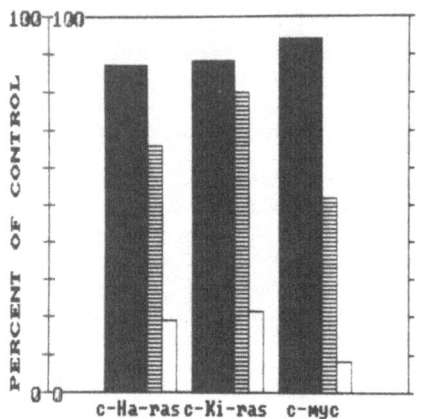

Fig. 3. Effect of SAM on proto-oncogene expression in NN. The rats were subjected to the initiation/selection/PB treatments as described in the legend of Fig. 1. SAM (384 μmol/Kg,day) injection was started at the 24th week after starting the 2-AAF treatment and was continued for 2 (■), 7 (▦), or 60 (□) days. Then the rats were killed. Total RNA was extracted and spotted on a nitrocellulose filter, and hybridized with oncogene-specific DNA. Radioactive spots were punched out and the radioactivity was evaluated in liquid scintillation computer. Absolute values for 10 μg of RNA from untreated controls were: 13,249 ± 1,628, for c-Ha-ras, 16,852 ± 1,516, for c-Ki-ras, and 18,877 ± 2,305 for c-myc. Results (means of 4 experiments) are expressed as percent of controls. The SD was lower than 10% of the means. "t"-Test: SAM vs control: c-Ha-ras, at 2, 7, and 60 days, and c-Ki-ras and c-myc, at 7 and 60 days, at least P < 0.05.

CONCLUSIONS

A good deal of evidence in favor of a SAM antipromotion and antiprogression effect during DENA-induced rat liver carcinogenesis exists. At least two different mechanisms may explain this SAM effect. High SAM doses reconstitute the liver SAM pool, otherwise low in the liver during hepatocarcinogenesis promotion and in NN, and cause accumulation of 5'-MTA, an inhibitor of polyamine synthesis and growth, in these tissues [19-21]. Polyamine synthesis is high in the liver and in NN, during hepatocarcinogenesis promotion [20]. 5'-MTA accumulation could thus explain the inhibitory effect of

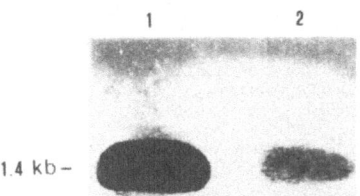

Fig. 4. Effect of SAM on the constitutive expression of the c-Ha-ras gene in NN. The same conditions as in Fig. 3. RNA was isolated from NN of rats treated 60 days with SAM (384 μmol/Kg,day) or their controls, electrophoresed (10 μg) on denaturating formaldehyde agarose gels, transferred to nitrocellulose membranes, and hybridized with a specific DNA probe. Lane 1, control; lane 2, SAM-treated.

high SAM doses. Low SAM doses cause: (a) a significant decrease in the growth of EAF coupled with a great increase in phenotypic reversion of EAF; (b) a near complete recovery of 5-mC content of liver DNA, otherwise reduced during the development of EAF, and of nodule DNA. All of these effects are reversed by 5-AzaC, while no interference of 5-AzaC with the effect of high SAM doses can be observed. This points to an antipromoting effect of low SAM doses and to a role of DNA undermethylation in the SAM antipromoting effect.

Inhibition of c-Ha-ras, c-Ki-ras and c-myc proto-oncogene expression by SAM has been observed in NN by using high SAM doses for a relatively short period of time (7 days). During this treatment no detectable nodule regression occurs. Thus, in our experimental conditions inhibition of gene expression precedes NN regression. The presence of activated c-Ha-ras, c-Ki-ras and c-myc proto-oncogenes has been widely described in experimental and human tumors[79-84], including hepatocarcinomas[85-87]. Proto-oncogene activation has also been found in liver nodules[88-90] as well as in early papillomas which develop during skin carcinogenesis promotion[91,92]. DNA from skin papillomas induces foci of transformed NIH 3T3 cells[91]. A point mutation in the 61st codon of one allele of the c-Ha-ras gene has been described in papillomas[92]. In addition, c-Ha-ras, c-Ki-ras, and c-myc proto-oncogenes, or some DNA sites near these oncogenes, are hypomethylated in tumors[93-96], and in nodules[89].

Reproducibility and specificity of proto-oncogene activation in neoplasms has been thought to be suggestive of a significant etiological role[80]. Interestingly, Thorgeirsson et al.[97] have recently reported that transfection of v-raf, v-Ha-ras and, in particular, a combination of v-raf and c-myc oncogenes are efficient in inducing the development of GGT-positive foci of altered cells in cultured rat liver epithelial cells. Focal cells induced the development of rapidly growing tumors, after transplantation into nude mice. It should be noted, however, that an elevation in proto-oncogene transcripts does not demonstrate per se transforming activity. An elevation in c-Ha-ras, c-Ki-ras and c-myc transcripts, coincident with the peak of DNA synthesis, occurs in liver after partial hepatectomy[98,99]. Mammalian ras genes acquire transforming-inducing properties by single point mutations within their coding sequences (see ref. 100 for review). myc Oncogenes have been implicated in tumors as genes which are altered by chromosome rearrangements or DNA amplification[101]. To our knowledge it has not yet been verified if these changes occur in liver preneoplastic nodules. It can be thus suggested that enhanced expression of the proto-oncogenes in liver nodules is linked to active proliferation of preneoplastic and neoplastic tissue. Indeed, a theme underlaying many researches is that proto-oncogenes control the molecular events that enable cells to grow[102,103].

The maintenance of a high methylating environment, by injecting exogenous SAM, during the development of preneoplastic tissue, causes DNA methylation, inhibition of gene expression, as early as after a 7-day SAM treatment, and, at least after 56-60 days, inhibition of nodule growth and progression. Preliminary experiments, in which nodule DNA has been analyzed for methylation using isoschizomeric enzymes HpaII/MspI and ^{32}P-labeled c-myc probe, showed hypomethylation of c-myc gene. A 7-day treatment with 384 µmol/Kg,day of SAM caused the disappearance of 0.38 and 0.87 Kb bands and the appearance of three bands of 0.63, 1.10 and 2.88 Kb in the HpaII digested DNA, which indicates the appearance of some methylated sites in the proto-oncogene DNA (Feo et al., unpublished results). Thus, our results envisage the possibility of modulating gene methylation and expression by maintaining a high SAM/SAH ratio in liver cells during hepatocarcinogenesis promotion.

ACKNOWLEDGEMENTS

Supported by CNR, Prog. Fin. Oncol., Assoc. It. Ric. Cancro, and MPI.

REFERENCES

1. A. K. Verma, B. G. Shapas, , M. H. Rice and R. K. Boutwell, Correlation of the inhibition by retinoids of the tumor promoter induced mouse epidermal ornithine decarboxylase activity and of skin tumor promotion, Cancer Res. 39:419 (1979).

2. S. M. Fisher, A. J. P. Klein-Szanto, L. A. Adams and T. J. Slaga, The first stage and complete promoting activity of retinoic acid but not the analog RO-10-9359, Carcinogenesis 6:575 (1985).

3. K. Chida, H. Hashiba, M. Fukushima, T. Suda and T. Kuroki, Inhibition of tumor promotion in mouse skin by $1\alpha,25$-dihydroxyvitamin D_3, Cancer Res. 45:5426 (1985).

4. J. T. Chan and H. S. Black, The mitigating effect of dietary antioxidants on chemically-induced carcinogenesis, Experientia 34:110 (1978).

5. T. J. Slaga, S. M. Fisher, C. E. Weeks, K. Nelson, Mamrack, M. and A. J. P. Klein-Szanto, Specificity and mechanism(s) of promoter inhibitors in multistage promotion, in: "Carcinogenesis: A Comprehensive Survey", T. J. Slaga, ed., Raven Press, New York (1980).

6. M. Takigawa, A. K. Verma, R. C. Simsiman, and R. K. Boutwell, Tumor promotion: Inhibition of 12-O-tetradecanoylphorbol-13-acetate mouse skin tumor formation by the irreversible inhibitor of ornithine decarboxylase α-difluoromethylornithine, Biochem. Biophys. Res. Commun. 105:969 (1982).

7. C. E. Weeks, A. L. Hermann, F. R. Nelson and T. J. Slaga, α-Difluoromethylornithine, an irreversible inhibitor of ornithine decarboxylase, inhibits tumor promoter-induced polyamine accumulation and carcinogenesis in mouse skin, Proc. Natl. Acad. Sci. USA, 79:6028 (1982).

8. R. G. Weeks, A. K. Verma and R. K. Boutwell, Inhibition by putrescine of the induction of epidermal ornithine decarboxylase activity and tumor promotion caused by 12-O-tetradecanoylphorbol-13-acetate, Carcer Res. 40:4013 (1980).

9. T. Narisawa, M. Satoh, M. Sano and T. Takahashi, Inhibition of initiation and promotion by N-methylnitrosourea-induced colon carcinogenesis in rats by non-steroid anti-inflammatory agent indomethacin, Carcinogenesis 44:1225 (1983).

10. W. Troll, A. Klassen and A. Janoff, Tumorigenesis in mouse skin: Inhibition by synthetic inhibitors of proteases, Science 169:1211 (1979).

11 S. N. Fischer, G. D. Mills and T. J. Slaga, Inhibition of mouse skin tumor promotion by several inhibitors of arachidonic acid metabolism, Carcinogenesis 3:1243 (1982).

12. T. Nakadate, S. Yamamoto, E. Aizu and R. Kato, Inhibition of 12-O-tetradecanoylphorbol-13-acetate-induced tumor promotion and epidermal ornithine decarboxylase activity in mouse skin by palmitoylcarnitine, Cancer Res. 46:1589 (1986).

13. L. L. Pashko, R. J. Rovito, J. R. Williams, S. L. Sobel and A. G. Schwartz, Dehydroepiandrosterone (DHEA) and 3-ßmethylandrost-5-en-17-one: inhibitors of 7,12-dimethylbenz(a)anthracene (DMBA)-initiated and 12-O-tetradecanoylphorbol-13-acetate (TPA)-promoted skin papilloma formation in mice, Carcinogenesis 5:463 (1984).

14. M. A. Moore, N. Thamavit, A. Tsuda, K. Sato, A. Ichihara and N. Ito, Modifying influence of dehydroepiandrosterone on the development of dehydroxy-di-n-propylnitrosamine-initiated lesions in the thyroid, lung and liver of F344 rats, Carcinogenesis 7:311 (1986).

15. R. Garcea, L. Daino, R. Pascale, S. Frassetto, P. Cozzolino. M. E. Ruggiu and F. Feo, Inhibition by dehydroepiandrosterone of liver preneoplastic foci formation in rats after initiation-selection of experimental carcinogenesis, Toxicol. Pathol. 15:164 (1987).

16. F. Feo, R. Garcea, L. Daino, S. Frassetto, P. Cozzolino, M.E Ruggiu,

G. Vannini, R. Pascale, L. Lenzerini, M. M. Simile and M. Puddu, Reversal by ribo- and deoxyribo-nucleosides of dehydroepiandrosterone-induced inhibition of enzyme altered foci in the liver of rats subjected to the initiation-selection process of experimental carcinogenesis, Carcinogenesis, in press.

17. F. Feo, Synthesis and accumulation of polyamines in N-nitrosodiethylamine-induced preneoplastic liver foci, Food Chem. Toxicol. 23:866 (1985).

18. F. Feo, R. Garcea, L. Daino, R. Pascale, L. Pirisi, S. Frassetto and M. E. Ruggiu, Early stimulation of polyamine biosynthesis during promotion by phenobarbital of diethylnitrosamine-induced rat liver carcinogenesis. The effect of variations of the S-adenosyl-L-methionine cellular pool, Carcinogenesis 6:1713 (1985).

19. F. Feo, R. Garcea, R. Pascale, L. Pirisi, L. Daino and A. Donaera, The variations of S-adenosyl-L-methionine content modulate hepatocyte growth during phenobarbital promotion of diethylnitrosamine-induced rat liver carcinogenesis. Toxicol. Pathol. 15:109 (1987).

20. R. Garcea, R. Pascale, L. Daino, S. Frassetto, P. Cozzolino, M. E. Ruggiu, M. G. Vannini, L. Gaspa and F. Feo, Variations of ornithine decarboxylase activity and S-adenosyl-L-methionine and 5'-methylthioadenosine contents during the development of diethylnitrosamine-induced liver hyperplastic nodules and hepatocellular carcinoma, Carcinogenesis 8:653 (1987).

21. F. Feo, R. Garcea, L, Daino and R. Pascale, Mechanism of the inhibition of liver hepatocarcinogenesis formation by S-adenosyl-L-methionine, in: "Experimental Hepatocarcinogenesis", M. Roberfroid and V. Préat, eds., Plenum Press, New York (1988) in press.

22. J. A. Stekol, E. I. Anderson and S. Weiss, S-Adenosyl-L-methionine in the synthesis of choline, creatine and cystine in vivo and in vitro, J. Biol. Chem. 233:425 (1958).

23. V. Zappia, P. Galletti, M. Porcelli, G. Ruggiero and A. Andreana, Uptake of adenosylmethionine and related sulfur compounds by isolated rat liver, FEBS Lett. 90:331 (1978).

24. C. Pezzoli, G. Stramentinoli, M. Galli-Kienle and E. Pfaff, Uptake and metabolism of S-adenosyl-L-methionine by isolated rat hepatocytes, Biochem. Biophys. Res. Commun. 85:1031 (1978).

25. J. Z. Farooqui, H. W. Lee, S. Kim and W. K. Paik, Studies on compartmentation of S-adenosyl-L-methionine in Saccharomyces cerevisiae and isolated rat hepatocytes. Biochim.. Bipophys. Acta 757:342 (1983).

26. P. Giulidori, M. Galli-Kienle, E. Catto and G. Stramentinoli, Transmethylation, transsulfuration, and aminopropylation reactions of S-adenosylmethionine, J. Biol. Chem. 259:4205 (1984).

27. M. A. Engstrom and N. J. Benevenga, Rates of oxidation of the methionine and S-adenosylmethionine methylcarbons in isolated rat hepatocytes, J. Nutr. 117:1820 (1987).

28. M. Frezza, G. Pozzato, L. Chiesa, G. Stramentinoli and C. DiPadova, Reversal of intrahepatic cholestasis of pregnancy in women after high dose S-adenosyl-L-methionine administration, Hepatology 4:274 (1984).

29. B. Lombardi and H. Shinozuka, Enhamcement of 2-acetylaminofluorene liver carcinogenesis in rats fed a choline-devoid diet, Int. J. Cancer 23:565 (1979).

30. A. E. Rogers, G, Lehnart and G, Morrison, Influence of dietary content of lipotropes and lipid on aflatoxin B1, N-2-fluorenylacetamide, 1,2-dimethylhidrazine carcinogenesis in rats, Cancer Res. 40:2802 (1980).

31. P. M. Newberne, J. L. V. deCarnago and A, J, Clark, Choline deficiency, partial hepatectomy, and liver tumors in rats and mice, Toxicol. Pathol. 2:95 (1982).

32. Y. B. Mikol, K. L. Hoover, D. Creasia and L. A. Poirier, Hepatocarcinogenesis in rats fed methyl-deficient amino acid-defined diets,

Carcinogenesis 4:1619 (1983).

33. A. K. Ghoshal and E. Farber, The induction of liver cancer by dietary deficiency of choline and methionine without added carcinogens, Carcinogenesis 5:1367 (1984).

34. S. Yokoyama, M. A. Sells, T. V. Reddy and B. Lombardi, Hepatocarcinogenic and promoting action of a choline-devoid diet in the rat, Cancer Res. 45:2834 (1985).

35. Y. B. Mikol and L. A. Poirier, An inverse relationship between hepatic ornithine decarboxylase and S-adenosylmethionine in rats, Cancer Lett. 13:195 (1981).

36. N. Shivapurkar and L. A. Poirier, Tissue levels of S-adenosylmethionine and S-adenosylhomocysteine in rats fed methyl-deficient, aminoacid-defined diets for one to five weeks, Carcinogenesis 4:1051 (1983).

37. N. Shivapurkar and L. A. Poirier, Decreased levels of S-adenosyl-methionine in the livers of rats fed phenobarbital and DDT, Carcinogenesis 3:589 (1982).

38. C. Peraino, R. J. M. Fry, E. Staffeldt and N. E. Kisielewsky, Effects of varying the exposure to phenobarbital on its enhancement of 2-acetylaminofluorene-induced hepatic tumorigenesis in the rat, Cancer Res. 33:1701 (1973).

39. T. Kitagawa and H. Sugano, Enhancing effect of phenobarbital on the development of enzyme-altered islands and hepatocellular carcinomas initiated by 3-methyl-4-(dimethylamino)azobenzene or diethylnitrosamine, Gann 69:678 (1978).

40. S. Yanagi, K. Sasaki and N. Yamamoto, Induction by phenobarbital of ornithine decarboxylase activity in rat liver after initiation with diethylnitrosamine, Cancer Lett. 12:87 (1981).

41. D. B. Solt and E. Farber, New principle for the analysis of chemical carcinogenesis, Nature 263:702 (1976).

42. H. G. Williams-Ashman, A. E. Pegg and D. H. Lockwood, Mechanisms and regulation of polyamine and putrescine biosynthesis in male genital glands and other tissues of mammals, Adv. Enzyme. Regul. 7:291 (1972).

43. R. L. Pajula and A. Raina, Methylthioadenosine, a potent inhibitor of spermine synthase from bovine brain, FEBS Lett. 99:383 (1979).

44. A. E. Pegg, R. T. Borchardt and J. K. Coward, Effects of inhibitors of spermidine and spermine synthesis on polyamine concentration and growth of transformed mouse fibroblasts, Biochem. J. 194:79 (1981).

45. G. Scalabrino, M. E. Ferioli and R. Candiani, New Insights into the regulation of S-adenosyl-L-methionine decarboxylase in normal and regenerating rat liver (this volume).

46. M. Tatematsu, Y. Nagamine and E. Farber, Redifferentiation as a basis for remodeling of carcinogen-induced hepatocyte nodules to normal appearing liver, Cancer Res. 43:5049 (1983).

47. M. A. Moore, H. J. Hacker and P. Bannasch, Phenotypic instability in focal and nodular lesions induced in a short term system in the rat liver, Carcinogenesis 5:595 (1983).

48. N. Bursch, B. Laner, I. Timmermann-Trosiener, G. Bartel, J. Schuppler and R. Schulte-Hermann, Controlled death (apoptosis) of normal and putative preneoplastic cell in rat liver following withdrawal of tumor promoters, Carcinogenesis 5:453 (1984).

49. C. Peraino, E. F. Staffeldt, B. A. Carnes, V. A. Ludeman, J. A. Blomquist and S. D. Vesselinovitch, Characterization of histochemically detectable altered hepatocyte foci and their relationship to hepatic tumorigenesis in rats treated once with diethylnitrosamine or benzo(a)-pyrene within one day after birth, Cancer Res. 44:3340 (1984).

50. R. Schulte-Hermann, W. Bursch, L. Fesus and B. Kraupp, Cell death by apoptosis in normal, preneoplastic and neoplastic tissue (this volume).

51. R. Schulte-Hermann, Tumor promotion in the liver, Arch. Toxicol. 57:

147 (1985).

52. E. Farber, Sequential events in chemical carcinogenesis, in: "Cancer: A Comprehensive Treatise", Plenum Press, New York (1982).

53. E. Farber and R. Cameron, The sequential analysis of cancer development, Adv. Cancer Res. 31:125 (1980).

54. Z. Brada, N. H. Altman, M. Hill and S. Bulba, The effect of methionine on the progression of hepatocellular carcinoma induced by ethionine, Res. Commun. Chem. Pathol. Pharmacol. 38:157 (1982).

55. Z. Brada, J. Hillova, M. Hill, N. H. Altman and S. Bulba, Effect of methionine on development of benzopyrene (BP) induced sarcomas, Proc. AACR, Abstr. No. 478, Cancer Res. 27:121 (1986).

56. E. Wainfan and M. Dizik, Suppression by methionine and choline of oncofetal patterns of liver tRNA methyltrasferase activities in carcinogen-treated rats, Carcinogenesis 8:615 (1987).

57. N. Shivapurkar, K. L. Hoover and L. A. Poirier, Effect of methionine and choline on liver tumor promotion by phenobarbital and DDT in diethyl nitrosamine-initiated rats, Carcinogenesis 5:547 (1986).

58. J. D. Finkelstein and J. J. Martin, Methionine metabolism in mammals. Adaptation to methionine excess, J. Biol. Chem. 261:1582 (1986).

59. D. R. Hoffman, D. W. Marion, W. E. Cornatzer and J. A. Duerre, S-adenosylmethionine and S-adenosylhomocysteine metabolism in isolated rat liver, J. Biol. Chem. 255:10822 (1980).

60. J. Hillova, M. Hill, J. Belehradek Jr., R. Mariage-Sanson and Z. Brada, Loss of the oncogene from human H-ras-1 transfected NIH/3T3 cells grown in the presence of excess methionine, J. Natl. Cancer Inst. 77:721 (1986).

61. K. Oden and S. Clarke, S-adenosyl-L-methionine synthase from human erythrocytes: role in the regulation of cellular S-adenosylmethionine levels, Biochemistry 22:2978 (1983).

62. C. Matsumoto, Y. Suma and K. Tsukada, Changes in the activities of S-adenosylmethionine synthetase isozymes from rat liver with dietary methionine, J. Biochem. 25:287 (1984).

63. V. Zappia, C. R. Zydek-Cwick and F. Schlenk, The simplicity of S-adenosylmethionine derivatives in methyl transfer reactions, J. Biol. Chem. 244:4499 (1969).

64. S. J. Kerr, Competing methyltransferase systems, J. Biol. Chem. 247:4248 (1972).

65. P. K. Chiang and G. L. Cantoni, Perturbation of biochemical transmethylations by 3-deazaadenosine in vivo, Biochem. Pharmacol. 28:1897 (1979).

66. M. J. Wilson, N. Shivapurkar and L. A. Poirier, Hypomethylation of hepatic nuclear DNA in rats fed with a carcinogenic methyl-deficient diet, Biochem. J. 218:987 (1984).

67. L. A. Poirier, M. J. Wilson and N. Shivapurkar, Carcinogenesis and EDNA hypomethylation in methyl-deficient animals, in: "Biological Methylations and Drug Design", R. T. Borchardt, C. R. Creveling and P. M. Veland eds., Humana Press, Clifton (1986).

68. J. Locker, T. V. Reddy and B. Lombardi, DNA methylation and hepatocarcinogenesis in rats fed a choline-deficient diet, Carcinogenesis 7:1309 (1986).

69. E. Wainfan, M. Dizik, M. Hluboky and M. E. Balis, Altered t-RNA methylation in rats and mice fed lipotrope-deficient diets, Carcinogenesis 7:473 (1986).

70. D. R. Hoffman, J. A. Hanig and W. E. Cornatzer, Effects of a methyl-deficient diet on rat liver phosphatidylcholine biosynthesis, Can. J. Biochem. 59:543 (1981).

71. R. Pascale, L. Pirisi, L. Daino, S. Zanetti, E. Satta, E. Bartoli, and F. Feo, Role of phosphatidylethanolamine methylation in the synthesis of phosphatidylcholine by hepatocytes isolated from choline-deficient rats, FEBS Lett. 145:293 (1982).

72. P. A. Jones and S. M. Taylor, Cellular differentiation, cytidine ana-

logs and DNA methylation, <u>Cell</u> 20:85 (1980).

73. F. Ceusot, G. Acs and S. K. Christman, Inhibition of DNA methylation and induction of erythroleukemia cell differentiation by 5-aza-cytidine and 5-deoxycytidine, <u>J. Biol. Chem.</u> 257:2041 (1982).

74. A. Cihak, J. Vesely and J. Skoda, Azapyrimidine nucleosides: metabolism and inhibitory mechanisms, <u>Adv. Enzyme Regul.</u> 24:335 (1985).

75. A. Cavaliere, A. Bufalari and R. Vitali, 5-Azacytidine carcinogenesis in Balb/c mice, <u>Cancer Lett.</u> 37:51 (1987).

76. L. P. Adams and R. H. Burdon, "Molecular Biology of DNA Methylation", Springer-Verlag, New York (1985).

77. R. M. Hoffman, Altered methionine metabolism and transmethylation in Cancer, <u>Antitumor Res.</u> 5:1 (1985).

78. W. Doerfler, The effect of DNA methylation on DNA-protein interactions and on the regulation of gene expression, in: "Oncogenes and Growth Control" P. Kahn and T. Graf. eds., Springer-Verlag, Berlin (1986).

79. S. Sukumar, V. Notario, D. Martin-Zanca and D. Barbacid, Induction of mammary carcinomas in rats by nitrosomethylurea involves malignant activation of Ha-<u>ras</u>-1 locus by single point mutations, <u>Nature</u>, 306:658 (1983).

80. G. M. Cooper and M.-A. Lane, Cellular transforming genes and oncogenesis, <u>Biochim. Biophys. Acta</u> 738:9 (1984).

81. T. Tachira, K. Hayashi, M. Ochiai, N. Tsuchida, M. Nagao and T. Sugimura, Structure of the c-Ki-<u>ras</u> gene in a rat fibrosarcoma induced by 1,8-dinitropyrene, <u>Molec. Cell. Biol.</u> 6:1349 (1986).

82. S. Sukumar, A. Perantoni, C. Reed, J. R. Rice and M. L. Wenk, Activated K-<u>ras</u> and N-<u>ras</u> oncogenes in primary renal mesenchymal tumors induced in F344 rats by methyl(methoxymethyl)nitrosamine, <u>Molec. Cell. Biol.</u> 6:2716 (1986).

83. H. D. Preisler, A. J. Kinniburgh, G. Wei-Dong and S. Khan, Expression of the protoncogenes c-<u>myc</u>, c-<u>fos</u> and c-<u>fms</u> in acute myelocytic leukemia at diagnosis and in remission, <u>Cancer Res.</u> 47:874 (1987)

84. R. Oscadir, R. Sanuda, M. Cruz, A. M. Graef and P. Cariglio, High correlation between molecular alterations of c-<u>myc</u> oncogene and carcinoma of uterine cervix, <u>Cancer Res.</u> 47:4173 (1987).

85. R. W. Wiseman, S. J. Stowers, E. C. Miller, M. W. Anderson and J. A. Miller, Activating mutations of the c-H-<u>ras</u> protoncogene in chemically induced hepatomas of the male B6C3F1 mouse, <u>Proc. Natl. Acad. Sci. USA</u> 83:5025 (1986).

86. B. E. Huber and S. S. Thorgeirsson, Analysis of c-<u>myc</u> expression in a human hepatoma cell line, <u>Cancer Res.</u> 47:3414 (1987).

87. G. J. Cota and J.-F Chiu, The expression of oncogenes and liver-specific genes in Morris hepatomas, <u>Biochem. Biophys. Res. Commun.</u> 143:624 (1987).

88. D. Corcos, N. Defer, M. Raymondjean, M. Paris, B. Corral, L. Tichonicky, J. Kruh, D. Glaise, A. Saulnier and C. Guguen-Guillouzo, Correlated increase of the expression of the c-<u>ras</u> genes in chem-ically induced hepatocarcinomas, <u>Biochem. Biophys. Res. Commun.</u> 122:259 (1984).

89. A. Antony, P. M. Rao, S. Rajalakshmi and D. S. R. Sarma, Hypometh-ylation of DNA during early stages of chemical carcinogenesis, Proc. AACR, Abstr. No. 412, <u>Cancer Res.</u> 28:104 (1987).

90. M. Kirsch-Volders, Cytogenetic and genetic alterations during hepato-carcinogenesis. Proceedings of the European Meeting on Experimental Hepatocarcinognesis, Abstracts (SPA, Belgium, May, 1987).

91. A. Balmain, M. Ramsden, G. T. Bowden and J. Smith, Activation of the mouse cellular Harvey-<u>ras</u> genes in chemically induced benign skin papillomas, <u>Nature</u> 307:658 (1984).

92. J. C. Pelling, S. M. Fisher, B. Neader, J. Strawhecker and L. Schweickert, Elevated expression and point mutation of the Ha-<u>ras</u> proto-oncogene in mouse skin tumors promoted by benzoylperoxide and other promoting agents, <u>Carcinogenesis</u> 8:1481 (1987).

93. A. P. Freinberg and B. Vogelstein, Hypomethylation of ras oncogenes in primary human cancers, Biochem. Biophys. Res. Commun. 198:47 (1983).

94. M. S. C. Cheah, C. D. Wallace and R. M. Hoffman, Hypomethylation of DNA in human cancer cells: a site-specific change in the c-myc oncogene, J. Natl. Cancer Inst. 73:1057 (1984).

95. M. Ramsden, G. Cole, J. Smith and A. Balmain, Differential methylation of c-H-ras gene in normal mouse cells during skin tumour progression, EMBO J. 4:1449 (1985).

96. L. A. Poirier, Methyl insufficiency in carcinogenesis, this volume.

97. S. S. Thorgeirsson, S. H. Garfield, B. E. Huber and R. P. Evarts, Cellular and molecular aspects of chemical carcinogenesis, Proceedings of the Fourth Sardinian International Meeting: Models and Mechanisms in Chemical Carcinogenesis (Alghero, Italy, October 1987), Abstracts book, La Celere, Alghero (1987).

98. R. Makino, K. Hayashi and T. Sugimura, c-myc Transcript is induced in rat liver at a very early stage of regeneration of by cycloheximide treatment, Nature 310:697 (1984).

99. M. Goyette, Ch. J. Petropoulos, P. R. Shank and N. Fausto, Expression of a cellular oncogene during liver regeneration, Science, 219:510 (1983).

100. M. Barbacid, ras Genes, Ann. Rev. Biochem. 56:79 (1987).

101. G. M. Cooper, Cellular oncogenes and Cancer, Clin. Physiol. Biochem. 5:122 (1987).

102. P. Kahn and Th. Graf (eds.) "Oncogenes and Growth Control", Springer Verlag, Berlin (1986).

103. B. I. Weinstein, Growth factors, oncogenes and multistage carcinogenesis, J. Cell. Biochem. 33:231 (1987).

DEVELOPMENT AND APPLICATION OF AN IN VIVO MEDIUM-TERM BIOASSAY SYSTEM
FOR THE SCREENING OF HEPATOCARCINOGENS AND INHIBITING AGENTS OF
HEPATOCARCINOGENESIS

Nobuyuki Ito, Hiroyuki Tsuda, Masae Tatematsu,
Katsumi Imaida, Masataka Kagawa, Atsushi Nakamura,
Shuji Yamaguchi and Mamoru Mutai

First Department of Pathology, Nagoya City University Medical
School, 1 Kawasumi, Mizuho-cho, Mizuho-ku, Nagoya 467, Japan

INTRODUCTION

The number of compounds which have been introduced into our environment
is now far beyond our capacity to perform costly long-term carcinogenicity
tests. For the purpose of performing mass-screening of compounds, in vitro
short-term tests such as the Salmonella/microsome assay (Ames' test) have
been developed [1,2], and a variety of carcinogenic compounds have been shown to
be mutagenic with an apparent good correlation [3-6]. Recently, however, it has
been shown that the mutagenicity results have not always correlated well to
carcinogenicity [4]. Therefore, the development of an appropriate in vivo
medium-term assay system which could bridge the gap between mutagenicity and
carcinogenicity tests has become an urgent necessity. Use of rat liver
utilizing the two-step concept for such a screening assay has the particular
advantage of relatively easy quantitative detection of preneoplastic enzyme
altered foci [7-10]. Recent studies in this laboratory indicated that immunohistochemically demonstrated GST-P+ foci show good conformity to previously
used hyperplastic nodules or γ-glutamyl transpeptidase-positive foci with the
advantage of having far less background hepatocyte staining due to non-specific induction [11-15].

The present report summarizes the development and application of this
medium-term model for the assay of 132 compounds and their relation to
available Ames' test and long-term carcinogenicity findings.

MATERIALS AND METHODS

A total of 5190 6-week-old male F344 rats (Charles River Japan Inc.,
Atsugi, Japan) were maintained on basal diet (Oriental M, Oriental Yeast Co.,
Tokyo) and housed in plastic cages in an air-conditioned room at 25°C.
Chemicals used were purest grades available.

Experiment I

The rats (100 rats) were initially treated with DEN (200 mg/kg, i.p.),
given basal diet for the following 2 weeks, and then divided into four
groups. Groups 1-3 were given basal diet containing 0.02% 2-AAF (group 1),
0.05% PB (group 2) or 2.0% BHA (group 3), respectively, for 6 weeks. Group 4

425

Table 1. Positive rate in the different category of compounds (%)

Test compound	Mutagen	Nonmutagen	Unknown	Total
Liver carcinogen	13/14 (92.9)	13/15 (86.7)	0/0	26/29 (89.7)
Nonliver carcinogen	1/14 (7.1)	1/6 (16.7)	0/1 (0)	2/21 (9.5)
Noncarcinogen	0/5 (0)	0/18 (0)	0/2 (0)	0/25 (0)
Unknown	1/8 (12.5)	7/25 (28)	4/22 (18.2)	12/55 (21.8)

was given basal diet only. All animals were subjected to two-thirds partial hepatectomy (PH) at the end of week 3. Three to four animals from each group were sacrificed 1, 2, 3 and 4 days and 1, 2, 3 and 5 weeks after PH. [^{3}H]-Thymidine was injected intraperitoneally at a dose of 1 μCi/g body weight 1 h before sacrifice. The liver was excised and 2-3 mm thick sections were fixed in ice-cold acetone for immunohistochemical demonstration of GST-P and autoradiography. Antibody against GST-P was raised as described previously[16]. The avidin-biotin-peroxidase complex (ABC) method described by Hsu et al.[17] was used to determine the localization of GST-P in the liver using Vectastain ABC Kit, PK-4001 (Vector Laboratories Inc., Burlingame, California, USA). The numbers and areas of GST-P$^+$ foci more than 0.2 mm in diameter were measured using an image processor[11-14]. The labeling indices (LI) of 2000 GST-P$^+$ and negative hepatocytes were determined.

Experiment II

For the several chemicals evaluated, a total of 4969 rats were divided into 3 groups. Group 1 was given a single intraperitoneal injection of DEN (200 mg/kg) dissolved in 0.9% NaCl to initiate hepatocarcinogenesis. After 2 weeks on basal diet, they received one of the test compounds administered either in the basal diet (D), drinking water (W), or by intraperitoneal (i.p.) or intravenous (i.v.) injection at the concentrations given in Table 1. Animals were subjected to PH at week 3 (total of 2454 rats). Group 2 was given DEN and PH in the same manner as for group 1 but without administration of any test compound (total 606 rats). Group 3 animals were injected with 0.9% NaCl instead of DEN solution and then subjected to administration of test compound and partial hepatectomy (total of 1909 rats, not listed in Table 1). Rats in each group were killed for examination at week 8 (Fig. 1). Results from 29 series of experiments using this regimen are summarized in this presentation, including results previously published.

Immunohistochemical staining and quantitative analysis for GST-P$^+$ foci were done as in Experiment I. The results were assessed by comparing the values between groups 1 (DEN → test compounds) and 2 (DEN alone). Group 3 served to evaluate the assay for the potential to induce GST-P$^+$ foci without prior DEN (Fig. 1). Scoring of the results was made on the basis of the difference of Student's P-values between groups 1 and 2: positive was considered an increase at $P < 0.05$ in either number or area of foci.

Experiment III

A total of 121 rats were divided into 3 groups in a similar protocol as in Experiment II. Clofibrate (0.3% in diet), a liver carcinogen[18], was given as the test compound, but was administered continuously for 30 weeks. Seven to 27 rats were killed at each of weeks 3, 8, 20 and 32 after treatment with DEN. Liver sections were prepared as in Experiment II using serial sections for GST-P staining and H&E staining.

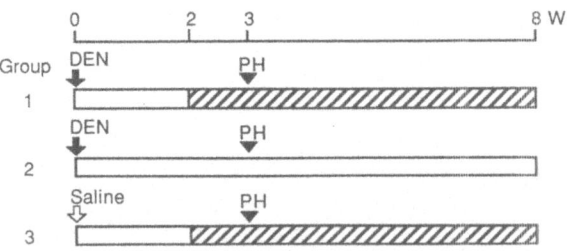

Fig. 1. Assay method. Group 1, DEN → test compound; group 2, DEN alone; group 3, test compound alone. All rats were subjected to partial hepatectomy at week 3 and killed at week 8.

RESULTS

Experiment I

2-AAF (group 1) and PB (group 2) significantly enhanced the induction of GST-P$^+$ lesions, and BHA (group 3) significantly inhibited their development (Fig. 2).

The sequential changes in LI of the cells in GST-P$^+$ foci and surrounding hepatocytes are presented in Figs. 3 and 4. In group 1, 2-AAF almost completely inhibited proliferation of surrounding hepatocytes, whereas GST-P$^+$ cells in the foci showed a high proliferative activity for at least 1 week after PH. In groups 2-4, the LI of the surrounding hepatocytes were higher than those of GST-P$^+$ foci cells on day 1, then quickly decreased to resting values. In contrast, the LI of cells in the foci began to show slightly higher values than the surrounding liver beginning on day 2.

Experiment II

The summarized results with regard to the categories of positive compounds are shown in Table 1. Most of chemicals which enhanced the induction of GST-P$^+$ foci have known carcinogenic potential, and almost all include the liver as one of their principal target organs (26 of 29 test compounds). Twenty-one of 40 positive compounds in this assay system are not mutagenic. GST-P$^+$ foci were also induced in group 3 rats given potent hepatocarcinogens. However, these values were far less than for group 1. No false-positive results were recorded; none of the tested chemicals without carcinogenic activity (25 compounds) enhanced the induction of GST-P$^+$ foci. Eleven of the 22 chemicals showing an inhibitory effect are antioxidants.

Experiment III

The total number of GST-P$^+$ and GST-P$^-$ foci at week 8 were significantly less in rats fed clofibrate after DEN than after DEN alone (Fig. 5). However, the values showed a gradual increase and became significantly elevated at week 32. Furthermore it was found that the increase in the total number of foci was due to GST-P$^-$ foci, which could be recognized in H&E stained foci. Similar results were also observed for the values for area of foci.

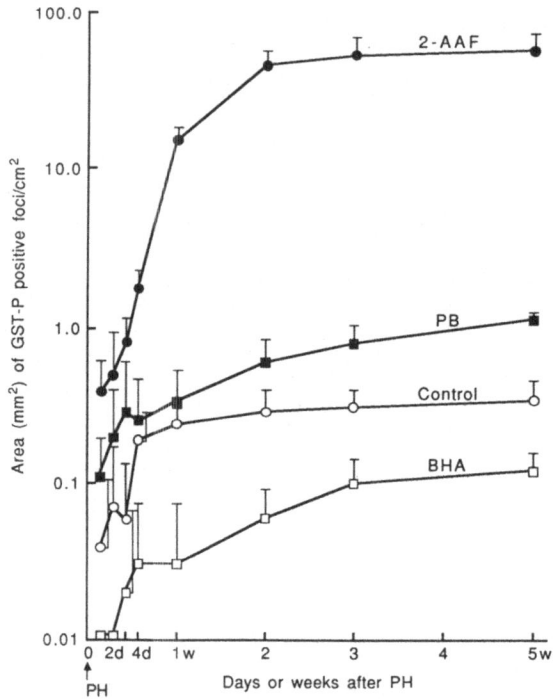

Fig. 2. Sequential changes in the areas of GST-P⁺ foci in rats fed 2-AAF
(●), PB (■), BHA (□), and basal diet (control,○). Mean ± SD.

DISCUSSION

        In experiment I, 2-AAF clearly inhibited the proliferation of surround-
ing hepatocytes whereas the cells in foci had considerably higher LI. This
observation is probably related to the differential toxic effect of 2-AAF on
foci cells compared to normal hepatocytes[10-19]. PB also inhibited the proliferation of surrounding hepatocytes to a much lesser extent than 2-AAF, but,
unlike in previous observations[20], PB did not cause any stimulation of proliferation of foci cells. In contrast, BHA did not show any inhibitory effect on surrounding hepatocytes, allowing them to proliferate continuously,
but inhibited the proliferation of hepatocytes in foci.

        In a large series of assays of various compounds with known hepatocarc-
inogenic effects or its inhibition, it is clear that the results obtained
using the present medium-term bioassay model show a positive association with
carcinogenicity but without obvious correlation to the results of Salmonella
mutagenicity testing[1-6] (Table 1). As summarized 15 mutagenic and 21 non-
mutagenic carcinogens were successfully detected as positive compounds in this
assay.

        It is especially noteworthy that 26 of 29 (89.7%) hepatocarcinogens gave
a positive result, irrespective of their mutagenicity, leaving only three
compounds, 4,4'-diaminodiphenylmethane, di(2-ethylhexyl)phthalate and
clofibrate as false-negatives[19,21]. As expected, positive results for non-
liver carcinogens were not as clear as for hepatocarcinogens. Most importantly is the fact that none of the compounds currently considered as
non-carcinogens proved positive in this assay. In other words, the bioassay
system did not generate false-positive results.

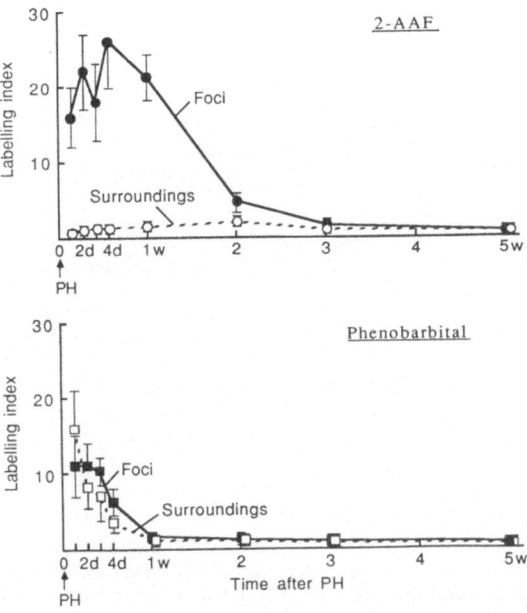

Fig. 3. Sequential changes in the LI of GST-P$^+$ foci and surrounding hepatocytes in rats fed 2-AAF or phenobarbital.

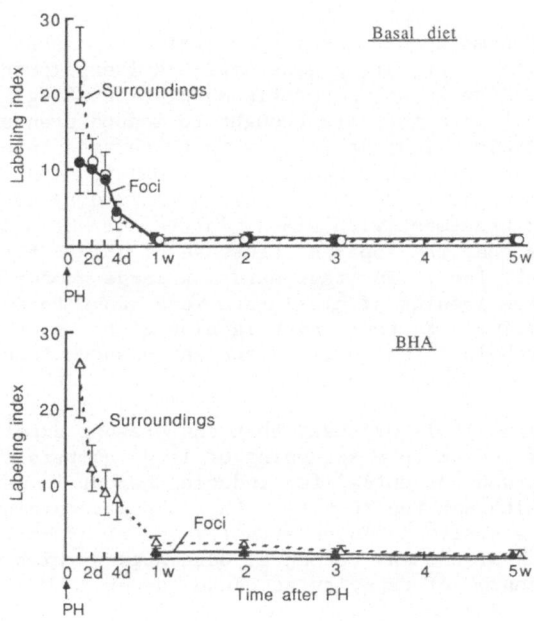

Fig. 4. Sequential changes in the LI of GST-P$^+$ foci and surrounding hepatocytes in rats fed basal diet or BHA.

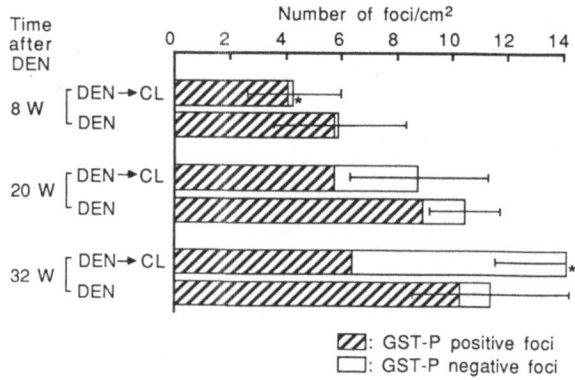

Fig. 5. Sequential changes in the number/cm^2 of GST-P$^+$ and GST-P$^-$ foci in rats given clofibrate (CL). P < 0.05 (*), compared to the groups given DEN alone. Bar represents mean ± S.D.

This assay system also possesses the advantage that detection of inhibitory agents for hepatocarcinogenesis is possible. For example, a number of antioxidants (11 of 22 inhibitory agents), mostly naturally occurring, brought about a reduction of GST-P$^+$ foci induction. BHA, ethoxyquin and α-tocopherol were previously reported as inhibitory agents for liver or colon carcinogenesis in long-term experiments[22,23]. However, since some antioxidant species, such as BHA and sodium ascorbate, are also carcinogenic or exert promoting effects on other organs, investigations using protocols based on initiating agents with a wide spectrum of organ effects, such as N-methyl-N-nitrosourea, is necessary to clarify their potential as environmental hazards[24,25].

Among the non-mutagenic carcinogens which demonstrated negative results in this study, the liver carcinogens di(2-ethylhexyl)phthalate and clofibrate, which are also known as peroxisome proliferating agents, are particularly interesting since they are thought to induce preneoplastic lesions which are phenotypically at variance with those associated with other carcinogens[19,20]. To clarify the apparent discrepancy, quantitative analyses of foci positive or negative for GST-P were performed by continuous feeding of clofibrate after treatment with DEN as in the current model. The results clearly indicate that the apparent decrease of GST-P$^+$ foci was caused by an increase of GST-P$^-$ foci. In other words, a large number of foci cells induced during the feeding of clofibrate were shown to be negative for GST-P rather than positive. Further investigation of the specific enzyme phenotype of such foci may help to solve this apparent disadvantage for use of our model system.

In conclusion, it is proposed that the present experimental protocol is advantageous for medium-term screening of large numbers of chemicals which may possess hazardous potential for inducing cancer in man, particularly liver cancer. Although the results with non-hepatocarcinogens were less than satisfactory, the system could nevertheless be of practical use because approximately 60% of the environmental carcinogens which were rated to have "sufficient evidence" of capacity to cause cancer in man listed by the IARC[26] are hepatocarcinogens in rats.

ACKNOWLEDGEMENTS

This work was supported in part by Grants-in-Aid for Cancer Research from the Ministry of Education, Science and Culture of Japan, the Ministry of Health and Welfare and the program for a Comprehensive 10 Year Strategy for Cancer Control, Japan. The authors would like to express their gratitude to Dr. Kiyomi Sato of Second Department of Biochemistry, Hirosaki University School of Medicine for providing us with a generous amount of GST-P antibody.

REFERENCES

1. B. N. Ames, J. McCann and E. Yamasaki, Methods for detecting carcinogens and mutagens with the Salmonella/mammalian microsome mutagenicity test, Mutat. Res. 32:347 (1975).
2. IARC, "Long-term and Short-term Assays for Carcinogens: A Critical Appraisal", IARC Sci. Publ., No. 83, Lyon, (1986).
3. S. Haworth, T. Lawlor, K. Morlelmans, W. Speak and E. Zeiger, Salmonella mutagenicity test results for 250 chemicals, Environ. Mutagenesis, Suppl. 1:3 (1983).
4. L. E. Kier, D. J. Brunsick, A. E. Auletta, E. S. Von Halle, M. M. Brown, V. F. Simmon, V. Dunkel, J. McCann, K. Morlelmans, M. Prival, T. K. Rao and V. Ray, The Salmonella typhimurium/mammalian microsomal assay. A Report of the U.S. Environmental Protection Agency Gene-tox Program, Mutat. Res. 168:69 (1986).
5. S. Rinkus and M. S. Legator, Chemical characterization of 465 known or suspected carcinogens and their correlation with mutagenic activity in the Salmonella typhimurium system, Cancer Res. 39:3289 (1979).
6. J. McCann, E. Choi, E. Yamasaki and B. N. Ames, Detection of carcinogens as mutagens in the Salmonella/microsome test: assay of 300 chemicals, Proc. Nat. Acad. Sci. USA 72:5135 (1975).
7. G. M. Williams, Phenotypic properties of preneoplastic rat liver lesions and applications to detection of carcinogens and tumor promoters, Toxicol. Pathol. 10:3 (1982).
8. P. Bannasch, Commentary: Preneoplastic lesions as end points in carcinogenicity testing, I. Hepatic neoplasia, Carcinogenesis 7:689 (1986).
9. E. Farber and R. Cameron, The sequential analysis of cancer development, Adv. Cancer Res. 31:125 (1980).
10. D. Solt and E. Farber, New principle for the analysis of chemical carcinogenesis, Nature (Lond.) 263:701 (1976).
11. N. Ito, H. Tsuda, R. Hasegawa and K. Imaida, Sequential observation of pathomorphologic alterations in preneoplastic lesions during the promoting stage of hepatocarcinogenesis and the development of short-term test system for hepatopromoters and hepatocarcinogens, Toxicol. Pathol. 10:37 (1982).
12. M. Tatematsu, R. Hasegawa, K. Imaida, H. Tsuda and N. Ito, Survey of various chemicals for initiating and promoting activities in a short-term in vivo system based on generation of hyperplastic liver nodules in rats, Carcinogenesis 4:381 (1983).
13. H. Tsuda, R. Hasegawa, K. Imaida, T. Masui, M. A. Moore and N.Ito, Modifying potential of thirty-one chemicals on the short-term development of γ-glutamyl transpeptidase-positive foci in diethylnitrosamine-initiated rat liver, Gann 75:876 (1984).
14. K. Sato, A. Kitahara, K. Satoh, T. Ishikawa, M. Tatematsu and N. Ito, The placental form of glutathione S-transferase as a new marker protein for preneoplasia in rat chemical hepatocarcinogenesis, Gann. 75:199 (1984).
15. M. Tatematsu, Y. Mera, N. Ito, K. Satoh and K. Sato, Relative merits of immunohistochemical demonstrations of placental, A, B, and C forms of glutathione S-transferase and histochemical demonstration of γ-glutamyl transferase as markers of altered foci during liver carcinogenesis in rats, Carcinogenesis 6:1621 (1985).
16. K. Satoh, A. Kitahara, Y. Soma, Y. Inaba, I. Hayama and K. Sato,

Purification, induction and distribution of placental glutathione transferase: a new marker enzyme for preneoplastic cells in rat chemical carcinogenesis, Proc. Nat. Acad. Sci. USA 82:3964 (1985).

17. S. M. Hsu, L. Raine and H. Fanger, Use of avidin-biotin-peroxidase complex (ABC) in immunoperoxidase techniques: A comparison between ABC and unlabeled antibody (PAP) procedures, J. Histochem. Cytochem. 29:577 (1981).

18. J. K. Reddy and S. A. Qureshi, Tumorigenicity of the hypolipidemic peroxisome proliferator ethyl-α-p-chlorophenoxyisobutyrate (clofibrate) in rats, Brit. J. Cancer 40:476 (1979).

19. D. B. Solt, A. Medline and E.Farber, Rapid emergence of carcinogen-induced hyperplastic lesions in a new model for the sequential analysis of liver carcinogenesis, Am. J. Pathol. 88:595 (1977).

20. R. Schulte-Hermann, G. Ohde, J. Schuppler and I. Timmermann-Trosiener, Enhanced proliferation of putative preneoplastic cells in rat liver following treatment with tumor promoters phenobarbital, hexachlorocyclohexane, steroid compounds, and nafenopin, Cancer Res. 41:2556 (1981).

21. W. M. Kluwe, J. K. Haseman, J. F. Douglas and J. E. Huff, The carcinogenicity of dietary di(2-ethylhexyl)phthalate (DEHP) in Fischer 344 rats and B6C3F$_1$ mice, J. Toxicol. Environ. Health 10:797 (1982).

22. M. G. Cook and P. McNamara, Effect of dietary vitamin E on dimethylhydrazine-induced colonic tumors in mice, Cancer Res. 40:1329 (1980).

23. H. Tsuda, T. Sakata, T. Masui, K. Imaida and N. Ito, Modifying effects of butylated hydroxyanisole, ethoxyquin and acetaminophen on induction of neoplastic lesions in rat liver and kidney initiated by N-ethyl-N-hydroxyethylnitrosamine, Carcinogenesis 5:525 (1984).

24. H. Tsuda, S. Fukushima, K. Imaida, Y. Kurata and N. Ito, Organ-specific promoting effect of phenobarbital and saccharin in induction of thyroid, liver, and urinary bladder tumors in rats after initiation with N-nitrosomethylurea, Cancer Res. 43:3292 (1983).

25. H. Tsuda, T. Sakata, T. Shirai, Y. Kurata, S. Tamano and N. Ito, Modification of N-methyl-N-nitrosourea initiated carcinogenesis in the rat by subsequent treatment with antioxidants, phenobarbital and ethinyl estradiol, Cancer Lett. 24:19 (1984).

26. IARC, "IARC Monograph on the Evaluation of the Carcinogenic Risk of Chemicals to Humans", Suppl. 4 (1982).

SECTION III

REGULATION OF CELL GROWTH IN CHEMICAL CARCINOGENESIS

NUCLEAR ALTERATIONS IN LIVER CARCINOGENESIS: THE ROLE OF NON-POLYPLOIDIZING GROWTH

Per O. Seglen, Per E. Schwarze and Gunnar Saeter

Department of Tissue Culture
Institute for Cancer Research
The Norwegian Radium Hospital
Montebello, 0310 Oslo 3, Norway

INTRODUCTION

Polyploidization is a characteristic and dominant aspect of rat liver growth. As the animal matures, an increasing number of proliferating diploid hepatocytes undergo nuclear mitosis without cell division, resulting in the formation of tetraploid cells with two diploid nuclei. In the next cell cycle these binucleated cells form a single mitotic spindle and divide, yielding mononucleated tetraploid progeny. The latter may in turn undergo ordinary cell division, probably several times, or take one or two further steps along the polyploidization pathway, generating first binucleated, then mononucleated octoploid cells[1,2].

Polyploidization is generally regarded as an irreversible process[3,4], which implies that a polyploid hepatocyte cannot serve as precursor (stem cell) for a clonal cell population that includes diploid cells. All hepatocellular tumors appear to be made up at least partially by diploid tumor cells; most are in fact predominantly diploid[5-10]. We have therefore suggested that liver tumors originate from diploid precursor cells, and that any treatment which reduces the extent of polyploidization in liver tissue may promote liver carcinogenesis by expanding the total diploid hepatocyte polulation, including the tumor precursor cell population. In addition, diploid hepatocytes would presumably be more susceptible towards mutagenic carcinogens than polyploid hepatocytes, since the quantitative dominance of normal alleles and suppressor genes in the latter would tend to prevent the expression of oncogenic mutations. Mutation-dependent carcinogenic progression would therefore be favoured by the maintenance of diploidy in the precursor populations[9-10].

A previous study of rat liver carcinogenesis, employing diethylnitrosamine (DEN) as initiator and 2-acetylaminofluorene (AAF) as promotor, demonstrated an increase in the fraction of diploid hepatocytes already in the preneoplastic liver[8]. Similar observations have been made with other carcinogen treatment regimens[5,6,11-14], indicating that induction of non-polyploidizing growth may play a role during the early stages of liver carcinogenesis. The present report summarizes our recent work on ploidy development during liver carcinogenesis in Wistar Kyoto rats, with particular emphasis on the promoting action of AAF. Use of the inbred Kyoto strain has permitted

435

transplantation experiments which both illuminate the role of ploidy and the mechanism of action of tumor promoters.

METHODS

An Effective Initiation-Promotion Model of Liver Carcinogenesis

Administration of DEN, 50 mg/kg, to young rats (70-80 g) 24 h after 2/3 partial hepatectomy (PH) is a very effective initiation regimen[15]. About one-half of Kyoto rats subjected to such initiation developed hepatocellular carcinoma within eight months even without promotion. With an additional 4 wk administration of dietary AAF (0.02%, starting one wk after PH), a promoting dose that by itself generated no carcinomas within this time period, 90% of the animals contracted cancer, and the incidence of carcinomas (per liver) increased threefold. A promoting effect of AAF was also evident at the nodule stage (4 mo.) and in the preneoplastic liver (atypical hepatocytes at 6 wk) (Table 1.).

Collagenase-isolation of Hepatocytes and Tumors as a Method for Quantitative Analysis of Liver Carcinogenesis

Dispersion of the liver by two-step collagenase perfusion converts the normal parenchyma to isolated hepatocytes with an efficiency of 100%[16]. The hepatocytes can be purified by differential centrifugation without any change in ploidy distributions or functional properties, while endothelial, connective, biliary epithelial and oval cells are removed[17]. Microscope counting of individual hepatocytes in the suspensions permits precise quantitation of cytochemically detectable phenotypic changes (e.g. γ-glutamyltranspeptidase -GGT- expression) and binuclearity; the suspensions are furthermore suitable for flow-cytometric analysis of DNA and protein[8]. Neoplastic nodules and hepatocellular carcinomas are not completely dissociated by the primary collagenase treatment; they can therefore be isolated by filtration, counted and weighed. By slicing and secondary collagenase treatment, however, isolated parenchymal cells can be prepared (albeit at low yield) even from the tumors.

Transplantation of Hepatocytes from Tumors and Preneoplastic Livers

Since Kyoto rats are an inbred strain, we can transplant hepatocytes and tumor cells from one animal to another. We have developed a procedure for injection of cell suspensions into a portal tributary vein immediately after PH; this results in the permanent retention of some 20% of the injected cells in the recipient liver[18]. Normal transplanted hepatocytes produced no tumors, whereas hepatocytes from preneoplastic (carcinogen-treated) livers, or from hepatocellular carcinomas, produced tumors in the host animals with high efficiency.

Flow-cytometric DNA Analysis and Calculation of Absolute Cell Numbers within the Hepatocytic Ploidy Classes

Hepatocellular and nuclear ploidy distributions were assessed by flow-cytometric DNA analysis of mithramycin-stained whole hepatocytes and propidium iodide-stained hepatocyte nuclei (Fig. 1)[8,19]. While the nuclear fraction values could be used directly, the cellular values included aggregates and binucleated cells, accounting for the difference between histograms A and B in Fig. 1. Counting by fluorescence microscopy provided total binuclearity values without distinguishing between tetraploid and octoploid binucleates; however, the ploidy analysis indicated that the relative rate of binucleation was four times higher among the diploid than among the tetraploid cells. Including this ratio in an extensive set of equations, all ploidy classes (2N, 2x2N, 4N, 2x4N and 8N) could eventually be resolved. By measuring liver wet weights, and deducing the relationship between hepatocellu-

436

Table 1. Promotion of phenotypic alterations and tumorigenesis by 2-acetylaminofluorene (AAF) at various times after initiation with diethylnitrosamine (DEN)

Type of lesion	DEN only	DEN + AAF
% GGT-positive hepatocytes		
6 wk	0.0 ± 0.0 (2)	29.1 ± 3.8 (4)[a]
4 mo	0.0 ± 0.0 (2)	15.2 ± 4.3 (6)[b]
No. of nodules (≧ 1mm)/liver		
4 mo	1.4 ± 0.5 (5)	14 ± 3 (11)[b]
8 mo	69 ± 45 (5)	50 ± 14 (6)
No. of hepatocellular carcinomas/liver		
4 mo	0.0 ± 0.0 (5)	0.4 ± 0.2 (11)[b]
8 mo	0.8 ± 0.3 (6)	2.4 ± 0.4 (8)[b]

[a] $P < 0.001$; [b] $P < 0.02$ for DEN+AAF vs. DEN according to the t-test. AAF (0.02%) was administered in the diet for 4 weeks, beginning one week after initiation (partial hepatectomy + DEN). Each value is the mean ± S.E. of the no. of animals given in parentheses.

larity, cell size and ploidy[8,16,17], absolute cell numbers within each ploidy class could be calculated.

RESULTS AND DISCUSSION

AAF Promotion of Liver Carcinogenesis by a Non-cytotoxic Mechanism

The ability of AAF to promote liver carcinogenesis (which appears to be independent of its initiating properties) is most often utilized according to the Solt-Farber protocol, i.e. by inducing regenerative growth during AAF administration subsequent to initiation by some other carcinogen (e.g. DEN given at a necrotizing dose)[20]. It has been suggested that AAF in that model promotes carcinogenesis by selective cytotoxicity, suppressing the regenerative growth of normal hepatocytes while permitting the outgrowth of abnormal cells rendered AAF-resistant by the initiating carcinogen[21].

In our model AAF exhibited, independently of liver regeneration (administration was for 1-5 wk after PH), promoting effects at several stages of carcinogenesis as indicated by the early appearance of phenotypically abnormal (GGT-positive) cells, the accelerated outgrowth of neoplastic nodules, and the increased incidence of hepatocellular carcinoma (Table 1). AAF was apparently not hepatotoxic, since liver growth proceeded at a normal rate in its presence, even without prior initiation[10]. Livers from AAF-treated rats thus had the same size, DNA and protein content as untreated livers (Table 2). AAF promotion in our model therefore must be due to some mechanism other than differential cytotoxicity.

Induction of a Non-polyploidizing, Regeneration-like Liver Growth Pattern by AAF

The previously reported changes in hepatocellular ploidy distributions during early carcinogenesis[8] were evident also in Kyoto rats, as a marked

437

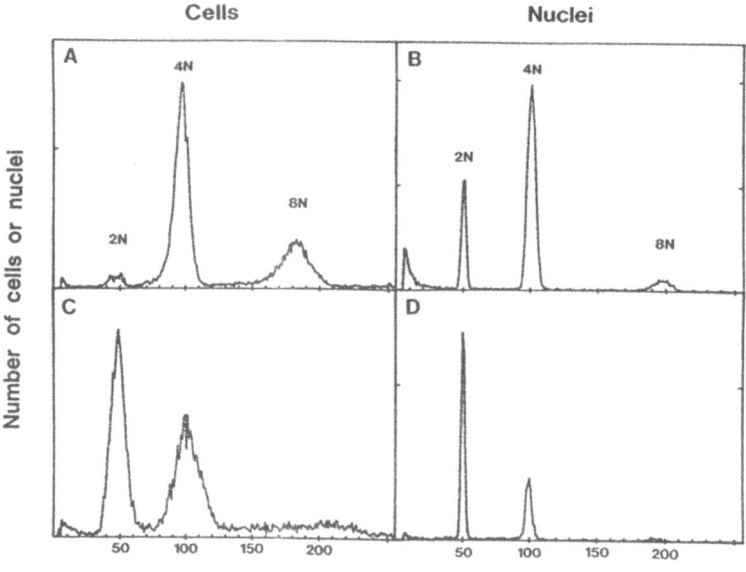

Fig. 1. Cellular and nuclear ploidy distributions of hepatocytes from normal
and carcinogen-treated Wistar Kyoto rats. DNA histograms were recorded by flow cytometry of whole cells (A,C) and nuclei (B,D) from
a normal (PH) control liver (A,B) or from the preneoplastic liver of
a carcinogen-treated (PH + DEN + AAF) animal (C,D), 6 wk after PH.

increase in the fraction of diploid hepatocytes and a virtual disappearance
of octoploid cells and nuclei 6 wk after initiation of treatment with DEN +
AAF (Fig. 1). Since these relative distributions did not provide information
about the direction or magnitude of changes in absolute cell numbers, the
latter were calculated on the basis of flow cytometric data, binuclearity
counts and liver weight determinations, and followed during growth and development of normal and carcinogen-treated rats.

Normal development was characterized by progressive polyploidization,
observed as a decrease in the number of diploid hepatocytes (Fig. 2A) and an

Table 2. Lack of AAF hepatotoxicity in Wistar Kyoto rats

	Liver size and composition	
	- AAF	+ AAF
Liver wet weight (g)	9.74 ± 0.35	9.73 ± 0.35
Dry weight (mg/g wet wt.)	250.4 ± 16.3	242.7 ± 8.3
Protein (mg/g wet wt.)	189.1 ± 10.5	173.6 ± 5.6
DNA (mg/g wet wt.)	1.62 ± 0.07	1.57 ± 0.02

Rats were given dietary AAF (0.02 %) 1-5 wk after partial
hepatectomy; livers were sampled at 8 wk. Each value is
the mean ± S.E. of three animals.

increase in all other ploidy classes, exemplified here by the binucleated cells (Fig. 2B). Initiation with DEN did not affect this pattern materially. AAF, on the other hand, induced a dramatic change in the growth pattern both with and without prior initiation: instead of decreasing, the number of diploid hepatocytes increased markedly (Fig. 2A), while the formation of binucleated cells was virtually arrested (Fig 2B).

Examination of the total ploidy distributions (Table 3.) made it evident that all types of mononucleated hepatocytes (2N, 4N and 8N) increased in number at a similar rate in the presence of AAF, while binucleated cells (2x2N and 2x4N) did not. This would suggest that liver cell proliferation in the presence of AAF proceeded at an undiminished rate, with all ploidy classes participating, but that no new polyploidization (binucleation) took place. The same pattern - equivalent increases in all mononucleated cells; no binucleation - was observed during the initial week of regenerative growth (Table 3). AAF would thus seem to effect a switch, in all hepatocytes, from the normal polyploidizing, developmental growth mode to a non-polyploidizing, regeneration-like mode. The normal rate of binucleation was rapidy restored following AAF withdrawal[10], suggesting that the switch was reversible.

The ability to induce non-polyploidizing growth in normal liver may be relatively unique to AAF. Phenobarbital (PB), which promoted primary carcinogenesis as effectively as did AAF (our unpublished results), did not alter the ploidy distribution of normal hepatocellular nuclei (Table 4). Previous studies have similarly indicated that PB may stimulate normal liver growth without interfering with polyploidization[22].

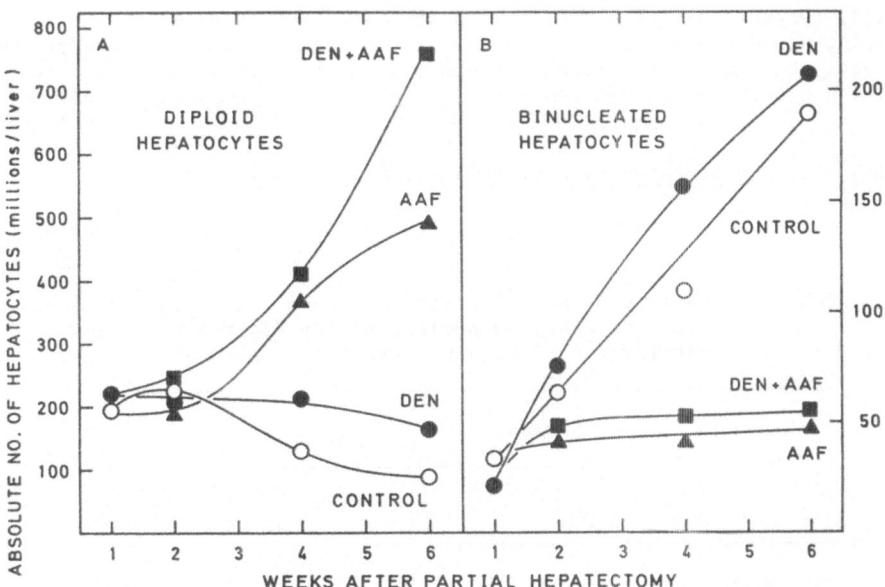

Fig. 2. Effects of AAF and DEN on absolute numbers of diploid (A) and binucleated (B) rat hepatocytes during developmental liver growth. Young rats were subjected to PH only (◯), PH + DEN (●), PH + AAF (▲) or PH + DEN + AAF (■) AAF (0.02 %) was given from 1 to 5 wk after PH. Absolute cell numbers (per liver) were calculated on the basis of flow-cytometric ploidy analyses of cells and nuclei, binuclearity counts and liver weight determinations.

Table 3. Net change in cell number (per liver) within each hepatocytic ploidy class during regeneration (0-1 wk after PH) and developmental growth (1-6 wk after PH)

	Change in cell number (millions/liver)					Fold increase (x)				
	2N	2x2N	4N	2x4N	8N	2N	2x2N	4N	2x4N	8N
Regeneration	170	4	225	0	17	8.1	1.2	7.1	1.0	6.7
Development										
PH control	-105	61	169	96	60	0.5	3.4	1.6	12.3	4.1
+ DEN	- 56	108	221	78	69	0.7	7.5	2.0	20.6	2.7
+ AAF	297	10	420	4	36	2.5	1.4	2.6	1.4	2.8
+ DEN + AAF	537	30	292	4	14	3.4	2.8	2.4	2.0	1.4

Constitutive Non-polyploidizing Proliferation of DEN-initiated Hepatocytes?

In DEN-pretreated rats, the AAF-induced proliferation of diploid hepatocytes tended to proceed at a higher rate than in the other mononucleated cell classes, and at a higher rate than with AAF alone (Fig. 2, Table 3). The fully carcinogen-treated animal thus maintained a higher fraction of diploid hepatocytes at all times. Eleven weeks after AAF withdrawal the fraction of diploid cells in rats treated with DEN + AAF was twice as high as in rats treated only with AAF, the percentage values being 24.2 ± 2.5 (9) and 12.3 ± 1.8 (11), respectively ($P < 0.02$). (In addition, tumors developing at this time point would contain predominantly diploid cells). This difference could conceivably be due to the progeny of DEN-initiated cells in which non-polyploidizing growth had become constitutive, possibly secondarily to AAF-induction.

Predominantly Diploid Phenotype of Hepatocellular Tumors

The relevance of non-polyploidizing growth to liver carcinogenesis

Table 4. Effects of phenobarbital (PB) and 2-acetylamino-fluorene (AAF) on nuclear ploidy distributions in rat liver 6-15 wk after start of treatment

	Nuclear ploidy (% of total)		
	2 N	4 N	8 N
Untreated control	21.8 ± 2.4 (9)	69.6 ± 1.7	8.6 ± 0.8
PB (continuously)	21.8 ± 1.4 (5)	69.5 ± 1.4	8.8 ± 1.0
AAF (4 weeks)	43.8 ± 2.7 (11)	52.2 ± 2.3	4.1 ± 0.4

Treatment was begun one week after partial hepatectomy and continued for 4 weeks (AAF) or until the time of sacrifice (PB). Each value is the mean ± S.E. of the no. of animals given in parentheses.

440

is particularly indicated by the fact that all tumors generated in our model are predominantly diploid[8-10]. Both hepatocellular carcinomas and large neoplastic nodules contained, on average, about 80% diploid parenchymal cells. There were few binucleated cells in the tumors; nuclear and cellular ploidy distributions were therefore relatively similar, in contrast to the situation among normal hepatocytes (Table 5).

Smaller nodules (< 2.5 mm in diameter) were, on average, significantly less diploid than larger nodules harvested at the same time (Table 5). Since tumor development in this model proceeds with a certain degree of synchrony, tumor size at a given time point would be expected to correlate with tumor growth rate. The data would therefore seem to suggest a negative correlation between the rate of growth and the rate of polyploidization (= differentiation?).

The development of diploid tumors was not restricted to our standard AAF promotion protocol. Hepatocellular carcinomas generated in the absence of promotion (DEN only), with PB promotion, or according to the Solt-Farber protocol were all predominantly diploid (Table 6). The diploidy of PB-promoted tumors, albeit somewhat lower than in the other groups, is of particular interest, since very slow-growing, PB-promoted tumors were reported to contain a majority of polyploid cells[23]. The present results suggest that the high degree of polyploidization in those tumors may be associated with their low growth rate rather than with PB promotion.

Transplanted Hepatocytes from Preneoplastic Liver Yield Diploid Host Tumors

Hepatocytes from preneoplastic livers of carcinogen-treated animals, transplanted to syngeneic recipients and promoted with PB[18], developed into hepatocellular carcinomas in the majority of the host animals within three months. Carcinomas were also formed in the absence of secondary promotion, but more slowly (no carcinomas at 3 mo., but 40% of the animals had cancer at 8-9 mo.). Both promoted and non-promoted tumors were predominantly diploid (Table 6). Secondary promotion with AAF + PB (sequential treatment) gave the same result as with PB alone. These transplantation experiments show that non-polyploidizing growth is a constitutive property of developing tumors, not dependent on carcinogen-induced alterations in the surrounding tissue.

Stimulation of Tumor Growth by AAF and Phenobarbital

Since PB apparently does not interfere with hepatocytic polyploidization, its promoting effect would probably be to accelerate the outgrowth of constitutively non-polyploidizing cells. To see if such growth stimulation could extend to established tumors, cells were isolated from primary hepatocellular carcinomas (induced by DEN + AAF) and transplanted to syngeneic hosts. As shown in Table 7, multiple tumors developed in the majority of the rats within two months. Both the number of detectable tumors and their total mass was significantly enhanced by treatment of the host rats with either PB or AAF. The ability of the promoters to stimulate the growth of already established carcinoma cells suggests that at least part of their promoting effect in primary carcinogenesis may simply be due to a growth factor-like action upon tumorigenic precursor cells.

Both PB[22] and an AAF analogue with particularly low toxicity, 4-AAF[24], have been reported to function as stimulators of normal liver growth. The response of tumors and possible tumor precursor cells towards these growth promoters may therefore represent a general hyperproliferative ability that could be triggered even by phsiological growth stimuli, and thus account for the outgrowth of tumors even in the absence of promotion. Hyperproliferation seems to be a characteristic feature of the hepatocellular tumor phenotype, established already at initiation[25].

441

Table. 5. Predominance of diploid cells and nuclei in
           carcinogen-induced hepatocellular tumors

	% Diploid cells	% Diploid nuclei
Normal liver	8.2 ± 1.4 (8)	21.8 ± 2.4 (9)
Nodules <2.5 mm	64.3 ± 3.0 (2)	65.8 ± 2.4 (16)[a]
Nodules >2.5 mm	80.5 (1)	81.4 ± 2.4 (13)[a]
Carcinomas	79.3 ± 2.8 (4)	83.0 ± 2.3 (15)

[a] P < 0.001 for difference between large and small nodules. All animals received complete carcinogen treatment
(PH + DEN + AAF). Nodules were harvested at 4 months;
carcinomas at 4-8 months. Each value is the mean ± S.E.
of the no. of tumors given in parentheses.

Possible Significance of Non-polyploidizing Growth in Liver Carcinogenesis

    Since AAF could stimulate the growth of tumors that were already di-
ploid, its growth-promoting properties would appear to be at least partially

Table 6. Effect of different promoting regimens on nuclear ploidy
          distributions in hepatocellular carcinomas developing during
          primary carcinogenesis or after transplantation of hepa-
          tocytes from preneoplastic livers

	Nuclear ploidy (% of total)		
	2 N	4 N	8 N
Normal liver, 4-8 mo. after PH	21.8 ± 2.4 (9)	69.6 ± 1.7	8.6 ± 0.8
Primary hepatocellular carcinoma			
No promotion, 8 mo.	88.8 ± 1.6 (3)	10.4 ± 1.4	0.8 ± 0.2
AAF promotion, 4-8 mo.	33.0 ± 2.3 (15)	15.1 ± 2.0	1.9 ± 0.3
PB promotion, 4 mo.	73.2 ± 4.3 (3)	24.0 ± 4.6	2.8 ± 0.3
Solt-Farber model, 4-8 mo.	74.1 ± 5-3 (4)	23.1 ± 4.7	2.8 ± 1.8
Post-transplantation carcinoma			
No sec. promotion, 8-10 mo.	90.6 ± 0.7 (11)	9.1 ± 0.6	0.3 ± 0.1
PB sec. promotion, 2-10 mo.	77.1 ± 2.4 (41)	21.3 ± 2.2	1.6 ± 0.3
AAF+PB sec. promot., 5 mo.	81.6 ± 2.3 (20)	17.2 ± 2.1	1.3 ± 0.3

Primary carcinogenesis was initiated with PH + DEN (50 mg/kg) except in
the Solt-Farber model, where DEN alone (200 mg/kg) was used. In the
latter model AAF was administered for 2 wk with PH after 1 wk of AAF;
in other regimens AAF (0.02%) was given for 4 wk both as primary and
secondary promoter. PB (0.04%), when given, was administered continuously for up to 4 mo. both in primary and secondary promotion. Transplant recipients were subjected to PH and given an intraportal injection of hepatocytes isolated from carcinogen-treated (PH + DEN + AAF)
donors 8 wk after initiation, or, in the last experiment, from PH +
DEN-treated donors 1 wk after initiation. Each value is the mean ±
S.E. of the no. of tumors given in parentheses.

Table 7. Effect of tumor promoters (PB, AAF) on outgrowth of
secondary (transplanted) hepatocellular carcinomas

Secondary treatment	Tumor-bearing animals		Promoted	No. of tumors per liver (all rats included)	Total tumor mass per liver (g)
	No.	%			
None (PH)	13/21	62	-	9.5 ± 6.3	31 ± 17[b]
+ AAF	18/20	90	15/20	42.0 ± 12.6[a]	398 ± 159[b]
+ PB	19/19	100	17/19	101.8 ± 35.1[b]	1608 ± 721[a]

[a] $P < 0.05$; [b] $P < 0.02$ for significance of AAF or PB treatment.
Host rats received intraportal transplants consisting of 500,000
donor cells from collagenase-dissociated primary hepatocellular
carcinomas, isolated 8 months after initiation of carcinogen
treatment (PH + DEN + AAF). The recipients were subjected to PH
immediately before transplantation, and were then given AAF for
4 weeks or PB until the time of sacrifice (2 months after transplantation). Promotion with AAF or PB was regarded positive if
both the number of tumors and the total tumor mass per liver was
increased relative to untreated rats receiving the same donor
cell suspension.

independent of its ability to induce a non-polyploidizing growth pattern. In
very early carcinogenesis, the combination of both of these AAF-effects may
provide a particularly effective promotion of initiated, hyperresponsive
cells. PB, which stimulates growth by a mechanism not involving ploidy alteration, would presumably be a more effective promoter after the establishment of non-polyploidizing growth. Such a dual requirement may be reflected
in the synergistic promotion of primary liver carcinogenesis upon sequential
administration of AAF and PB [26]. However, it cannot be excluded that the two
promoters may simply act as synergistic growth stimulators by mechanisms that
are unrelated to ploidy, and that non-polyploidizing growth conversely can
become established independently of promotion. Our results in fact show that
DEN-initiated tumors are predominantly diploid regardless of whether they are
promoted by AAF, PB or not promoted at all. The non-polyploidizing phenotype
may therefore be present already as a result of initiation, and would
furthermore seem to be strongly favoured during tumor development.

The probable importance of a large diploid cell fraction in terms of
stem cell maintenance and carcinogenic (mutational) progression risk has already been pointed out [8-10]. Even more importantly, the negative correlation
between tumor size and polyploidy suggests that non-polyploidizing growth may
be intrinsically more rapid than polyploidizing growth (cf. also regeneration) and thus represent a tumor advantage. Whether the reduced rate of
polyploidization is a cause or a consequence of rapid growth remains to be
established.

ACKNOWLEDGEMENT

This work has been generously supported by the Norwegian Cancer Society.

REFERENCES

1. M. Alfert and I. I. Geschwind, The development of polysomaty in rat

liver, Exp. Cell Res. 15:230 (1958).

2. C. Nadal and F. Zajdela, Polyploïdie somatique dans le foie de rat. I. Le role des cellules binuclées dans la genèse des cellules poly-ploïdes, Exp. Cell Res. 42:99 (1966).

3. R. Carriere, The growth of liver parenchymal nuclei and its endocrine regulation, Int. Rev. Cytol. 25:201 (1969).

4. W. Y. Brodsky and I. V. Uryvaeva, Cell polyploidy: its relation to tissue growth and function, Int. Rev. Cytol. 50:275 (1977).

5. A. Simard, G. Cousineau and R. Daoust, Variations in the cell cycle during azo dye hepatocarcinogenesis, J. Natl. Cancer Inst. 41:1257 (1968).

6. F. F. Becker, R. A. Fox, K. M. Klein and S. R. Wolman, Chromosome patterns in rat hepatocytes during N-2-fluorenylacetamide carcino-genesis, J. Natl. Cancer Inst. 46:1261 (1971).

7. G. E. Neal, H. M. Godoy, D. J. Judah and W. H. Butler, Some effects of acute and chronic dosing with aflatoxin B on rat liver nuclei, Cancer Res. 36:1771 (1976).

8. P. E. Schwarze, E. O. Pettersen, M. C. Shoaib and P. O. Seglen, Emergence of a polulation of small, diploid hepatocytes during hepatocarcinogenesis, Carcinogenesis 5:1267 (1984).

9. P. O. Seglen, P. E. Schwarze and G. Saeter, Changes in cellular ploidy and autophagic responsiveness during rat liver carcinogesis, Toxicol. Pathol. 14:342 (1986).

10. P. O. Seglen, G. Saeter and P. E. Schwarze, Nuclear alterations during hepatocarcinogenesis: promotion by 2-acetylaminofluorine, in: "Experimental Hepatocarcinogenesis", V. Préat and M. Roberfroid, eds., Plenum Press, London, (1988).

11. J. H. Holzner, T. Barka and H. Popper, Changes in deoxyribonucleic acid content of rat liver cells during ethionine intoxication, J. Natl. Cancer Inst. 23:1215 (1959).

12. C. C. Irving, J. A. Roszell and J. L. Fredi, Effects of chronic feeding of 2-acetylaminofluorene on nuclear populations in rat liver, Adv. Enz. Reg. 16:365 (1977).

13. J. Styles, B. M. Elliott, P. A. Lefevre, M. Robinson, N. Pritchard, D. Hart and J. Ashby, Irreversible depression in the ratio of tetraploid:diploid liver nuclei in rats treated with 3'-methyl-4-dimethylaminoazobenzene (3'M), Carcinogenesis 6:21 (1985).

14. A. Deleener, P. Castelain, V. Préat, J. de Gerlache, H. Alexandre and M. Kirsch-Volders, Changes in nucleolar transcriptional activity and nuclear DNA content during the first steps of rat hepatocarcino-genesis, Carcinogenesis 8:195 (1987).

15. E. Scherer and P. Emmelot, Kinetics of induction and growth of pre-cancerous liver-cell foci, and liver tumor formation by diethyl-nitrosamine in the rat, Eur. J. Cancer 11:689 (1975).

16. P. O. Seglen, Preparation of isolated rat liver cells, Methods Cell Biol. 13:29 (1976).

17. P. E. Schwarze, E. O. Pettersen, H. Tolleshaug and P. O. Seglen, Isolation of carcinogen-induced diploid rat hepatocytes by cen-trifugal elutriation, Cancer Res. 46:4732 (1986).

18. G. Saeter, P. E. Schwarze, J. Nesland and P. O. Seglen, Transplan-tation of preneoplastic rat hepatocytes by intraportal injection, Toxicol. Pathol. 15:78 (1987).

19. P. E. Schwarze, E. O. Pettersen and P. O. Seglen, Characterization of hepatocytes from carcinogen-treated rats by two-parametric flow cytometry, Carcinogenesis 7:171 (1986).

20. D. B. Solt, A. Medline and E. Farber, Rapid emergence of carcinogen-induced hyperplastic lesions in a new model for the sequential analysis of liver carcinogenesis, Am. J. Pathol. 88:595 (1977).

21. E. Farber and D. S. R. Sarma, Hepatocarcinogenesis: a dynamic cel-lular perspective, Lab. Invest. 56:4 (1987).

22. R. Schulte-Hermann, R. Thom, I. Schlicht and W. Koransky, Zahl und

Ploidiegrad der Zellkerne der Leber unter dem Einfluss körperfremder Stoffe, Naunyn-Schmiedebergs Arch. Pharmakol. Exp. Pathol. 261:42 (1968).

23. M. Sarafoff, H. M. Rabes and P. Dörmer, Correlation between ploidy and initiation probability determined by DNA cytophotometry in individual altered hepatic foci, Carcinogenesis 7:1191 (1986).

24. J. Ashby, P. A. Lefevre, B. Burlinson and B. Beije, Potent mitogenic activity of 4-acetylaminofluorene to the rat liver, Mutation Res. 72:271 (1986).

25. C. Peraino, E. F. Staffeldt, B. A. Carnes, V. A. Ludeman, J. A. Blomquist and S. D. Vesselinovitch, Characterization of histochemi-cally detectable altered hepatocyte foci and their relationship to hepatic tumorigenesis in rats treated once with diethylnitrosamine or benzo(a)pyrene within one day after birth, Cancer Res. 44:3340 (1984).

26. V. Préat, M. Lans, J. de Gerlache, H. Taper and M. Roberfroid, Com-parison of the biological effects of phenobarbital and nafenopin on rat hepatocarcinogenesis, Jpn. J. Cancer Res. 77:629 (1986).

GROWTH CONTROL OF HEPATOCYTES IN PRIMARY CULTURE AND ITS ABERRATION IN

CARCINOGENESIS

Akira Ichihara

Institute for Enzyme Research, University of Tokushima
Tokushima 770, Japan

INTRODUCTION

Liver has diverse functions, and these functions are mostly carried out by mature parenchymal hepatocytes. These cells are usually quiescent in vivo, but after injury of the liver or partial hepatectomy, they grow actively, and when the original mass of liver is restored, they stop growing. Therefore, problems of central interest in hepatology are how growth of hepatocytes is initiated, how it is stopped, and how hepatocytes function during liver regeneration. These problems are important not only in cell biology, but also in hepatic surgery and carcinogenesis.

For studies on these problems, a simple, clear experimental system is desirable. Primary cultures of hepatocytes have been shown to be the most suitable for this purpose, because these cells express most liver functions and are regulated by various hormones in the same way as they are in vivo[1-4]. In primary cultures, differentiated hepatocytes are quiescent, even in the presence of serum, but they can grow actively when appropriate mitogens are added. This paper mainly describes recent work in our laboratory on the mechanism regulating hepatocyte growth in vitro. Many important findings in other laboratories have been omitted because space is limited, and the reader should refer to review articles on these findings[1,2].

PLATELET-DERIVED REGULATORS OF HEPATOCYTE GROWTH

Mature hepatocytes have been shown to proliferate in vitro when insulin and epidermal growth factor (EGF) are added to the medium. However, the serum of rats after partial hepatectomy contains a potent mitogen, and this blood-born mitogen has been thought to be a genuine hepatocyte growth factor (HGF). Recently, this HGF was purified from rat platelets and found to be a heterodimer protein with sub-units of 34 and 69 KD[5,6]. HGF had an additive effect with insulin and EGF in stimulating DNA synthesis of primary cultured rat hepatocytes. There have been several reports on serum growth factors for hepatocytes (see ref. 1, p.1) and although these factors have not been well characterized, they are likely to be similar to HGF released into the serum. Moreover, rat platelets also contain factors that inhibit HGF (platelet-derived growth inhibitors, PDGIs)[7]. These PDGIs were separated into two factors, α and ß, on Biogel P-60. PDGI-ß was identified as transforming growth factor-ß (TGF-ß)[8], which also stimulates colony formation of NRK49F cells in soft agar, but PDGI-α did not show this activity. These PDGIs are present in latent forms in a platelet extract, and are activated by treatment with 6M urea or 1M acetic acid[9]. Therefore, these PDGIs are bound to a masking protein (MP) and usually do not inhibit hepatocyte growth. But under as yet unknown physiological conditions, they are dissociated from the MP and exert their inhibitory action. This MP was also purified and found to be a glycoprotein consisting of a tetramer of 46 KD. MP is a double-negative regulatory protein for hepatocyte growth (Fig. 1). Platelet-derived growth factor (PDGF) did not stimulate hepatocyte growth and HGF did not stimulate growth of mesenchymal cells.

EFFECTS OF CELL-CELL AND CELL-MATRIX CONTACTS: EFFECT OF TISSUE ARCHITECTURE ON HEPATOCYTE GROWTH

Hepatocytes in liver lobules are in contact not only with each other, but also with non-parenchymal liver cells (endo- and epi-thelial cells) and with the extracellular matrix (collagen, proteoglycan, fibronectin and laminin). DNA synthesis of hepatocytes is markedly influenced by contact with neighboring hepatocytes[10]: at high cell density, cell-cell contact inhibits DNA synthesis, with enhancement of functions of mature liver. Co-culture with liver epithelial cells (putative immature bile duct cells) also has a similar effect (see ref. 1, p.259), while co-culture with endothelial cells stimulates DNA synthesis of hepatocytes[11]. These effects of other cells are mediated by the membrane fractions of these respective cells. Similarly, hepatocytes cultured on collagen IV or fibronectin showed higher synthesis of DNA when humoral mitogen was added. Some proteoglycan also stimulated expression of liver-specific characters, such as albumin synthesis and gap-junction formation (see ref. 1, p.225).

These results suggest that, not only a specific signal from membranes in contact with hepatocytes, but also the hepatocyte shape, controlled in general by surrounding cells and the matrix, may be important for regulation of liver cell growth and expression of functions. Indeed, the cells become smaller and more compact in these conditions than in control cultures, in which they spread out very much.

Based on these findings, several modifications of culture conditions were introduced, such as coating the dishes wit. biomatrix and co-culture of the cells with non-parenchymal liver cells, which increased the functions of

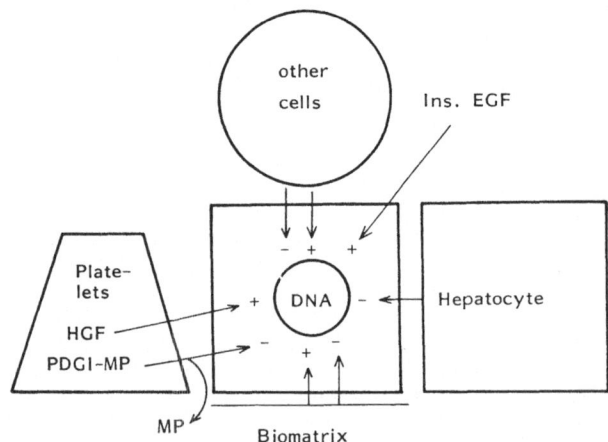

Fig. 1. Effects of microenvironment on hepatocyte growth

the hepatocytes and prolonged their survival in culture (for over one month).

AUTONOMOUS GROWTH AND DIFFERENTIATION OF EARLY NEONATAL HEPATOCYTES

In contrast to adult hepatocytes, early neonatal hepatocytes showed marked proliferation in culture without added serum or hormones. A stimulatory factor was found in the medium of 0-day-old rat hepatocytes[12]. This conditioned medium also stimulated growth of 3T3 cells, but not DNA synthesis of adult hepatocytes. The secretion of this growth factor and the autonomous growth of neonatal hepatocytes decrease rapidly within one week after birth. In contrast to their active growth, neonatal hepatocytes are immature in ability to express liver-specific functions, but acquire this ability as their growth decreases during culture. For differentiation in vitro, neonatal hepatocytes require contact with adult hepatocytes. We showed by immunostaining of tryptophan oxygenase (TO) that in culture autonomous growth of neonatal hepatocytes and acquirement of TO were segregated, and that contact with adult hepatocytes resulted in cascade spread of expression of differentiated characters and decrease of autonomous growth[13]. During this work on the development of expression of differentiated characters by neonatal hepatocytes, we found that expression of albumin by neonatal hepatocytes occurs from a very early stage, even by cells of rats at birth, whereas expression of TO does not occur until two weeks after birth[14]. These expressions are regulated by the amounts of the respective mRNAs. Therefore, it is better to use TO than albumin as a marker of differentiation in studies on terminal differentiation. In fact, many hepatoma cells still retain the ability to express albumin, but no cell line has been found to express TO or show its induction by hormones[15]. Similarly, expression of tyrosine aminotransferase (TAT) has often been used as a differentiated liver character, but this marker is also unsuitable for use in studies on differentiation.

MECHANISMS OF LIVER REGENERATION AND HEPATOCARCINOGENESIS

These results on growth of hepatocytes in primary culture showed that platelets play a central role in liver regeneration. Mature hepatocytes in vivo are in the typical G_0 state, the cells being quiescent and expressing differentiated characters under tight contact with each other and with non-parenchymal materials (Fig. 1). However, in injured liver, their contact may be loosened and they may proceed to the G_1 stage (Fig. 2). At this stage, platelets aggregate at the site of injury and secrete HGF to stimulate DNA synthesis of the hepatocytes. PDGIs are secreted similarly, and released from their complex with MP to exert fine regulation of cell growth. During the cell cycle, expression of differentiated functions is suppressed by the loose contact between cells. However, after one or two cell cycles, cell contact becomes tight, cell growth is suppressed, and the cells move back into G_0 state. PDGI may also participate in this process. Insulin and EGF may also participate, but HGF must be the most important mitogen for hepatocyte growth, because it is secreted at high concentration at a specific site. This situation is also favorable for growth of non-parenchymal cells, because platelets also secrete PDGF, which stimulates growth of non-parenchymal cells, but not of hepatocytes. Therefore, HGF and PDGF may stimulate growth of the whole liver in a concerted way.

It should be emphasized that, besides hormonal factors, many insoluble factors, such as the membranes of hepatocytes and non-parenchymal cells (epithelial and endothelial cells) and the extracellular matrix, regulate hepatocyte growth. These factors have specific negative and positive effects on growth and expression of liver functions (Fig. 1). In general their effects on growth and differentiation are reciprocal: when growth is enhanced, differentiation is suppressed, and vice versa. The effects of these factors

449

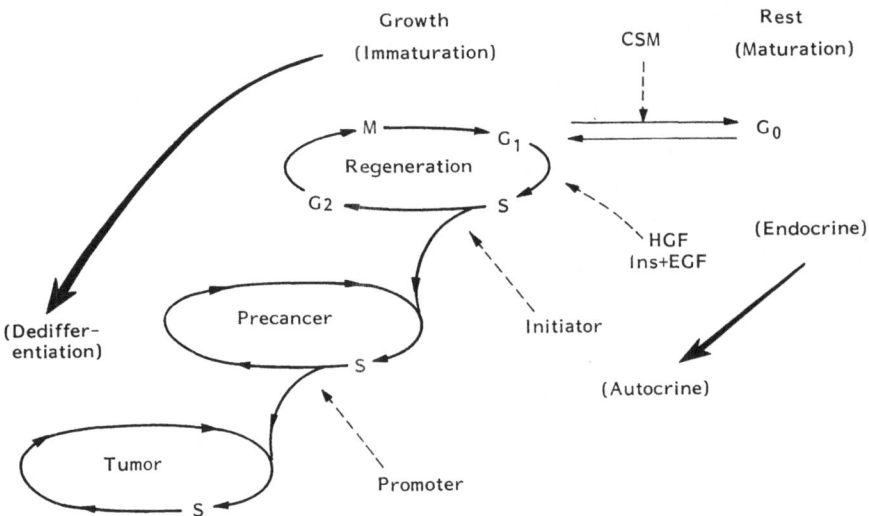

Fig. 2. Diagramatic representation of hepatocyte growth in liver regeneration and carcinogenesis.

provide some idea of the significance of the architecture of the liver for the function and regeneration of the liver. Not only hepatocytes, but also many other cells and the matrix influence hepatocyte growth, and hepatocytes influence the growth of other cells.

The mechanism of hepatocarcinogenesis can be understood in the light of this mechanism of growth regulation. During the normal cell cycle of hepatocyte growth, if there is some carcinogenic condition, such as viral infection or a carcinogen that modifies DNA, hepatocytes may lose both their normal response to environmental growth-regulating factors and their differentiated characters. Thus they will not be able to return to the G_0 state, and will continue to grow with HGF (Fig. 2).

This state may correspond to that of putative hyperplastic nodules. We found that in this precancerous state, glucose 6-phosphate dehydrogenase increased, while TO was lost[16,17], and that this dehydrogenase was a good marker of cell growth[10]. As these irreversibly altered cells continue to grow, they progress to a more malignant state in which they lose differentiated functions, such as the expressions of albumin and TAT, and become typical hepatoma cells, showing no contact-inhibition of growth and no expression of differentiated characters[18].

This state corresponds to that of Morris or Reuber hepatomas, which show various spectra of phenotypes. At this stage, the cells lose HGF-dependent growth, perhaps lose HGF receptor, and acquire autonomous growth. This autonomous growth somewhat resembles that of neonatal hepatocytes[12]. Recently, Isom's group obtained transformed hepatocytes that had acquired immortalized growth, but retained various differentiated characters and closely resembled some hepatoma cells with differentiated characters[19]. Therefore, there is a close relation among in vivo carcinogenesis, retrogression of development, and phenotypic changes in culture[15].

REFERENCES

1. A. Guillouzo and C. Guguen-Guillouzo, eds., Isolated and Cultured

Hepatocytes, John Libbey Eurotext, London (1986).

2. T. Nakamura and A. Ichihara, Control of growth and expression of differentiated functions of mature hepatocytes in primary culture, Cell Struct. Funct. 10:1 (1986).

3. T. Nakamura, S. Niimi, K. Nawa, C. Noda, A. Ichihara,Y. Takagi, M. Anai and Y. Sakaki, Multihormonal regulation of transcription of the tryptophan 2,3-dioxygenase gene in primary cultures of adult rat hepatocytes with special reference to the presence of a transciptional protein mediating the action of glucocorticoids, J. Biol. Chem. 262:727 (1987).

4. K. Nawa, T. Nakamura, A. Kumatori, C. Noda and A. Ichihara, Gluco-corticoid-dependent expression of the albumin gene in adult rat hepatocytes, J. Biol. Chem. 261:16883 (1986).

5. T. Nakamura, H. Teramoto and A. Ichihara, Purification and character-ization of a growth factor from rat platelets for mature paren-chymal hepatocytes in primary culture, Proc. Natl. Acad. Sci. USA 83:6489 (1986).

6. T. Nakamura, K. Nawa, A. Ichihara, N. Kaise and T. Nishino, Large scale purification and subunit structure of hepatocyte growth factor from rat platelets, FEBS Lett. 224:311 (1987).

7. T. Nakamura, H. Teramoto, Y. Tomita and A. Ichihara, Two types of growth inhibitor in rat platelets for primary cultured rat hepato-cytes, Biochem. Biophys. Res. Commun. 134:755 (1986).

8. T. Nakamura, Y. Tomita, R. Hirai, K. Yamaoka, K. Kaji and A. Ichihara, Inhibitory effect of transforming growth factor-ß on DNA synthesis of adult rat hepatocytes in primary culture, Biochem. Biophys. Res. Commun. 133:1042 (1985).

9. T. Nakamura, T. Kitazawa and A. Ichihara, Partial purification and characterization of masking protein for ß-type transforming growth factor from rat platelets, Biochem. Biophys. Res. Commun. 141:176 (1986).

10. T. Nakamura, K. Yoshimoto, Y. Nakayama, Y. Tomita and A. Ichihara, Reciprocal modulation of growth and differentiated functions of mature rat hepatocytes in primary culture by cell-cell contact and cell membranes, Proc. Natl. Acad. Sci. USA 80:7229 (1983).

11. S. Shimaoka, T. Nakamura and A. Ichihara, Stimulation of growth of primary cultured adult rat hepatocytes by co-culture with nonparen-chymal liver cells, Exp. Cell Res. 172:228 (1987).

12. T. Nakamura, T. Fujii and A. Ichihara, Autocrine mechanism of growth of neonatal rat hepatocytes in primary culture, J. Biochem. 103:313 (1988).

13. T. Nakamura, M. Nagao and A. Ichihara, In vitro induction of terminal differentiation of neonatal rat hepatocytes by direct contact with adult rat hepatocytes, Exp. Cell Res. 169:1 (1987).

14. M. Nagao, T. Nakamura and A. Ichihara, Developmental control of gene expression of tryptophan-2,3-dioxygenase in neonatal rat liver, Biochim. Biophys. Acta 867:179 (1986).

15. A. Ichihara, Relation of the characteristics of liver cells during culture, differentiation and carcinogenesis, in: "Control Mechanisms in Cancer", W. E. Criss, T. Ono, and J. R. Sabine, eds., Raven Press, New York, (1976).

16. M. A. Moore, T. Nakamura, A. Shirai, A. Ichihara and N. Ito, Immuno-histochemical demonstration of increased glucose 6-phosphate de-hydrogenase in preneoplastic and neoplastic lesions induced by propylnitrosamine in F344 rats and Syrian hamsters, Jpn. J. Cancer Res. (Gann) 77:131 (1986).

17. M. A. Moore, T. Nakamura, A. Ichihara and N. Ito, Immunohistochemically demonstrated suppression of tryptophan oxygenase, a marker of liver differentiation, within putative preneoplastic rat liver lesions, Carcinogenesis 7:1393 (1986).

18. T. Nakamura, Y. Nakayama, H. Teramoto, K. Nawa and A. Ichihara, Loss of

reciprocal modulation of growth and liver function of hepatoma cells in culture by contact with cells or cell membranes, <u>Proc.</u> <u>Natl.</u> <u>Acad.</u> <u>Sci.</u> <u>USA</u> 81:6398 (1984).

19. C. Woodworth, T. Scott, and H. C. Isom, Transformation of rat hepatocytes by transfection with simian virus 40 DNA to yield proliferating differentiating cells, <u>Cancer</u> <u>Res.</u> 46:4018 (1986).

CONSTITUTIVE AND INDUCED EXPRESSION OF HEAT SHOCK PROTEINS DURING

LIVER CARCINOGENESIS AND IN HEPATOMAS

Gaetano Cairo, Emilia Rappocciolo, Lorenza Tacchini,
Lidia Bardella, Luisa Schiaffonati and Aldo Bernelli-
Zazzera

Istituto di Patologia Generale dell'Università degli Studi
Centro di Studio sulla Patologia Cellulare del CNR
Via Mangiagalli 31, Milano, Italy

INTRODUCTION

Exposure of cells to heat or other stresses[1] induces in the chromosomes the appearance of a new set of puffs[2] and causes changes in gene structure and regulation that finally lead to the activation of synthesis of a distinct group of proteins. These proteins, known historically as heat-shock proteins (hsps), are now often referred to as stress proteins[3-5]. The best evolutionarily conserved heat-shock protein, hsp 70, is coded by a multigene family whose members respond in different ways to temperature and are subject to different regulatory mechanisms; the heat-inducible and constitutive proteins have sometimes been confused[4]. At least one of the members of the hsp 70 group is also growth-regulated[6-7] and the activity of the corresponding gene is induced by viral and cellular oncogenes[8,10]. In previous studies we have examined the synthesis of hsp in injured liver cells[11] and in some hepatomas of different growth rates[12], and have shown that in these tumours hsp 89 and hsp 70 are expressed constitutively, while induction by heat, at least in the case of hsp 70, is progressively reduced from the slow to the fast-growing hepatomas[13]. Later on we became aware of the fact that hsc 73, the major heat-shock related protein, rather than hsp 70 itself, is the main constitutively synthesized protein in unstressed cells; moreover recent data showed that hsc 73 is expressed at higher level in transformed cells than in their non-transformed counterparts[14]. In the present paper we report the results of an investigation with a probe for hsc 73, which became recently available to us, on the constitutive expression of hsc 73 gene in liver cells submitted to a multistep carcinogenic treatment and in transplantable hepatomas, both solid and in ascitic form, of various growth rates. The main purpose of the investigation was to see if the constitutive expression of hsc 73 is an event occurring during the carcinogenic treatment, presumably associated with initiation and/or promotion, or a relatively late outcome during the growth of the established tumour, presumably associated with tumor progression.

MATERIALS AND METHODS

Multistep liver carcinogenesis was induced according to the resistant-hepatocyte model[15] in Fisher 344 rats. Rats were utilized 16 h after partial hepatectomy (PH), performed after 1 week of feeding a diet containing acetyl-amino fluorene (AAF) started 15 days after a single intraperitoneal treatment of diethylnitrosamine (DEN; 200 mg/Kg).

Transplantable hepatomas of the Morris series, defined by the time between transplantation (t.t.), were the slow-growing 9618A, t.t 60-90 days and the fast growing 3924A, t.t. 20-25 days. Tumor-bearing rats were used at half the average, t.t. of the tumors. AH-130 Yoshida-ascites hepatoma, t.t. 10-12 days, was used 5-6 days after transplantation, in the middle of the logarithmic phase of growth. When AH-130 cells were injected subcutaneously in the back their growth rate was remarkably reduced: a sizeable nodule appeared after 10-13 days and the tumor was taken for the experiment after 20-25 days when the nodule reached a diameter of about 1 cm. RNA was extracted from livers, solid tumors and ascites cells with guanidium thiocyanate[16] and subjected to: 1) functional analysis in a cell-free, mRNA dependent translation system. The proteins synthesized by this system were then separated by 2D electrophoresis and characterized according to their molecular weight (Mw) and isoelectric point[11]; 2) Northern blot analysis[17] with the hsc 73 cDNA kindly provided by Dr. H. Pelham, MRC Laboratory of Molecular Biology, Cambridge, U.K. Southern blot analysis, DNA extraction, digestion and hybridization with the hsc 73 cDNA were performed according to Cairo et al.[18]

RESULTS AND DISCUSSION

Two-dimensional electrophoresis of proteins synthesized by reticulocyte cell-free systems supplemented with exogenous RNAs reveals that liver cells of hyperthermic rats contain two stress proteins with molecular masses of 68-73 Kd, and pI of 5.8-6.45. The more acidic polypeptide has the higher molecular weight (hsc 73), is barely detectable before heat-shock and is only moderately induced by hyperthermia: the second and more basic hsp, on the contrary, is highly inducible upon heat-shock (Fig. 1, a and b). In the present work we focused our attention on hsc 73 which is also barely detectable in liver subjected to DEN + AAF + PH, and in the slow-growing 9618A Morris hepatoma.

On the contrary, hsc 73 is well visible in AH-130 hepatoma, both in ascitic and in solid form and is highly expressed in the fast-growing 3924A Morris hepatoma (Fig. 1, c, d, e). Therefore a functionally competent mRNA for hsc 73 seems to be constitutively expressed at appreciable levels only in established tumors and in particular in the solid tumor with the highest growth rate. These results are confirmed by Northern blot analysis which allows a more direct quantitation of hsc 73 mRNA. Fig. 2 shows that the levels of expression of hsc 73 mRNA are very low in control liver, in liver subjected to the carcinogenic treatment and in the slow-growing transplantable hepatoma. Due to the weakness of the signal the small increase observable in the carcinogen-treated liver cannot be taken as significant. Expression of the hsc 73 mRNA is very strong in the fast-growing 3924A Morris hepatoma, much stronger than in its slow-growing counterpart. The steady state level of hsc 73 mRNA is also very high in AH-130 ascites tumor. Differences in physical character and in nature between the latter tumor and the hepatomas of the Morris series make scarcely reliable any direct comparison between them. But comparison of AH-130 in ascites form with the same cells that grow as a solid tumor, and show a clearly reduced growth rate, seems to suggest that expression of hsc 73 is in some way connected to the rate of growth of tumors. It has recently been shown that the levels of hsc 73 mRNA are about five-fold higher in rapidly-growing tissue-culture cells than in cells whose growth has been arrested by serum starvation[19]. Our results do not give evidence of a constitutive expression of hsc 73 mRNA in a tissue subjected to a complete carcinogenic treatment, destined to originate a tumor after suitable lag phase. However, the technique used in the present study can reveal very small amounts of specific mRNA in a cell population, but does not exclude

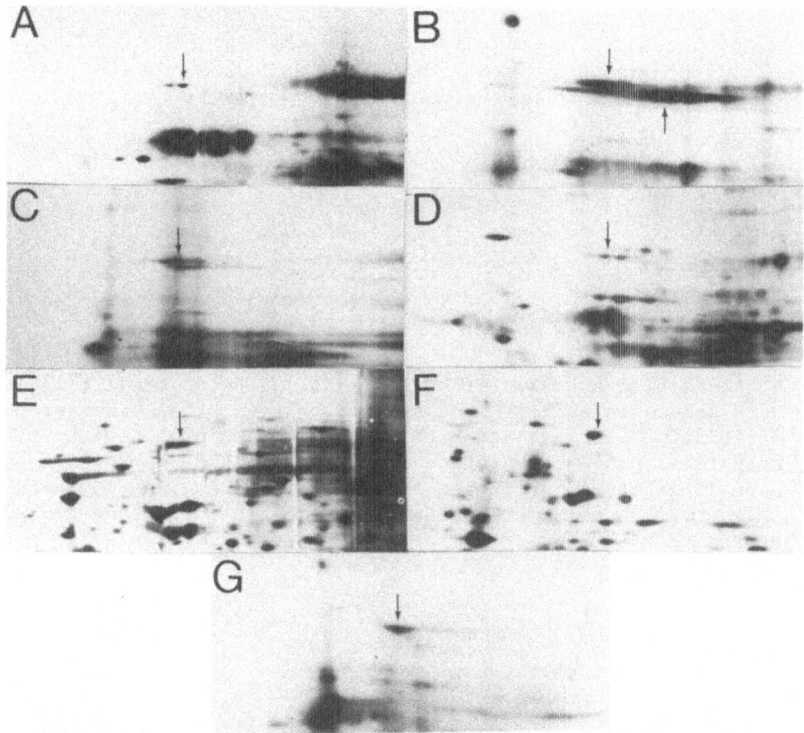

Fig. 1. Fluorograms of 2-D gel electrophoresis of the translational products of total RNA. A) Liver from normothermic control rat. B) Liver from hyperthermic control rat. C) Liver from normothermic rat subjected to DEN + AAF + PH. D) 9618A slow-growing Morris hepatoma. E) AH-130 hepatoma in ascitic form. F) AH-130 hepatoma in solid form. G) 3924A fast-growing Morris hepatoma. Downward arrows indicate the migration position of hsc 73. Upward arrow indicates the migration position of the heat-inducible hsp 70.

that the constitutive expression of hsc 73 mRNA would occur in an extremely limited number of cells, destined to become a visible tumor later on.

The use of _in situ_ hybridization will hopefully permit to overcome this difficulty in the future. The technique of Southern blotting can reveal overall changes of gene structure, and justify further work. Analysis of DNA extracted from 3924A hepatoma, the fast growing tumor with a strong constitutive expression of hsc 73 mRNA, does not show appreciable changes in the overall gene structure with any of the enzymes tested (Fig. 3). Therefore the high level of expression of hsc 73 in the latter tumor is most probably dependent on the increased transcription and/or increased stability of a message transcribed from a gene of an essentially normal structure.

ACKNOWLEDGEMENTS

We thank Prof. T. Galeotti for providing the transplantable Morris hepatomas, Mr. V. Albini for technical assistance and Ms. M. G. Bombonato for typing the manuscript. The work was supported by grant 85.02027.44 of C. N. R. Progetto Finalizzato Oncologia.

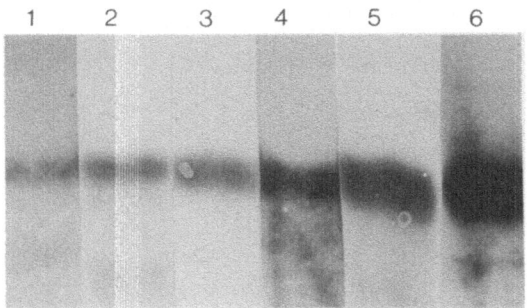

Fig. 2. Northern blot analysis of hsc 73 mRNA. 30 μg of total cellular RNA
were loaded in each lane of the gel that was then processed as described in Materials and Methods section and probed with the nick
translated hsc 73 cDNA. Lane 1: rat liver, Lane 2: 9618A hepatoma,
Lane 3: liver of carcinogen-treated rat, Lane 4: AH-130 hepatoma
solid form, Lane 5: AH-130 hepatoma ascitic form, Lane 6: 3924A
hepatoma.

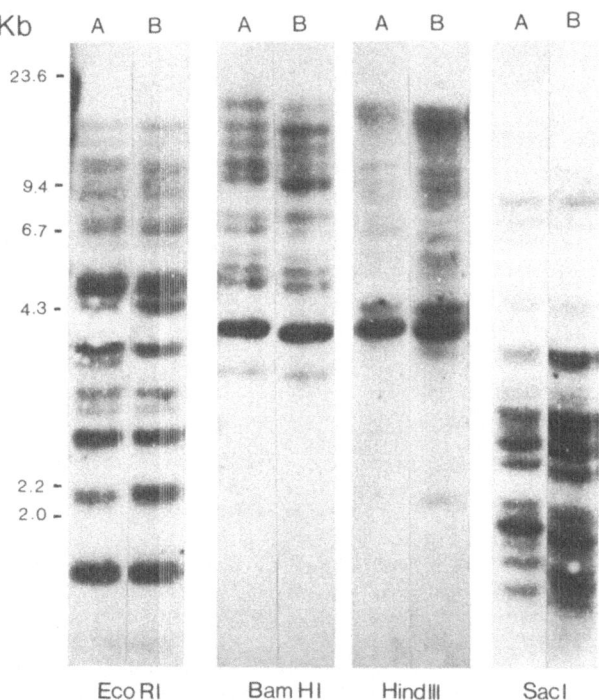

Fig. 3. Southern blot analysis of hsc 73 in liver and 3924A Morris hepatoma.
10 μg of genomic DNA from rat liver (A) and 3924A hepatoma (B) were
digested with the indicated restriction enzymes, separated on agarose gels, blotted to filters and hybridized to the nick translated
hsc 73 cDNA. The migration position of DNA fragments of DNA digested
with Hind III and run in a parallel slot of the gel is indicated on
the left.

REFERENCES

1. M. J. Schlesinger, M. Ashburner and A. Tissieres, "Heat shock. From Bacteria to Man", Cold Spring Harbor Laboratory, Cold Spring Harbor (1982).
2. F. Ritossa, A new puffing pattern induced by temperature shock and DNP in Drosophilia, Experientia 18:571 (1962).
3. R. H. Burdon, Heat shock and the heat shock proteins, Biochem. J. 240:313 (1986).
4. S. Lindquist, The heat shock response, Ann. Rev. Biochem. 55:1151 (1986).
5. J. R. Subjeck and T-T. Shyy, Stress protein systems of mammalian cells, Am. J. Physiol. 250:c1 (1986).
6. BN. J. Wu and R. I. Morimoto, Transcription of the human hsp 70 gene is induced by serum stimulation, Proc. Natl. Acad. Sci. USA 82:6070 (1985).
7. H. T. Kao, O. Capasso, N. Heintz and J. R. Nevins, Cell cycle control of the human hsp 70 gene: implications for the role of a cellular E1A-like function, Mol. Cell. Biol. 5:628 (1985).
8. J. R. Nevins, Induction of the synthesis of 70.000 dalton mammalian heat shock protein by the adenovirus E1A gene product, Cell 29:913 (1982).
9. M. J. Imperiale, H. T. Kao, L. T. Feldman, J. R. Nevins and S. Strickland, Common control of the heat shock gene and early adenovirus genes: evidence for a cellular E1A-like activity, Mol. Cell. Biol. 4:867 (1984).
10. R. E. Kingston, A. S. Baldwin, Jr. and P. A. Sharp, Regulation of heat shock gene 70 expression by c-myc, Nature 312:280 (1984).
11. G. Cairo, L. Bardella, L. Schiaffonati and A. Bernelli-Zazzera, Synthesis of heat shock proteins in rat liver after ischemia and hyperthermia, Hepatology 5:357 (1985).
12. L. Bardella, L. Schiaffonati, G. Cairo and A. Bernelli-Zazzera, Synthesis of heat-shock proteins in liver and hepatomas of different growth rate, Biochem. Soc. Trans. 14:974 (1986).
13. L. Bardella, L. Schiaffonati, G. Cairo and A. Bernelli-Zazzera, Heat-shock proteins and mRNAs in liver and hepatoma, Br. J. Cancer 55:643 (1987).
14. O. Pinhasi-Kimhi, D. Michalovitz, A. Ben-Zeev and M. Oren, Specific interaction between the p53 cellular tumor antigene and major heat shock proteins, Nature 320:182 (1986).
15. D. B. Solt and E. Farber, New Principle for the analysis of chemical carcinogenesis, Nature 263:701 (1976).
16. Y. Raymond and G. C. Shore, The precursor for carbamyl phosphate synthetase is transported to mitochondria via a cytosolic route, J. Biol. Chem. 19:9335 (1979).
17. G. Cairo, L. Bardella, L. Schiaffonati, P. Arosio, S. Levi and A. Bernelli-Zazzera, Multiple mechanisms of iron-induced ferritin synthesis in HeLa cells, Biochem. Biophis. Res. Commun. 133:314 (1985).
18. G. Cairo, P. Vezzoni, L. Bardella, E. Rappocciolo, S. Levi, P. Arosio and A. Bernelli-Zazzera, Regulation of ferritin synthesis in malignant and non-malignant lymphoid cells, Biochem. Biophys. Res. Commun. 139:652 (1986).
19. P. K. Sorger and H. R. B. Pelham, Cloning and expression of a gene incoding hsc 73, the major hsp 70-like protein in unstressed rat cells EMBO J. 6:993 (1987).

REGULATION OF S-ADENOSYL-L-METHIONINE DECARBOXYLASE IN NORMAL AND REGENERATING RAT LIVER BY ADENOSINE-CONTAINING MOLECULES

Giuseppe Scalabrino, Maria Elena Ferioli and Rossella Candiani

Institute of General Pathology and C.N.R. Centre
for Research in Cell Pathology, University of Milan
Via Mangiagalli, 31, 20133 - Milano (Italy)

INTRODUCTION

The pathway for polyamine biosynthesis in its entirety thought to play a pivotal role in the regulation of the proliferation of eukaryotic cells, simultaneously provides the cells with two types of factors that regulate the growth and the differentiation of mammalian cells in opposite directions. These are the polyamines, which stimulate cell growth and differentiation, and the sulphur-containing purine nucleoside 5'-deoxy-5'methylthioadenosine (MTA), which strongly inhibits growth and differentiation of mammalian cells[1-5]. Therefore, the study of the key points in the polyamine biosynthetic pathway seems to be of paramount importance for better understanding the mechanisms regulating the proliferation of eukaryotic cells.

S-Adenosyl-L-methionine decarboxylase (E.C. 4.1.1.50) (SAMD) is one of the two regulatory enzymes in the biosynthetic pathway of polyamines, the other being ornithine decarboxylase (ODC)[2]. SAMD provides decarboxylated S-adenosyl-L-methionine (Dec-SAM), which is used as an aminopropyl donor substrate and is needed by spermidine- and spermine-synthases[2]. Mammalian SAMD recognizes putrescine as a physiological positive factor and Dec-SAM, spermidine and spermine as physiological negative regulatory factors[2,6-8]. SAMD activity like ODC activity, can be easily induced by a wide variety of hormonal, developmental and cell growth-related stimuli, including liver regeneration following partial hepatectomy of the rat[2]. Conversely, various inhibitors of eukaryotic SAMD activity have been shown to have an antiproliferative effect on several cell lines[9-11]. SAMD has been the subject of several recent review[6-8].

In the experiments reported here we compared the effects of several nucleoside-containing molecules, some produced in different steps of the polyamine biosynthetic pathway, i.e., Dec-SAM and MTA[2], and some not physiological, i.e., 5'-isobutylthioadenosine (SIBA) and 2-chloro-2'-deoxyadenosine (CdA), on the activity of SAMD purified from normal rat liver and of SAMD purified from regenerating rat liver. Our main aim was to see whether or not the regulation of regenerating liver SAMD by these molecules (chiefly Dec-SAM and MTA) might differ from the regulation of SAMD from normal liver by the same molecules. A previous and preliminary in vivo study in our laboratory indicated that it might be, since we observed differences between normal and regenerating liver of the rat in the degree of the SAMD inhibition in response to the administration of MTA and of Dec-SAM[12].

We also had another ancillary goal of the present experiments. It was to investigate whether or not MTA and SIBA, whose strong antiproliferative effects[13,14], are due also to inhibition of spermidine- and spermine-synthases[15,16], might inhibit another step in the polyamine biosynthetic pathway, namely SAMD. In addition, it is not known at present whether or not the antiproliferative effects of CdA[17-19] might involve any step of the polyamine biosynthetic pathway.

MATERIALS AND METHODS

Animals

Adult male non-inbred rats of the Sprague-Dawley strain, weighing about 200 g, were obtained from Charles River, Calco, Italy and were used in this work. All rats were housed as previously described[20].

Treatments of Animals

Two-thirds partial hepatectomies were performed under light ether anesthesia by the method of Higgins and Anderson[21].

Tissue Preparation

The rats were killed by decapitation, the livers were rapidly removed and weighed. For purification of SAMDs, the normal and regenerating livers were homogenized in 1 vol. of a medium whose composition has been published by Seyfried et al.[22].

SAMD Purification

In preliminary experiments, whose results are not reported here, on the kinetics of hepatic SAMD activity after partial hepatectomy, we observed that the peak of the enzyme activity occurred at 96 h after operation. This result is in keeping with what reported by Sturman and Gaull[23]. Therefore, we chose that time to take the livers to purified SAMD from regenerating rat liver. Before beginning the purification of the SAMD from regenerating liver, the induction of SAMD activity was checked in the 20,000 x g supernatant, measuring SAMD activity as described later. The procedure for SAMD purification was essentially that reported by Seyfried et al.[22], except that the MGBG-Sepharose chromatography was omitted. The final enzyme preparation from normal liver had a specific activity of about 400 pmol/mg protein and that from regenerating liver a specific activity of about 850 pmol/mg protein, under the standard assay conditions. The final preparations used in our experiments represented purifications of about 60-fold. All the purified SAMDs were stored for a short time (at maximum 1 week) at 0-2°C at a concentration of 4-8 mg of protein/ml, with practically no loss of enzyme activity.

Assay of SAMD Activity

Putrescine-activated SAMD activity was assayed by measuring $^{14}CO_2$ released from carboxyl-labeled substrate. The experimental conditions were as published previously[20]. The composition of the standard incubation mixture was that routinely used in our laboratory except that putrescine was 1.0 mM (final)[20]. Each sample contained an aliquot of enzyme preparation corresponding to about 300 µg protein. All the assays were done in triplicate. The results are expressed as pmoles of CO_2 liberated per mg protein per 30 min incubation. The K_m values of the two SAMDs in the absence of any inhibitor were also determined in an incubation mixture which differed from the above only in having a higher putrescine concentration (2.5 mM, final). The higher putrescine concentration in the incubation mixture did not modify the values K_ms to any appreciable extent (results not reported here). To obtain two classical plots of enzyme kinetics, i.e., the double-reciprocal Lineweaver-Burk plot (in which 1/V versus 1/S is plotted)[24] and the Dixon plot (in which 1/V versus [I] is plotted[25,26]) for each compound for each SAMD, enzyme activity was determined at a constant putrescine concentration (1.0 mM, final) with six different concentrations of S-adenosyl-L-methionine (SAM) (ranging from 0.02 mM to 0.4 mM, final concentration) in the absence (control) and in the presence of the compound tested at different fixed concentrations. The final concentration for each compound were: 0.01 mM, 0.05 mM and 0.1 mM for Dec-SAM; 1.0 mM and 1.5 mM for MTA; 2.5 mM, 3.5 mM and 5.0 mM for CdA; 0.5 mM, 1.0 mM and 3.0 mM for SIBA. To obtain the two plots for each compound for each SAMD, the data of four to six experiments were collected. From the intersection point of the straight lines on the ordinate or on the abscissa in the Lineweaver-Burk plots[24], we established the type of SAMD inhibition for each compound. The Dixon plots allowed us to calculate directly the value of the inhibition constant (K_i) for each compound for each SAMD from the intersection point of the star of the straight lines[25,26].

Protein Determination

Protein concentrations were estimated by the method of Geiger and Bessman[27], using crystalline serum albumin as standard.

Statistical Analyses

The K_i values were determined as described by Todhunter[28], using linear regression analyses to determine the slopes of the straight lines and a parallelism test[29,30]. The estimates of the kinetics parameters were calculated by the method of last-squares for a star of straight lines[29,30]. We have developed a program for a personal computer that is available on request. In each statistical test used, an alpha-level of 0.05 or less was the limit of statistical significance.

RESULTS

The SAMD from normal and regenerating rat liver have very similar values of K_m for SAM in the presence of putrescine. The K_m value for SAMD from normal liver was 25 μM, which is in accordance with the values reported by other investigators for putrescine-activated SAMD from the rat liver and other mammalian organs[31,32]. The value of K_m for SAMD from regenerating liver was 33 μM, which did not differ significantly from that observed for the other SAMD.

The in vitro effects of several nucleoside-containing compounds structurally related to SAM on the activities of the two SAMDs were determined. The type of inhibition for each compound and the value of inhibition constant (K_i) for each compound for each SAMD are reported in Table 1.

MTA is a non-competitive inhibitor of both the hepatic SAMDs and the values of K_is for the two SAMDs significantly differ. The K_i value for SAMD from normal liver is significantly lower than that for SAMD from regenerating liver. MTA is actually a more powerful SAMD inhibitor in normal liver than in regenerating liver.

We confirmed that Dec-SAM is a potent competitive inhibitor for SAMD from normal liver (as earlier reported[31]) and also for SAMD from regenerating liver. However, the K_i value for Dec-SAM for regenerating liver SAMD is significantly lower than that for normal liver SAMD. Parenthetically, the K_i value for Dec-SAM for normal liver SAMD is in good agreement with the value found by others[31]. Dec-SAM is therefore a more potent SAMD inhibitor in regenerating rat liver than in normal rat liver. The Dixon plots for Dec-SAM

461

Table 1. Type of inhibition and inhibition constants for
SAMD from normal and regenerating rat liver of
some adenosine-containing molecules

Compound[a]	Type of inhibition[b]	Inhibition constant (K_i)[c] for	
		SAMD from normal liver	SAMD from regenerating liver
M T A	non-competitive	0.6 mM	1.2 mM*
Dec-SAM	competitive	7.6 µM	3.3 µM*
CdA	competitive	4.7 mM	3.0 mM*
S I B A	no inhibition		

[a]Each compound was added to the SAMD assays at the concentrations indicated in the Materials and Methods section.
[b]Type of inhibition was established on the double-reciprocal Lineweaver-Burk plots[24]. [c]The K_i values were calculated as described under Statistical analyses paragraph. K_i values with * are different from K_i values for normal liver SAMD at $P \leq 0.05$.

for each of the two SAMDs are presented in Fig. 1, as paradigms of all the Dixon plots for all the compounds (not shown). It should be noted here that the concentrations that significantly inhibited both the SAMD activities were of the order of the millimolar for MTA and of the order of micromolar for Dec-SAM. This clearly means that Dec-SAM is a much more potent inhibitor of hepatic SAMD activities than MTA. Two other compounds, containing adenosine like those above but not formed in the polyamine biosynthetic pathway, were evaluated as inhibitors of the activities of the SAMDs. These molecules are CdA and SIBA. SIBA was shown to have no effect at all on either of the SAMDs (Table 1). On the contrary, CdA is a new competitive inhibitor of both SAMDs, with greater inhibition of SAMD from regenerating rat liver (the K_i value for this last enzyme is significantly lower than the K_i value for SAMD from normal rat liver) (Table 1).

DISCUSSION

The present results indicate that the sensitivity of SAMD prepared from regenerating rat liver (at that time after partial hepatectomy when hepatic SAMD activity reaches its peak) to some inhibitory nucleoside-containing compounds substantially differs from the sensitivity of normal rat liver SAMD to the same compounds. We have demonstrated that normal liver SAMD is inhibited more by MTA, whereas regenerating liver SAMD is inhibited more by Dec-SAM and CdA. The present results with MTA and Dec-SAM are in complete agreement with our previous results from in vivo experiments[12]. Although Dec-SAM is a well known in vitro SAMD inhibitor[31], it is a new finding that MTA and CdA are also SAMD inhibitors. By analogy with other drugs that inhibit SAMD activity and have antiproliferative effects[9,10], it is conceivable that one mechanism of CdA inhibition of mammalian cell proliferation is through inhibition of SAMD activity.

Again, it is also a new finding that the intensity of the negative regulation of hepatic SAMD by these compounds differs according to the proliferative or non proliferative status of the organ.

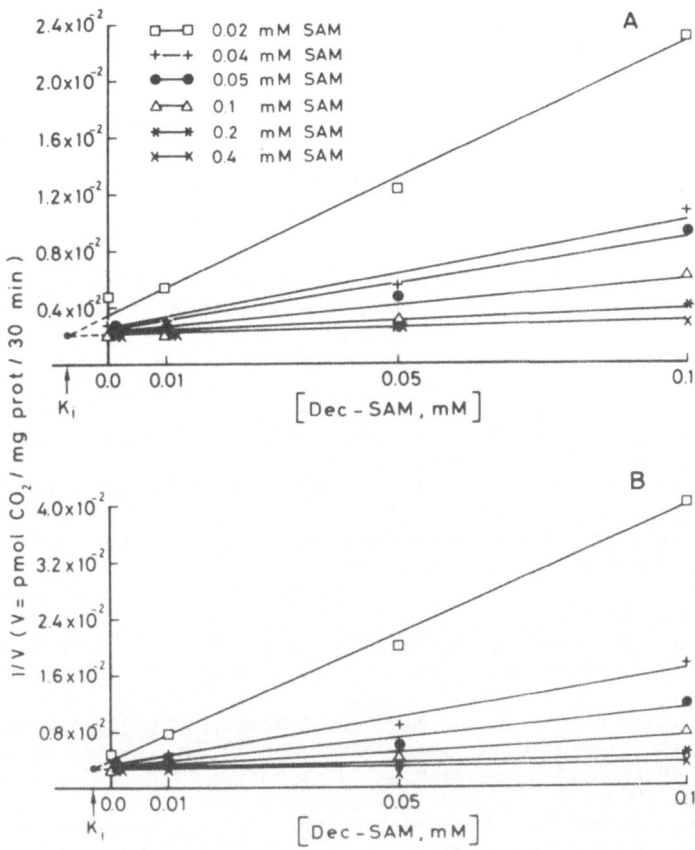

Fig. 1. Dixon plot of the effects of the different substrate (S-adenosylmethionine, SAM) concentrations on the activity of S-adenosylmethionine decarboxylase (SAMD) purified from normal rat liver (Panel
A) or from regenerating rat liver (Panel B), in the absence and in
the presence of different concentrations of decarboxylated SAM
(Dec-SAM). The enzyme activity was determined as described under
Materials and Methods section. The reciprocals were plotted as
described under Materials and methods.

At this stage of the research, although the greater sensitivity of regenerating liver SAMD to Dec-SAM than that of normal liver SAMD is not easy
to explain, it seems to be of particular interest. Until now, in fact, it
has been demonstrated that both putrescine-activated SAMD and putrescine-
insensitive SAMD from various organisms are inhibited similarly by the product of the reaction, i.e., Dec-SAM [33].

All these results raise the possibility that there are two different
forms of SAMD, one in quiescent and one in proliferating liver of the rat.
It is exceedingly difficult at this stage of the research to know whether or
not there are two forms of SAMD. On the one hand, we have demonstrated that
the two SAMDs have very similar values of K_m for SAM, but on the other hand
they also have clear and definite differences in sensitivity to the negative
regulatory influences of some nucleoside-containing compounds. Different
forms of SAMD have been demonstrated so far only in different organs (i.e.,
liver and skeletal muscles) of the rat [34,35]. It remains to be established
whether or not these differences between normal and regenerating rat liver in

SAMD regulation are partially or totally cancelled after the complete restoration of the liver mass. In a previous study[36], we demonstrated that rat
liver maintains abnormally high levels of SAMD activity (but not of ODC activity) even at those times after partial hepatectomy when the reconstitution
of liver mass has been long completed, suggesting that certain derangements
in hepatic SAMD regulation do not go away in "post-regeneration" rat liver.

ACKNOWLEDGEMENTS

This study was supported by grant No. 104348.44.8402849 of the Finalized
Project "Oncology", Consiglio Nazionale delle Ricerche (C.N.R.), Rome, Italy.

REFERENCES

1. H. G. Williams-Ashman and Z. N. Canellakis, Polyamines in mammalian
   biology and medicine, Perspect. Biol. Med. 22:421 (1979).
2. G. Scalabrino and M.E. Ferioli, Polyamines in mammalian tumors, Part I,
   Adv. Cancer Res. 35:151 (1981).
3. O. Heby, Role of polyamines in the control of cell proliferation and
   differentiation, Differentiation 19:1 (1981).
4. A. E. Pegg and P. P. McCann, Polyamine metabolism and function, Am. J.
   Physiol. 243:C212 (1982).
5. H. G. Williams-Ashman, J. Seidenfeld and P. Galletti, Trends in
   biochemical pharmacology of 5'-deoxy-5'-methylthioadenosine,
   Biochem. Pharmacol. 31:277 (1982).
6. H. G. Williams-Ashman and A. E. Pegg, Aminopropyl group transfers in
   polyamine biosynthesis, in: "Polyamines in Biology and Medicine",
   D. R. Morris and L. J. Marton, eds., Marcel Dekker, New York,
   (1981).
7. A. E. Pegg, S-adenosylmethionine decarboxylase: a brief review, Cell
   Biochem. Funct. 2:11 (1984).
8. C. W. Tabor and H. Tabor, Methionine adenosyltransferase (S-adenosylmethionine synthetase) and S-adenosylmethionine decarboxylase, Adv.
   Enzymol. 56:251 (1984).
9. G. Scalabrino and M. E. Ferioli, Polyamines in mammalian tumors. Part
   II., Adv. Cancer Res. 36:1 (1982).
10. E. Hölttä, H. Korpela and T. Hovi, Several inhibitors of ornithine and
    adenosylmethionine decarboxylase may also have antiproliferative
    effects unrelated to polyamine depletion, Biochim. Biophys. Acta
    677:90 (1981).
11. A. E. Pegg, Inhibition of S-adenosylmethionine decarboxylase, Meth.
    Enzymol. 94:239 (1983).
12. G. Scalabrino and M. E. Ferioli, New insights into the regulation of
    S-adenosyl-L-methionine decarboxylase by adenosine-containing
    molecules in rat liver, in: "Biomedical Studies of Natural Poly-
    amines", C. M. Caldarera, C. Clô and C. Guarnieri, eds., CLUEB,
    Bologna, (1986).
13. R. W. Wolford, M. R. Macdonald, B. Zehfus, T. J. Rogers and A. J.
    Ferro, Effect of 5'-methylthioadenosine and its analogs on murine
    lymphoid cell proliferation, Cancer Res. 41:3035 (1981).
14. M. W. White, M. K. Riscoe and A. J. Ferro, The comparative effects of
    5'-methylthioadenosine and some of its analogs on cells containing,
    and deficient in, 5'-methylthioadenosine phosphorylase, Biochim.
    Biophys. Acta 762:405 (1983).
15. A. E. Pegg, R. J. Borchardt and J. K. Coward, Effects of inhibitors of
    spermidine and spermine synthesis on polyamine concentrations and
    growth of transformed mouse fibroblasts, Biochem. J. 194:79 (1981).
16. H. Hibasami, M. Tanaka and J. Nagai, Inhibition of aminopropyltransfer-
    ases by 5'-S-isobutyl-5'-deoxyadenosine in vitro, Biochem.

Pharmacol. 31:1649 (1982).

17. J. Uberti, J. J. Lightbody and R. M. Johnson, The effect of nucleosides and deoxycoformycin on adenosine and deoxyadenosine inhibition of human lymphocyte activation, J. Immunol. 123:189 (1979).

18. D. A. Carson, D. B. Wasson, R. Taetle and A. Yu, Specific toxicity of 2-chlorodeoxyadenosine toward resting and proliferating human lymphocytes, Blood 62:737 (1983).

19. D. A. Carson, D. B. Wasson and E. Beutler, Antileukemic and immunosuppressive activity of 2-chloro-2'-deoxyadenosine, Proc. Natl. Acad. Sci. USA 81:2232 (1984).

20. G. Scalabrino, M. E. Ferioli, R. Nebuloni and F. Fraschini, Effects of pinealectomy on the circadian rhythms of polyamine biosynthetic decarboxylases and tyrosine aminotransferase in different organs of the rat, Endocrinology 104:377 (1979).

21. G. M. Higgins and R. M. Anderson, Experimental pathology of the liver. I. Restoration of the liver of the white rat following partial surgycal removal, Arch. Path. 12:186 (1931).

22. C. E. Seyfried, O. E. Oleinik, Y. L. Degen, K. Resing and D.R. Morris, Purification, properties and regulation of the level of bovine S-adenosyl methionine decarboxylase during lymphocyte mitogenesis, Biochim. Biophys. Acta 716:169 (1982).

23. J. A. Sturman and G. E. Gaull, Changes in subcellular distribution of S-adenosylmethionine decarboxylase in regenerating and in developing rat liver, Biochim. Biophys. Acta 428:70 (1976).

24. H. Lineweaver and D. Burk, The determination of enzyme dissociation constants, J. Am. Chem. Soc. 56:658 (1934).

25. M. Dixon, The determination of enzyme inhibitor constants, Biochem. J. 55:170 (1953).

26. A. Cornish-Bowden, A simple graphical method for determining the inhibition constants of mixed, uncompetitive and non-competitive inhibitors, Biochem. J. 137:143 (1974).

27. P. J. Geiger and S. P. Bessman, Protein determination by Lowry's method in the presence of sulphydryl reagents, Anal. Biochem. 49:467 (1972).

28. J. A. Todhunter, Reversible enzyme inhibition, Meth. Enzymol. 63:383 (1979).

29. E. J. Williams, "Regression Analysis", J. Wiley, New York (1959).

30. C. Daniel and F. S. Wood, "Fitting Equations to Data", J. Wiley, New York (1971).

31. B. Yamanoha and K. Samejima, Inhibition of S-adenosylmethionine decarboxylase from rat liver by synthetic decarboxylated S-adenosylmethionine and its analogs, Chem. Pharm. Bull. 28:2232 (1980).

32. R. Porta, C. Esposito and G. Della Pietra, S-adenosylmethionine decarboxylase from human placenta, Int. J. Biochem. 8:347 (1977).

33. H. Pösö, R. Sinervirta, J. J. Himberg and J. Jänne, Putrescine-insensitive S-adenosyl-L-methionine decarboxylase from Tetrahymena pyriformis, Acta Chem. Scand. B 29:932 (1975).

34. H. Pösö and A. E. Pegg, Differences between tissues in response of S-adenosylmethionine decarboxylase to administration of polyamines, Biochem. J. 200:629 (1981).

35. H. Pösö and A. E. Pegg, Comparison of S-adenosylmethionine decarboxylase from rat liver and muscle, Biochemistry 21:3116 (1982).

36. M. E. Ferioli and G. Scalabrino, Permanently decreased hepatic levels of 5'-deoxy-5'-methylthioadenosine during regeneration of and chemical carcinogenesis in rat liver, J. Natl. Cancer Inst. 76:1217 (1986).

DERANGEMENTS OF CATIONIC AMINO ACID TRANSPORT IN FIBROBLASTS

FROM HUMAN DESMOID TUMOR

Franca A. Nucci, Ovidio Bussolati, Valeria Dall'Asta,
Giancarlo Gazzola, Roberto Giardini[1] and Guido G. Guidotti

Istituto di Patologia Generale, Università di Parma
Parma (Italy) and [1]Divisione Anatomia Patologica e
Citologia, Istituto Nazionale per lo Studio e la Cura
dei Tumori, Milano (Italy)

INTRODUCTION

An increased rate of amino acid uptake by System A has been considered a consistent feature of tumor cells[1]. This inference is based on the following facts: (a) enhancements in the transport rates of site A-reactive amino acids are among the early events associated with cell transformation in vitro[2]; (b) increased uptake of some amino acid substrates of System A has been reported for virus- and chemically-transformed cell lines[3,4] and (c) rat fibroblasts made tumorigenic by ras transfection exhibit a somewhat faster uptake of 2-(methyl)aminoisobutyric acid (a transport-specific substrate of System A) than non-transfected cells[5]. However, tumor cell lines have been described that do not exhibit this metabolic feature[6,7] and little is known on amino acid transport changes in cells from spontaneously occurring tumors[8]. We therefore devised experiments to investigate the amino acid transport in cells from a spontaneous human tumor (desmoid tumor fibroblasts) as compared with their normal counterpart (skin-derived fibroblasts). To avoid variability among individuals for transport, the comparison was made between cell cultures obtained from the same donor. The present study concerns the six transport systems for amino acids so far characterized in human fibroblasts[9-13].

MATERIALS AND METHODS

Explants of surgically excised extra-abdominal desmoid tumors and of skin biopsies obtained from 3 female patients (Z. R., S. V., M. L., respectively 8-, 19- and 42-years old) were cultured in Dulbecco's modified Eagle's medium (D-MEM) supplemented with 20% fetal bovine serum (FBS). Fibroblasts derived from desmoid tumors (DTF) and skin biopsies (HF) were routinely grown in the same medium supplemented with 15% FBS, penicillin (100 units/ml) and streptomycin (100 µg/ml). Culture conditions were: pH 7.4; atmosphere 5% CO_2 in air; temperature 37°C.

For the study of growth kinetics, cells were seeded in D-MEM onto 12-well trays (COSTAR) and cultured for 11 days with medium changes every 72 h. At daily intervals, cells were harvested in phosphate buffered saline

(PBS) containing trypsin (0.0115%) and EDTA (0.02%) and counted with a Coulter Counter ZM.

The assay of transport activity was performed on cells (HF and DTF) cultured in 24-well trays (COSTAR) and used while still far from confluence (4-5 x 10^4 cells/sq cm). The culture medium was renewed every 72 h and always 24 h before the experiment. Amino acid inward transport was assayed as described by Gazzola et al.[14]. At the beginning of the experiment, cell monolayers were washed twice in Earle's balanced salt solution (EBSS) and preincubated for appropriate periods in the same solution supplemented with 10% dialyzed FBS. Amino acid uptake was measured at 37°C in EBSS for 10 sec (initial entry rate) under discriminating conditions for each individual transport system[9-13]. Cells were extracted in 0.25 ml ethanol. Scintillation fluid (2.5 ml) was added to cell extracts and radioactivity was measured with a liquid scintillation spectrometer Packard 460C.

Transport activity was expressed as nmoles or μmoles/ml of cell water/min. Cell water was estimated by measuring the distribution space of 3-O-methyl-D-glucose and expressed as μl/mg of protein. Kinetic analysis was performed with a BASIC program applying Marquardt's algorithm, a general procedure for least squares estimation of non linear parameters. The equation used was[13]:

$$v = \frac{V_{max} [S]}{K_m + [S]} \frac{1 - e^{-K_D}}{K_D} + [S](1 - e^{-K_D})$$

Membrane potential (E_m) was estimated by a null-point method, using the transmembrane distributions of the lipophilic cation tetraphenylphosphonium (TPP^+) and of the amino acid L-arginine as E_m-sensitive probes[15]. Intracellular [K^+] was measured with an atomic absorption spectrophotometer as described[15] and approached 170 mM both in HF and DTF.

Fetal bovine serum (FBS), growth medium and antibiotics were purchased from Flow Laboratories. L-[5-^{3}H]proline (30 Ci/mmol); L-[2,3-^{3}H] aspartic acid (15 Ci/mmol); L-[2,3-^{3}H]serine (35 Ci/mmol); L-[4,5-^{3}H] leucine (61 Ci/mmol) and tetra[^{3}H]phenylphosphonium bromide (24 Ci/mmol) were from Amersham; L-[1-^{14}C]-glutamic acid (55.3 Ci/mol) and 3-O-methyl-D-[U-^{14}C]glucose (329 Ci/mol) were obtained from New England Nuclear. The source of all the other chemicals was Sigma.

RESULTS

Morphology and Growth Features of Desmoid Tumor Fibroblasts in Culture

HF and DTF did not exhibit striking differences when grown in culture. Both cell types grew as monolayers without evidence of helter-skelter piling-up. However, DTF showed a less regular alignement in bundles. The most prevalent cells in DTF cultures were fusiform with a minority of quadrangular cells with ill-defined cytoplasmic processes.

HF and DTF cultures showed a comparable growth dependence on serum concentration in D-MEM (not shown). Cell growth kinetics of HF and DTF are presented in Fig. 1. The growth rate of DTF was slightly lower than that exhibited by HF, the doubling time of cell population being 55 h for DTF and 49 h for HF. Cell density at confluence was 20 to 30% lower in DTF cultures.

Activity of Amino Acid Transport Systems in HF and DTF

Both HF and DTF are endowed with the six transport systems for amino

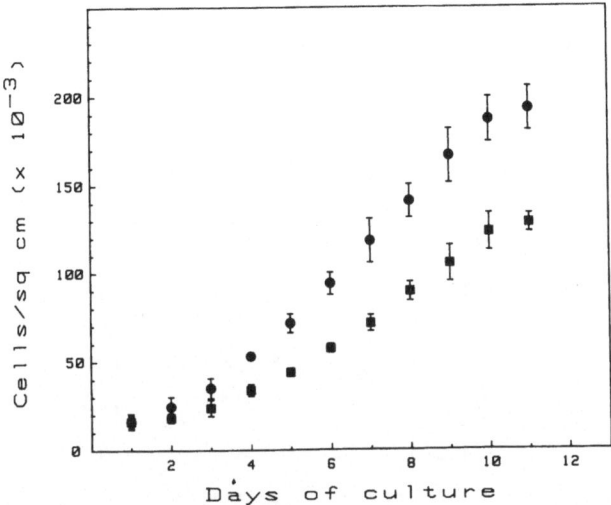

Fig. 1. Cell growth kinetics of HF and DTF. Skin-derived (●) and tumor-
derived fibroblasts (■) were from the same patient, Z. R.. After 7
passages in vitro, cells were plated at an initial density of 10^4
cells/sq cm in 12-well trays. Cells from either strain were
harvested and counted daily for an 11-day period. The points are
means of triplicate determinations with bars indicating S.D. when
greater than size of point.

acids, formerly characterized in human fibroblasts[9-13]. Operation features
(Na^+-dependence, sensitivity to pH, membrane potential-dependence, substrate
recognition properties, specific kinetic behavior) of each system were remarkably similar in HF and DTF (not shown). The same was true for trans-ef-
fects due to the presence of amino acids in the intracellular compartment
(Fig. 2). As expected for a release from trans-inhibition, the activities of
Systems A and X_{AG}^- increased slightly as the cells were depleted. The activities of Systems ASC, L, y^+ and x_c^- decreased with cell depletion as the trans-
stimulation by internal amino acids declined progressively. The basal activity of the various agencies, as estimated after a 90-min depletion of the
intracellular amino acid pool[9], was not consistently different in HF and DTF,
except for System y^+. The comparison between HF and DTF from the same donor
showed that basal transport activity of System y^+ was always reduced by about
50% in tumor cells (Fig. 2). The kinetic analysis presented in Fig. 3 shows
that in both cell types the inward transport of L-arginine (a y^+ site-reac-
tive substrate) can be described by the sum of a non saturable process (diffusion) and a saturable component (System y^+). The diffusion coefficient
(K_D) was not significantly different in HF and DTF and the decrease in the
uptake of L-arginine by DTF was fully attributable to a lowered activity of
System y^+. This change was dependent upon a decrease in transport V_{max} , in
the absence of a substantial modification in the concentration for half-maxi-
mal transport (K_m).

Membrane Potential in HF and DTF

Data given in Fig. 4 indicate that the null point, as measured with the
distribution ratio of TPP^+, is slightly higher for DTF than for HF (13.5 and
9.2 mM external $[K^+]$, respectively). These values correspond to a membrane
potential of -77 mV for HF and -67 mV for DTF (Table 1, left column). Comparable changes of E_m were obtained with L-arginine distribution ratios either

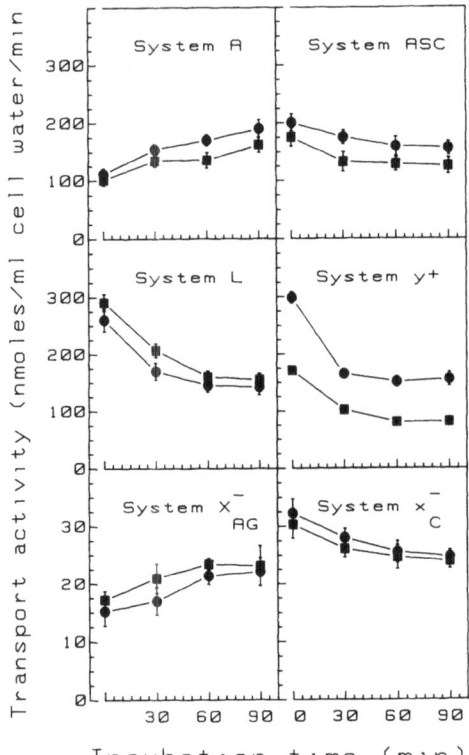

Fig. 2. Changes in amino acid transport during depletion of the internal amino acid pool in HF and DTF. Skin-derived (●) and tumor-derived fibroblasts (■) were from the same patient, S. V.. Cells were incubated for 90 min at 37°C in EBSS containing 10% FBS. At successive intervals cells were washed and uptake of labeled amino acids was measured under conditions approaching initial entry rates (see Materials and Methods). The activity of the various transport systems was assessed with the following amino acids: 0.1 mM L-proline in EBSS for System A; 0.01 mM L-serine in EBSS containing 5 mM unlabeled MeAIB for System ASC; 0.01 mM L-leucine in Na^+-free EBSS for System L; 0.01 mM L-aspartate in EBSS for System X_{AG}^-; 0.05 mM L-glutamate in Na^+-free EBSS for System x_c^- and 0.02 mM L-arginine in Na^+-free EBSS for System y^+. The points are means of triplicate determinations with <u>bars</u> indicating S. D. when greater than size of <u>point</u>.

exploited in null point experiments or directly substituted in the Nernst equation (Table 1, <u>middle</u> and <u>right</u> <u>columns</u>).

DISCUSSION

The results presented above indicate that the basal activity of the transport systems for neutral amino acids (Systems A, ASC and L) and for anionic amino acids (Systems X_{AG}^- and x_c^-) does not change appreciably in desmoid tumor fibroblasts as compared with normal human fibroblasts obtained from the same donor. This observation supports the conclusion formerly attained for neutral amino acids in experiments with normal and neoplastic fibroblasts derived from different individuals[8] and extends it to anionic amino acids. In contrast, all the DTF strains studied showed a consistent decrease in the basal transport activity for cationic amino acids (System y^+).

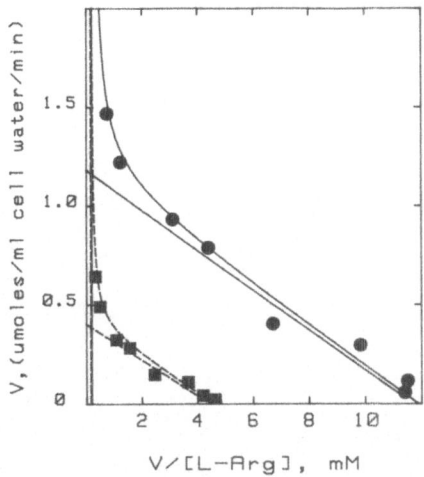

Fig. 3. Kinetic behavior of L-arginine transport in HF and DTF. Skin-derived (●) and tumor-derived fibroblasts (■) from the same patient, S. V., were depleted of intracellular amino acids for 90 min (see Fig. 2). Transport was measured for 10 sec over a range of 0.005-2 mM L-arginine in Na^+-free EBSS. Entry rates of the amino acid are plotted against the ratio of velocity to medium concentration according to an Eadie-Hofstee graphical representation. The experimental data were best fitted by the sum of a single rectangular hyperbola and a linear (non-saturable) component (see Equation under Materials and Methods). Solid lines, L-arginine transport in HF; dashed lines, L-arginine transport in DTF. Kinetic parameters are: (a) for HF, V_{max} = 1.180 ± 0.102 µmoles/ml cell water/min, K_m = 0.101 ± 0.013 mM, K_D = 0.195 ± 0.027 min^{-1}; (b) for DTF, V_{max} = 0.405 ± 0.061 µmoles/ml cell water/min, K_m = 0.095 ± 0.008 mM, K_D = 0.160 ± 0.035 min^{-1}.

This change was independent of the growth rate of the cultures and still detectable in cells made quiescent by serum withdrawal (not shown). System y^+ operates as a facilitated diffusion agency driven by the membrane potential[15]. Therefore, the decreased transport activity of this system in DTF might be secondary to an impaired ability of the tumor cells to maintain the membrane potential at a normal value. Indeed, membrane potential was slightly lower in DTF than in HF (Table 1). However, the activity of System y^+ is directly proportional to E_m (unpublished results) and, therefore, the membrane potential change in DTF is inadequate to account in full for the lowered amino

Table 1. Membrane Potential in HF and DTF

Method Probe	Membrane potential (E_m), mV		
	Null point TPP$^+$	Null point L-Arg	Distribution ratio L-Arg
HF	- 77	- 74	- 73
DTF	- 67	- 63	- 63

471

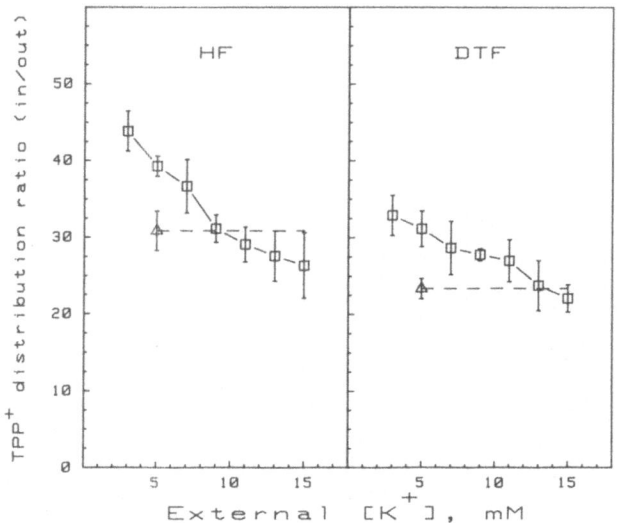

Fig. 4. Null-point experiment with TPP$^+$ as a probe of E_m. HF (<u>left</u> <u>panel</u>)
and DTF (<u>right</u> <u>panel</u>) from the same patient, S. V., were depleted of
intracellular amino acids for 90 min. Cells were then incubated for
60 min with ^{3}H-TPP$^+$ (0.2 μM, 0.5 μCi/ml) in EBSS containing 50 nM
valinomycin at 5 mM [K$^+$] (▲) or 50 μM valinomycin at variable [K$^+$]
(■). The points are means of triplicate determinations with <u>bars</u>
indicating S.D..

acid transport. Fig. 3 shows that the decreased activity of System y$^+$ in DTF
can be ascribed to a lowered transport V_{max}. This suggests that a change in
the capacity of the System underlies the transport derangement.

The present study proves that an increased amino acid transport is not a
constant feature of spontaneously occurring human tumors. The desmoid tumor,
however, is not representative of a frank malignancy, being at the borderline
between a stubbornly infiltrative fibroblastic proliferation and a low-grade
fibrosarcoma. It will be of interest to extend these studies to sarcoma-de-
rived fibroblasts.

ACKNOWLEDGEMENTS

The authors are indebted to Professor Franco Rilke, Istituto Nazionale
per lo Studio e la Cura dei Tumori, Milano, for supplying the primary
explants. This study has been aided by Lega Italiana per la Lotta contro i
Tumori, Sezione di Parma, and by Ministero della Pubblica Istruzione, Rome.

REFERENCES

1. K. J. Isselbacher, Increased uptake of amino acids and 2-deoxyglucose
     by virus-transformed cells in culture, <u>Proc</u>. <u>Natl</u>. <u>Acad</u>. <u>Sci</u>. <u>USA</u>
     69:585 (1972).
2. J. F. Perdue, Transport across serum-stimulated and virus-transformed
     cell membranes, <u>in</u>: "Virus-transformed Cell Membrane", C. Nicolau,
     ed., Academic Press, London (1980).
3. A. F. Borghetti, G. Piedimonte, M. Tramacere, A. Severini, P.
     Ghiringhelli and G. G. Guidotti, Cell density and amino acid trans-

port in 3T3, SV3T3, and SV3T3 revertant cells, $\underline{J}$. $\underline{Cell}$. $\underline{Physiol}$. 105:39 (1980).

4. P. Boerner and M. H. Saier Jr., Growth regulation and amino acid transport in epithelial cells: influence of culture conditions and transformation on A, ASC and L transport activities, $\underline{J}$. $\underline{Cell}$. $\underline{Physiol}$. 113:240 (1982).

5. E. Racker, R. J. Resnick and R. Feldman, Glycolysis and methylaminoisobutyrate uptake in rat-1 cells transfected with ras or myc oncogenes, $\underline{Proc}$. $\underline{Natl}$. $\underline{Acad}$. $\underline{Sci}$. $\underline{USA}$ 82:3535 (1985).

6. M. K. Patterson Jr., P. J. Birckbichler, E. Conway and G. R. Orr, Amino acid and hexose transport of normal and Simian Virus 40-transformed human cells, $\underline{Cancer}$ $\underline{Res}$. 36:394 (1976).

7. G. B. Segel, A. M. Tometsko and M. A. Lichtman, Y^{+}- and L-System amino acid transport in normal and chronic lymphocytic leukemia lymphocytes: photoinhibition by fluoronitrophenylazide, $\underline{Arch}$. $\underline{Biochem}$. $\underline{Biophys}$. 242:347 (1985).

8. G. C. Gazzola, V. Dall'Asta, R. Franchi-Gazzola, O. Bussolati, N. Longo and G. G. Guidotti, Amino acid transport in normal and neoplastic cultured human fibroblasts, $\underline{in}$: "Cell Membranes and Cancer", T. Galeotti et al., eds., Elsevier Science Publishers B. V., Amsterdam (1985).

9. G. C. Gazzola, V. Dall'Asta and G. G. Guidotti, The transport of neutral amino acids in cultured human fibroblasts, $\underline{J}$. $\underline{Biol}$. $\underline{Chem}$. 255:929 (1980).

10. S. Bannai and E. Kitamura, Transport interaction of L-cystine and L-glutamate in human diploid fibroblasts in culture, $\underline{J}$. $\underline{Biol}$. $\underline{Chem}$. 255:2372 (1980).

11. M. F. White, G. C. Gazzola and H. N. Christensen, Cationic amino acid transport into cultured animal cells, $\underline{J}$. $\underline{Biol}$. $\underline{Chem}$. 257:4443 (1982).

12. R. Franchi-Gazzola, G. C. Gazzola, V. Dall'Asta and G. G. Guidotti, The transport of alanine, serine and cysteine in cultured human fibroblasts, $\underline{J}$. $\underline{Biol}$. $\underline{Chem}$. 257:9582 (1982).

13. V. Dall'Asta, G. C. Gazzola, R. Franchi-Gazzola, O. Bussolati, N. Longo and G. G. Guidotti, Pathways of L-glutamic acid transport in cultured human fibroblasts, $\underline{J}$. $\underline{Biol}$. $\underline{Chem}$. 258:6371 (1983).

14. G. C. Gazzola, V. Dall'Asta, R. Franchi-Gazzola and M. F. White, The cluster-tray method for rapid measurement of solute fluxes in adherent cultured cells, $\underline{Anal}$. $\underline{Biochem}$. 115:368 (1981).

15. O. Bussolati, P. C. Laris, F. A. Nucci, V. Dall'Asta, N. Longo, G. G. Guidotti and G. C. Gazzola, Dependence of L-arginine accumulation on membrane potential in cultured human fibroblasts, $\underline{Am}$. $\underline{J}$. $\underline{Physiol}$. 253:C391 (1987).

LIPIDS AS EFFECTORS AND MEDIATORS IN GROWTH CONTROL OF ASCITES TUMOR CELLS

Eberhard W. Haeffner[1], Joanna B. Strosznajder[2] and Claudia J. K. Hoffmann[1]

[1]Institute of Cell and Tumor Biology
German Cancer Research Center
D-6900 Heidelberg, FRG
[2]Department of Neurochemistry
Medical Research Center
Polish Academy of Sciences
00-784 Warsaw, Poland

INTRODUCTION

The growth requirement of some cell types for preformed cholesterol, normally supplied by the serum, has been established[1-5]. Cholesterol, which is a major component in the plasma membrane of eukaryotes, modulates bilayer fluidity[6], and affects the activity of membrane-bound enzymes[7]. It can also modulate e.g. the characteristics of the beta-adrenergic receptor[8].

Brennemann et al.[9] have shown that ascites tumor cells take up major amounts of cholesterol from the ascites fluid, and that their cholesterol de novo synthesis is rather low. Using the same cell line Haeffner et al.[10] have demonstrated that cultured ascites cells need cholesterol for growth which cannot be replaced by precursors of the biosynthetic pathway or by some growth factor-like substances, and that this compound affects the properties of the plasma membrane in several aspects. It was further shown that the cholesterol biosynthetic pathway is blocked at the stage of lanosterol conversion to cholesterol. The same defect has also been observed for kidney cells[11], human blood leukocytes[12] and human lymphocytes and granulocytes[13].

By what mechanism does cholesterol stimulate cell proliferation? Is it its properties as a necessary membrane constituent and modulator of membrane function? Does it act as a mitogen-like substance similar to phosphatidic acid[14], coupled to a receptor system which transmits its signals through the breakdown of inositol phospholipids. In a recent study on human erythrocytes M'Zali and Giraud[15] have observed that cholesterol can indeed affect the activity of phosphatidylinositol kinases and phosphodiesterases. In our system we have recently measured the phosphodiesterase activity acting against phosphatidyl-4-monophosphate (PIP) and phosphatidyl-4,5-biphosphate (PIP_2) and the PIP-specific kinase under cholesterol growth-promoting conditions. These enzymes are stimulated 2 to 4-fold supporting previous observations that the growth factor-like action of cholesterol may be explained by its modulation of enzyme activities involved in the synthesis and breakdown of polyphosphoinositides which is followed by a cascade of reactions ultimately leading to the initiation of DNA synthesis and cell proliferation.

MATERIALS AND METHODS

Hyperdiploid ascites tumor cells, type Karzel, were cultivated in minimum essential medium with Earle's salt, 1% HEPES, pH 7.4, and horse serum. Preparation of the cholesterol cocktail[16] and the cell culture studies were performed according to Haeffner et al.[10]

Plasma membranes were prepared and characterized from a 12.000 x g pellet by gel filtration on Sephacryl S-1000[17,18]. Incorporation of [^{3}H]-thymidine into DNA and of [^{14}C] acetate into the sterol fraction was performed as described by Haeffner et al.[10]. Lipids were extracted according to Folch et al.[19] and quantitation of cholesterol was done by gas-liquid chromatography[20]. The water-soluble inositol metabolites were separated by anion exchange chromatography[21]. The PI, PIP and PIP$_2$-specific phosphodiesterase was measured essentially as described by Wikiel and Strosznajder[22], and the PI and PIP kinase according to the procedure of Garret and Redman[23].

RESULTS

Initial studies about the relationship between serum concentration and ascites cell growth have shown that these cells can be cultivated at reduced serum concentrations (from 10% to 5%) without any significant reduction in growth properties. However, a reduction of the serum concentration below 5% results in a retardation and finally cessation of cell proliferation. Cells with 0.5% serum in the medium do not grow at all, whereas cells at 3% serum proliferate until about 5 days. Addition of some selected well-known growth factors such as insulin, transferrin and prostaglandins, either alone or in combination, have no stimulating effect. However, when cholesterol is added in a concentration of 2.50 mg/100 ml of medium, then we observe some growth in the case of 0.5% serum and nearly optimal growth in the case of 3% serum concentration. In the latter case, cells can be kept as long-term cultures (Fig. 1a). Supplementation of the medium with linoleic acid alone which is added in combination with cholesterol, does not stimulate cell growth at a substantial rate, which has also been found for normal and virus-infected chicken embryo fibroblasts[24]. Furthermore, reduction of the linoleic acid concentration to one-tenth of the amount normally added does not have any effect on cell growth. Also, the substitution of linoleic by other natural and unnatural fatty acids like oleic, palmitoleic, or linolelaidic acid, has no influence on the growth properties. The growth stimulation is specific for cholesterol. This has been investigated by replacing cholesterol with various precursors of the cholesterol-biosynthetic pathway such as mevalonic acid in mM concentrations, squalene and lanosterol. Neither of these precursors can imitate the growth-stimulating effect of cholesterol, and it has also been shown that the rate of proliferation is dependent upon the cholesterol concentration in the medium (Fig. 1b).

We have also measured macromolecule synthesis by thymidine incorporation into DNA in relation to the cholesterol concentration in the medium which is shown in Fig. 2. Saturation is obtained at about 50 µg of cholesterol per ml of medium. Table 1 gives a summary of the effect of other stimuli such as insulin, phorbol ester (TPA), lanosterol and a diglycerid analog, 1-oleoyl-2-acetyl-glycerol (OAG), on the rate of thymidine incorporation. DNA distribution within the cell cycle of cells cultured at 3% serum without and in the presence of cholesterol reveals an accumulation of cells in the mitotic and G$_2$ phase under the non-growth-promoting conditions.

The feedback control of cholesterol synthesis is known to be a complex phenomenon. We have measured the rate of cholesterol de novo synthesis from labeled acetate of cells grown in 3% serum medium either with or without cholesterol added to the medium. The results (Fig. 3) indicate that the rate

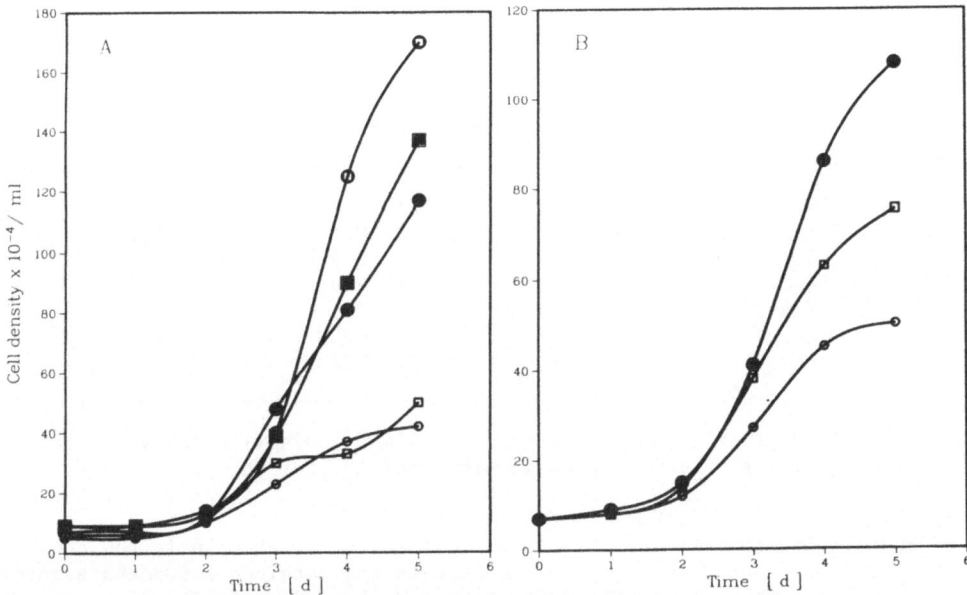

Fig. 1.   A) Growth curves of ascites cells. A, medium with 10% serum (O—O);
with 3% serum plus (■—■) and minus (O—O) cholesterol; the same as
before but with 0.25 instead of 2.5μg fatty acid (●—●); with 3%
serum plus either insulin, transferrin or prostaglandin (1 to 4
μg/10⁵ cells) (□—□). B) Growth curves of ascites cells cultured
at 3% serum concentration plus increasing amounts of cholesterol,
0.4 mg (O—O), 1.25 mg (□—□), and 2.5 mg (●—●) per 100 ml of medium.

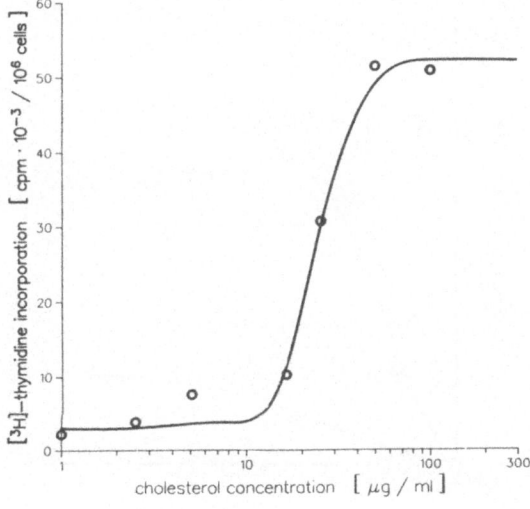

Fig. 2.   Incorporation of [³H]-thymidine into DNA in relation to the cholesterol concentration in the medium. Cells were first cultured in
low serum (24 h), then stimulated with cholesterol for another
24 h prior to thymidine labeling (30 min).

Table 1. Stimulation of DNA synthesis in quiescent ascites tumor cells

Condition	[^{3}H]-thymidine incorporation cpm x $10^3/10^6$ cells	Stimulation (fold)
2% serum	3.07 ± 1.6	1.0
+ cholesterol (25 µg/ml)	47.84 ± 20.8	15.6
+ OAG (20 µg/ml)	28.36 ± 2.1	9.2
+ insulin (2.5 µg/ml)	3.90 ± 2.8	1.3
+ TPA (5 x 10^{-8})	3.35 ± 2.1	1.1
+ lanosterol (25 µg/ml)	0.11 ± 0.03	0.04
10% serum	51.75 ± 6.2	16.9

Mean ± SD of at least 2 experiments. OAG = 1-oleoyl-2-acetyl-glycerol; TPA = 12-O-tetradecanoylphorbol-13-acetate.

of acetate incorporation is depressed by a factor of about 6 in the cholesterol-supplemented cells. Similar results are obtained for acetate incorporation into the entire sterol fraction and the radioactivity distribution between cholesterol precursors and the products of the branched pathways, ubiquinone and dolichol, is shown in Fig. 4. Unlike in liver, the major labeled products are found to be lanosterol, followed by squalene, and the

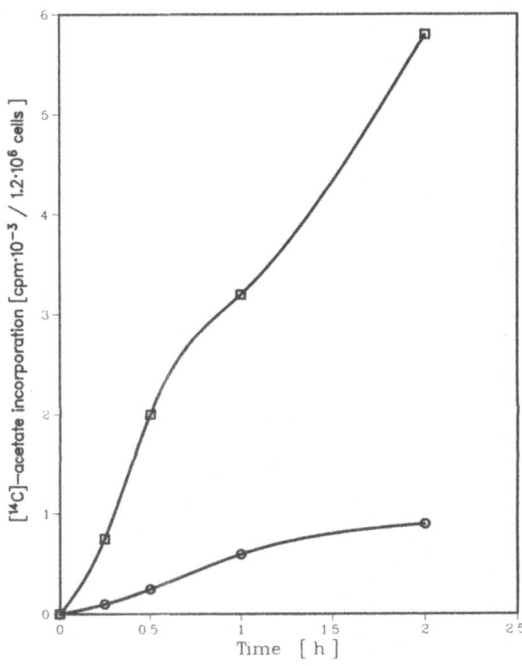

Fig. 3. Synthesis of cholesterol from labeled [^{14}C]acetate of cells cultivated at 3% serum concentration in the presence (O—O) and absence (□-□) of cholesterol. The curves represent the mean of 3 individual experiment.

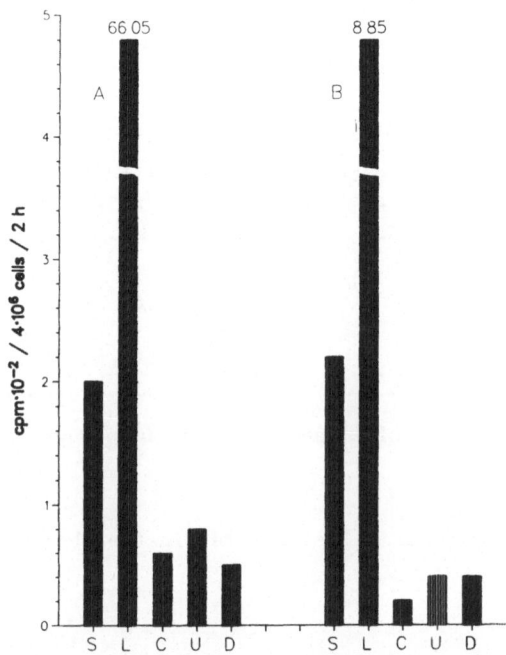

Fig. 4. [^{14}C]Acetate incorporation into the non-saponifiable lipid fraction. A, at 3% serum concentration without and B, in the presence
of cholesterol; 66.05 and 8.85 x 10^2, total incorporated radioactivity in cpm. S, squalene; L, lanosterol; C, cholesterol; U,
ubiquinone and D, dolichol.

lanosterol-labeling is about 7-fold higher in the cholesterol-depleted compared to the cholesterol-rich cells. From these data, it becomes clear that
a block exists in the sterol synthesis pathway at the stage of lanosterol
conversion to cholesterol. This also explains why neither mevalonic acid nor
squalene nor lanosterol added to the culture medium have a growth-stimulating
effect. The data in Fig. 4 further show that the main compound of the isoprenoid pathway does not seem to be cholesterol but ubiquinone. Surprisingly
radioactive-labeled squalene is about the same under both growth conditions.

Cholesterol is mainly localized in the plasma membrane, and therefore
its content should be influenced by changing the cholesterol concentration in
the culture medium (Table 2). Measurements of membrane fluidity with the
fluorescence depolarization technique reveal no significant changes between
the cholesterol-poor and cholesterol-rich membranes.

Nevertheless, since cholesterol is known to modulate the activity of
membrane-bound enzymes, we have investigated several enzymes which are involved in the synthesis and degradation of polyphosphoinositides. These compounds yield upon hydrolysis two second messenger molecules, 1,2-diacylgly-
cerol and inositol-1,4,5-trisphosphate. Table 3 gives the data for PI and
PIP kinase activity without and in the presence of extra cholesterol, and
Table 4 shows the data for the phosphodiesterase activity measured with all
inositol phospholipids as substrates. Significant stimulations of the PIP
kinase and the PIP and PIP$_2$-specific phophodiesterase are found under
cholesterol-supplemented conditions. For the latter enzyme a more than two-
fold stimulation is also observed with GTP present in the reaction mixture.
From these data it should further be expected that the metabolites of the
polyphosphoinositides, specially IP$_3$, are affected, which is shown in

Table 2. Cholesterol and phospholipid content, and the C/P molar ratio of the plasma membranes of ascites cells grown under normal and reduced serum concentrations (3%) without and with cholesterol supplementation

Condition	Cholesterol (μg/mg protein)	Phospholipid (μg/mg protein)	C/P (M)
10% serum	19.6	151.8	0.29
3% serum	9.0	108.0	0.18
+ cholesterol (25 μg/1 ml)	25.2	256.0	0.25

Data represent the average of at least 2 experiments with duplicate analyses.

Table 5. About twice as much IP$_3$ is detected in the cholesterol-stimulated cells.

DISCUSSION

Our results clearly demonstrate that ascites tumor cells need preformed cholesterol for growth which normally is supplied by the serum. The observation that cells at 0.5% serum plus exogenous cholesterol in the medium grow only to some extent indicates that other growth-promoting factors become limiting. These factors may be different from those growth factors which we have added to the medium and which have been shown to induce proliferation in ascites cells[25] and in others[26]. It has been known that the physical state of the lipid-protein matrix of the plasma membrane undergoes profound changes after stimulation of cell division and that the changes are correlated in one way or another to the presence of cholesterol[27,28].

The regulation of cholesterol biosynthesis and the homeostatic control of the cholesterol level are very complex mechanisms, still far from being

Table 3. Plasma membrane-bound PI and PIP kinase activity in non-stimulated and cholesterol-stimulated ascites cells

Condition	Kinase Activity (cpm/min/mg protein)	
	PI	PIP
2% serum	869 ± 76.7[a] (3)	220.7 ± 42.1 (3)
2% serum + cholesterol (25 μg/ml)	793 ± 146.7 (3)	705.0 ± 57.2 (3)

[a]Mean ± SD with number of experiments in parentheses.

480

Table 4. Plasma membrane-bound phospholipase C activity acting on PI, PIP and PIP_3 in non-stimulated and cholesterol-stimulated cells

Condition	Phospholipase C (nmol/min/mg protein)		
	PI	PIP	PIP_2
2% serum	2.34 ± 0.28 [a] (3)	3.31 ± 0.58 (3)	15.67 ± 2.09 (10)
2% serum + cholesterol (25 µg/ml)	2.26 ± 0.18 (3)	5.95 ± 0.36 (3)	26.26 ± 3.0 (8)

[a] Mean ± SD with number of experiments in parentheses.

resolved. For liver and also some hepatomas[29], it is known that the ß-hydroxymethylglutaryl coenzyme A reductase is the key enzyme in the sterol synthesis pathway which is controlled through a multivalent feedback mechanism, involving cholesterol itself and possibly also a nonsterol compound[30]. With respect to the involvement of cholesterol, there are at least 3 independently operating negative-feedback mechanisms, which have been described by Sabine[31]. Taking into account the low cholesterol-synthetic capacity of these ascites cells, feedback control by endogenous cholesterol can almost be excluded. It also remains to be investigated whether or not a nonsterol compound plays any role in this respect, or whether a second rate-limiting step exists besides the ß-hydroxymethylglutaryl coenzyme A reductase controlled reaction, since the amount of radioactive squalene is about the same both in the presence and absence of cholesterol.

Cultivation of our ascites cells under cholesterol-rich and -depleted conditions cause significant changes in the content of this molecule as well as in the C/P ratio of the isolated plasma membranes. The total lipid phosphorus content is also affected, but the relative proportions of the individual phospholipids are not altered to a major extent between the two cell cultures (data not shown). In spite of some changes in the C/P ratio between the plasma membranes of the cholesterol-rich and cholesterol-poor cells, we

Table 5. Radioactive labeling of inositol-1,4,5-trisphosphate (IP) of growth-stimulated ascites cells

Condition	IP_3 cpm/5 x 10^6 cells	Stimulation (fold)
2% serum	348 ± 109 [a]	1.0
+ cholesterol (25 µg/ml)	691 ± 160	2.0
+ OAG (20 µg/ml)	576 ± 217	1.7
10% serum	2255 ± 525	6.5

[a] Mean ± SD of 3 to 5 experiments.

could not find any significant alterations in membrane lipid fluidity. There is some tendency for a rigidifying effect of the extra cholesterol in the higher temperature range beginning at about 30°C. These results indicate that these membranes apparently have the ability to control their fluidity, e.g. by releasing phospholipids in the case of a low cholesterol supplementation and vice versa. With respect to membrane lipid fluidity which was measured on the entire membrane, it cannot be excluded that differences do exist between the outer or inner monolayers since cholesterol is asymmetrically distributed between the two layers[32]. If this turns out to be the case, it can explain the stimulatory effect of cholesterol upon the PIP kinase and the phosphodiesterase.

ACKNOWLEDGEMENTS

This work was supported in part by a grant from the Deutsche Forschungsgemeinschaft (Ha 666/2-1). J. B. S. as a guest scientist was supported by a stipend from the German Cancer Research Center.

REFERENCES

1. R. Holmes, J. Helms and G. Mercer, Cholesterol requirement of primary diploid human fibroblasts, J. Cell Biol. 42:262 (1969).
2. L. Kahane and S. Razin, Cholesterol-phosphatidylcholine dispersions as donors of cholesterol to mycoplasma membranes, Biochim. Biophys. Acta 471:32 (1977).
3. D. Monard, M. Rentsch, Y. Schuerch-Rathgeb and R. M. Lindsay, Morphological differentiation of neuroblastoma cells in medium supplemented with delipidated serum, Proc. Natl. Acad. Sci. USA 74:3893 (1977).
4. J. M. Odriozola, E. Waitzkin, T. L. Smith and K. Bloch, Sterol requirement of mycoplasma capricolum, Proc. Natl. Acad. Sci. USA 75:4107 (1978).
5. R. E. Ostlund, Jr. and J. W. Yang, Effect of cholesterol and growth factors on the proliferation of cultured human skin fibroblasts, Exp. Cell Res. 161:509 (1985).
6. E. Oldfield and D. Chapman, Dynamics of lipids in membranes: Heterogeneity and the role of cholesterol, FEBS Lett. 23:285 (1972).
7. P. L. Yeagle, Cholesterol modulation of $(Na^+ + K^+)$-ATPase ATP hydrolyzing activity in the human erythrocyte, Biochim. Biophys. Acta 727:39 (1983).
8. P. J. Scarpace, S. J. O'Connor and I. B. Abrass, Cholesterol modulation of beta-adrenergic receptor characteristics, Biochim. Biophys Acta 845:520 (1985).
9. D. E. Brennemann, L. McGee and A. A. Spector, Cholesterol metabolism in the Ehrlich ascites tumor, Cancer Res. 34:2605 (1974).
10. E. W. Haeffner, C. J. K. Hoffmann, M. Stoehr and H. Scherf, Cholesterol induced growth stimulation, cell aggregation and membrane properties of ascites tumor cells in culture, Cancer Res. 44:2668 (1984).
11. K. H. Hellstrom, M. D. Sipperstein, L. A. Bricker and L. J. Luby, Studies of the in vivo metabolism of mevalonic acid in the normal rat, J. Clin. Invest. 52:1303 (1973).
12. P. Burns, I. R. Welshman, J. Edmond and A. A. Spector, Evidence for rate-limiting steps in sterol synthesis beyond 3-hydroxy-3-methylglutaryl-coenzyme A reductase in human leukocytes, Biochim. Biophys. Acta 572:345 (1979).
13. P. Burns, I. R. Welshman, T. J. Scallen and A. A. Spector, Mechanism of defective sterol synthesis in human leukocytes, Biochim. Biophys. Acta 713:519 (1982).

14. W. H. Moolenar, W. Kruijer, B. C. Tilly, I. Verlaan, A. J. Bierman and S. W. de Laat, Growth factor-like action of phosphatidic acid, Nature (Lond.) 323:171 (1986).

15. H. M'Zali and F. Giraud, Phosphoinositide reorganization in human erythrocyte membrane upon cholesterol depletion, Biochem. J. 234:13 (1986).

16. L. M. Corwin, I. P. Humphrey and J. Shloss, Effect of lipids on the expression of cell transformation, Exp. Cell Res. 108:341 (1977).

17. E. W. Haeffner, K. Kolbe, D. Schroeter and N. Paweletz, Plasma membrane heterogeneity in ascites tumor cells. Isolation of a light and a heavy membrane fraction of the glycogen-free Ehrlich-Lettr substrain, Biochim. Biophys. Acta 603:306 (1980).

18. E. W. Haeffner, A. Holl and D. Schroeter, Preparation of two plasma membrane fractions from ascites tumor cells by gel chromatography on Sephacryl S-1000, J. Chromatogr. 382:107 (1986).

19. J. Folch, M. Lees and G. H. Sloane-Stanley, A simple method for the isolation and purification of total lipids from animal tissues, J. Biol. Chem. 226:497 (1957).

20. E. W. Haeffner and C. J. K. Hoffmann, Direct quantitation of free cholesterol from total serum lipid extracts by computer-assisted gas liquid chromatography, J. Chromatogr. 228:268 (1982).

21. M. J. Berridge, R. M. C. Dawson, C. P. Downes, J. P. Heslop and R. F. Irvine, Changes in the levels of inositol phosphates after agonist-dependent hydrolysis of membrane phosphoinositides, Biochem. J. 212:473 (1983).

22. H. Wikiel and J. Strosznajder, Phospatidylinositol degradation in ischemic brain specifically activated by synaptosomal enzymes, FEBS Lett. 216:57 (1987)

23. R. J. B. Garrett and C. M. Redman, Localization of enzymes involved in polyphosphoinositide metabolism on the cytoplasmic surface of the human erythrocyte membrane, Biochim. Biophys. Acta 382:58 (1975).

24. H. A. Hale, J. E. Pessin, F. Palmer, M. J. Weber, and M. Glaser, Modification of the lipid composition of normal and Rous sarcoma virus-infected cells, J. Biol. Chem. 252:6190 (1977).

25. W. Lehmann, H. Graetz, M. Schütt, and P. Langen, Antagonistic effects of insulin and a negative growth regulator from ascites fluid on the growth of Ehrlich ascites carcinoma cells in vitro, Exp. Cell Res. 119:396 (1979).

26. D. Barnes and G. H. Sato, Growth of human mammary tumor cell line in a serum-free medium, Nature (Lond.) 281:388 (1979).

27. M. M. Burger, Cell surfaces in neoplastic transformation, in: "Current Topics in Cellular Regulation", B. L. Horecker and E. R. Stadtman, eds., Academic Press, New York (1971).

28. K. Bloch, Sterol structure and membrane function, in: "Current Topics in Cellular Regulation, Biological Cycles", R. W. Estabrook and P. Srere, eds., Academic Press, New York (1981).

29. J. Avigan, C. D. Williams, and J. P. Blass, Regulation of sterol synthesis in human skin fibroblast cultures, Biochim. Biophys. Acta 218:381 (1970).

30. M. S. Brown and J. L. Goldstein, Multivalent feed-back regulation of HMG-CoA reductase, a control mechanism coordinating isoprenoid synthesis and cell growth, J. Lipid Res. 21:505 (1980).

31. J. R. Sabine, Defective control of cholesterol synthesis and the development of liver cancer: a review, in: " Tumor Lipids: Biochemistry and Metabolism", R. Wood, ed., American Oil Chemists' Society, Champaign (1973).

32. F. Schroeder, Use of fluorescent sterol to probe the transbilayer distribution of sterols in biological membranes, FEBS Lett. 135:127 (1981).

SUSCEPTIBILITY OF GRC-BEARING RATS TO DEN AND ITS RELATIONSHIP TO THE HMP PATHWAY

Mona F. Melhem, Kalipatnapu N. Rao, Heinz W. Kunz and
Thomas J. Gill III

Department of Pathology, University of Pittsburgh School
of Medicine, Pittsburgh, PA 15261, and the Veterans
Administration Medical Center, Pittsburgh, PA 15240

INTRODUCTION

The genetic factors underlining the susceptibility to cancer have been examined in a number of experimental[1] and clinical[2] settings, and there is substantial evidence for the existence of such factors. Studies in both rats[3] and mice[4] have provided evidence that genes linked to the major histocompatibility complex (MHC) play an important role in the susceptibility to cancer following exposure to chemical carcinogens. Previously, we[3] showed that two strains that differ in their MHC's (RT1) and in the presence of the MHC-linked growth and reproduction complex (grc) differed in their susceptibility to the chemical carcinogen N-2-acetylaminofluorene (AAF). The R10 ($RT1.A^n B^l D^l E^-$ grc) and BY1 ($RT1.A^l B^l D^l E^-$ grc) strains were highly susceptible to AAF, whereas the BI ($A^n B^a D^a E^u$ grc$^+$) and BY2 ($A^u B^u D^u E^u$ grc$^+$) strains were not. These findings indicated that genes in, or linked to, the MHC were involved in the increased susceptibility to preneoplastic changes and suggest that the presence of the grc is critical in determining this susceptibility.

Biochemical studies[5] suggested that changes in cholesterol synthesis, DNA synthesis and the hexose monophosphate (HMP) pathway, and the critical enzymes involved therein: 3-hydroxy-3-methyl-glutaryl coenzyme A (HMG-CoA) reductase, glucose-6-phosphate dehydrogenase (G6PD), and 6-phosphogluconate dehydrogenase (6PGD), may be important elements in the differential susceptibility of the two strains. The more susceptible R10 strain had an increased cholesterol synthesis, increased DNA synthesis and stimulation of the HMP pathway compared to the less susceptible BI strain. An extensive body of work from our laboratory and by others has demonstrated enhanced de novo cholesterogenesis and HMP pathway activity in fetal tissues, regenerating normal pancreas and liver[5,6], lead-induced hyperplasia[7],preneoplastic foci in the liver[8,9], and malignant tissues[5,10]. Since NADPH generated through the HMP pathway is utilized for cholesterol and DNA synthesis, G6PD appears to play an important role in cell proliferation[11]. Indeed, increased levels of G6PD correlate with an increased incidence of cancer in experimental animals[3]. On the other hand, inhibition of G6PD activity was shown to reduce susceptibility to carcinogens in experimental animals[12], and G6PD deficiency may offer protection against the development of cancer in humans[13,14].

The present study was undertaken to expand our initial observations

485

about the role of the grc in the differential susceptibility of rats to chemical carcinogens and to characterize various inbred strains of rats for their potential use in further investigation of the biochemical basis of carcinogenesis. The specific objectives were four-fold. First, a carcinogen other than AAF was used in order to test the generality of the difference in susceptibility between grc-bearing rats and their wild type counterparts (grc^+). The system chosen was induction by diethylnitrosamine (DEN) and feeding either a choline-deficient (CD) or choline-supplemented (CS) diet, since the combination of a single necrogenic dose of DEN and a CD diet is an effective way of inducing preneoplastic foci, nodules and hepatomas in rats[15]. Second, we wanted to determine whether the effect of the grc could be identified more clearly by comparing strains that had the same MHC but differed at the grc. The strains so identified were R16 (A^a B^a D^a E^- grc) and ACP (A^a B^a D^a E^- grc^+). Third, a variety of strains were analyzed in order to find one that would have the same biochemical level of activity in the HMP pathway as the R16 strain but did not carry the grc in order to use it for comparison in the induction of hepatocarcinoma in the R16 strain: the YO strain was identified as the appropriate one. Finally, the effect of MHC-linked genes on G6PD and 6PGD activities was tested using MHC-congenic strains in which the donor and background strains differed in the level of activities of these enzymes.

MATERIALS AND METHODS

Animals

Male rats of 26 strains and 4 to 60 weeks of age from our colony at the University of Pittsburgh School of Medicine were used. For the hepatocarcinogenesis experiment four strains of rat were used: R16, which is a recombinant (A^a B^a D^a E^- grc) between R10 (A^n B^ℓ D^ℓ E^- grc) and ACP (A^a B^a D^a E^- grc^+); ACP, which is its normal counterpart; and BN and YO which have tissue G6PD activities comparable to the ACP and R16 strains, respectively.

Two types of experiment were conducted. In the first experiment, a group of rats was fed laboratory chow throughout the experiment (9 months) to evaluate the response of different strains to the same carcinogen without any promotion. In the second experiment, the rats were fed laboratory chow until they were placed on a choline-deficient diet (CD) two weeks after initiation with diethylnitrosamine (DEN), in order to evaluate their response to a promoting regimen. They had free access to food and water and were housed in an air-conditioned room with a 12 h light (7 a.m. to 7 p.m.) and dark (7 p.m. to 7 a.m.) cycle. At various times, the animals were weighed, anesthetized with fluorothane and sacrificed (at 9 a.m.) by bleeding through the abdominal aorta.

Glucose-6-Phosphate Dehydrogenase and 6-Phosphogluconate Dehydrogenase Assays

Blood was collected in ethylenediaminetetraacetate (EDTA), and the red blood cells (RBC) were separated immediately[16]. Livers were resected, washed, weighed and homogenized in 9 volumes of cold isotonic saline in a Polytron PT 10 Sonicator at setting 7 for 30 sec. The homogenates were centrifuged for 60 min at 100,000 x g, and the supernatants were used immediately for the assay of G6PD and 6PGD as described[17]. The units of activity in all cases were expressed as nmoles of pyrimidine nucleotide produced per min, and the specific activities were expressed as units per 10^{10} RBC or per mg protein (liver).

Induction and Promotion of Hepatocarcinogenesis

Induction of preneoplastic and neoplastic lesions in the liver was investigated after exposure of rats to a single necrogenic dose of DEN[13]. Six to eight week old R16, ACP, BN and YO male rats were injected intraperitoneally (i.p.) with a single dose of DEN (200 mg/kg of body wt.). After two weeks, the rats were divided into two groups: one was fed a choline supplemented (CS) diet, and the other was fed a choline-deficient (CD) diet[18]. The animals were sacrificed at different time intervals and complete autopsies were performed. Blocks of liver tissue and tumors were prepared for histological examination (hematoxylin-eosin, H & E) and for histochemical staining with gamma-glutamyltranspeptidase (GGT)[19]. Livers were examined microscopically for cirrhosis, nodules, fatty infiltrate and tumors. Sections were scored for GGT-positive foci, bile duct proliferation, disruption of hepatic architecture and degree of cellular atypia using a scale of 0 to ++++ [19].

Morphometric Measurements

Foci of GGT-positive hepatocytes were readily distinguishable in liver sections histochemically as discrete areas of intensely orange-brown cells on a blue hematoxylin background. The areas of the sections and of the foci were measured by tracing their perimeters with the transverse arm of an electronic digitizer/planimeter interfaced with a computer by means of an acoustic coupler (Numonics Corp., Lansdale, PA). The area of the liver section was measured at 5x and that of the foci at 40x. The results were expressed as the diameter (mean ± SD) of the foci in nm and as their number (mean ± SD) per cm^2 of liver section.

Other Procedures

Total protein was estimated by the method of Lowry et al.[20] using bovine serum albumin as the standard. Statistical analysis of the data was performed using an analysis of variance, and the difference between means was considered significant if $P < 0.05$ [21].

RESULTS

A variety of inbred and congenic strains were analyzed for G6PD levels in the liver and red blood cells and for 6PDG levels in the liver in order to identify those strains that would be useful for comparison with the R16 and ACP strains (Fig. 1, Table 1). There was approximately a four-fold variation in liver G6PD activity levels and a three-fold variation in red blood cell

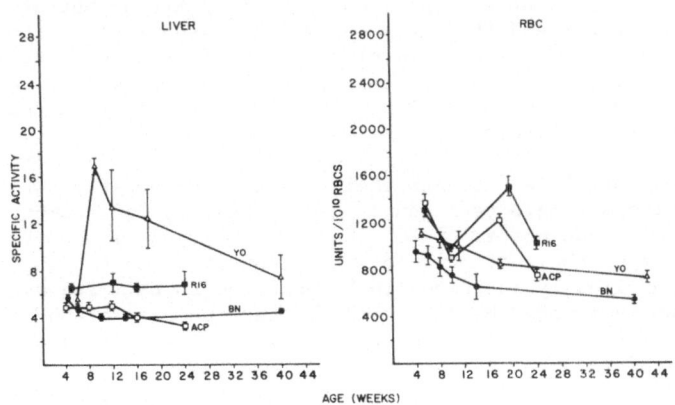

Fig. 1. The specific activity of G6PD in the liver and red blood cells of the R16, ACP, YO and BN strains as a function of age.

Table 1. G6PD and 6PGD activities in liver and red blood cells of 12 to 16 week old male rats [a]

Strain	Body weight (g)	G6PD [b]		6PGD [b]
		Liver	RBC	Liver
OKA	249 ± 4	3.8 ± 0.1	996 ± 17	13.6 ± 0.3
BN	265 ± 7	4.0 ± 0.2	791 ± 61	11.7 ± 0.3
ACP	231 ± 3	4.5 ± 0.3	1333 ± 51	19.5 ± 1.2
DA	278 ± 5	4.6 ± 0.3	779 ± 36	17.5 ± 1.6
WF	267 ± 5	5.3 ± 0.3	820 ± 30	16.9 ± 0.8
F344	372 ± 31	6.0 ± 1.4	961 ± 10	15.4 ± 2.4
BUF	427 ± 64	6.3 ± 0.6	658 ± 38	18.2 ± 1.4
KGH	281 ± 2	6.8 ± 0.3	560 ± 36	13.9 ± 1.3
WKA	310 ± 19	6.9 ± 1.3	680 ± 41	14.6 ± 0.7
PVG	299 ± 4	6.9 ± 0.1	968 ± 60	20.8 ± 0.8
R16	190 ± 7	8.3 ± 0.8	1323 ± 80	16.9 ± 0.7
MR	284 ± 11	8.9 ± 0.6	684 ± 30	20.7 ± 1.7
NBR	268 ± 8	9.1 ± 0.7	852 ± 6	21.4 ± 1.4
LEW	403 ± 5	9.9 ± 1.0	388 ± 40	18.8 ± 2.4
TAL/K	292 ± 18	10.4 ± 1.1	494 ± 18	20.1 ± 0.6
AUG	259 ± 1	13.8 ± 1.6	1189 ± 46	39.4 ± 3.3
YO	325 ± 4	16.3 ± 1.8	704 ± 117	28.4 ± 2.4

[a] Each value is the mean ± SEM of 7-12 animals. [b] Values are in units as defined under Materials and Methods. G6PD, glucose-6-phosphate dehydrogenase, and 6PDG, 6 phosphogluconate dehydrogenase.

G6PD activity levels and liver 6PGD activity levels when the strains were assayed at 12-16 weeks of age. The differences in the G6PD activity levels among the strains persisted with age in all of the tissues studied (liver and red blood cells) (Table 2). No significant G6PD deficiency was observed in any strain.

The G6PD levels that were optimal for comparison with R16 and ACP were found in the YO and BN strains, and they were studied in greater detail. The growth patterns of all four strains showed the same trends but at different levels: the body weight and the liver weight increased rapidly for the first twelve weeks and then increased more slowly. The fastest growing strain was YO, and the ACP strain grew more rapidly than the R16 strain. The relative growth of the liver in the YO and R16 strains was greater than that in the BN and ACP strains, respectively.

The response of R16 and YO and their normal counterparts ACP and BN at 9 months following a single necrogenic dose of DEN without promotion is presented in Table 3. The number of GGT-positive foci per cm^2 and the area of the foci per area of the section were significantly increased in R16 compared to ACP and was marginally increased in YO compared to BN. The average diameters of all the foci were the same. The number and size of GGT-positive foci in R16 and ACP after initiation with a single intraperitoneal dose of DEN and 9 weeks of CD promotion are presented in Table 4. The R16 strain consistently showed a significantly greater number of GGT-positive foci, larger foci area/section area and increased diameter of foci both with promotion (CD) and without promotion (CS) when compared to the ACP strain.

488

Table 2. Effect of age on the G6PD and 6PGD activities in liver and red blood cells of male rats from MHC-congenic strains [a]

Age	4-9 weeks			12-16 weeks			24 weeks		
Strain	G6PD [b]		6PGD [b]	G6PD		6PGD	G6PD		6PGD
	RBC	liver	liver	RBC	liver	liver	RBC	liver	liver
BN	909 ± 35	4.9 ± 0.2	14.1 ± 0.5	791 ± 61	4.0 ± 0.2	11.7 ± 0.3	510 ± 25	4.5 ± 0.1	11.8 ± 0.1
YO	1113 ± 54	11.2 ± 2.6	21.7 ± 2.1	704 ± 117	16.3 ± 1.8	28.4 ± 1.6	806 ± 32	14.5 ± 3.4	26.3 ± 1.8
BN.1U(YO)	814 ± 87	5.1 ± 0.3	14.9 ± 1.6	501 ± 41	3.5 ± 0.2	11.2 ± 0.2	620 ± 40	3.1 ± 0.1	10.6 ± 0.4
KGH	905 ± 93	8.1 ± 0.7	19.1 ± 2.1	560 ± 36	6.8 ± 0.3	13.9 ± 1.3	446 ± 79	6.0 ± 0.7	11.9 ± 0.6
BN.1K(KGH)	694 ± 26	4.2 ± 0.2	13.1 ± 0.8	624 ± 52	3.6 ± 0.3	10.3 ± 0.5	520 ± 52	4.6 ± 0.3	12.4 ± 0.3

[a] Each value is the mean ± SEM of 7-12 animals. [b] Values are in units as defined under Materials and Methods. G6PD, glucose-6-phosphate dehydrogenase, and 6PGD, 6-phosphogluconate dehydrogenase.

Table 3. Interstrain difference by number and size of GGT positive foci in the liver at 9 months after DEN[a] initiation. Rats fed laboratory chow

Strain	Number of foci per cm^2	Foci area per section area	Average diameter (nm)
R16	49.3 ± 22.3[b]	2.95 ± 1.3[b]	265 ± 2
ACP	2.4 ± 0.5[b]	0.16 ± 0.02[b]	282 ± 13
YO	6.6 ± 4.5[c]	0.40 ± 0.20	278 ± 30
BN	2.7 ± 1.0[c]	0.10 ± 0.05[c]	241 ± 32

[a]Values are mean ± SEM for 4-6 animals. [b]Significantly different from R16 (P < 0.05). [c]Significantly different from YO (P < 0.05).

The morphologic and histologic findings are summarized in Table 5. The R16 strain showed severe cellular atypia, cirrhosis and hyperplastic nodules and finally hepatoma when fed a CD diet for 6 months. This response is markedly different from that of its normal counterpart, ACP. On the other hand, when fed a CS diet, the R16 strain showed only severe cellular atypia when compared to ACP.

DISCUSSION

In previous studies[3] we showed that rats carrying the grc genes (R10) were highly susceptible to the induction of GGT-positive foci in the liver after feeding a diet containing 0.02% N-2-acetylaminofluorene (AAF) for 4 weeks. When compared with the normal grc[+] rats, they showed elevated G6PD activity, de novo cholesterogenesis and DNA synthesis in the liver.

G6PD is a rate limiting enzyme in the HMP pathway of glucose metabolism and plays an important role in cell proliferation[8]. Elevated G6PD activities

Table 4. Number and size of GGT foci in the livers of R16 and ACP rats given DEN and placed on a CD or a CS diet for 9 weeks[a]

Strain	Diet	Number of foci per cm^2	Foci area per section area	Average diameter (nm)
R16	CD[b]	12.9 ± 5.5	0.55 ± 0.17	231 ± 14
ACP	CD	4.8 ± 0.4[c]	0.15 ± 0.05[c]	195 ± 6[c]
R16	CS[b]	5.1 ± 1.1	0.23 ± 0.1	227 ± 22
ACP	CS	1.1 ± 0.3[c]	0.02 ± 0.01[c]	158 ± 12[c]

[a]Values are mean ± SEM for five animals each. [b]CD, choline-deficient; and CS, choline-supplemented. [c]Significantly different from R16 (P < 0.05).

Table 5. Morphologic assessment of R16 and ACP with CD or CS diets at 6 months after DEN[a]

Strain	Diet[b]	Fatty infiltrate	Disruption of lobular architecture	Bile duct proliferation	Cellular atypia	Hyperplastic nodules	Oncocytic nodules	Cirrhosis	Hepatocellular carcinoma
R16	CD	Severe	+++	++	++++	4/6	3/6	4/6	2/6
ACP	CD	Minimal	+	+	±	0	-	-	-
R16	CS	Patchy	+++	+	++++	0	-	-	-
ACP	CS	Minimal to moderate	0	+	+	0	-	-[c]	-

[a] Each group contained 6 animals. Histological characteristics are scored on a scale of 0 to ++++ on liver sections stained with hematoxylin and eosin. [b] CD, choline-deficient; and CS, choline-supplemented. [c] Mild portal fibrosis was present.

491

were shown to occur in rapidly growing normal, as well as in preneoplastic and neoplastic tissues. Inhibition of G6PD activity has been shown to inhibit neoplastic growth[12]. Surveying several strains of rats disclosed a number of strains with high G6PD activity in the liver that appeared to correlate with the degree of cell proliferation. Since cell proliferation is a prerequisite for neoplastic growth[9], animals with high G6PD levels and a higher rate of cell proliferation should be more susceptible to cancer. Indeed, both the YO and R16 strains showed more GGT-positive foci in the liver even without promotion. The CD diet promotes the development of hepatocellular carcinoma[19], and this diet was used to test the role of the grc in susceptibility to cancer. On a CD diet, hepatocellular carcinoma developed 24 weeks after initiation in the R16 strain, while the ACP strain was free of tumor.

The enhanced susceptibility of grc-bearing rats to the development of preneoplastic hepatocellular nodules is now firmly established with two different carcinogens, AAF and DEN. The biochemical pathways that may be involved in the development of cancer are the HMP pathway and the synthetic pathway for cholesterol. The question being actively pursued involves the role of grc in modulating these two metabolic pathways. Another aspect of the work is the relationship between embryogenesis and carcinogenesis and the role of grc genes in growth and development[22].

ACKNOWLEDGEMENTS

This work was supported by grants from the National Institutes of Health (CA 18659 and HD 08662) and from the National Dairy Council of America

REFERENCES

1. H. C. Pitot, Neoplasia: A somatic mutation or heritable change in cytoplasmic membranes, J. Natl. Cancer Inst. 53:905 (1974).
2. W. T. London, Primary hepatocellular carcinoma: Etiology, pathogenesis and prevention, Hum. Pathol. 12:1085 (1981).
3. K. N. Rao, H. Shinozuka, H. W. Kunz and T. J. Gill III, Enhanced susceptibility to a chemical carcinogen in rats carrying MHC-linked genes influencing development (GRC), Int. J. Cancer 34:113 (1984).
4. D. W. Nebert and N. M. Jensen, The ah locus: Genetic regulation of the metabolism of carcinogens, drugs, and other environmental chemicals by cytochrome p450-mediated monoxygenase, CRC Crit. Rev. Biochem. 6:401 (1979).
5. K. N. Rao, S. Kottapally and H. Shinozuka, Acinar cell carcinoma of rat pancreas. Mechanism of deregulation of cholesterol metabolism, Toxicol. Path. 12:62 (1984).
6. K. N. Rao, S. Kottapally and H. Shinozuka, Lipid composition and 3-hydroxy-3 methylglutaryl CoA reductase activity of acinar cell carcinoma of rat pancreas, Biochim. Biophys. Acta 759:74 (1983).
7. P. Pani, S. Dessi, K. N. Rao, B. Batetta and E. Laconi, Changes in serum and hepatic cholesterol in lead induced liver hyperplasia, Toxicol. Pathol. 12:162 (1984).
8. G. M. Ledda-Columbano, A. Columbano, S. Dessi, P. Coni, C. Chiodino and P. Pani, Enhancement of cholesterol synthesis and pentose phosphate pathway activity in proliferative hepatocyte nodules, Carcinogenesis 6:1371 (1985).
9. R. Schulte-Herman, Tumor promotion in the liver, Arch. Toxicol. 57:147 (1985).
10. H. J. Hacker, M. A. Moore, D. Mayer and P. Bannash, Correlative histochemistry of some enzymes of carbohydrate metabolism in preneoplastic and neoplastic lesions in the rat liver, Carcinogenesis

3:1265 (1982).

11. K. N. Rao, Regulatory aspects of cholesterol metabolism in cells with different degrees of replication, Toxicol. Pathol. 14:430 (1986).

12. A. G. Schwartz, Inhibition of spontaneous breast cancer formation in female C3H(A^{vy}/a) mice by long-term treatment with dehydroepiandrosterone, Cancer Res. 39:1129 (1979).

13. P. Beaconsfield, R. Rainsbury and G. Kalton, Glucose-6-phosphate dehydrogenase deficiency and the incidence of cancer, Oncologia 19:11 (1965).

14. S. N. Naik and D. E. Anderson, The association between glucose-6-phosphate dehydrogenase deficiency and cancer in American Negroes, Oncology 25:356 (1971).

15. H. Shinozuka and B. Lombardi, Synergistic effect of a choline-devoid diet and phenobarbital in promoting the emergence of foci of gamma-glutamyltranspeptidase-positive hepatocytes in the liver of carcinogen-treated rats, Cancer Res. 40:3846 (1980).

16. K. Betke, E. Keutler, G. J. Brewer, H. N. Kirkman, L. Luzzatto, A. G. Motulsky, B. Ramst and M. Simiscalco, Standarization procedures for the study of G6PD, Report, WHO Scientific Group, WHO Tech. Rep. Ser. No. 366 (1967).

17. Worthington Enzyme Manual, Worthington biochemical Corporation, Freehold, New Jersey, pp. 212 (1979).

18. K. N. Rao, J. Tuma and B. Lombardi, Acute hemorrhagic pancreatic necrosis in mice: Intraparenchymal activation of zymogens and other enzyme changes in pancreas and serum, Gastroenterology 70:720 (1976).

19. H. Shinozuka, M. A. Sells, S. L. Katyal, S. Sell and B. Lombardi, Effects of choline devoid diet on the emergence of gamma-glutamyl transpeptidase positive foci in the liver of carcinogen treated rats, Cancer Res. 39:2515 (1979).

20. O. H. Lowry, N. J. Rosebrough, A. L. Farr and R. J. Randall, Protein measurement with folin phenol reagent, J. Biol. Chem. 193:265 (1951).

21. R. G. D. Steel and J. H. Torrie, "Principles and procedures of statistics: a biomedical approach", 2nd ed., pp. 137, McGraw Hill, New York, (1980).

22. T. J. Gill III, The borderland of embryogenesis and carcinogenesis. Major histocompatibility complex-linked genes affecting development and their possible relationship to the development of cancer, Biochim. Biophys. Acta 738:93 (1984).

LIPOPROTEINS, CELL PROLIFERATION AND CANCER

Kalipatnapu N. Rao, Hoda F. Gabriel, Elhamey D. Eskander
and Mona F. Melhem

Department of Pathology, School of Medicine, University of
Pittsburgh, Pittsburgh, PA 15261, and The Veterans
Administration Medical Center, Pittsburgh, PA 15240

INTRODUCTION

In experimental animals, diets high in polyunsaturated fatty acids (USF)
greatly promote tumorigenesis relative to the same animals fed saturated fat
(SF)[1-6]. It appears that the tumor-promoting properties of a high fat diet
are more a function of fatty acid composition than of fat content per se or
total caloric intake[5]. Various mechanisms that were examined to explain the
promotion of tumorigenesis by USF diet include: alterations in hormone levels, membrane fluidity, intracellular communication, prostaglandins, protein
kinases, immune system and cell proliferation[5]. In spite of studies by cancer biologists and nutritionists, the exact mechanism(s) by which USF diets
promote tumorigenesis is not well understood. It is likely that the tumor
promoting properties of high USF diets may be related to their capacity to
eliminate bile acids, stimulate de novo cholesterogenesis and decrease serum
cholesterol ester levels[7]. Moreover, in various experimental tumor model
systems[3-5] and in epidemiological studies[8], the serum cholesterol and triglyceride levels were found to be reduced. However, a cause and effect
relationship between low serum low density lipoproteins (LDL) and promotion
of cancer has not been explored thoroughly. Nonetheless, it is widely accepted that there is a cause and effect relationship between serum LDL and
coronary heart disease (CHD) and that a reduction in serum LDL will greatly
reduce the risk of CHD[9].

We present evidence suggesting that serum lipoproteins not only transport cholesterol, phospholipids and triglycerides in the blood, facilitate
movement of lipids between various tissues, and regulate lipid synthesis and
catabolism, but also regulate immune function and cell proliferation. We
suggest that reduced serum LDL levels may provide the mitogenic signal to
promote tumorigenesis.

LIPOPROTEINS AND IMMUNE FUNCTION

Dietary lipids influence immune function as well as tumorigenesis, and
by modifying the diet it is possible to alter the progression of lymphoproliferative disorders as well as tumorigenesis[10]. Relative to SF diets, diets
high in USF are immunosuppressive[11-14]. Feeding USF diets results in the loss
of an animal's ability to reject skin allografts[15-19]. USF diets are used as

adjuvants for immunosuppression in renal transplant patients[13,16,20,21] and in patients suffering from multiple sclerosis[22]. These diets also suppress delayed hypersensitivity reactions in cancer patients[23]. Guinea pigs challenged with an antigen and fed USF diets show lower antibody titres than those fed saturated fat[24].

The immunosuppression by USF diets can be linked to changes in serum lipoproteins. Indeed various studies indicate that chylomicrons, very low density lipoproteins (VLDL), intermediate density lipoproteins (IDL) and LDL inhibit lymphocyte proliferation in increasing order and high density lipoproteins (HDL) had no effect[25,26]. The regulation of lymphocyte proliferation by LDL has been studied extensively[25,27-32], and these studies showed that LDL decreases proliferation of phytohemagglutinin (PHA)-stimulated lymphocytes by first inhibiting de novo cholesterogenesis and subsequently DNA synthesis[32]. LDL was found to affect clonal expansion but had no effect on subsequent events such as differentiation and maturation[32].

The immune response to antigenic stimulation needs the cooperation of macrophages, T and B lymphocytes, and production of various differentiation and maturation factors, including interleukin-1 and interleukin-2. Activation of B-lymphocytes by antigen has two consequences: proliferation that results in clonal expansion and differentiation and maturation that results in antibody-secreting cells. In the first situation, the B-lymphocyte goes from the resting G_0 phase to the S phase. While activation by an antigen results from the direct contact between B and T cells, which is MHC restricted, the activation of B cells by T cell soluble factors does not require direct cell contact. Overall, the sequence of events results in a series of clonal expansion and differentiation of lymphocytes to plasma cells that synthesize immunoglobins[23]. The inhibition of mitosis of PHA-activated lymphocytes by LDL suggests that LDL plays an important role in controlling clonal expansion so that the activated lymphocytes can further differentiate and mature[27-32]. At this stage, it can only be speculated that a decrease in serum LDL results in uncontrolled lymphocyte proliferation which consequently suppresses their differentiation and maturation. Preliminary experiments from our laboratory suggest that a drastic reduction in serum lipoproteins in experimental animals results in lymphoproliferative disorders, enhanced susceptibility to infection, and development of cancer.

LIPOPROTEINS AND CELL PROLIFERATION

We have shown that cholesterol synthesis plays an essential role in cell proliferation[33-36], and synchrony between de novo cholesterogenesis and DNA synthesis has been suggested[34-37]. Since NADPH generated through the hexose monophosphate pathway (HMP-pathway) is important for DNA, RNA and cholesterol synthesis, the HMP-pathway also plays a major role in cell proliferation[6,34]. In PHA-stimulated lymphocytes, de novo cholesterogenesis happens much earlier than DNA synthesis[38], and DNA synthesis and mitosis of these stimulated lymphocytes can be arrested by blocking cholesterol biosynthesis with LDL[32]. Our studies have shown that there are elevated cholesterol levels, enhanced cholesterogenesis, and stimulation of HMP-pathway in all rapidly proliferating normal as well as neoplastic cells[33,34]. The rates of de novo cholesterogenesis and the activity of the HMP-pathway decrease as the cells cease to proliferate. In addition, the activity of acyl CoA: cholesterol acyltransferase increases during cell proliferation to convert newly synthesized cholesterol to cholesterol esters. Accompanying these changes, there is a decrease in serum cholesterol esters, and lecithin: cholesterol acyltransferase (LCAT) activity levels in serum[39]. Thus, these findings indicate that enhanced de novo cholesterogenesis and reduced influx of serum cholesterol esters into the cell are required during the cell cycle if the cells are to grow and divide[3-6].

All nucleated cells in the body have genomic information for the synthesis of cholesterol[40], and all the tissues that have been examined show feed back control of cholesterol synthesis[41]. In addition, extrahepatic rat tissues have functioning high affinity LDL receptors, and the rates of endogenous cholesterol synthesis in a number of tissues can be increased by lowering plasma lipoprotein levels. Intravenous infusion of LDL was shown to reduce the rates of cholesterol synthesis in some of the tissues examined towards control values[42]. There is evidence to suggest that extrahepatic tissues take up mostly LDL cholesterol[43]. Since preneoplastic as well as neoplastic tissues exhibit loss of feed back control of cholesterol synthesis, such deregulated cholesterol metabolism may be an adaptive phenomenon to achieve continuous cell growth[40]. This view is supported by the observation that fetal and regenerating normal tissues express LDL receptors[44], whereas they appear to be reduced in tumor tissues[34,36]. These observations suggest that influx of circulating cholesterol esters into the cell is reduced by lowering serum LDL during normal cell proliferation and by down regulation of LDL receptors in cancer cells in vivo[36-39].

Phenobarbital (PB)[45] and a choline deficient (CD) diet[46] and diets high in USF[1-6] all promote cancer and reduce serum LDL levels. Indeed, feeding a choline deficient diet alone was shown to cause cancer in rats . Moreover, rats carrying genes linked to the major histocompatibility complex (MHC) that control growth and development (grc) have reduced serum lipoproteins and enhanced de novo cholesterogenesis and DNA synthesis in the liver. These rats were found to be unusually susceptible to chemical carcinogens[6]. A decrease in serum LDL levels leads to a reduced influx of circulating cholesterol esters which removes the inhibition of 3-hydroxy-3-methylglutaryl coenzyme A (HMG-CoA) reductase leading to enhanced de novo cholesterogenesis, DNA synthesis and cell proliferation[35,36]. Since cell proliferation is a prerequisite for the development of cancer[46], a drastic reduction in serum LDL promotes cancer[35,36].

LIPOPROTEINS AND CANCER

Even though, PB[45], CD diet[46] and USF diets[1-6] reduce serum lipoproteins and promote cancer in experimental animals, they are also known to affect many other metabolic pathways. For this reason, it can be argued that a reduction in serum LDL may not be the primary mitogenic signal. We have solved this problem by utilizing a 4% cholestyramine diet and reduced serum lipoproteins. Such reductions in serum lipoproteins resulted in enhanced 7,12-dimethylbenzanthracene (DMBA) induced breast cancer in female Wistar rats[35].

Fifty-day-old Wistar rats were initiated with a single intragastric dose of 5 mg of DMBA in 0.8 ml corn oil. Control rats received the vehicle alone. After 7 days on laboratory chow the rats were fed a control diet containing 5% corn oil (American Institute of Nutrition, AIN) prepared according to Berri et al.[48]. The experimental diets consisted of 4% cholestyramine resin in AIN diet (CHST), a 2% corn oil plus 18% coconut oil diet (SF), a 20% corn oil diet (USF), and a 4% CHST in 20% corn oil diet (USF + CHST). The rats were sacrificed at the end of 100 and 200 days. A detailed histological evaluation of grossly normal mammary tissue as well as any tumor mass was done. Lipids in serum, liver, breast tissue and tumors were analyzed[6]. In addition, serum LCAT levels[39] and cholesterol biosynthesis in breast tissue and tumors[34] were measured.

The length of feeding has a significant effect on serum lipid parameters. At the end of 200 days, the total lipids (TL) decreased significantly in rats fed CHST when compared to the rats fed the AIN diet. Similarly the USF and USF + CHST groups had a significant decrease in serum TL when

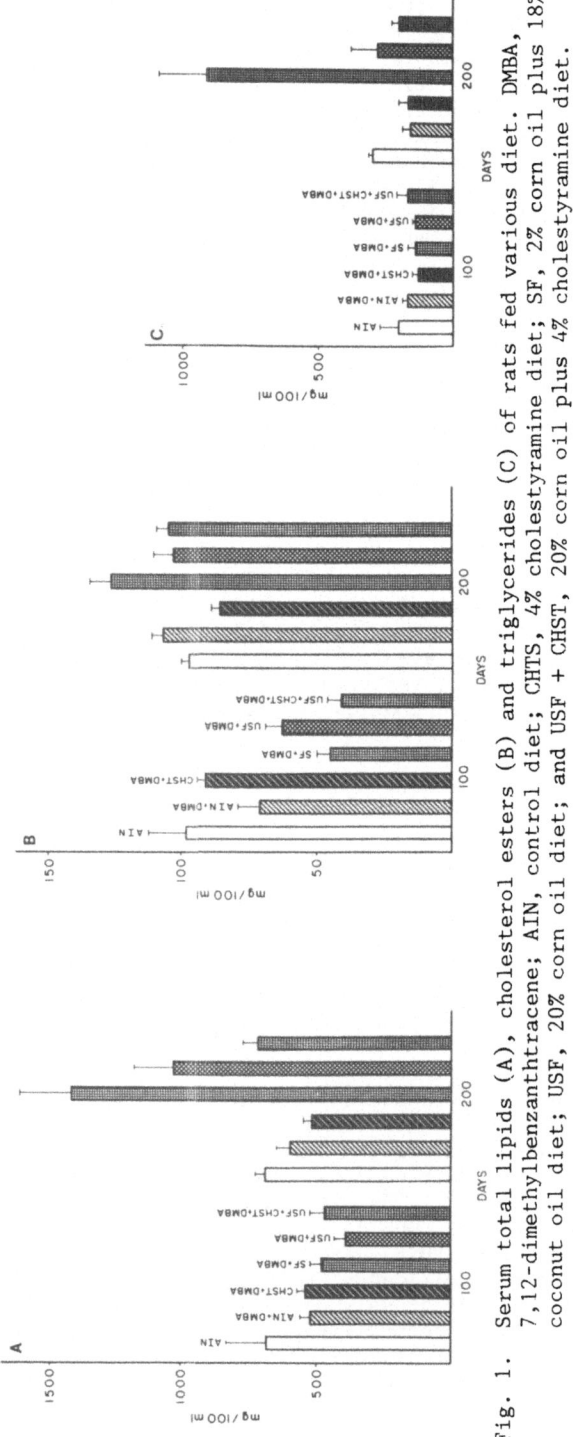

Fig. 1. Serum total lipids (A), cholesterol esters (B) and triglycerides (C) of rats fed various diet. DMBA, 7,12-dimethylbenzanthtracene; AIN, control diet; CHTS, 4% cholestyramine diet; SF, 2% corn oil plus 18% coconut oil diet; USF, 20% corn oil diet; and USF + CHST, 20% corn oil plus 4% cholestyramine diet.

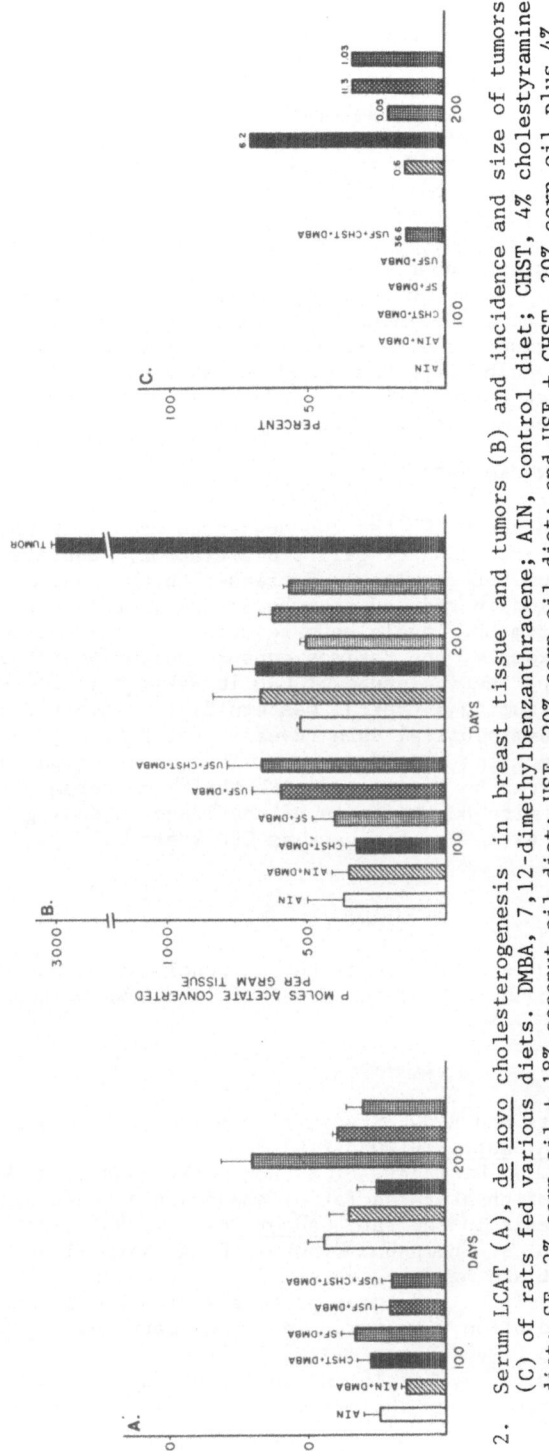

Fig. 2. Serum LCAT (A), de novo cholesterogenesis in breast tissue and tumors (B) and incidence and size of tumors (C) of rats fed various diets. DMBA, 7,12-dimethylbenzanthracene; AIN, control diet; CHST, 4% cholestyramine diet; SF, 2% corn oil + 18% coconut oil diet; USF, 20% corn oil diet; and USF + CHST, 20% corn oil plus 4% cholestyramine diet. The average weight of the tumors in grams is indicated on top of each panel in figure C.

compared to the rats fed the SF diet (Fig. 1A). Similar decreases were seen in cholesterol esters (Fig. 1B), triglycerides (Fig. 1C), and serum LCAT (Fig. 2A) levels. At the same time, the de novo cholesterogenesis showed a significant increase in breast tissues of USF and USF + CHST groups at 100 days and in CHST, USF and USF + CHST groups at the end of 200 days. Tumors in rats fed any diet showed a several-fold increase in cholesterol synthesis when compared to apparently normal breast tissue (Fig. 2B). The percent incidence of tumors and the size of adenocarcinomas increased many-fold in rats fed CHST, USF and USF + CHST diets when compared to rats fed AIN and SF groups. It is of interest to note that a combination of USF and CHST gave an aggressively growing tumor in one rat as early as 100 days (Fig. 2C). Both CHST and USF diets are known to reduce serum lipoproteins, eliminate bile acids and enhance de novo cholesterogenesis[7,49]. Since dietary CHST is not absorbed through the gastrointestinal tract and cannot reach the target tissue, it can be concluded that a reduction in serum LDL stimulates cells to enhanced de novo cholesterogenesis and DNA synthesis. These results support our contention that a reduction in serum LDL provides the mitogenic signal resulting in the promotion of cancer.

RELEVANCE TO HUMAN CANCERS

A decrease in serum LDL has two consequences: 1) suppression of immune function and 2) stimulation of cell proliferation. Besides these functions, a decrease in serum LDL results in decreased thromboxane A_2 synthesis in platelets[50] resulting in reduced thrombosis and increased bleeding. All these effects of low serum LDL levels have relevance to the development of cancer in humans. For example, the Eskimos consume polyunsaturated fats derived from fish oils, and the incidence of CHD in Eskimos is low when compared to the population consuming a typical western diet. However, a critical evaluation of the epidemiological data reveals that Eskimos have a higher prevalence of infections and of cancer. Thus, reduced CHD and increased cancer prevalence can be correlated with their low serum lipoprotein levels[51]. Low serum LDL may also be the cause of the longer bleeding times in Eskimos for which records extend as far back as 500 years[52].

ACKNOWLEDGEMENTS

This research is supported by the National Dairy Council. We thank Drs. T. J. Gill and H. W Kunz for thoughtful reviews of the manuscript.

REFERENCES

1. E. A. Alcantara and E. W. Speckman, Diet, Nutrition and Cancer, Am. J. Clin. Nutr. 29:1035 (1976).
2. B. D. Roebuck, J. D. Yager, D. S. Longnecker and S. A. Willpone, Promotion of unsaturated fat of azaserine induced pancreatic carcinogenesis in the rat, Cancer Res. 41:3961 (1981).
3. L. A. Cohen, D. A. Thompson, K. Choi, R. A. Karmali and D. P. Rose, Dietary fat and mammary cancer II. Modulation of serum and tumor lipid composition and tumor prostaglandins by different dietary fats: Association with tumor incidence patterns, J. Natl. Cancer Inst. 77:43 (1986).
4. S. A. Broitman, J. J. Vitale, E. Vavrousek-Jakuba and L. S. Gottlieb, Polyunsaturated fat, cholesterol and large bowel tumorigenesis, Cancer 40:2455 (1977).
5. C. W. Welsch, Host factors affecting the growth of carcinogen induced rat mammary carcinomas: A review and tribute to Charles Brenton Higgins, Cancer Res. 45:3415 (1985).
6. K. N. Rao, H. Shinozuka, H. W. Kunz and T. J. Gill III, Enhanced

500

susceptibility to a chemical carcinogen in rats carrying MHC-linked genes influencing development (grc), Int. J. Cancer 34:113 (1984).

7. C. S. Ramesha, R. Paul and J. Ganguly, Effect of dietary unsaturated oils on the biosynthesis of cholesterol and on biliary and fecal excretion of cholesterol and bile acids in rats, J. Nutr. 110:2149 (1980).

8. S. Graham, Dietary factors in the prevention of cancer, Transplantation Proceedings 16:392 (1984).

9. B. M. Rifkind, The lipid research clinics coronary primary prevention trial, Drugs Suppl. 1, 53 (1986).

10. J. J. Vitale and S. A. Broitman, Lipids and immune function, Cancer Res. 41:3706 (1981).

11. G. M. Kollmorgen, W. A. Sansing, A. A. Lehman, G. Fischer, R. E. Longley, S. S. Alexander, M. M. King and P. B. McCay, Inhibition of lymphocyte function in rats fed high fat diets, Cancer Res. 39:3458 (1979).

12. W. L. Kos, R. M. Loria, M. J. Snodgrass, D. Cohen, T. G. Thorpe and A. M. Kaplan, Inhibition of host resistance by nutritional hypercholesterolemia, Infect. Immun. 26:658 (1979).

13. M. I. McHugh, R. Wilkinson, R. W. Elliott, E. J. Field, P. Dewar, R. R. Hall, R. M. R. Taylor and P. R. Uldall, Immunosuppression with polyunsaturated fatty acids in renal transplantation, Transplantation 24:263 (1977).

14. C. J. Meade and J. Mertin, Fatty acids and immunity, Adv. Lipid Res. 16:127 (1978).

15. C. J. Meade and J. Mertin, The mechanism of immunoinhibition by arachidonic acid and linoleic acid: effects on the lymphoid and reticuloendothelial systems, Inf. Archs. Allergy Appl. Immunol. 51:2 (1976).

16. J. Mertin, Effects of polyunsaturated fatty acids on skin allograft survival and primary and secondary cytotoxic response in mice, Transplantation 21:1 (1976).

17. J. Mertin and R. Hunt, Influence of polyunsaturated fatty acids on survival of skin allografts and tumor incidence in mice, Proc. Natl. Acad. Sci. USA 73:928 (1976).

18. J. Ring, J. Seifert, J. Mertin and W. Brendel, Prolongation of skin allografts in rats by treatment with linoleic acid, Lancet 2:1331 (1974).

19. P. R. Uldall, R. Wilkinson, M. I. McHugh, E. J. Field, B. K. Shenton, K. Baxby and R. M. R. Taylor, Linoleic acid and transplantation, Lancet 2:128 (1975).

20. J. Mertin, Polyunsaturated fatty acids and cancer, Br. Med. J. 4:357 (1973).

21. P. R. Uldall, R. Wilkinson, M. I. McHugh, E. J. Field, B. K. Shenton, K. Baxby, R. M. R. Taylor and J. Swinney, Unsaturated fatty acids and renal transplantation, Lancet 2:514 (1974).

22. J. H. D. Miller, K. J. Zilkha, M. J. S. Langman, H. Play-wright, A. D. Smith, J. Belin and R. H. S. Thompson, Double blind trial of linoleate supplementation of the diet in multiple sclerosis, Br. Med. J. 1:765 (1973).

23. G. B. Otto, W. Dias da Silva, W. Gotze and I. Mota, Fundamentals of immunology, Springer-Verlag, New York, (1986)

24. J. V. Friend, S. O. Lock, M. I. Gurr and W. E. Parish, Effect of different dietary lipids on immune responses of Hartley strain guinea pigs, Int. Archs. Allergy Appl. Immunol. 62:292 (1980).

25. J. H. Morse, L. D. Witte and D. S. Goodman, Inhibition of lymphocyte proliferation stimulated by lectins and allogenic cells by normal plasma lipoproteins, J. Exp. Med. 146:1791 (1977).

26. C. C. Waddel, O. D. Taunton and J. J. Twomey, Inhibition of lymphoproliferation by hyperlipoproteinemic-plasma, J. Clin. Invest. 58:950 (1976).

27. L. K. Curtiss, D. H. Deheer and T. S. Edgington, Influence of the immunoregulatory serum lipoprotein LDL-In on the in vivo proliferation and differentiation of antigen-binding and antibody-secreting lymphocytes during a primary immune response, Cellular Immunol. 49:1 (1980).

28. D. Y. Hui and J. A. K. Harmony, Inhibition of low density lipoproteins of mitogen-stimulated cyclic nucleotide production by lymphocytes, J. Biol. Chem. 255:1413 (1980).

29. D. Y. Hui and J. A. K. Harmony, Inhibition of Ca^{2+} accumulation in mitogen-activated lymphocytes: Role of membrane-bound plasma lipoproteins, Proc. Natl. Acad. Sci. USA 77:4764 (1980).

30. F. Ito, Y. Takii, J. Suzuki and Y. Masamune, Reversible inhibition by human serum lipoproteins of cell proliferation, J. Cellular Physiol. 113:1 (1982).

31. M. Macy, Y. Okano, A. D. Cardin, E. M. Avila and J. A. K. Harmony, Suppression of lymphocyte activation by plasma lipoproteins, Cancer Res. (suppl.) 43:2496 (1983).

32. J. A. Cuthbert and P. E. Lipsky, Immunoregulation by low density lipoproteins in man: low density lipoprotein inhibits mitogen stimulated human lymphocyte proliferation after initial activation, J. Lipid Res. 24:1512 (1983).

33. K. N. Rao, S. Kottapally and H. Shinozuka, Lipid composition and 3-hydroxy-3-methylglutaryl-CoA reductase activity of acinar cell carcinoma of rat pancreas, Biochim. Biophys. Acta 759:74 (1983).

34. K. N. Rao, S. Kottapally and H. Shinozuka, Acinar cell carcinoma of rat pancreas: Mechanism of deregulation of cholesterol metabolism Toxicol. Pathol. 12:62 (1984).

35. M. F. Melhem, H. F. Gabriel, E. D. Eskander and K. N. Rao, Cholestyramine promotes 7,12-dimethylbenzanthracene induced mammary cancer in Wistar rats, Br. J. Cancer 56:45 (1987).

36. K. N. Rao, Regulatory aspects of cholesterol metabolism in cells with different degrees of replication, Toxicol. Pathol. 14:430 (1986).

37. M. D. Siperstein, Role of cholesterogenesis and isoprenoid synthesis in DNA replication and cell growth, J. Lipid Res. 25:1462 (1984).

38. H. W. Chen, H. J. Heiniger and A. A. Kandutsch, Relationship between sterol synthesis and DNA synthesis in phytohemagglutinin-stimulated mouse lymphocytes, Proc. Natl. Acad. Sci. USA 72:1950 (1975).

39. K. N. Rao, S. Kottapally, E. D. Eskander, H. Shinozuka, S. Dessi and P. Pani, Acinar cell carcinoma of rat pancreas: Regulation of cholesterol esterification, Br. J. Cancer 54:305 (1986).

40. P. S. Coleman and B. B. Lavietes, Membrane cholesterol and tumorigenesis, CRC Critical Rev. Biochem. 11:341 (1981).

41. A. Swann, M. H. Wiley and M. D. Siperstein, Tissue distribution of cholesterol feedback control in guinea pig, J. Lipid Res. 16:360 (1975).

42. J. M. Anderson and J. M. Dietschy, Regulation of sterol synthesis in 15 tissues of rat: II Role of rat and human high and low density plasma lipoproteins and of rat chylomicron remnants, J. Biol. Chem. 252:3652 (1977).

43. M. S. Brown, P. T. Kovanen and J. L. Goldstein, Regulation of plasma cholesterol by lipoprotein receptors, Science 212:628 (1981).

44. M. J. Rudling and C. O. Peterson, LDL receptors in bovine tissues assayed as the heparin-sensitive binding of ^{125}I-labeled LDL in homogenates: relation between liver LDL receptors and serum cholesterol in the fetus and post term, Biochim. Biophys. Acta 836:96 (1985).

45. E. Farber and D. S. R. Sarma, Biology of disease. Hepatocarcinogenes: A dynamic cellular perspective, Lab. Invest. 56:4 (1987).

46. L. I. Giambarresi, S. L. Katyal and B. Lombardi, Promotion of liver carcinogenesis in the rat by a choline-devoid diet: Role of liver cell necrosis and regeneration, Br. J. Cancer 46:825 (1982).

47. N. Chandar, J. Amenta, J. C. Kandala and B. Lombardi, Liver cell turnover in rats fed a choline devoid diet, Carcinogenesis 8:669 (1987).

48. J. C. Bieri, G. S. Stoewsand, G. M. Briggs, R. W. Phillips, J. C. Woodward and J. J. Knapka, Report of the American Institute of Nutrition. Ad hoc committee on standards for nutritional studies, J. Nutr. 107:1340 (1977).

49. N. B. Myant, The biology of cholesterol and related steroids, William Heineman Medical Books Inc. London, (1981).

50. T. I. Pynadath and A. Z. Haghighi, Inhibition of thromboxane A_2 synthesis in rats treated with phenobarbital, Prostaglandins Leukotrienes Med. 28:61 (1987).

51. W. E. M. Lands, Fish and human health, Academic Press Inc. New York, (1986).

52. H. O. Bang and J. Dyerberg, The bleeding tendency in Greenland Eskimos, Danish Med. Bull. 27:202 (1980).

A MECHANISTIC ASSOCIATION BETWEEN CHOLESTEROL METABOLISM AND CELL PROLIFERATION

Sandra Dessì, Barbara Batetta, Caterina Chiodino
and Paolo Pani

Istituto di Farmacologia e Patologia Biochimica
Università degli Studi di Cagliari
Via Porcell 4, 09124 Cagliari, Italy

Cholesterol is one of the major components of animal cell plasma membranes. It plays a vital role in the maintenance of cell integrity and in the regulation of membrane fluidity; it regulates more or less directly many cellular functions including cell growth, in normal and malignant processes[1,2].

It has been repeatedly shown by our laboratory that constant changes of cholesterol metabolism occur during different types of cellular proliferation irrespectively of the proliferative stimuli[3-8].

Increase of cholesterol synthesis was observed not only in liver, a major organ of cholesterol production, but also in kidney, an organ devoid of any cholesterol synthesizing activity under normal resting conditions[3,4]. These observations were made following a single administration of lead nitrate a potent mitogen for liver[9] and kidney[4]. The increase of cholesterol synthesis was also observed in other types of cellular proliferation, during liver regeneration after partial surgical hepatectomy[5], in liver hyperplasia occuring in diabetic rats given insulin and in fasted-refed rats[6,7], and in hepatic nodules during rat liver carcinogenesis[8]. In most of these models the increase of cholesterol synthesis preceded and accompanied DNA synthesis.

A direct association between cholesterol and DNA synthesis has been proposed by several investigators in the last decade. Compactin, mevinolin and other inhibitors of 3-hydroxy-methyl glutaryl CoA reductase (HMGCoA), the rate controlling enzyme of cholesterol synthesis, were shown to inhibit in vitro DNA synthesis[10-12].

The levels of HMGCoA reductase were found proportional to the degree of cell proliferation during fetal and neonatal development, in pancreatic regeneration after tissue loss as well as in tumors of different growth rates[13].

A role for cholesterol synthesis in regulating cell proliferation is also supported by the early report of Siperstein and Fagan on the loss of feedback inhibition on HMGCoA reductase in tumoral tissue[14] and by more recent data on the possible effect of mevalonic acid and/or other isoprenoid units in the initiating phases of DNA synthesis[15,16].

Despite the overwhelming data concerning intracellular cholesterol synthesis during cell growth, the role of circulating cholesterol during such process has been so far neglected.

Mammalian cells obtain cholesterol from exogenous source or through "de novo" synthesis, while a continuous exchange occurs between intracellular and plasma cholesterol pools in the whole organism; it is then predictable that the variation in the two pools can be reciprocally influenced.

In humans a possible relationship between circulating cholesterol and cancer has been extrapolated from studies designed to examine the association between cholesterol and cardiovascular diseases. Those studies were not specifically intended to measure cancer incidence and mortality. In most cases an inverse correlation between serum cholesterol levels and cancer incidence was found [17-20]. The results however were sometimes contradictory and not always consistent. In a clinical study on the relationship between colon cancer and serum cholesterol, Rose et al. observed that the initial levels of serum cholesterol in colon cancer patients were lower than expected [21]. However, the same Authors also reported that serum cholesterol levels were higher than in controls in patients with cancer of the stomach, pancreas, liver, bile ducts and rectum [22]. Similarly, low serum cholesterol levels were found by Kark et al. in lung cancer patients [23], while Stamler et al. observed higher serum cholesterol levels in lung cancer cases compared to controls [24]. Moreover, not always association was found between serum cholesterol levels and overall risk of death from cancer in epidemiological studies by Dier et al. [25]. The Authors reported an inverse correlation between serum cholesterol level and cancer death for sarcoma, leukemia and Hodgkin's disease but not for lung cancer, colon-rectal cancer, cancer of oral cavity, pancreatic cancer, or all other cancer combined. The lowering of serum cholesterol by a diet rich in polyunsaturated fatty acids resulted in a higher incidence of cancer death [26]. Similar results were obtained in epidemiological studies conducted in humans treated with hypolipidemic agents: the beneficial effect against hypercholesterolemia of clofibrate, the most widely used lipid-lowering drug in Europe and in USA, is counterbalanced by a probable higher risk of tumors of the gastrointestinal tract [27].

Experimental studies have shown that cholestyramine, another widely used hypolipidemic agent that differs from other drugs since it is not absorbed by the intestine, was associated with cancer of the gastrointestinal tract [28,29]. More recently, Melhem et al. have shown an increase in the incidence of 7,12-dimethylbenzanthracene (DMBA)-induced breast cancer in rats fed a 4% cholestyramine diet [30]. In the same study, the serum total lipids, cholesterol esters and triglycerides, decreased significantly in rats treated with DMBA and fed a 4% cholestyramine diet when compared with rats fed a control diet.

Although the clinical and the epidemiological results on cholesterol and cancer are often contradictory and experimental evidences are far from being conclusive, the cumulative data provide enough support to suggest that cellular proliferation and tumoral progression are characterized by peculiar changes of cholesterol metabolism not only in the proliferating tissues but also in the plasma compartment.

However, on the basis of the available findings, at least three important points need to be better considered:

1. Epidemiological data based on self-reported information need to be corrected, possibly by using more precise biological indicators.
2. If a correlation between plasma cholesterol levels and cancer incidence and mortality does exist it still remains to be established whether alterations of plasma cholesterol is the cause or whether it reflects the metabolic consequences of cancer.
3. Circulating cholesterol metabolism is too complex to be expressed

merely by plasma cholesterol levels. For this reason the different classes of lipoproteins in relation to the intracellular cholesterol synthesis and catabolism must be carefully investigated.

In our laboratory, intracellular and circulating cholesterol as well as VLDL, LDL, HDL_2 and HDL_3 lipoproteins were investigated in rats in two different models of cell proliferation: liver regeneration after partial surgical hepatectomy and liver hyperplasia induced by lead nitrate. These two different models were chosen since the first is a compensatory process subsequent to loss of parenchyma while the latter is a process characterized by a net tissutal gain with no apparent sign of liver injury. In both models the "de novo" cholesterol synthesis was measured in vivo by the incorporation of ^3H_2O into hepatic free and esterified cholesterol.

In liver regeneration after partial hepatectomy, no significant changes of labelling in free cholesterol and very high levels of labelled esterified cholesterol esters were observed (Fig. 1). The incorporation of tritiated water into free and esterified cholesterol increased after lead treatment, the percentage being higher in cholesterol esters than in free cholesterol (Fig. 2). A massive accumulation of cholesterol esters but not of free cholesterol was also found in both proliferating parenchymas (Fig. 3 and 4).

These data were further confirmed by a study on the regulation of cholesterol esterification in two different growing (fast and slow) pancreatic tumors transplanted in nude mice [31]. Both tumors showed an increase in the activity of acyl CoA:cholesterol acyltransferase (ACAT), the intracellular esterifying enzyme of free cholesterol, during the active tumoral growth phase. The enzymatic activity was significantly higher in fast than in slow

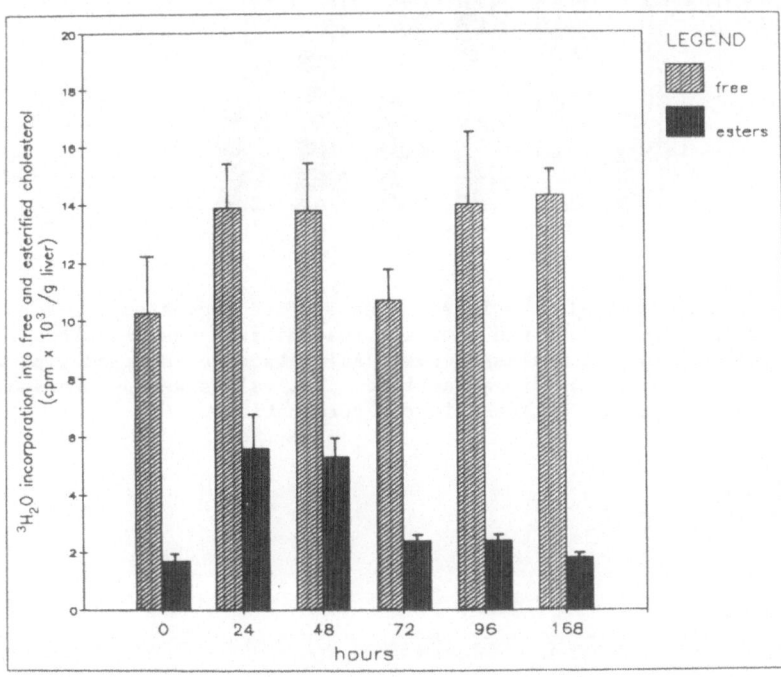

Fig. 1. ^3H_2O incorporated into free and esterified cholesterol in liver regeneration after partial hepatectomy. All rats were injected i.p. with 10 µmCi/100 g b.w. of ^3H_2O 6 h before sacrifice. The values were expressed as the means ± S.E. of at least four determinations.

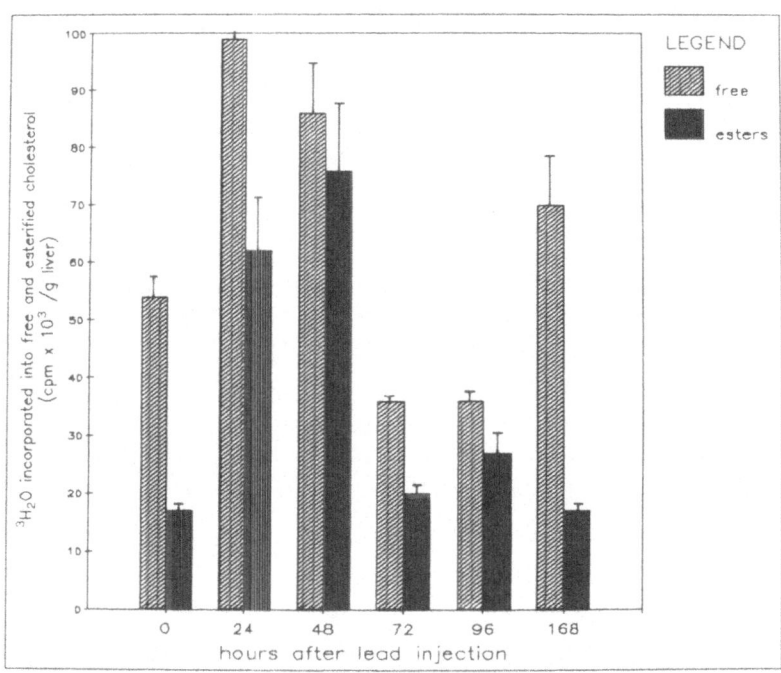

Fig. 2. ^3H_2O incorporated into free and esterified cholesterol after a single administration of lead nitrate. Lead nitrate was injected i.v. at the dose of 100 µmoles/Kg b.w.. All rats were injected i.p. with 40 mCi of ^3H_2O 12 h before sacrifice. The values were expressed as the means ± S.E. of at least four determinations.

growing tumors. These results suggest that the newly synthesized cholesterol during massive cell proliferation is immediately esterified in order to obtain sufficient cholesterol storage, probably for the biogenesis of new membranes. On the other hand, cholesterol enrichment of cellular membranes was shown in mitochondria and in microsomes isolated from Morris hepatoma[32].

Total plasma cholesterol levels did not change after partial hepatectomy (Fig. 5), while hypercholesterolemia was found in lead-treated rats (Fig. 6). In spite of this discrepancy a drastic decrease in cholesterol HDL levels (Fig. 5 and 6), accompanied by a reduced activity of lecithin:cholesterol acyl transferase (LCAT), the esterifying enzyme of free cholesterol in plasma, was observed in both experimental models[33,34]. A decrease of LCAT activity was also found in rats bearing pancreatic tumors[31].

The analysis of serum lipoproteins by HPLC showed a decrease in HDL_2 and a concomitant increase of HDL_3 (Fig. 7 and 8), thus resulting in a very strong decrease in HDL_2/HDL_3 ratio (Fig. 9). No apparent differences in VLDL and a slight increase in LDL were also observed (Fig. 7 and 8). A recovery of the normal lipoprotein pattern was achieved together with the retrieve of hyperplasia (Fig. 8), thus showing that the changes in cholesterol metabolism are synchronous with the different phases of cell proliferation.

Even if the interpretation of these data is rather difficult and needs further experimental support, we believe that cholesterol changes occurring in plasma are a consequence of intracellular cholesterol metabolism taking place in tissues during massive parenchymal proliferation. On the basis of our data some general conclusions can be drawn. It has been postulated that HDLs are involved in a process of reverse cholesterol transport[35]. According

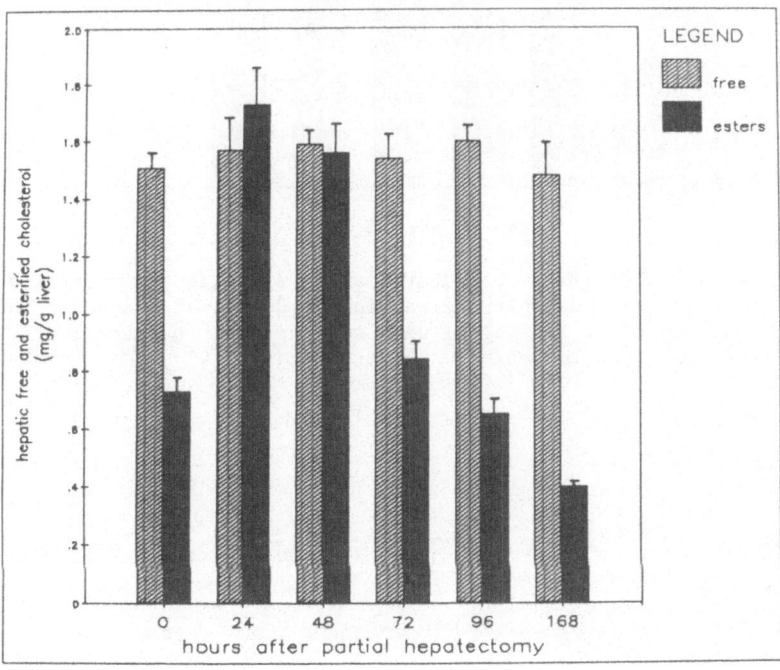

Fig. 3. Free and esterified cholesterol during liver regeneration after partial hepatectomy. The values were expressed as the means ± S.E. of at least four determinations.

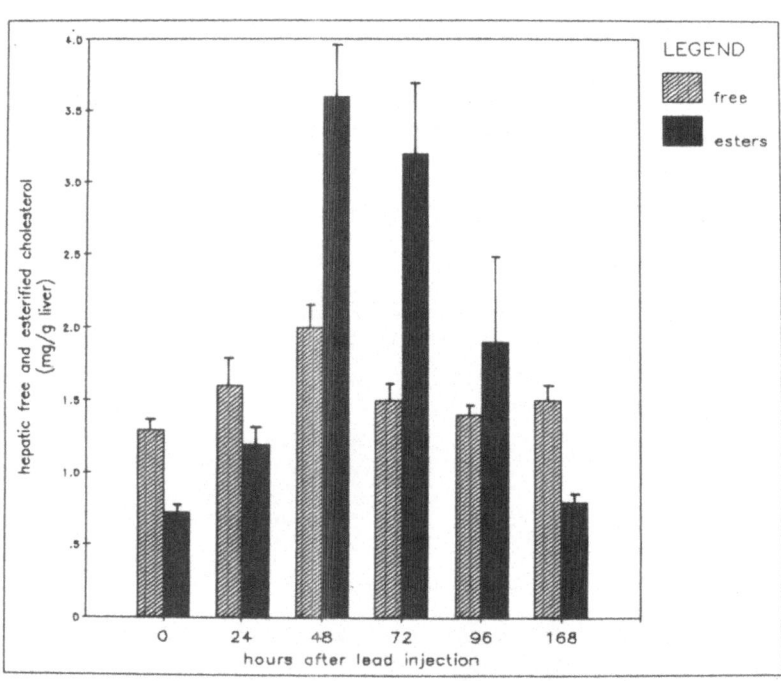

Fig. 4. Free and esterified cholesterol after a single administration of lead nitrate. Lead nitrate was injected i.v. at a dose of 100 μmoles/Kg b.w.. The values were expressed as the mean ± S.E. of at least four determinations.

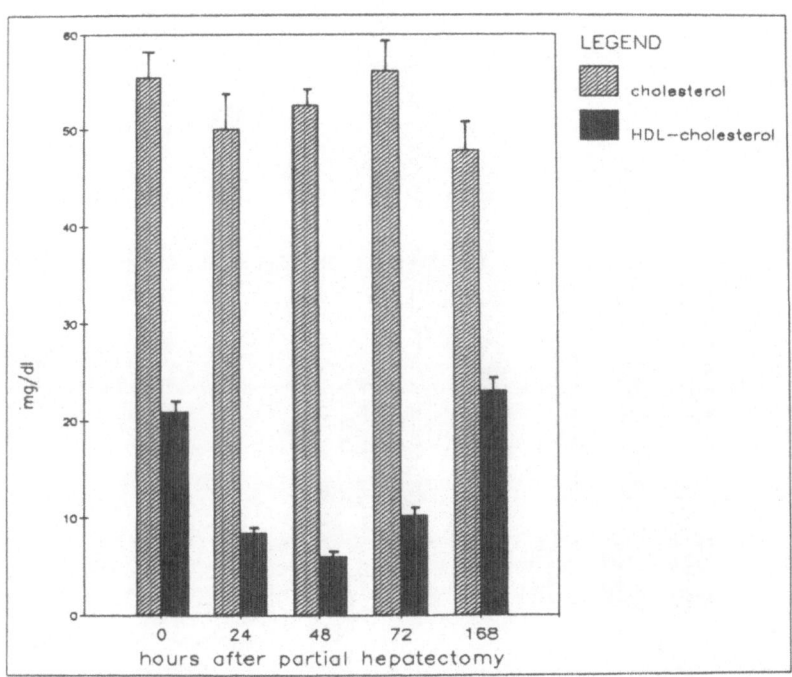

Fig. 5. Total and HDL serum cholesterol during liver regeneration after partial hepatectomy. The values were expressed as the mean ± S.E. of at least four determinations.

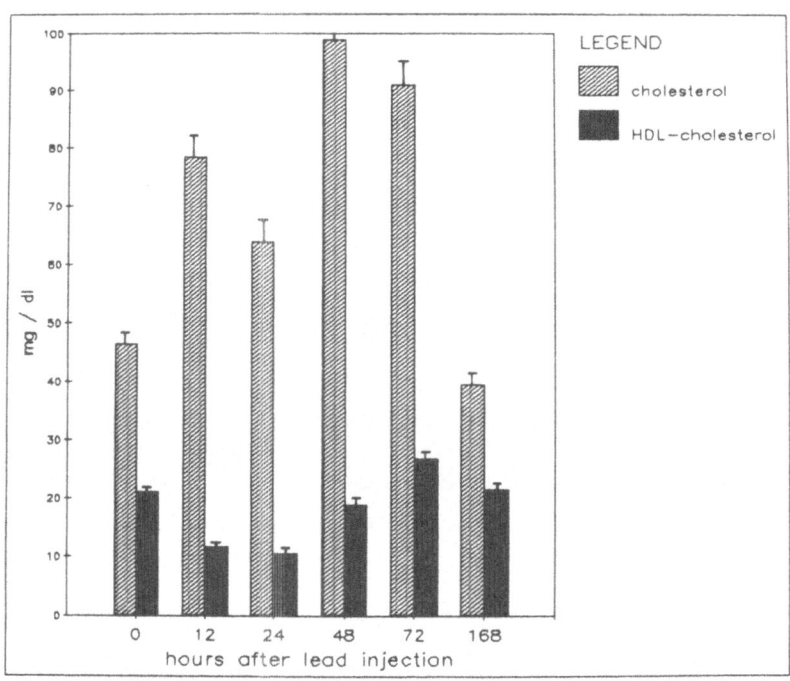

Fig. 6.   Total and HDL serum cholesterol after a single administration of
lead nitrate. Lead nitrate was injected i.v. at a dose of 100
μmoles/Kg b.w.. The values were expressed as the mean ± S.E. of at
least four determinations.

to this hypothesis, HDLs remove cholesterol from various tissues and other lipoproteins; cholesterol is then transported to the liver and excreted from the body. HDLs form a heterogenous group of particles which can be divided in subfractions, mainly HDL_2 and HDL_3, with different protein and lipid composition. It is assumed that HDL_3 are converted in HDL_2 in the plasma compartment by enrichment of cholesterol ester molecules, via LCAT reaction. Cell membranes and other lipoprotein fractions may be considered a potential source of free cholesterol for the conversion of HDL_3 to HDL_2, suggesting that HDL_2 are directly implied in reverse cholesterol transport [36]. The decrease in HDL_2 and LCAT activity and the concomitant increase in HDL_3 found under our experimental conditions indicate a decrease of circulating cholesterol catabolism, probably for a decreased conversion of HDL_3 to HDL_2 in the plasma compartment. This could be explained by the fact that the reserve cholesterol transport process may be inhibited if cholesterol is needed during massive cellular growth. A recent study on mice bearing a transplanted lymphoid tumor showed that the rate of cholesterol synthesis in the tumoral cells was extremely high when compared to various lymphoid cells in normal control mice [37]. In contrast to tumoral cells, serum cholesterol content decreased to very low levels in tumor-bearing mice. In the same work it was also shown that outgrowth of tumor cells was accompanied by a drastic alteration in the plasma lipoproteins: HDL became strongly reduced while LDL and VLDL were increased. A decrease in HDL was also found in mice bearing Ehrlich ascites tumor; however, in this model the total cholesterol levels were higher than normal [38]. More recently a decrease in HDL levels was observed in patients with mieloproliferative disorders [39] as well as in patients with hepatic metastases of colorectal cancer and with primary liver cancer [40]. Finally, a fall of cholesterol HDL has been reported under physiological con-

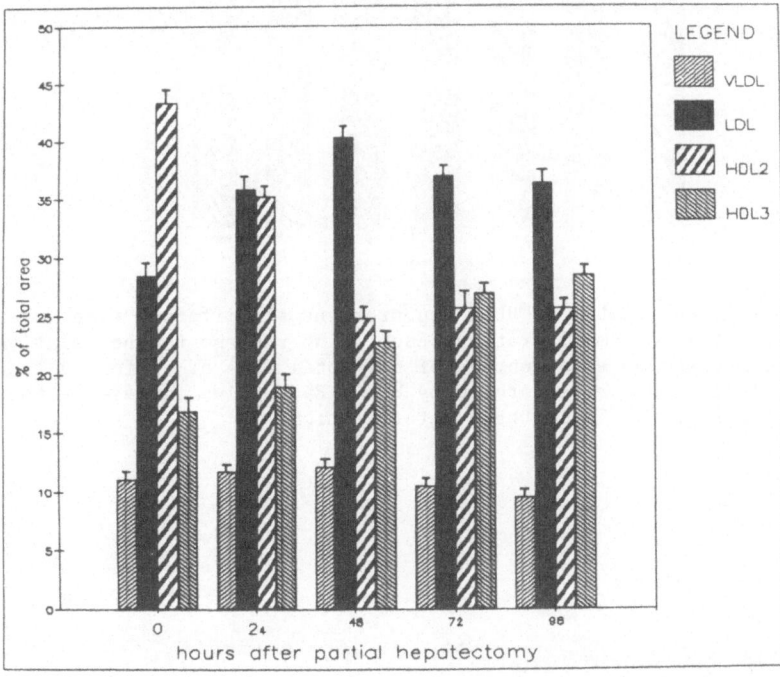

Fig. 7. VLDL, LDL, HDL_2 and HDL_3 lipoproteins at different time intervals after partial hepatectomy. Each point represents the relative area, expressed as a percentage of the total area of different fractions of lipoproteins separated by HPLC. Values were expressed as the mean ± S.E. of at least three determinations.

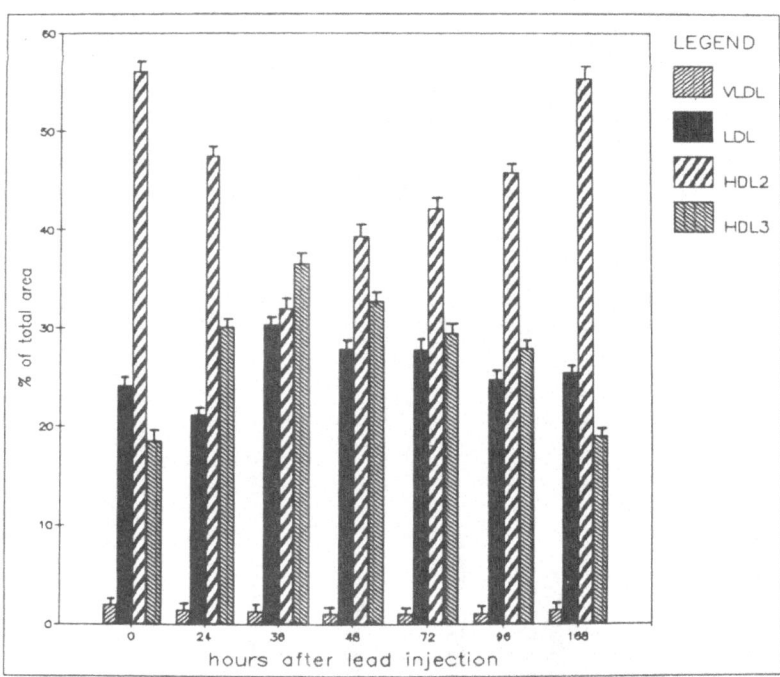

Fig. 8. VLDL, LDL, HDL$_2$ and HDL$_3$ lipoproteins at different time intervals after lead administration. Each point represents the relative area, expressed as a percentage of the total area of different fractions of lipoproteins separated by HPLC. Each value represents the mean ± S.E. of at least three determinations.

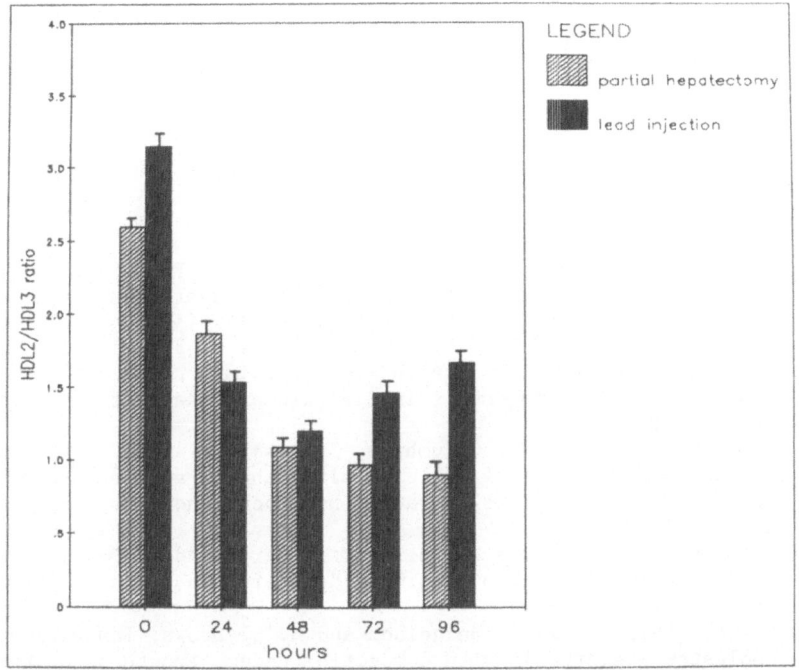

Fig. 9. HDL_2/HDL_3 ratios at different time intervals after partial and lead nitrate injection. Each value represents the mean ± S.E. of at least three determinations.

ditions of extended cellular growth as occurring during fetal and neonatal development[41,42].

In conclusion, although the available data are not sufficient to establish the exact mechanism underlying changes of lipoprotein metabolism and cellular proliferation, they offer the possibility to hypothesize that a decrease in HDL, mainly in HDL_2 fraction, rather than changes in serum cholesterol levels, may be a generalized phenomenon related to cell growth, in normal and malignant processes. On the basis of these considerations, we believe that it is also worthy to re-evaluate the epidemiological and experimental data on serum cholesterol levels and cancer.

ACKNOWLEDGEMENTS

This work was supported by funds from C.N.R. (Oncologia, No. 86.00510.44) Rome, Italy, A.I.R.C., Milan, Italy and N.A.T.O., Bruxelles, Belgium. Dr. C. Chiodino receives salary support through a "Regione Automoma della Sardegna" career development award.

REFERENCES

1. H. W. Chen, A. A. Kandutsch and H. J. Heiniger, The role of cholesterol in malignancy, Prog. exp. Tumor Res. 22:275 (1978).
2. P. S. Coleman and B. B. Lavietes, Membrane cholesterol, tumorigenesis, and the biochemical phenotype of neoplasia, CRC Crit. Rev. Biochem 11:341 (1981).
3. S. Dessì, B. Batetta, E. Laconi, C. Ennas and P. Pani, Hepatic choles-

terol in lead nitrate induced liver hyperplasia, Chem.-Biol. Interactions 48:271 (1984).

4. G. M. Ledda-Columbano, A. Columbano, S. Dessì, P. Coni, C. Chiodino, G. Faa and P. Pani, Hexose monophosphate shunt and cholesterogenesis in lead-induced kidney hyperplasia, Chem.-Biol. Interactions 62:209 (1987).

5. S. Dessì, C. Chiodino, B. Batetta, A. M. Fadda, C. Anchisi and P. Pani, Hepatic Glucose-6-Phosphate Dehydrogenase, cholesterogenesis, and serum lipoproteins in liver regeneration after partial hepatectomy, Exp. Mol. Pathol. 44:169 (1986).

6. S. Dessì, C. Chiodino, B. Batetta, E. Laconi, C. Ennas and P. Pani, Hexose monophosphate shunt and cholesterol synthesis in the diabetic and fasting states, Exp. Mol. Pathol. 43:177 (1985).

7. S. Dessì, C. Chiodino, B. Batetta, M. Armeni, M. F. Mulas and P. Pani, Comparative effects of insulin and refeeding on DNA synthesis, HMP shunt and cholesterogenesis in diabetic and fasted rats, Pathology 20:53 (1988).

8. G. M. Ledda-Columbano, A. Columbano, S. Dessì, P. Coni, C. Chiodino and P. Pani, Enhancement of cholesterol synthesis and pentose phosphate pathway activity in proliferating hepatocyte nodules, Carcinogenesis 6:1371 (1985).

9. A. Columbano, G. M. Ledda, P. Sirigu, T. Perra and P. Pani, Liver cell proliferation induced by a single dose of lead nitrate, Amer. J. Pathol. 110:83 (1983).

10. J. L. Goldstein, J. A. S. Helgelson and M. S. Brown, Inhibition of cholesterol synthesis with compactin renders growth of cultured cells dependent on the low density lipoprotein receptor, J. Biol. Chem. 254:5403 (1979).

11. A. A. Kandutsch and H. W. Chen, Consequences of sterol synthesis in cultured cells, J. Biol. Chem. 252:409 (1977).

12. M. S. Brown and J. L. Goldstein, Suppression of 3-hydroxy-3-methyl-glutaryl coenzyme A reductase activity and inhibition of growth of human fibroblasts by 7-ketocholesterol, J. Biol. Chem. 249:7306 (1974).

13. K. N. Rao, Regulatory aspects of cholesterol metabolism in cells with different degrees of replication, Toxicol. Pathol. 14:430 (1986).

14. M. D. Siperstein and V. M. Fagan, Deletion of the cholesterol negative feedback system in liver tumors, Cancer Res. 24:1108 (1964).

15. V. Quesney-Huneeus, M. H. Wiley and M. D. Siperstein, Essential role for mevalonate synthesis in DNA replication, Proc. Natl. Acad. Sci. USA 76:5056 (1979).

16. A. J. R. Habenicht, J. A. Glomset and R. R. Ross, Relation of cholesterol and mevalonic acid to the cell cycle in smooth muscle and Swiss 3T3 cells stimulated to divide by platelet-derived growth factor, J. Biol. Chem. 225:5134 (1980).

17. A. Kagan, D. L. McGee, K. Yano, G. G. Rhoads and A. Nomura, Serum cholesterol and mortality in a Japanese-American population, Amer. J. Epidemiol. 114:11 (1981).

18. D. Kozarevic, D. McGee, N. Vojvodic, T. Gordon, Z. Racic, W. Zukel and T. Dawber, Serum cholesterol and mortality. The Yugoslavia cardio-vascular disease study, Amer. J. Epidemiol. 114:21 (1981).

19. M. R. Garcia-Palmieri, P. D. Sorlie, R. Costas and R. J. Hawlik, An apparent inverse relationship between serum cholesterol and cancer mortality in Puerto Rico, Amer. J. Epidemiol. 114:29 (1981).

20. B. Peterson, E. Trell and N. H. Sternby, Low cholesterol level as risk factor for noncoronary death in middle-aged men, J. Amer. Med. Assoc. 245:2056 (1981).

21. G. Rose, H. Blackburn, A. Keys, H. L. Taylor, W. B. Kannel, O. Paul, O. Paul, D. D. Reid and J. Stamler, Colon cancer and blood-cholesterol, Lancet 1:181 (1974).

22. G. Rose and M. J. Shipley, Plasma lipids and mortality: a source of

error, Lancet 1:523 (1980).

23. J. D. Kark, A. H. Smith and C. G. Hames, The relationship of serum cholesterol to the incidence of cancer in Evans County, Georgia, J. Chronic. Dis. 33:311 (1980).

24. J. Stamler, D. M. Berkson, H. A. Linber, W. A. Miller, C. R. Soyugene, T. Tokish and T. Whipple, Does hypercholestorolemia increase risk of lung cancer in cigarette smokers?, Circulation Suppl. 6:188 (1968).

25. A. R. Dyer, J. Stamler, O. Paul, R. B. Shekelle, J. A. Schoenberger, D. M. Berkson, M. Lepper, P. Collette, S. Shekelle and H. A. Lindberg, Serum cholesterol and risk of death from cancer and other causes in three Chicago epidemiological studies, J. Cronic. Dis. 34:249 (1981).

26. M. L. Pearce and S. Dayton, Incidence of cancer in men on a diet high in polyunsaturated fat, Lancet. 1:464 (1971).

27. Committee on Principle Investigators, A cooperative trial in the primary prevention of ischaemic heart disease using clofibrate, Brit. Heart J. 1069-1118 (1978).

28. N. D. Nigro, N. Bhadrachari and C. Chomchai, A rat model for studying colon cancer. Effect of cholestyramine on induced tumors, Dis. Colon Rectum 16:438 (1973).

29. T. Asano, M. Pollard and D. C. Madsen, Effects of cholestyramine on 1,2-dimethylhydrazine-induced enteric carcinoma in germfree rats, Proc. Soc. Exp. Biol. Med. 150:780 (1975).

30. M. F. Melhem, H. F. Gabriel, E. D. Eskander and K. N. Rao, Cholestyramine promotes 7,12-dimethylbenzanthracene induced mammary cancer in Wistar rats, Brit. J. Cancer 56:45 (1987).

31. K. N. Rao, S. Kottapally, E. D. Eskander, S. Shinozuka, S. Dessì and P. Pani, Acinar cell carcinoma of rat pancreas: regulation of cholesterol esterification, Brit. J. Cancer 54:305 (1986).

32. F. Feo, R. A. Canuto, G. Bertone, R. Garcea and P. Pani, Cholesterol and phospholipid composition of mitochondria and microsomes isolated from Morris hepatoma 5123 and rat liver, FEBS Letters 33:229 (1973).

33. G. Fex and L. Wallinder, Liver and plasma cholesteryl esters metabolism after partial hepatectomy in the rat, Biochim. Biophys. Acta 316:91 (1973).

34. P. Pani, S. Dessì, K. N. Rao, B. Batetta and E. Laconi, Changes in serum and hepatic cholesterol in lead-induced liver hyperplasia, Toxicol. Pathol. 12:162 (1984).

35. J. Loeb and G. Davson, High density lipoprotein exhange reactions, Mol. Cell Biochem. 52:161 (1983).

36. S. Eisenberg, High density lipoprotein metabolism, J. Lipid Res. 25:1017 (1984).

37. W. J. Van Blitterswijk, J. Damen, H. Hilkman and J. De Widt, Alterations in biosynthesis and homeostasis of cholesterol and in lipoprotein patterns in mice bearing a transplanted lymphoid tumor, Biochim. Biophys. Acta 816:46 (1985).

38. D. E. Brenneman, S. N. Nathur and A. A. Spector, Characterization of the hyperlipemia in mice bearing the Ehrlich Ascites tumor, Europ. J. Cancer 11:225 (1975).

39. H. N. Ginsberg, L. Ngoc-Anh and H. S. Gilbert, Altered high density lipoprotein metabolism in patients with myeloproliferative disorders and hypocholesterolemia, Metabolism 35:878 (1986).

40. H. Hachem, G. Favre, G. Raynal, G. Blavy, P. Canal and G. Soula, Serum apolipoproteins AI, AII and B in hepatic metastases. Comparison with other liver diseases: hepatomas and cirrhosis, J. Clin. Chem. Clin. Biochem. 24:161 (1986).

41. J. Argiles and E. Herrera, Lipids and lipoproteins in maternal and fetus plasma in the rat, Biol. Neonate 39:37 (1981).

42. M. B. N. Johanson, Lipoproteins and lipids in fetal, neonatal and adult rat serum, Biol. Neonate 44:278 (1983).

MODIFYING INFLUENCE OF FASTING ON DNA SYNTHESIS, CHOLESTEROL METABOLISM

AND HMP SHUNT ENZYMES IN LIVER HYPERPLASIA INDUCED BY LEAD NITRATE

Sandra Dessì, Barbara Batetta, Donatella Pulisci, Annalisa
Carrucciu, Marina Armeni, Maria F. Mulas and Paolo Pani

Istituto di Farmacologia e Patologia Biochimica
Via Porcell 4
09124 Cagliari, Italy

INTRODUCTION

Previous studies by Columbano et al.[1] have shown that lead nitrate when
injected intravenously as a single dose to rats, induces an hyperplastic response in the liver as indicated by an increased DNA synthesis and by an enhanced mitotic index. Liver hyperplasia was accompanied by a stimulation of
cholesterol synthesis and hexose-monophosphate (HMP) shunt enzymes[2].

The pathophysiological link between DNA, cholesterol synthesis and HMP
shunt enzymes, is not completely known, but in a variety of experimental models, an increase of these metabolic pathways has been associated with hyperplasia[3-6].

If these biochemical changes have an essential role during cell proliferation, it is conceivable that the inhibition of these pathways could re-
strain the proliferative capacity of the organ following a mitogenic stimulus. In fact, dehydroepiandrosterone, a non-competitive inhibitor of glu-
cose-6-phosphate dehydrogenase (G6PD) has been shown to have a protective
effect against chemically induced tumors[7]. On the other hand, mevinolin,
compactin, and other inhibitors of cholesterol synthesis are able in vitro to
block DNA synthesis[8-10].

In vivo, however, the inhibition of cholesterol synthesis is not easily
obtained in rats[11]. Since fasting is a metabolic condition unfavorable to
growth, characterized by a very strong depression of both cholesterol synthesis and HMP shunt enzymes in the liver[12], aim of the present work was to
verify whether the hyperplastic response and the associated metabolic changes
can be modified by fasting.

For this purpose, DNA and cholesterol synthesis as well as G6PD and
6-phosphogluconate dehydrogenase (6PGD) activities were investigated in liver
of fasted rats treated with the hyperplastic agent lead nitrate.

In addition, since fasting is also characterized by a decrease in the
activities of the key glycolytic enzymes and by an increasing rate of the
opposing gluconeogenic enzymes, glucose-6-phosphatase (G6Pase) and pyruvate
kinase (PK) were also investigated in this study.

519

MATERIALS AND METHODS

Male Wistar rats weighing 200-250 g were used in these experiments. Lead nitrate (Carlo Erba, Milano, Italy), dissolved in distilled water, was injected i.v. under light ether anesthesia at the dose of 100 μmol/kg body weight. Control rats received a corresponding amount of distilled water.

The animals were divided into two groups (1 and 2). Group 1 was fed ad libitum and killed 36 h after lead injection. Group 2 was fasted 36 h before and 36 h after lead injection for a total period of 72 h of fasting; the rats of this group were sacrificed 36 h after treatment.

DNA and cholesterol synthesis were measured by determining the incorporation of ^{3}H-thymidine into DNA and ^{14}C-acetate into cholesterol respectively, as described previously [12].

For the determination of the enzymatic activities, liver homogenates were centrifuged at 9000 g for 20 min and the supernatants were used to assay G6PD and 6PGD [13], G6Pase [14] and PK [15].

RESULTS

The effect of lead nitrate on DNA synthesis in fed and fasted animals is presented in Fig. 1. An increase of DNA synthesis was observed in both groups, however the rate of ^{3}H-thymidine incorporation into DNA was very lower in fasted than in fed animals. An approximate 50% decrease of G6PD and 6PGD activities and a very strong depression of cholesterol synthesis caused by fasting is shown in Fig. 2. Lead is able to revert these depressions

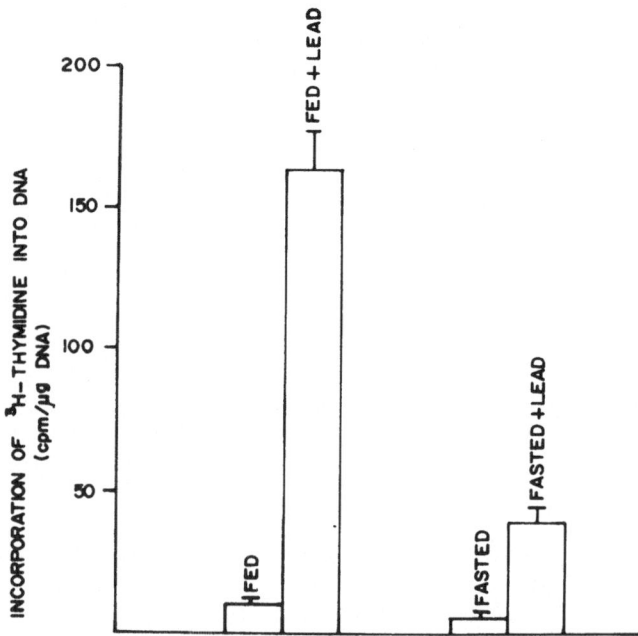

Fig. 1. Effect of lead nitrate on DNA synthesis in fed and fasted rats. Lead nitrate was injected intravenously at a dose of 100 μmol/kg body weight. Fed and fasted controls received distilled water. Vertical bars represent mean ± SE of at least 4 animals.

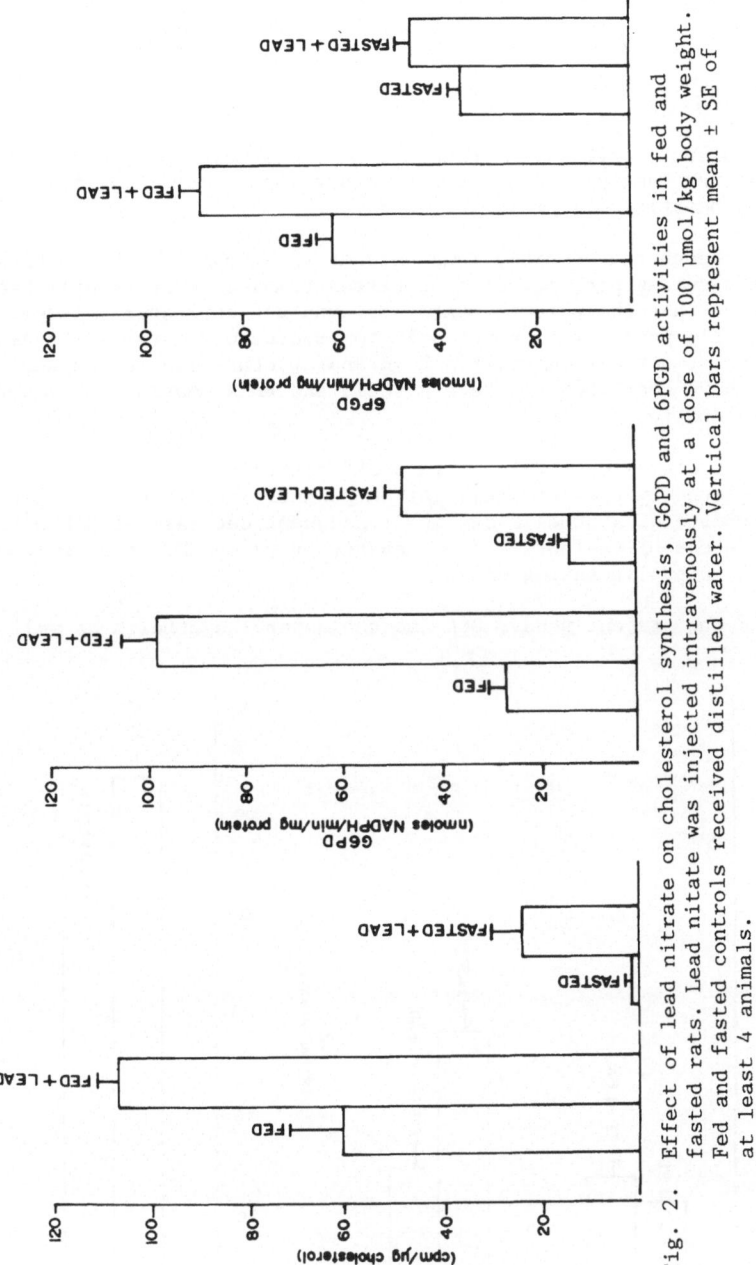

Fig. 2. Effect of lead nitrate on cholesterol synthesis, G6PD and 6PGD activities in fed and fasted rats. Lead nitate was injected intravenously at a dose of 100 μmol/kg body weight. Fed and fasted controls received distilled water. Vertical bars represent mean ± SE of at least 4 animals.

inducing both cholesterol synthesis and HMP shunt enzymes. Also in this case however the reached levels were lower in fasted than in fed animals (Fig. 2).

No change in G6Pase and a significant decrease in PK were found in lead-treated fed rats (Fig. 3), on the contrary a decrease in G6Pase and an increase in PK were observed in lead-treated fasted rats (Fig. 3).

DISCUSSION

A direct correlation between DNA, cholesterol synthesis and HMP shunt enzymes has been previously shown in our laboratory in different models involving hepatic cell proliferation[2-6].

Although the possible cause-effect relationship between these metabolic pathways during proliferating processes remains to be established, it was suggested that cholesterol synthesis, and not cholesterol itself, is needed during cell proliferation not only to produce cholesterol for new membrane biogenesis, but also mevalonic acid and/or other isoprenoid units that recently have been shown to have a rapid and an essential initiator function in DNA replication[9,10].

Cholesterol synthesis is an NADPH-consuming pathway and since HMP shunt is regulated by the intracellular NADPH/NADP ratio, the utilization of NADPH for cholesterol synthesis and the consequent decrease of NADPH/NADP ratio could be one of the possible mechanisms by which HMP shunt enzymes are induced in proliferating cells.

In the present study, DNA and cholesterol synthesis as well as G6PD,

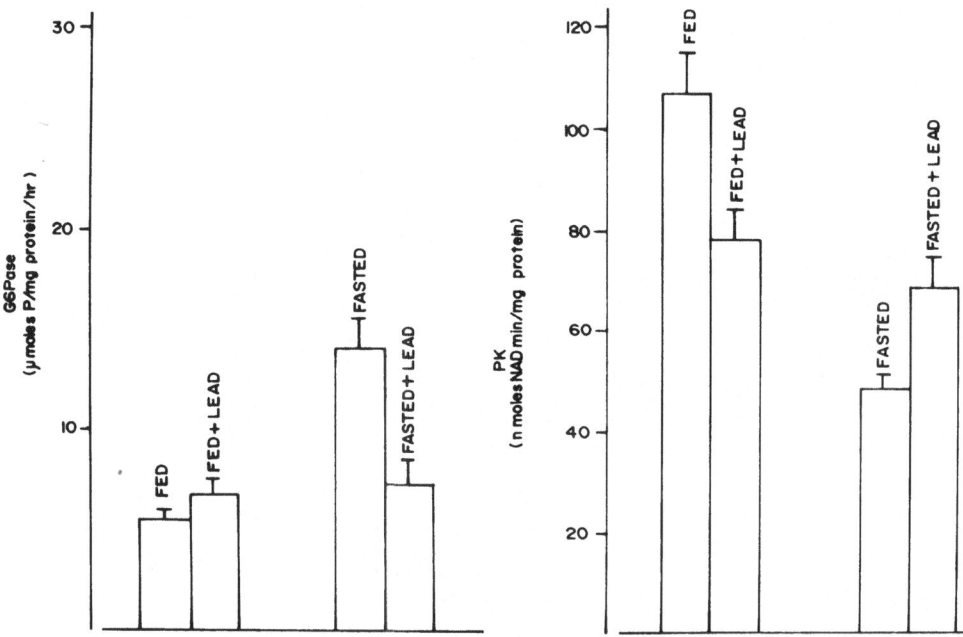

Fig. 3. Effect of lead nitrate on G6Pase and PK activities in fed and fasted rats. Lead nitrate was injected intravenously at a dose of 100 µmol/kg body weight. Fed and fasted controls received distilled water. Vertical bars represent mean ± SE of at least 4 animals.

6PDG, G6Pase and PK were investigated in liver of fasted rats treated with the potent mitogen lead nitrate.

Fasting was chosen as an experimental model characterized by a strong depression of those metabolic pathways that we suggest related to liver hyperplasia[3,12].

Our results show that even if at minor extent, hepatic cell proliferation induced by lead also occurs during fasting, and it is accompanied by a de-repression of cholesterol synthesis and HMP shunt enzyme activities which are inhibited during fasting. The levels of cholesterol synthesis and HMP shunt enzyme activities in lead-treated fasted rats, however, are lower than in fed rats.

These data suggest that cholesterol synthesis and HMP shunt pathway in proliferating processes respond only partially to the regulatory factors that normally control these pathways, and add new support for the concept that an endogenous source of newly synthesized cholesterol and a suitable increase of HMP shunt enzymes positively correlated to the degree of DNA synthesis are needed during cell proliferation.

Furthermore, the fact that G6Pase and PK activities are regulated differently during cell proliferation in fed and fasted conditions, while G6PD and 6PGD always increase, suggests a shift of glucose metabolism towards HMP shunt pathway during proliferating processes.

Since acetyl-CoA, the precursor of cholesterol synthesis, is formed from pyruvate, the possibility that the observed changes in PK activity in fed and fasted animals, may be related to acetyl-CoA production available for cholesterol synthesis must be also considered.

ACKNOWLEDGEMENTS

This work was supported by CNR Roma (No. 86.00510.44), AIRC, SARAS "Raffinerie Sarde" and by Assessorato Igiene e Sanità della Regione Autonoma della Sardegna (Cagliari, Italy).

REFERENCES

1. A. Columbano, G. M. Ledda, P. Sirigu, T. Perra and P. Pani, Liver cell proliferation induced by a single dose of lead nitrate, Amer. J. Pathol. 110:83 (1983).
2. S. Dessì, B. Batetta, E. Laconi, C. Ennas and P. Pani, Hepatic cholesterol in lead nitrate induced liver hyperplasia, Chem. Biol. Interactions 48:271 (1984).
3. S. Dessì, C. Chiodino, B. Batetta, E. Laconi, C. Ennas and P. Pani, Hexose-monophosphate shunt and cholesterol synthesis in the diabetic and fasting states, Exp. Mol. Pathol. 43:177 (1985).
4. S. Dessì, C. Chiodino, B. Batetta, A. M. Fadda, C. Anchisi and P. Pani Hepatic G6PD, cholesterogenesis and serum lipoproteins in liver regeneration after partial hepatectomy, Exp. Mol. Pathol. 44:169 (1986).
5. G. M. Ledda-Columbano, A. Columbano, S. Dessì, P. Coni, C. Chiodino and P. Pani, Enhancement of cholesterol synthesis and pentose phosphate pathway activity in proliferating hepatocyte nodules, Carcinogenesis 6:1371 (1985).
6. G. M. Ledda-Columbano, A. Columbano, S. Dessì, P. Coni, C. Chiodino, G. Faa and P. Pani, Hexose monophosphate shunt and cholesterogenesis in lead-induced kidney hyperplasia, Chem. Biol. Interactions

62:209 (1987).

7. A. G. Schwartz and R. H. Tonnen, Inhibition of 7,12-dimethyl-benz(a)-anthracene and urethane-induced lung tumor formation in A/J mice by long-term treatment with dehydroepiandrosterone, Carcinogenesis 2: 1335 (1981).

8. M. S. Brown and J. L Goldstein, Suppression of 3-hydroxy-3-methyl-glutaryl coenzyme A reductase activity and inhibition of growth of human fibroblasts by 7-ketocholesterol, J. Biol. Chem. 249:7306 (1974).

9. V. Q. Huneeus, M. H. Wiley and M. D. Siperstein, Essential role for mevalonate synthesis in DNA replication, Proc. Natl. Acad. Sci. 76:5056 (1979).

10. A. J. R. Habenicht, J. A. Glomset and R. Ross, Relation of cholesterol and mevalonic acid to the cell cycle in smooth muscle and swiss 3T3 cells stimulated to divide by platelet-derived growth factor, J. Biol. Chem. 255:5134 (1980).

11. A. Endo, Y. Tsujita, M. Kuroda and K. Tanzawa, Effects of ML-236B on cholesterol metabolism in mice and rats: lack of hypocholesterolemic activity in normal animals, Biochim. Biophys. Acta 575:266 (1979).

12. S. Dessi, C. Chiodino, B. Batetta, M. Armeni, M. F. Mulas and P. Pani, Comparative effect of insulin and refeeding on DNA synthesis, HMP shunt and cholesterogenesis in diabetic and fasted rats, Pathology 20:53 (1988).

13. G. E. Glock and P. McLean, Further studies on the properties and assay of glucose-6-phosphate dehydrogenase and 6-phosphogluconate dehydrogenase of rat liver, Biochem. J. 55:400 (1953).

14. R. C. Nordlie and W. J. Arion, Glucose-6-phosphatase, in: "Methods in Enzymology", W. A. Wood, ed., Academic Press, New York (1966).

15. T. Bücher and G. Pfleiderer, Pyruvate kinase from muscle, in: "Methods in Enzymology", S. P. Colowick and N. O. Kaplan, eds., Academic Press, New York (1955).

REGULATION OF PROTEIN DEGRADATION IN NEOPLASTIC CELLS

Franco M. Baccino, Riccardo Autelli, Gabriella Bonelli,
Paola Costelli, Ciro Isidoro and Luciana Tessitore

Department of Experimental Medicine and Oncology, General
Pathology Section, Torino University, Corso Raffaello 30
10125 Torino, Italy

GROWTH

Changes in the overall rate of cell protein degradation (PD) were included by Hershko and co-workers in 1971[1] among the elements of the "pleiotypic response", which was categorized as "positive" if associated with growth stimulation or "negative" if accompanying growth suppression. Protein turnover modulations were thus regarded as instrumental in effecting growth-phase transitions. In 1977 Warburton and Poole[2] analyzed the relationship between cell growth and PD rates in tissue cultures exposed to serum deprivation or other manipulations. Moreover, in 1977 Gunn et al.[3] reported that basal PD rates in various transformed cells were lower than in their normal counterparts, hypothesizing that a reduction in intracellular proteolysis could play a role in neoplastic growth. Since then, many methodological advances have been developed to analyze PD in tissues or cells, particularly in tissue cultures. Nevertheless, studies on the relationship between cell protein turnover rates and growth or neoplastic transformation have generally relied on experimental models not adequately defined in terms of growth properties and kinetics, not to mention the rather crude tools often used to elicit growth-phase transitions.

Both abrupt serum deprivation and gradual shift to confluency and/or quiescence in unchanged growth medium have been the experimental protocols most frequently adopted to effect on tissue cultures the shift that can be categorized as "emergence from proliferation" or growth-to-quiescence transition. Alternatively, "emergence from quiescence" or quiescence-to-growth transition has been investigated and most frequently brought about by replating cells or supplying cultures with fresh medium or serum[2,4]. With both kinds of approaches, however, far more than a single variable is affected, thereby precluding any univocal interpretation of results. It is impossible, indeed, to discriminate whether protein turnover shifts, whenever present, are inherent to the growth-phase transition or rather are merely concomitant phenomena due to changes in medium composition.

These limitations notwithstanding, there is no doubt that many mammalian cells, though not all of them, show a rather stringent control of overall cell PD versus the growth state, as established for tissue cultures and for cells or tissues in the animal as well[5-11]. In principle, such growth-associated modulations of PD appear quite reasonable, since a reduction in breakdown rates is an economical and very effective way to determine or enhance protein accumulation, i.e. growth. Of interest, the same strategy is adopted during the development. Many developing organisms or tissues, indeed, have been shown to turn over their own protein constituents at rates quite lower than those set in during the adult life[11].

The growth-related restriction in PD is most commonly associated with enhanced protein synthesis, which probably is an even more general process occurring in proliferating cells or tissues. Therefore, growth is quite often the result of a double, convergent regulation: of protein synthesis, which is enhanced, and of PD, which is reduced. In spite of many efforts, the precise nature of the factor(s) and mechanism(s) associating reduced PD with growth or coordinating the accelerated synthesis with restricted PD in growing cells remains elusive (cf. 9).

NEOPLASTIC TRANSFORMATION

A spectrum of abnormalities has been reported for protein catabolic properties in transformed or tumoral cells. Most frequent is a quantitative difference, since the overall rate of protein turnover as well as the amplitude of protein catabolic modulations in response to environmental stimuli are lower in neoplastic than in normal cells[3,5,6,8,12,13]. In this direction, Schwarze and Seglen[14] have reported that even hepatocytes isolated from the liver of rats under carcinogenic treatment with diethylnitrosamine are already deviated from the normal pattern. On this basis, it has been hypothesized that cells developing relaxations in growth control may benefit of restrictions in protein turnover rates, being thus provided with selective metabolic advantage for growth[3,13,15].

Gronostajski and Pardee[16] have addressed the issue by comparing overall proteolytic rates in different lines of Balb/c 3T3 cells. Four transformed lines were examined and, with respect to their normal counterpart, all exhibited lower rates of "basal" proteolysis as well as reduced proteolytic elevations in response to serum deprivation. However, differences among transformed lines did not correlate with the tumorigenic potential and were also higher than those between transformed and normal cells. It was thus concluded that there was "no consistent change in any rate of protein degradation due to transformation to tumorigenicity".

Whether tumorigenicity is the proper reference parameter in the above or similar studies remains to be assessed, however. For instance, it is not known whether tumorigenicity strictly correlates with proliferation or growth rates. Yet PD rates could well be related to rates of growth or cell birth (cf. 17). Unfortunately, such parameters are not evaluated in most investigations. The key issue, indeed, is whether for a given cell type the neoplastic transformation modifies PD and its regulation in such a way as to determine or favor an anabolic attitude. Generally speaking, it is still unsettled whether alterations in overall protein catabolism inhere to the neoplastic transformation or rather reflect differences in growth kinetic parameters between neoplastic and normal cell populations.

ASCITES HEPATOMA

In view of the many problems arising in tissue culture studies, not to mention the difficulties to make any safe extrapolation from such in vitro studies to neoplastic growth in the animal, we have developed an in vivo-in vitro model based on the use of an ascites hepatoma of the rat. The Yoshida AH-130 tumor has been analyzed in many details with respect to growth kinetics and PD regulations associated with growth-state transitions[7,18,19].

Growth Properties

Inoculating intraperitoneally 10^8 cells, derived from exponentially-growing tumors, the ascites hepatoma grows rapidly in rats for about a week, the growth fraction being close to 1 and the t_d about a day. Thereafter growth slows down and eventually subsides; the tumor shifts to a quasi-stationary state, wherein the population size remains approximately constant until animals die, about two weeks after transplantation.

Cell Protein Turnover

Protein metabolism has been investigated[18] by monitoring over three days the decay of specific and total cell protein radioactivity since 24 h after labeling proteins with ^{14}C-bicarbonate (at day 1 or day 8 of tumor growth for exponential and stationary tumors, respectively). The protein balance can be expressed in the form of an equation based on the fractional rates of protein accumulation (k_a, estimated from the actual protein content of tumors; apparent k'_a, calculated by difference between k_s and k_d), synthesis (apparent k_s, calculated from the specific protein radioactivity decay), and degradation (apparent k_d, from the total protein radioactivity decay):

$$k'_a = k_s - k_d$$

The experimental data, expressed as %/h, were as follows:

I. exponential tumors (days 2-4): $2.49 = 3.10 - 0.61$ ($k_a = 2.77$)

II. stationary tumors (days 9-11): $0.06 = 1.49 - 1.43$ ($k_a = 0.00$)

These results show that in AH-130 tumors the exponential growth is associated with high synthetic and low degradative rates, while cessation of growth is achieved by convergent reduction of synthesis and two-fold enhancement of breakdown (Fig. 1). The apparent average half-life of cell protein (slow-turnover pool) drops from 7-8 days in growing tumors to less than 2 days in growth-arrested tumors[18].

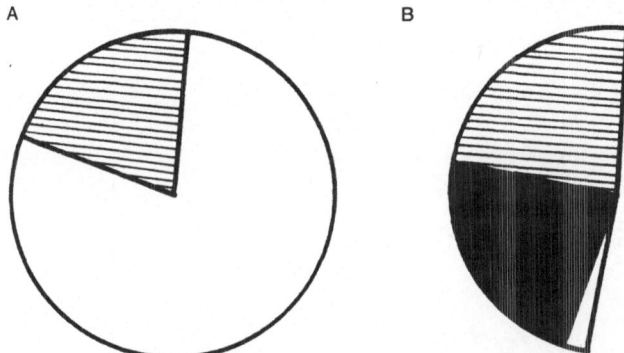

Fig. 1. Rates of apparent protein accumulation (k'_a), synthesis (k_s) and degradation (k_d) in cells from exponential (A) and stationary (B) Yoshida ascites hepatomas. All rates expressed as percentage of the protein synthesis rate in exponential tumor cells, the value for which corresponds to the circle in A; synthesis rate for stationary tumor cells corresponds to the whole sector in B. ☐ Protein accumulation rate; ▤ ammonia-resistant degradation rate; ■ ammonia-suppressible degradation rate (acidic vacuolar proteolysis).

<u>Lysosomal Proteolysis and Autophagy</u>

Proteolytic rates have also been measured <u>in vitro</u>, 24 h after labeling protein in the animal with ^{3}H-leucine, on AH-130 cells incubated in ascitic fluid from unlabeled paired tumors[18,19]. These <u>in vitro</u> rates are virtually identical to the degradation rates measured in the animal and thus allow moment-to-moment estimates of the latter. In addition, it has been possible to dissect <u>in vitro</u> the proteolytic mechanisms by using appropriate drugs. Energy metabolism inhibitors, such as rotenone, are quite effective in suppressing proteolysis in cells from growing tumors and, even more markedly, in those from stationary tumors. The residual activity is similar for cells in either growth state, indicating that energy-independent proteolytic processes do not contribute substantially to the growth-associated regulation of PD.

Virtually all of the elevation of PD in stationary tumor cells is due to activation of the lysosomal mechanism from a state of virtually complete suppression in growing tumor cells, as shown by using inhibitors of the acidic vacuolar pathway for PD (Figs. 1 and 2). Moreover, as shown in Fig. 2, inhibitors of the final hydrolytic step in this pathway, such as ammonia or leupeptin, are immediately effective on stationary tumor cells; in contrast,

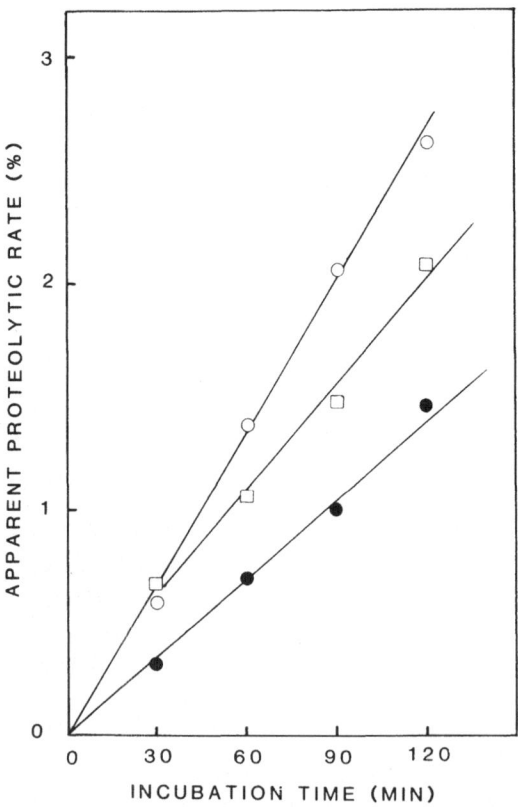

Fig. 2. Kinetics of the effects of lysosomal inhibitors on endogenous proteolysis in AH-130 cells. Prelabeled cells from stationary tumors were incubated in ascitic fluid in the absence (○) or in the presence of 40 mM ammonia (●) and 10 mM asparagine (□).

inhibitors of the early sequestrational or fusional steps, such as 3-methyl-adenine or cycloheximide, show a lag of about 30 min. Such difference is conceivably due to the time required for cells to break down those protein substrates they have already sequestered in the vacuolar system when exposed to the inhibitors.

In collaboration with U. Pfeifer [20], the issue has been further investigated by morphometric analysis, in order to quantify autophagy. The fractional cytoplasmic volume occupied by early autophagic vacuoles (those containing recognizable cytoplasmic structures) increases from 0.31×10^{-4} in exponential tumor cells to 1.37×10^{-4} in stationary tumors. At a first sight, these data appear consistent with the above biochemical measurements in pointing to an elevation of the autophagic pathway in stationary tumors. Nevertheless the two sets of data show a remarkable quantitative discrepancy. Assuming for autophagic vacuoles a life span of the same order of that estimated in the case of hepatocytes [21,22], the 4 to 5-fold increase in morphologically recognizable autophagic vacuoles would only account for a minor part of the elevation of acidic vacuolar proteolysis occurring in stationary tumor cells.

Triggers

In a search for signals or factor(s) that might be responsible for the elevation of PD in stationary AH-130 tumors, or for its reduction in exponential ones, the composition of the ascitic fluid has been first considered. Exponential tumor cells have been incubated in liquid from unlabeled stationary tumors and vice versa; both complementations result in only small alterations of proteolytic rates, far from accounting for the difference between tumors in the two growth states. Cells from exponential tumors have also been exposed to a complete stepdown, by incubation in purely saline media, or to classical inducers of autophagy, such as glucagon or cAMP; but the changes thus elicited are again quite narrow. All such data consistently show that PD in AH-130 cells is quite refractory to environmental factors or, what is more relevant to the present discussion, that growing tumor cells cannot be induced by a variety of environmental factors to elevate their PD up to the rates occurring in stationary tumors.

Free Amino Acids

Amino acids are thought to play a regulatory role on the lysosomal pathway for PD [23,24]. Therefore, the tumoral ascitic fluid and the blood plasma of tumor hosts have been analyzed for free amino acid concentrations [19]. Relevant to the present discussion are three main observations, all of which are a bit surprising, indeed, and rather seem to support the view that "in the whole animal amino acid concentrations are regulated and not regulators" [25].

The total amino acid concentration in the ascitic fluid is much higher for stationary than for exponential tumors. Therefore, depletion of amino acids cannot be responsible for the elevation of PD in the plateau phase of tumor growth.

On the whole, amino acid concentrations in the ascitic fluid from stationary tumors do not notably differ from those in normal blood plasma. The main exceptions are provided by Gln and Ser, the concentrations of which are considerably lower in the ascitic fluid. Since their concentrations are extremely low in the fluid from exponential tumors, it can be inferred that Gln and Ser are avidly consumed by tumor cells.

The amino acid concentration in blood plasma is somewhat reduced in correspondence with the initial phases of tumor growth, yet it increases subsequently, in spite of the markedly negative nitrogen balance of the host.

529

The increase is largely due to a marked elevation in the concentration of branched-chain amino acids, namely, Ile, Leu and Val. Such an increase is of usual observation in starved animals. In AH-130 hosts, however, glycemia does not vary significantly over the course of tumor growth. It can thus be excluded that we are dealing here with a simple situation of starvation (cf.18,26).

Hormones

The plasmatic levels of different hormones have been measured in AH-130 tumor-bearing rats. Insulin and glucagon show quite marked changes since day 4, the sign of the shifts being compatible with the induction of a catabolic state. Yet these hormones are virtually ineffective on AH-130 cells in vitro and, in addition, their changes are already evident in correspondence of the exponential phase of tumor growth. For the latter reason, the changes observed for thyroid hormones should also have no relevance, while attention should be paid to the altered levels of corticosterone. Neither thyroid hormones nor corticosterone have been tested in vitro so far.

Rather than on tumor cells, however, these hormonal changes are likely to play some role in the protein metabolic perturbations that develop in tumor hosts and lead to cachexia. An elevation of the host body temperature and an increase in the plasmatic level of triglycerides are also worth of being noted, since both changes might reflect an increased production of mediators such as interleukin-1 and cachetin[26]. In this regard, the tumor grows in rats which develop a markedly negative nitrogen balance since early after tumor transplantation. It has been shown that such metabolic perturbation mostly involves tissue protein breakdown rather than synthesis, at least in liver and gastrocnemius muscle[26]. Therefore, the net host-to-tumor flux of metabolites in the present experimental model does not depend on simple competition between host and tumor cells. A decisive role is also played by changes in tissue protein metabolism, which is probably driven into a catabolic state by factors, not clearly identified as yet, to be included among those which mediate the host response to the tumor.

Cell Turnover (cell loss)

A convergent modulation of protein synthesis and degradation is operated by many cells to regulate protein accumulation, and neoplastic cells may show defects in such regulations. Yet even highly-deviated cells such as those of the Yoshida ascites hepatoma cease growth by reducing protein synthesis and enhancing PD. To evaluate properly the latter change, it is worthy noting that AH-130 cells in the plateau phase do not reach a real quiescence, as indicated by thymidine incorporation or DNA radioactivity decay and also inferred from DNA distribution by flow cytometry[18]. In stationary tumors, rather than accumulating in G1 or in a putative G0, cells remain largely distributed all along the cycle, although with a significant decrease in the proportion of cells in S phase.

These data indicate that attainment of the stationary state in AH-130 tumors involves at least two phenomena, on the degradative side: I. an increase in protein turnover, via activation of the acidic vacuolar pathway for PD; and II. an increase in cell turnover, which means cell loss. The relative contributions of the two mechanisms remain to be assessed (cf. 17). While the acidic vacuolar proteolysis is undoubtedly activated, so far we don't know how much of it is related to, or even contributed for, by cell death. On one hand, the increase in PD might be instrumental in leading to cell death. On the other, dying cells or debris could be taken up and degraded by other cells; this heterophagic mechanism could closely mimic autophagy, both morphologically and biochemically.

CONCLUSIONS AND PERSPECTIVES

Growth-state transitions in the AH-130 tumor are associated with marked changes in PD. The total amino acid concentration in the ascitic fluid is increased in stationary tumors, rather than decreased; thus changes in amino acid concentration do not seem to play any role in causing tumors to cease growth and to enhance PD. As far as known, growth of the ascites hepatoma AH-130 only subsides when the oxygen concentration becomes critical for cells engaging in DNA duplication[27]. In this regard, recent observations by Pettersen et al.[28] on NHIK 3025 cells are of considerable interest. These cells, a human line derived from cervical carcinoma, do not show any PD regulation versus the growth rate[29], yet they cease growth and protein accumulation through a marked elevation of PD when exposed to acute hypoxia.

No environmental signal or factor that might directly account for the elevation of PD in stationary AH-130 tumors has been recognized so far. The acidic vacuolar proteolysis in cells from exponential tumors can not be activated substantially in vitro by means such as complementation with ascitic fluid from stationary tumors or incubation in saline or addition of cAMP. Nor the reciprocal hypothesis, that environmental agents could suppress PD down to the level occurring in exponential tumors, is supported by any finding in the present work.

Absence of evidence is not evidence of absence, yet AH-130 cells seem very little responsive to environmental stimuli, which are known to be very effective in elevating PD in cells of many types, hepatocytes included. In these tumor cells the pace for PD rates seems mostly controlled by intracellular rather than extracellular signals, which might be a further manifestation of the autonomous growth of the AH-130 tumor.

Even cells exhibiting extensive defects in growth regulation such as those of the AH-130 tumor somehow still maintain an option to emerge from the proliferative cycle and to adjust accordingly their protein turnover state. However, rather than immediate environmental stimuli, the growth state itself seems to be what determines the protein turnover state of such tumor cells. In this direction, we are presently evaluating whether the enhanced proteolysis in stationary tumors preferentially involves cells in a specific phase of the cell cycle or reflects a generalized or random change in all cells.

AKNOWLEDGEMENTS

Work supported in part by grants from the Associazione Nazionale per la Ricerca sul Cancro, Milano, the Ministero della Pubblica Istruzione, Roma, and the Consiglio Nazionale delle Ricerche, Roma (Progetto Finalizzato "Oncologia", contratto 87.01170.44; progetto bilaterale Italia-USA, CT 87.00039.04).

REFERENCES

1. A. Hershko, P. Mamont, R. Shields and G. M. Tomkins, Pleiotypic response, Nature 232:206 (1971).
2. M. J Warburton and B. Poole, Effect of medium composition on protein degradation and DNA synthesis in rat embryo fibroblasts, Proc. Natl. Acad. Sci. USA 74:2427 (1977).
3. J. M. Gunn, M. G. Clark, S. E. Knowles, M. F. Hopgood, and F. J. Ballard, Reduced rates of proteolysis in transformed cells, Nature 266:58 (1977)
4. A. M. Spanier, W. A. Clark jr., and R. Zak, Replacement perfusion of

cultured eucaryotic cells: a method for the accurate measurement of the rates of growth, protein synthesis, and protein turnover, J. Cell. Biochem. 26:67 (1984).

5. F. M. Baccino, M. Messina, M. Musi, and L. Tessitore, Regulation of cell protein turnover and proteinase activities in relation to growth in: "Recent trends in Chemical Carcinogenesis", P. Pani, F. Feo, and A. Columbano, eds., ESA, Cagliari (1982).

6. F. M. Baccino, L. Tessitore, and G. Bonelli, Control of protein degradation vs. growth phase in normal and neoplastic cells, Toxicol. Pathol. 12:281 (1984).

7. F. M. Baccino, L. Tessitore, G. Bonelli, and C. Isidoro, Protein turnover state of tumour cells and host tissues in an experimental model, Biomed. Biochim. Acta 45:1585 (1986).

8. F. J. Ballard, S. E Knowles, S. S. C. Wong, J. B. Bodner, C. M. Wood, and J. M. Gunn, Inhibition of protein breakdown in cultured cells is a consistent response to growth factors, FEBS Lett. 114:209 (1980).

9. F. J. Ballard, Regulation of protein breakdown by epidermal growth factor in A431 cells, Exp. Cell Res. 157:172 (1985).

10. F. J. Ballard, Regulation of intracellular protein breakdown with special reference to cultured cells, in: "Lysosomes: their Role in Protein Breakdown", H. Glaumann and F. J. Ballard, eds., Academic Press, London (1987).

11. O. A. Scornick and V. Botbol, Protein metabolism and liver growth, in: "Lysosomes: their Role in Protein Breakdown", H. Glaumann and F. J. Ballard, eds., Academic Press, London (1987).

12. S. M. Cockle and R. T. Dean, Derangement of regulation of protein degradation in transforming fibroblasts, Biosci. Rep. 2:107 (1982).

13. T. D. Lockwood and I. A. Minassian, Protein turnover and proliferation. Failure of SV-3T3 cells to increase lysosomal proteinases, increase protein degradation and cease net protein accumulation, Biochem. J. 206:251 (1982).

14. P. E. Schwartze and P. O. Seglen, Reduced autophagic activity, improved protein balance and enhanced in vitro survival of hepatocytes isolated from carcinogen-treated rats, Exp. Cell Res. 157:15 (1985).

15. A. Anastasi and R. T Dean, Regulation of protein degradation in fibroblasts, in: "Intracellular Protein Catabolism", E. A. Khairallah, J. S. Bond, and J. W. C. Bird, eds., Alan R. Liss, New York, (1985).

16. R. M. Gronostajski and A. B. Pardee, Protein degradation in 3T3 cells and tumorigenic transformed 3T3 cells, J. Cell. Physiol. 119:127 (1984).

17. J. S. Amenta, J. Mehta and F. M. Baccino, Proteolysis associated with thymidine-induced selective cell death in L-cell cultures, this book.

18. L. Tessitore, G. Bonelli, G. Cecchini, J. S. Amenta and F. M. Baccino, Regulation of protein turnover versus growth state: ascites hepatoma as a model for study both in the animal and in vitro, Arch. Biochem. Biophys. 255:372 (1987).

19. L. Tessitore, G. Bonelli, G. Cecchini, R. Autelli, J. S. Amenta and F. M. Baccino, Regulation of protein turnover vs. growth state. II. Studies on the mechanisms of initiation of acidic vacuolar proteolysis in cells of stationary ascitic hepatoma, Biochem. J. in press (1988).

20. U. Pfeifer, L. Tessitore, G. Bonelli and F. M. Baccino, Regulation of protein turnover versus growth state. III, Growth cessation is associated with activation of autophagy in Yoshida ascites hepatoma AH-130, submitted (1988).

21. J. Kovàcs, E. Fellniger, P. A. Kàrpàti, A. L. Kovàcs and L. Làzlò, The turnover of autophagic vacuoles: evaluation by quantitative electron microscopy, Biomed. Biochim. Acta 45:1543 (1986).

22. U. Pfeifer, Inhibition by insulin of the formation of autophagic vacuoles in rat liver, J. Cell Biol. 78:152 (1978).

23. G. E Mortimore and A. R. Poso, The lysosomal pathway of intracellular proteolysis in liver: regulation by amino acids, Adv. Enzymol. 25:257 (1986).

24. P. O. Seglen and P. B. Gordon, Amino acid control of autophagic sequestration and protein degradation in isolated rat hepatocytes, J. Cell Biol. 99:435 (1984).

25. J. C. Waterlow, Protein turnover with special reference to man, Quart. J. Exp. Physiol. 69:409 (1984).

26. L. Tessitore, G. Bonelli and F. M. Baccino, Early development of protein metabolic perturbations in the liver and skeletal muscle of tumour-bearing rats: a model system for cancer cachexia, Biochem. J. 241:153 (1987).

27. M. Olivotto, R. Caldini, M. Chevanne and M. G. Cipolleschi, The respiration-linked limiting step of tumor cell transition from the non-cycling to the cycling state: its inhibition by oxidizable substrates and its relationships to purine metabolism, J. Cell. Physiol. 116:149 (1983).

28. E. O. Pettersen, N. O. Juul and O. W. Ronning, Regulation of protein metabolism of human cells during and after acute hypoxia, Cancer Res. 46:4346 (1986).

29. Ø. W. Rønning, T. Lindmo, E. O. Petterson and P. O. Seglen, Effect of serum step-down on protein metabolism and proliferation kinetics of NHIK 3025 cells, J. Cell. Physiol. 107:47 (1981).

PROTEOLYSIS ASSOCIATED WITH THYMIDINE-INDUCED SELECTIVE

CELL DEATH IN L-CELL CULTURES

Joseph S. Amenta[1], Jardir Mehta[1] and Francesco M. Baccino[2]

[1]Department of Pathology, University of Pittsburgh School of
Medicine, Pittsburgh, PA 15261, USA and [2]Dipartimento di
Medicina ed Oncologia Sperimentale, Sezione di Patologia
Generale, Università di Torino, Corso Raffaello 30,
10125 Torino, Italy

INTRODUCTION

Previous studies in our laboratories have suggested that some established cell lines, both in vitro and in vivo, show an increased rate of proteolysis as they attain high density[1-5]. Unusual was the observation that this increased proteolysis appeared to be associated with a unique type of selective
cell death (SCD), while the remaining viable cells continued to show a low
rate of protein turnover. The exact relationship between this increased
proteolysis and SCD, however, has remained elusive.

In this report we have re-evaluated the relationship between proteolysis
and cell death in L-cell cultures, comparing the protein hydrolysis observed
in TdR(thymidine)-induced SCD with that produced by GCN (general cell necrosis) induced by chemical and physical agents. These experiments indicated
that L-cells under conditions that induce SCD hydrolyze about 40% of the cell
protein. GCN, in contrast, resulted in only minimal hydrolysis of cell protein. The proteolysis associated with SCD was inhibited by NH_4Cl, cycloheximide (CH), and was reduced when the number of viable cells remaining on the
monolayer was reduced. These observations suggested that the TdR-induced SCD
produced in L-cell cultures is apoptosis, which can be functionally distinguished from GCN by the hydrolysis of the protein released from the apoptotic
cell.

METHODS

Mouse L-cells were maintained in Eagle's minimum essential medium
(Gibco, Grand Island, NY, USA) supplemented with 10% (v/v) fetal-calf serum
(Gibco) in 75 cm flasks (Falcon Becton-Dickenson, Rutherford, NJ, USA). For
the experiments the cells were subcultured in 28 cm^2 Petri dishes (Falcon).
Penicillin, streptomycin, kanamycin and mycostatin were added to the medium
in all experiments. Cells were labeled for 3 days with ^{14}C-leucine (0.05
µCi/ml) and ^{3}H-TdR (0.10 µCi/ml), as previously described[3]. To clear the
acid soluble radioactivity, we placed all cultures in a nonlabeled chase
medium containing 4.0 mM leucine for 1 h before initiating the experiment.
For the freeze-thaw experiments cell monolayers were placed in a -20°C
freezer for 15 min, then rapidly thawed at room temperature; this procedure

was repeated twice more. Fresh chase medium was then added to each dish. For the experiments using a suspension of freeze-thawed cells on viable monolayer, the labeled L-cells were freeze-thawed x3, scraped into chase medium, and the suspension placed over a monolayer of subconfluent nonlabeled L-cells. The protein in the labeled cellular debris/viable monolayer protein approximated a ratio of approximately 1:2. For the experiments with inhibitors (TdR, NH$_4$Cl and CH), the chase medium was changed daily. Daily assays of cell protein, DNA and radioactivity in cells and medium were as previously described[1-6]. Calculations of fractional accumulation, turnover and synthesis rates for protein and DNA were based on the changes in mass, radioactivity and specific activity[1]. The fractional degradation of cell protein and of DNA were also determined from the accumulation of radioactivity in the medium for each 24 h period: total turnover for the culture was calculated, based on total medium radioactivity; degradation was then calculated, based on the TCA (trichloroacetic acid) soluble fraction.

RESULTS

General Cellular Necrosis

GCN was induced in L-cell cultures by treating the cultures with 10 mM NaCN. Most of the ^{14}C-labeled cell protein was recovered in the medium within 4 h, reflecting the rapid disintegration of the monolayer cells (Fig. 1A). By 24 h approximately 90% of the ^{14}C was recovered in the medium. Only a small fraction of this label, however, was hydrolyzed to TCA soluble form, reaching approximately 8.5% by 24 h. Labeled DNA was released more slowly into the medium (Fig. 1B), reflecting the loss of nuclear material. In contrast to the protein, hydrolysis of labeled DNA in the medium continued throughout the experimental period, reaching 74% by 24 h.

GCN was also produced by treating the cells with three freeze-thaw cycles over a period of approximately 1 h. As expected, a large fraction of the cell protein, 65%, was found in the medium directly after the freeze-thaw cycles (Fig. 2A). This increased to approximately 85% by 2 h and reached 90% in 24 h. During the freeze-thaw period, the acid soluble fraction increased from less than 1% to 4%. During the next 24 h a further increase of 4-5% in acid soluble radioactivity was observed. This pattern of secondary proteolysis was initially slightly more rapid than that observed previously in L-cells treated with NaCN, though not different at the end of 24 h. Freeze-thawing also released the labeled DNA more rapidly into the medium, along with a more rapid hydrolysis of the labeled DNA. Again at 24 h the effects of freeze-thawing and CN were exactly the same (Fig. 2B).

These data demonstrated that GCN was characterized by a small amount of secondary proteolysis, in the range of 8% over a 24 h period. Much of this appeared to occur shortly after cell death, while the cell structure was still partially intact. Secondary hydrolysis of DNA, however, was much higher, in the range of 60-70%, suggesting that high neutral DNAase activity existed in the medium-cell debris suspension.

We next considered the possibility that proteins in the cellular debris could be hydrolyzed by proteases in or from adjacent viable cells remaining in the monolayer. In this experiment, labeled L-cell cultures were killed by a triple freeze-thaw cycle and then added to the medium of nonlabeled L-cell cultures. The presence of viable cells increased slightly the hydrolysis of the protein debris from the freeze-thawed cells (Fig. 3A). This increase was observed only during the second phase of the experiment, extending from 3 to 25 h. Total hydrolysis reached at most 15% of the labeled protein during the 25 h experimental period. Hydrolysis of DNA was not affected by the presence of viable cells (Fig. 3B). Thus, even under circumstances designed to optim-

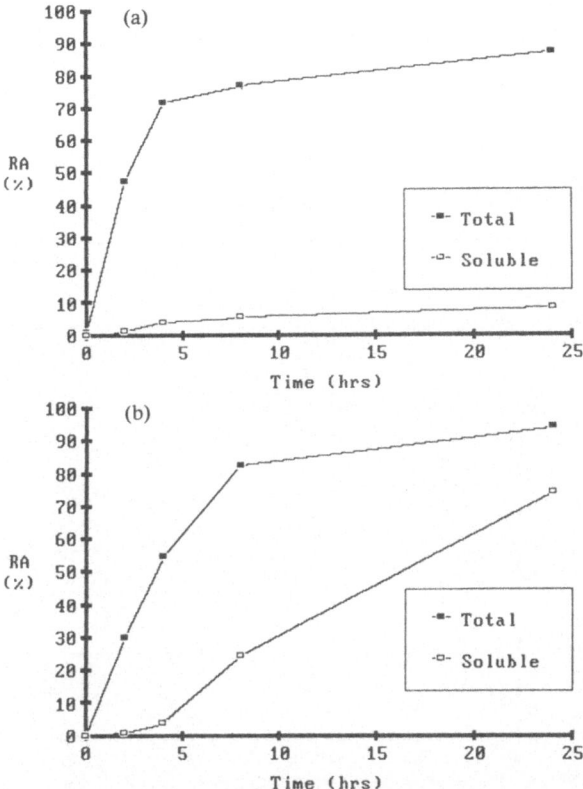

Fig. 1. Medium radioactivity from L-cells treated with 10 mM NaCN. L-cells
were labeled with leucine and TdR, as described in Methods Cultures
were washed x3 with chase medium, incubated for 1 h to clear soluble
radioactivity. Chase medium containing NaCN was added at TO. Total
and TCA soluble radioactivity was measured, as derived from the cell
protein (A) and DNA (B).

ize the degradation of proteins in GCN, we could produce no more than a 15%
hydrolysis of the total protein released into the medium. Under circumstances in which cells would be dying progressively over a 24 h period, we would
estimate that the hydrolysis rate would be significantly less.

Thymidine-induced Selective Cell Death

To effect a selective mechanism of cell death, we exposed cells to different concentrations of TdR. Concentrations between 1.0 and 10.0 mM TdR effected a substantial loss of labeled cell DNA, particularly after the first
24 h of the experiment (Fig. 4). As expected, the loss of labeled cell protein was also increased by TdR (data not shown).

We then studied in detail the effects of 4.0 mM TdR on cell growth and
proliferation. Protein accumulation was only slightly reduced by TdR during
the initial 24 h of the experiment (Fig. 5A). Beyond this point, however,
further accumulation was blocked completely, with net loss of protein during
the third day. DNA accumulation was blocked completely by 4 mM TdR on the
first day of the experiment, with net loss on the second and third day of the
experiment (Fig. 5B). The result of these differential effects on protein
and DNA accumulation was a 50% increase in cell size, measured by the

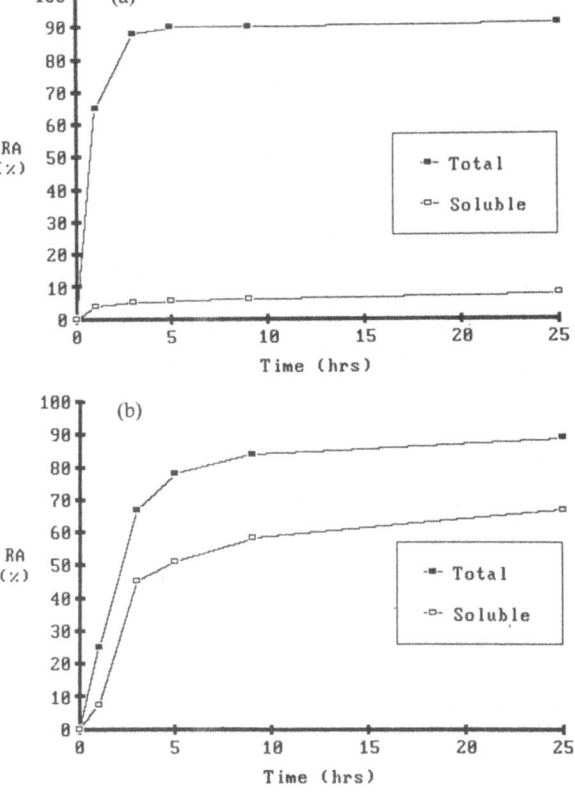

Fig. 2. Medium radioactivity from freeze-thawed L-cells. L-cells were
labeled with leucine and TdR, washed and cleared of soluble label,
as described in Methods. Monolayers were freeze-thawed x3 for a
period of 1 h. Chase medium was added. Total and TCA soluble
radioactivity was measured, as derived from cell protein (A) and
DNA (B).

protein/DNA ratio, with most of this change occurring on the first day of the
experiment (Fig. 5C).

A differential rate of protein and DNA synthesis was sufficient to
account for the increase in cell size induced by TdR. On the first day protein synthesis was decreased slightly by TdR (0.66/d versus 0.82/d for control), then dropping to about 1/2 of the control values on the next two days
(Fig. 6A). In contrast, TdR reduced the rate of DNA synthesis by 2/3 on all
days of the experiment (Fig. 6B). The ratio of the protein/DNA specific activities, demonstrated the rate of differential synthesis that was increasing
cell size. TdR clearly affected protein and DNA synthesis differently on the
first day of the experiment and resulted in cellular hypertrophy (Fig. 6C).

Consistent with the pilot experiment, TdR at the concentration of 4.0 mM
effected a significant cell death after 24 h of treatment, measured by release of ^{3}H-labeled DNA from the monolayer (Fig. 7A). This TdR-induced cell
loss was also reflected in the excess loss of ^{14}C-labeled protein from the
monolayer beyond the first day of the experiment (Fig. 7B). The ratio of
cell protein radioactivity/DNA radioactivity measured protein turnover in the
viable cells remaining on the monolayer. We could not detect any significant

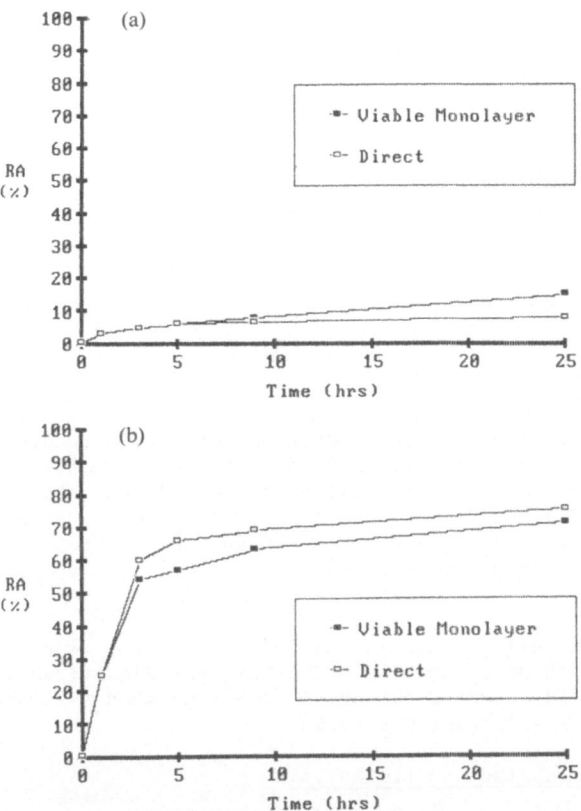

Fig. 3. Medium radioactivity from freeze-thawed L-cells on a viable mono-
         layer. Cell cultures were labeled and killed by freeze-thawing,
         as described in Fig. 2. Cell debris was scraped into chase medium
         and placed over a monolayer of nonlabeled L-cells. Total and tri-
         chloroacetic soluble radioactivity was measured, as derived from
         cell protein (A) and DNA (B).

effect of TdR on this ratio (Fig. 7C). Both controls and TdR-treated cells
that remained viable showed cell protein turnover rates that average approximately 0.13/d, consistent with previously reported values[6,7].

When the proteolysis was measured as the accumulation of TCA-soluble
radioactivity in the medium, however, the data indicated that, with TdR
treatment, proteolysis was significantly higher than 0.13/d measured in the
viable monolayer. Based on medium values, controls showed an expected proteolysis rate of 0.13/d, while the TdR-treated cell cultures showed a rate of
0.23/d on day 2 (Fig. 8B). Since approximately 18% of the cells died during
this 24 h period (Fig. 7B), this indicated that just about half of the protein released by cell death was hydrolyzed. This can be visualized by
comparing the data in Figs. 8A and 8B: of the excess TdR-induced radioactivity from the cell protein (Fig. 8A), approximately 40% of this was recovered in the acid soluble form (Fig. 8B). Thus, SCD results in a significantly higher fraction of protein hydrolysis than observed with GCN.

Two additional conclusions could be drawn from this experiment.
Firstly, a small, but consistent increase in proteolysis could be detected on
day 1, before we could detect any significant cell death, suggesting that the

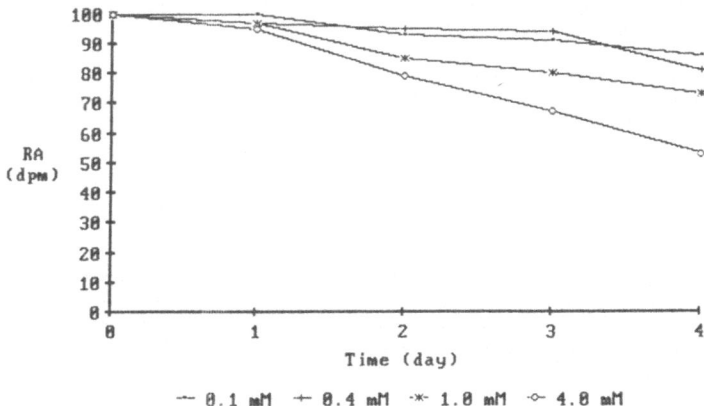

Fig. 4. Effect of different concentrations of TdR on recovery of 3H-labeled DNA in cell monolayer. Labeled L-cell cultures were treated with chase medium containing different TdR concentrations, as indicated in the figure. Data for 10 mM TdR followed the same curve as shown for the 4.0 mM TdR and is not shown. Cells were assayed daily for label remaining in cell DNA.

hydrolysis of the cell protein is initiated prior to release of labeled DNA. Secondly, the fraction of protein hydrolyzed was significantly less on the third day in the TdR-treated cultures, at a time when a number of residual viable cells was significantly reduced.

Effect of NH_4 on TdR-induced Proteolysis

To evaluate the role of vacuolar proteolysis in SCD, we studied the effect of NH_4Cl on the TdR-induced cell death and associated proteolysis. The proteolysis associated with TdR-induced cell death was largely inhibited by NH_4 (Fig. 9A left), approaching the level seen in the corresponding NH_4-treated controls (Fig. 9A right). As previously reported [3], NH_4 has a small effect on proteolysis in growing L-cells. We could detect a consistent effect of NH_4 neither on the release of total radioactivity from the labeled cell protein from cells treated with TdR (Fig. 9B left), nor on the release into the medium of ^3H from cell DNA (data not shown). These observations suggested that most of the proteolysis associated with TdR-induced cell death was occurring in the vacuolar system, either the vacuolar system of the dying cells and/or the vacuolar system of the viable cells remaining on the monolayer. However, NH_4Cl did not reduce the rate of SCD induced by TdR.

Selective Cell Death and Cycloheximide

Since CH appears to be an inhibitor of cell death induced by agents blocking DNA synthesis[8,9], we studied the effect of this agent on SCD, and, in particular, its effect on hydrolysis of cellular protein. Low concentrations of CH, sufficient to inhibit protein synthesis by 90%, however, also effect a progressive and selective cell loss, when applied to cultures of L-cells (unpublished observations), in contrast to the relative stability of CH-treated cultures of rat embryo fibroblasts[10]. Thus, in this experiment, we were able to evaluate the effects of an agent that inhibits SCD-apoptosis, while at the same time inducing a substantially different type of SCD.

Addition of CH, 0.02 µg/mL, to growing L-cell cultures immediately blocked the further accumulation of cell protein (Fig. 10A). This contrasted with the effect of TdR, which showed significant inhibition only after 24 h. The combination of agents, CH and TdR, showed that the CH effect was predomi-

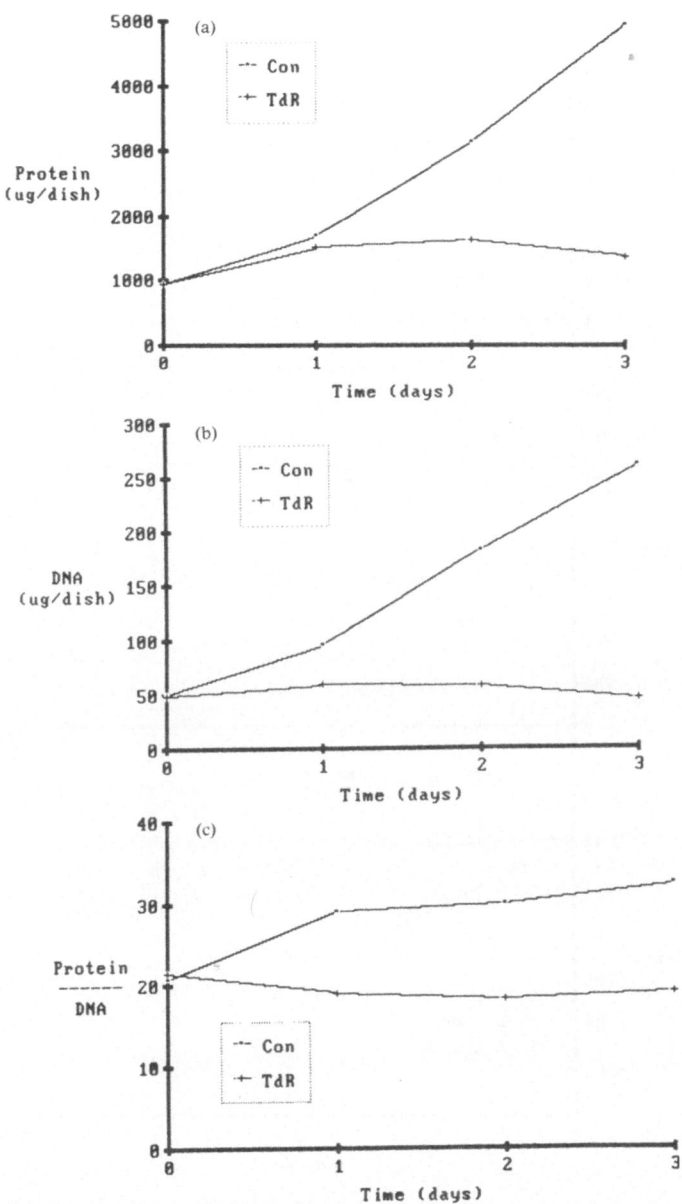

Fig. 5. Effect of TdR on accumulation of cell protein and DNA. L-cells were
         labeled, washed and cleared of soluble radioactivity, as described
         in Methods. At T0, cells were placed in chase medium containing 4.0
         mM TdR. Cultures were assayed daily for protein (A) and DNA (B).
         Ratio of protein/DNA is plotted in C.

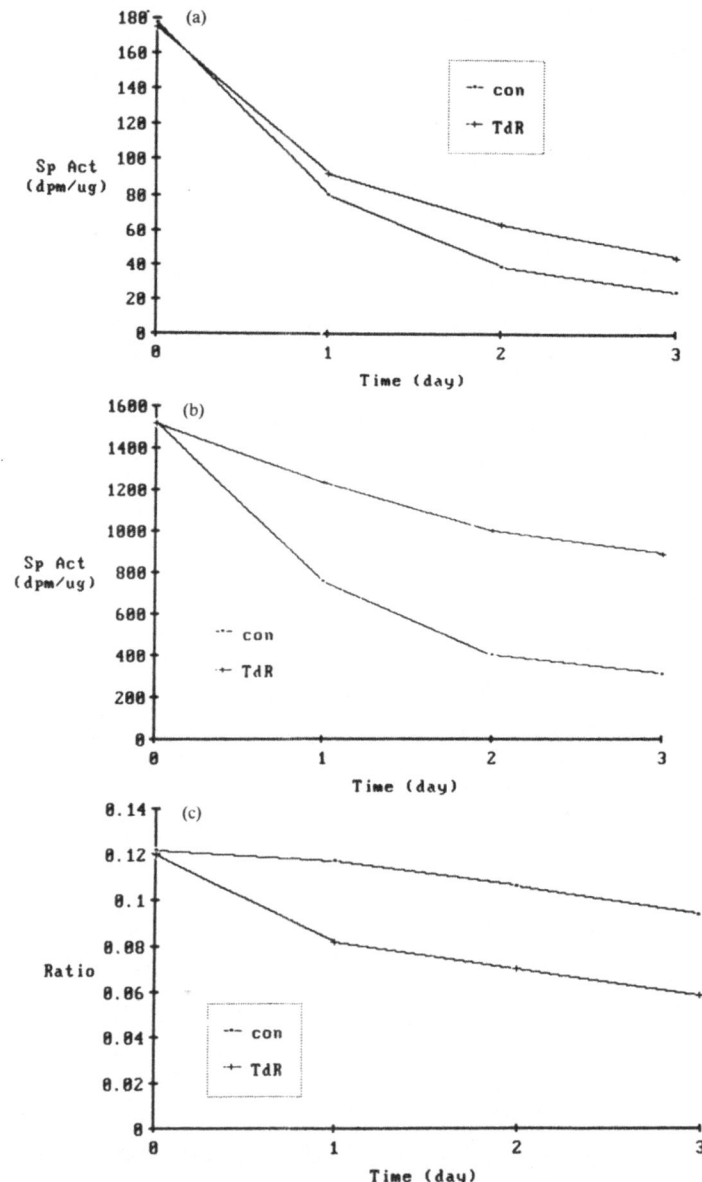

Fig. 6. Effect of TdR on synthesis of cell protein and DNA. Cells were treated and processed as described in 5. Radioactivity in cell protein and cell DNA was assayed daily. Radiospecific activity was calculated for both protein (A) and DNA (B). The change in radiospecific activity reflects protein synthesis. The plot of the ratios of the protein/DNA specific activity (C) reflects effect of unbalanced synthesis on cellular hypertrophy.

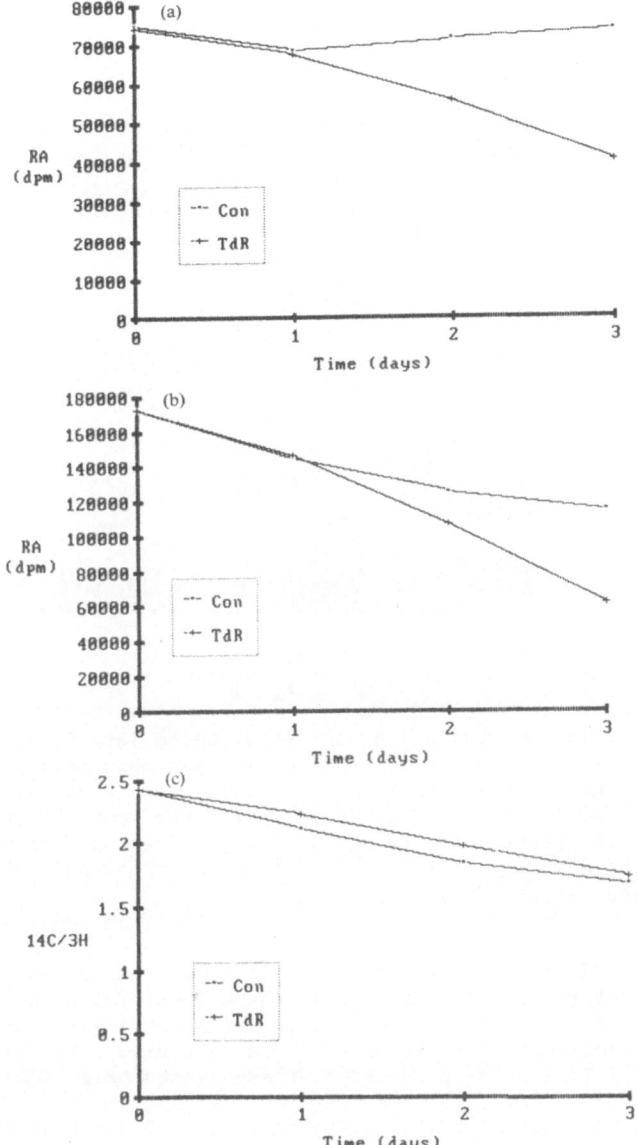

Fig. 7. Effect of TdR on turnover of cell protein and DNA. Cells were treated and processed as described in 5. Radioactivity in cell DNA (A)
and protein (B) was analyzed daily. The change in radioactivity in
the DNA reflects cell death and loss from the monolayer; the change
in radioactivity in cell protein reflects cell death and protein
turnover. The plot of the ratios of the protein/DNA (C) radioactiv-
ities indicates the rate of turnover in the viable cells recovered
in the monolayer. Differences in values between control and TdR-
treated in C were not significant.

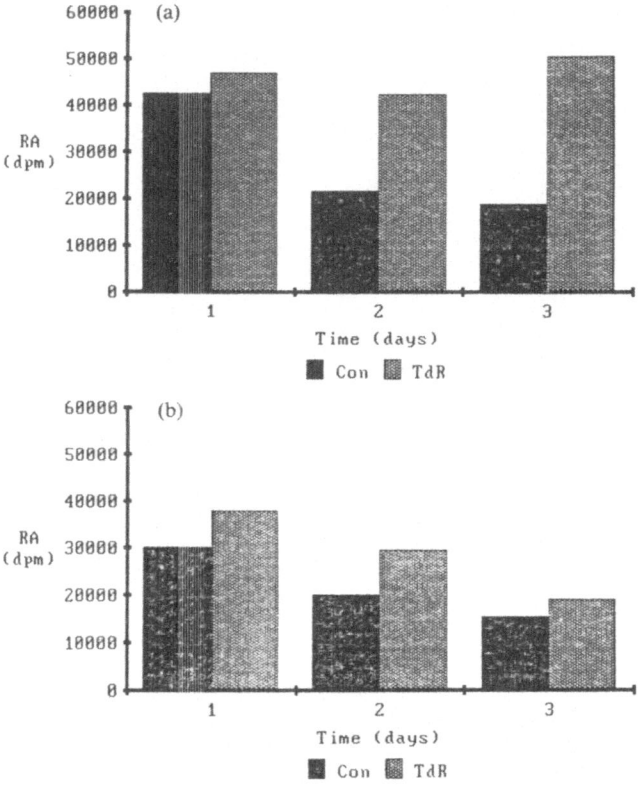

Fig. 8. Effect of TdR on medium radioactivity. Cells were treated and processed as described in 5. Chase medium was changed daily. The total
(A) and the TCA-soluble (B) radioactivity were assayed, as described
in Methods. The rates of proteolysis for the controls averaged
0.13/d. The proteolysis rates for the TdR-treated were 0.16, 0.23,
0.21/d on days 1-3. Differences between controls and TdR-treated
were significant (P < 0.05).

nant. Similarly, CH immediately effected cell loss from the monolayer, as
measured by loss of radioactive DNA, while the effect of TdR on cell loss was
delayed for 24 h (Fig. 10B). The combination of agents again showed that the
CH effect was predominant. The release of radioactivity from cell protein
reflected the cell loss rates produced by these agents (Fig. 10C).

Analysis of the medium radioactivity demonstrated striking differences
in proteolysis rates. TdR, as expected, showed increased rates of proteolysis, starting from day 1 (Fig. 11A). CH, however, appeared to inhibit
proteolysis, even though associated with some cell death from the first day.
The combination of agents, CH and TdR, acted like CH alone, inhibiting the
TdR induced proteolysis entirely. That both agents were effecting cell loss
and death was clear from the amount of total radioactivity released into the
medium, particularly on days 2 and 3 (Fig. 11B). Virtually all of this
increase was in the form of acid insoluble radioactivity in the cultures
treated with CH. These observations suggested that agents known to inhibit
apoptosis will also inhibit the proteolysis associated with this SCD. CH
alone, however, will induce a different type of SCD that is not associated
with hydrolysis of cell protein.

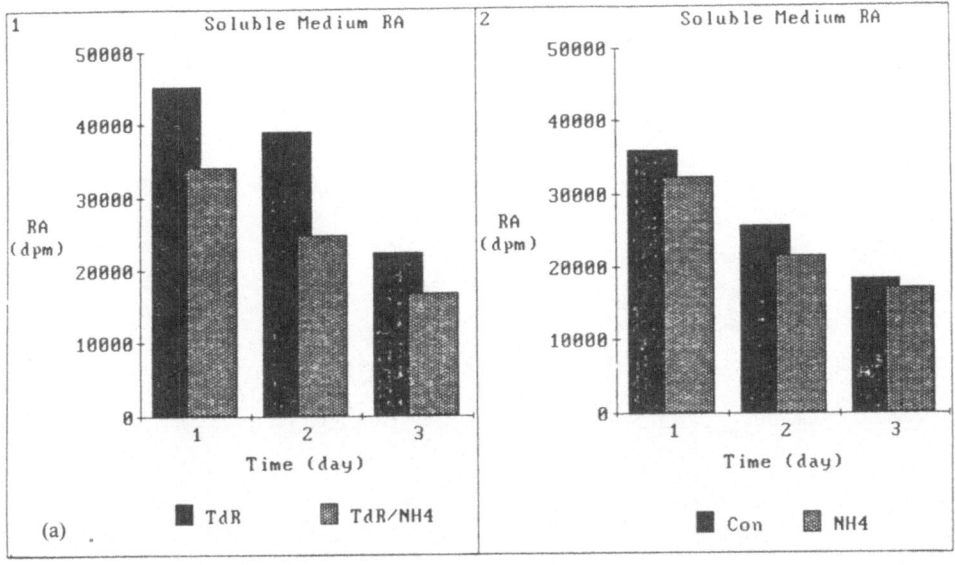

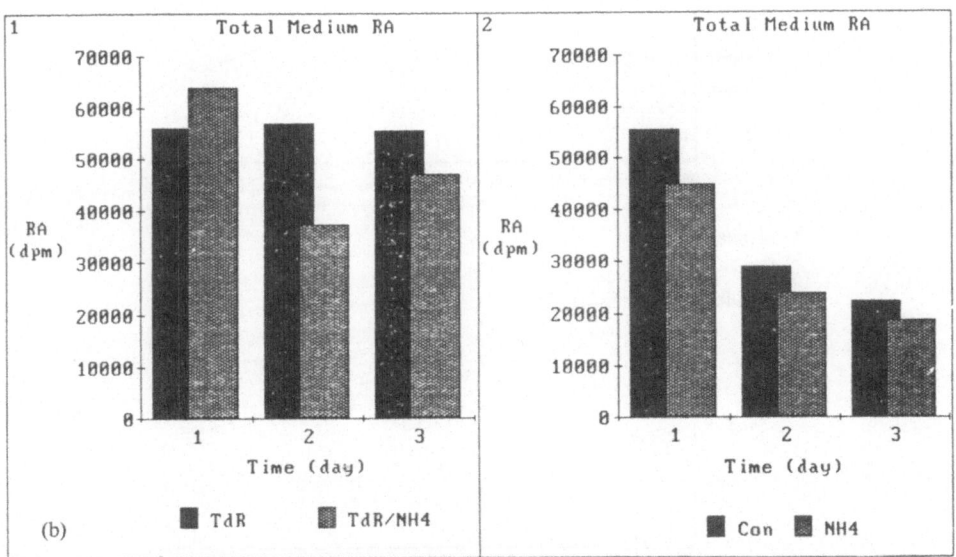

Fig. 9. Inhibition of TdR-induced proteolysis by NH₄Cl. Cells were labeled and processed as described in 5. Controls (Con) were placed in chase medium containing only 4 mM leucine; chase medium for TdR-treated (TdR) group contained leucine and 4mM TdR; medium for NH₄Cl (NH₄) group contained leucine and 10mM NH₄Cl; medium for combination (TdR/NH₄) group contained leucine, TdR and NH₄. Media were changed daily. Assays of TCA-soluble (A) and total (B) radioactivity were performed daily. For TCA-soluble group, all values for the NH₄-treated cells were significantly lower than their corresponding controls. Release of acid insoluble was calculated as the difference between the total and the TCA-soluble radioactivity for each medium (data not shown). None of these differences were significant.

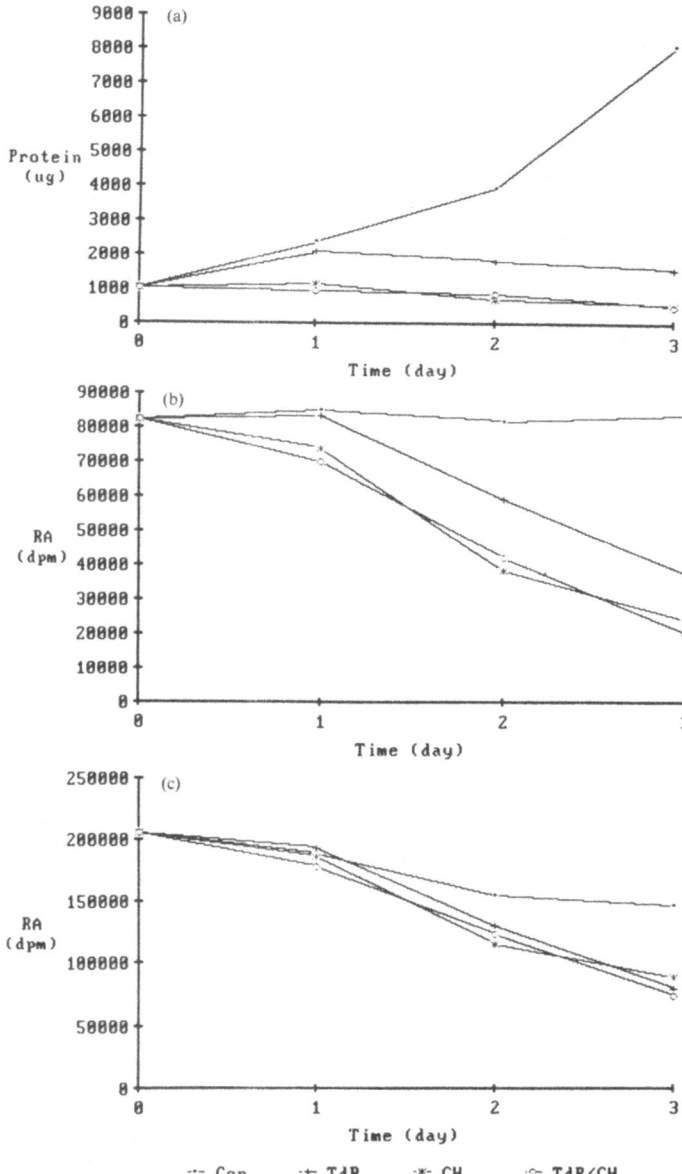

Fig. 10. Inhibition of TdR-induced proteolysis by CH: cell analyses. Cells were labeled and processed as described in 5. Controls (Con) were placed in chase medium containing only 4 mM leucine; chase medium for TdR-treated (TdR) group contained leucine and 4 mM TdR; medium for CH-treated (CH) group contained leucine and CH (0.02 μg/mL); medium for combination (TdR/CH) group contained leucine, TdR and CH. Media were changed daily. Cells were assayed for protein (A), radioactivity in DNA (B) and radioactivity in protein (C).

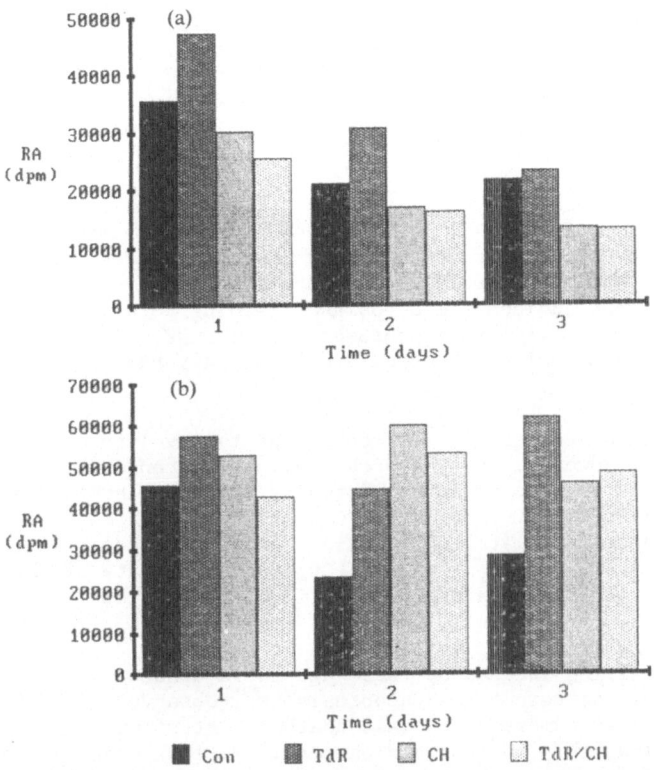

Fig. 11.   Inhibition of TdR-induced proteolysis by CH; medium analyses. Experimental conditions and treatment groups as described in 10.
Assays of TCA-soluble (A) and total (B) radioactivity were performed daily. For TCA-soluble group, all values for the CH-treated
cells were significantly lower than their corresponding controls
(P < 0.01); there was no difference between the CH and the CH/TdR
groups.

DISCUSSION

Selective Cell Death

        SCD induced by agents inhibiting DNA synthesis has been extensively
studied, since this phenonemon forms the basis for the chemotherapy of neoplastic diseases[11].   Cells selected for death are generally cells actively
involved in DNA synthesis.   Since DNA synthesis is almost totally inhibited,
while protein synthesis is only minimally affected by these agents, cell
protein mass increases relative to the DNA mass by 50-100%[11,12]. Our results
with high concentrations of TdR support this hypothesis, showing that protein
synthesis was only minimally decreased during the first day of the experiment
while DNA synthesis was severely compromised.   Protein turnover was not affected.   Thus, the cellular enlargement produced by TdR was due entirely to an
unbalanced synthesis of protein and DNA.

        After an 18-24 h period of cellular hypertrophy, cell death becomes
evident in cultures treated with DNA synthesis inhibitors[11,12,13] .   The biochemical mechanism by which these chemotherapeutic agents kill cells,
however, remains unknown.   Almost certainly cell death is not an inevitable
consequence of the DNA synthesis inhibition, since removal of the inhibitor

547

prior to 18 h will for some cell lines completely prevent cell killing[12]. Further, protein synthesis is required to express this process leading to cell death, since CH treatment will also prevent cell killing[8]. These observations suggest that cell death induced by these agents is not a direct consequence of inhibition of DNA synthesis, but requires a significant period for further metabolic reactions, one of which may be protein synthesis and increased cell mass.

The mechanism of cell killing has been related to the unbalanced growth of the cell. One hypothesis proposes that the hydrolases accumulate in these enlarging cells. These hydrolases are then released from the segregating compartments in the nucleus and cytoplasm, leading to a lytic cell death[14,15]. Unbalanced growth has also been proposed as a trigger to activate a high rate of autophagy in some cells[16]. Unregulated autophagy has been shown to also lead to cell death[17,18].

The problem of cell killing by these agents involves much more than unbalanced growth, however. Hydroxyurea, while simultaneouly killing cells in intestinal crypts, has no effect on normally quiescent cells that have been induced to proliferate[19]. Yet, after cessation of treatment with chemical agents that induce cellular proliferation, these same cells will show SCD as the organ regresses in size[20,21]. Cell type and the state of the cell prior to the effect of the agent are obviously important factors in determining whether or not SCD will occur.

The cell killing induced by these agents is associated with a distinctive morphological pattern termed apoptosis[22,23], observed in both solid tissues and ascites tumors. The cell death in intestinal crypts described above also follows this morphologic pattern. Nuclear changes involve fragmentation and margination of chromatin[23,24], followed by release of cellular cytoplasmic constituents in membrane-bound bodies, termed apoptotic bodies. The morphology of apoptosis is associated with a rapid cleavage of DNA at internucleosomal sites, apparently by an activated endogenous endonuclease[24]. Again, agents blocking protein synthesis were shown to inhibit this process. Hydroxyurea produces an increase in a number of cellular hydrolases, including DNAase II[14,15], again associated with DNA fragmentation. The hypothesis has been proposed that the mechanism of cell death involves the activation of endonucleases, which inititiate DNA injury and triggers the apoptotic response and SCD. While it has been suggested that DNA fragmentation may be a primary event in the process, it is clear from our data that cell death induced by virtually any agent results in rapid DNA fragmentation. It remains to be seen whether the pattern of fragmentation of DNA in SCD-apoptosis is any different from that seen in GCN.

Whatever the mechanism of cell death produced by these inhibitors of DNA synthesis, morphologic and biochemical evidence strongly suggests that these cytotoxic agents producing SCD-apoptosis should also be associated with significant hydrolysis of cellular protein. A number of hypotheses have been proposed that involve hydrolysis of cellular protein: a lytic cell death from release of hydrolases into the cytoplasm, an autophagic cell death, and finally a homophagic process in which adjacent viable cells degrade apoptotic bodies.

Our data support the idea that SCD-apoptosis is characterized by increased hydrolysis of cellular protein. Approximately 40% of the released cellular protein is rapidly degraded with cell death, far more than would be expected with GCN. Further, this cell death-associated proteolysis appears to start while the cell is enlarging during the first 24 h of the experiment, is inhibited by NH_4Cl, and decreases at a time when cell death is highest and few viable cells remain in the culture. These observations are most consistent with the hypothesis that the SCD in these cultures is apoptosis, involvving release of apoptotic bodies and homophagy, rather than autophagy or lytic cell death. We have obtained similar data in studies on Yoshida ascites tumor: a high rate of cell death and vacuolar proteolysis was seen as these ascites tumors reached high density[5].

The phenomenon of proteolysis associated with SCD may have been complicating the interpretation of many experiments studying protein turnover in L-cells [16,25,26,27,28]. Studies in our laboratory have suggested a basis for many of these apparently descrepant observations: the observed protein turnover in some L-cell cultures may be due to a high rate of proteolysis occurring in nonproliferating subfraction of the cells on the monolayer, whereas the rapidly proliferating cells showed relatively low rates of protein turnover [1,2,6,7]. Protein turnover in growing L-cells is at a relatively low rate, in the range of 0.12/d [6], consistent with data reported here. Though our initial estimates were biased downward by the residual labeled protein that remained on the monolayer[1], it still appears that L-cells have an unusually low rate of protein turnover in monolayer culture, about 30-50% that usually observed in fibroblasts cultures [29]. The relatively large amount of acid soluble radioactivity in the medium derived from cell protein reflects the high rate of proteolysis associated with the SCD that occurs in these cultures. Thus, the critical functional difference between GCN and SCD-apoptosis may be that the latter process efficiently recaptures the macromolecular constituents in the dying cells, degrades them to elementary units and makes these units available for immediate reutilization by proliferating viable cells.

ACKNOWLEDGEMENTS

Work supported in part (F.M.B.) by Consiglio Nazionale delle Ricerche (Progetto Finalizzato Oncologia, contratto 86.00297.44; programma bilaterale Italia-USA CT 86.00012.04), Ministero della Pubblica Istruzione, and Associazione Italiana per la Ricerca sul Cancro.

REFERENCES

1. J. S. Amenta and M. J. Sargus, Mechanisms of protein degradation growing and non-growing L-cell cultures, Biochem. J. 182:847 (1979).
2. J. S. Amenta and S. C. Brocher, Evidence for heterogeneity of protein-turnover states in cultured cells, Biochem. J. 190:673 (1980).
3. J. S. Amenta, S. C. Brocher, J. Mehta, D. Manjunath and F. M. Baccino, Evidence for a special relationship between proteolysis and single cell necrosis, Toxicol. Pathol. 14:335 (1986).
4. F. M. Baccino, L. Tessitore and G. Bonelli, Control of protein degradation and growth phase in normal and neoplastic cells, Toxicol. Pathol. 12:281 (1984).
5. L. Tessitore, G. Bonelli, G. Cecchini, J. S. Amenta and F. M. Baccino, Regulation of protein turnover versus growth state: ascites hepatoma as a model for studies both in the animal and in vitro, Arch. Biochem. Biophys. 255:372 (1987).
6. J. A. Silverman, J. Mehta, S. Brocher and J. S. Amenta, Analytical errors in measuring radioactivity in cell proteins and their effect on estimates of protein turnover in L cells, Biochem. J. 226:361 (1985).
7. J. S. Amenta and J. A. Silverman, Analytical errors in measuring radioactivity in cell protein and their effect of estimates of protein synthesis, in: "Intracellular Protein Catabolism", I. E. Khairrallah, J. F. Bond and J. W. C. Bird, eds., Alan R. Liss, New York, (1985).
8. M. W. Lieberman, R. S. Verbin, M. Landay, H. Liang, E. Farber, T. -N. Lee and R. Starr, A probable role for protein synthesis in

intestinal epithelial cell damage induced in vivo by cytosine arabinoside, nitrogen mustard, or X-radiation, Cancer Res. 30:942 (1970).

9. A. H. Wyllie, Cell death: a new classification separating apoptosis from necrosis, in: "Cell Death in Biology and Pathology", I. D. Bowen and R. A. Lochsin, eds., Chapman and Hall, London, (1981).

10. J. S. Amenta, M. J. Sargus and F. M. Baccino, Inhibition of basal protein degradation in rat embryo fibroblasts by cycloheximide: Correlation with activities of lysosomal proteases, J. Cell. Pathol. 97:267 (1978).

11. D. W. Ross, The nature of unbalanced cell growth caused by cytotoxic agents, Virchows Arch. Cell Path. 37:225 (1981).

12. S. E. Pfeiffer and L. J. Tolmach, Inhibition of DNA synthesis in HeLa cells by hydroxyurea, Cancer Res. 27:124 (1967).

13. R. E. Bennet, M. W. Harrison, C. J. Bishop, J. Searle and J. F. R. Kerr, The role of apoptosis in atrophy of the small gut mucosa produced by repeated administration of cytosine arabinoside, J. Pathol. 142:259 (1984).

14. J. Malec, W. M. Przybyszewski, M. Grabarczyk, E. Sitarska and B. Czartoryska, Mechanism of unbalanced growth-induced cell damage. I. A probable role for hydrolytic enzymes in synthesis, Chem.-Biol. Inter. 57:315 (1986).

15. J. Sawecka, B. Golos and J. Malec, Mechanism of unbalanced growth-induced cell damage.II. A probable relationship between unbalanced growth, DNA breakage and cell death, Chem.-Biol. Inter. 60:47 (1986).

16. J. Sparkuhl and R. Sheinin, Protein synthesis and degradation during expression of the temperature-sensitive defect in ts A1S9 mouse L-cells, J. Cell. Physiol. 105:247 (1980).

17. P. E. Schwarze and P.O. Seglen, Protein metabolism and survival of rat hepatocytes in early culture, Exp. Cell Res. 130:185 (1980).

18. P. E. Schwarze and P.O. Seglen, Reduced autophagic activity, improved protein balance and enhanced in vitro survival of hepatocytes isolated from carcinogen-treated rats, Exp. Cell Res. 157:15 (1985).

19. E. Farber and R.Baserga, Differential effects of hydroxyurea on survival of proliferating cells in vivo, Cancer Res. 29:136 (1969).

20. A. Columbano, G. M. Ledda-Columbano, P. M. Rao, S. Rajalakshmi and D. S. R. Sarma, Occurrence of cell death (apoptosis) in preneoplastic and neoplastic liver cells, Amer. J. Pathol. 116:441 (1984).

21. W. Bursch, H. S. Taper, B. Lauer and R. Schulte-Hermann, Quantitative histological and histochemical studies on the occurrence and stages of controlled cell death (apoptosis) during regression of rat liver hyperplasia, Virchows Arch. Cell Path. 50:153 (1985).

22. J. Searle, T. A. Sawson, P. J. Abbott, B. Harmon and J. F. R. Kerr, Electron-microscope study of the mode of cell death induced by cancer-chemotherapeutic agents in populations of proliferating normal and neoplastic cells, J. Pathol. 116:129 (1975).

23. A. H. Wyllie, J. F. R. Kerr and A. R. Currie, Cell death: the significance of apoptosis, Int. Rev. Cytol. 68:251 (1980).

24. A. H. Wyllie, R. G. Morris, A. L. Smith and D. Dunlop, Chromatin cleavage in apoptosis: association with condensed chromatin morphology and dependence on macromolecular synthesis, J. Pathol. 142:67 (1984).

25. D. W. King, K. G. Bensch and R. B. Hill, State of dynamic equilibrium in protein of mammalian cells, Science 131:106 (1960).

26. L. Warren and M. C. Glick, Membranes of animal cells. II. The metabolism and turnover of the surface membrane, J. Cell Biol. 37:729 (1968).

27. H. C. Jordan and P. A. Schmidt, Constant protein turnover in mammalian cells during logarithmic growth, Biochem. Biophys. Res. Commun. 4:313 (1961).

28. G. B. Gordon, Lipid accumulation in the stationary phase of strain L cells in suspension culture, Lab. Invest. 36:114 (1977).

29. J. S. Amenta, F. M. Baccino and M. J. Sargus, Cell protein degradation in cultured rat embryo fibroblasts. Suppression by vinblastine of the enhanced proteolysis by serum-deficient media, Biochim. Biophys. Acta 451:511 (1976).

ROLE OF INTERMEDIATE FILAMENTS AND UBIQUITIN

IN PROTEIN CATABOLISM IN TUMOR CELLS

Rupert Earl, Heather Mangiapane, Daniel Tuckwell,
Simon Dawson, Michael Landon, Michael Billet, James
Lowe and R. John Mayer

Department of Biochemistry, University of
Nottingham Medical School, Queen's Medical Centre
Nottingham, NG7 2UH, U.K.

INTRODUCTION

Coordinate stimulation of protein synthesis and inhibition of protein catabolism generally follows the addition of hormones or growth factors to cells bearing the requisite receptors. This anabolic effect may occur constitutively in transformed cells during exponential phases of growth. Cessation of growth in certain tumor cell lines, e.g. at high density, can be achieved by a large increase in protein degradation together with a significant decrease in protein synthesis. It is even possible that tumor cell death is accompanied by high protein catabolic rates mediated by the acid vesicular system[1].

Clearly, to understand the relationship between protein synthesis and degradation in growth regulation, we must understand the mechanisms involved in both processes. While much is understood about the mechanism of protein synthesis, little is known about the mechanisms of intracellular protein catabolism. The relationship between synthesis and degradation of proteins is contained in the notion of a protein turnover cycle with orthograde transport of newly synthesized proteins to functional sites and retrograde transport of proteins to degradative sites[2]. There are two major sites of protein catabolism. A cytosolic degradation system which may involve protein targeting via ubiquitination or involve multicatalytic large proteases[3] and also the acidic vesicular (lysosomal) degradation system. The results presented here will demonstrate a molecular biological mechanism by which proteins are fed into the acid vesicular system in rat hepatoma cells and a newly discovered novel role of ubiquitination in human astrocytomas.

METHODS

Methods for culturing Morris hepatoma 7288c cells (HTC cells) are described in 4. The procedures for the preparation of Sendai virus, reconstituted Sendai virus envelopes (RSVE), virus-cell fusion and microinjections are described in 4, 5, 9. The methods for the preparation of fluorescein-conjugated virus and biotinylated virus are described in 6. The procedure for carrying out immunohistochemistry of cytoskeletal proteins is described in 6.

RESULTS

Degradation Systems in HTC Cells

We have previously shown that Sendai virus membrane haemagglutinin (HN) and fusogen (F) proteins are good probes for the degradation of one of the two groups of HTC cell plasma membrane proteins, i.e. the glycoprotein sub-group having $t_{\frac{1}{2}}$'s of 60-90h[4]. The HN and F protein probes are degraded like HTC surface glycoproteins in the acid vesicular system[4]. Therefore, HTC cells have an active protein degradation system involving the acid vesicle system which degrades plasma membrane[4] and other endogenous proteins[7]. The results in Fig. 1 show that microinjected ^{125}I-human serum albumin-methotrex-ate complex is rapidly degraded to amino acids ($t_{\frac{1}{2}}$, 35h) by a mechanism which is not susceptible to lysosomotropic agents (not shown) indicating that albu-min degradation is achieved by a cytosolic degradation system as we have shown in fibroblasts[8]. Fig. 1 also shows that a significant proportion (42%) of [125I]-human serum albumin-methotrexate is not degraded but ejected from the cell: a similar proportion of HN and F proteins are ejected intact from HTC cells[6]. The combined data show that both cytosolic and acid vesicular degradation systems exist in HTC cells.

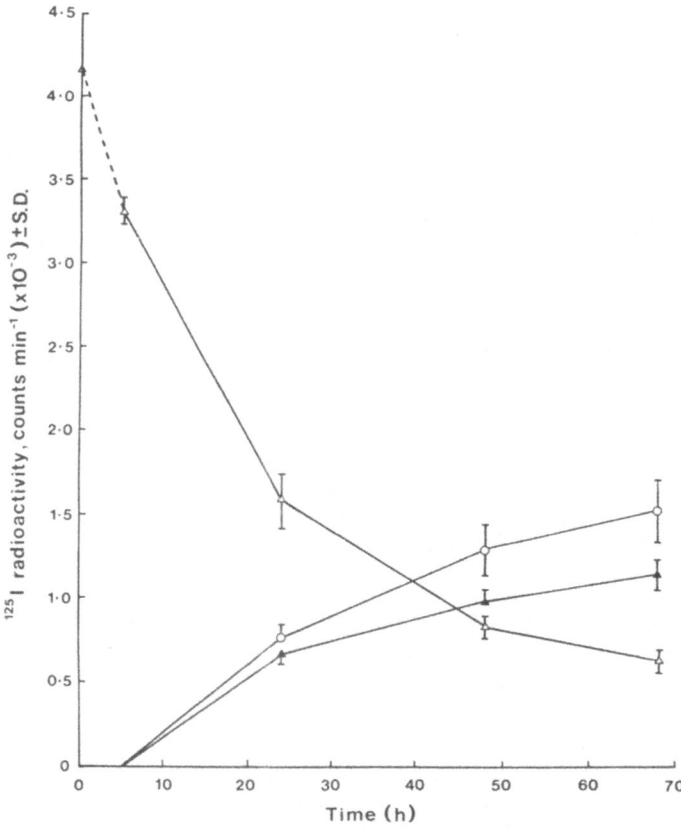

Fig. 1.   Fate of microinjected ^{125}I human serum albumin-methotrexate in HTC cells. Loss of radioactivity from cells (△), appearance of trichlo-roacetic acid-soluble (O) and insoluble (▲) radioactivity in the culture medium. Albumin was microinjected by RSVE-cell fusion[4,5].

Mechanism of Viral Membrane Protein Sequestration in HTC Cells

We have previously shown that Sendai HN and F proteins are rapidly sequestered (<5h) into a nuclear-intermediate filament (vimentin) fraction after transplantation into HTC cells[5]. The nuclear fraction containing viral HN and F proteins is enriched in multivesicular bodies[6]. Fig. 2 shows the behavior of cells in response to fusion with RSVE containing biotinylated HN and F proteins as detected with streptavidin-peroxidase. After RSVE-cell fusion the cytoplasm retracts as shown by phase microscopy (Fig. 2a) so that cells have a rounded appearance. However, fluorescence microscopy shows that the cell membrane does not retract: cells retain their triangular appearance as illustrated by distribution of actin stress-fibers (Fig. 3e) and microtubules (Fig. 3f). After transplantation into the plasma membrane, Sendai HN and F proteins are rapidly pulled into the cell to form small perinuclear inclusions (<2h) which then appear (<5h) as large perinuclear dense bodies (Fig. 2a) as detected by streptavidin-peroxidase. Subsequently, the cytoplasm extends again (over 2-8h) so that cells assume their normal triangular appearance by phase (Fig. 2b) or fluorescence (Fig. 3e) microscopy with large perinuclear viral protein inclusions (Fig. 2b, arrow). Perinuclear sequestration of viral proteins is achieved by reversible vimentin intermediate collapse filament collapse as detected with anti-vimentin immunofluorescence, i.e. with rounded cells showing collapsed filaments 1-2 h after RSVE-cell fu-

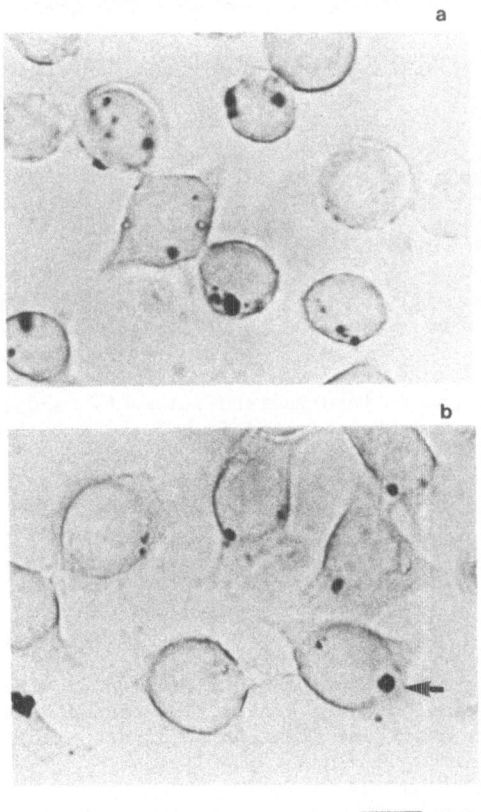

10μm

Fig. 2. Fate of biotinylated Sendai haemagglutin (HN) and fusogen (F) in HTC cells. Biotinylated-viral proteins were detected by streptavidinperoxidase reaction product[6]. Cells were stained at 5 h (a) and 18 h (b) after RSVE-cell fusion.

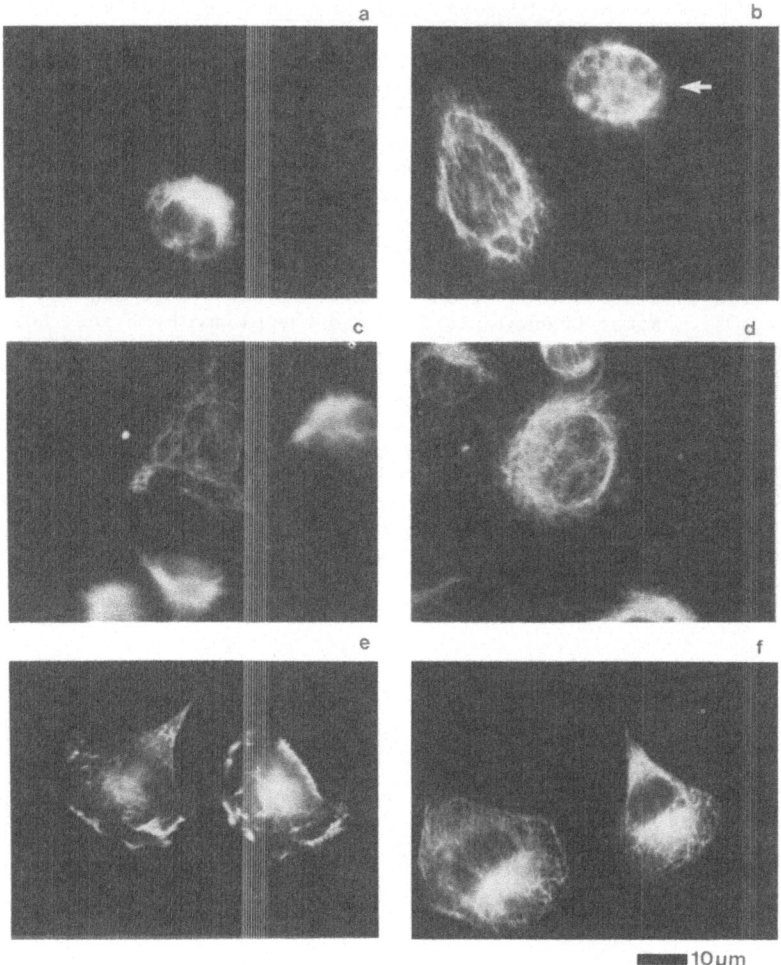

Fig. 3. Response of vimentin intermediate filaments (a-d); actin stress-
fibers (e) and microtubules (f) in HTC cells to heat-shock. (a)
Immediately after heat-shock (1 h at 43°C); (b) 2 h later; (c) 8 h
later (d) 13 h later. Cytoskeletal staining was by immunofluorescence (a-d, f) or TRITC-phalloidin staining (e).

sion (similar to Fig. 3a); extending intermediate filaments 2-8 h after fusion (similar to Fig. 3b and 3d respectively) and extended filaments in tri-
angular-shaped cells 13 h after fusion (similar to Fig. 3c). No change in
actin-stress fibres (Fig. 3e) or microtubules (Fig. 3f) occur to accompany
the changes in distribution of intermediate filaments. The kinetics of intermediate filament extension after contraction caused by virus are very
similar to the kinetics of filament extension after cellular heat-shock which
is shown in Figure 3a-d.

The rapid sequestration of viral transmembranous HN and F proteins into
perinuclear inclusions in HTC cells is therefore caused by specific reversible intermediate filament collapse with concomitant cytoplasmic retraction
before viral protein degradation in the acid vesicle system[4,5] or protein
ejection in vesicular form from the cell[6] . Viral HN and F proteins serve as
excellent probes of the mechanism of degradation of HTC cell surface glyco-

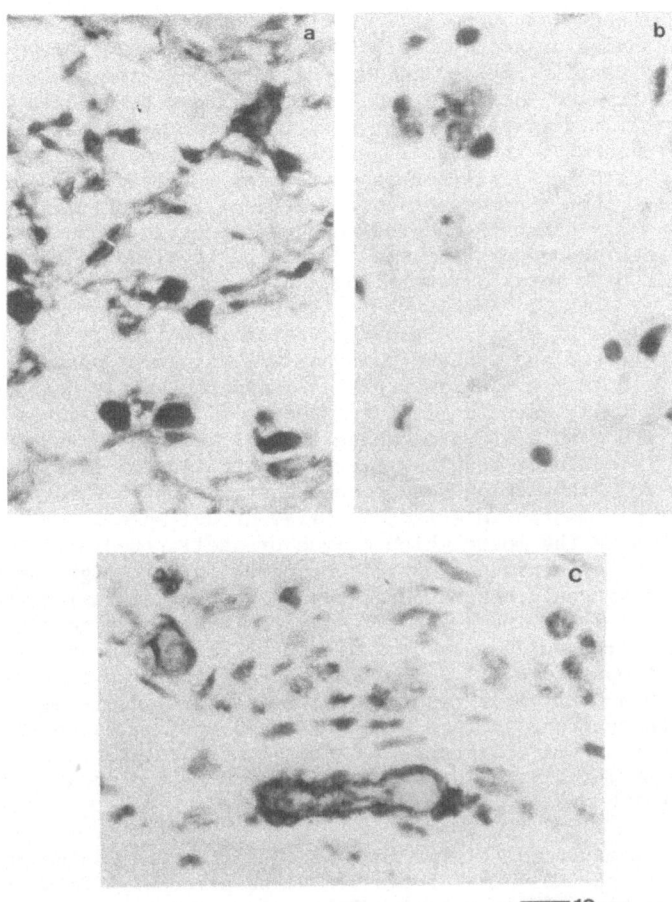

━━━10μm

Fig. 4. Immunoperoxidase staining of Rosenthal fibres in human astrocytomas
stained with anti-GFAP (a) and anti-ubiquitin (b,c). Immunochemical
methods as in 13.

proteins[4]. Therefore, a similar mechanism of internalization, processing and
degradation of HTC plasma membrane glycoproteins presumably occurs.

Intermediate Filaments and Ubiquitination in Tumor Cells

Inspection of the literature shows that vimentin intermediate filament
inclusion bodies, often in juxtanuclear positions, are common, e.g. in temperature-sensitive Rous sarcoma virus transformed cells at the non-permissive
temperature[10] or in Epstein-Barr transformed lymphocytes[11]. Intermediate
filament inclusions (Rosenthal fibres) are commonly observed in astrocytomas.
Fig. 4 shows that Rosenthal fibres contain abundant glial fibrillary acid
protein (GFAP, Fig. 4a) and are ubiquitinated (Fig. 4b) as detected by immunoperoxidase staining. Fig. 4c shows the amorphous central region of a
Rosenthal fibre and the intense ubiquitination of the periphery of the body.

DISCUSSION

Transplantation of Sendai HN and F[4,5] and microinjection of [^{125}I]-human serum albumin-methotrexate (Fig. 1) clearly demonstrate that HTC cells possess both acid vesicular and cytosolic degradation systems respectively. Interestingly, the data in Fig. 1 and in 6 also demonstrate that intact [^{125}I]-human serum albumin and intact HN and F proteins[6] can be ejected from the cell. The most likely possibility is that the viral membrane proteins[6] or albumin (Fig. 1) enter multivesicular bodies[6] which fuse with the cell membrane to eject vesicular viral membrane proteins or vesicular albumin from the tumor cells. Such a process is of importance for the immune response to tumor cells in that intact fragmented oncogenic protein products (or inappropriately expressed endogenous proteins) could be involved in the generation of the cytotoxic or humoral immune responses.

Transplantation of viral HN and F proteins into the HTC cell surface invokes rapid reversible cytoplasmic withdrawal driven by reversible intermediate filament collapse. Viral proteins are subsequently degraded or ejected from the cell. Reversible intermediate filament collapse is also caused by heat-shock (Figs. 3,a-d) which implies that the phenomenon is generally related to cellular insults, i.e. sequestration of foreign or endogenous proteins for "inspection" and either destruction or ejection from the cell. Reversible vimentin intermediate filament collapse in cells of mesenchymal origin may be the motor which drives organellar redistribution to eliminate unwanted or noxious agents from the cells. It might be expected that such a process would become deranged or inactivated during the cell transformation. This has been observed with a temperature sensitive mutant of Rous sarcoma virus which only causes intermediate filament collapse at the non-permissive temperature[10]. Furthermore, transformation of human lymphocytes with Epstein-Barr virus results in the association of the transforming latent infection membrane protein with collapsed perinuclear filament[11]. During virally induced cell transformation some transforming viral gene products may immobilize or inactivate intermediate filaments to prevent degradation or ejection of viral proteins.

In human astrocytomas (Fig.4) intermediate filament inclusions (GFAP) are frequently found (Rosenthal fibres). We have recently found that Rosenthal fibres are ubiquitinated[12]. This raises the possibility that ubiquitination is acting as a marker for cytosolic degradation[3] of filaments or proteins trapped by collapsed filaments (cf. 10,11): alternatively, ubiquitination could be a signal for protein incorporation into multivesicular bodies for eventual degradation or ejection from the cell. As indicated for virally transformed cells[10,11] the filament inclusions in human astrocytomas may represent a "frustrated" attempt at filament-related "abnormal" protein processing. Such a process is not confined to astrocytomas being observed in Alzheimer plaques[13], granulovacuolar degeneration, Parkinsonian Lewy bodies (containing neurofilaments) and alcoholic liver Mallory bodies (containing cytokeratins): each case being an example of a chronically stressed cell[12].

ACKNOWLEDGEMENT

We would like to thank the MRC and Wellcome Trust for generous financial support and Mrs. M. A. Spooner for typing the manuscript.

REFERENCES

1. L. Tessitore, G. Bonelli, C. Cecchini, J. S. Amenta and F. M. Baccino, Regulation of protein turnover versus growth state: Ascites hepatoma as a model for studies both in the animal and in vivo, Arch. Biochem. Biophys. 255:372 (1987).
2. F. J. Doherty and R. J. Mayer, Intracellular protein catabolism: state

of the art, FEBS Letts. 198:181 (1986).

3. M. Rechsteiner, Ubiquitin mediated pathways for intracellular proteo-
    lysis, Ann. Rev. Cell Biol. 3:1 (1987).

4. T. T. Earl, E. E. Billet, I. M. Hunneyball and R. J. Mayer, Sendai-
    viral HN & F glycoproteins as probes of plasma-membrane protein
    catabolism in HTC cells, Biochem. J. 241:801 (1987).

5. R. T. Earl, E. H. Mangiapane, E. E. Billet and R. J. Mayer, A putative
    protein sequestration site involving intermediate filaments for
    protein degradation by autophagy, Biochem. J., 241:809 (1987).

6. A. J. Ingram, S. Dawson, R. T. Earl, H. E. Mangiapane and R. J. Mayer,
    Reversible intermediate filament collapse with concomitant revers-
    ible cytoplasmic retraction results in rapid perinuclear sequestra-
    tion of viral membrane proteins before degradation, Biochem. J.
    submitted.

7. S. M. Russell, J. S. Amenta and R. J. Mayer, Degradation of proteins in
    rat liver mitochondrial outer membrane transplanted into different
    cell types, Biochem. J. 220:489 (1984).

8. M. J. Gaskell, P. C. Heinrich and R. J. Mayer, Mechanisms of intracel-
    lular protein catabolism. Intracellular fate of microinjected poly-
    peptides translated in vitro, Biochem. J. 241:817 (1987).

9. F. J. Doherty, J. A. Wassell and R. J. Mayer, A putative protein-se-
    questration site involving intermediate filaments for protein deg-
    radation by autophagy. Studies with microinjected purified gly-
    colytic enzymes in 3T3-L1 cells, Biochem. J. 241:793 (1987).

10. E. H. Ball and S. J. Singer, Mitochondria are associated with micro-
    tubules and not with intermediate filaments in cultured fibroblasts,
    Proc. Natl. Acad. Sci. 79:123 (1982).

11. D. Liebowitz, R. Kopan, E. Fuchs, J. Sample and E. Kief, An Epstein-
    Barr virus transforming protein associates with vimentin in lympho-
    cytes, Mol. Cell Biol. 7:2299 (1987).

12. J. Lowe, A. Blanchard, K. Morrell, G. Lennox, L. Reynolds, M. Billet,
    M. Landon and R. J. Mayer, Ubiquitin is a common factor in inter-
    mediate filament inclusion bodies of diverse type in man, J. Pathol.
    submitted.

13. V. A. Fried, H. T. Smith, E. Hilderbrandt and K. Weiner, Ubiquitin has
    intrinsic proteolytic activity: Implications for cellular regula-
    tion, Proc. Natl. Acad. Sci. 84:3685 (1987).

SECTION IV

NUTRITIONAL CARCINOGENESIS AND ANTICARCINOGENESIS

THE CHOLINE-DEVOID DIET MODEL OF HEPATOCARCINOGENESIS IN THE RAT

Benito Lombardi

University of Pittsburgh School of Medicine
Department of Pathology, Pittsburgh, PA 15261

INTRODUCTION

The notion that a choline-devoid (CD) diet is hepatocarcinogenic in the rat has already gone through two historical phases, and is at the beginning of its third (for a detailed account, see ref. 1). In 1946, Copeland and Salmon published the first of a series of papers showing the development of hepatocellular carcinomas (HCCs) in rats chronically fed a CD diet[2,3]. The finding attracted much attention at the time, since the modalities of the experiments involved no addition to the diet of, or treatment of the animals with, chemical carcinogens. However, subsequent studies by Newberne et al.[4,5] cast doubts on whether the diet, and the diet alone, was responsible for the genesis of the tumors, and attributed the latter, instead, to a likely contamination of the diet with aflatoxin B1, a newly discovered[6] and most potent hepatocarcinogen in the rat[7]. In the last few years, the question was reopened by a repetition of the original findings of Copeland and Salmon in three different laboratories[8-11]. In these instances, both diets and rat's environment were scrutinized for relevant contamination with chemical carcinogens, with negative results. At the present time, therefore, the conclusion seems unavoidable that CD diets are indeed hepatocarcinogenic, and that the genesis of the tumors resides in effects of these diets on rat liver. However, at least one primary issue awaits resolution, before the CD-diet model of hepatocarcinogenesis can be fully categorized; that is whether the diets act as complete carcinogens, able to initiate de novo liver cells, as well as to promote their evolution to cancer; or whether they merely promote the evolution to cancer of endogenous initiated cells.

Comprehensive reviews of the pathology of dietary choline-deficiency, and of the interactions of CD or lipotrope-deficient diets with chemical carcinogens, have been recently made[1,12,13]. This presentation, therefore, will be confined to a consideration of the CD-diet model of hepatocarcinogenesis, even though much of what can be said at the present time, from a point of view of underlying mechanism(s), is either tentative or only of a speculative nature. In addition, only a few of the effects of CD diets on rat liver were selected for consideration.

THE MODEL

Choline

Choline is a quaternary ammonium compound widely present in natural foods and food products[14]. It was described in 1849 by Strecker[15], but it was only in the early 1930s that its role as a dietary factor was clarified, when Best and associates discovered that it is required to maintain the lipid homeostasis of rat liver[16,17]. Rats fed CD diets were shown to accumulate fat in the liver, and to readily remove it after restoration of choline to their diet. The term "lipotropic phenomenon" was coined to refer to these effects of choline[18]. The seminal work by Best and associates was instrumental in leading later to the discovery by others of similar effects by methionine, and of the role of folic acid and of vitamin B12 in the biosynthesis of methyl groups. These four compounds are therefore often referred to as "lipotropes" or "lipotropic factors". Choline acts also as donor of methyl groups (after oxidation to betaine), as a precursor of choline-containing phospholipids, and of the neurotansmitter acetyl-choline (for detailed historical and scientific accounts, see 1, 14).

A condition of dietary choline-deficiency has not been identified in humans, even though use of choline has recently been advocated in the therapy of certain neurological disorders[19]. Such a condition, however, readily occurs in several animal species, in which it results in a variety of pathological processes involving the kidney, the cardiovascular system, the pancreas, the immune system, and most notably the liver (see 1, 12, 14). For this reason, it has been a good source of experimental models of human diseases, models which have been intermittently under active investigation during the past 50 years.

Choline-devoid Diets

Rats synthesize choline endogenously via transmethylation of phosphatidylethanolamine to phosphatidylcholine (lecithin), S-adenosylmethionine acting as the donor of methyl-groups[20]; choline is then obtained from lecithin catabolism[21]. To achieve a dietary deficiency of choline, therefore, CD diets need to be not only as free as possible of choline, but to provide also a less than adequate supply of methionine. Both requirements are best met by formulating the diets with mixtures of synthetic aminoacids, as was done in a few cases[1,22]. This approach, though, has the disadvantage of cost, especially for experiments of long duration. Hence the more frequent use of methionine-poor natural proteins, the source and content of which represent perhaps the most significant differences between the variety of refined CD-diets employed, through the years, by different investigators (see 14). Refined diets too have disadvantages, which include presence of residual choline in some of the purified proteins; less flexibility in setting the dietary level of methionine; and some batch-to-batch variability in the methionine (and choline) content of the purified proteins, as supplied by commercial sources.

Table 1 shows the composition of the refined CD-diet originally formulated by Young et al. (basal B diet[23]), that has been extensively used in this and other laboratories. This diet contains about 6 mg of choline and 1.7 g of methionine per kg, or about 0.05 and 11.39 mmoles of labile methyl groups/kg, respectively[10]. It is therefore essentially devoid of choline, and relatively deficient in methionine, since it supplies about 0.15% of the choline, and about 50% of the methionine requirements of young, growing rats[24]. This diet provides an adequate amount (15%) of proteins and of essential amino acids, L(-)cystine being added as a supplemental source of organic sulfur. The vitamin mixture includes folic acid and vitamin B12, the two other lipotropic factors. The fat content is high (15%), but can and has been varied greatly, since the total content and the ratio of saturated to unsaturated fat both modulate certain consequences of dietary choline-deficiency in the rat[14,25]. The remainder of the diet consists of carbohydrates, fiber and a mineral mixture. A control, choline-supplemented (CS) diet is prepared by replacing equivalent amounts of sucrose with choline chloride (0.2-0.8%). These CD and CS diets are well taken by rats and afford them

Table 1. Refined choline-devoid (CD) and choline-supplemented (CS) diets

Component (g/kg)	CD diet	CS diet
Alcohol-extracted peanut meal	120.00	120.00
Water-washed soya protein	80.00	80.00
Casein, vitamin-free	10.00	10.00
L-Cystine	2.00	2.00
Sucrose	389.00	381.00
Corn starch	100.00	100.00
Dextrin	100.00	100.00
Alphacel	10.00	10.00
Corn oil	50.00	50.00
Hydrogenated fat	100.00	100.00
Vitamin mixture	10.00	10.00
Salt mixture	29.00	29.00
Choline-chloride	0.00	8.00

equally good growth, both being important requisites, since induction of choline deficiency depends on diet intake and the growth rate of the animals[14]. Both diets are available commercially.

There appear to be no marked differences in the sensitivity of different strains of rats to CD diets[1,14]. The earliest responses, though, may be more severe in strains having a more rapid natural-rate of growth. On the other hand, males are distinctly more sensitive than females, a difference that , to my knowledge, has not been accounted for. The choline and methionine requirements of rats are greatest in the postweanling period, and diminish appreciably after active growth ceases[14]. Rats should therefore be ideally placed under experimentation at weanling. In this case, though, an early loss of animals occurs, due to onset of a fatal hemorrhagic-kidney syndrome[26]. The latter does not ensue if feeding is begun when the animals have grown to about 100 g of body weight, and this is the compromise choice frequently made by investigators interested in pathological consequences other than the kidney syndrome.

THE LATENCY PERIOD OF TUMOR DEVELOPMENT

Selected Liver Responses

Lecithin and membrane changes. The effects of CD diets on rat liver stem from alterations in the biosynthesis of lecithins that lead, on one hand, to widespread changes in the structure and functions of the membranous organelles of hepatocytes, and, on the other, to a faulty synthesis and secretion of serum lipoproteins (27-29, and see 1, 12, 14). The transmethylation pathway of lecithin biosynthesis has been already alluded to. Quantitatively more important, though, is the Kennedy's pathway, which proceeds via conversion of choline to phosphorylcholine, CDP-choline and phosphatidylcholine[30]. This pathway utilizes and requires a sufficient availability of free choline, and is the one primarily affected by the dietary deficiency. Within a few days of feeding a CD diet, a marked decrease in the lecithin content of the liver ensues, due to depletion in microsomal, mitochondrial, and other cellular membranes. As a consequence, the overall phospholipid composition of the membranes changes, as indicated by a lowering of the ratio of lecithins to phosphatidylethanolamines, the second major membrane-phospholipid constituents. These structural changes are accompanied by distortions in the architecture of the endoplasmic reticulum, mitochondria and Golgi apparatus of hepatocytes[31]. Integrated functions of the organelles, such as intracellular transport and secretion of albumin and serum lipoproteins, become affected[32,33], as are membrane-bound enzymes; among the latter, especially those that require a specific lecithin-microenvironment for their activity, such as cytochrome P-450 and associated drug-metabolizing enzymes, or the mitochondrial beta-hydroxybutyrate dehydrogenase[1,12]. Significant changes are introduced also in the species heterogeneity of hepatic lecithins. The Kennedy's pathway synthesizes species containing prevalently saturated fatty acids, while species originating via the transmethylation pathway are particularly rich in arachidonic acid[34]. In rats fed a CD-diet, the activity of the latter pathway increases, in a likely but abortive attempt to compensate for the overall reduced synthesis of lecithins. The net result, instead, is an abnormally high content of arachidonic acid in the membrane lecithins[35,36].

A majority of the above described changes were observed in rats fed CD diets for relatively short periods of time; whether they persist in chronically fed rats, or how they may evolve, has not been determined.

Fatty liver. Development of a fatty liver is another early, and the most conspicuous consequence of the quantitative and qualitative changes induced by CD diets in hepatic lecithins[27-29]. The changes lead to an impaired conjugation and secretion of VLD lipoproteins, the physiological vehicles that transport fat (triacylglycerols) out of the liver into the blood circulation[27]. As a result, fat accumulates in the hepatocytes[37-39]. The accumulation is initially in the form of small cytoplasmic droplets, but the latter steadily enlarge and fuse, as the accumulation progresses, and soon completely fill the cells. The plasma membrane next ruptures; the fat within contiguous hepatocytes merges; and large extracellular collections (fatty cysts, lipodiastemata) are formed, a process which was first described and then studied in detail by Hartroft[40]. Characteristic of CD diets is the extent of the induced fat-accumulation, which can reach 30-40% of the liver weight[14]. Under conditions used in this laboratory, accumulation of fat was found to peak at 25-30% of liver weight within 2-3 weeks[41]. The far content declined thereafter, and at the end of a 16-month feeding-period it was essentially the same as that in rats fed a CS diet[42]. While initially (after the peak), the decline may be underlined by the diet effects on the turnover of liver cells (see below), later on it is probably accounted for by the relationships between choline, methionine and rat's growth[14]. During active growth, methionine is in all likelihood used mostly for protein synthesis; on the other hand, as growth slows down and then ceases, methionine becomes increasingly available for choline (and VLD lipoproteins) synthesis, and the accumulated fat is slowly removed from the liver. A transition of the liver from "fatty" to "nonfatty" has frequently been noted in rats chronically fed CD diets, a transition which seems to be marked also by increasing amounts of glycogen in hepatocytes[1-4,10].

Among the great number and variety of agents that can induce a fatty liver in the rat, formation of fatty cysts is almost unique to dietary choline-deficiency. The importance of this process resides in the fact that it results in hepatocyte death, upon rupture of their plasma membrane if not before[43,44].

Liver cell death and proliferation. Whether CD diets are hepatonecrogenic has long been debated in the literature, since single hepatocytes or cluster of only a few cells are involved, which, in a liver heavily infiltrated and distorted by fat, are not easily recognizable as dead cells in histopathological examinations[45]. On the other hand, several investigators noted, mostly in autoradiographic studies, high numbers of proliferating cells, which could be best accounted for only as being compensatory to liver

cell-death (see 1, 41). In one such study, MacDonald estimated that a CD
diet reduces the life-span of liver cells to 26-60 days, from 191-453 days in
control rats, life-span being defined as the time needed by a cell to either
divide or die[46] Clear evidence that CD diets are hepatonecrogenic was recently provided by experiments in which liver DNA was (pre)labeled, before
placing rats on the diet, and cell death was assessed from the loss of radioactivity in total liver-DNA[41,47,48]. The loss was as much as 50% after a
3-week feeding period, indicating that about that many original cells had
died during that time interval. In one of these studies[41], estimates were
made of cell half-life, and of the fractional rates of cell death, proliferation and growth. Liver cells were estimated to die at a rate of about 4%/day
during an initial 8-week feeding period, and of about 1%/day during a second
8-week period; corresponding $t_{\frac{1}{2}}$ were about 17 and 75 days, respectively. In
contrast, in rats fed a CS control diet the $t_{\frac{1}{2}}$ was 292-818 days during the
entire 16-week period. At the end of the latter, less than 10% of the original (pre)labeled cells remained in the liver of rats fed the CD diet. Thus,
there can be no doubt left that cell death is a major consequence of this
diet, since it has such a profound effect on the turnover of rat liver cells,
and leads to an almost complete renewal of the organ within an initial
16-week period. In the same study, however, further evidence was obtained
that CD diets have also a primary mitogenic action vis à vis liver cells[1,49],
besides those involved in the replacement of dead cells, and the normal
accrual of the organ.

In a follow-up experiment (Fig. 1), the extent of cell proliferation
([^{3}H]-thymidine incorporation into DNA) was assessed in rats fed a CD diet
for 3-month incremental-periods, up to 16 months[42]. Liver-cell proliferation
was found to be high after 3-6 months of feeding, and to decline steadily

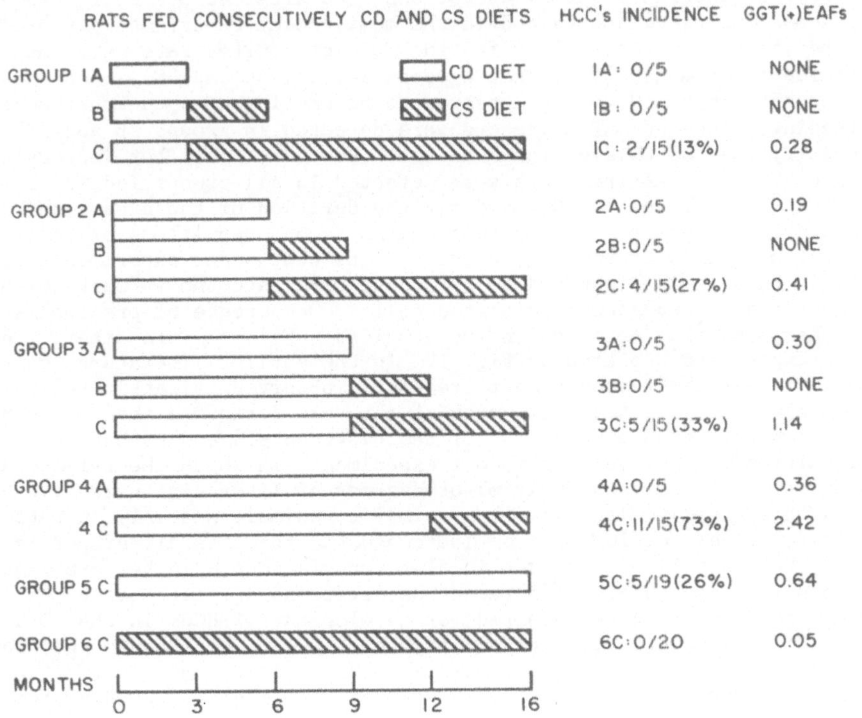

Fig. 1. Effect of consecutive feeding of CD and CS diets on the incidence of
        HCC and GGT(+)EAF.

thereafter, attaining control levels by the end of the experiment. In the
same experiment, other groups of rats were initially fed the CD diet for
3-month incremental-periods, up to 12 months, followed by feeding a CS diet
for 3 months, or for the duration of the experiment. The shorter feeding
with the control diet was performed to probe whether a CD diet causes any
change(s) that is not reversible upon restoring choline to the diet; feeding
for the duration of the experiment had instead the primary objective of assessing the minimum length of CD-diet feeding that is required for tumor induction. Subsequent feeding the CS diet promptly removed any accumulated
fat from the liver. On the other hand, cell proliferation was reduced but
still at elevated levels after 3 months of CS diet feeding, with further significant reductions occurring when the latter was fed for the duration of the
experiment. Thus, either restoration of choline to the rat's diet does not
abolish completely liver cell death, and hence proliferation; or mitogenic
stimuli generated by a CD diet persist for at least 3 months after its
discontinuation. In this context, it appears indeed arguable[48] whether
accumulation of fat and formation of fatty cysts[41,50] can alone account for the
rate of liver-cell death observed in rats fed CD diets.

Preneoplastic lesions. Only very limited information is available as to
the nature and type of preneoplastic lesions that might arise before development of HCCs in rats fed CD diets. Small numbers of foci of enzyme-altered
hepatocytes (EAFs) have been observed. However, only one marker, gammaglu-
tamyltranspeptidase (GGT), has been used for their detection in the studies
so far reported. Given the unusual aspects of the CD-diet model, and the
insufficient reliability of this marker in at least another model[51], further
studies with multiple markers will be needed to adequately assess the presence and properties of this type of lesion. In one of the published studies[52], a mean of 5.0 EAF-transects/cm^2 of section were noted after 15 months
of feeding. In a second study[10], EAFs were observed at mean transect numbers
of 1.24/cm^2 of section after 24 weeks, and 1.08 after 48 weeks of feeding.
Fig. 1 shows the results obtained in the most recent experiment performed in
this laboratory. In rats solely fed the CD diet, GGT(+) EAFs were seen only
in rats fed for 6 months or longer (groups 2A, 3A, 4A and 5C). EAFs that
developed after 6 and 9 months appeared to be contingent upon continuous CD
diet feeding, since no EAF-transects were detected in groups 2B and 3B,
subsequently fed the CS diet for 3 months. However, small but increasing
numbers of GGT (+) EAF-transects were detected in all groups fed for 3-12
months the CD diet, and the CS diet for the duration of the experiment.
Taken together, these results suggest that a second population of EAFs, not
contingent upon CD diet feeding, arose in rats kept under experimentation for
over 9-12 months. As discussed below, these EAFs have the earmarks of having
been generated by endogenous initiated cells. Occurrence of preneoplastic
nodules has been also observed in two studies[8,9], but not in another[10], nor in
the stop-experiment depicted in Fig. 1. In the earlier literature, cirrhosis
of the liver was the response most frequently observed in rats chronically
fed CD diets (see 1). Actually, much of the work performed then on such rats
was directed toward an elucidation of the histogenesis and pathogenesis of
this condition[45]. However, in recent experiments in which the refined diet
of Young et al.[23] was used, only minor degrees of liver cirrhosis have been
noted, with a great variability from animal to animal, and only in rats fed
the diet for a year or longer[9,10]. The reason(s) for this difference is not
apparent, but most likely resides in diet formulations or other experimental
conditions. At any rate, these recent observations suggest that onset of an
overt cirrhosis may not be required for development of HCCs in the CD-diet
model. CD diets have a striking ability to stimulate or enhance the proliferation of the nonparenchymal epithelial (NPE) cells of rat liver[53,54], and
increasing numbers of these cells are seen as the length of CD-diet feeding
increases[2,10]. Recently, fairly conclusive evidence has been presented that
these cells include a phenotype(s) which, under carcinogenic conditions, has
the ability to differentiate into hepatocytes and to generate HCCs[55-60]; a few

of the resulting tumors were hepatoblastomas[61], indicating that the phenotype(s) has the potential of a stem cell[62], rather than merely of an epithelial progenitor cell committed to hepatocyte differentiation[63]. NPE cells, therefore, deserve consideration as potential progenitors of preneoplastic and neoplastic lesions also in this model of hepatocarcinogenesis.

THE TUMORS AND TUMOR INCIDENCES

The liver tumors induced by CD diets are mostly well- to moderately well-differentiated HCCs, of the trabecular, adenomatous, and mixed types[1-5,8-10]. Cholangiocarcinomas have been also noted, underscoring an involvement of NPE cells. In several instances, the tumors produced both local and distal metastases. Tumor incidences appear to bear a direct relationship to the length of the experiments. No tumors have been observed in rats fed control CS diets.

In the most recent experiment performed in this laboratory, tumor incidences were as shown in Figure 1. No tumors developed in rats fed the CS diet throughout, or in rats killed after 12 months or less of experimentation. The incidence was 26% in rats fed the CD diet throughout, but almost 3 times as large in rats fed the control diet for the last 4 months. This result clearly indicates that occurrence of a late event(s) is critical in this model, for the progression of any preneoplastic lesion to cancer; an event(s), whose occurrence is either hindered by the lack of choline in the diet, or is actually mediated by choline restoration to the diet[64]. The occurrence coincides with the late transition of the liver from "fatty" to "nonfatty", a transition which is undoubtedly facilitated by feeding the control rather than CD diet. The tumor incidence was 13% in rats fed the CD diet for 3 months, and the control diet for the duration of the experiment, while it rose to 26% and 33% in rats initially fed the CD diet for 6 or 9 months. An unnecessary but fruitful period of promotion appears thus to be provided by the CD diet. Two of the 15 rats fed initially the CD diet for 3 months developed carcinomas. This result could be statistically irrelevant, were it not for a similar finding previously made in rats fed a methyl-deficient diet for 15 weeks, and laboratory chow for the next 37 weeks[65]. It is apparent, therefore, that a relatively short 3-4 month exposure to these diets is sufficient to yield carcinomas, and that initiated cells must be present in the liver within this time. It may or not be coincidental that this is also the time during which the liver undergoes the most drastic changes in cell turnover. It is apparent, also, that during the same time CD diets must create local tissue and cellular conditions, able to mediate the evolution of initiated cells, and their neoplastic transformation, even in the absence of subsequent active promotion.

As to the origins of the initiated cells, given the lack of evidence that chemical-carcinogen contaminants are involved in this model, only two choices remain: either CD diets initiate de novo liver cells, or endogenous cells are the precursors of the carcinomas. Plausible means whereby CD diets could effect initiation de novo are considered below. The case of endogenous initiated cells (EICs) will be instead argued here. This is undoubtedly a difficult task at the present time, since so little is known about the biological nature, properties and significance of EICs, beyond the fact that they do exist, and that they have the constitutive ability to evolve, under natural conditions, into preneoplastic and neoplastic lesions[66-80]. Cells presumed to be EICs have been detected in the liver of rats as young as 6 weeks[81]; they may therefore be present in this organ, and possibly in others, since birth if not before. Exposure of the mothers or pups to environmental chemical agents or to ionizing radiation; unknown endogenous factors; or a high rate of fetal and perinatal cell division in many tissues, could be postulated as possible causes of their initiation[77,82,83].

EICs appear to have extremely low rates of amplification and evolution, since under natural conditions they become overtly manifest as EAFs, adenomas, or carcinomas, only late in the life of the rat [66-70,76,78,79] . They have however been shown to respond positively to promoters [72,74,75], albeit apparently not as strongly as do cells initiated by chemical carcinogens. Conjecturally, it could be that EICs are constitutively endowed also with a poor ability to metabolize and hence respond to xenobiotics, as the NPE cells of the liver are[84]. With some notable exceptions, EICs are usually dealt with, in much of the pertinent literature, as being no more than background noise in the process of hepatocarcinogenesis, or are outrightly dismissed as being of no significance or consequence. Such a position may be indeed correct. However, it would seem to be at least premature, since the whole phenomenology of EICs has not as yet been subjected to close scrutiny and experimental analysis. At any rate, it seems hardly probable that EICs would remain completely unaffected by an intervening carcinogenic treatment, be the latter in the form of a CD diet or of a chemical carcinogen. Actually, some of the results obtained in the above-described stop-experiment appear to strongly implicate EICs as the likely progenitors of the observed HCCs: small but increasing numbers of GGT(+)EAFs emerged in rats fed the control CS diet for 4-13 months before tumor detection; EAFs emerged late in the treatment, at a time overlapping the natural emergence of EICs in the liver of aging rats; the number of EAFs appeared to be directly related to the length of previous CD-diet feeding, and was the largest in the rat group with the highest incidence of HCCs; HCCs developed only in those groups that were kept under experimentation for 16 months, and, thus, while tumor incidences varied greatly in the tumor-bearing groups, the latency period for tumor development did not; presence of EICs has been reported also in rat pancreas[85-88], and while EAFs and carcinomas arose in the liver, acidophilic atypical acinar-cell nodules, adenomas and carcinomas in situ developed in the pancreas of the same animals[89]. The last, of course, are not new findings, since not only HCCs, but also carcinomas of the pancreas, lung and mammary gland, as well as hemangioendotheliomas and sarcomas, have been shown to develop in rats chronically fed a CD diet[2-4]. In this respect, therefore, the CD-diet model of "hepatocarcinogenesis" is actually a misnomer. At the same time, though, the probability that the diet may have the ability to initiate cells de novo, in such a variety of tissues, appears to be remote, at least at the present time. In the case of the liver, on the other hand, one could visualize how EICs, after a 3-4 month period of diet-mediated amplification, could receive a second and diet-related hit, rather than the first (initiation), and thus be able to fully evolve to HCCs even after subsequent feeding with control diets. An involvement of EICs, in the genesis of HCCs, is indicated also by the results recently obtained by Newberne et al. in mice fed a CD diet[90].

PLAUSIBLE MECHANISMS WHEREBY CD DIETS COULD LEAD TO A NEOPLASTIC TRANSFORMATION OF LIVER CELLS

The Mitogenic Milieu Created by CD Diets in Rat Liver

Feeding a CD diet can be viewed as a "nutritional partial-hepatectomy" that mimics, in several respects, events accompanying a surgical partial-hepatectomy (PH) in rats. As is in the case of the latter, the diet-induced cell-loss triggers cell proliferation and a compensatory regeneration. After a PH, proliferation of liver cells appears to be mediated more by paracrine or autocrine than endocrine factors[91], the cells losing for example surface receptors for epidermal growth factor[92,93]. At the same time, endogenous growth-modulating factors such as HSS (hepatic stimulator substance[94]), which is normally quiescent in young-adult and adult-rat liver, becomes very active, while an expressed chalone becomes inactive[95]. Similar cytological and humorohomeostatic changes have been shown to occur in rats fed a CD diet[96-98]. However, there is at least one major and significant difference between the two conditions: in the case of a PH, the changes are episodic, acute and transient, while in rats fed a CD diet they persist in various degrees for as long as the diet is fed, and some of them even after its discontinuation. Thus, when CD diets are fed for periods of 6-9 months or longer, a highly abnormal condition arises in the liver, since its cells, normally fairly stable in adult rats, are instead forced to divide and renew themselves, repeatedly. Such a condition would appear to constitute a highly favorable ground for the occurrence of alteration(s) in the processes of cell division, growth or differentiation, that could contribute to a neoplastic transformation of cells.

As postulated by Potter and by Scherer[99,100], promoters must create local conditions having at least two effects: I) to achieve an amplification of the finite number of cells that are initiated, as well as as that of their progenies; and II) to bring to fruition[101], or lead to, further genomic alteration(s) in one cell, among the progenies of initiated cells, conferring to it the neoplastic phenotype. The mitogenic milieu created in the liver by CD diets could very well fulfill the second requirement; by forcing cells to continuously divide, for example, it could increase the frequency with which mutational errors in DNA replication occur[82,83]. It obviously fulfills the first requirement, whether the diets are acting alone[1-5,8-10], or in the context of a chemical hepatocarcinogenesis model[1,12,13]. At any rate, there seems to be no doubt that it is the primary factor underlying the promoting action of these diets[102]. Reservations to this conclusion have been expressed[13,103], since inclusion of phenobarbital (PHB) in a CD diet leads to an inhibition of the diet-induced increases of [^{3}H]-thymidine incorporation into liver DNA, and of the mitotic index; yet, it results in a synergistic promotion of the emergence of diethylnitrosamine-initiated cells, as EAFs[13]. However, subsequent results obtained in this laboratory[104] have shown that the principal effect of PHB is to prevent instead, to a large extent, the hepatonecrogenic action of a CD diet; furthermore, induction of tumors by the diet was found to be not enhanced by PHB inclusion in it[105].

The forced renewal of liver cells, though, poses also questions that might require attention when considering the actions of CD diets vis à vis hepatocarcinogenesis in the rat. For example, within the overall population of hepatocytes are all cells sensitive to the necrogenic action of these diets, or is there a subpopulation, however small, that is constitutively refractory. In the former case, clonal evolution of initiated cells and their progenies would depend not only on the frequency with which relevant events occur[99,100], but also the frequency with which evolving cells escape eventual death. In the same case, relevant events might have to occur repeatedly with time, and the search for them might be more fruitful if made on livers obtained after medium- or long-, rather than short-term feeding-periods, as indicated by some of the results obtained in the above described stop-experiment. Histological analyses provide only an ambiguous answer to this question. Scattered hepatocytes free of fat, the latter being a good index of response to the diets, are consistently seen even in the most severe of the fatty livers induced by CD diets. Similarly, in studies in which these diets have been used as a promoter of chemical hepatocarcinogenesis, the observation has been repeatedly made that cells, within emerging EAFs, are either completely or relatively free of fat, in contrast to hepatocytes in the surrounding parenchyma[10]. However, it is difficult to ascertain, from both observations, whether they pertain to cells with a native or acquired resistance, or to hepatocytes which, having just divided, have as yet not had a sufficient exposure-time to respond, visibly, to the lack of choline in the diets. Assessment of the relevant importance of some of the diet's effects may also be hampered by the underlying high background of liver-cell death and proliferation, which forces the need to evaluate whether, or to what extent, the effect is merely a manifestation of ongoing cell death, or of the presence of a sizeable number of regenerating cells, rather than an index of ongoing cell transformation. It seems finally likely that other agents, shown to modulate the actions of CD diets[25], may do so, in part or completely, by modifying the underlying rates of liver cell death and regeneration, as is the case of PHB.

Peroxidation of Liver Lipids

Much attention has been recently given to the possibility that CD diets may lead to peroxidation of liver lipids. This is indeed an event which, if real, could go a long way in accounting for the hepatocarcinogenicity of these diets, since the generated lipoperoxide-radicals could contribute to liver-cell death[13], on one hand, and be potentially genotoxic[106], on the other. Results indicating an ongoing peroxidation of hepatic lipids have been presented [1,13,107,108]. The results consist of detection of diene conjugates, by means of the spectrum difference between total lipids of liver nuclei, mitochondria and microsomes[109] from rats fed either a CD or a CS diet, for a few hours, days or weeks. This is a methodological approach which, in similar studies of other experimental conditions, has consistently proven to be reliable and completely satisfactory[39,110]. In this particular instance, though, it may overlook the possibility that the source of the diene conjugates, detected in total lipids, may not be peroxidation of liver lipids, but rancid fat(ty acids) present in the diets; indeed, use of difference spectra does not allow to test whether diene conjugates are present in the lipids of rats fed the control diet. In a series of analyses just completed in this laboratory[111], absolute spectra were taken[112] of total lipids, neutral lipids and phospholipids, extracted either from liver microsomes, or intestinal mucosa homogenates of rats fed refined CD or CS diets. Diene conjugates were detected in the total and neutral lipids, but not in the phospholipids of both tissue preparations, and irrespective of which diet was fed to the rats. Were a CD diet to indeed trigger lipoperoxidation, it is hard to visualize how the latter could leave unscathed the phospholipids, especially those of liver microsomes, whose lecithins are so rich in arachidonic acid. On the other hand, the finding of diene conjugates in the intestinal mucosa neutral lipids, of rats fed both diets, clearly implicates dietary fat as their source. Indeed, as dietary fat is stored in adipose tissue[113], diene conjugates were detected also in the total lipids of the latter, and again in rats fed either diet. Furthermore, diene conjugates were detected not only in the fat of the diets themselves, but also in the layer that floats upon centrifugation of liver homogenates from rats fed a CD diet; this is the fat, mostly of dietary origin, that has accumulated in the liver. The last finding could perhaps account for some of the discrepancies in previously reported results [108,109,114,115], concerning the intracellular site(s) of the presumed ongoing peroxidation of liver lipids. The amount of fat that floats is a direct function of the centrifugal force that is applied to homogenates; the higher the force, the more fat-free is the resulting sediment. Widely different forces are used in the preparation of subcellular organelles such as microsomes or nuclei; moreover, differences in the forces used by different laboratories, in the preparation of one and the same organelle, or in washing procedures, could result in different degrees of dietary fat "contamination" of the organelles, before extraction of their total lipids. At any rate, the above results underscore at least the need for a critical reassessment of the overall methodologies used in determining whether CD diets induce an active peroxidation of hepatic membrane-lipids. Whether effects ascribed to the inclusion in the diet of additional antioxidants[1,114] may reflect an inhibition of dietary fat rancidity, rather than of an active lipoperoxidation, may also need to be reevaluated. It seems then likely that other organs that utilize fatty acids, mobilized from the adipose tissue, may also show diene conjugates in their total lipids[116].

The initial results obtained in this laboratory pertain to rats fed the diets for up to two weeks; it remains to be established, therefore, whether CD diets fed for longer periods of time can lead to peroxidation of the phospholipids of hepatic microsomes, or of other hepatocyte-organelles. Cellular metabolism involves a variety of oxidation-reduction reactions that produce radical intermediates or metabolites capable of initiating lipid peroxidation[117]. It seems therefore conceivable, at least, that more chronic alterations caused by the diet, in the structure and functions of hepatocyte organelles, could result in an increased oxidative stress in the cells, due either to an increased metabolic production of radicals, or to a decrease in the enzymatic, antioxidant or radical-scavenger, cellular defenses. The same results raise also other questions that will have to be addressed, such as what are the fate and disposition of rancid fat in the liver and other organs. A great variety of electrophilic and toxic compounds has been shown to result from active peroxidation of membrane lipids[110]. Thus, to the extent that similar products may be generated from cellular breakdown of rancid fatty acids, they could still contribute to liver-cell death, and, conceivably, lead to cell-initiation de novo not only in the liver, but also in other organs, such as the pancreas. Obviously, much work lies ahead, before clarification of whether autooxidized or peroxidized fatty acids play any role in the CD-diet model of carcinogenesis.

Methionine Versus Choline Dietary Deficiency, and Gene Undermethylation

The question of the relative methionine-deficiency of refined CD diets, and of the extent to which it contributes to the pathology induced by them, has long attracted the attention of investigators [1,14,22]. It has been however difficult to study, and still is, largely, unresolved. Diets formulated with synthetic amino acids, and devoid of choline and/or methionine, have been used in some of these studies. These diets reproduce much if not all the pathology of choline dietary deficiency[1,22]; however, it seems probable that those devoid of methionine may induce also additional and unique responses, especially when fed to weanling rats. Refined CD and CS diets usually contain the same level of methionine. Since no overt pathology is observed in rats fed the control diet, the lack of choline, rather than the relative deficiency of methionine, would appear to be the critical factor. The growth rate of rats fed a refined CS diet[23] is nearly optimal, and that of rats fed the corresponding CD diet is equally good. These diets, therefore, seem to satisfy fairly adequately the overall methionine needs of the animals, in particular that for (liver)protein synthesis[32,33,118]. Yet, decreased levels of free methionine have been found in the liver of rats fed a refined CD-diet , and of S-adenosyl methionine (SAM) in that of rats fed a synthetic aminoacid-diet[22]. The latter could obviously be a more critical consequence, since so many metabolic reactions require a methylation step, with SAM as the methyl-group donor[119]. However, it is not known whether or to what extent either reduction is due to insufficient availability of methionine, defects in its activation to SAM, and/or an increased metabolic utilization of both methionine and SAM, as in the case of the transmethylation pathway of lecithin biosynthesis.

Undermethylation of liver t-RNA and DNA has been shown to occur in rats fed for 6 months aminoacid-formulated diets, devoid of both methionine and choline (see 21). In contrast, no undermethylation of DNA was seen in rats fed a refined CD-diet for up to 12 months, even though it was present later on, just before the development of HCCs, and in some of the tumors as well[120]. When the methylation state of the alphafetoprotein gene of the same tissues was analyzed, slight changes in its overall methylation, but not of its methylation pattern, were observed in the livers, while both were affected in the tumors[121]. It seems therefore possible that alterations in gene methylation may mark or contribute to a neoplastic transformation[122,123] of liver cells in this carcinogenesis model, also in view of some of the results obtained in the stop-experiment (Fig. 1). However, given the late occurrence of the observed changes, it seems unlikely that their cause can be ascribed to an undersupply of labile methyl groups of dietary origin. Recent evidence indicates that epimutations may result from gene undermethylation[124]. It may therefore be worthy of exploration whether gene undermethylation, or changes in the intragene methylation-pattern, rather than those of liver total DNA, occur earlier in rats fed a refined or aminoacid-formulated diet.

Proto-oncogene Activation to Oncogenes

Initial studies have been made in rats fed refined CD or CS diets for 14 months[125]. K-ras, H-ras and N-ras transcripts were found to be elevated in 7 HCCs that developed in CD diet fed rats, but not in the liver from which they arose, nor in nontumor-bearing livers of CD or CS diet fed rats. Structural alterations of the ras genes were not detected by Southern-blot analyses. Since ras oncogenes generally differ from ras proto-oncogenes by point mutations[126,127], and elevated H-ras and K-ras transcripts occur after a PH[128], it seems likely that the observed elevations may simply reflect the high rate of proliferation of the tumoral cells.

A mutation in the coding domain of the gene, leading to expression of an abnormal product, is one way in which activation of proto-oncogenes to oncogenes occurs; another is by expression of a normal product, but either in abnormal quantities, or at inappropriate times during the life span of a cell at cancer risk[129,130]. Involvement of proto-oncogene products in the processes of cell division, growth or differentiation has been shown[130-133], and, as already mentioned, a transiently elevated expression of some of them occurs after a PH[128]. Given the analogies between the latter, and the effects of a CD diet on the turnover of rat liver cells, it seems likely that expression of certain proto-oncogenes may increase, early in the liver of rats fed the diet; that it may persist elevated, for as long as the diet is fed; and/or that the elevation may occur at inappropriate time(s), in an organ whose cells are proliferatively quite stable for much of the adult life of the rat. It is worthy of note, also, that transforming-growth-factor beta, which can be considered as a proto-oncogene product[132], is also involved in cell regenerative processes[134].

CONCLUSIONS

It is widely accepted today that neoplastic transformation of cells is a multistage process[135], transition of one stage to the next being mediated by the occurrence of genomic alterations[100,136,137] ranging from single point mutations to large scale chromosomal rearrangements[131]. The CD diet model of hepatocarcinogenesis, therefore, poses a considerably greater challenge than models of chemical carcinogenesis, in trying to understand the nature of those genomic events, and how a diet, and diet alone, can generate them. The challenge, of course, goes beyond the question of whether CD diets act as complete carcinogens, or merely as promoters. At the same time, though, the model seems to offer some almost unique opportunities, such as that of probing deeper into the biology of endogenous initiated cells, in both liver and pancreas, at least, biology whose implications could extend to chemical-carcinogenesis models as well. It could serve as a "control" to the latters, and help to clarify whether any phenomenology seen in them reflects toxicological, rather than the carcinogenic properties of the chemicals used.

The CD-diet model, moreover, may be more attuned to the human experience than many of the chemical hepatocarcinogenesis models presently under study. In humans, a high association exists between HCC, on one hand, and liver cirrhosis or viral hepatitides on the other[138,139]. The latter entities are characterized by a chronic condition of heightened liver cell turnover[140,141], much alike that seen in rats fed CD diets; in the viral diseases, it is still not clear whether insertion and expression of the viral genome, or hepatocyte turnover, is the crucial determinant of HCC development[142].

Finally, CD diets and their hepatic consequences could be construed as being an example of severe malnutrition; and malnutrition, in contrast to undernutrition, knows no boundaries within mankind's societies. On the other hand, one indeed wonders how high the probability or frequency are that humans, barring catastrophies, may undergo single or multiple exposures to chemical carcinogens, at dosages translated into a human scale from those employed in many carcinogenesis models[143]. The CD-diet model of hepatocarcinogenesis, therefore, would seem to deserve a wider-than-present adoption for study, not only in view of the challenges and opportunities it offers, but also because only its bare outlines have been so far recognized.

ACKNOWLEDGEMENTS

Thanks are expressed to Mrs. M. L. Rotz for her assistance in the manuscript's preparation. The work performed in the author's laboratory was supported in part by grants from the National Institutes of Health (CA 23449) and the American Cancer Society (BC 471).

REFERENCES

1. P. M. Newberne, Lipotropic Factors and Oncogenesis, in: "Essential Nutrients in Carcinogenesis", L. A. Poirier, P. M. Newberne and M. W. Pariza, eds., Plenum Press, New York (1986).
2. D. H. Copeland and W. D. Salmon, The occurrence of neoplasms in the liver, lungs and other tissues of rats as a result of prolonged choline deficiency, Am. J. Pathol. 22:1059 (1946).
3. W. D. Salmon and D. H. Copeland, Liver carcinoma and related lesions in chronic choline deficiency, Ann. N. Y. Acad. Sci. 57:664 (1954).
4. W. D. Salmon and P. M. Newberne, Occurrence of hepatomas in rats fed diets containing peanut meal as a major source of protein, Cancer Res. 23:571 (1963).
5. P. M. Newberne, Biological activity of the aflatoxins in the domestic and laboratory animals, U. S. Fish Wildlife Serv. Res. Rep. 70:131 (1967).
6. K. Sargeant, R. B. A. Carnaghan and R. Allcroft, Toxic products in groundnuts: chemistry and origin, Chem. Ind. 53:55 (1963).
7. G. N. Wogan, A. S. Paglialunga and P. M. Newberne, Carcinogenic effects of low dietary levels of aflatoxin B in rats, Food Cosmet. Toxicol. 12:681 (1974).
8. Y. B. Mikol, K. L. Hoover, D. Creasia and L. A. Poirier, Hepatocarcinogenesis in rats fed methyl-deficient, amino acid-defined diets, Carcinogenesis 4:1619 (1983).
9. A. K. Ghoshal and E. Farber, The induction of liver cancer by a dietary deficiency of choline and methionine without added carcinogens, Carcinogenesis 5:1367 (1984).
10. S. Yokoyama, M. A. Sells, T. V. Reddy and B. Lombardi, Hepatocarcinogenic and promoting action of a choline-devoid diet in the rat, Cancer Res. 45:2834 (1985).
11. R. C. Gupta, J. Locker and B. Lombardi, ^{32}P-Postlabeling analysis of liver DNA-adducts in rats chronically fed a choline-devoid diet, Carcinogenesis 8:187 (1987).
12. H. Shinozuka and S. L. Katyal, Pathology of choline deficiency, in: "Nutritional Pathology", H. Sidransky, ed., Marcel Dekker, New York (1985).
13. H. Shinozuka, S. L. Katyal and M. I. R. Perera, Choline deficiency and chemical carcinogenesis, in: "Essential Nutrients in Carcinogenesis", L. A. Poirier, P. M. Newberne and M. W. Pariza, eds., Plenum Press, New York (1986).
14. C. C. Lucas and J. J. Ridout, Fatty livers and lipotropic phenomena,

<u>Progr</u>. <u>Chem</u>. <u>Fats</u> <u>Other</u> <u>Lipids</u> 10:1 (1967).

15. A. Srecker, Beobachtungen uber die Galle ver shiedener Thiere, <u>Ann.</u> <u>Chim.</u> 70:149 (1849).

16. C. H. Best and M. E. Huntsman, The effects of the components of lecithins upon deposition of fat in the liver, <u>J.</u> <u>Physiol.</u> 75:405 (1932).

17. C. H. Best and J. H. Ridout, Choline as a dietary factor, <u>Ann.</u> <u>Rev.</u> <u>Biochem.</u> 8:349 (1939).

18. C. H. Best, M. E. Huntsman and J. H. Ridout, The "lipotropic" effect of proteins, <u>Nature</u> 135:821 (1935).

19. S. H. Zeisel, Dietary choline: biochemistry, physiology, and pharmacology, <u>Annu.</u> <u>Rev.</u> <u>Nutr.</u> 1:95 (1981).

20. J. A. Bremer and D. M. Greenberg, Methyl transferring enzyme system of microsomes in the biosynthesis of lecithin (phosphatidyl choline), <u>Bioch.</u> <u>Biophys.</u> <u>Acta</u> 46:205 (1961).

21. A. Contardi and A. Ercoli, Uber die enzymatische spaltung der Lecithine und Lysolecithine, <u>Biochem.</u> <u>Z.</u> 261:275 (1933).

22. L. A. Poirier, The role of methionine in Carcinogenesis, <u>in</u>: "Essential Nutrients in Carcinogenesis", L. A. Poirier, P. M. Newberne and M. W. Pariza, eds., Plenum Press, New York (1986).

23. R. J. Young, C. C. Lucas, J. M. Patterson and C. H. Best, Lipotropic dose-response studies in rats: comparison of choline, betain, and methionine, <u>Canad.</u> <u>J.</u> <u>Biochem.</u> <u>Physiol.</u> 34:713 (1956).

24 Nutrient requirements of laboratory animals, <u>Natl.</u> <u>Res.</u> <u>Council</u> 10:7 (1978).

25. H. Shinozuka, S. L. Katyal and M. I. R. Perera, Fat, lipotropes, hypolipidemic agents and liver cancer, <u>in</u>: "Dietary Fat and Cancer", C.Ip., D. F. Birt, A. E. Rogers and C. Mettline, eds., Alan R. Liss Inc., New York (1986).

26. W. H. Griffith and N. J. Wade, Choline metabolism. 1. The occurrence and prevention of hemorrhagic degeneration in young rats on a low choline diet, <u>J.</u> <u>Biol.</u> <u>Chem.</u> 131:567 (1939).

27. R. E. Olson, Scientific contributions of Wendel H. Griffith to our understanding of the function of choline, <u>Fed.</u> <u>Proc.</u> 30:131 (1971).

28. B. Lombardi, Effects of choline deficiency on rat hepatocytes, <u>Fed.</u> <u>Proc.</u> 30:139 (1971).

29. A. Kuksis and S. Mookerjea, Choline, <u>Nutr.</u> <u>Rev.</u> 36:201 (1978).

30. E. P. Kennedy and S. B. Weiss, The function of cytidine coenzymes in the biosynthesis of phospholipids, <u>J.</u> <u>Biol.</u> <u>Chem.</u> 222:193 (1956).

31. C. Bruni and D. M. Hegsted, Effects of choline-deficient diets on the rat hepatocyte. Electron microscopic observations, <u>Am.</u> <u>J.</u> <u>Path.</u> 61:413 (1970).

32. B. Lombardi and A. Oler, Choline deficiency fatty liver. Protein synthesis and release, <u>Lab.</u> <u>Invest.</u> 17:308 (1967).

33. A. Oler and B. Lombardi, Further studies on a defect in the intracellular transport and secretion of proteins by the liver of choline-deficient rats, <u>J.</u> <u>Biol.</u> <u>Chem.</u> 245:1282 (1970).

34. W. E. M. Lands, Lipid metabolism, <u>Ann.</u> <u>Rev.</u> <u>Biochem.</u> 34:313 (1965).

35. R. L. Lyman, S. M. Hopkins, G. Sheehan and J. Tinoco, Effects of estradiol and testosterone on the incorporation and distribution of [Me- C] methionine methyl in rat liver lecithins, <u>Bioch.</u> <u>Biophys.</u> <u>Acta</u> 152:197 (1968).

36. B. Lombardi, P. Pani, F. F. Schlunk and S-H. Chen, Labeling of liver and plasma lecithins after injection of 1-2-^{14}C-2-Dimethyl-amino-ethanol and ^{14}C-L-Methionine-Methyl to choline deficient rats, <u>Lipids</u> 4:67 (1969).

37. B. Lombardi, Considerations on the pathogenesis of fatty liver, <u>Lab.</u> <u>Inv.</u> 15:1 (1966).

38. K. J. Isselbacher and D. H. Alpers, Fatty liver, biochemical and clinical aspects, <u>in</u>: "Disease of Liver", L. Schiff, ed., Lippincott, Philadelphia (1969).

39. R. O. Recknagel and E. A. Glende, Carbon tetrachloride hepatoxicity: An example of lethal cleavage, CRC Crit. Rev. Toxicol. 2:263 (1973).

40. W. S. Hartroft, Histological studies on fatty infiltration of the liver in choline-deficient rats, in: "Liver Disease", S. Scherlock and G. E. W. Wolstenholme, eds., J. & A. Churchill, London (1951).

41. N. Chandar, J. Amenta, J. C. Kandala and B. Lombardi, Liver cell turnover in rats fed a choline-devoid diet, Carcinogenesis, 8:669 (1987).

42. N. Chandar and B. Lombardi, Liver cell proliferation, and incidence of hepatocellular carcinomas, in rats fed consecutively a choline-devoid and choline-supplemented diet, submitted for publication.

43. J. L. Farber, Reactions of the liver to injury: Necrosis, in: "Toxic Injury of the Liver, Part A", E. Farber and M. M. Fisher, eds., Marcel Dekker, New York (1979).

44. B. F. Trump, E. M. McDowell and Arstila, A. V., in: "Principles of Pathobiology", R. B. Hill and M. F. LaVia, eds., Oxford University Press (1980).

45. W. S. Hartroft, The sequence of pathological events in the development of experimental fatty livers and cirrhosis, Ann. N. Y. Acad. Sci. 57:633 (1954).

46. R. A. MacDonald, "Lifespan" of liver cells. Autoradiographic study using tritiated thymidine in normal, cirrhotic, and partially hepatectomized rats, Arch. Intern. Med. 107:79 (1961).

47. L. K. Giambarresi, S. L. Katyal and B. Lombardi, Promotion of liver carcinogenesis in the rat by a choline-devoid diet: role of liver cell necrosis and regeneration, Br. J. Cancer 46:825 (1982).

48. A. K. Ghoshal, M. Ahluwalia and E. Farber, The rapid induction of liver cell death in rats fed a choline-deficient methionine-low diet, Am. J. Pathol. 113:309 (1983).

49. K. N. Rao, Regulatory aspects of cholesterol metabolism in cells with different degrees of replication, Toxicol. Path. 14:430 (1986).

50. A. E. Rogers and R. A. MacDonald, Hepatic vasculature and cell proliferation in experimental cirrhosis, Lab. Invest. 14:1710 (1965).

51. H. P. Glauert, D. Beer, M. S. Rao, M. Schwartz, Y-D. Xu, T. L. Goldsworthy, J. Coloma and H. Pitot, Induction of altered hepatic foci in rats by the administration of hypolipidemic peroxisome proliferators above or following a single dose of diethylnitrosamine, Cancer Res. 46:4601 (1986).

52. A. K. Ghoshal and M. Farber, Induction of liver cancer by a diet deficient in choline and methionine (CMD), Proc. Am. Assoc. Cancer Res. 24:98 (1983).

53. J. W. Grisham and E. A. Porta, Origin and fate of proliferated hepatic ductal cells in the rat: Electron microscopic and autoradiographic studies, Exp. Mol. Path. 3:242 (1964).

54. H. Shinozuka, B. Lombardi, S. Sell and R. M. Iammarino, Early histological and functional alterations of ethionine liver carcinogenesis in rats fed a choline-deficient diet, Cancer Res. 38:1092 (1978).

55. Y. Inaoka, Significance of the so-called oval cell proliferation during azo-dye hepatocarcinogenesis, Gann 58:355 (1967).

56. J. M. Grisham, Cell types in long-term propagable cultures of rat liver, Ann. N. Y. Acad. Sci. 349:128 (1980).

57. H. Yoshimura, R. Harris, S. Yokoyama, S. Takahashi, M. A. Sells, S. F. Pan and B. Lombardi, Anaplastic carcinomas in nude mice and in original donor strain rats inoculated with cultured oval cells Am. J. Path. 110:322 (1983).

58. N. T. Hayner, L. Braun, P. Yaswen, M. Brooks and N. Fausto, Isozyme profiles of oval cells, parenchymal cells, and biliary cells isolated by centrifugal elutriation from normal and preneoplastic livers, Cancer Res. 44:332 (1984).

59. M-S. Tsao, J. D. Smith, K. G. Nelson and J. W. Grisham, A diploid

epithelial cell line from normal adult rat liver with phenotypic
properties of oval cells, Exp. Cell Res. 154:38 (1984).

60. L. Braun, M. Goyette, P. Yaswen, N. L. Thompson and N. Fausto, Growth
in culture and tumorigenicity after transfection with ras oncogene
of liver epithelial cells from carcinogen-treated rats, Cancer
Res. 47:4116 (1987).

61. M-S. Tsao and J. W. Grisham, Hepatocarcinomas, cholangiocarcinomas, and
hepatoblastomas produced by chemically transformed cultured rat
liver epithelial cells. A light and electron-microscopic analysis,
Am. J. Path. 127:168 (1987).

62. J. W. Grisham, S. B. Thal and A. Nagel, Cellular derivation of continuously cultured epithelial cells from normal rat liver, in: "Gem
Expression and Carcinogenesis in Cultured Liver Cells", L. E.
Gershcenson and E. Brad Thompson, eds., Academic Press, New York
(1975).

63. B. Lombardi, On the nature, properties, and significance of oval cells,
in: "Recent Trends in Chemical Carcinogenesis", P. Pani, F. Feo and
A. Columbano, eds., ESA, Cagliari (1982).

64. J. C. Linnell, M. J. Wilson, Y. B. Mikol and L. A. Poirier, Tissue
distribution of methylcobalamin in rats fed amino acid-defined
methyl-deficient diets, J. Nutrit. 113:124 (1983).

65. K. L. Hoover, P. H. Lynch and L. A. Poirier, Profound post-initiation
enhancement by short-term severe methionine, choline, vitamin B_{12},
and folate deficiency of hepatocarcinogenesis in F344 rats given a
single low-dose diethylnitrosamine injection, JNCI 73:1327 (1984).

66. D. G. Goodman, J. M. Ward, R. A. Squire, K. C. Chu and M. S. Linhart,
Neoplastic and nonneoplastic lesions in aging F344 rats, Toxicol.
Appl. Pharmacol. 48:237 (1979).

67. M. Pollard and P. H. Luckert, Spontaneous liver tumors in aged germ
free Wistar rats, Lab. Anim. Sci. 29:74 (1979).

68. D. G. Goodman, J. M. Ward and R. A. Squire, Neoplastic and nonneoplastic lesions in aging Osborne Mendel rats, Toxicol. Appl. Pharmacol.
55:433 (1980).

69. J. M. Ward, Morphology of foci of altered hepatocytes and naturally
occurring hepatocellular tumors in F344 rats, Virchows Arch. Pathol.
Anat. 390:339 (1981).

70. K. Ogawa, T. Onoe and M. Tachenchi, Spontaneous occurrence of gamma-
glutamyltranspeptidase positive hepatocytic foci in 105-week old
Wistar and 72-week-old Fischer 344 male rats, JNCI 67:407 (1981).

71. G. M. William, Liver carcinogenesis: the role for some chemicals of an
epigenetic mechanism of liver-tumor promotion involving modification
of the cell membrane, Food Cosmet. Toxicol. 19:577 (1981).

72. R. Schulte-Hermann, G. Ohde, J. Schuppler and I. Timmermann-Trosiener,
Enhanced proliferation of putative preneoplastic cells in rat liver
following treatment with the tumor promoters phenobarbital, hexachlorocyclohexane, steroid compounds, and nafenopin, Cancer Res.
41:2556 (1981).

73. P. M. Newberne, Assessment of the hepatocarcinogenic potential of
chemicals: Response of the liver, in: "Toxicology of the Liver",
G. L. Plaa and W. R. Hewitt, eds., Raven Press, New York (1982).

74. R. Schulte-Hermann, I. Timmerman-Trosiener and J. Schuppler, Promotion
of spontaneous preneoplastic cells in rat liver as a possible
explanation of tumor production by non-mutagenic compounds, Cancer
Res. 43:2644 (1983).

75. J. M. Ward, Increased susceptibility of livers of aged F344/NCr rats to
the effects of phenobarbital on the incidence, morphology and histochemistry of hepatocellular foci and neoplasms, J. Natl. Cancer
Inst. 71:815 (1983).

76. H. A. Solleveld, J. K. Haseman and E. E. McConnell, Natural history of
weight gain, survival and neoplasia in the F344 rat, J. Natl. Cancer
Inst. 72:929 (1984).

77. T. Goldswothy, H. A. Campbell and H. C. Pitot, The natural history and dose-response characteristics of enzyme-altered foci in rat liver following phenobarbital and diethylnitrosamine administration, Carcinogenesis 5:67 (1984).

78. J. A. Popp, B. H. Scartichini and L. K. Garvey, Quantitative evaluation of hepatic foci of cellular alteration occurring spontaneously in Fisher-344 rats, Fund. Appl. Toxicol. 5:314 (1985).

79. G. Bode, F. Hartig, G. Hebold and H. Czerwek, Incidence of spontaneous tumors in laboratory rats, Exp. Path. 28:235 (1985).

80. T. Tanaka, H. Mori and G. M. Williams, Enhancement of dimethylnitrosamine-initiated hepatocarcinogenesis in hamsters by subsequent administration of carbon tetrachloride but not phenobarbital or p,p'-dichlorodiphenyltrichloroethane, Carcinogenesis 8:1171 (1987).

81. M. A. Moore, K. Nakagawa, K. Satoh, T. Ishikawa and K. Sato, Single GST-P positive cells-putative initiated hepatocytes, Carcinogenesis 8:483 (1987).

82. R. J. Monnat, Jr. and L. A. Loeb, Mechanisms of neoplastic transformation, Cancer Invest. 1:175 (1983).

83. W. Den Otter, J. W. Koten and D. J. Kerkinderen, Carcinogenesis Revisited, Cancer Invest. 5:69 (1987).

84. G. M. Ledda, M. A. Sells, S. Yokoyama and B. Lombardi, Metabolic properties of isolated rat liver cell preparations enriched in epithelial cells other than hepatocytes, Int. J. Cancer 31:231. (1983).

85. D. S. Longnecker, J. French, E. Hyde, H. S. Lilja and J. Yager, Jr., Effect of age on nodule induction by azaserine and DNA synthesis in rat pancreas, J. Natl. Cancer Inst. 58:1769 (1977).

86. D. A. Banas, Evaluations of pancreas lesions in corn oil vehicle control rats: Comparison of quantity of pancreatic tissue examined to the number of proliferative lesions. Report submitted by Experimental Pathology Laboratories, Inc., to the National Toxicology Program, May 19, 1983.

87. G. A. Boorman and S. L. Eustis, Proliferative lesions of the exocrine pancreas in male F344/n rats, Environ. Health Perspect. 56:213 (1984).

88. S. L. Eustis and G. A. Boorman, Proliferative lesions of the exocrine pancreas: relationship to corn oil gavage in the National Toxicology Program, J. Natl. Cancer Inst. 75:1067 (1985).

89. D. S. Longnecker, N. Chandar, B. D. Roebuck, D. G. Sheahan and B. Lombardi, Preneoplastic and neoplastic lesions in the pancreas of rats fed choline-devoid or choline-supplemented diets, with or without phenobarbital, in preparation.

90. P. M. Newberne, J. L. V. deCarmago and A. J. Clark, Choline deficiency, partial hepatectomy, and liver tumors in rats and mice, Toxicol. Pathol. 2:95 (1982).

91. M. B. Sporn and G. J. Todaro, Autocrine secretion and malignant transformation of cells, New Engl. J. Med. 303:878 (1980).

92. H. S. Earp and E. J. O'Keefe, Epidermal growth factor receptor number decreases during rat liver regeneration, J. Clin. Invest. 67:1580 (1981).

93. A. Francavilla, P. Ove, L. Polimeno, C. Sciascia, M. L. Coetzee and T. E. Starzl, Epidermal growth factor and proliferation of rat hepatocytes in primary culture isolated at different times after partial hepatectomy, Cancer Res. 46:1318 (1986).

94. D. R. LaBreque and N. R. Bachur, Hepatic stimulator substance: Physiochemical characteristics and specificity, Am. J. Physiol. 242:G281 (1982).

95. T. P. Iype and J. B. McMahon, Hepatic proliferation inhibitor, Mol. Cell. Biochem. 59:57 (1984).

96. B. Lombardi, P. Ove and T. V. Reddy, Endogenous hepatic growth-modulating factors and effects of a choline-devoid diet and of phenobarbi-

tal on hepatocarcinogenesis in the rat, <u>Nutr. Cancer</u> 7:145 (1985).

97. J. M. Betschart, M. A. Virji, M. I. R. Perera and H. Shinozuka, Alterations in hepatocyte insulin receptors in rats fed a choline-deficient diet, <u>Cancer Res.</u> 46:4425 (1986).

98. C. Gupta, A. Hattori, J. M. Betschart, M. A. Virji and H. Shinozuka, Inhibition of EGF binding in rat hepatocytes by liver tumor promoters, <u>Cancer Res.</u> 28:173 (1987).

99. V. R. Potter, Initiation and promotion in cancer formation: the importance of studies on intercellular communication, <u>Yale J. Biol. Med.</u> 53:367 (1980).

100. E. Scherer, Neoplastic progression in experimental hepatocarcinogenesis, <u>Bioch. Biophys. Acta</u> 738:219 (1984).

101. P. C. Nowell, The clonal evolution of tumor cell populations, <u>Science</u> 194:23 (1976).

102. A. W. Pound and L. J. McGuire, Repeated partial hepatectomy as a promoting stimulus for carcinogenic response of liver to nitrosamine in rats, <u>Br. J. Cancer</u> 37:585 (1978).

103. P. M. Newberne and H. E. Rogers, Labile methyl groups and the promotion of cancer, <u>Ann. Rev. Nutrit.</u> 6:407 (1986).

104. T. V. Reddy, N. Chandar and B. Lombardi, in preparation.

105. N. Chander and B. Lombardi, in preparation.

106. T. H. Rushmore, E. Farber, A. K. Ghoshal, S. Parodi, M. Pala and M. Taningher, A choline-devoid diet, carcinogenic in the rat, induces DNA damage and repair, <u>Carcinogenesis</u> 7:1677 (1986).

107. G. Ugazio, L. Gabriel and E. Burdinor, Osservazioni sperimentali sui lipidi accumulati nel fegato di ratto alimentato con dieta colino priva, <u>Sperimentale</u> 117:1 (1967).

108. T. H. Rushmore, Y. P. Linn, E. Farber and A. K. Ghoshal, Rapid lipid peroxidation in the nuclear fraction of rat liver induced by diet deficient in choline and methionine, <u>Cancer Lett.</u> 24:251 (1984).

109. A. Ghoshal, T. H. Rushmore, Y. P. Linn and E. Farber, Early detection of lipid peroxidation in the hepatic nuclei of rats fed a diet deficient in choline and methionine (CMD), <u>Proc. Am. Cancer Res.</u> 25:94 (1984).

110. M. Comporti, Lipid peroxidation and cellular damage in toxic liver injury, <u>Lab Invest.</u> 53:599 (1985).

111. S. Banni, F. Corongiu and B. Lombardi, in preparation.

112. F. P. Corongiu, G. Poli, M. U. Dianzani, K. V. Cheeseman and T. F. Slater, Lipid peroxidation and molecular damage to polyunsaturated fatty acids in rat liver. Recognition of two classes of hydroperoxides found under conditions in vivo, <u>Chem. Biol. Interactions</u> 59:147 (1986).

113. R. J. Havel, Metabolism of lipids in chylomicrons and very low density lipoproteins, <u>in</u>: "Adipose Tissue", A. E. Renold and G. F. Cahill, Jr., eds., American Physiological Society, Washington (1965).

114. M. I. R. Perera, A. J. Demetris, S. L. Katyal and H. Shinozuka, Lipid peroxidation of liver microsome membranes induced by choline-deficient diets and its relationship to the diet-induced promotion of the induction of gamma-glutamyltranspeptidase-positive foci, <u>Cancer Res.</u> 45:2533 (1985).

115. M. A. Bansk and H. Shinozuka, Methapyrilene induces lipid peroxidation in the nuclear fraction of rat liver, <u>Cancer Res.</u> 28:90 (1987).

116. A. J. Monserrat, A. K. Ghoshal, W. S. Hartroft and E. A. Porta, Lipoperoxidation in the pathogenesis of renal necrosis in choline-deficient rats, <u>Am. J. Path.</u> 55:163 (1969).

117. A. Sevanian and P. Hochstein, Mechanisms and consequences of lipid peroxidation in biological systems, <u>Ann. Rev. Nutrit.</u> 5:365 (1985).

118. H. Sidransky and E. Verney, Influence of ethionine on choline-deficiency fatty liver, <u>J. Nutrit.</u> 97:419 (1969).

119. E. Usdiu, R. T. Borchardt and C. R. Creveling, eds., "Transmethylation", Elsevier, New York, (1979).

120. J. Locker, T. V. Reddy and B. Lombardi, DNA methylation and hepato-carcinogenesis in rats fed a choline-devoid diet, Carcinogenesis 7:1309 (1986).

121. J. Locker, S. Hunt and B. Lombardi, Alpha-fetoprotein gene methylation and hepatocarcinogenesis in rats fed a choline-devoid diet, Carcinogenesis 8:241 (1986).

122. R. Holliday, A new theory of carcinogenesis, Br. J. Cancer 40:513 (1979).

123. R. M. Hoffman, Altered methionine metabolism, DNA methylation and oncogene expression in carcinogenesis, Biochim. Biophys. Acta 738:49 (1983).

124. R. Holliday, The inheritance of epigenetic defects, Science 238:163 (1987).

125. N. Chandar, B. Lombardi, W. Schultz and J. Locker, Analysis of ras genes and linked viral sequences in rat hepatocarcinogenesis, Am. J. Path. in press.

126. R. Muller and I. Verma, Expression of cellular oncogenes, Curr. Topics Microbiol. Immunol. 112:73 (1984).

127. T. Tanaka, D. J. Slamon, H. Battifora and M. J. Cline, Expression of 21 ras oncoproteins in human cancers, Cancer Res. 46:1465 (1986).

128. N. Fausto and P. R. Shank, Oncogene expression in liver regeneration and hepatocarcinogenesis, Hepatology (Baltimore) 3:1016 (1983).

129. J. M. Bishop, Cellular oncogenes and retroviruses, Annu. Rev. Biochem. 53:301 (1983).

130. J. L. Marx, Oncogene action probed, Science 237:602 (1987).

131. P. K. Vogt, The Seventeenth International Symposium of the Princess Takamatsu Cancer Research Fund: Oncogenes and Cancer, GANN 78:529 (1987).

132. B. I. Weinstein, Growth factors, oncogenes and multistage carcinogenesis, J. Cell. Biochem. 33:213 (1987).

133. V. Chiarugi, M. Ruggiero and F. Porciatti, Oncogenes and transmembrane cell signaling, Cancer Inv. 5:215 (1987).

134. M. B. Sporn, A. B. Roberts, L. M. Wakefield and R. K. Assoian, Transforming growth factor-beta: biological function and chemical structure, Science 233:532 (1986).

135. E. Farber, The multistep nature of cancer development, Cancer Res. 44:4217 (1984).

136. S. H. Yuspa, T. Ben, H. Hennings and U. Lichti, Divergent responses in epidermal basal cells exposed to the tumor promoter 12-0-tetra-decanoylphorbal-13-acetate, Cancer Res. 42:2344 (1982).

137. S. Yokoyama and B. Lombardi, Stage dependent enhanced induction of hepatocellular carcinomas in rats administered a second dose of diethylnitrosamine, Cancer Lett. 25:171 (1985).

138. S. N. Zaman, W. M. Melia, R. D. Johnson, B. L. Portman, P. J. Johnson and R. Williams, Risk factors in development of hepatocellular carcinoma in cirrhosis: prospective study of 613 patients, Lancet 1:1357 (9185).

139. K. Okuda, I. Fujimoto, A. Hanai and Y. Urano, Changing incidence of hepatocellular carcinoma in Japan, Cancer Res. 47:4967 (1987).

140. P. P. Anthony, K. G. Ishak, N. C. Nayak, H. E. Pulsen, P. J. Schever and L. H. Sobin, The morphology of cirrhosis: definition, nomenclature, and classification, Bull. WHO 55:521 (1977).

141. A. Baggenstoss, R. D. Soloway, W. H. J. Summerskill, L. R. Elveback and L. J. Schoenfield, Chronic active liver disease. The range of histologic lesions, their response to treatment and evolution, Hum. Pathol. 3:183 (1972).

142. D. Ganem and H. E. Varmus, The molecular biology of the hepatitis B virus, Ann. Rev. Biochem. 56:651 (1987).

143. D. E. Koshland, Jr., Immortality and risk assessment, Science 236:241 (1987).

METHYL INSUFFICIENCY IN CARCINOGENESIS

Lionel A. Poirier

Division of Comparative Toxicology
National Center for Toxicological Research
Jefferson, Arkansas 72079 USA

INTRODUCTION

In 1946 Copeland and Salmon found that the chronic administration of a choline-deficient diet to rats led to the formation of liver cancer[1]. Thus began one of the most prolonged and controversial areas of study in the field of chemical carcinogenesis. At first, such results were not regarded as surprising, for at the time other dietary deficiencies had also been shown to enhance carcinogenesis[2,3]. However, subsequent studies showed that the peanut meal diets used to obtain the choline deficiency were in fact contaminated with aflatoxin and that the original strain of rats used by Salmon's group appeared to undergo genetic drift such that the original results in tumor formation by choline deficiency alone could no longer be repeated[4-6]. Although the original observations became doubted, they did serve to focus interests of other groups on the possible role of labile methyl group deficiency in carcinogenesis. Thus, Farber et al.[7] showed that ethionine, an antagonist of the labile methyl group donor methionine also caused liver cancer in rats. Together, methionine and choline constitute the chief dietary sources of methyl groups. The Millers[2] postulated that a labile methyl group insufficiency might by involved in hepatocarcinogenesis by the monomethyl aminoazo dyes. However, from 1960 to 1975, with the exception of the group of P. Newberne[6], little research was conducted on the possible etiological role of dietary methyl donors in carcinogenesis. In the early 1970s, this group showed that several biochemical effects of hepatocarcinogens could be enhanced by the administration of diets marginally deficient in lipotropes, including methionine and choline[8]. Also, in the late 1970s, the group of Lombardi and Shinozuka showed that choline-devoid diets enhanced the activities of several hepatocarcinogens[9-11] (Table 1).

TUMOR FORMATION

In 1983 Mikol et al.[16] showed that an amino acid-defined diet lacking the methyl donors methionine and choline not only promoted liver tumor formation in rats given a single initiating dose of diethylnitrosamine, but even induced liver carcinomas in the absence of such initiation. Similar observations were made by Ghoshal and Farber[17] and by Yokoyama et al.[18] who used choline-deficient diets containing alcohol-extracted peanut meal. Thus began a marked resurgence of interest in the role of physiological methyl donors in carcinogenesis. A series of studies were undertaken to extend the observa-

Table 1. Hepatocarcinogens whose activities are inhibited by methyl donors or enhanced by methyl deficiency

Hepatocarcinogen	Species	References
2-Acetylaminofluorene	Rat	6,9,11,12
Ethionine	Rat	11,12,13
Diethylnitrosamine	Rat, Mouse	6,11,12,14
Aflatoxin B_1	Rat	6,11,12
Phenobarbital	Rat, Mouse	12,15
N,N-Dimethyl-4-aminoazobenzene	Rat	2,12
Dibutylnitrosamine	Rat	6,11
Azaserine	Rat	10,11

tions on the liver tumor promoting and hepatocarcinogenic activities of dietary methyl deprivation. The hepatocarcinogens and liver tumor promoters whose activity has been shown to be enhanced by dietary deficiencies of methionine and/or choline or to be suppressed by increased administration of these compounds are listed in Table 1. This table also shows that the effects of methyl donors on the activity of hepatocarcinogens can also be seen in mice. Later studies showed that B6C3F1 mice were also sensitive to the hepatocarcinogenic effects of methionine and choline deficiency [14].

Consistent with the hepatocarcinogenicity of dietary methyl deprivation are its effects on the preneoplastic lesions commonly associated with such activity. For example, the formation of γ-glutamyltranspeptidase-positive foci, commonly seen as an early lesion in liver tumor formation in rats, is enhanced in both initiated and uninitiated animals fed a diet severely deficient in methionine and choline [19]. Also, the serum levels of α-fetoprotein, which are often elevated during the chronic administration of hepatocarcinogens [20], are markedly elevated in rats receiving a methionine- and choline-deficient diet [21]. Finally, subchronic treatment with either ethionine-containing or methyl-deficient diet led to strong elevations in the hepatic levels of the enzyme ornithine decarboxylase (ODC) [22]. The rises in liver ODC were shown to be proportional to the decrease noted in the hepatic contents of S-adenosylmethionine (SAM), the chief physiological methyl donor [22].

S-ADENOSYLMETHIONINE

The biological effects of dietary methyl deprivation in rodents served to increase emphasis on SAM insufficiency as a possible etiologic factor in carcinogenesis. SAM is the methyl donor for a host of biochemical methylation reactions, including DNA and RNA methylations, as well as for the methylation of phosphatidylethanolamine to form phosphatidylcholine. In addition, SAM regulates the distribution of folate cofactors, is the source of the aminopropyl groups of the polyamines spermidine and spermine, and serves as a cofactor in the vitamin B_{12}-dependent pathway of methionine biosynthesis [23]. Its product in methylation reactions, S-adenosylhomocysteine (SAH), is a potent inhibitor of methylases [24,25].

Similarly, the ethyl analog of SAM, S-adenosylethionine (SAE), the major metabolite of the methionine antagonist ethionine, is a potent inhibitor of methylase reactions [13]. Thus, in cells and tissues a hypomethylating environment may be produced by decreasing the levels of SAM or by increasing those of SAE and SAH [25]. The chronic feeding of lipotrope-deficient or amino acid-defined, methyl-deficient diets to rodents decreases the levels of SAM in

Table 2. Hepatocarcinogens and liver tumor promoters suppressing the bioavailability of hepatic S-adenosylmethionine (SAM) during chronic administration

Hepatocarcinogen	Species	References
2-Acetylaminofluorene	Rat	12,26
Diethylnitrosamine	Rat	28
Ethionine	Rat, Mouse, Hamster	12,13,29
Phenobarbital	Rat	12
DDT	Rat	12

their livers[12,26,27]. The chronic administration of several hepatocarcinogens can also effect such changes. These are listed in Table 2. The carcinogenic activity of most is inhibited by dietary administration of either methionine or choline (Table 1). Recent studies by Feo et al. have produced strong evidence that the direct administration of SAM in vivo can suppress the preneoplastic changes in the livers of rats undergoing an initiation/promotion regimen[30].

HYPOMETHYLATION AND DIFFERENTIATION

One expected consequence of a SAM insufficiency in vivo is hypomethylation of macromolecules. Such indeed is the case of undermethylation of phospholipids, which has been observed in the livers of rats fed methyl-deficient or ethionine-containing diets[24]. Undermethylation of tRNA has also been demonstrated in the livers of rats fed methyl-deficient or ethionine-containing diets[31,32]. Of possibly greater significance to the process of carcinogenesis on cell differentiation is DNA hypomethylation. The role of 5-methyldeoxycytidine (5-MC) on gene expression and cell differentiation has been widely investigated[33,35]. The chronic administration of an amino acid-defined diet deficient in both methionine and choline to rats for 22 weeks led to a 15% drop in the proportion of 5-MC residues in liver DNA[36]. Similarly, the administration of 0.3% ethionine to rats for 10 weeks led to a drop of 8% in such DNA methylation[37]. The C3H mouse, on the other hand, appeared to be insensitive to both the DNA hypomethylating and the hepatocarcinogenic effects of deprivation of both methionine and choline[27]. Thus, to the limited degree studied, a correlation can be seen between DNA hypomethylation in the livers of methyl-deficient animals and the subsequent formation of hepatocellular carcinomas. Cell culture studies extended such correlations. Rat liver epithelial cells could be transformed to tumor cells capable of giving rise to tumors when implanted into isologous hosts, following long-term treatment in culture by the methylation inhibitors ethionine, SAE and 3-deazaadenosine[38,39]. The latter compound, which inhibits methylation through the accumulation of SAH, produced a 40% decrease in the extent of DNA methylation under the transforming conditions[39].

CELL TRANSFECTION AND GENE HYPOMETHYLATION

The transfecting activity in NIH 3T3 cells and the extent of oncogene hypomethylation was investigated with the DNA obtained from the tumors and preneoplastic livers of methyl-deficient rats and from the grossly normal livers of control rats fed a methyl-adequate diet. Both the control and the methyl-deficient groups were subdivided into uninitiated and those receiving a single initiating dose of 20 mg/kg diethylnitrosamine. The results are

Table 3. Cell transfection and c-H-ras and c-K-ras gene hypomethylation with DNA from the livers and liver tumors in rats fed methyl-deficient diets [a]

Diet group	Liver tissue	DEN Initiation	Transfecting Activity [b]	Hypomethylation	
				c-H-ras	c-K-ras
Control	Normal	-	0/4	0/4	0/3
		+	0/3	0/3	0/3
Methyl-deficient	Preneoplastic [c]	-	0/4	4/4	2/2
		+	0/6	6/6	5/5
	Neoplastic [d]	-	0/6	4/4	3/3
		+	4/22	21/21	13/13

[a] Male, weanling F-344 rats were given a single initiating dose of diethylnitrosamine (DEN, 20 mg/kg) and subsequently placed on an amino acid-defined diet deficient in methionine and/or choline for 18 months[40,41].
[b] Towards NIH 3T3 cells. [c] Including neoplastic nodules. [d] Hepatocellular carcinomas.

summarized in Table 3. The results show that the DNA from 4 of 28 hepatocellular carcinomas examined could transform NIH 3T3 cells in vitro; such transfecting activity was found to be accompanied by amplification of the c-H-ras oncogene in the transfected cells [40]. All of the transfecting DNA samples were obtained from initiated animals, and none from the six liver carcinomas found in rats simply fed the methyl-deficient diet. No activated c-K-ras or c-myc oncogenes were found in the transfectants. The extent of hypomethylation of the c-K-ras and c-H-ras oncogenes was studied using an isoschizomeric pair of restriction endonucleases MspI and HpaII followed by gel electrophoresis. As shown in Table 3, no hypomethylation of either gene was seen in the livers from animals fed the control diet[41]. However, both genes were undermethylated in all livers, both neoplastic and preneoplastic, obtained from animals fed the methyl-deficient diets. DEN initiation exerted no discernible effect on the extent of methylation of either gene. The results suggest that oncogene hypomethylation may be a predisposing lesion to subsequent cell transformation. In specific in vitro systems, DNA methylation has been shown to diminish the ability of viruses and oncogenes to transform cells[42-44].

In summary, evidence has been presented showing that chronic dietary methyl insufficiency leads to the formation of hepatocellular carcinomas in male F-344 rats and B6C3F1 mice. The administration of these diets also produces a hypomethylating environment in the livers of such animals, with a concomitant undermethylation of hepatic DNA. Chronic hepatocarcinogen administration also produces each of these changes. Consistent with the hypomethylation of total liver DNA seen in methyl-deficient rats is the hypomethylation of the c-K-ras and c-H-ras oncogenes in the liver DNA of such animals. If increased gene expression is the consequence of such hypomethylation, a mechanism is provided by which dietary methyl deprivation may lead to tumor formation.

REFERENCES

1. D. H. Copeland and W. D. Salmon, The occurrence of neoplasms in the liver, lungs, and other tissues of rats as a result of prolonged choline deficiency, Am. J. Pathol. 22:1059 (1946).
2. J. A. Miller and E. C. Miller, The carcinogenic aminoazo dyes, Adv. Cancer Res. 1:339 (1953).
3. R. C. Moon and R. G. Mehta, Anticarcinogenic effects of retinoids in animals, in: "Essential Nutrients in Carcinogenesis", L. A. Poirier, P. M. Newberne, and M. W. Pariza, eds., Plenum Press, New York (1986).
4. W. D. Salmon and P. M. Newberne, Occurrence of hepatomas in rats fed diets containing peanut meal as a major source of protein, Cancer Res. 23:571 (1963).
5. P. M. Newberne, W. W. Carlton and G. N. Wogan, Hepatomas in rats and hepatorenal injury induced by peanut meal in Aspergillus flavus extract, Pathol. Vet. 1:105 (1964).
6. P. M. Newberne, Lipotropic factors and oncogenesis, in: "Essential Nutrients in Carcinogenesis", L. A. Poirier, P. M. Newberne, and M. W. Pariza, eds., Plenum Press, New York (1986).
7. E. Farber, Carcinoma of the liver in rats fed ethionine, Arch. Pathol. 66:445 (1956).
8. A. E. Rogers and P. M. Newberne, Lipotrope deficiency in experimental carcinogenesis, Nutr. Cancer 2:104 (1980).
9. B. Lombardi and H. Shinozuka, Enhancement of 2-AAF liver carcinogenesis in rats fed a choline-deficient diet, Int. J. Cancer 23:565 (1979).
10. H. Shinozuka, S. L. Katyal and B. Lombardi, Azaserine carcinogenesis: Organ susceptibility change in rats fed a diet devoid of choline, Int. J. Cancer 22:36 (1978).
11. H. Shinozuka, S. L. Katyal and M. I. R. Perera, Choline Deficiency and Chemical Carcinogenesis, in: "Essential Nutrients in Carcinogenesis", L. A. Poirier, P. M. Newberne, and M. W. Pariza, eds., Plenum Press, New York (1986).
12. L. A. Poirier, The role of Methionine in Carcinogenesis In Vivo, in: "Essential Nutrients in Carcinogenesis", L. A. Poirier, P. M. Newberne, and M. W. Pariza, eds., Plenum Press, New York (1986).
13. E. Farber, Ethionine carcinogenesis, Adv. Cancer Res. 7:383 (1963).
14. L. A. Poirier and K. H. Hoover, Liver tumor formation in male B6C3F1 mice fed methyl-deficient, amino acid-defined diets with and without diethylnitrosamine initiation, Proc. Am. Assoc. Cancer Res. 27:129 (1986).
15. N. Shivapurkar, K. L. Hoover and L. A. Poirier, Effect of methionine and choline on liver tumor promotion by phenobarbital and DDT in diethylnitrosamine-initiated rats, Carcinogenesis 7:547 (1986).
16. Y. B. Mikol, K. L. Hoover, D. Creasia and L. A. Poirier, Hepato-carcinogenesis in rats fed methyl-deficient, amino acid-defined diets, Carcinogenesis 4:1619 (1983).
17. A. K. Ghoshal and E. Farber, The induction of liver cancer by dietary deficiency of choline and methionine without added carcinogens, Carcinogenesis 5:1367 (1984).
18. S. Yokoyama, M. A. Sells, T. V. Reddy and B. Lombardi, Hepatocarcino-genic and promoting action of a choline-devoid diet in the rat, Cancer Res. 45:2834 (1985).
19. K. L. Hoover, P. Lynch and L. A. Poirier, Profound postinitiation enhancement by short-term methionine, choline, vitamin B_{12}, and folate deficiency of hepatocarcinogenesis in F-344 rats given a single low dose of diethylnitrosamine, J. Natl. Cancer. Inst. 73:1327 (1984).
20. L. Belanger, P. Baril and M. Guertin, Oncodevelopmental and hormonal regulation of alpha-fetoprotein gene expression, Adv. Enzyme Regulation 21:73 (1983).

21. H. Gourdeau, Ph.D. Thesis, Université de Laval, Quebec (1986).
22. Y. B. Mikol and L. A. Poirier, An inverse correlation between hepatic ornithine decarboxylase and S-adenosylmethionine in rats, Cancer Letters 13:195 (1981).
23. S. H. Mudd and H. L. Levy, Disorders of Transsulfuration, in: " The Metabolic Basis of Inherited Disease", J. B. Stansbury, J. B. Wyngaarden, D. S. Frederickson, J. L. Goldstein, and M. S. Brown, eds., McGraw-Hill, New York (1983).
24. D. R. Hoffman, J. A. Hanig and W. E. Cornatzer, Effects of a methyl-deficient diet on rat liver phosphatidylcholine biosynthesis, Can. J. Biochem. 59:543 (1981).
25. P. K. Chiang and G. L. Cantoni, Perturbations of biochemical trans-methylations by 3-deazaadenosine, Biochem. Pharmacol. 28:1897 (1979).
26. L. A. Poirier, P. H. Grantham, and A. E. Rogers, The effects of a mar-ginally lipotrope-deficient diet on the hepatic levels of S-adeno-sylmethionine and on the urinary metabolites of 2-acetylaminofluo-rene in rats, Cancer Res. 37:744 (1977).
27. N. S. Shivapurkar, M. J. Wilson, K. L. Hoover, Y. B. Mikol, D. Creasia and L. A. Poirier, Hepatic DNA methylation and liver tumor formation in male C3H mice fed methionine- and choline-deficient diets, J. Natl. Cancer Inst. 77:213 (1986).
28. Y. S. S. Buehring, L. A. Poirier and E. L. R. Stoksad, Folate defi-ciency in the livers of diethylnitrosamine-treated rats, Cancer Res. 36:2775 (1976).
29. Z. Brada, S. Bulba and N. H. Altman, The influence of DL-methionine on the metabolism of S-adenosylethionine in rats chronically treated with DL-ethionine, Cancer Res. 36:1573 (1976).
30. F. Feo, R. Garcea, L. Daino, P. Pascale, S. Frassetto, P. Cozzolino, M. G. Vannini, M. E. Ruggiu, M. M. Simile and M. Puddu, S-adenosyl-L-methionine antipromotion and antiprogression effect in hepatocar-cinogenesis. Its association with inhibition of gene expression, this volume.
31. E. Wainfain, M. L. Moller, F. A. Maschio and M. E. Balis, Time-depend-ent ethionine-induced changes in rat liver transfer RNA methylation, Cancer Res. 35:2830 (1977).
32. E. Wainfain, M. Dizik, M. Hluboky and M. E. Balis, Altered tRNA meth-ylation in rats and mice fed lipotrope-deficient diets, Carcinogenesis 7:473 (1986).
33. D. N. Cooper, Eukaryotic DNA methylation, Hum. Genet. 64:315 (1983).
34. A. D. Riggs and P. A Jones, 5-Methylcytosine, gene regulation, and cancer, Adv. Cancer Res. 40:1 (1983).
35. P. A. Jones, DNA methylation and cancer, Cancer Res. 46:461 (1986).
36. M. J. Wilson, N. Shivapurkar and L. A. Poirier, Hypomethylation of hepatic nuclear DNA in rats fed with a carcinogenic methyl-deficient diet, Biochem. J. 218:987 (1984).
37. N. Shivapurkar, M. J. Wilson and L. A. Poirier, Hypomethylation of DNA in ethionine-fed rats, Carcinogenesis 5:989 (1984).
38. J. D. Brown, M. J. Wilson and L. A. Poirier, Neoplastic conversion of rat liver epithelial cells in culture by ethionine and S-adenosyl-ethionine, Carcinogenesis 4:173 (1983).
39. M. J. Wilson, R. M. Bare, E. D. Kwiecinski and L. A. Poirier, 3-Deaza-adenosine induces transformation and DNA hypomethylation in rat liver cells in culture, Proc. Am. Assoc. Cancer Res. 26:506 (1985).
40. M. J. Wilson, L. A. Poirier, K. J. Dunn, D. O. Halverson, L. A. Eader and D. G. Blair, Activation of the c-H-ras oncogene in hepatocellu-lar carcinomas initiated with diethylnitrosamine and promoted with methyl-deficient diets, Fed. Proc. 45:1707 (1986).
41. M. R. Bhave, M. J. Wilson and L. A. Poirier, c-H-ras and c-K-ras gene hypomethylation in the livers and hepatomas of rats fed methyl-defi-cient amino acid-defined diets, Carcinogenesis in press (1988).

42. M. L. McGeady, C. Jhappan, R. Ascione and G. F. Vande Woude, In vitro methylation of specific regions of the cloned Maloney sarcoma virus genome inhibits its transforming activity, Mol. Cell. Biol. 3:305 (1983).

43. J. Groffen, N. Heiserdamp, G. Blennerhassett and J. R. Stephenson, Regulation of viral and cellular oncogene expression by cytosine methylation, Virology 126:213 (1983).

44. B. A. Christy and G. A. Scangos, In vitro methylation of bovine papillomavirus alters its ability to transform mouse cells, Mol. Cell. Biol. 6:2910 (1986).

CHOLINE DEFICIENCY, LIPID PEROXIDATION, LIVER CELL SURFACE

RECEPTOR ALTERATIONS AND LIVER TUMOR PROMOTION

Hishashi Shinozuka, Chhanda Gupta, Atsuo Hattori,
James M. Betschart and Mohamed A. Virji

Department of Pathology, University of
Pittsburgh, School of Medicine
Pittsburgh, PA 15261 USA

INTRODUCTION

The two-stage model of carcinogenesis, originally demonstrated in the skin of mice and rabbits, has been extended to several other organ systems including the liver[1,2]. The importance of tumor promotion in the genesis of human cancers has been recognized[3], and the better understanding of the mechanism of tumor promotion would be critical in the formulation of the overall strategies for the prevention of human cancers. Considerable insights have been gained for the mechanisms of action of phorbol esters, the classical and perhaps most intensively studied skin tumor promoter[4]. Studies of liver carcinogenesis during the past 15 years have identified a number of promoters with diverse properties[1], but the mechanisms of their action remain elusive.

Phenobarbital (PB) and a CD diet serve as two prototype liver tumor promoting regimens, the former being the well known pharmacological agent and demonstrated as a liver tumor promoter over 15 years ago[5], the latter being the well known nutritional regimen which modifies structure and functions of liver cells and was demonstrated as an efficient liver tumor promoting regimen[6]. Ugazio et al.[7] and Wilson et al.[8] demonstrated lipid peroxidation of liver cells in rats fed a CD diet and linked it to the possible mechanism of the diet-induced fatty liver. Since there is some evidence to indicate that free radicals and enhanced lipid peroxidation may be involved in the promotion of tumor induction in some organs[9], we extended the original studies of Ugazio et al. and demonstrated lipid peroxidation in microsomal membrane lipids of liver cells in rats fed a CD diet[10]. With the use of antioxidants or hypolipidemic agents and by modifying the dietary fat components of a CD diet, it was shown that the extent of CD diet-induced lipid peroxidation was positively correlated to its promoting activity[11]. Rushmore and his associates[12] demonstrated lipid peroxidatin in the nuclear fraction of hepatocytes of rats fed a CD diet.

A causal role for lipid peroxidation in various pathological processes has not been definitively established[13]. A number of pathological consequences can be envisioned resulting from cellular lipid peroxidation which may be relevant to the tumor promotion (Fig. 1). Each area listed in Fig.1 has been under investigation in several different laboratories, and we focused our attention on the cell membrane receptor alterations and possible

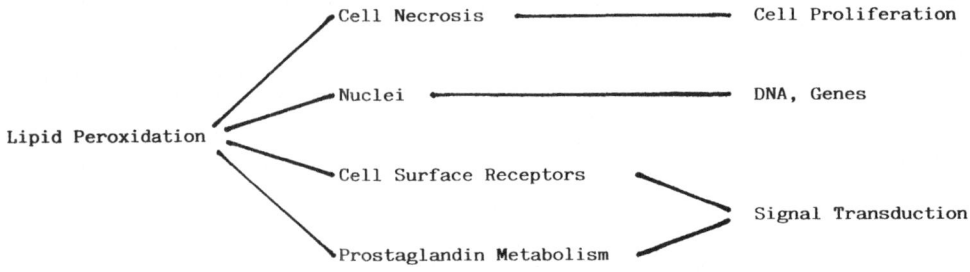

Fig. 1. Possible pathological consequences of lipid peroxidation.

changes in prostaglandin metabolism. Recently considerable attention has been focused on the possible involvement of cell surface receptors for growth factors in malignant transformation of cells[14]. Equally, the possible involvement of altered prostaglandin metabolism has been implicated in tumor promotion[15]. Changes in both cell surface receptors and prostaglandin metabolism may have profound effects on cellular signal transduction and its aberation may trigger abnormal cell growth related to tumor promotion. In this paper, we will summarize our recent experimental data on the changes in rat hepatocyte surface receptors, and discuss the possible significance of these changes.

HEPATOCYTE INSULIN RECEPTORS IN RATS FED A PB DIET

In a recent report from our laboratory, the alterations in hepatocyte receptors for insulin and cellular glycogen biosynthetic response were demonstrated in rats following feeding of a CD diet[16]. In order to determine whether the alterations in plasma membrane insulin receptors and insulin-mediated glycogen synthesis are unique to the CD diet and whether similar effects can be induced by other known liver tumor promoters, effects of feeding PB on hepatocyte insulin receptors were examined. Male Sprague-Dawley rats, weighing 175-200 gm at the beginning of the experiments were fed the basal diet (Dyets Inc., Bethelhem, PA) or the same diet containing 0.06% PB for 4-5 weeks. Hepatocytes were isolated by the collagenase perfusion techniques according to the method of Seglen[17] and the binding of ^{125}I-labeled insulin (200-250 mCi/mg) was assayed as described previously[16]. The competition curves and Scatchard plots for the control hepatocytes and hepatocytes of rats fed PB for 4-5 weeks are presented in Fig. 2. The number of high affinity insulin receptors per cell and Kd for the control hepatocytes were respectively 183,000 ± 19,000 and 15.3 ± 2.5 nM (n = 8), while those for PB-treated hepatocytes were 47,000 ± 5,000 and 2.8 ± 0.3 (n = 5). Both the number of insulin receptors per cell and the Kd were significantly reduced in the PB-treated hepatocytes compared to the controls.

The glycogen biosynthetic response for control and PB-treated hepatocytes was measured as described previously[16], and Fig. 3a depicts the incorporation of radiolabeled glucose into glycogen in the presence and absence of insulin. The basal amount of ^{14}C-glucose incorporated into glycogen in PB-treated hepatocytes was 2.5 ± 0.2 nmol/10^6 cells/60 min which was not significantly different from the basal rate of the control hepatocytes. The percent change in the glycogen synthesis (-6%) from the basal rate in the presence of insulin in the PB-treated hepatocytes was not significantly different either. The control hepatocytes, however, did respond to insulin stimulation of glycogen synthesis (+44%). Thus, hepatocytes from rats fed a PB diet did not respond to insulin.

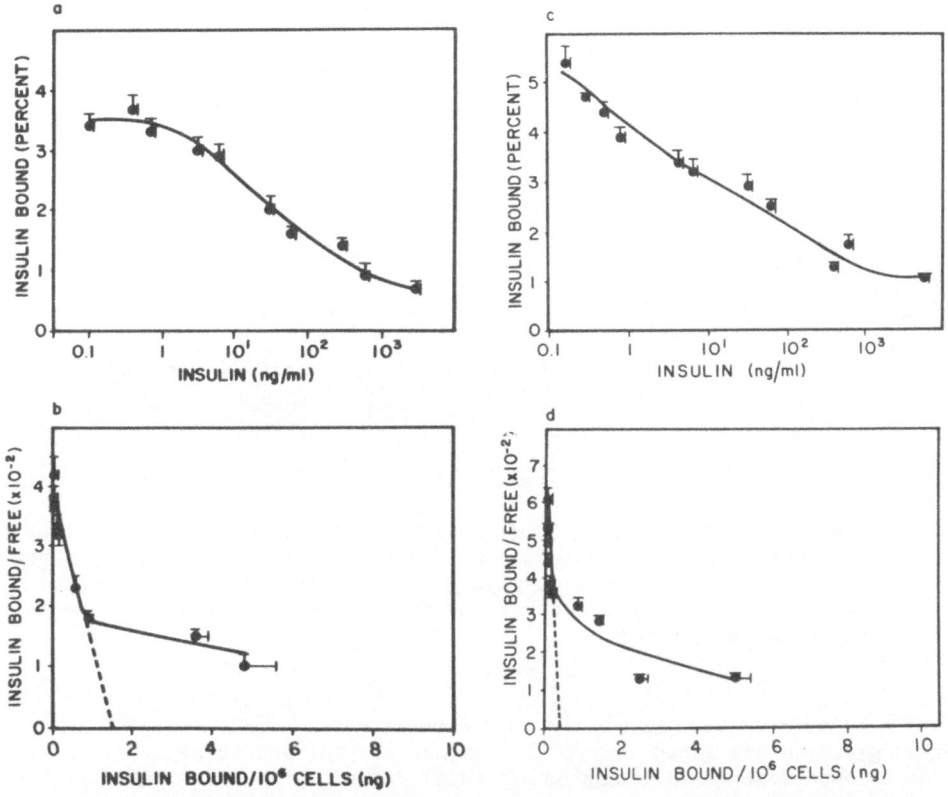

Fig. 2. Insulin binding to hepatocytes of the control (a, b) and phenobarbi-
tal-treated (c, d) rats. a, c: percentage bound with increasing concentration; b, d: Scatchard plot of the competitive binding data.

HEPATOCYTE GLUCAGON RECEPTORS IN RATS FED A PB OR A CD DIET

At tracer concentration of 300-400 pg, the binding of ^{125}I-labeled glucagon to hepatocytes plateaued at 5 min, and remained constant up to 15 min
at 30°C (data not shown). All experiments for the glucagon binding shown
were carried out with 8 min incubation and with 1.5-2.0 x 10[6] cells/ml. Competition curves and Scatchard plots derived from the glucagon binding for the
control hepatocytes and PB-treated hepatocytes are shown in Fig. 4. Although
the Kd and number of high affinity glucagon receptors/cell were similar in
both the control and PB-treated hepatocytes, the Scatchard plot for glucagon
binding in the hepatocytes of PB-treated rats revealed the presence of only
the high affinity receptors as indicated by the rectilinear plot, unlike the
control hepatocytes. There is evidence to indicate that glucagon receptors
on rat hepatocyte plasma membrane are a homogenous group of receptors with
similar binding affinity[18], and that the curvilinear nature of such a plot is
due to a two step dissociation process of the bound glucagon receptor complex
rather than a noncooperative type of binding, or two class of receptors with
high and low affinity binding. PB treatment is apparently altering the asso-
ciation-dissociation process of glucagon receptor complex. Similar analyses
of the glucagon binding between hepatocytes of rats fed a CS or a CD diet
showed that neither the Kd nor number of high affinity glucagon receptors
were significantly different between the two groups.

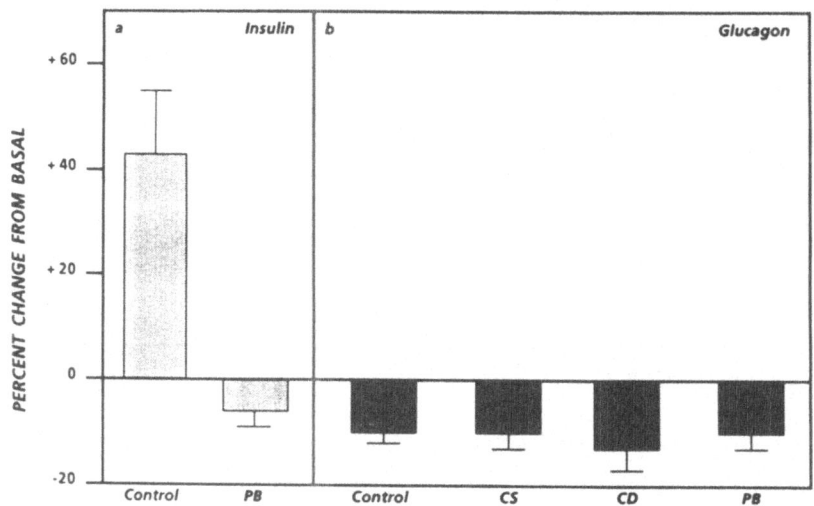

Fig. 3. The incorporation of U-^{14}C glucose from glycogen in hepatocytes of the control and PB-treated rats in response to insulin (a), and the loss of U-^{14}C glucose from glycogen in hepatocytes of the control and CS, CD and PB-treated rats in response to glucagon (b). The bars represent mean ± SEM.

The effects of glucagon on the disappearance of ^{14}C glucose from pre-labeled glycogen are shown in Fig. 3b for the control and PB hepatocytes and CS and CD hepatocytes. Glucagon enhanced glycogen degradation in all groups to the same magnitude.

HEPATOCYTE EGF RECEPTORS IN RATS FED A PB OR A CD DIET

In order to determine whether the hepatocyte membrane receptor altera-tions induced by liver tumor promoters are limited to the peptide hormone receptors or affect receptors for other types of peptides, we examined the binding of EGF to hepatocytes of rats fed a CD diet, PB diet and CD + PB diet. The EGF receptor assays were based on the method of Carpenter[19] and the assays were carried out with freshly prepared hepatocytes, at 4°C and at 37°C. The former represents the surface binding and the latter the net intra-cellular binding of EGF. Fig. 5 shows the results of the effects of CS, CD, CS + PB and CD + PB diets on the number of EGF receptors. In our preliminary experiments, we found a lower number of EGF receptors on hepato-cytes of rats fed a CS diet than those on hepatocytes of rats fed a commer-cial purina chow. The differences in the composition of these two diets may account for this effect. Feeding a CD or PB diet for 3 days produced no sig-nificant effect. After 10 days feeding, a CD diet and PB induced 53% and 60% decreases respectively of the surface receptor number. While PB induced no further reduction in the receptor number after 28 days, rats fed a CD diet after 28 days showed a slight recovery in the EGF receptor number. The com-bination of a CD and PB diet produced a greater decrease in the receptor num-ber than that observed with either the PB or CD diet alone. A significant decrease was evident even at 3 days of feeding the combined diet when neither of the diets alone had any effect on the binding. The effect of the experi-mental diets on Kd of EGF receptor determined at 4°C is shown in Table 1. All the experimental diets, either alone or in combination produced a 50% de-crease in the Kd at 10 and 28 days. At 3 days, the changes in Kd were appar-ent only in the hepatocytes of PB and CD + PB-treated rats. The intracellu-

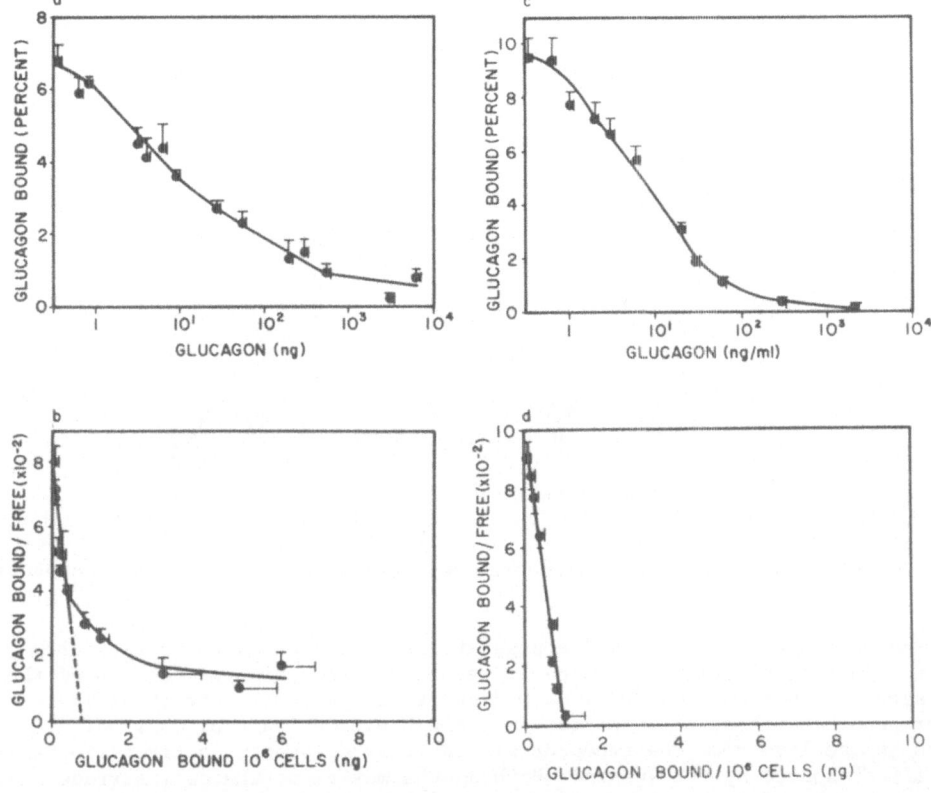

Fig. 4. Glucagon binding to hepatocytes of the control (a, b) and phenobarbital treated (c, d) rats. a, c: percentage bound with increasing glucagon concentration; b, d: Scatchard plots.

lar binding of EGF as assayed at 37°C showed that both PB and CD + PB diet decreased the binding activity similar to the results of surface receptors. However, a CD diet alone induced no significant changes in the intracellular binding as compared to the control CS diet.

DISCUSSION AND PERSPECTIVE

The results presented here together with the findings of our earlier study[16] clearly show that two prototype liver tumor promoters, a CD diet and PB, induced a significant and selective alterations in the properties of hepatocyte surface receptors for a specific hormone and a growth factor. The findings are summarized in Table 2. Both PB and a CD diet induced the changes in insulin and EGF receptors but their effects on glucagon receptors were negligible. The insulin and EGF receptors have a common feature that both receptors are associated with tyrosin kinase activity, and the receptors are autophosphorylated preferentially at the tyrosin residue upon ligand-receptor interaction[20]. Glucagon receptor mediated actions involve activation of the adenylate cyclase-cyclic AMP system. Furthermore, when a PB and CD diet were combined, the decrease in the EGF receptor number was accentuated over that produced by either of the diets alone. In our recent studies, similar augmenting effect of the combined diet was observed in the decrease of the number of hepatocyte insulin receptors. Earlier we demonstrated that the combination of a PB and CD diet exerted a stronger promoting action than

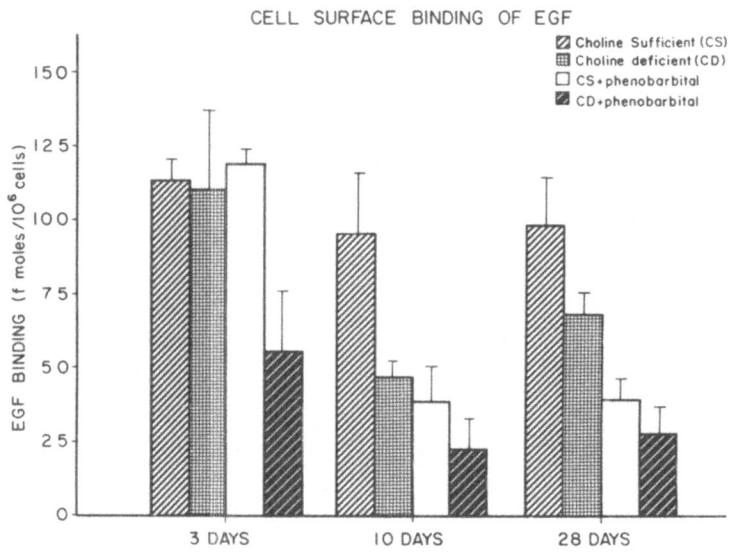

Fig. 5. Effects of various experimental regimens on cell surface EGF receptors.

each agent alone [21]. The findings suggest that the overall changes of the hepatocyte receptors may be, in part, related to the promoting action of the two agents. Even though a PB and CD diet exert the common effect of liver tumor promotion, they often differ in their biological and biochemical effects on the liver [22]. The receptor alterations represent the first biochemical effect shared by the agents. Both agents may alter the cell surface receptors by changing the phospholipid composition of the cell membranes or by altering receptor production. Alternatively the receptor changes induced by a CD diet may be the pathological consequences of peroxidative damage of cell membrane lipids [10].

An increasing number of reports indicate that changes in cell surface receptors for various growth factors are associated with cellular malignant

Table 1. Effect of a CS, CD, CS + PB and CD + PB diets on dissociation constant of EGF receptor

Treatment	Dissociation constant of EGF receptor (nM)		
	3 days	10 days	28 days
Chow	1.01 ± 0.12	1.72 ± 0.10	0.97 ± 0.15
Choline sufficient	0.87 ± 0.18	0.81 ± 0.11	0.80 ± 0.22
Choline deficient	0.85 ± 0.21	0.52 ± 0.13	0.47 ± 0.26
Choline sufficient plus phenobarbital	0.46 ± 0.13	0.44 ± 0.09	0.47 ± 0.08
Choline deficient plus phenobarbital	0.49 ± 0.14	0.43 ± 0.11	0.40 ± 0.12

Data represent mean ± SD. Significant differences (P < 0.001) exist between the control (CS) and the treatment groups by ANOVA test.

Table 2. Hepatocyte surface receptor alterations by liver tumor promoters

Promoter	Cell surface receptors					
	Insulin		Glucagon		EGF	
	No.	Kd.	No.	Kd.	No.	Kd.
Choline deficient diet	↓	↓	-	-	↓	↓
Phenobarbital	↓	↓	-	-	↓	↓
Choline deficiency plus phenobarbital	↓↓	↓	-	-	↓↓	↓↓

-: No changes; ↓: decrease; ↓↓: decrease over ↓.

transformation [14]. The structural homologies between certain cellular onco-genes and cell surface receptors, including EGF and insulin receptors, have been reported [23]. Thus, the receptor alterations may be coupled with altered control of oncogene expression. The decrease in receptors for several lig-ands have been demonstrated in the liver after exposure to carcinogens [24] and to liver tumor promoters [25,26]. EGF binding is decreased in several trans-plantable hepatomas [20] and the progressive decrease in EGF binding in normal liver, preneoplastic nodules and hepatomas during hepatocarcinogenesis has been reported [27]. These observations suggest that the decrease in insulin and EGF binding induced by liver tumor promoters may alter the control of liver cell growth. A number of peptides other than insulin and EGF have been im-plicated to participate in the process [28]. Several liver cell specific growth factors have been demonstrated [29] which may also exert their effects through the receptor mediated events. More recently the transforming growth factor ß has been shown to be involved in the control of liver cell proliferation [30]. It is conceivable that changes in any one of the cell surface receptors in-volved in liver cell growth may lead to disturbances of the homeostatic growth control of the liver cells, and such changes may play a critical role in liver tumor promotion. In-depth analyses of the alterations of hepatocyte surface receptors may be a fruitful avenue to explore the mechanism of liver tumor promotion.

ACKNOWLEDGEMENTS

The study was supported by Grants (CA 26556 and CA 40062) awarded by the National Institute of Health, Department of Health and Human Services and a grant for the American Institute for Cancer Research. We gratefully acknowl-edge the technical assistance of Lynn A. Witkowski and Mary Joe Dimasi and the typing of the manuscript by Marie Gardner.

REFERENCES

1. T. L. Goldworthy, M. H. Hanegan and H. C. Pitot, Models of hepato-carcinogenesis in the rat - contrast and comparison, CRC Crit. Rev. Toxicol. 17:61 (1986).
2. E. Farber and D. S. R. Sarma, Hepatocarcinogenesis: A dynamic cellular

perspective, Lab. Invest. 56:4 (1987).

3. N. E. Day, Epidemiological data and multistage carcinogenesis, in: "Models, Mechanisms and Etiology of Tumour Promotion", M. Borzsonyi, K. Lapis, N. E. Day, and H. Yamasaki, eds., Int. Agency for Research on Cancer, Lyon (1984).

4. T. J. Slaga, Mechanisms involved in two-stage carcinogenesis in mouse skin, in: "Mechanisms of Tumor Promotion", T. J. Slaga, ed., CRC Press, Boca Raton, FL (1984).

5. C. Peraino, R. J. M. Fry and E. Staffeldt, Reduction and enhancement by phenobarbital of hepatocarcinogenesis induced in the rat by 2-acetylaminofluorene, Cancer Res. 31:1506 (1971).

6. S. Takahashi, B. Lombardi and H. Shinozuka, Progression of carcinogen induced foci of γ-glutamyltranspeptidase positive hepatocytes to hepatomas in rats fed a choline deficient diet, Int. J. Cancer 29: 445 (1982).

7. G. Ugazio, L. Gabriel and E. Bardinor, Osservazioni sperimentale sui lipidi accumulati nel fegato di ratto alimentato con dieta colina priva, Lo Sperimentale 117:1 (1967).

8. R. B. Wilson, N. S. Kula, P. M. Newberne and M. W. Conner, Vascular damage and lipid peroxidation in choline deficient rats, Exp. Mol. Path. 18:357 (1973).

9. P. A. Cerutti, Prooxidant states and tumor promotion, Science 227:375 (1985).

10. M. I. R. Perera, A. J. Demetris, S. L. Katyal and H. Shinozuka, Lipid peroxidation of liver microsome membrane induced by choline-deficient diets and its relationship to the diet-induced promotion of the induction of γ-glutamyltranspeptidase-positive foci, Cancer Res. 45:2533 (1985).

11. M. I. R. Perera, J. M. Betschart, M. A. Virji, S. L. Katyal and H. Shinozuka, Free radical injury and liver tumor promotion, Toxicol. Path. 15:51 (1987).

12. T. H. Rushmore, Y. P. Lim, E. Farber and A. K. Ghoshal, Rapid lipid peroxidation in the nuclear fraction of rat liver induced by a diet of deficient in choline and methionine, Cancer Lett. 24:251 (1984).

13. D. L. Tribble, T. Y. Aw and D. P. Jones, The pathophysiological significance of lipid peroxidation in oxidative cell injury, Hepatology 7:377 (1987).

14. A. S. Goustin, E. B. Leof, G. D. Shipley and H. L. Moses, Growth factors and cancer, Cancer Res. 46:1015 (1986).

15. S. M. Fischer, The role of prostaglandins in tumor promotion, in: "Mechanisms of Tumor Promotion", T. J. Slaga, ed., CRC Press, Boca Raton, FL (1984).

16. J. M. Beschart, M. A. Virji, M. I. R. Perera and H. Shinozuka, Alterations in hepatocyte insulin receptors in rats fed a choline-deficient diet, Cancer Res. 46:4425 (1986).

17. P. O. Seglen, Preparation of isolated rat liver cells, Meth. Cell Biol. 13:29 (1976).

18. E. M. Horwitz, W. T. Jenkins, N. M. Hoosein and R. S. Gurd, Kinetic identification of a two-stage glucagon receptor system in isolated hepatocytes, Interconversion of homogenous receptors, J. Biol. Chem. 260:9307 (1985).

19. G. Carpenter, Binding assay for epidermal growth factor, Meth. Enzymol. 109:101 (1985).

20. P. J. Blackshear, R. A. Nemenoff and J. Auruch, Characteristics of insulin and epidermal growth factor stimulation of receptor autophosphorylation in detergent extracts of rat liver and transplantable rat hepatomas, Endocrinology 14:141 (1984).

21. H. Shinozuka and B. Lombardi, Synergistic effect of a choline-devoid diet and phenobarbital in promoting the emergence of foci of -glutamyltranspeptidase positive hepatocytes in the liver of carcinogen-treated rats, Cancer Res. 40:3846 (1980).

22. H. Shinozuka, A. J. Demetris, S. L. Katyla and M. I. R. Perera, Promotion of liver carcinogenesis: Interactions of barbiturates and a choline-deficient diet, in: "Xenobiotic Metabolism: Nutritional Effects", J. W. Finley and D. E. Schwass, eds., American Chemical Society, Washington DC (1985).

23. G. Carpenter, Receptors for epidermal growth factor and other poly-peptide mitogens, Ann. Rev. Biochem. 56:881 (1987).

24. B. I. Carr, A. Roitman, D. L. Hwang, G. Barseghian and A. Lev-Ran, Effects of diethylnitrosamine on hepatic receptor binding and auto-phosphosylation of epidermal growth factor and insulin in rats, J. Nat. Cancer Inst. 77:219 (1986).

25. R. A. Evarts, E. R. Marsden and S. S. Thorgeirsson, Modulation of asialoglycoprotein receptor levels in rat liver by phenobarbital treatment, Carcinogenesis 6:1767 (1985).

26. D. L. Hwang, A. Roitman, A. Lev-Ran and B. I. Carr, Chronic treatment with phenobarbital decreases the expression of rat liver EGF and insulin receptors, Biochem. Biophys. Res. Comm. 135:501 (1986).

27. L. Harris, V. Préat and E. Farber, Patterns of ligand binding to normal, regenerating, preneoplastic and neoplastic rat hepatocytes, Cancer Res. 47:3954 (1987).

28. N. R. L. Bucher, U. Patel and S. Cohen, Hormonal factors and liver growth, Adv. Enzym. Reg. 16:205 (1978).

29. F. J. Thaler and G. Michalopoulos, Hepatopoietin A. partial character-ization and trypsin activation of a hepatocyte growth factor, Cancer Res. 45:2545 (1985).

30. B. I. Carr, I. Hayashi, E. L. Branum and H. L. Moses, Inhibition of DNA in rat hepatocytes by platelet-derived type transforming growth factor, Cancer Res. 46:2330 (1986).

MECHANISM OF INITIATION OF LIVER CARCINOGENESIS BY DIETARY IMBALANCE

Amiya K. Ghoshal, Danny Ghazarian, Amit Ghoshal, Thomas
Rushmore and Emmanuel Farber

Department of Pathology, Medical Sciences Building
University of Toronto, Toronto, Ontario, Canada M5S 1A8

INTRODUCTION

The role of diet in the modulation of carcinogenic process is known for
several years. However, most of the emphasis was on how different macro and
micro-constituents of diet can alter the biology of the system with respect
to different carcinogens and modify the carcinogenic potential of the latter
either by enhancing or diminishing the carcinogenicity to different organs.
Recently, we and two other laboratories[1-5] have shown that about 50% of rats
develop hepatocellular carcinoma when fed a diet which is devoid of choline
and low in methionine (CD) without any added carcinogen. The role of choline
deficiency as a promoter in the carcinogenic process has been established[6-7].
The recent demonstration that CD alone without any added carcinogens caused
liver cancer strongly suggests that CD both initiates and promotes, i.e. it
is a carcinogenic regimen. This observation raised a puzzling question as to
how the absence of something can cause initiation. A hypothesis has been
proposed [8] based on the following observations: (a) early hepatic nuclear lipid peroxidation which occurs in choline deficiency[9] is followed closely by
(b) DNA alteration[10] and (c) liver cell proliferation[11]. These may represent
the chain of events leading to initiation, promotion and eventually to cancer
in this model where no known carcinogen is added.

CHOLINE DEFICIENT DIET AS A COMPLETE CARCINOGEN

The observation that rats exposed to a CD diet without any added carcinogen develop a 100% incidence of hepatic nodules and 50% incidence of
hepatocellular carcinoma[4] suggests that CD is a complete carcinogen regimen.
The ability of the CD diet as a promoter when liver cells are initiated by a
chemical carcinogen is well established[6,7]. This observation is to be expected, since we now know that CD is a complete carcinogen.

CD diet can initiate rat liver [12]. The generation of initiated hepatocytes was monitored by the appearance of γ-GT positive foci with the imposi-
tion of a strong selection procedure for resistant hepatocytes, a phenomenon
induced by many chemical carcinogens[13]. These resistant hepatocytes are
precursors of hepatocyte nodules and these in turn are at least one site of
origin for hepatocellular carcinoma [14-15]. However, unlike many hepatocarcinogens, it takes about 10 weeks of exposure for a CD diet to induce initiated
hepatocytes [12].

601

INITIATION WITHOUT A CARCINOGEN

The question that remains is how initiation can occur without the active participation of a carcinogen. There is evidence[16-17] that free radicals can cause DNA damage which coupled with proliferation can lead to initiation. Therefore it was logical to look for the generation of free radicals in the liver of rats on CD diet.

Evidence for Free Radical Involvement in CD Model

Diene Conjugates in Lipids. Current views favour the thesis that lipid peroxidation is due to the genesis of free radicals. We have demonstrated that hepatic nuclear lipid peroxidation can be detected within 24 h after the exposure to a choline deficient diet (Fig. 1).

4-Hydroxyalkenals. Generation of 4-hydroxyalkenals from peroxidized lipid in livers have been demonstrated in carbon tetrachloride treated rats and in liver preparations with in vitro forced peroxidation[18,19]. Fig. 2 shows that within 2 days after the exposure to the CD diet, hydroxyalkenals can be detected histochemically in the rat liver using Schiff reagent. In contrast, livers from rats on the CS diet (control) for up to 5 days did not show any Schiff positive staining. 4-Hydroxyalkenals were also detected biochemically, 48 h after the exposure to the CD diet.

Free Radical Trapping

The next series of experiments was designed to provide additional evidence about the generation of free radicals on exposure to the CD diet. We argued that if free radicals can be trapped by a free radical trapping agent, e.g., N-tert-butyl-α-phenylnitrone (PBN), then such an agent should also prevent CD-induced nuclear lipid peroxidation. Free radicals generated from CCl_4 can be trapped by PBN[20]. Accordingly, PBN prevented CCl_4 induced microsomal lipid peroxidation[21]. If CD diet causes free radical generation in the liver then challenging the rat with PBN would be expected to inhibit lipid peroxidation. The results shown in Fig. 3 clearly indicate that PBN, when given before the start of lipid peroxidation in the rat fed the CD diet totally prevented lipid peroxidation. Further generation of lipid peroxidation can

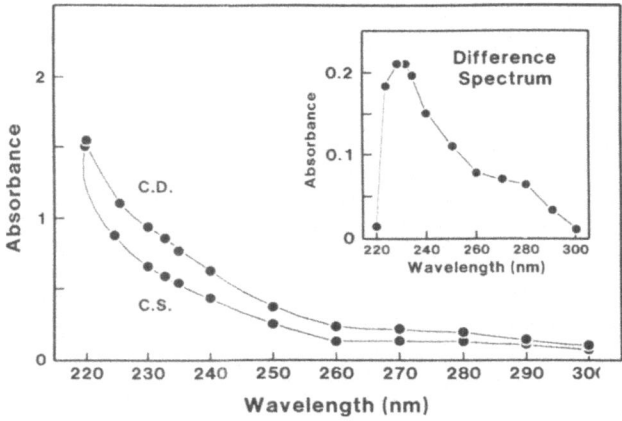

Fig. 1. Ultraviolet absorption spectra and difference spectrum (inset) of nuclear lipids recovered from the livers of rats exposed to the CD or CS diets for 24 h. The difference spectrum (inset) was obtained by subtracting the values at each wavelength for control (CS) from the values for experimental (CD).

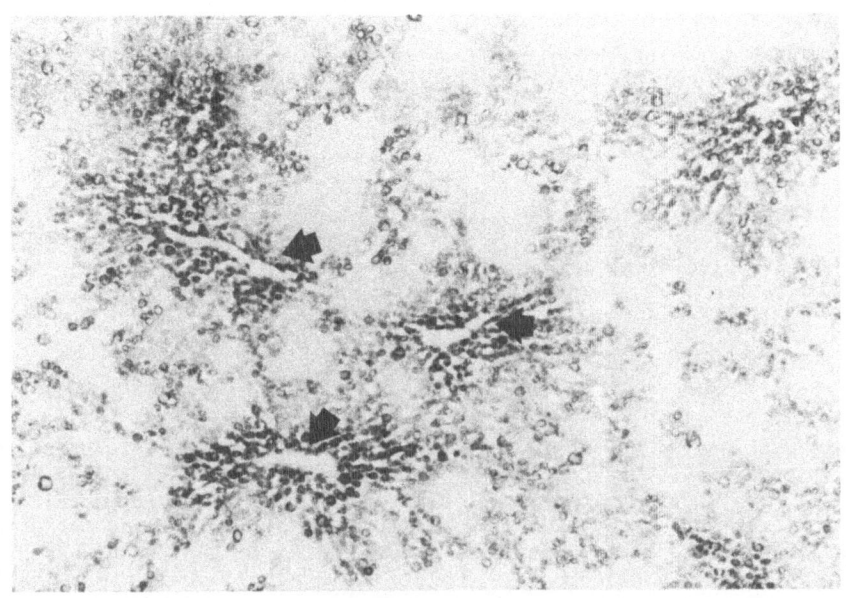

Fig. 2. Shiff-positive staining (for hydroxy alkenals) in hepatocytes in zones 2 and 3 in a rat exposed to CD diet for 2 days. The terminal hepatic vein (central vein) is indicated by arrows, x 25.

also be stopped when the rats are challenged with PBN while peroxidation is in progress [21]. Interestingly, PBN did not inhibit CD induced hepatic lipid accumulation.

NATURE OF FREE RADICALS IN CD INDUCED LIPID PEROXIDATION

Having established that CD induces free radical mediated lipid peroxidation, it became important to determine the mechanism by which CD increases the free radical status in hepatocytes.

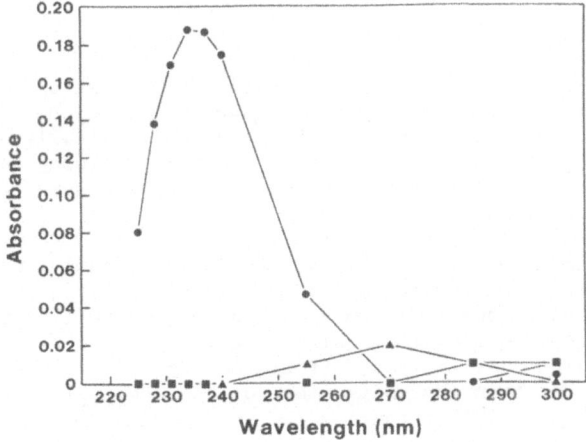

Fig. 3. The prevention of nuclear lipid peroxidation in rats exposed to CD diet by PBN. (●—●) CD, killed at 50 h; (■—■) CD + PBN at 12 and at 24 h and killed at 50 h; (▲—▲) CD, killed at 12 h.

Table 1. Glutathione and free radical metabolism related enzymes in CD induced in rat liver

Glutathione	increased significantly (2 to 3 times)
Glutathione peroxidase	No change
Glutathione reductase	No change
Superoxide dismutase	No change
Catalase	No change
DT-diaphorase	No change

There are several ways by which a free radical induced lipid peroxidation might be achieved: (a) a decrease in the free radical scavenging systems in the cell, (b) excessive generation of free radicals, e.g., superoxide, hydroxy radicals, etc., (c) excessive susceptibility of membranes for attack by normally generated free radicals or a combination of all the above three.

STATUS OF FREE RADICAL SCAVENGING ENZYMES

We measured glutathione peroxidase, glutathione-S-transferase, superoxide dismutase, catalase and DT-diaphorase and could not detect any significant change in the level of these enzymes in the liver of rats fed the CD diet when compared with animals on the CS diet[22]. See also Table 1.

GLUTATHIONE STATUS

Glutathione has been implicated in protecting the hepatocytes against reactive oxygen species and free radicals[23]. Glutathione levels both in the cytosol and whole liver homogenate from 1 to 8 days after exposure to CD diet were determined and it was found that glutathione content does not decrease, if at all it increases steadily from 1 to 8 days following exposure to CD diet[22]. (See also Table 1).

DISCUSSION

The major question which emerges from this presentation is how a diet which is devoid of choline (CD) can cause liver cancer without any known participation of carcinogen. The diet used was nutritionally adequate in respect to growth of animals. The CD diet exposure to young adult rats does not cause any major alterations to any organs excepting some very early changes and eventual cancer in the liver.

Emanating from here were three key questions. Firstly, does the process of initiation by this CD regimen take a totally novel pathway or follow the same pattern as present day paradigm accepted for chemical carcinogenesis? Secondly, what is the mechanism of initiation in this model in the absence of any exogenous carcinogens? Thirdly, if the CD regimen follows the same pattern of initiation then why unlike most of the chemical carcinogens does it take about 10 weeks for initiation?

Our presentation amply justifies the hypothesis that this CD model follows the same pattern of DNA damage, cell proliferation and initiation. The observation that free radicals generated _in vivo_ by metabolic imbalance

Table 2. Experiments suggesting "spontaneous" existence
of initiated hepatocytes

Observations	References
γ-GT positive hepatocytes in zone 1 - exaggerated by exposure to phenobarbital. No preneoplastic nodules or cancer.	Kitagawa et al. Gann 71:362 (1980).
Old animals or chronic treatment with phenobarbital, 1-2 cm^2 γ-GT positive foci without any appearance of preneoplastic nodules or liver cancer.	Ogawa et al. J. Natl. Cancer Inst. 67:407 (1981). Schulte-Hermann et al. Cancer Res. 43:837 (1981).
Six per cent incidence of hepatocellular nodules or carcinoma in aged rats. Phenobarbital treatment did not increase the incidence.	J. M. Ward, J. Natl. Cancer Inst. 71:815 (1983).

the biological effects of which is of great significance. The generation of free radicals in the liver due to CD diet is detected by indirect evidence, which includes (a) lipid peroxidation in the nuclei, (b) the genesis of hydroxyalkenals in association with lipid peroxidation, and (c) prevention of lipid peroxidation by the free radical trapping agent PBN. Subsequent to lipid peroxidation, DNA alterations can be detected[10]. The present challenge is that neither the genesis and nature of free radicals nor the cause of DNA alterations are known at present in this CD model. The important question to answer is how lipid peroxidation is initiated in this model without any apparent exogenous agent. Three possibilities by which CD diet might cause lipid peroxidation have been considered. (a) An increased generation of free radicals, (b) increased susceptibility of membranes to oxidative damage or (c) a decreased availability of free radical scavenger and enzymes. As shown in this study the status of various enzymes detoxifying oxygen radicals remains virtually unaltered during the period when active lipid peroxidation in the nuclei was in progress. Thus the third possibility that the increased susceptibility of hepatic nuclear lipid peroxidation in rats fed a CD diet is due to a deficiency of oxygen radical detoxifying machinery may be eliminated at least at the very early stages. Other two possibilities (a) and (b) as mentioned above are being investigated at present.

Thus, it is clear that choline deficiency is a complete carcinogen with respect to liver. The possibility of "spontaneous" origin of initiated hepatocytes has been suggested from several laboratories. This suggestion has prompted some investigators to believe that liver cancer arising from a CD diet may be due to promotion of preexistent initiated hepatocytes. Table 2 shows that some altered hepatocytes may exist in the very old animals but promoting agents e.g. phenobarbital treatment could not promote them further to hepatocyte nodules or to hepatocellular carcinoma. In our laboratories when rats not exposed to initiating treatments were kept on promoting agents like 1% orotic acid or 0.05% phenobarbital for nearly two years no cancer was observed in the liver. Furthermore, when a strong promoting regimen like 0.02% 2-acetylaminofluorene coupled with partial hepatectomy[13] was applied to rats without preinitiation by any carcinogen, no hepatocellular carcinoma developed within two years.

Recently, we kept rats on a CD diet for 11 weeks (the time when liver cells are initiated by CD diet, see ref. 12), then CD diet was replaced by laboratory chow diet (No. 5001) for more than two years. Out of 15 rats not a single rat developed hepatocellular carcinoma. It should be pointed out that when rats are initiated with one of several carcinogens, feeding a CD diet for a period of 5 weeks as a promoting procedure generate many γ-GT positive foci [24,25]. Thus, the studies on the presumed presence of initiated hepatocytes without prior exposure to a carcinogen seem to remain speculative.

Another important observation that merits consideration is the long exposure of 10 to 11 weeks required for CD to initiate hepatocarcinogenesis in spite of very early occurrences of DNA alterations and cell proliferation. Since exposure to a CD diet results in extensive liver cell necrosis, it could be that initiated hepatocytes are eliminated by CD-induced necrosis. It may be noted that CD is not the only treatment which takes a long time to initiate. There are other instances e.g. ethionine, aflatoxin B_1 etc. Ethionine inhibits cell proliferation very effectively. Probably even the initiated cells are also prevented to proliferate for a long time. Aflatoxin B_1 on the other hand is a very potent hepatotoxic agent. Possibly, due to its toxicity it takes 10-12 weeks of multiple exposures for B_1 to induce initiated hepatocytes. It is known that 50% of the liver cells die due to CD diet within 1 to 2 weeks [11,26]. It is not unreasonable to suggest that like B_1 cell death may be a determinant of delayed initiation by CD. Experiments are in progress to test this hypothesis.

ACKNOWLEDGEMENTS

The authors would like to express their sincere thanks to Lori Cutler for her excellent secretarial help. This work was supported by grants from the U. S. Public Health Service Research grant CA 41577, from the National Cancer Institute of Canada and the Medical Research Council of Canada (MT 5994). T. Rushmore was supported by a PHS National Research Service Award F-32 CA 07806 from the National Cancer Institute, NIH (USA).

REFERENCES

1. A. K. Ghoshal and E. Farber, Induction of liver cancer by a diet deficient in choline and methionine (CMD), Proc. Am. Assoc. Cancer Res. 24:98 (1983).
2. L. A. Poirier, Y. B. Mikol, K. Hoover and D. Creasia, Liver tumor formation in rats fed methyl-deficient amino acid-defined diets with and without diethylnitrosamine initiation, Proc. Am. Assoc. Cancer Res. 24:97 (1983).
3. Y. B. Mikol, K. L. Hoover, D. Creasia and L. A. Poirier, Hepatocarcinogenesis in rats fed methyl-deficient, amino acid defined diets, Carcinogenesis 4:1619 (1983).
4. A. K. Ghoshal and E. Farber, The induction of liver cancer by dietary deficiency of choline and methionine without added carcinogens, Carcinogenesis 5:1367 (1984).
5. S. Yokoyama, M. A. Sells, T. V. Reddy and B. Lombardi, Hepatocarcinogenesis and promoting action of a choline-devoid diet in the rat, Cancer Res. 45:2834 (1985).
6. M. A. Sells, S. L. Katyal, S. Sell, H. Shinozuka and B. Lombardi, Induction of foci of altered, γ-glutamyltranspeptidase-positive hepatocytes in carcinogen-treated rats fed a choline-deficient diet,

Br. J. Cancer 40:274 (1979).

7. B. Lombardi and H. Shinozuka, Enhancement of 2-acetylaminofluorene liver carcinogenesis in rats fed a cholinr-devoid diet, Int. J. Cancer 23:565 (1979).

8. A. K. Ghoshal, D. S. R. Sarma and E. Farber, Ethionine in the analysis of the possible separate roles of methionine and choline deficiences in carcinogenesis, in: "Essential Nutrients in Carcinogenesis", L. A. Poirier, P. M. Newberne, and M. W. Pariza, eds., Plenum Press, New York (1986)

9. T. H. Rushmore, Y. P Lim, E. Farber and A. K. Ghoshal, Rapid lipid peroxidation in the nuclear fraction of rat liver induced by a diet deficient in choline and methionine, Cancer Lett. 24:251 (1984).

10. T. H. Rushmore, E. Farber, A. K. Ghoshal, S. Parodi, M. Pala and M. Taningher, A choline-devoid diet, carcinogenic in the rat, induces DNA damage and repair, Carcinogenesis 7:1677 (1986).

11. A. K. Ghoshal, M. Ahluwalia and E. Farber, The rapid induction of liver cell death in rats fed a choline-deficient methionir a-low diet, Am. J. Pathol. 113:309 (1983).

12. A. K. Ghoshal, T. H. Rushmore and E. Farber, Initiation of carcinogenesis by a dietary deficiency of choline in the absence of added carcinogens, Cancer Lett. 36:289 (1987).

13. H. Tsuda, G. Lee and E. Farber, Induction of resistant hepatocytes as a new principle for a possible short-term in vivo test for carcinogens, Cancer Res. 40:1157 (1980).

14. D. B. Solt, E. Cayama, H. Tsuda, K. Enomoto, G. Lee and E. Farber, Promotion of liver cancer development by brief exposure to dietary 2-acetylaminofluorene plus partial hepatectomy or carbon tetrachloride, Cancer Res. 43:188 (1983).

15. D. B. Solt, A. Medline and E. Farber, Rapid emergence of carcinogen-induced hyperplastic lesions in a new model for the sequential analysis of liver carcinogenesis, Am. J. Pathol. 88:595 (1977).

16. S. Inouye, Site-specific cleavage of double-strand DNA by hydroperoxide of linoleic acid, FEBS Lett. 172:231 (1984).

17. T. J. Slaga, A. J. R. Klein-Szanto, L. L. Triplett, L. P. Yotti and J. E. Trosko, Skin tumor-promoting activity of bezoyl peroxide, a widely used free radical-generating compound, Science 213:1023 (1981).

18. A. Benedetti, G. Malvaldi, R. Fulceri and M. Comporti, Loss of lipid peroxidation as a histochemical marker for preneoplastic hepatocellular foci in rats, Cancer Res. 44:5712 (1984).

19. H. Esterbauer, Aldehydic products of lipid peroxidation, in: "Free Radicals, Lipid Peroxidation and Cancer, D. C. H. McBrien and T. C. Slater, eds., Academic Press, London (1982).

20. E. K. Lai, P. B. McKay, T. Noguchi and K-L. Fong, In vivo spin-trapping of trichloromethyl radicals formed from CCl_4, Biochem. Pharmacol. 28:2231 (1979).

21. A. K. Ghoshal, V. Subrahmanyan, T. H. Rushmore and E. Farber, The reversal of nuclear lipid peroxidation by a spin trap agent "PBN" in rats on a choline deficient diet, Proc. Am. Assoc. Cancer Res. 28:161 (1987).

22. A. Ghoshal, A. K. Ghoshal, T. H. Rushmore and E. Farber, Glutathione and enzymes related to free radical metabolism in liver of rats fed a choline deficient diet, Proc. Am. Assoc. Cancer Res. 28:162 (1987).

23. A. Meister, Glutathione, in: "The Liver: Biology and Pathology", L. A. Arias, H. Popper, B. Schachter, and E. A. Shefritz, eds., Raven Press, New York (1982).

24. A. Columbano, S. Rajalakshmi and D. S. R. Sarma, Requirement of cell proliferation for the induction of presumptive preneoplastic lesions in rat liver by a single dose of 1,2-dimethylhydrazine, Chem. Biol. Interactions 32:347 (1980).

25. H. Shinozuka and B. Lombardi, Synergistic effect of a choline-devoid

diet and phenobarbital in promoting the emergence of foci of γ-glutamyl-transpeptidase-positive hepatocytes in the liver of carcinogen treated rats, Cancer Res. 40:3846 (1980).

26. L. I. Giambarresi, S. L. Katyal and B. Lombardi, Promotion of liver carcinogenesis in the rat by a choline-devoid diet: Role of liver cell necrosis and regeneration, Br. J. Cancer 46:825 (1982).

MODULATION BY POLYUNSATURATED LIPID DIETS OF NITROSAMINE-INDUCED

CANCERS IN RATS: EFFECTS ON PROOXIDANT STATE AND DRUG METABOLISM

Helmut Bartsch, Eino Hietanen, Markku Ahotupa,
Anne-Marie Camus and Jean-Claude Béréziat

International Agency for Research on Cancer, F-69372 Lyon
Cedex 8, France and Department of Physiology, University
of Turku, SF-20520 Turku, Finland (M.A.)

INTRODUCTION

Dietary lipids are associated with increased risks for certain types of cancer, especially those of the breast and colon. This enhancing effect has been documented in humans by epidemiological studies[1,2,3], as well as by studies in experimental animals [4,5,6,7,8]. Although the association between dietary lipids and carcinogenesis has been known for a long time, the molecular mechanism by which lipids modulate tumor development remains largely unknown.

We have studied the modulating effects of diets with various fat contents on nitrosamine-induced cancers in rats by monitoring simultaneously indicators of lipid peroxidation (LPO), antioxidant defence, drug-metabolizing enzymes, membrane composition and blood chemistry. These parameters were measured during the course and at the end of experiments, in different organs, blood, urine and exhaled air. The tumor yields in different treatment groups were related to the biochemical changes measured[9]. A brief account of the major findings is given below.

MATERIALS AND METHODS

Animals, Diets and Carcinogen Treatment

Experiment I: Male Wistar rats (n = 144) were taken into the experiment after weaning at the age of three weeks. Low fat (LF) and high fat (HF) diets were purchased from ICN Nutritional Biochemicals (Cleveland, OH, USA) as 0 and 45% fat diets. During the first eight weeks of the feeding period, the HF diet contained 45% (w/w); for the rest of the experiment, it consisted of 30% (w/w) polyunsaturated fat (cotton-seed oil). This diet was prepared by mixing the 45% and 0% fat diets. The LF diet was totally fat-free for the first 20 weeks; thereafter, to avoid symptoms of essential fatty acid deficiency, 2% sunflower oil was added. In order to ensure an equal energy intake for rats in the two groups, they were pair-fed. Food intakes and weekly weight gain were measured for the duration of the experiment. After two weeks on the diets, the groups were divided and one half was given N-nitrosodiethylamine (NDEA) (3 mg/kg bw) intragastrically on five days per week for ten weeks; the control group received an equal volume of 0.9% NaCl. Some rats in each group were killed at 4, 12, 30 and 40 weeks after the beginning of feeding, and samples were taken from all organs and tissues for histological analysis and biochemical assays.

Experiment II: Groups of male Wistar rats (n = 73) were fed after weaning diets containing 2%, 12.5% or 25% by weight of either saturated fat (SF, lard) or polyunsaturated fatty acids (PUFA, sunflower oil). The diets, prepared by Usine d'Alimentation Rationnelle (Villemoisson, France), contained no added antioxidants; details of its composition are given elsewhere[9]. Possible auto-oxidation of dietary lipids was checked routinely by analyzing the malondialdehyde (MDA) content. One group of rats on the 25% polyunsaturated fat diet was also given indomethacin in the diet (50 mg/kg w/w). Another group of rats was fed standard laboratory chow (IARC diet). After ten weeks, each dietary group was divided, and one (n = 44) then received intragastric doses of N-nitrosodimethylamine (NDMA, 200 µg/rat) dissolved in 1 ml of water on 5 days per week for 30 weeks, a dose expected to produce hepatocellular tumors in about half of the animals[10]. Rats in all groups grew normally, and no difference was seen in weight gain between the groups.

For short-term studies on LPO, male Wistar rats (6-10 weeks old) were used. NDMA was dissolved in water prior to intraperitoneal administration at the doses indicated in the text.

For measurement of oxidative stress at the tissue level, rats were killed by decapitation 10-60 min or 24 h after dosing. The liver was removed and hepatic tissue fractions prepared by routine procedures immediately after sacrifice. All analyses of LPO were carried out on freshly prepared tissue fractions. Lipids were extracted from tissue specimens by the conventional chloroform-methanol procedure.

Lipid Peroxidation (LPO)

The extracted lipid fraction was used to estimate the levels of diene conjugation[11,12] and fluorescent products of LPO[13]. The amount of thiobarbituric acid (TBA)-reactive material was measured in the S9 fraction of liver[14]. NADPH-stimulated, lucigenin-dependent chemiluminescence in microsomal fractions was measured using a Wallac 1251 Luminometer[15,16].

The rate of ethane exhalation of individual rats was measured as described by Wendel and Dumelin[17]. Superoxide dismutase (SOD) activity (Cu/Zn form) was assayed as described by Flohe and Otting[18].

Cytochrome P450 concentration and drug-metabolizing enzymes were determined using standard procedures, similar to those reported by Hietanen et al.[19]. The production of NADPH by the hexose monophosphate shunt (HMS), which is the sum of glucose-6-phosphate dehydrogenase and 6-phosphogluconate dehydrogenase activities, was determined as described by Clock and McLean[20]. The fatty acid compostion of liver microsomal phospholipids was analyzed by gas chromatography as described by Norred and Wade[21].

RESULTS AND DISCUSSION

Major biochemical changes and tumor outcomes in different treatment groups in experiment I are summarized in Table 1.

NDEA-induced Tumors in Rats Fed HF and LF Diets

In NDEA-treated groups, there were more tumors in rats on the LF (18/23) than in those on the HF (13/22). However, the level of fat in the diet strongly affected the tissue distribution of tumors: rats on the HF diet had almost exclusively liver tumors (12 hepatocellular carcinomas, 1 hemangiosarcoma), while only 9/23 rats on the LF diet did so ($P < 0.1$); LF animals had several extrahepatic tumors (eight of the nervous system and seven in the kidney) some animals having multiple tumors.

Monooxygenase Activities

Monooxygenase activities were measured using six different xenobiotic and endogenous substrates. Changes related to dietary fat were seen mainly at early time points and not in all monooxygenase activities: ethoxycoumarin O-deethylase, testosterone 6ß-hydrolase and NDMA demethylase activities were increased by the HF diet. Testosterone 7α- and 16α-hydroxylase activities were higher in rats of the LF group, and aryl hydrocarbon hydrolase and ethoxyresorufin O-deethylase showed no diet-related alteration.

NADPH Production by the HMS in Liver

NADPH production by the HMS in liver cytosol was affected dramatically

Table 1. Major biochemical changes induced by LF and diets with and without NDEA treatment, with tumor outcome

	HF/O[a] vs LF/O	HF/NDEA vs LF/NDEA	HF/NDEA vs HF/O	LF/NDEA vs LF/O
Total no. of tumors	-	↓	↑	↑
Liver tumors	-	↑	↑	↑
Extrahepatic tumors	-	↓	↑	↑
Plasma[b]				
Total cholesterol	↓	∅	∅	↓
Free cholesterol	↓	∅	↓	↓
Triglyceride	↓	∅	↓	↓
Glutathione-peroxidase	∅	∅	↑	∅
Exhaled ethane	↑	∅	↑	↑
Liver				
Cytochrome P450[b]	↑	↑	∅	∅
Monooxygenases	↑	↑	∅	∅
NADPH production	↓	↓	∅	∅
SOD	↑	↑	∅	∅
MDA	↑	↑	∅	∅
Chemiluminescence	↑	↑	∅	∅
PUFA	↑	↑	∅	∅
SF	↓	↓	∅	∅
Extrahepatic organs[b]				
Renal glutathione	↓	↓	∅	∅

[a] O, no treatment. [b] Data not discussed in the text. ↓ Decrease; ↑ increase; ∅ no change.

by the level of dietary fat: up to 12 weeks' feeding, the amount of NADPH produced was about 20 times higher in the LF than in the HF group, while the cytosolic protein concentration remained unaffected. After this period, the activity of the HMS in the LF group decreased, but a diet-related difference of about 2.5-3-fold remained at the end of the 40-week feeding period. NDEA treatment had no effect of NADPH production by the HMS. As rats in the LF and HF groups were fed equicaloric diets, different carbohydrate intakes cannot explain these large differences. Thus, it appears that the fat intake also controls HMS activity in the liver; however, it is not known whether such differences affect tumor outcome. In red blood cells, the amount of NADPH is important in controlling (indirectly, through the action of gluta-thione reductase) tissue levels of antioxidant defence, and an increased cel-lular prooxidant state has been linked to processes of carcinogenesis[22].

Membrane Composition and Lipid Peroxidation (LPO)

When comparing the fatty acid compostion of phospholipids in liver microsomes of rats fed HF and LF diets (without NDEA treatment), significant diet-related differences were found: the HF diet enhanced the relative amount of PUFA. The relative amounts of palmitic, stearic and linolenic acids were smaller in the HF group, and oleic, linoleic and arachidonic acid levels were higher than in the LF group.

LPO, as measured by TBA in hepatic microsomes, was increased several fold by the HF diet. No difference was found between NDEA-treated and un-treated animals. Similarly, measurement of chemiluminescence in liver micro-somes revealed an increase in lipid peroxidation by the HF diet (Table 2). In both the above assays, the microsomal lipid peroxidation was estimated after stimulation by ADP-Fe^{++} and/or NADPH. Thus, the results indicate that liver microsomal membranes of rats on a HF diet are more prone to LPO.

Exhaled ethane was used as a measure of whole body LPO after 40 weeks on diet (Table 2). Ethane exhalation was elevated significantly in the HF/O group over that in the LF/O group. In both groups, NDEA treatment markedly enhanced ethane exhalation. On an individual basis, ethane exhalation was similar in rats with liver and extrahepatic tumors. On a group basis, ethane production was closely related to the prior NDEA administration.

The level of Cu/Zn SOD activity was higher in animals on the HF diet up to 20 weeks of age. Between 20 and 45 weeks, however, SOD activity in the HF group declined by about 50%. No such decrease was seen in the LF group.

The data summarized in Table 1 (experiment I) might explain both the higher rate of hepatic tumors in the HF/NDEA vs the LF/NDEA group, as well as the higher yield of extrahepatic tumors in LF/NDEA vs HF/NDEA. (I) The high-er hepatic monooxygenase activities in the HF/NDEA and HF/O groups in com-parison with LF/NDEA and LF/O groups suggest enhanced oxidative metabolism of NDEA, leading to more DNA damage. (II) Conversely, in the LF/NDEA group, higher levels of NDEA could circulate in the blood, as the first-pass clear-ance in the liver would be less effective; thus, NDEA would reach extrahepat-ic organs at higher concentrations. (III) As a result of the higher content of PUFA in phospholipids in liver-cell membranes, LPO is greatly enhanced in liver microsomes by the HF diet, and further by NDEA. The increased cellular prooxidant state linked to promotion/progression steps in carcinogenesis[22] could explain the increased yield of liver tumors in the HF group.

In experiment II, NDMA produced mainly hepatic hemangiosarcomas. Up to 12.5% fat in the diet resulted in similar incidences of tumors, whether the diet contained SF or PUFA. However, when the dietary fat content was 25%, the group of rats on PUFA had more liver tumors (12/15) than the group on the

Table 2. Ethane exhalation in vivo, chemiluminescence and MDA production
         (TBA-reactive material) of hepatic microsomes (NADPH stimulat-
         ed) of rats after 40 weeks on various fat diets with and with-
         out NDEA treatment

| Diet/treatment[a] | Ethane exhaled[b] (nmol/kg b.w. x h) | NADPH stimulated | |
		Chemiluminescence[c] (x 10^3 mV/nmol cyt. P450)	MDA production [μmol/20 min x (g wet wt)$^{-1}$]
LF/O	1.05 ± 0.30	6 960 ± 590	2.2 ± 0.2
LF/NDEA	3.63 ± 0.63 **	7 070 ± 1 690	3.2 ± 0.6
HF/O	2.20 ± 0.36 *	7 660 ± 680	12.9 ± 0.7
HF/NDEA	3.76 ± 0.30 **	9 560 ± 1 420	12.5 ± 2.3
HF/NDEA (tumor)		28 670 ± 9 040	7.2 ± 2.4

[a]O, not treatment. [b]Statistical analyses are shown as compared to respective dietary group without NDEA: **, P < 0.01; *, significantly higher
(P < 0.05) than respective LF/O group. [c]Under optimal conditions in 0.1 M
Tris buffer, pH 7.4, in the presence of 0.25 mM (final conc.) lucigenin,
0.2 mM NADPH, and 0.25 - 0.5 mg microsomal protein; peak values were
recorded. Mean ± SEM from 4-6 animals.

SF diet (5/12). In the group fed 25% PUFA, indomethacin at 50 mg/kg diet resulted in a decreased tumor incidence (9/14).

Ethane exhalation in different groups of rats was monitored ten, 20-27
and 43 weeks after the start of feeding (Fig. 1). When the amount of SF or
PUFA dietary fat was raised from 2% to 12.5%, the rate of ethane exhalation
increased. A further increase in the level of dietary fat did not, however,
augment ethane production: animals receiving either 12.5% or 25% fat in the
diet exhaled at similar rates. With the exception of rats on 2% dietary fat,
animals consuming PUFA produced more ethane than those on SF diets. Within
all dietary groups, administration of NDMA elevated ethane production by two
to four times, and the increase brought about by NDMA was more pronounced
than the variation in dietary lipid content. This increased LPO by NDMA was
confirmed in short-term experiments (see below, 23, 24).

The presence of 0.005% indomethacin in the 25% PUFA diet diminished
ethane production to less than that seen with 2% fat (Fig. 1). Moreover, in
rats treated with NDMA, indomethacin suppressed the increase in ethane exhalation.

Induction of LPO by NDMA was investigated further. A single intraperitoneal administration of NDMA to rats increased the amount of ethane in exhaled air in a dose-related manner (Fig. 2), and a dose of 6.7 μmol/kg (the
lowest dose tested) was sufficient to enhance ethane exhalation. The rate of
ethane exhalation increased rapidly during the first 60 min after single administration of NDMA, and remained elevated for several days (Fig. 2B).

As reported previously[23,24], NDMA treatment rapidly enhanced LPO in liver
tissue. The greatest increase was seen in the amount of diene conjugation
and fluorescent products. These two measures of LPO, like microsomal NADPH-
stimulated chemiluminescence, showed highest levels 20 min after dosing,
whereas the production of TBA-reactive material was maximal 60 min after the

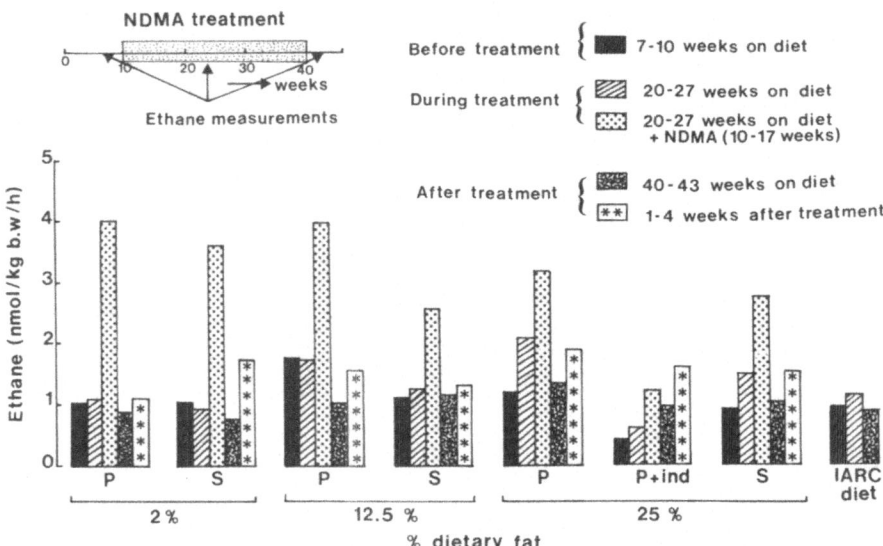

Fig. 1. Ethane production in rats on diets with different fat contents,
with or without NDMA; mean (n = 3-5 rats). P, polyunsaturated fat;
S, saturated fat; ind, indomethacin.

dose. The values for fluorescent and TBA-reactive products were still above
control levels 24 h after dosing.

It was found earlier[24] that in isolated rat hepatocytes lucigenin-dependent chemiluminescence and hydrogen peroxidase release are increased by
micromolar concentration of NDMA. NADPH-stimulated ethane production by hepatic microsomes (from untreated rats) was also increased in the presence of
2-8 μM NDMA.

Two further carcinogenic nitrosamines were tested for their ability to
induce oxidative stress in vivo. N-nitrosodiethanolamine, which is hepatocarcinogenic to rats, enhanced ethane exhalation and LPO in the liver, although when compared to NDMA, a higher dose was required. N-Nitrosomethylbenzylamine which is an esophageal but not a liver carcinogen in rats, had no
effect on LPO. Thus, the hepatocarcinogenic nitrosamines induce LPO in the
target tissue, and this effect is not related to the acute toxicities of
these compounds[23].

In summary, our results indicate, for the first time, that the amount
and composition of dietary lipids modify the oxidative state in experimental
animals in vivo, as measured by LPO.

We found that the rate of LPO in vivo is raised more readily by PUFA
than by SF in the diet. This finding is not unexpected, since only PUFA can
serve as substrates in LPO reactions. Experimental studies have indicated
generally that PUFA are more effective than fats containing mainly SF in
enhancing chemically induced tumors[25,5,26], as we found in experiment II. The
finding that the amount of LPO was similar in rats receiving 12.5% and 25%
fat in the diet was unexpected.

Administration of the hepatocarcinogens NDMA and NDEA at relatively low
doses also markedly enhanced ethane production, independently of the level of
fat in the diet. It has been demonstrated that the superoxide anion and hydrogen peroxide form during normal cytochrome P450-mediated metabolism of

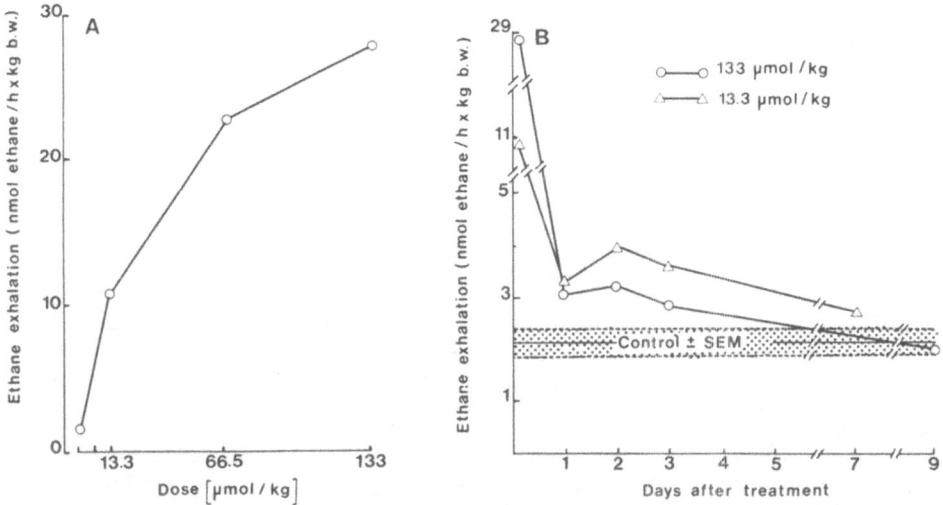

Fig. 2. A, rate of ethane exhalation during the first 60 min after intraperitoneal administration of different doses of NDMA to rats. Each
point represents the mean of two to three rats. B, rate of ethane
exhalation by rats up to 9 days after a single intraperitoneal dose
of NDMA. Each point represents the mean of two rats.

xenobiotics[27]. The results presented here suggest that reactive species generated during carcinogen metabolism increase the peroxidation of cellular
lipids. The possible consequences of this effect in carcinogenesis remain to
be investigated.

A conspicuous finding in the present study was the suppression of LPO by
indomethacin in vivo at doses that are comparable to those used for therapy
in humans. The results cannot, however, elucidate the mechanism by which
indomethacin blocks peroxidation induced by dietary lipid and NDMA. It may
represent a further link between prostaglandin synthesis and LPO, since it is
well known that products of LPO regulate prostaglandin synthesis[28].

In conclusion, our results suggest that dietary lipids regulate the carcinogenesis process in a complex way, depending on both the quantity and the
quality of dietary fats, and especially the degree of saturation. Under our
experimental conditions, dietary fats appear to be strong modulators of nitrosamine-induced tumors in rats, affecting both the incidence and distribution of target organs. Of all the parameters examined, exhaled ethane appeared to be most closely related to the risk of developing tumors on a group
basis. It may thus be useful to study its predictive power for cancer development longitudinally.

ACKNOWLEDGEMENTS

The authors wish to thank Miss Y. Granjard for secretarial assistance
and Mr. E. Heseltine for editing the manuscript.

REFERENCES

1. B. Armstrong and R. Doll, Environmental factors and cancer incidence
and mortality in different countries, with special reference to

dietary practices, _Int_. _J_. _Cancer_ 15:617 (1975).

2. K. K. Carroll, Lipids and carcinogenesis, _J_. _Environ_. _Pathol_. _Toxicol_. 3:253 (1980).

3. P. Correa, Epidemiological correlations between diet and cancer frequency, _Cancer Res_. 41:3685 (1981).

4. L. A. Hillyard and S. Abraham, Effect of dietary polyunsaturated fatty acids on growth of mammary adenocarcinomas in mice and rats, _Cancer Res_. 39:4430 (1979).

5. K. K. Carroll and M. B. Davidson, The role of lipids in tumorigenesis, _in_: "Molecular Interrelations of Nutrition and Cancer", M. S. Arnott, J. Van Eys and Y.-M. Wang, eds., Raven Press, New York (1982).

6. C. A. Carter, R. J. Milholland, W. Shea and M. M. Ip, Effect of prostaglandin synthetase inhibitor indomethacin on 7,12-dimethylbenz(a)anthracene-induced mammary tumorigenesis in rats fed different levels of fat, _Cancer Res_. 43:3559 (1983).

7. A. F. Watson and E. Mellanby, Tar cancer in mice. II: The condition of the skin when modified by external treatment or diet, as a factor in influencing the cancerous reaction, _Br_. _J_. _Exp_. _Pathol_. 11:311 (1930).

8. A. Tannenbaum, Nutrition and cancer, _in_: "The Physiopathology of Cancer", 2nd ed., F. Homburger, ed., Hoeber-Harper, New York (1959).

9. E. Hietanen, M. Ahotupa, J.-C. Béréziat, V. Bussacchini, A.-M. Camus and H. Bartsch, Elevated lipid peroxidation in rats induded by dietary lipids and _N_-nitrosodimethylamine and its inhibition by indomethacin monitored via ethane exhalation, _Toxicol_. _Pathol_. 15:93 (1987).

10. W. Lijinsky and M. D. Reuber, Carcinogenesis in rats by nitrosodimethylamine and other nitrosomethylalkylamines at low doses, _Cancer Lett_. 22:83 (1984).

11. F. Corongiu, M. Lai and A. Milia, Carbon tetrachloride, bromotrichloromethane and ethanol acute intoxication, _Biochem_. _J_. 212:625 (1983).

12. F. Corongiu, S. Vargiolu, A. Milia, K. Cheeseman and T. Slater, Antioxidants and lipid peroxidation: _in vivo_ studies, _Biochem_. _Pharmacol_. 34:397 (1985).

13. C. Dillard and L. Tappel, Fluorescent damage products of lipid peroxidation, _Meth_. _Enzymol_. 105:337 (1984).

14. R. P. Bird and H. H. Draper, Comparative studies on different methods of malonaldehyde determination, _Meth_. _Enzymol_. 105:299 (1984).

15. J. R. Wright, R. C. Rumbaugh, H. D. Colby and P. R. Miles, The relationship between chemiluminescence and lipid peroxidation in rat hepatic microsomes, _Arch_. _Biochem_. _Biophys_. 193:344 (1979).

16. E. Cadenas and H. Sies, Low-level chemiluminescence as an indicator of singlet-molecular oxygen in biological systems, _Meth_. _Enzymol_. 105:221 (1984).

17. A. Wendel and E. E. Dumelin, Hydrocarbon exhalation, _Meth_. _Enzymol_. 77:10 (1981).

18. L. Flohe and F. Otting, Superoxide dismutase assays, _Meth_. _Enzymol_. 105:93 (1984).

19. E. Hietanen, C. Malaveille, A.-M. Camus, J.-C. Béréziat, G. Brun, M. Castegnaro, J. Michelon, J. R. Idle and H. Bartsch, Interstrain comparison of hepatic and renal microsomal carcinogen metabolism and liver S9-mediated mutagenicity in DA and Lewis rats phenotyped as poor and extensive metabolizers of debrisoquine, _Drug Metab_. _Dispos_. 14:118 (1986).

20. G. E. Glock and P. McLean, Further studies on the properties and assay of glucose-6-phosphate dehydrogenase and 6-phosphogluconate dehydrogenase of rat liver, _Biochem_. _J_. 55:400 (1953).

21. W. P. Norred and A. E. Wade, Dietary fatty-acid-induced alterations of hepatic microsomal drug metabolism, _Biochem_. _Pharmacol_. 238:292 (1972).

22. P. A. Cerutti, Prooxidant states and tumor promotion, Science 227:375 (1985).

23. M. Ahotupa, J.-C. Béréziat, V. Bussacchini-Griot, A.-M. Camus and H. Bartsch, Lipid peroxidation induced by N-nitrosodimethylamine (NDMA) in rats in vivo and in isolated hepatocytes, Free Rad. Res. Comms. 3:285 (1987).

24. M. Ahotupa, V. Bussacchini-Griot, J.-C. Béréziat, A.-M. Camus and H. Bartsch, Rapid oxidative stress induced by N-nitrosamines, Biochem. Biophys. Res. Commun. 146:1047 (1987).

25. J. A. Miller, B. E. Kline, H. P. Rush and C. A. Baumann, The effect of certain lipids on the carcinogenicity of p-dimethylaminoazobenzene, Cancer Res. 4:756 (1944).

26. K. K. Carroll and H. T. Khor, Effects of level and type of dietary fat on incidence of mammary tumors induced in female Sprague-Dawley rats by 7,12-dimethylbenz(a)anthracene, Lipids 6:415 (1971).

27. A. G. Hildebrandt, C. Bergs, G. Heinemeyer, E. Schlede, J. Roots, P. Abbas-Ali and A. Schmoldt, Studies on the mechanism of stimulation of microsomal H_2O_2 formation and benzo(a)pyrene hydroxylation by substrates and flavone, Adv. Exp. Med. Biol. 136A:179 (1982).

28. W. E. M. Lands, R. J. Kulmacz and P. J. Marshall, Lipid peroxide actions in the regulation of prostaglandin biosynthesis, in: "Free Radicals in Biology", vol. VI, W. A. Pryor, ed., Academic Press, New York (1984).

LIPIDS AND COLONIC CARCINOGENESIS: FACT OR ARTEFACT ?

Myriam Wilpart and Marcel Roberfroid

Laboratoire de Biochimie Toxicologique et Cancérologique
U.C.L. 73.69, B-1200 Brussels, Belgium

INTRODUCTION

Life style and dietary habits play an important role in the causation and development of a number of major human cancers[1]. This conclusion is partly supported by evidence from epidemiological and laboratory animal studies. Investigators have attempted to study the mechanisms by which diet may influence carcinogenesis and to examine the ability of nutrients, food components or non-nutritive food additive components to enhance or to inhibit carcinogenesis. Cancer of the colon is one of the most common tumors observed in the affluent western populations[2] for which the relationship between epidemiological and laboratory findings and an overall assessment of the influence of diet on carcinogenesis is not straight forward.

THEORY OF CARCINOGENESIS: CONCEPT OF MODULATION[3]

The natural history of cancer development begins with the initiating treatment. If initiation is necessary and sufficient, the process progresses to malignancy even in the absence of further exogenous treatment.

That progress can be modified by modulating treatments. If these treatments shorten the latency period before the appearance of the malignancy, they are said to induce a positive modulation, part of which is equivalent to what has been called up to now a "promotion". Such a positive modulation may be the consequence of dietary imbalances of surgery or of repeated doses of initiator. If these treatments increase the latency period for the appearance of malignancy, they are said to cause a negative modulation which is the objective of cancer chemoprevention. Such a negative modulation can result from administrating free radical scavengers or antioxidants or by rebalancing dietary habits such as e.g. by increasing fiber and/or reducing fat intakes. The present report shows experimental evidence for a positive modulation of colonic carcinogenesis by dietary lipids.

DIETARY FAT AND COLON CANCER: RESULTS FROM ANIMAL STUDIES

Critical Evaluation

During the last decade, many reports have been concerned with the influence of dietary lipids on chemically initiated experimental carcinogenesis both in rats and in mice. However, interpretation of the data is difficult because they have been provided by various experimental protocols in which different fatty diets were administered to rats belonging to various strains treated by carcinogens differing in nature and doses. Knowing the differences between rat strains with regard to the metabolism of the carcinogens it would be expected that such variable protocols would lead to important variations in the tumor-modulating effect of fat. Moreover, in most cases, it was not stated whether injected dose of carcinogen was adjusted for the increase in weight of animals that occurs with aging. In some studies, the total dose of the injected carcinogen (mg/kg/rat) was so high that a cancer incidence of 100% was observed in the control group of treated rats. Consequently, it was thus not possible to detect any influence of high fat diets on tumor incidence.

The use of poorly defined diets of varying caloric density caused different intakes of other nutrients which might have produced experimental artefacts[4,5]. The number of tumorigenic parameters analyzed is often limited; indeed, only the number of rats bearing tumors and the number of tumors/rat are given in details. Few or none of the reports show the curves of cumulative incidence. Moreover, they do not indicate where the tumors were located they do not give their volume and they do not classify them according to histological or morphological criteria. Moreover, the terminology of the tumorigenic parameters is usually confusing. For example, frequency is defined as the number of animals bearing tumors for some authors whereas it is the number of tumors/rat for others[4]. The meaning of this last parameter is quite variable; in some papers, the mean number of tumors/rat, but in others it is the mean number of tumors/rat bearing tumors. Moreover, it is often not possible to know if this value refers to benign + malignant tumors or only malignant carcinomas.

In order to fully evaluate the influence of lipids on experimental colon carcinogenesis, the following parameters need to be analyzed concomitantly[6]:

Incidence (%): ° number of rats bearing tumors/total number of rats.

Intensity or frequency : ° number of tumors per rat bearing tumors and/or tumor-bearing rats with multiple tumors (%).

% of metastasis: ° number of rats with metastasis/number of rats with tumors.

Location of tumors: ° number or % of tumors in small bowel, large bowel, different segments of large bowel (caecum + ascending, transverse, descending).

Tumoral volume expressed in cm^3/rat and/or tumor size: expressed in mm. Moreover the histological nature of the tumors should be indicated.

Despite such limitations, the influence of the nature of fat as well as that of the amount of fat have been analyzed.

Influence of the Nature of Fat

Table 1 summarizes the results of experiments published between 1975 and 1986 on the influence of the nature of fat on colonic carcinogenesis, studies chosen for their partly compatible experimental protocol.

The protocols used mainly differed by the strain of rat as well as the nature of fat. The animal treated was almost exclusively the rat and the carcinogen used 1,2-dimethylhydrazine or its metabolite azoxymethane. Fat was most frequently given before, during and after exposure to the carcinogen. Although this protocol may be analogous to the human situation, it does not allow conclusions to be drawn concerning the influence of fat intake on the initiation or promotion of colon carcinogenesis.

Analysis of the results shown in Table 1 demonstrates the discrepancy of these data. Indeed, in experiment 2, corn oil-vegetal oil, induces a similar effect to that observed with animal fats; on the other hand, in experiment 3, corn oil has a positive modulating effect significantly greater than that of high saturated lipids (like animal fats).

In conclusion, although a great number of data are available on the relation between nature of fat in diet and colon cancer, no clear conclusion can be drawn from their analysis. The lack of consistent findings may have arisen from methodological differences between the studies. The question of whether the effect of dietary fat on carcinogenesis is due to the specific action of fat or to an associated caloric effect has been raised several times[10].

Although the studies which have been done addressed the question of the role of caloric intake versus fat intake in carcinogenesis, the observed effects might have been due to a reduction in body weight or to changes in other dietary components such as non-nutritive fiber, protein and micronutrients. A more systematic approach is required which would test the varying levels of caloric restriction while keeping the fat, micronutrient and non-nutritive fiber intakes constant. That kind of approach will separate the effect of fat, micronutrients and fiber from that of total calories[10].

Influence of the Amount of Dietary Fat on Colon Carcinogenesis

Up to now, it has been demonstrated that corn oil or beef fat content in diets given to rats after exposure to AOM increased either incidence and/or intensity of colonic tumors (Table 2).

In Reddy's study[10], the positive modulating activity of corn oil on colonic carcinogenesis was only observed when the amount was increased from 13.6% to 23.5% of food intake. No difference in colon tumor incidence was seen between 5 and 13.6% corn oil diet groups, suggesting a threshold effect. This observation is of first importance for cancer prevention and must therefore be confirmed for other fats. Systematic dose effect relationship studies are necessary.

It is the reason why we studied the influence on intestinal tumorigenesis of high dietary fat (10-15-20% wt./wt. of a mixture corn oil-palm oil 50:50 containing equal proportions of saturated, monounsaturated and polyunsaturated fatty acids) given after a chronic treatment of male Wistar rats with 1,2-dimethylhydrazine (DMH).

A diet containing 5% of such a fat mixture was given as a control. The experimental diets were isocalorically balanced to be equivalent with respect to the nutrient to calorie ratio (Table 3).

Preliminary experiments (data not shown) during which various parameters such as food consumption, body weight, fecal outflow, fecal concentrations of water, Na^+, K^+ and free fatty acids were measured, has given us confidence that giving such isocaloric diets allows the animals to live in normal physiological conditions. The dosage schedule (30 mg/kg/rat of 1,2-dimethylhydrazine injected subcutaneously under chlorhydrate form once per week during 15 weeks) was chosen to produce tumors in 50% of the control animals allowing assessment of enhancing effects on tumors incidence and intensity as well as tumor yield.

Table 4 shows the influence of the administration of these high fatty

Table 1. Influence of nature of fat on colonic carcinogenesis

Study	Treated animals [a]	Diet: lipids		Period administ.[b]	Point of sacrifices[c]	Results	References
		Nature	Amount (wt/wt)				
1.	DMH-2HCl M Sprague-Dawley	Safflower oil Coconut oil	20%	BE+E+PE	35	Enhancing effect of safflower oil	7
2.	DMH-2HCl and MNU M Sprague-Dawley	Beef fat Corn oil Crisco	24%	BE+E+PE	55	No difference	5, 8
3.	AOM F-F344 rats	Low trans fat intermediate trans fat, high trans fat corn oil	23.5%	PE	37	Enhancing effect of corn oil	9

[a] Carcinogen used: DMH-2HCl: chlorhydrate of 1,2-dimethylhydrazine; MNU: methylnitrosourea; AOM: azoxymethane; sex of rats: M = male, F = female. [b] Period of administration: BE = before exposure; E = exposure; PE = post exposure to carcinogen. [c] In weeks from first injection.

Table 2. Influence of amount of fat on colonic carcinogenesis

Study	Treated animals [a]	Diet: lipids		Period administ. [b]	Results	References
		Nature	Amount (wt/wt)			
1.	AOM F-F344 rats	Corn oil	5;13.6;23.5	PE	Promoting effect at 23.5%	9
2.	AOM M Sprague-Dawley rats	Purina chow Beef fat	5-35	PE	Promoting effect at 35%	10
3.	AOM M Sprague-Dawley rats	Beef fat	5-30	PE	Promoting effect at 30%	4

[a]Carcinogen used: AOM: azoxymethane; sex rat: M = male; F = female. [b]PE: post exposure to carcinogen.

Table 3. Diet composition

Component	Amount (g/100g) in function of the % of the lipidic fraction	
	5	20
Caseine	25	31
Corn starch	50	22.5
Corn oil	2.5	10
Palm oil	2.5	10
Cellulose	2	6
Mineral complex	7	9.2
Vitamin complex	1	1.3
Sipernatt [a]	10	10
Caloric density (Kcal/g)	3.5	3.9

[a]Added to allow oil compression.

diets on the tumorigenic parameters in DMH treated rats. In group 5/5, 54% of the rats had intestinal tumors. In group 5/20, that percentage was significantly (P < 0.01) increased up to 82%; that increase was mainly due to the fact that the number of rats with colonic carcinomas was going up from 41% (group 5/5) to 59% (group 5/20). Although the incidence of intestinal and large bowel carcinogenesis was increased in rats fed the 20% lipids diet during the promotional phase, the intestinal and colonic tumor yields expressed as the number of tumors/rat bearing tumors were not modified. Indeed the mean number of intestinal and colonic tumors/rat bearing tumors was 1.5 in rats of group 5/5 whatever the localisation was and respectively 1.8 and 1.5 in animals of group 5/20. The mean tumoral volume as well as the percentage of rats with metastasis were significantly (P < 0.01) higher in rats of group 5/20 than in those of group 5/5. With regard to the localisation of colonic tumors, the fatty treatment changed the relative distribution between the transverse and the descending fragment since in group 5/20 most (71%) of the colonic carcinomas were found in the transverse whereas the ratio transverse/ descending was 2:3 (transverse, 32%; descending 56% in group 5/5). All these data demonstrated a modulating effect of the high fatty diet on DMH initiated colon carcinogenesis.

Nevertheless, increasing the amount of fat from 5% to 10% and 15% failed to enhance colonic carcinogenesis. Indeed, incidence, intensity, mean tumoral volume as well as number of rats with metastasis did not differ in the animals of the first three experimental groups (5/5-5/10-5/15).

In conclusion, increasing the amount of fat from 5% (wt/wt) up to 15% did not influence intestinal carcinogenesis but increasing that content from 15% to 20% induced a positive modulation of colonic carcinogenesis in rats treated with DMH.

In addition to the experiment described above, we applied the following protocol during which rats were exposed alternatively to 3 weeks of low fat diet and to 3 weeks of high fat diet for a total period of 18 weeks. Those regimens were given after the exposure to the carcinogen.

The data obtained showed a dramatic increase of the incidence and intensity of the colonic carcinogenesis. Indeed, 100% of the rats had colonic

Table 4. Influence of the administration of high fatty diets on the tumorigenic parameters in DMH treated rats

Tumorigenic parameters	Experimental groups			
	5/5	5/10	5/15	5/20
Number of rats in experiment	56	56	56	56
% of survival rats [a]	98	96	100	82*
Incidence [b] of intestinal carci-nogenesis (small + large bowel)	30/55 (54%)	32/54 (59%)	34/56 (60%)	38/46* (82%)
Incidence of colonic carci-nogenesis	23/55 (41%)	26/54 (49%)	27/56 (48%)	27/46* (59%)
Intensity [c] of intestinal carci-nogenesis	1.5 ± 0.2	1.6 ± 0.2	1.2 ± 0.1	1.8 ± 0.4
Intensity of colonic carcino-genesis	1.5 ± 0.1	1.5 ± 0.1	1.2 ± 0.1	1.5 ± 0.4
Mean colonic tumoral volume (cm^3/rat)	0.5	0.4	0.6	1.5*
% of rats with metastasis [d]	15	18	20	74*
Localisation of colonic tumors:				
transverse	11/34 (32%)	18/42 (43%)	18/32 (56%)	29/41 (71%)
descending	19/34 (56%)	20/42 (48%)	12/32 (37%)	10/41 (24%)

[a] % Survival rats at first sacrifices. [b] Incidence defined as number of rats bearing tumors/total number of rats. [c] Intensity defined as number of tumors/rat bearing tumors. [d] Number of rats with metastasis/number of rats bearing tumors. * Value significantly ($P < 0.01$) different from that obtained for group 5/5.

tumors with metastasis. Moreover the intensity of colonic carcinogenesis expressed by the number of tumors/rat bearing tumors was 3.2 ± 0.3.

CONCLUSIONS

Although a great number of data are available on the relation between
nature of fat in diet and colon cancer, no clear conclusions can be drawn

from their analysis. The lack of consistent findings may have arisen from methodological differences between the studies.

To date available reports demonstrate a promoting effect of increasing dietary fat content on colon carcinogenesis. Nevertheless, a threshold may exist for that effect, an observation of first importance for human cancer prevention. Repeated sequential changes in dietary fat content dramatically reinforce the positive modulating effect of dietary fat an observation which is also of first importance for cancer prevention.

But since the consequence of such feeding protocols might be repeated changes in metabolic homeostasis of the animal, it is worth asking the question if the positive modulating effect of fat on colon carcinogenesis is a fact or not simply an artefact due to such changes in metabolism.

REFERENCES

1. B. S. Reddy, Nutritional aspects of colon cancer, in: "Progress in Food and Nutrition Science", B. S. Reddy and L. A. Cohen, eds., Pergamon Press Ltd (USA), vol. 9 (1986).
2. D. M. Jensen, Colon cancer epidemiology, in: "Experimental Colon Carcinogenesis", H. Autrup and G. M. Williams, eds., CRC Press, Boca Raton (1983).
3. M. B. Roberfroid, From normal cell to cancer an overview introducing the concept of modulation of carcinogenesis, in: "Concepts and Theories in Carcinogenesis", A. P. Maskens, P. Ebbesen, and A. Burny, eds., Elsevier Amsterdam (1986).
4. A. W. Bull, B. K. Soullier, P. S. Wilson, M. T. Hayden and N. D. Nigro, Promotion of azoxymethane-induced intestinal cancer by high fat diet in rats, Cancer Res. 39:4956 (1979).
5. K. M. Nauss, M. Lockniskar and P. M. Newberne, Effects of alterations in the quality and quantity of dietary fat on 1,2-dimethylhydrazine-induced colon tumorigenesis in rats, Cancer Res. 43:4083 (1983).
6. M. Wilpart, Dietary fats and fiber and experimental colon carcinogenesis: a critical review of published evidences, in: "Causation and Prevention of Colorectal Cancer", J. Faivre and M. J Hill, eds., Elsevier, Amsterdam (1987).
7. S. A. Broitman, J. Vitale, E. Varrousek-Jakula and L. S. Gottlich, Polyunsaturated fat: cholesterol and large bowel tumorigenesis, Cancer 40:2455 (1977).
8. K. M. Nauss, M. Lockniskar, D. Sondergaard and P. M. Newberne, Lack of effect of dietary fat on N-nitrosomethylurea induced colon tumori-genesis in rats, Carcinogenesis 5:255 (1984).
9. K. M. Watanabe, B. S. Reddy, C. Q. Wong and J. H. Weisburger, Effect of undegraded carrageenan on colon carcinogenesis in F 344 rats treated with azoxymethane or methylnitrosourea, Cancer Res. 38:4427 (1978).
10. B. S. Reddy, Dietary fat and cancer: specific action or caloric effect, J. Nutr. 116:1132 (1986).
11. N. D. Nigro, D. V. Singh, R. L. Campbell and M. S. Pack, Effects of dietary fat on intestinal tumour formation by azoxymethane in rats, J. Natl. Cancer Inst. 54:439 (1972).

626

SECTION V

MECHANISMS IN HUMAN CARCINOGENESIS

MECHANISMS OF ENDOGENOUS NITROSATION

Steven R. Tannenbaum

Massachusetts Institute of Technology
Department of Applied Biological Sciences
77 Massachusetts Avenue, 56-311 Cambridge
MA 02139, U.S.A.

INTRODUCTION

Substantial evidence has been assembled suggesting the possible health risk from exposure to N-nitroso compounds. Sander[1] originally proposed the concept of endogenous intragastric nitrosation, a process capable of generating carcinogenic N-nitroso compounds in humans. It has also been demonstrated that N-nitroso compounds are formed endogenously in animals given nitrite and a suitable amine[2,3]. However, the initial research on endogenous synthesis of N-nitroso compounds in humans was complicated by the presence of numerous artifacts of the analytical methods and the collection procedures. A method proposed by Ohshima and Bartsch[4] was the first potentially suitable procedure for estimating daily human exposure to endogenously formed N-nitroso compounds. The monitoring of urinary levels of N-nitrosoproline (NPRO) after dosing with nitrate and proline was utilized in a human volunteer without adverse biological effects. Since this publication, a number of other investigators have used this procedure to demonstrate the endogenous formation of nitrosoproline as well as other nitrosoamino acids in humans.

A summary of some mechanisms of nitrosation in biological systems is presented in Table 1. Endogenous nitrosation occurs via homogeneous chemical reaction or via cell mediated or catalyzed reaction. The source of nitrosating agent for homogeneous chemical reaction may be atmospheric NO_2 (lipophilic nitrosation) or nitrite formed in saliva or gastric juice. In the latter case, nitrosation will most probably be catalyzed by the presence of chloride and thiocyanate in the acidic environment of the stomach. In the case of NO_2, the reaction occurs indirectly through the preliminary interaction of NO with unsaturated lipids.

Cell mediated nitrosation may occur via bacteria or macrophages. Bacteria catalyze the nitrosation of amines via an enzyme catalyzed process in a neutral pH environment. Possible anatomical sites for this type of reaction include areas where bacteria are normally present in large numbers, e.g., the mouth or the intestinal tract, or areas where bacteria are present in large numbers due to a specific pathological condition, e.g., the hypochlorhydric stomach or the infected bladder.

Another category of cell-mediated nitrosation occurs in macrophages. When macrophages are elicited in vivo with thioglycolate and stimulated in

Table 1. Mechanisms of nitrosation in biological systems

Nitrosating agent	Site	Comments	Ref.
Nitrite	stomach	acid and thiocyanate catalyzed	5
NOX (gas)	lungs, skin	reacts via products with unsaturated lipids	32, 33
nitrite esters	all	phosphate catalyzed at neutral pH	34
Macrophages	possibly lung, intestine	requires infection and/or inflammation	19
Bacteria	oral cavity, GI tract	species specific	24, 25

vitro with E. coli lipopolysaccharide, copious amounts of nitrite and nitrate are formed. Amines present in the incubation medium are nitrosated by trapping a reactive nitrosating species prior to the formation of nitrite and nitrate. These results suggest that immunostimulated macrophages may be capable of nitrosamine formation under phsiological conditions and be related to the pathology of the inflammatory state.

The measurement of these various types of endogenous nitrosation in both experimental animals and in humans poses a challenge to the investigator because there is no suitable method for estimating the extent of cell-mediated nitrosation in the intact animal. In this paper I will briefly review the state of knowledge of different mechanisms of nitrosation in biological systems and their possible significance for cancer etiology.

Acid-Catalyzed Nitrosation

A general scheme for acid-catalyzed nitrosation is shown in Fig. 1. The relevance of this scheme for nitrosation of various classes of nitrogen compounds has been reviewed [5] , so this subject will only receive brief attention here.

The relative concentrations of the various nitrosating species depend on the pH and on the concentrations of halogens and pseudohalogens. In general, for a typical dialkylamine in a straightforward chemical situation, the maximum initial rate for nitrosation will occur at about pH 3.4, resulting from a combination of increased concentrations of nitrosating species, and decreased concentration of the free amine (the protonated amine does not react) as the pH is lowered. Catalysts such as thiocyanate generally shift the position of the pH maximum and cause the rate to fall off more slowly as the pH decreases. Ascorbic acid, under anaerobic conditions, can usually react with N_2O_3, $H_2NO_2^+$, and NOX with rates higher in each case than the corresponding nitrosation rates for amides or dialkyl amines. Ascorbic acid can therefore generally inhibit the in vitro nitrosation of these classes of compounds[6,7]. This property has been exploited to prevent nitrosamine formation in foods[8], and to some extent, to inhibit in vivo nitrosation[9,10,11].

Ascorbic acid thus appears to have potential importance as an in vivo nitrite scavenger. This potential, however, has yet to be completely realized; not only are the reactions that occur in live organisms more complex than might have been expected, but the in vitro model systems themselves, as noted earlier, are not straightforward. The equilibria shown in Figs. 1 and 2, for example, involve several nitrosating species that can react with

$$NO_2^- + H^+ \rightleftharpoons HNO_2$$

$$HNO_2 + H^+ \rightleftharpoons H_2NO_2^+ \rightleftharpoons H_2O + NO^+$$

$$2HNO_2 \rightleftharpoons N_2O_3 + H_2O$$

$$HNO_2 + HX \rightleftharpoons NOX + H_2O$$

Fig. 1. Chemistry of acid-catalyzed nitrosation

ascorbic acid to form NO, which does not nitrosate amines. Under anaerobic conditions these reactions can then exhaust the nitrosating capacity of the system. Oxygen, however, can react with NO to form N_2O_3 and N_2O_4, both of which are capable of nitrosation. Therefore, under the aerobic conditions of the stomach, ascorbic acid may be reduced in effectiveness as a nitrosation inhibitor [12,13,14].

Ohshima and Bartsch, in 1981, developed the "nitrosoproline test" for endogenous nitrosation, in which proline and/or nitrate can be given to humans and the resulting N-nitrosoproline can be detected in the urine[4]. N-nitrosoproline is not metabolized in humans and is consequently nontoxic and noncarcinogenic. Proline is a naturally occurring amino acid and can thus be administered safely to humans. Nitrate is nontoxic to humans, but can be reduced endogenously by bacteria to nitrite. Levels of urinary nitrosoproline are then assumed to be related to endogenous nitrosation, and experiments based on this test have been carried out in many laboratories. Although the results have often been difficult to interpret because of confounding factors, one immutable conclusion has emerged, i.e., virtually all humans tested thus far have been found to have N-nitrosoproline in their urine, and it is undoubtedly of endogenous origin.

In the original form of the test, and the one subject to the least number of variables, the system is swamped with nitrate and proline so that variations in the amount of nitrosoproline are dependent on other factors. Among the most important variables are the temporal relationships among the reactants and catalysts in the reaction compartment (presumably the stomach). The administered nitrate must be distributed throughout the body, taken up into saliva, converted to nitrite, and swallowed, prior to the reaction. Individual differences in rates of nitrate distribution, rates of conversion of nitrate to nitrite, and other factors such as rates of stomach emptying will be important determinants of nitrosoproline yield. These and additional parameters, e.g., concentrations of inhibitors and catalysts, will finally determine the overall amount of excreted N-nitrosoproline.

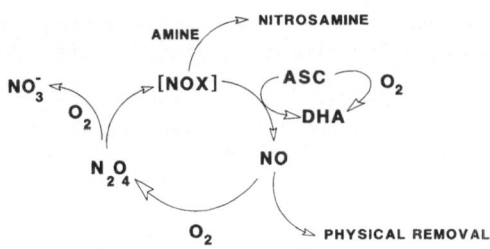

Fig. 2. Inhibition of nitrosation by ascorbic acid.

Despite the inherent complexity of the nitrosoproline test, it has been used successfully to demonstrate partial in vivo inhibition of endogenous nitrosation by ascorbic acid in humans. In some early experiments, ascorbic acid was simply administered prior to or concurrently with proline or proline and nitrate; the yield of N-nitrosoproline was lower in the presence of vitamin C [10]. In other experiments, the effect of ascorbic acid on baseline nitrosoproline levels was studied. In these cases it was found that nitroso-proline is apparently synthesized in at least two physiological compartments, one that utilizes an endogenous nitrosating agent and is unaffected by ascorbic acid, and another that can utilize exogenous nitrite and in which ascorbic acid can inhibit nitrosation. The latter compartment is probably the stomach; the nature of the former is unknown, but could conceivably be related to the macrophage-induced synthesis of nitrite and N-nitroso compounds which will be discussed in the next section.

Macrophages

In a recent series of papers, we have elaborated the mechanisms of nitrate biosynthesis in mammalian animals and cells[15,16]. As a result of the experiments in cell culture it became apparent that stimulated macrophages produced nitrite and nitrate in a constant ratio (approximately 3:2 M/M). In whole animal experiments nitrite could not be observed, because of immediate oxidation of nitrite to nitrate by oxyhemoglobin. However, it is reasonable to assume that 60 % of the nitrate formed by endogenous processes in vivo was originally formed as nitrite. This could amount to 300 μmol per day of nitrite in the average infection-free person[17] and up to 2 or more mmole per day in the case of viral or bacterial infection[18].

By any standard, this is a large quantity of nitrite. If this nitrite participated in nitrosation reactions prior to conversion to nitrate, even a small yield of N-nitroso compound would represent a significant addition to the existing body burden. In this connection, it is not unreasonable to speculate that the basal, nongastric synthesis of N-nitrosoproline in man is related to nitrite synthesis by macrophages. We have estimated this basal level to be on the order of 20-30 nmol/day[15], which would be approximately 10^{-4} of the amount of nitite.

Our results clearly show that stimulated macrophages, during synthesis of NO_2^- and NO_3^-, will also produce N-nitrosamines from a variety of secondary amines[19]. Also, and importantly, the formation of these nitrosamines does not appear to be due to an acid-catalyzed solution reaction, since nitrosamines are formed only when the secondary amine is added to the cell culture medium during the time that the cells are being stimulated. The control reactions produce only background levels of NMOR. In addition, when NO_2^- is added to the medium at a concentration similar to that expected at the end of a 48h synthesis, and MOR is added to non-stimulated cells, no formation of NMOR above background levels is observed. The implication then is that MOR or other amines are acting to trap out some intermediate in the pathway from the precursor to the final products, NO_2^- and NO_3^-.

Table 2 shows the results of nitrosation of various amines by the cell line RAW 264. What emerges from these studies is first that nitrosation of structurally diverse amines occurs and that there is a pK_a dependence in which the lower pK_a values correspond to higher yields of nitrosamine. This pK_a dependence is analogous to that for bacteria-mediated nitrosation[14] and for chemical nitrosation with NO_2^--derived nitrosating agents[8]. Other factors are certainly involved as evidenced by the the fact that diethylamine (DEA) and dibutylamine (DBA) which have the same pK_a differ by 5-fold in the amount of nitrosamine produced suggesting a lipophilic effect.

It is well known that leukocytes in general, and macrophages in particular are capable of actively concentrating ascorbic acid from plasma to intra-

632

Table 2.　N-Nitrosation of amines by RAW 264 [a]

Amine	Nitrite (μM)	Nitrosamines (nM)	pKa
DEA	57	4	10.7
DBA	77	23	10.7
MBA	60	255	9.5
MOR	60	1680	8.7

[a] Incubations were carried out for 72 h and contained the listed amine (5 mM), LPS (10 μg/ml) and IFN (500 U/ml).

cellular concentrations in the mM range [20].　The question we asked was whether ascorbic acid played a role in the biosynthesis of nitrite and nitrate and the concomittant formation of nitrosamines.

The experiments were conducted with RAW 24 macrophages as described in our published work [19].　These cells normally grow in the complete absence of ascorbic acid, but actively concentrate ascorbate to the levels shown in Table 3.　Experiments with a wide range of ascorbate concentrations in the medium revealed that formation of $NO_2^- + NO_3^-$ was enhanced throughout the range up to and beyond normal levels in human serum.　What was most interesting was that inhibition of nitrosamine formation is found at concentrations of ascorbate that correspond to normal physiological levels.　Therefore, high intake of ascorbic acid enhances NO_2^- formation but inhibits nitrosamine formation, a condition that best fits the needs of the host.

From an etiological point of view it is not immediately obvious what the relevance of the synthesis of nitrosamines by macrophages is to human cancer. However, a long-standing problem in cancer etiology is the role of infection and inflammation.　I have no intention of reviewing this vast field of literature, and the effects of inflammation certainly include the critical steps of cell proliferation in the process of carcinogenesis.　However, generation of nitrite and nitrosamines by macrophages may also play a role in special cases.　For example, coeliac disease carries a higher (80-100 fold) risk of non-Hodgkin's lymphoma compared to the general population [21].　Inflammatory bowel disease produces detectable colonic nitrite as has been shown by rectal dialysis experiments [22].　I would like to conclude this discussion by speculating that there is a connection between these observations.　We are currently trying to devise a test of this hypothesis.

Table 3. Ascorbate levels in RAW 24 macrophages

Ascorbate in medium, μM	Ascorbate in cell, mM
5	0.5
50	2.9
120	3.2
200	5.0

Bacteria

The concept that bacteria might participate in nitrosation of amines (i.e. formation of an NNO bond to yield the N-nitroso derivative, NOX) has been considered to some extent in most hypotheses concerning the possible role of endogenous nitrosation in cancer etiology[23-27]. As recently as 1981, it was felt that the major contribution of bacteria to nitrosation was the reduction of nitrate to nitrite, resulting in an increase in the concentrations of nitrosating species[28]. More recently, however, several groups have shown that bacteria can in fact participate directly in the nitrosation of amines, and this area has consequently gained renewed interest[24-25].

A number of bacterial genera, including Neisseria, Pseudomonas, Escherichia, Klebsiella, Proteus, Alcaligenes and Bacillus, have been investigated from several experimental perspectives and reported to have some ability to catalyze nitrosation[24,25,29,30]. This catalysis is generally believed to be an anaerobic process which occurs in intact resting cells. The mechanism of nitrosation of amines by bacteria is currently under investigation in several laboratories, and may be related to the process of denitrification. While the potential for this process has been demonstrated _in vitro_ there is as yet no conclusive demonstration of its occurrence _in vivo_. Therefore I will briefly discuss some possible sites and consequences of this mechanism of nitrosation.

Approximately ten years ago we demonstrated that nitrosamines can form in saliva at neutral pH[31]. This strongly suggests that nitrosamine formation can proceed in the presence of the proper strains of bacteria. Calmels and coworkers[24] have shown that the rate of nitrosation is proportional to the concentrations of amine, nitrite and bacteria in the following manner: direclty proportional to concentration of bacteria,, double-reciprocal for nitrite and amine. The apparent K_m values for amine and nitrite were in the mM range. Under actual concentrations in the body, which would be μM the kinetics would therefore be first-order in amine, nitrite and bacteria.

Anatomical compartments which are likely to contain bacteria, amines and nitrite at concentrations yielding ng (or greater) quantities of nitrosamines include the oral cavity, hypochlorhydric stomach, intestinal tract and the infected bladder. While each of these sites has the potential for measurable rates of nitrosamine synthesis we do not necessarily have a good idea of which compounds to look for. Furthermore, nitrosamines synthesized in any particular compartment may be organ-specific for any tissue, including or excluding the site of synthesis.

In the light of all of this uncertainty, bacterial nitrosation remains a mechanism in search of a disease, and we must develop methods suitable for estimating amounts and types of compounds synthesized via this pathway.

CONCLUSION

The ultimate question remains whether or not endogenous nitrosation represents a human health risk and whether or not this risk can be significantly lowered. There is currently no direct answer to either aspect of this question. The uncertainty arises largely from the facts that human exposure to nitrosamines, whether endogenous of exogenous, generally involves chronic low-dose situations and is superimposed on countless other factors (e.g., exposure to other types of toxic substances or to modifiers of metabolism) that may enhance, inhibit, or mask the effects of the N-nitroso compounds. N-nitroso compounds have, nonetheless, been implicated in several human epidemiological situations, including elevated risk toward gastric cancer in some well-defined geographical areas. Fig. 3 shows schematically some of the characteristics of gastric cancer, along with some etiological hypotheses, based on available epidemiological and biochemical evidence. In virtually all cases, a well-defined progressive change in stomach physiology and morphology from chronic gastritis through atrophic gastric to apparently precancerous intestinal metaplasia is observed. As the stomach begins increasingly to resemble the intestine, the pH rises, leading to conditions that support or favor increased bacterial growth. This in turn leads to higher levels of gastric nitrate from reduction of nitrate by some of these bacteria. This could conceivably result in increased formation of N-nitroso compounds from amines present in the stomach either via the diet or from endogenous synthesis. If this is indeed the case, then it is also conceivable that the process could be interrupted at this point by the ingestion of ascorbic acid.

As noted earlier in this report, however, the interactions of ascorbic acid with nitrosating agents in the stomachs are far from straightforward, and the clinical effectiveness of vitamin C in reducing a real human cancer risk still remains to be demonstrated.

With regard to cell-mediated nitrosation, it remains to be demonstrated that this is an important mechanism for endogenous formation of N-nitroso compounds. If this mechanism can be demonstrated in animals (and possibly in humans) then attempts can be made to link this mechanism with the etiology of a carcinogenic process. Methods for inhibiting cell-mediated nitrosation must be developed, and then tested in vivo.

Therefore there is much work remaining in the investigation of whether endogenous nitrosation plays a role in human disease.

ACKNOWLEDGEMENTS

This investigation was supported by PHS grant number CA26731, awarded by the National Cancer Institute, DHHS and by the American Meat Institute.

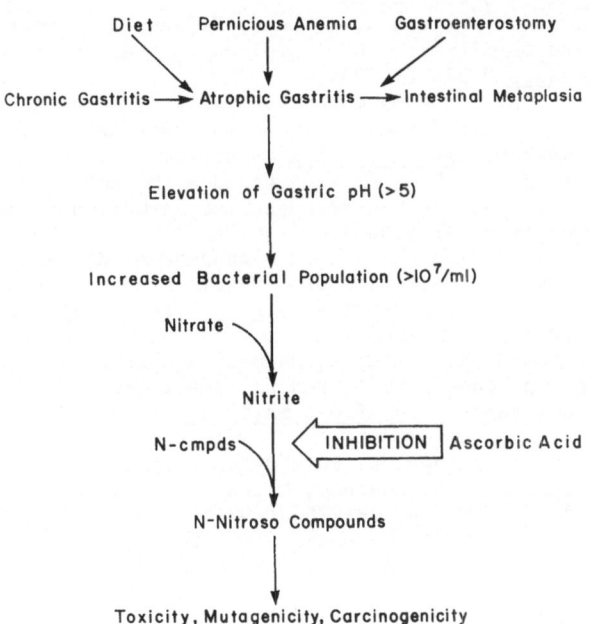

Fig. 3. Model for gastric cancer etiology.

REFERENCES

1. J. Sander, Kann nittit in der menschlichen Nahrung Ursache einer Krebsentstehung durch Nitrosaminbildung sein? Arch. Hyg. Bakteriol. 151:22 (1967).

2. M. Greenblatt and S. S. Mirvish, Dose-response studies with concurrent administration of piperazine and sodium nitrite to strain A mice, J. Natl. Cancer Inst. 50:119 (1973).

3. W. Lijinsky and H. W. Taylor, Feeding tests in rats on mixtures of nitrite with secondary and tertiary amines of environmental significance, Food Cosmet. Toxicol. 34:255 (1977).

4. H. Ohshima and H. Bartsch, Quantitative estimation of endogenous nitrosation in humans by monitoring N-nitrosoproline excreted in the urine, Cancer Res. 41:3658 (1981).

5. S. S. Mirvish, Formation of N-nitroso compounds: chemistry, kinetics and in vivo occurrence, Toxicol. Appl. Pharmacol. 31:1325 (1975).

6. S. S. Mirvish, L. Wallcave, M. Eagen and P. Shubik, Ascorbate-nitrite reaction: Possible means of blocking the formation of carcinogenic N-nitroso compounds, Science 177:65 (1972).

7. M. C. Archer, S. R. Tannenbaum, T. Y. Fan and M Weisman, The reactions of nitrite with ascorbate and its relation to nitrosamine formation, J. Natl. Cancer Inst. 154:1203 (1975).

8. W. Fiddler, J. W. Pensabene, E. G. Piotrowski, R. C. Doerr and A. E. Wasserman, Use of sodium ascorbate or erythorbate to inhibit formation of N-nitrosodimethlyamine in frankfurters, J. Food Sci. 38:1084 (1973).

9. H. Bartsch and R. Montesano, Relevance of nitrosamines to human cancer, Carcinogenesis 5:1381 (1984).

10. D. A. Wagner, D. E. G. Shuker, C. Bilmazes, M. Obiedzinski, I. Baker, V. R. Young and S. R. Tannenbaum, Effect of vitamins C and E on endogenous synthesis of N-nitrosoamino acids in humans: precursor-product studies with ^{15}N-nitrate, Cancer Res. 45:6519 (1985).

11. C. Lintas, J. Fox, S. R. Tannenbaum and P. M. Newberne, In vivo stability of nitrite and nitrosamine formation in the dog stomach. Effect of nitrite and amine concentration and of ascorbic acid, Carcinogenesis 3:161 (1982).

12. W. R. Licht, J. G. Fox and W. M. Deen, Effects of ascorbic acid and thiocyanate on nitrosation of proline in the dog stomach, Carcinogenesis 9:373 (1988).

13. W. R. Licht, S. R. Tannenbaum and W. M. Deen, Use of ascorbic acid to inhibit nitrosation: Kinetic and Mass transfer considerations for an in vitro system, Carcinogenesis in press.

14. C. D. Leaf, A. J. Vecchio, D. A. Roe and W. M. Hotchkiss, Influence of ascorbic acid dose on N-nitrosoproline formation in humans, Carcinogenesis 8:791 (1987).

15. D. A. Wagner, V. R. Young and S. R. Tannenbaum, Mammalian nitrate biosynthesis: incorporation of [^{15}N]-ammonium into nitrate is enhanced by endotoxin treatment, Proc. Natl. Acad. Sci. U.S.A. 80:4518 (1983).

16. D. J. Stuehr and M. A. Marletta, Mammalian nitrate biosynthesis: mouse macrophages produce nitrite and nitrate in response to Escherichia coli lipopolysaccharide, Proc. Natl. Acad. Sci. U.S.A. 82:7738 (1985).

17. L. C. Green, D. A. Wagner, K. Ruiz deLuzuriaga, W. Rand, N. Istfan, V. R. Young and S. R. Tannenbaum, Nitrate biosynthesis in man, Proc. Natl. Acad. Sci. U.S.A. 78:7764 (1981).

18. D. A. Wagner and S. R. Tannenbaum, Enhancement of nitrate biosynthesis by E. coli lipopolysaccharide, in: "Nitrosamines and Human Cancer", P. N. Magee, ed., Banbury Report No. 12, Cold Spring Harbor Labs, New York (1982).

19. M. Miwa, D. J. Stuehr, M. A. Marletta, J. S. Wishnok and S. R.

Tannenbaum, Nitrosation of amines by stimulated macrophages, Carcinogenesis 8:955 (1987).

20. U. Moser, Uptake of ascorbic acid by leukocytes, Ann. N.Y. Acad. Sci. 498:200 (1987).

21. J. R. G. Nash, E. Gradwell and D. W. Day, Large-cell intestinal lymphoma occurring in coeliac disease: morphological and immuno-histochemical features, Histopathology 10:195 (1986).

22. W. E. W. Roediger, M. J. Lawson, S. H. Nance and B. C. Radcliffe, Detectable colonic nitrite levels in inflammatory bowel disease - mucosal or bacterial malfunction?, Digestion 35:199 (1986).

23. R. Bockler, H. Meyer and P. Schlag, An experimental study on bacterial colonization, nitrite and nitrosamine production in the operated stomach, J. Cancer Res. Clin. Oncol. 105:62 (1983).

24. S. Calmels, H. Ohshima, P. Vincent, A. M. Gounot and H. Bartsch, Screening of microorganisms for nitrosation catalysis at pH 7 and kinetic studies on nitrosamine formation from secondary amines by E. coli strains, Carcinogenesis 6:911 (1985).

25. S. Calmels, H. Ohshima, H. Rosenkranz, E. McCoy and H. Bartsch, Bio-chemical studies on catalysis of nitrosation by bacteria, Carcinogenesis 8:1085 (1987).

26. G. DeBernadis, S. Guadagni, M. A. Pistoria, G. Amicucci, C. Masci, C. Herfath, A. Agnifili and M. Carboni, Gastric juice, nitrite and bac-teria in gastroduodenal disease and resected stomach, Tumori 96:231 (1983).

27. P. I. Reed, P. L. R. Smith, K. Haines, F. R. House and C. L. Walters, Gastric juice N-nitrosamines in health and gastrointestinal disease, Lancet II:550 (1981).

28. D. Ralt and S. R. Tannenbaum, Role of bacteria in nitrosamine forma-tion, N-nitroso compounds, Adv. Chem. 174:159 (1981).

29. S. A. Leach, A. R. Cook, B. C. Challis, M. J. Hill and M. H. Thompson, Bacterially mediated N-nitrosation reactions and endogenous forma-tion of N-nitroso compounds, Proc. of the Ninth International Meeting on N-Nitrosamines, Baden, Austria, September, 1986 (1987).

30. H. Ohshima, S. Calmels, B. Pignatelli, P. Vincent and H. Bartsch, N-Nitrosamine formation in urinary tract infections, Proc. of the Ninth International Meeting on N-nitrosamines, Baden, Austria, September, 1986 (1987).

31. S. R. Tannenbaum, M. C. Archer, J. S. Wishnok and W. W. Bishop, Nitro-samine formation in saliva, J. Natl. Cancer Inst. 60:251 (1978).

32. S. S. Mirvish, J. P. Sams and P. Issenberg, The nitrosating agent in mice exposed to nitrogen dioxide: improved extraction method and localization in the skin, Cancer Res. 43:2550 (1983).

33. H. D. Ross, J. Henion, J. G. Babish and J. H. Hotchkiss, Nitrosating agents from the reaction between methyl oleate and dinitrogen trioxide: Identification and mutagenicity, Food Chemistry 23:207 (1987).

34. R. Dabora, M. Molina, V. Ng, J. S. Wishnok and S. R. Tannenbaum, in: "N-Nitroso Compounds: Occurrence, Biological Effects and Relevance to Human Cancer", I. K. O'Neill, R. C. vonBorstel, C. T. Miller, J. Long and H. Bartsch, eds., IARC Scientific Publication No. 57, International Agency for Research on Cancer: Lyon, France (1984).

THE ROLE OF N-NITROSO COMPOUNDS IN HUMAN CANCER

William Lijinsky

NCI-Frederick Cancer Research Facility
BRI-Basic Research Program
Frederick, MD 21701

INTRODUCTION

Only a small proportion of human cancers appear to be caused by exposure to carcinogenic agents of known nature, for example radiation and industrial chemicals. For the remainder, it is probable that no single cause is prominent, but that combinations of many carcinogenic agents, probably mainly at low concentrations, are responsible. Apart from the use of tobacco, which seems to be related to several types of cancer, especially lung cancer, there are no reasonable explanations for the most common types of cancer, including stomach, liver (in the non-industrial world), colon, nervous system, breast, uterus and cervix, prostate and pancreas; esophagus and bladder cancer are also unknown in this regard, but they seem to be particularly common in certain locations and among certain groups.

A lot of information has been obtained about a few types of carcinogen, such as polycyclic aromatic compounds and aromatic amines, but these are probably not involved in most of the common types of human cancer, the exception being the skin cancer which is related to exposure of workers to coal tar and petroleum products. While the carcinogenic hydrocarbons are very potent inducers of skin cancer, in animals or humans, they are very weakly carcinogenic for other types of cancer, such as lung, and they have not been shown convincingly to induce cancer of other organs. Aromatic amines have induced tumors of the bladder and some other organs, in a quite species-specific manner, but large doses are required to induce cancer. Nevertheless, they are probable contributors to human cancer risk, but the types of cancer are not known. Aflatoxins and other carcinogenic mycotoxins almost certainly play a role in cancer in the non-industrial countries, and types of cooking seemingly related to particular types of cancer prevalent in certain communities, have been related to polyheterocyclic aromatic amines formed by pyrolysis. Again, the carcinogenic potencies of these compounds seems too low to consider them sole inducers of these cancers. Instead it seems likely that they contribute initiating activity, but the action of other substances is needed for cancer formation and progression.

Experimental studies of many types of carcinogen have been extensive during the past 50 years or so, and many exotic compounds have been found to possess carcinogenic activity. This is most interesting intellectually and provides fascinating studies of mechanisms of cancer induction, but the practical relevance of these compounds in understanding human cancer is small.

639

N-Nitroso compounds are exceptional carcinogens in that they are easily formed, are water soluble and soluble in organic solvents, so that they diffuse readily through the skin or other epithelia, and many of them are volatile. The precursors of nitrosamines are amines and nitrosating agents. The amines can be secondary or tertiary, the latter usually reacting more slowly than the former. Nitrosating agents include inorganic or organic nitrites, nitrogen oxides and some N-nitroso compounds; formation of nitrosamines from the latter is often called transnitrosation. The precursors of N-nitroso compounds are very common, both amines and nitrosating agents, so that human exposure to N-nitroso compounds is not rare, particularly at low concentrations.

Because nitrites are used to preserve and color meats and fish, nitrosamines are usually present in such cured foods. The concentrations vary from a few parts per billion to approaching 100 parts per billion, in such foods as sausages and cooked bacon. Cooked bacon contains nitrosopyrrolidine formed by decarboxylation of nitrosoproline or by other processes that have been suggested[1]. The most common nitrosamine in the environment, nitrosodimethylamine, is also found in beer and in some distilled spirits, the concentrations being higher in some than others. It seems that the source of this nitrosamine is reaction of certain alkloids, including the tertiary amines gramine and hordenine, in the grain with nitrogen oxides in the burning fuel used to heat it[2]. The nitrosamine concentration in the beer and spirits ranged from a few parts per billion to 50 parts per billion. The higher figures in the beer have been reduced considerably by changing manufacturing procedures. Before the changes were made, it is likely that large numbers of people had considerable exposure to nitrosodimethylamine in beer (much more so than in spirits) because beer is drunk in such large quantities. That might have been one of the major sources of exposure of humans to nitrosamines. There are lesser quantities of nitrosamines in cheese and in smoked fish.

Although many of these exposures are small, because the concentrations of nitrosamines are low, they tend to be continuous over a major portion of the life of many humans, so that the cumulative dose can be appreciable. Even small doses of nitrosamines can be important since they are among the most potent carcinogens known, and they are effective in all species in which they have been tested. For example, in Table 1. are listed several nitrosamines which have been evaluated in dose-response experiments, together with the lowest concentrations and the smallest cumulative doses which induced a significant tumor incidence.

There are often considerable differences between species in the types of tumor induced by particular N-nitroso compounds, so that it can be quite incorrect to assume that an organ or cell type susceptible in the rat or hamster to a particular carcinogenic N-nitroso compound would also be susceptible in the human. There seem to be characteristic tumors induced in a particular species by N-nitroso compounds, for example, rats and hamsters (Table 2). While this makes epidemiological studies of the response to exposure to nitrosamines difficult, it also suggests that any exposure to a N-nitroso compound is likely to increase the overall cancer risk of people exposed. Several nitrosamines have induced significant incidences of tumors in animals such as rats after administration of doses of 2 to 5 milligrams per kilogram body weight during the 2 year lifetime of those species, which would translate to approximately 5 milligrams per year in a human to provide the same risk, assuming that humans are no more susceptible than rats or other laboratory animal. For example, the incidence of liver tumors in female rats given nitrosomorpholine in drinking water increases from 6% at the lowest dose, 3 mg/kg, to 50% at 100 mg/kg (Fig. 1).

The most likely sub-group of N-nitroso compounds contributing to human

Table 1. Lowest effective doses of nitrosamines in rats

Compound	Concentration in water (ppm)	Total dose mg/kg	% rats with tumors
Nitrosodiethylamine	0.45	5	40
	0.45	20	90
Nitrosoheptamethyleneimine	1	25	50
Nitrosomethylphenylethyl-amine	0.4	10	40
Nitroso-1,2,3,6-tetrahydro-pyridine	1	10	50
Dinitrosohomopiperazine	1	33	20
Nitrosomethyl-n-butylamine	6.3	30	100
Nitrosomethylcyclohexylamine	5	50	100
Nitrosomorpholine	0.07	3	6
Nitrosodiethanolamine	28	1,100	26

exposure are the nitrosamines, which require metabolic activation. There is no good evidence of human exposure to nitrosoalkylureas or other nitrosamides, which are directly acting carcinogens and mutagens, and it is difficult to propose likely sources of such compounds. However, there are many well established sources of nitrosamines. The number of nitrosamines that have been found in the environment is quite small, especially in comparison with the approximately 300 compounds in this category that have been tested for car-

Table 2. Target organs of N-nitroso compounds

Rats (n = 158)		Hamsters (n = 51)	
Liver	41%[a]	Forestomach	53%
Esophagus	39%	Liver	51%
Forestomach	30%	Nasal Mucosa	43%
Nasal Mucosa	30%	Lung	29%
Lung	28%	Pancreas	27%
Tongue	19%	Spleen	20%
Nervous System	9%	Trachea	18%
Bladder	9%	Uterus	8%
Mammary	8%	Bladder	6%
Uterus	8%		
Colon	8%		
Trachea	7%		
Mesothelioma	7%		
Thymus Lymphoma	6%		
Kidney	6%		
Thyroid	5%		
Intestine	5%		
Zymbal Gland	5%		
Duodenum	3%		
Glandular Stomach	2%		

[a]Proportion of compounds tested inducing tumors in these organs.

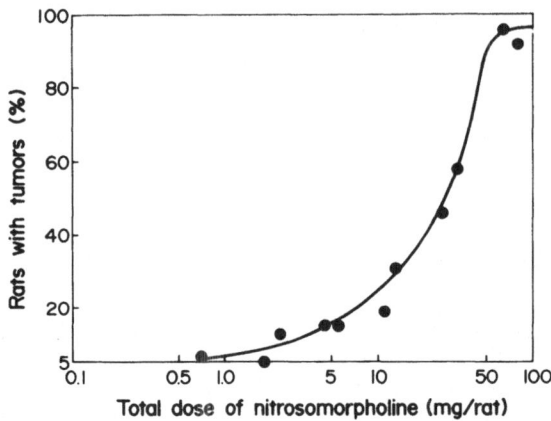

Fig. 1. Increase in incidence of treatment-related tumors in female rats
receiving increased doses of nitrosomorpholine in drinking water.

cinogenic activity, of which perhaps 85% have been carcinogenic[3]. Most of
those relevant to human exposure are among the most potent of carcinogenic
nitrosamines, although the remainder show a considerable range of carcinogenic activity, and induce in animals a variety of tumors, including some of the
common human tumors.

Apart from the occurrence of nitrosamines in some foods, there are other
major sources of human exposure to nitrosamines. Principal among these are
tobacco and tobacco products, including tobacco smoke, both active and pas-
sive[4]. Concentrations of the several nitrosamines typical of tobacco are
many orders of magnitude greater than those found in food, and the former
might be responsible in major part for the several human cancers that have
been associated with use of tobacco, such as lung, oral cavity, bladder and
perhaps others. The source of most of the nitrosamines in tobacco seems to
be nicotine, although nitrosodimethylamine and nitrosopyrrolidine are also
present. Nitrosamines derived from nicotine are nitrosonornicotine and NNK
(nitrosomethylamino-3-pyridyl-butanone), and these are commonly present in
tobacco smoke[4], as well as in tobacco itself, which is frequently used as
snuff or chewing tobacco. Other nitrosamines analogous to nitrosonornicotine
are the piperidine derivatives nitrosoanabasine and nitrosoanatabine, but
these are weak carcinogens or non-carcinogenic.

Other nitrosamines associated with tobacco, although not derived from
tobacco alkaloids, are nitrosodiethanolamine and nitrosomorpholine, the
latter arising from the packaging of smokeless tobacco, the former from the
use of tobacco on which agricultural chemicals containing diethanolamine
salts had been applied. NNK and nitrosomorpholine are particularly potent
carcinogens, inducing a variety of tumors in rats, including lung tumors in
the case of NNK[5]. Nitrosonornicotine is somewhat less potent than the two
mentioned, and appears only to induce tumors of the esophagus and nasal
mucosa in rats, but it is an important carcinogenic constituent of tobacco
and tobacco smoke because it is present in large concentrations. Through
these carcinogenic constituents the contribution of tobacco use to the
occurrence of many kinds of cancer in humans must be considered significant.
Since it was found that unburnt, but processed, tobacco contains high concentrations of several carcinogenic nitrosamines it becomes clearer why the oral
cancer which is quite common in India and other areas is associated with the
habit of keeping a mixture of tobacco and lime in the mouth; in these places

smoking of tobacco is not common. The lime facilitates the release of nitrosonornicotine and NNK (and other tobacco nitrosamines) which are bases.

Another major source of human exposure to N-nitroso compounds at what must be considered high concentrations is various industrial processes and particular types of factory. Several surveys have shown that certain nitrosamines occur commonly in rubber factories[6], leather works and machine shops. The latter, through the use of modern synthetic cutting oils, might represent the greatest exposure of humans to nitrosamines, since nitrosodiethanolamine is a common and large constituent of these cutting oils. Concentrations of nitrosodiethanolamine as high as 3% have been found, although most cutting oils contain substantially lower concentrations. The source of the nitrosodiethanolamine in cutting oils, like that in various cosmetic preparations, is the emulsifier triethanolamine, which invariably contains the easily nitrosated diethanolamine as a component to react with the nitrite salts; triethanolamine also reacts with nitrites[7]. The concentrations in the cosmetics are very much smaller than in cutting oils. Smaller concentrations of several other carcinogenic nitrosamines have been found in cutting oils and in toilet preparations, including nitrosobis-(2-hydroxypropyl)amine[8] (derived from triisopropanolamine), nitrosomethyl-n-tetradecylamine and nitrosomethyl-n-dodecylamine[9]. The last two compounds have induced bladder tumors in rats[10].

Nitrosodiethanolamine and other nitrosamines readily penetrate skin[11], and nitrosamines, like many other types of carcinogen, act systemically to induce tumors in specific target organs, regardless of the route of administration. There are differences in response to carcinogenic nitrosamines between different dose rates, which makes prediction of target organs in man virtually impossible. The common nitrosamine nitrosodimethylamine, for example, induces only tumors of the liver when given to rats in drinking water at a wide range of concentrations. However, when small doses are given by gavage twice a week to rats tumors of the lung, kidney and nasal mucosa are induced, as well as liver tumors; only 50% of the rats had liver tumors in that experiment. In Syrian hamsters, on the other hand, nitrosodimethylamine given by gavage induces 100% incidence of liver tumors, but not tumors of lung or kidney and few of the nasal mucosa[12]. This pattern is not common with other types of carcinogen.

Exposure to nitrosamines in factories in which rubber is manufactured or processed can be considerable, and there are a number of nitrosamines of concern. Not only are many nitrosatable amines, both secondary and tertiary, used in rubber manufacture, but the unstable nitrosamine, nitrosodiphenylamine, is also used and is a good nitrosating agent, in addition to being a weak bladder carcinogen itself[13]. There is, of course, exposure to nitrosodiphenylamine in those factories in which it is used. But of far greater concern is the reaction of nitrosodiphenylamine with a number of derivatives of morpholine, dimethylamine, diethylamine and di-n-butylamine to form the N-nitroso derivatives of those amines. Nitrogen oxides are also plentiful in rubber factories and these "nitrous gases" can also form nitrosamines from the accelerators, a number of which are tertiary amines[14]. So much nitrosamine has been found in some rubber factories that they occur as deposits on floors and other surfaces. It is an inescapable consequence of the process of rubber manufacture that, not only are workers in the factories exposed to nitrosamines, but nitrosamines are present in rubber products such as tires or nipples for baby bottles, two products about which concern has been raised in the past few years. Even though the exposure of the general population to nitrosamines from this source is relatively small, because of the low concentrations, the potency of the carcinogenic action of nitrosamines makes such exposures important.

Another source of exposure to N-nitroso compounds, which might be as important as any other, is due to the chemistry of formation of N-nitroso compounds. They are formed, as has been known for more than a century, by reaction of secondary or tertiary amines with nitrosating agents, which include nitrogen oxides and nitrites, particularly inorganic nitrites. As has been stated earlier, sodium nitrite has long been used for preserving and curing meat and fish. Saltpeter (potassium nitrate) - used since antiquity for this purpose - is effective because the nitrate is reduced bacterially or chemically to nitrite, the active agent. Similar bacterial reduction of nitrate in saliva to nitrite occurs in the mouth, resulting in a continuous flow of nitrite at low concentrations into the stomach of people. These reactions are important sources of human exposure to nitrosamines because of the large variety of nitrosatable amines ingested by humans as drugs and medicines[15], residues of agricultural chemicals[16] and components of food itself. Nitroso derivatives of several insecticides are potent carcinogens[17]. There is always nitrate in body fluids derived from plants in the diet or, as recently shown by Marletta and others, formed _in vivo_ by oxidation in macrophages[18]. For formation of N-nitroso compounds, nitrite in food is more effective than nitrite in saliva because the concentration provided by the former is greater and more heterogeneous; the rate of nitrosation is proportional to the square of the nitrite concentration. Both sources of nitrite are important, however. The rate of reaction of most tertiary amines with nitrous acid is much slower than that of secondary amines[19], although yields of nitrosamines can be appreciable if reaction times are long, as they can be in the stomach, where the essential acid environment is present to facilitate the reaction. The presence or absence of accelerators of nitrosation, such as carbonyl compounds and thiocyanate or halide ions, and of inhibitors of nitrosation, such as ascorbic acid or tocopherols, modulate the yields of nitrosamines from these reactions. However, the effects of the complex mixtures that might be present at any time in the stomach of a person are very difficult to model, so that any estimates of probable nitrosamine yields can be wildly inaccurate.

One way out of the dilemma is the method used by Ohshima and Bartsch[20] to measure the formation of a model nitrosamine, nitrosoproline, after administration of a large dose of proline. Proline is readily nitrosated and nitrosoproline is excreted in the urine and can be measured; nitrosoproline is not carcinogenic and is not, as far as is known, metabolized. Unfortunately, there are two impediments to the acceptance of such a procedure as an accurate measure of nitrosation potential. One is that the yield of nitrosoproline that is measured is exceedingly small; the second is that there is a back-ground level of nitrosoproline excretion that is variable, and that seems not to be inhibited by administration of ascorbic acid, whereas the quantity of newly formed nitrosoproline is reduced by ascorbic acid administration. It is possible that the recently studied formation and excretion of nitrosothioproline will provide a better measure of nitrosating potential[21], since the rate of nitrosation of thioproline is much greater than that of proline, and the yields are higher.

There is considerable experimental evidence that formation of nitrosamines _in vivo_ can be a significant cause of tumor induction, apart from the chemical demonstration that the reactions can take place. A variety of secondary and tertiary amines has been administered to experimental animals together with nitrite, and formation of the expected nitrosamine has been measured in the gastric juice. Rats, rabbits, Guinea pigs and dogs have been used, as well as humans. In addition, chronic feeding of mixtures of amines and nitrite to rats and mice has led to induction of tumors expected to arise from the particular nitrosamine that is formed. The amines that have given positive results in such experiments are listed in Table 3, together with the nitrosamine expected and the type of tumor induced. Many other nitrosatable amines have given marginal or null tumor effects in these experiments and it must be assumed that, under the conditions used, insufficient nitroso compound was produced to induce tumors within the lifetime of the animal, although it can be assumed that formation of the N-nitroso compound predicted

Table 3. Carcinogenicity of amines fed with nitrite to rats

	Amine Concentration (ppm)	Nitrite Concentration (ppm)	Nitrosamine Formed	% Tumors Induced
Allantoin	2000	2000	?	Forestomach 30
Aminopyrine	250	250	Nitrosodimethylamine	Liver 97
Chlorpheniramine	1000	2000	Nitrosodimethylamine	Liver 58
Dimethyldodecyla-mine-n-oxide	1000	2000	Nitrosodimethylamine	Liver 42
Diphenhydramine	2000	2000	Nitrosodimethylamine	Liver 46
Disulfiram	1000	2000	Nitrosodiethylamine	Esophagus 55
Heptamethylene-imine	2000	2000	Nitrosoheptamethyl-eneimine	Lung 53 Esophagus 77
Methylbenzylamine	500	500	Nitrosomethylbenzyl-amine	Esophagus 100
Morpholine	500	500	Nitrosomorpholine	Liver 100
Oxytetracycline	1000	1000	Nitrosodimethylamine	Liver 20
Thiram	500	2000	Nitrosodimethylamine	Nasal Mucosa 75
Tolbutamide	1000	2000	?	Forestomach 25 Liver 25

from our chemical knowledge did occur. However, the differences in response in these experiments between one amine and another does enable us to rate the risk of ingestion of one amine compared with another. For example, the large tumorigenic response of rats to feeding of the tertiary amine aminopyrine (pyramidon)[22] illustrates a much greater risk to humans than is posed by the tertiary amine chlorpromazine or the secondary amine chlordiazepoxide, which did not produce a significant tumorigenic response when fed to rats together with nitrite[23].

In conclusion, because of the great carcinogenic potency of many N-nitroso compounds, and their broad carcinogenic activity across many species and organs, the exposure of humans to these compounds, exogenously or endogenously formed, must be considered to contribute to our carcinogenic risk. Even though the concentrations are often small, they are quite variable from one group of people to another, depending on diet, habits and occupation, and can be continuous over a lifetime of 50 years or more. Although no specific human cancer has yet been associated with exposure to N-nitroso compounds, other than oral cancer in users of smokeless tobacco[24] it is precisely the fact that exposure to them at some level is almost universal that inhibits the application of the usual epidemiological procedures to this problem. It is important that we attempt to develop a more accurate method of assessing the seriousness of their potential for adverse health effects.

ACKNOWLEDGEMENTS

Research sponsored by the National Cancer Intitute, DHHS, under contract No. NO1-CO-74101 with Bionetics Research, Inc. The contents of this publication do not necessarily reflect the views or policies of the Department of Health and Human Services, nor does mention of trade names, commercial products, or organizations imply endorsement by the U.S. Government.

REFERENCES

1. E. T. Huxel, R. A. Scanlan and L. M. Libbey, Formation of N-nitrosopyrrolidine from pyrrolidine ring compounds at elevated temperatures, J. Agric. Food Chem. 22:698 (1974).
2. M. M. Mangino and R. A. Scanlan, Rapid formation of N-nitrosodimethylamine from gramine, a naturally occurring precursor in barley malt, in: "N-Nitroso Compounds: Occurrence, Biological Effects and Relevance to Human Cancer", IARC Scientific Publications, No. 57 (1984).
3. R. Preussmann and B. W. Steward, N-Nitroso carcinogens, in: "Chemical Carcinogens", C. E. Searle, ed., American Chemical Society Monograph, No. 182 (1984).
4. D. Hoffmann, K. D. Brunnemann, J. D. Adams and S. S. Hecht, Formation and analysis of N-nitrosamines in tobacco products and their endogenous formation in consumers, in: "N-Nitroso Compounds: Occurrence, Biological Effects and Relevance to Human Cancer", IARC Scientific Publications, No. 57 (1984).
5. S. S. Hecht, A. Castonguay, F. L. Chung and D. Hoffmann, Carcinogenicity and metabolic activation of tobacco-specific nitrosamines: current status and future prospects, in: "N-Nitroso Compounds: Occurrence, Biological Effects and Relevance to Human Cancer", IARC Scientific Publications, No. 57 (1984).
6. J. M. Fajen, D. P. Rounbehler and D. H. Fine, Summary report on N-ni-

trosamines in the factory environment, in: "N-Nitroso Compounds: Occurrence and Biological Effects, IARC Scientific Publications, No. 41 (1982).

7. W. Lijinsky, L. Keefer, E. Conrad and R. Van de Bogart, The nitrosation of tertiary amines and some biologic implications, J. Natl. Cancer Inst. 49:1239 (1972).

8. P. Issenberg, E. E. Conrad, J. W. Nielsen, D. A. Klein and S. E. Miller, Determination of N-nitrosobis-(2-hydroxypropyl)amine in environmental samples, in: "N-Nitroso Compounds: Occurrence, Biological Effects and Relevance to Human Cancer", IARC Scientific Publications, No. 57 (1984).

9. S. S. Hecht, J. B. Morrison and J. A. Wenninger, N-Nitroso-N-methyl-dodecylamine and N-nitroso-N-methyltetradecylamine in hair care products, Fd. Chem. Toxicol. 20:165 (1982).

10. W. Lijinsky, J. Saavedra and M. D. Reuber, Induction of carcinogenesis in Fischer rats by methylalkylnitrosamines, Cancer Res. 41:1288 (1981).

11. W. Lijinsky, A. M. Losikoff and E. P. Sansone, N-Nitrosodiethanolamine and N-nitrosomorpholine readily penetrate rat skin, J. Natl. Cancer Inst. 66:125 (1981).

12. W. Lijinsky, R. M. Kovatch and C. W. Riggs, Carcinogenesis by nitroso-dialkylamines and azoxylkanes given by gavage to rats and hamsters, Cancer Res. 47:3968 (1987).

13. R. H. Cardy, W. Lijinsky and P. Hildebrandt, Neoplastic and nonneoplastic urinary bladder lesions induced in Fischer 344 rats and B6C3F1 hybrid mice by N-nitrosodiphenylamine, Ecotoxicology and Environ. Safety 3:29 (1979).

14. B. Spiegelhalder and R. Preussmann, Nitrosamines and rubber, in: "N-Nitroso Compounds: Occurrence and Biological Effects", IARC Scientific Publications, No. 41 (1982).

15. W. Lijinsky, Reaction of drugs with nitrous acid as a source of carcinogenic nitrosamines, Cancer Res. 34:255 (1974).

16. R. Elespuru and W. Lijinsky, The formation of carcinogenic N-nitroso compounds from nitrite and some types of agricultural chemicals, Fd. Cosmet. Toxicol. 11:807 (1973).

17. W. Lijinsky and D. Schmähl, Carcinogenicity of N-nitroso derivatives of N-methylcarbamate insecticides in rats, Ecotoxicology and Environ. Safety 2:413 (1978).

18. D. J. Stuehr and M. A. Marletta, Synthesis of nitrite and nitrate in murine macrophage cell lines, Cancer Res. 47:5590 (1987).

19. A. R. Jones, W. Lijinsky and G. M. Singer, Steric effects in the nitrosation of piperidines, Cancer Res. 34:1079 (1974).

20. H. Ohshima and H. Bartsch, Quantitative estimation of endogenous nitrosation in man by monitoring N-nitrosoproline excreted in the urine, Cancer Res. 41:3658 (1981).

21. M. Tsuda, T. Kakizoe, T. Hirayama and T. Sugimura, New type of N-nitro-samino acids, N-nitroso-L-thioproline and N-nitroso-L-methylthiopro-lines, found in human urine as major N-nitroso compounds, in: "N-Nitroso Compounds: Occurrence, Biological Effects and Relevance to Human Cancer", IARC Scientific Publications, No. 57 (1984).

22. H. W. Taylor and W. Lijinsky, Tumor induction in rats by feeding amino-pyrine or oxytetracycline with nitrite, Intern. J. Cancer 16:211 (1975).

23. W. Lijinsky and H. W. Taylor, Nitrosamines and their precursors in food, Cold Spring Harbor Symposium on the Origins of Human Cancer, Book C (1977).

24. D. M. Winn, Tobacco chewing and snuff dipping: An association with human cancer, in: "N-Nitroso Compounds: Occurrence, Biological Effects and Relevance to Human Cancer", IARC Scientific Publications, No. 57 (1984).

MODELS FOR COMBINED ACTION OF ALCOHOL AND TOBACCO ON RISK OF CANCER:

WHAT DO WE REALLY KNOW FROM EPIDEMIOLOGICAL STUDIES ?

Jacques Estève and Albert J. Tuyns

International Agency for Research on Cancer
150, cours Albert-Thomas, 69372 Lyon Cedex 08, France

INTRODUCTION

There is a considerable amount of literature on the subject of interaction or combined effect of etiological factors on the risk of cancer[1-7]. It is, however, still uncertain which model - multiplicative or additive - is adequate for describing the combined effect of alcohol and tobacco on the risk of esophageal, laryngeal and pharyngeal cancer[8-15], even if there is convincing evidence that each factor increases "independently" the risk of these cancers. Even though many analyses of epidemiological data are based on the implicit assumption of a multiplicative model and a good fit is often obtained, the size of most studies does not permit the testing of interaction in the conventional way[16,17]. Therefore, the use of such statistical tools for inferring on the underlying biological process would be misleading. The complete understanding of the effect of alcohol on the risk of cancer is in addition complicated by the lack of carcinogenicity of ethanol or alcoholic beverages in animal experiments. As a consequence, there are divergent opinions on the role played by alcohol in increasing the risk of specific cancers among non-smokers. The latter issue has furthermore been confused with the problem of demonstrating a synergistic effect of the two factors because almost tautologically alcohol and tobacco would have a greater effect in combination than alone if the former had no effect in non-smokers. The present paper will deal only with the problem of distinguishing between additive and multiplicative joint effect of the two factors by reanalyzing the data of several epidemiological studies on oral, esophageal, laryngeal and hypopharyngeal cancers with a method recently proposed by Breslow and Storer[8].

MATERIALS AND METHODS

The data of study on oral cancer[18] was taken from a paper by Rothman[4] dealing with the estimation of synergy.

The data on esophageal cancer are pooled data from two similar studies carried out in Calvados[19,20] and in Ille-et-Vilaine[21], the latter being the example taken by Breslow and Storer[8] to demonstrate that both additive and multiplicative models could describe equally well the risk of esophageal cancer.

The data on laryngeal and hypopharyngeal cancers came from an international study[22] carried out in France, Italy, Spain and Switzerland to examine, among other things, the combined effect of alcohol and tobacco. Alcoholic beverages are more widely consumed in these countries than in Anglo-Saxon populations and are in addition often consumed in large quantities by light and non-smokers. These features gave us a better opportunity to study these factors alone and in combination.

The statistical analysis is based on the fitting of a family of relative risk functions having as particular cases additive and multiplicative models[8]. More explicitly, the model to be fitted is described by the following equations linking the probability of disease (d) to the relative risk (R):

$$\text{logit } (\text{pr}(d \mid X)) = \alpha + \log R(X) \qquad (E1)$$

$$\log R(X) = \frac{(1 + X\beta)^{\lambda} - 1}{\lambda} \; ; \lambda \neq 0 \qquad (E2)$$

$$R(X) = 1 + X\beta \; ; \lambda = 0 \qquad (E3)$$

where α is a stratum parameter (age and center of study in this instance), X is a vector of binary covariates characterizing the categories of alcohol and tobacco consumption, and β are parameters to be estimated and which will describe the relative risk R(X) as a function of the class of consumption of alcohol and tobacco. When $\lambda = 1$ the model is multiplicative, and when $\lambda = 0$ the model described by equation (E3) is obviously the additive model. Let R_{ij} be the relative risk for smoking "i" and drinking "j"; the multiplicative model predicts that $R_{ij} = R_{i1} * R_{1j}$ and the additive model $R_{ij} = R_{i1} + R_{1j} - 1$. When λ is comprised between 0 and 1 the relative risks R_{i1} and R_{1j} combine in a way which is intermediate between additive and multiplicative. When λ is outside this interval the model is subadditive ($\lambda < 0$) or supramultiplicative ($\lambda > 1$). The maximum likelihood method was used to estimate λ. The tests of hypotheses $\lambda = 1$ and $\lambda = 0$ were performed by calculating the difference in deviance of the best fitting model and of the model with a specified λ. The latter statistics is distributed as chi-squared with one degree of freedom.

RESULTS

Esophageal Cancer

A total of 903 cases is obtained when pooling the data from Calvados and Ille-et-Vilaine, and the evidence is no longer favoring the additive model as suggested by a previous estimate[8]; the present estimate is 0.52 and the additive model would be rejected by the test $\lambda = 0$ ($\chi^2 = 6.36$) whereas the multiplicative model is compatible with the observed data ($\chi^2 = 1.78$). The risks for alcohol and tobacco at the lower level of the other factor implied by the various models are shown in Table 1. The observed combined effect of the two factors and the number of cases in each consumption category is shown in Table 2. The fitted combined effect may be obtained from equations (E2) and (E3).

Laryngeal Cancer

A total of 727 cases of cancer of the endolarynx was analyzed and the fit of the multiplicative model is almost perfect ($\lambda = 0.93$). The test of the additive model would reject it very strongly ($\chi^2 = 26.1$). Table 3 gives the relative risks for alcohol and tobacco implied by the various models and the observed combined effect may be found elsewhere[21].

650

Table 1. Relative risks of esophageal cancer implied by various models

Model	Alcohol (g/day)[a]				Tobacco (g/day)[b]			
	0-39	40-79	80-119	120+	0-9	10-19	20-29	30+
Multiplicative ($\lambda = 1$)	1	2.9	6.7	24.7	1	1.5	1.8	2.9
Best fitting ($\lambda = 0.52$)	1	3.6	8.4	33.8	1	1.9	2.3	4.1
Additive ($\lambda = 0$)	1	4.7	11.9	50.3	1	2.5	3.2	5.8

[a]Associated with a consumption of less than 10 g of tobacco/day. [b]Associated with a consumption of less than 40 g of alcohol/day.

Hypopharyngeal Cancer

A total of 409 cases of cancer of the hypopharynx and epilarynx was analyzed and the fit is again in agreement with a multiplicative model ($\lambda = 0.78$; $\chi^2 = 0.51$ for the test of $\lambda = 1$). On the contrary, the additive model would be strongly rejected ($\chi^2 = 47.4$). Table 4 gives the relative risks implied by the various models and the observed combined effect may be found elsewhere [22].

Oral Cancer

A total of 483 cases of cancer of the mouth and pharynx was analyzed using only two age strata as given by Rothman[4]. A slight residual confound-

Table 2. Observed joint effect of alcohol and tobacco on the risk of esophageal cancer (number of cases

Alcohol (g/day)	Tobacco (g/day)			
	0-9	10-19	20-29	30+
0-39	1 (26)	2.9 (30)	2.7 (17)	4.4 (17)
40-79	5.1 (92)	4.8 (58)	7.7 (52)	10.7 (34)
80-119	8.1 (54)	15.5 (87)	18.5 (57)	23.8 (38)
120+	41.5 (79)	51.8 (110)	43.0 (86)	150.8 (66)

Table 3. Relative risks of laryngeal cancer implied by various models

Model	Alcohol (g/day)[a]				Tobacco (g/day)[b]			
	0-40	41-80	81-120	121+	0-7	8-15	16-25	26+
Multiplicative ($\lambda = 1$)	1	1.1	1.8	2.7	1	4.5	9.3	11.1
Best fitting ($\lambda = 0.93$)	1	1.1	1.8	2.8	1	4.6	9.5	11.4
Additive ($\lambda = 0$)	1	1.5	3.8	8.8	1	6.8	15.0	18.9

[a]Associated with a consumption of less than 8 g of tobacco/day. [b]Associated with a consumption of less than 41 g of alcohol/day.

ing may have occurred which can explain the minor differences from the results provided in the above reference. A multiplicative model gives a good description of the data although the estimate of λ suggests a supramultiplicative effect ($\lambda = 3.37$; $X^2 = 1.90$ for the test of $\lambda = 1$). The additive model would again be rejected ($X^2 = 11.19$). Table 5 gives the risks for alcohol and tobacco implied by the various models and Table 6 gives the observed combined effect as calculated from the data available in Rothman[4].

DISCUSSION

This study of the joint effect of alcohol and tobacco suffers from a lack of individuals with "pure" exposure to the factors under consideration. In particular, one might think more appropriate to avoid mixing those not exposed with those with light exposure. This has not been possible for practi-

Table 4. Relative risks of hypopharynx implied by various models

Model	Alcohol (g/day)[a]				Tobacco (g/day)[b]			
	0-40	41-80	81-120	121+	0-7	8-15	16-25	26+
Multiplicative ($\lambda = 1$)	1	2.2	4.6	10.2	1	4.9	7.2	7.3
Best fitting ($\lambda = 0.78$)	1	2.6	6.1	14.6	1	6.1	9.4	9.4
Additive ($\lambda = 0$)	1	5.2	21.6	71.0	1	11.6	24.0	17.5

[a]Associated with consumption of less than 8 g of tobacco/day. [b]Associated with a consumption of less than 41 g of alcohol/day.

Table 5. Relative risk of oral[a] cancer implied by various models

Model	Alcohol (oz/day)[b]				Tobacco (g/day)[c]			
	0	0.1-0.3	0.4-1.5	1.6+	0	1-19	20-39	40+
Best fitting ($\lambda = 3.37$)	1	1.2	1.6	2.1	1	1.5	2.1	2.6
Multiplicative ($\lambda = 1$)	1	1.4	2.4	4.2	1	1.8	2.9	4.2
Additive ($\lambda = 0$)	1	1.4	3.6	8.2	1	1.8	3.1	5.6

[a]Mouth and pharynx, see 18; [b] and no tobacco; [c] and no alcohol.

cal reasons. In the French study of Ille-et-Vilaine the non-smoker cannot be separated from the light smoker, and it was thought useful to have the same approach for all sites examined. On the contrary, the data from Keller and Terris[18] was not published with sufficient details to permit the creation of a category of light smokers. Despite the fact that the choice of the baseline may have influenced the results of the analysis, our results clearly show that a joint effect of alcohol and tobacco on the risk of cancer of the aerodigestive tract is multiplicative rather than additive. The previously reported uncertainties[8,13] were mainly caused by the small size of the studies which have been used to attempt to discriminate between the two models. The observed joint effect in larger studies is clearly incompatible with the large risk at low doses that would imply an additive model: the lack of fit becomes significant because a sufficient number of cases is then available

Table 6. Observed joint effect of alcohol and tobacco on the risk of oral[a] cancer (number of cases)

Alcohol oz/day	Tobacco (g/day)			
	0	1-19	20-39	40+
0	1 (10)	1.5 (11)	1.3 (13)	3.9 (9)
0.1-0.3	0.9 (7)	1.8 (16)	3.2 (50)	3.2 (16)
0.4-1.5	1.2 (4)	4.2 (18)	4.8 (60)	7.4 (27)
1.6+	2.3 (5)	4.0 (21)	9.4 (125)	13.2 (91)

[a]Mouth and pharynx, see 18.

in these low dose exposure categories. It is therefore not surprising that esophageal cancer, which is the site where the effect of alcohol at low dose of tobacco is the largest, is also the site for which the best fitting model is further away from the multiplicative one; this latter model could not, however, be ruled out on the basis of the present data.

The method of statistical analysis used in this paper describes a family of models of risk combination which has mainly an operational purpose and which has no particular biological interpretation. Other families could have been used[23,24]. Its main interest is to describe synergism with one parameter and to enable one-degree of freedom testing. From that point of view it appears to be more informative than the index proposed by Rothman[4]. As a matter of fact, Walker and Rothman[25] proposed later an approach which in principle is equivalent to the present one in simpler situations.

There has been a long debate at the end of the seventies on the meaning of interaction and on what should be understood by independent causes of disease[1-7]. Rothman et al.[5] proposed four contexts in which the interaction should be evaluated. The statistical interaction is the best defined concept and using this concept in the present situation one may say that there is no significant departure from the multiplicative model for the joint effect of alcohol and tobacco on the risk of cancer of the upper aerodigestive tract. On the contrary, there is significant departure from the additive model. The practical consequence is that the relative risk for alcohol adjusted for tobacco is a sufficient summary for describing the effect of alcohol, and it is worth noting that, given a multiplicative model, the knowledge of this relative risk also permits the calculation of the proportion of cases attributable to alcohol exposure. On the contrary, the excess risks for alcohol and tobacco at low dose of the other factor are uninformative. Following Blot and Day[1] it may be added that the rejection of the additive model implies synergism, which means that the excess incidence due to alcohol is far greater among heavy smokers than among light smokers. The latter concept is called public health interaction by Rothman et al.[5]

The understanding of the mechanism by which alcohol and tobacco act together to produce cancer is probably not considerably improved by the present analysis; in other words, a better knowledge of the interaction of the two factors does not improve our knowledge of their biological interaction. This latter rather fuzzy concept might be useful in the framework of initiation/promotion experiment or in the interpretation of epidemiological observation with the help of the multistage model: it has often been said that etiological factors acting interchangeably in the same step of the process would correspond to additivity, whereas those acting on different steps would combine multiplicatively. However, rejecting the hypothesis of a common step of action on the sole ground of non-additivity would be valid only if the two factors would act with the same linear dose effect relationship[2]. The evidence accumulated up to now would not support such a simple mechanism and would suggest, on the contrary, that both factors act on several steps of a possible multistage process.

REFERENCES

1. W. J. Blot and N. E. Day, Synergism and interaction: Are they equivalent?, Am. J. Epidemiol. 110:99 (1979).
2. L. L. Kupper and M. D. Hogan, Interaction in epidemiologic studies, Am. J. Epidemiol. 108:447 (9178).
3. O. S. Miettinen, Causal and preventive interdependence. Elementary principles, Scand. J. Work Environ. Health 8:159 (1982).
4. K. J. Rothman, The estimation of synergy or antagonism, Am. J. Epidemiol. 103:506 (1976).

5. K. J. Rothman, S. Greenland and A. M. Walker, Concepts of interaction, Am. J. Epidemiol. 112:467 (1980).

6. R. Saracci, Interaction and synergism, Am. J. Epidemiol. 112:465 (1980).

7. S. D. Walter and T. R. Holford, Additive, multiplicative, and other models for disease risks, Am. J. Epidemiol. 108:341 (1978).

8. N. E. Breslow and B. E. Storer, General relative risk functions for case-control studies, Am. J. Epidemiol. 122:149 (1985).

9. S. Graham, H. Dayal, T. Rohrer, M. Swanson, H. Sultz, D. Shedd and S. Fischman, Dentition, diet, tobacco, and alcohol in the epidemiology of oral cancer, J. Nat. Cancer Inst. 59:1611 (1977).

10. J. Olsen, S. Sabreo and U. Fasting, Interaction of alcohol and tobacco as risk factors in cancer of the laryngeal region, J. Epidemiol. Commun. Health 39:165 (1985).

11. J. Olsen, S. Sabreo and J. Ipsen, Effect of combined alcohol and tobacco exposure risk of cancer of the hypopharynx, J. Epidemiol. Commun. Health 39:304 (1985).

12. K. Rothman and A. Keller, The effect of joint exposure to alcohol and tobacco on risk of cancer of the mouth and pharynx, J. Chron. Dis. 25:711 (1972).

13. S. D. Walter and M. Iwane, re: "Interaction of alcohol and tobacco in laryngeal cancer" (Letter to the Editor), Am. J. Epidemiol. 117:639 (1983).

14. R. T. Zagraniski, J. L. Kelsey and S. D. Walter, Occupational risk factors for laryngeal carcinoma: Connecticut, 1975-1980, Am. J. Epidemiol. 124:67 (1986).

15. W. D. Flanders and K. J. Rothman, Interaction of alcohol and tobacco in laryngeal cancer, Am. J. Epidemiol. 115:371 (1982).

16. S. Greenland, Tests for interaction in epidemiologic studies: A review and a study of power, Stat. Med. 2:243 (1983).

17. P. G. Smith and N. E. Day, The design of case-control studies: the influence of confounding and interaction effects, Int. J. Epidemiol. 13:356 (1984).

18. A. Z. Keller and M. Terris, The association of alcohol and tobacco with cancer of mouth and pharynx, Am. J. Publ. Health 55:1578 (1965).

19. A. J. Tuyns and M. X. Hu, Changing smoking patterns in the département of Calvados (France), Brit. J. Addict. 77:167 (1982).

20. A. J. Tuyns, E. Riboli, G. Doornbos and G. Péquignot, Diet and esophageal cancer in Calvados (France), Nutr. Cancer 9:81 (9187).

21. A. J. Tuyns, G. Péquignot and O. M. Jensen, Le cancer de l'oesophage en Ille-et-Vilaine en fonction des niveaux de consommation d'alcool et de tabac. Des risques qui se multiplient, Bull. Cancer 64:45 (1977).

22. A. J. Tuyns, J. Estève, L. Raymond, F. Berrino, E. Benhamou, F. Blanchet, P. Boffetta, P. Crosignani, A. Del Moral, W. Lehmann, F. Merletti, G. Péquignot, E. Riboli, H. Sancho-Garnier, B. Terracini, A. Zubiri and L. Zubiri, Cancer of the larynx/hypopharynx, tobacco and alcohol, Int. J. Cancer in press (1987).

23. F. J. Aranda-Ordaz, An extension of the proportional-hazards model for grouped data, Biometrics 39:109 (1983).

24. D. C. Thomas, General relative-risk models for survival and matched case-control analysis, Biometrics 37:673 (1981).

25. A. M. Walker and K. J. Rothman, Models of varying parametric form in case-referent studies, Am. J. Epidemiol. 115:129 (1982).

EPIDEMIOLOGICAL APPROACH TO THE INVESTIGATION OF MECHANISMS OF ACTION

IN HUMAN BLADDER CARCINOGENESIS

Paolo Vineis and Benedetto Terracini

Servizio di Epidemiologia dei Tumori, Dipartimento di
Scienze Biomediche e Oncologia Umana, Ospedale Maggiore
e Università di Torino
Via Santena 7, 10126 Torino, Italy

INTRODUCTION

In 1972, Doll et al.[1] suggested that the concentration of 2-naphthyl-amine in cigarette smoke was similar to concentrations found in coal plants, i.e. a setting where an elevated risk of bladder cancer was identified. Sub-sequent chemical analyses of tobacco smoke[2] showed that several aromatic amines were present, with higher concentrations in black (air-cured) tobacco. This suggested that in countries where prevalently black tobacco is smoked the relative risk for bladder cancer could be higher than in countries, such as the U. K. and U. S.. where blond (flue-cured) tobacco is used.

Such hypothesis has been confirmed by a study in Italy and one in Argen-tina[3,4] showing that black tobacco was associated with risks of bladder cancer 2-3 times higher than blond tobacco.

Further evidence was given by biochemical studies. Urines of black to-bacco smokers were slightly more mutagenic than urines of blond tobacco smok-ers in a pilot study in Italy (paper in preparation); in a German study, a single subject switching from blond to black cigarettes showed higher urine mutagenicity when smoking the latter[5]. Independently, the exceptionally high mutagenic activity in the urine of a single smoker was attributed to the iso-lation of an aromatic amine, 2-amino-7-naphthol (a metabolite of 2-naphthyl-amine)[6]. Finally, the blood of black tobacco smokers was reported to contain a higher concentration of adducts formed by 4-aminobiphenyl with hemoglobin (288 pg/g),in comparison with blond tobacco smokers (175 pg/g) and non-smok-ers (51 pg/g)[7]. Interestingly, the levels of adducts among Italian blond to-bacco smokers was similar to the levels found in an investigation of American cigarette smokers[8].

TEMPORAL ASPECTS OF BLADDER CARCINOGENESIS

The different composition of black and blond tobacco smoke suggests the possibility that various stages of action in a multi-stage carcinogenic pro-cess may be differently affected by the two types of tobacco. In a recently published and wide study in the U. S., where blond tobacco is almost exclu-sively smoked, the risks following cessation of smoking clearly suggested that late-stage carcinogens were present[9]. This was consistent with a number

Table 1. Male smokers of black tobacco: role of time variables in bladder cancer risk (duration and cessation)

		Case/Controls	Or [a]	95% CI	Trend X^2
Duration	1-19	6/29	1.0		
	20-39	68/77	3.4	(1.1-10.0)	2.9 (p=0.08)
	40+	81/45	6.7	(1.7-26.1)	
Intensity	1-14	50/77	1.0		
	15-29	93/62	1.8	(1.1-3.0)	
	30+	12/12	1.6	(0.6-4.0)	
Cessation	Current	115/72	1.0		
	<3 years	10/15	0.46	(0.19-1.15)	
	3-9 years	11/25	0.32	(0.14-0.75)	0.03 (p=0.87)
	10+ years	19/39	0.59	(0.25-1.38)	

Model X^2 50.36 (13 d.f.)

[a]Logistic regression estimates; models including age at diagnosis/interview

of previous investigations[10]. The American study, in addition, was properly analyzed for the role of age at start of smoking, a variable which can indicate an early stage of action of the exposure[11]. However, in that study age at start was not clearly associated with the relative risk, although the suggestion of a decrease in the risk with increasing age at start was present. The previous literature on this subject is very sparse[10].

The major problem with the analysis of time aspects is collinearity between variables, particularly with a case-control design. Age at start, for instance, does not vary when duration, age at diagnosis and time since stopping are fixed[12].

We have analyzed a case-control study in Torino for the role of time variables in black and in blond tobacco smokers, respectively. Since there is evidence that a greater content in 4-aminobiphenyl and, possibly, other aromatic amines might explain the greater bladder carcinogenicity of black tobacco, the latter could be associated with a more evident effect of age at start. In fact, aromatic amines are mutagenic[13] and covalently bind with macromolecules[8], thus suggesting a potential early-stage (initiating?) activity.

Details on the study design and further analyses are reported elsewhere[3,14].

AGE AT START AND TIME SINCE CESSATION: SMOKING OF BLACK OR BLOND TOBACCO

Duration, age at start, age at diagnosis and time since cessation cannot be analyzed in a same statistical model. We have chosen, therefore, to analyze separately the effects of duration and, respectively, age at start, adjusting for age at diagnosis and time since cessation. Table 1 shows the logistic regression model including duration, and Table 2 the model including age at start for male smokers of black tobacco for most of their smoking lives (155 cases and 151 controls). A clearcut effect of cessation is evident, par-

Table 2. Male smokers of black tobacco: role of time variables in bladder
cancer risk (age at start and cessation)

		Case/Controls	Or[a]	95% CI	Trend X^2
Age at start	25+	6/16	1.0		
	21-24	16/15	2.4	(0.7-8.7)	
	17-20	63/63	2.3	(0.8-6.9)	2.9 (p=0.08)
	<17	70/57	2.8	(0.9-8.3)	
Intensity	<15	50/77	1.0		
	15-29	93/62	1.9	(1.1-3.3)	
	30+	12/12	1.5	(0.6-4.0)	
Cessation	Current	115/72	1.0		
	<3	10/15	0.45	(0.18-1.11)	
	3-9	11/25	0.27	(0.11-0.61)	8.9 (p=0.003)
	10+	19/39	0.27	(0.14-0.53)	

Model X^2 45.86 (14 d.f.)

[a]Logistic regression estimates; models including age at diagnosis/interview

ticularly when age at start is included. The differences between duration
and age at start are due to the different categories which have been used;
both suggest the presence of an early stage effect, although the role of age
at start is not statistically significant.

Since only 27 male cases and 76 controls smoked blond tobacco for most
of their smoking lives, and the distribution by categories of age at start was
sparse, it is premature to draw any conclusion in this group of smokers. The
effect of cessation, however, is clear, with a relative risk of 0.36 (95% c.1.
0.1-1.2) among former smokers compared to current smokers.

A comparison is possible between those who smoked black tobacco throughout life (87 cases and 109 controls) and those who switched from black to
blond (Table 3). Apparently it makes no difference stopping smoking black
tobacco or switching to blond before stopping; a very slight difference is
suggested, among current smokers, between those who never quit black tobacco
and those who switched to blond.

A possible interpretation of these data, which require confirmation, is
that black tobacco acts at early stages of carcinogenesis; the role of blond
tobacco at the same stages cannot be solved because of paucity of data; both
types of tobacco seem to have a late stage effect, suggested by a decrease in
the relative risk after smoking cessation.

CONCLUSIONS AND PERSPECTIVES

In the case of smoking and lung cancer, both an early and a late stage
of action have been proposed[10]. The literature on bladder cancer is limited,
at least as far as the effect of age at start is concerned, by the paucity of
data and methodological problems. Previous studies on the subject either did
not adjust the estimates concerning time variables for the confounding role of
other tobacco-related variables or did not consider properly the strong

Table 3. Odds ratios and 95% confidence intervals according to duration and cessation among smokers of black tobacco throughout life and among smokers switching from black to blond tobacco, males[a]

Duration	Black tobacco throughout life		Black tobacco switching to blond[b]	
	Current	Former	Former black Current blond	Formerly black, later blond, then quitting
1-19	2.1 (0.2-22.9)	0.8 (0.2-3.0)	3.3 (1.0-10.3)	1.3 (0.1-12.6)
20+	6.6 (3.5-12.6)	2.2 (1.0-4.9)	5.7 (3.3-10.0)	2.6 (1.2-5.6)
All durations	6.2 (3.3-11.7)	2.1 (1.1-4.0)	5.5 (3.2-9.5)	2.5 (1.3-5.0)
Cases/Controls	65/47	22/62	151/127	30/45

[a]All odds ratios are age-adjusted, reference category: non-smokers.
[b]Including 68/42 subjects smoking mostly black, 13/14 smoking mostly blond tobacco, and 100/116 mostly mixed smokers

collinearity between time variables.

An apparent early-stage effect of black tobacco is suggested by the present analysis, and is coherent with the biologic properties of aromatic amines. Future studies should include a sufficient number of subjects who quit temporarily and then started smoking again, in order to disentangle effectively between age at start and duration. In addition, biochemical studies are warranted to confirm the relevance to bladder carcinogenicity of the aromatic amines found in tobacco smoke. A particularly interesting question is whether the level of adducts formed by aromatic amines with hemoglobin (and, possibly, DNA) is influenced by the phenotype for N-acetyltransferase; this has been suggested by a small investigation [15] and is consistent with the hypothesis that slow acetylators exposed to aromatic amines are at higher risk of bladder cancer than fast acetylators [16].

REFERENCES

1. R. Doll, H. P. Vessey, R. W. R. Beasely, A. R. Buckley, E. C. Fear, R. E. W. Fisher, E. T. Gammon, W. Gunn, G. O. Hughes, K. Lee and B. Norman-Smith, Mortality of gas-workers: Final report of a prospective study, Br. J. Industr. Med. 29:394 (1972).
2. C. Patrianakos and O. Hoffman, Chemical studies of tobacco smoke, LXIV. On the analysis of aromatic amines in cigarette smoke, J. Anal. Chem. 3:150 (1979).
3. P. Vineis, J. Esteve and B. Terracini, Bladder cancer and smoking in males: types of cigarettes, age at start, effect of stopping and interaction with occupation, Int. J. Cancer 34:165 (1984).
4. J. Iscovich, R. Castelleto, J. Esteve, N. Munoz, R. Colanzi, A. Coronel, I. Deamezola, V. Tassi and A. Arslan, Tobacco smoking, occupational exposure, and bladder cancer in Argentina, Int. J. Cancer in press.
5. E. Mohtashamipur, K. Norpoth and F. Lieder, Urinary excretion of

mutagens in smokers of cigarettes with various tar and nicotine yelds, black tobacco, and cigars, Cancer Lett., 34:103-112.

6. T. H. Connor, V. M. S. Ramanujam, J. B. Ward, Jr. and M. S. Legator, The identification and characterization of a urinary mutagen resulting from cigarette smoke, Mutat. Res. 113:161 (1983).

7. M. S. Bryant, P. Vineis, P. L. Skipper and S. R. Tannenbaum, Hemoglobin adducts of 4-aminobiphenyl: correlation with relative risks for bladder cancer in cigarette smokers, submitted for publication.

8. S. R. Tannenbaum, M. S. Bryant, P. L. Skipper and M. McLure, Hemoglobin adducts of tobacco-related aromatic amines: application to molecular epidemiology, Banbury Report 23:63 (1986).

9. P. Hartge, D. Silverman, R. Hoover, C. Schairer, R. Altman, D. Austin, K. Cantor, M. Child, C. Key, L. Marrett, T. J. Mason, J. W. Meigs, M. H. Myers, A. Narayana, J. W. Sullivan, G. M. Swanson, D. Thomas and D. West, Changing cigarette habits and bladder cancer risk: a case-control study, J. Natl. Cancer Inst. 78:1119 (1987).

10. IARC Monographs on the Evaluation of the Carcinogenic Risk of Chemicals to Humans, Vol. 38, Tobacco smoking, International Agency for Research on Cancer, Lyon (1986).

11. N. E. Day, Epidemiological data and multistage carcinogenesis, in: "Models, Mechanisms and Etiology of Tumor Promotion", M. Borzsonyi, K. Lapin, N. E. Day and H. Yamasaki, eds., IARC Sci. Publ. No. 56, International Agency for Research on Cancer, Lyon (1985).

12. P. Vineis and J. Esteve, Temporal aspects of bladder carcinogenesis, Toxicol. Pathol. in press (1987).

13. IARC Monographs on the Evaluation of the Carcinogenic Risk of Chemicals to Humans, Suppl. 4 (volumes 1-29), Chemicals, Industrial Processes and Industries Associated with Cancer in Humans, Lyon (1982).

14. P. Vineis, J. Esteve, P. Hartge, R. Hoover, D. T. Silverman and B. Terracini, Cigarette-induced bladder cancer: the effects of timing and type of tobacco, Cancer Res. in press

15. J. Lewalter and V. Korallus, Blood protein conjugates and acetylation of aromatic amines, Int. Arch. Occup. Environ. Health 56:176 (1985).

16. R. Cartwright, R. A. Ahmad, D. Barham-Hall, R. W. Glashan, H. J. Rogers, E. Higgins and M. A. Kahn, Role of N-acetyltransferase phenotypes in bladder carcinogenesis: a pharmacogenetic epidemiologic approach to bladder cancer, Lancet 2:842 (1982).

CONCLUDING REMARKS

Peter Bannasch

Institut fur Experimentelle Pathologie
Deutsches Krebsforschungszentrum
6900 Heidelberg, Im Neuenheimer Feld 280, FRG

The Fourth Sardinian International Meeting on Models and Mechanisms in
Chemical Carcinogenesis was a great challenge for all participants. In
rather a tight and exhausting schedule, recent advances in carcinogenesis research were presented in numerous reviews, short oral presentations and posters. Both the oral presentations and the posters stimulated a lively and
fruitful discussion. It is virtually impossible to summarize the plethora of
information given during the five days of the meeting within a few minutes.
I will, therefore, pick out some aspects which I felt to be of particular interest. This selection will certainly be biased by my specific interests in
some research areas and my ignorance in others.

How do chemical carcinogens interact with their target cells, and how do
they induce the persistent intracellular changes which eventually lead to
neoplastic cell transformation? Many contributions dealt with these crucial
questions and provided interesting new information. It is now well established that the majority of chemical carcinogens require metabolic activation, and it has been shown that the cytochrome P-450 enzyme system plays an
important role in this process. Elegant new tools for studying the cytochrome P-450 enzymes have been developed recently. Monoclonal antibodies to
the cytochrome P-450 enzymes, which inhibit the specific individual isoenzymes, can be used for reaction phenotyping of this enzyme system and may
help to further analyse the mechanisms of metabolic activation of chemical
carcinogens. Another promising approach to studying this mechanism is the
production of viral vectors by inserting P-450 DNA coding sequences into a
vaccinia virus.

Examples for the formation of reactive metabolites from chemical carcinogens and their subsequent interaction with chromosomal DNA were given for
a number of compounds such as aromatic amines, nitroaromatic hydrocarbons,
polycyclic hydrocarbons and some chemicals which appear to be of particular
interest as environmental pollutants (alipathic halocompounds, methyl halides). The importance of species differences and alternative metabolic pathways in the activation of chemical carcinogens has been emphasized, especially with respect to the extrapolation from data obtained in animals to the
situation in man. Detoxicating metabolic processes mediated, for example, by
the glutathione system, and DNA repair processes may counteract the formation
and accumulation of DNA-adducts and hence exert "antigenotoxic" and anticarcinogenic effects. It is evident, however, that there are still many gaps in
our understanding of the possible role of DNA-adducts in initiation of chemi-

663

cal carcinogenesis. Thus, in vitro studies of the methylation of calf thymus DNA by three different N-nitrosoalkylamines revealed that the reactive intermediates did not react with DNA in a manner which led to any specificity of DNA methylation. The results rather suggested that methylation of DNA occurred via a common diazonium ion entermediate. It is still unclear whether alkylation due to chemical carcinogens and subsequent repair of DNA are distributed in the genome in a random or non-random fashion. In combination with modern methods of molecular genetics, antibodies to DNA-adducts may be used as powerful tools to further elucidate the role of DNA-alterations induced by chemical carcinogens and their potential significance for preventive measures. Interestingly enough, in contrast to the surrounding liver tissue, preneoplastic hepatic foci produced in rats by continuous administration of AAF did not react with antibodies to AAF-DNA-adducts.

In this context, I would like to stress that the popular phrasing "DNA-adducts (or other phenomena) are necessary but not sufficient" for neoplastic cell transformation is problematic. Whereas this may be true, as long as we do not understand the whole process of carcinogenesis, we have to keep in mind that such changes may be irrelevant. Some other possibilities of primary DNA-alterations like DNA-hypomethylation or oncogene activation were discussed, and it was explicitly stated that changes in genetic regulation may eventually turn out to be more important than structural differences in genes.

As to the induction of tumours by compounds which have not been shown to interact with DNA, it has frequently been postulated that "spontaneously" initiated cell populations are always present and may give rise to tumours under the additional exogenous influence of chemicals usually called promoters. This speculation can neither be proven nor disproven at present. I wonder, however, why only so-called promoters but not carcinogens per se should use such "spontaneously" initiated cells as targets? Are we able to identify "initiators" at all under these conditions? It was mentioned just in time during our discussion that there is most probably no animal available which has not been exposed to exogenous carcinogens.

The developmental stages of carcinogenesis induced by chemicals have been studied in considerable detail in a number of tissues such as the liver, skin and kidney. In liver and kidney, characteristic phenotypic cellular changes were detected which apparently follow an ordered sequence during neoplastic development. In comparison to the frequently studied role of cell proliferation for carcinogenesis, the significance of cell death for this process has been neglected for a long time. At this meeting, several speakers emphasized the importance of a better understanding of the balance between cell death and cell proliferation during tumour development. The challenging idea has been proposed that hyperplasia due to mitogens is biologically different from hyperplasia due to necrosis. The concept of "apoptosis" as a genetically programmed cell death is also attractive. However, so far no specific morphological or biochemical criteria have been described which would allow us to clearly distinguish "apoptosis" from cell death due to other factors such as toxic effects or haemodynamic alterations.

Many laboratories earlier showed that partial hepatectomy enhances hepatocarcinogenesis induced by various chemicals. Regenerative cell proliferation and "fixation" of carcinogen-induced DNA lesions were usually made responsible for this effect. This interpretation is questioned, however, by the observation reported at this meeting that partial hepatectomy performed ten weeks before administration of the carcinogen apparently also increased the incidence of liver tumours. Decades ago pathologists postulated that "hyperplasiogenic tumours" may appear in certain tissues, particularly in endocrine organs. Today, it is difficult to accept that cell proliferation as such can lead to neoplasia since some persistent cellular alterations appear to be a prerequisite to neoplastic transformation. During hepatocarcinogenesis, early appearing phenotypic cellular changes are not associated with an appreciable increase in cell proliferation but there is a slow progression to phenotypic alterations characterized by a high cell turnover.

In some models of hepatocarcinogenesis, diploid cell populations appear to be the precursors of hepatocellular tumours but this observation should not be generalized, neither for the liver nor for other tissues. There is no doubt that tetraploid, polyploid and aneuploid preneoplastic and neoplastic lesions may occur under other experimental conditions or in man. The controversial discussion on the significance of diploid "oval" cells for hepatocarcinogenesis goes back to the early fifties when a paper coauthored by the Millers appeared implicating that the oval cells might be precursors of hepatocytes and hepatoma cells. In the meantime, many workers have tried to substantiate this idea. However, as far as I can see, no convincing evidence has been provided up to date which would prove this postulated transition. On the contrary, in a number of experimental models it has clearly been shown that hepatocellular tumours arise without any preceding proliferation of oval cells. The majority of oval cells, which appear early after repeated administration of sublethal doses of hepatocarcinogens, die later on. There is little doubt that those which survive may progress to cholangiofibrosis and to cholangiocellular tumours, but this histogenetic relation is frequently neglected in the ongoing debate on the origin and fate of the oval cell.

I was most impressed by the large number of contributions dealing with metabolic aberrations in preneoplastic and neoplastic lesions produced or promoted by chemical carcinogens. There is no time to go into the details, but I would like to list some of the topics discussed:

- alterations in drug metabolism

- relation of oncogene activation to metabolic aberrations

- changes in carbohydrate metabolism concerning glycosyltransferases, nucleotide sugars, glycogen, the hexose monophosphate shunt and glycolysis

- alterations of lipid metabolism during tumour promotion

- cholesterol metabolism and cell proliferation

- changes in signal transduction during neoplastic transformation including the role of proteinkinase C in tumour promotion

- anticarcinogenesis induced by dehydroepiandrosterone

A central position in the discussion on metabolic aberrations during neoplastic transformation has been given to the key enzyme of the hexose monophosphate shunt, the glucose 6-phosphate dehydrogenase (G6PDH), and the possible relation of the anticarcinogenic effect of dehydroepiandrosterone, which has been shown earlier to be an inhibitor of the G6PDH, to the activity of this metabolic pathway.

The influence of nutrition (choline deficiency, administration of orotic acid, caloric restriction, dietary fat) on chemical carcinogenesis described by many speakers, was in line with the concept that metabolic aberrations are closely related to neoplastic transformation.

In the last session of the meeting, some chemical risk factors for human cancer were discussed, including nitrosamines which have been shown to induce cancer of nearly all sites in various animal species, endogenous nitrosation processes, alcohol and tobacco. It was evident from the discussion that considerable progress has been made in this research area.

We have certainly heard many excellent presentations, but the most perfect presentation was that of Franco Maggio Ormezowski playing the Sonata for solo cello composed by Zoltan Kodaly. I am convinced that I can speak on behalf of all participants when I thank Professor Feo for this unforgettable concert and ask him to convey our gratitude to Mr. and Mrs. Ormezowski.

In my guide book for tourists I found two remarkable Sardinian sayings:

- "Sa domu est minore, su coru est mannu"
  This means: "His house is small but his heart is great"

Having experienced your most generous hospitality and your friendship, we can confirm this saying.

The second saying is more complicated and may need some explanation:

- "Furat chie venit da'e su mare" which means "He who comes over the sea will always steal something from our country".

I am afraid that this saying is also true. When we share with you the beauty of this island, when we accept your warm hospitality, when we drink your delicious wines and eat your food, and when we, after all, also take with us some of the original scientific ideas developed in your laboratories, then we are undoubtedly thieves. However, I hope that we can compensate to some extent what we have taken away with us by bringing you the best possible present scientists can give, namely the results which came out of hard working and thinking for many years, frequently for a lifetime. Thus, our meeting may be of mutual advantage. Many thanks to you, Francesco, to your wife and to your staff who had been working so efficiently to make the days in Alghero as pleasant and fruitful as possible for all of us.

668

Calcium (continued)
  cycling, 146, 149
  homeostasis, 154
  -Mg++-dependent DNase, 225
  pumps, 156
  transport, 147, 149
Calmodulin, 229
  -dependent enzymes 185
Cancer(s), 389, 609, 642
  of mouth, 651
    colon, 619
    oral cavity, 506
    rectum, 506
    bile ducts, 506
    liver, 506
    pancreas, 506
    stomach, 506
Carbocation, 140
Carbamylphosphate, 318
Carbon tetrachloride, 53, 143,
      147, 263, 333, 378, 392,
      394, 602
Carbonyl compounds, 644
Carcinogen-DNA adducts, 20, 104-
      108, 122, 123, 125, 194,
      195, 224-248, 333, 378,
      385, 436-441, 443, 657,
      660
Carcinoma(s)
  in situ, 570
  of liver, 338
    lung, 180
    pancreas, 338
Catalase, 604
Catecholamines, 220
CD diet, see choline-devoid diet
Cell(s), 180, 224, 340
  death, 154, 168, 225, 263, 265,
      275-278, 289, 409, 531,
      544, 547, 548, 553, 567,
      572
  differentiation, 585
  division, 574
  growth, 525
  loss, 167, 168, 170, 540, 544
  proliferation, 168, 188, 211,
      245, 264, 270, 278, 293,
      384, 385, 395, 475, 567,
      604, 633
  surface receptors, see
      receptor(s)
  turnover, 256, 530
Cellular growth, 513
  control, 187
Cellular hypertrophy, 538
Cellular necrosis, 210
Cellular oncogenes, see onco-
      gene(s)
Cellular proliferation, 515
Cervical cancer, 177
Cervical neoplasia, 176

CH, see cycloheximide
Chalone, 571
Chicken embryo fibroblasts, 476
Chlorinated hydrocarbons, 93, 287
Chloroacetaldehyde, 93
Chloroxirane, 93
Cholangiocarcinomas, 569
Cholesterol, 4, 333, 475-482, 490, 495,
      497, 499, 505, 506, 513
  esters, 495-497, 507
  synthesis, 362, 476, 485, 496, 497,
      499, 519, 520, 522
Cholestyramine, 497, 506
Choline, 390, 411, 583-585, 563-566,
      569, 572, 583-585
  deficiency, 565, 566, 583, 601, 605
  devoid diet, 195, 236, 238, 338, 413,
      486, 566-575, 583, 591-595, 602,
      604, 605
Chromatid exchanges, 222
Chromophobic cell tumors, 209, 213
Chromophobic epitheliomas, 210
Chromophobic lesions, 210, 211
Chromosomal aberrations, 222
Chromosomal damage, 223
Chromosomal fragile sites, 176
Chromosome rearrangements, 417
Chronic lymphocyte leukemia, 177
Chrysene, 37, 43, 67, 123
Chrysene-1,2-diol-3,4-oxide(s), 37
Chylomicrons, 496
Cigarette smoke, 17, 137, 657
  condensate, 75
Cimetidine, 76
Ciprofibrate, 105, 106
Cirrhosis, 120, 143, 489, 568
Clastogenic factors, 223
Clear cell tumors, 209, 214
Clear renal epitheliomas, 213
Clofibrate, 323-327, 331, 426-430,506
Clonal nodules, 167
Coal plants, 657
Collagens
  type I, 179
  type III, 179
Colon, 17, 362, 609, 639
  cancer, 621, 624
  carcinogenesis, 188, 620, 621, 626
Colonic carcinogenesis, see colon car-
      cinogenesis
Colonic tumor, see colon cancer
Colorectal cancer, 506, 513
Compactin, 505, 519
Conjugated dienes, 148
Contact inhibition, 340, 450
Conversion, 217-219, 222-224, 227-229
Corticosterone, 530
CPA, see cyproterone acetate
Cyclic adducts, 244
Cyclic AMP-dependent protein
      kinase, 187

Gene(s) (continued)
mutation, 223, 387
ornithine decarboxylase, 222
rearrangement, 387
suppressor, 222
structure, 453
T antigen, 180
thymidine kinase, 10
translocation, 223, 387
Genetic polymorphisms, 125
Genital dysplasias, 177
GGT, see gamma-glutamyltranspeptidase
Glucagon, 350, 530, 593, 594
receptors, see receptor(s)
Glucocorticoids, see glucocorticosteroids
Glucocorticosteroid(s), 225, 228,
264
Glucose-6-phosphatase, 104, 212,
213, 523
Glucose 6-phosphate dehydrogenase,
76, 211-213, 324, 325, 333,
362, 364, 365, 369-371,
399-402, 408, 450, 485-489,
519, 522, 523, 610
deficiency, 369, 372
deficient cells, 370, 388, 391
inhibitors, 365
Glucuronic acid, 137
Glutamate-pyruvate transaminase,
158
Glutathione, 46, 48, 53, 604
peroxidase, 160, 604
reductase, 612
-S-transferase(s) (P-form), 53,
67, 69, 71, 94, 99, 153,
260, 283, 284, 293, 346,
350, 352, 399, 400-402, 426
-S-transferase-P positive foci,
425-427
hepatocytes, 297
nodule(s), 293, 294, 297
Glyceraldehyde-3-phosphate dehydrogenase, 211-213
Glycogen
phosphorylase, 212, 231
storage cells, 212
synthase, 212, 213
Glycoproteins, 341
Glycosyltransferase, 338, 339
GnTase, I, III, IV, V, see
N-acetyl-glucosaminyl-
transferases
grc Gene(s), see gene(s)
grc, see growth reproduction
complex
Growth, 574
anchorage-independent, 314, 340
factor(s), 179-181, 185, 217,
441, 476, 480, 553, 591,

Growth (continued)
factor(s) (continued) 597
receptors, see receptor(s)
rate, 441
regulation, 553
reproduction complex, 485-490
GSH, see reduced glutathione
depletion, see reduced glutathione
depletion
peroxidase, see glutathione peroxidase
-S-transferase, see glutathione-S-
transferase
Guanine, 115, 125, 371
methylation, 111
Guanine-8-arylamine(s), 125
DNA adducts, 120
Guanylate cyclase, 351

H2-blockers, 76
H-301 line, 275
H-ras oncogene, see oncogene(s)
Haloaldehydes, 53
Haloalkanes, 93, 99
Haloethane, 54
Halogenated hydrocarbons, 53
Hamster
embryonic fibroblasts, 362
kidney tumor cell line, 275
HDL(2,3), see high density lipopro-
tein(s)
Heat-shock, 453, 556, 558
proteins, 453
HeLa cells, 176, 177
Hemagiosarcoma, 611
Hemangioendotheliomas, 570
Hemorrhagic-kidney syndrome, 565
Hepatic foci, 288
Hepatic hemangiosarcomas, 612
Hepatic microsomes, 612
Hepatic nodule(s), 235, 239, 288, 505,
601
Hepatitis, 263
Hepatoblastomas, 569
Hepatocarcinoma, see hepatocellular
carcinoma(s)
Hepatocarcinogenesis promotion, 371,
407, 416
Hepatocellular carcinoma(s), 167, 203,
207, 251, 259, 317, 333, 345,
348, 350, 352, 391, 394, 406-
408, 410, 411, 417, 436, 437,
441, 486, 489, 513, 553, 563,
575, 585, 601, 611
Hepatocyte
foci, 251-257, 260
growth factor, 447, 449
receptor, see receptor(s)
primary cultures, 447
proliferation, 206
nodules, 169, 259
Hepatoma, see hepatocellular

PK, see pyruvate kinase
PKC, see protein kinase C
PKC-catalyzed protein phosphoryla-
    tion, 221
Plasmamembrane(s), 348, 350, 475,
    476, 479, 505, 591, 593
Platelet-derived growth factor,
    188, 217, 448, 449
Platelet-derived growth inhibitors,
    447
Polyamine(s), 220, 319, 497, 408,
    416
  biosynthesis, 459
Polycyclic aromatic compound(s),
    639
Polycyclic aromatic hydrocarbon(s),
    3, 18, 20, 25, 43, 67, 122,
    137, 138
Polymorphonuclear leukocyte(s),
    223, 371
Polyoma, 174
  virus, 173
Polyphosphoinositide(s), 479
Polyploidization, 435, 438, 439,
    441
Polyunsaturated fatty acid(s), 495,
    506, 610, 612-614
Portacaval shunt, 390-394
Portacaval transposition, 391-394
Preneoplastic cell(s), 180, 409
Preneoplastic foci, 270, 333, 362
Preneoplastic intestinal metapla-
    sia, 635
Preneoplastic lesions, 108, 213,
    243, 247, 372, 377, 390,
    399, 568
Preneoplastic liver(s), 435, 436,
    441
Preneoplastic nodules, 259, 348,
    372, 352, 417, 450
Preneoplastic tissue(s), 272, 275,
    410
Preneoplastic tubular lesions, 209,
    210
Primary liver cancer, 513
Progesterone, 4
Programmed cell death, see
    apoptosis
Progression, 129, 167, 217, 219,
    227, 235, 281, 337, 372,
    395, 453, 612
Promoter(s), 185, 219, 228, 238,
    270, 301, 304, 435, 441,
    570, 571, 594, 601
Promoting
  action, 596
  effect, 224, 391
  regimen, 244
Promotion, 129, 167, 180, 188, 195,
    198, 217, 218, 220, 222,
    227, 228, 229, 235, 236,

Promotion (continued)
    238, 247, 281, 282, 287, 301,
    308, 317, 337, 362, 364, 372,
    377, 318, 371, 437, 441, 443,
    453, 495, 601, 605, 612, 619,
    654
Prostacyclins, 365
Prostaglandin(s), 4, 8, 220, 365, 592
  E2, 228
  F2 , 223, 228
  synthesis, 615
Protease(s), 220, 553
  inhibitors, 407
Protein
  catabolism, see protein degradation
  degradation, 525, 526, 528-531, 535-
    537, 539, 540, 544, 549, 553,
    554
  kinase-C, 185-188, 220, 301
  kinase(s), 185, 495
  synthesis, 526, 530, 540, 547, 548,
    553
  thiol(s), 154, 156, 158
  turnover, 526, 530, 535, 539, 547,
    549, 553
Proteolysis, see protein degradation
Protooncogene(s), 176, 185, 217, 248,
    574
  c-abl, 177
  c-fos, 222, 319
  c-H-ras, see C-Ha-ras
  c-Ha-ras, 115, 181, 187, 220, 319,
    415, 417, 574, 586
  c-K-ras, see c-Ki-ras
  c-Ki-ras, 415, 417, 586
  c-myc, 177, 222, 319, 345, 415, 476,
    586
  c-sis, 177
PUFA, see polyunsaturated fatty acid(s)
Putrescine, 307, 461
Pyruvate kinase, 323-325, 327, 333, 519,
    523

Quinones, 88

R10 rat strain, 485
R16 rat strain, 486-488
Radical(s), see free radical(s)
    scavengers, see free radical(s)
    scavengers
Raf oncogene, see oncogene(s)
Ranitidine, 76-83
Ras gene, see oncogene(s)
Ras protooncogene(s), see protoonco-
    gene(s)
Rat 6 fibroblast cell line, 187
Reactive oxygen, 246
  species, 75, 143
Receptor(s), 243, 341
  beta-adrenergic, 475
  cell-surface, 592